Helpful Geometric Formulas for Solving A

4. Triangle of base a and height h:	2
5. Cube of side s:	volume $= s^3$ surface area $= 6s^2$
6. Rectangle of length l, width w, and height h:	volume $= lwh$ surface area $= 2lw + 2lh + 2wh$
7. Right circular cylinder of radius r and height h:	volume $= \pi r^2 h$ lateral surface area $= 2\pi rh$ total surface area $= 2\pi rh + 2\pi r^2$
8. Sphere of radius r:	volume $= \dfrac{4}{3}\pi r^3$ surface area $= 4\pi r^2$

Fundamental Properties of Logarithms

$a^{\log_a x} = x, \quad x > 0$	$\log_a a^x = x, \quad x$ any real number
$\log_a(xy) = \log_a x + \log_a y$	$\log_a x^y = y \log_a x$
$\log_a\left(\dfrac{x}{y}\right) = \log_a x - \log_a y$	$\log_a x = \dfrac{\log_b x}{\log_b a}$

Instructor's Edition

Precalculus and Its Applications

Larry Joel Goldstein

Wm. C. Brown Publishers

Dubuque, Iowa•Melbourne, Australia•Oxford, England

In memory of Emil Grosswald—
Teacher, Colleague, Friend

On the cover . . .
  Upright shell by Stewart Dickson © 1993 All Rights Reserved

$X = 2*(1 - E\hat{\ }(u/(6*Pi)))*Cos[u]*Cos[v/2]\hat{\ }2;$

$Y = 2*(-1 + E\hat{\ }(u/(6*Pi)))*Cos[v/2]\hat{\ }2*Sin[u];$

$Z = 1 - E\hat{\ }(u/(3*Pi)) - Sin[v] + E\hat{\ }(u/(6*Pi))*Sin[v];$

ParametricPlot3D[{X, Y, Z}, {u, 0, 6 Pi},
    {v, 0, 2 Pi},
    PlotPoints → {96, 20},
    ViewPoint → {1.0992, 3.2001, 0.0255}, Boxed→False,
    PlotRange → All, Axes → False,

LightSources →
{{{0.7071, 0, 0.7071}, RGBColor[1, 0, 0]},
 {{0.5773, 0.5773, 0.5773}, RGBColor[0, 1, 0]},
 {{0, 0.7071, 0.7071}, RGBColor[0, 0, 1]}},
 RenderAll→False

Book Team

Editor *Paula-Christy Heighton*
Developmental Editor *Theresa Grutz*
Publishing Services Coordinator *Julie Avery Kennedy*

Wm. C. Brown Publishers
A Division of Wm. C. Brown Communications, Inc.

Vice President and General Manager *Beverly Kolz*
Vice President, Publisher *Earl McPeek*
Vice President, Director of Sales and Marketing *Virginia S. Moffat*
National Sales Manager *Douglas J. DiNardo*
Marketing Manager *Julie Joyce Keck*
Advertising Manager *Janelle Keeffer*
Director of Production *Colleen A. Yonda*
Publishing Services Manager *Karen J. Slaght*
Permissions/Records Manager *Connie Allendorf*

Wm. C. Brown Communications, Inc.

President and Chief Executive Officer *G. Franklin Lewis*
Corporate Senior Vice President, President of WCB Manufacturing *Roger Meyer*
Corporate Senior Vice President and Chief Financial Officer *Robert Chesterman*

Copyediting and production by Publication Services, Inc.

The credits section for this book begins on page 639 and is considered an extension of the copyright page.

PREFACE

Precalculus plays a very significant role in the undergraduate mathematics curriculum. On one hand, it provides preparation and background for calculus and other higher-level mathematical courses. On the other hand, it serves as a basic mathematics requirement in many core curricula.

In recent years, the course has confronted many problems and changes, some unique to the subject itself and others reflecting the general turmoil and uncertainty in the undergraduate mathematics curriculum. Students, always a heterogeneous group in this course, today exhibit a wider range of mathematical preparation and maturity. This brings into question the proper level for the course and the content that should be included. New technology has been developed and cries out to be integrated into the traditional curriculum. New teaching methodology is being discussed and applied. New supplements are required to enhance student learning and to ease the burdens of a teaching staff that has more students and fewer resources.

It is with these problems and changes in mind that I have written the present text. Developed over a period of five years, the text presents a modern approach to a traditional curriculum. Some of the features of the text are:

1. **Standard content**   The text covers what has become the standard curriculum for the precalculus course for college students.
2. **Mathematical precision**   I have attempted to incorporate a high degree of mathematical precision into the discussion while omitting much of the formalism and forbidding language of many earlier books.
3. **Training for calculus**   Above all else, a precalculus book must prepare students for the standard calculus course, in terms of both conceptual development and technical expertise. I have kept this primary goal in mind in designing the approach, exercises, and level of the text. In addition, I have included many discussions of topics that point directly to calculus.
4. **A readable book**   I have attempted to write a book that a student can read. The writing style is relaxed and informative without being condescending.
5. **Motivation**   Throughout, I have included motivation for the mathematical concepts so that students will see their logical necessity even before formal definitions and proofs are offered.
6. **Emphasis on graphical analysis**   For applying precalculus topics to other disciplines and to higher mathematics, graphical analysis is one of the most important skills required. Throughout the book, we have emphasized understanding graphs and applying them to solve problems.

7. **Applications**   Mathematics does not exist in a vacuum. Mathematical ideas are often suggested by physical problems, and mathematical abstractions describe real-world phenomena. Throughout the book, the connection between precalculus and its applications is a major theme. There are hundreds of applications, many using real-life data. The applications cover many fields students are likely to study.

8. **Problem solving**   One of the principal goals of an introductory mathematics course is to teach problem solving. To include this thread in the text, I have included strategies for solving many types of problems and have illustrated the strategies at work. I have also included cautions about common errors.

9. **Examples**   Worked examples form the body of any precalculus text. This book is no exception. I have included a large number of examples that illustrate algebraic techniques, problem-solving strategies, and applications. I have attempted to make the examples "talk" to the student as a live instructor might.

10. **Exercise sets**   The exercise sets are probably the most important part of any mathematics textbook. For this text, I have included voluminous exercise sets that test and develop a large number of skills, both technical and problem-solving. Innovative optional technology exercises allow students to apply graphing calculators and computers to solve algebra and trigonometry problems. Application problems describe how algebra is used in other fields. Essay problems require students to develop their verbal skills and demonstrate understanding of the material by describing mathematical ideas and procedures in their own words.

11. **Technology**   In addition to the technology exercises described above, the text contains discussions of technology use in precalculus. Included in the technology line-up are scientific calculators, graphing calculators, spreadsheets, graphing software, and symbolic manipulation packages. Graphic calculator examples are contained throughout the book. Available with the text is a self-contained, window-based graphing utility that emulates a graphing calculator (better than the original) and may be used on student computers, in computer labs, or for class demonstration. This software is available in IBM and Macintosh versions.

12. **Mathematical history**   I have indulged my interest in mathematical history by including boxes throughout the text that describe how algebra and trigonometry came to be and some of the interesting personalities who developed the subject.

13. **Photographs**   I have included a large number of photographs illustrating applications of precalculus topics. I want students to see what exciting subjects these are and how they relate to the world they live in.

14. **Four-color format**   A functional use of four colors enhances the text and graphics.

■ SUPPLEMENTS

A modern college textbook requires an extensive set of supplements. I was fortunate to be closely involved in writing and overseeing many of these supplements.

For the Instructor

Instructor's Edition   You are reading the Instructor's Edition of this text. We have prepared this special edition for your convenience. The Instructor's Edition contains an additional appendix with the answers to all of the even-numbered exercises. This makes answering student questions about them easier and more convenient. You can find this additional appendix by looking for the pages with the blue highlighted edges at the end of the Instructor's Edition. The student edition contains only the odd exercise answers and is 48 pages shorter than the edition you are holding.

Instructor's Solutions Manual   This contains detailed solutions to all the exercises in the text.

Test Bank   This is a printed test bank containing about 100 questions per chapter. The questions are of varying levels of difficulty and are in multiple-choice, free response, matching, and fill-in-the-blank formats. Also included are two prepared tests for each chapter along with their answer sheets.

Computer Test Generator   This test generator allows you to select test questions from any of the test questions in the printed test bank. The questions may be selected manually or randomly by the computer subject to criteria you control. There are also algorithmically generated questions that provide a virtually limitless supply of questions. Graphics are incorporated and the test generator supports graphing and special math characters on a wide selection of printers. This is available in IBM-compatible and Macintosh versions. A key factor in choosing this system to accompany the text was its high degree of user-friendliness.

Video Tapes   Ten hours of video tapes, covering the major topics in each chapter, are available. Each concept is introduced with a real-world problem and is followed by careful explanation and worked out examples using computer-generated graphics. These videos can be used in the math lab for remediation or even in the classroom to motivate or enhance the lecture.

For the Student

Student Solutions Manual   This contains detailed solutions to one third of the text exercises. All of the solutions are from the odd-numbered exercises.

Graphing Utility Guide   This contains numerous short laboratory-type graphing calculator explorations and exercises. Many topics in algebra and trigonometry are covered. Keystroke instructions are provided in separate appendices for TI, HP, Casio, and Sharp calculators. Keystrokes are kept to a minimum in the utility guide, so this manual can be used with any graphing utility. The pages are perforated, so students can turn in their assignments.

The Plotter   This software is for graphing and analyzing functions. This software simulates a graphing calculator on a PC. You may use it to do the technology exercises even if you don't have a graphics calculator. A manual is included that describes operations and includes student exercises. The software is menu-driven and has an easy-to-use window interface. The high-quality graphics can also be used for classroom presentation and demonstration. Students who go on to calculus classes will want to keep the software for future use.

Interactive Tutorial Software   Interactive tutorial software that follows each section of the book provides numerous exercises and diagnostic feedback. It also contains a glossary of mathematical terms and a context-sensitive help system. The software will also monitor student progress and test scores.

ACKNOWLEDGMENTS

Reviewers   We reviewed three drafts of the manuscript with more than 25 different professors and instructors. Their suggestions and comments helped shape the text as well as point out errors or inconsistencies. I would like to thank the following individuals who reviewed all or part of the manuscript.

Greg M. Dotseth
University of Northern Iowa

Dr. Cynthia Snead

Daniel Martinez
California State University, Long Beach

Toni Kasper
Borough of Manhattan Community College, CUNY

Bill Miller
Central Michigan University

Rudy Radulovich
Florida Community College at Jacksonville

Anthony Vance
Austin Community College

Adrian Riskin
Northern Arizona University

Bill Tomhave
Concordia College

Dr. Mahbobeh Vezvaei
Kent State University

Sandra Wray-McAfee
University of Michigan–Dearborn

Jon D. Weerts
Triton College

Janice Roe Kilpatrick
University of Toledo

Dr. William Stout
Salve Regina University

Judy Barclay
Cuesta College

Ben P. Bockstege
Broward Community College

David Collingwood
University of Washington

John Gosselin
University of Georgia

Delois Salter-Jones
Pensacola Junior College

Nancy Jim Poxon

Accuracy At the very beginning of this textbook project, the publisher and I decided that the most important thing we could do was to ensure that these texts are accurate. As a longtime classroom teacher and user of textbooks I am well aware of the frustrations caused by errors in text or in the answer section. Therefore, throughout the writing and production of this series we have single-mindedly pursued a process designed to eliminate errors. I would like to especially thank Sandra Beken, Linda Murphy, Carol Hay, and Nancy Nickerson for their assistance in reading the page proofs and checking the answers to the exercises.

Typesetting was done from my computer disks, minimizing the chance of errors being introduced at this stage. The typeset manuscript was exhaustively checked; the writing, examples, figures, and the answers appendix were checked by me. As the author, I feel completely confident that this textbook is as error-free as possible, and that it will serve you well in this regard.

My association with Wm. C. Brown has been very exciting and rewarding. The professionalism and concern for quality of everyone there has been most appreciated. I would like to especially thank Theresa Grutz, Paula Heighton, Julie Keck, and Earl McPeek for their kind encouragement and support.

Thank you, everyone, for your help and unfailing conscientiousness. I have really enjoyed writing this text and hope you enjoy using it. Please let me hear from you with suggestions for improvements and additions to future editions.

Dr. Larry Joel Goldstein
Brookeville, Maryland

CONTENTS

INDEX OF APPLICATIONS

CHAPTER 1

ALGEBRAIC PRELIMINARIES

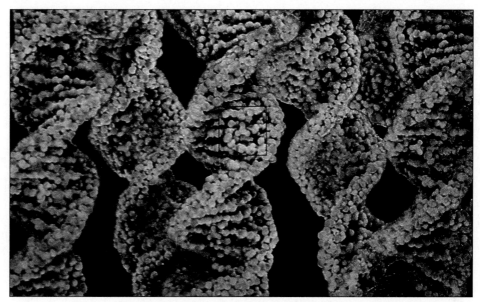

A close-up of the genetic material DNA. The molecules shown here are 10^{-8} meters in diameter. (From Powers of Ten *by Phillip Morrison, Phyllis Morrison, and The Office of Charles and Ray Eames,* Scientific American Library, *1982.)*

O ver the last 3000 years, algebra has evolved into a major field that is a prerequisite to learning and applying many branches of mathematics, including calculus, statistics, and numerical analysis, to mention but a few. These branches are, in turn, the foundation for solving problems in almost all fields in which quantitative problems arise: business, biology, chemistry, physics, law, sociology, psychology, medicine, and computer science. As the British philosopher Alfred North Whitehead put it, "Algebra is the intellectual instrument for rendering clear the quantitative aspects of the world."

This book provides a thorough look at algebra as a prerequisite to other mathematics courses, especially calculus, and to other fields. In order to enhance your perspective on the subject, we will attempt to keep both the history and the applications of the subject in view.

1.1 THE REAL NUMBER SYSTEM

The real number system is the foundation upon which elementary algebra rests. In this section, we summarize the most elementary facts about the real numbers as a basis for our subsequent development of algebra.

Real Numbers

Arithmetic and geometric problems involve numbers like the following:

$$5, \quad -18, \quad 1.782, \quad \tfrac{1}{3}, \quad \tfrac{7}{5}, \quad \pi, \quad \sqrt{2}$$

These numbers are examples of **real numbers**.

A real number can be written using decimal notation. A real number whose decimal expression requires only a finite number of digits is called a **terminating decimal**. Here are some examples of terminating decimals:

$$-58.354, \quad 12, \quad 0.0000031$$

A decimal expression that contains an infinite number of nonzero digits is called a **nonterminating decimal**. Here are some examples of nonterminating decimals:

$$\tfrac{1}{3} = 0.3333333\ldots, \qquad \sqrt{2} = 1.4142135\ldots$$

The three dots (...) indicate that the decimal expression continues indefinitely. A decimal expression that has a repeating pattern after some point, such as $12.247247247\ldots$, is called a **repeating decimal**. We indicate the pattern by a line over it: $12.\overline{247}$. A decimal expression that has no repeating pattern, such as that of $\sqrt{2}$ given above, is called a **nonrepeating decimal**.

An **integer** is a real number in which all digits to the right of the decimal point are 0. Here are some examples of integers:

$$5, \quad 177, \quad -3, \quad 0, \quad -300$$

A **rational number** is a real number that can be written as the quotient of two integers a/b, where b is not zero. Here are some examples of rational numbers:

$$\tfrac{5}{3}, \quad \tfrac{117}{89}, \quad \tfrac{45}{7}, \quad \tfrac{5}{1} = 5$$

It can be proven that every rational number is either a repeating decimal or a terminating decimal. The repeating decimal representation of a rational number may be determined by long division. Moreover, every repeating decimal represents a rational number. You may determine the rational number using a computation like the following: Consider the repeating decimal $x = 12.\overline{247}$. Then

$$1000x = 12{,}247.\overline{247}$$

$$x = 12.\overline{247}$$
$$999x = 12{,}235$$
$$x = \frac{12{,}235}{999}$$
$$12.\overline{247} = \frac{12{,}235}{999}$$

An integer a is a rational number because it can be written as the quotient $a/1$.

A number that is not rational is called **irrational**. Examples of irrational numbers are $\sqrt{2}$, $\sqrt[3]{5}$, and π.

Sets

A **set** is a collection of objects. The collection of all real numbers is an example of a set. Sets are customarily denoted using capital letters, such as A, B, or C. We will reserve the notation **R** to mean the set of real numbers.

An object in a set is called an **element** of the set and is said to **belong** to the set. Thus, $\frac{1}{3}$ and -3.75 are elements of the set **R**. If every element of the set A is also an element of the set B, then we say that A is a **subset** of B. For example, let A denote the set consisting of the numbers 0 and 1, and B the set consisting of all real numbers between 0 and 1 inclusive. Then A is a subset of B. We write this symbolically as $A \subset B$.

Because every integer is also a real number, the set of integers is a subset of the set of real numbers. Because every integer is also a rational number, the set of integers is a subset of the set of rational numbers. The interrelationships between the sets of integers, rational numbers, and real numbers are illustrated in Figure 1.

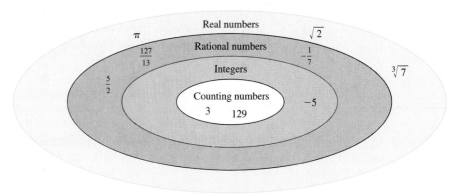

Figure 1

Throughout this book, unless we explicitly say otherwise, all sets are subsets of the real numbers.

One way of describing a set is to list its elements within braces. For instance, the set A consisting of the integers 1, 2, 3, 4, and 5 can be written as

$$A = \{1, 2, 3, 4, 5\}$$

Another way of describing sets is to use *set builder notation*, which lists the conditions for an element to belong to the set. Using this notation, for example, the set of nonzero real numbers x may be written as

$$\{x \mid x \neq 0\} \qquad \text{or} \qquad \{x : x \neq 0\}$$

This notation is read "The set of all x such that x is not equal to 0."

The Real Number Line

It is possible to represent the real number system geometrically using a line, as shown in Figure 2. Each point on the number line corresponds to exactly one real number, and each real number corresponds to exactly one point on the number line. The number corresponding to a point is called the **coordinate** of the point. We say that there is a **one-to-one correspondence** between the real number system and the number line. Figure 3 shows the point on the number line corresponding to a real number a, which is between 3 and 4.

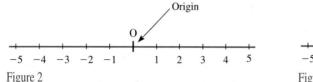

Figure 2

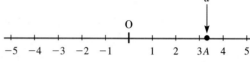

Figure 3

Inequalities

Real numbers have a natural ordering that allows any two numbers to be compared with one another. Let a and b be real numbers. We say that a is **less than** b, denoted $a < b$, provided that $b - a$ is positive. If a is less than b, we also say that b is **greater than** a, denoted $b > a$. The symbols $<$ and $>$ are called **inequality symbols**, and statements involving them are called **inequalities**.

Because $5 - 1 = 4$ is positive, $-0.5 - (-1) = 0.5$ is positive and $\frac{1}{3} - 0.3 = 0.0\overline{3}$ is positive, the following inequalities are true statements:

$$1 < 5, \quad -1 < -0.5 \quad \tfrac{1}{3} > 0.3$$

If a is either less than b or equal to b, then we write $a \le b$ ("a is less than or equal to b"). Similarly, if a is either greater than b or equal to b, then we write $a \ge b$ ("a is greater than or equal to b").

Thus, for example, the following inequalities are true:

$$3 \le 3, \quad -5 \le 1, \quad 4 \ge 4$$

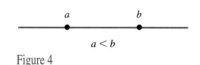

Figure 4

The number line is useful for illustrating inequalities geometrically. If a and b are real numbers, then $a < b$ is true if and only if the point with coordinate a lies to the left of the point with coordinate b (see Figure 4).

The fundamental rules for manipulating inequalities are summarized in the following.

PROPERTIES OF INEQUALITIES

Let a, b, and c be real numbers. Then the following properties of inequalities hold.

1. If $a < b$ and $b < c$, then $a < c$.
2. If $a < b$ and c is any real number, then $a + c < b + c$.
3. If $a < b$ and c is any real number, then $a - c < b - c$.
4. If $a < b$ and c is positive, then $ac < bc$.
5. If $a < b$ and c is negative, then $ac > bc$. That is, when an inequality is multiplied by a negative number, it is necessary to reverse the inequality sign.

We will omit the proofs of these properties.

Inequalities may be treated very much like equations. You may add or subtract the same quantity from both sides, and you may multiply or divide both sides by a positive number. However, if you multiply or divide an inequality by a *negative* number, then you must reverse the direction of the inequality.

The above laws of inequalities were stated for the inequality symbol $<$. However, throughout their statement you may replace $<$ with any of the other inequality symbols $>$, $\leq$, or $\geq$. (In Property 5, where two opposite symbols are used, you may replace the first symbol by any of the inequality symbols and the other by its opposite. That is, if the first symbol is replaced by $\leq$, then the second is replaced by $\geq$.)

Absolute Value

In many applications, it is necessary to consider the magnitude but not the sign of a real number. For dealing with such applications, it is convenient to define the notion of absolute value.

Definition 1 **Absolute Value**	Let a be a real number. The **absolute value** of a, denoted $|a|$, is the real number defined by the formula $$|a| = \begin{cases} a & \text{if } a \geq 0 \\ -a & \text{if } a < 0 \end{cases}$$

➢ **EXAMPLE 1**
Calculating Absolute Values

Calculate the following:

1. $\left| -\frac{3}{8} \right|$

2. $|2|$

3. $\left| \frac{33}{12} - \pi \right|$

Solution

1. Because $-\frac{3}{8}$ is negative, we have

$$\left| -\tfrac{3}{8} \right| = -(-\tfrac{3}{8}) = \tfrac{3}{8}$$

2. Because 2 is positive, we have

$$|2| = 2$$

3. We have

$$\tfrac{33}{12} - \pi = 2.75 - 3.14159\ldots = -0.39159\ldots$$

which is less than 0. Therefore, we have

$$\left| \tfrac{33}{12} - \pi \right| = -(\tfrac{33}{12} - \pi)$$
$$= \pi - \tfrac{33}{12}$$
$$\approx 0.39159$$

The following theorem summarizes some of the most useful properties of absolute values.

PROPERTIES OF ABSOLUTE VALUE

Let a and b be real numbers.

1. The absolute value of the additive inverse of a number is the same as the absolute value of the number. That is,
$$|-a| = |a|$$

2. The absolute value of a number is nonnegative. That is,
$$|a| \geq 0$$

3. The absolute value of a product equals the product of the absolute values. That is,
$$|ab| = |a||b|$$

4. The absolute value of a quotient equals the quotient of the absolute values. That is,
$$\left|\frac{a}{b}\right| = \frac{|a|}{|b|} \quad (b \neq 0)$$

5. $|a^2| = a^2$

Proof   To prove Property 1, first assume that $a > 0$. Then $-a < 0$, so that
$$|-a| = -(-a) = a = |a|$$
If $a = 0$, then
$$|-0| = 0 = |0|$$
If $a < 0$, then $-a > 0$, and
$$|-a| = -a = |a|$$
The proofs of Properties 2 through 5 are left as exercises.   ◆

Distances on the Number Line

Using absolute values, we can define the distance between points on the number line as follows:

Definition 2
Distance on the Number Line

Suppose that A and B are points on the number line with corresponding coordinates a and b. The distance between A and B, denoted $d(A, B)$, is defined by the formula
$$d(A, B) = |a - b|$$

➤ **EXAMPLE 2**
Calculating Distances

Find the distance $d(A, B)$ between the points A and B on the number line, whose respective coordinates a and b are given as follows:

1. $a = 1, b = 9$
2. $a = -1, b = -5$

Solution

In each case, the value of $d(A, B)$ is derived using a simple absolute value calculation. Note that the value of the distance corresponds with our intuitive notion of distance as illustrated in the accompanying figures (Figures 5 and 6).

1. $d(A, B) = |a - b| = |1 - 9| = |-8| = 8$

2. $d(A, B) = |a - b| = |-1 - (-5)| = |4| = 4$

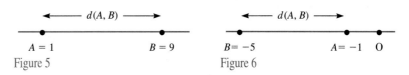

Figure 5   Figure 6

Suppose that A is any point on the number line. We can compute the distance of A to the origin in terms of the coordinate a of A. The origin corresponds to the number 0, so that we have the formula

$$d(A, 0) = |a - 0| = |a|$$

Exercise 1.1

Consider the following numbers:

$$24, \quad \tfrac{2}{3}, \quad -1.8, \quad 2\pi, \quad 0, \quad 93\%,$$

$$\sqrt[5]{32}, \quad \sqrt{5}, \quad \sqrt{16}, \quad \sqrt{\tfrac{1}{4}}, \quad -234,$$

$$-\tfrac{16}{7}, \quad 0.0035$$

1. List the numbers that are integers.

2. List the numbers that are rational.

3. List the numbers that are irrational.

4. List the numbers that are real.

Find decimal notation for each of the following.

5. $\frac{2}{7}$

6. $\frac{45}{11}$

7. $\frac{23}{12}$

8. $\frac{34}{9}$

Calculate the following.

9. $\left|\frac{1}{4}\right|$

10. $|5 - 3|$

11. $|4 + 7|$

12. $|-5 + 0|$

13. $|-7 - 3|$

14. $|-8 + 2|$

15. $|3.28 - 3.28|$

16. $\left|-\frac{2}{3}\right|$

17. $|\sqrt{10}|$

18. $|\sqrt{10} - \pi|$

19. $|2 - \sqrt{5}|$

20. $|\pi - \sqrt{10}|$

Fill in the blank with the correct symbol ($<$, $>$, or $=$).

21. $|4| + |-6|$ _____ $|4 + (-6)|$

22. $|7| - |9|$ _____ $|7 - 9|$

23. $|-8| - |17|$ _____ $|-8 - 17|$

24. $|-24| + |-18|$ _____ $|-24 + (-18)|$

Find the distances between points A and B on the number line for the given coordinates a and b.

25. $a = 7, b = 10$

26. $a = -13, b = 6$

27. $a = \frac{2}{5}, b = \frac{3}{4}$

28. $a = 0, b = -89$

29. $a = -6.7, b = -34.2$

30. $a = \sqrt{2}, b = \pi$

31. Write set builder notation for the set of all negative real numbers.

32. Write set builder notation for the set of all real numbers that are not equal to 2.

33. Write set builder notation for the set of all real numbers that are not less than -3.

34. Write set builder notation for the set of all positive numbers greater than or equal to -7.

Let a and b be real numbers. Prove each of the following.

35. $|-a| = |a|$ (Hint: Consider two cases: $a \geq 0$ and $a < 0$.)

36. $|ab| = |a||b|$

37. $\left|\dfrac{a}{b}\right| = \dfrac{|a|}{|b|}$

38. For what values of x is $|x| - x = 0$?

39. Give an example to disprove that $a|b + c| = |ab + ac|$.

1.2 ALGEBRAIC EXPRESSIONS

In this section, we review the basic facts about exponents and radicals, polynomials, factoring, and rational expressions.

Exponents and Radicals

If n is a positive integer, then we write a^n to mean a product of n factors of a. That is, we have

$$a^1 = a$$
$$a^2 = a \cdot a$$
$$a^3 = a \cdot a \cdot a$$
$$a^4 = a \cdot a \cdot a \cdot a$$

The expression a^n is read **a to the nth power** or simply **a to the nth.** Computing a^n is called **raising a to the nth power.** In an expression a^n, a is called the **base,** and n is called the **exponent.**

Definition 1
Principal nth Root

Let n be a positive integer and a be a real number. The **principal nth root** of a, denoted $\sqrt[n]{a}$, is defined as follows:

1. If a is positive, then $\sqrt[n]{a}$ is the positive real number whose nth power is a, that is,

$$(\sqrt[n]{a})^n = a, \qquad \sqrt[n]{a} > 0$$

2. If a is negative and n is odd, then $\sqrt[n]{a}$ is the negative number whose nth power is a, that is,

$$(\sqrt[n]{a})^n = a, \qquad \sqrt[n]{a} < 0$$

3. $\sqrt[n]{0} = 0$

If $n = 2$, we write $\sqrt{a}$ instead of $\sqrt[2]{a}$. In this case, $\sqrt{a}$ is called the **principal square root of a** or, for short, the **square root of a.**

For example, we have

$$\sqrt{4} = 2, \qquad \sqrt{\tfrac{1}{16}} = \tfrac{1}{4}, \qquad \sqrt{0} = 0$$

$\sqrt{-1}$ is undefined in the real number system.

In the three examples above, the values of $\sqrt{a}$ are rational numbers. However, for many rational numbers a, the value of $\sqrt{a}$ is irrational. For example, it can be proved that $\sqrt{2}$ is irrational. The decimal expression of this real number is nonterminating and nonrepeating, and begins

$$\sqrt{2} = 1.414213\ldots$$

We have

$$\sqrt[3]{8} = 2, \quad \sqrt[4]{\tfrac{1}{16}} = \tfrac{1}{2}, \quad \sqrt[3]{-\tfrac{8}{27}} = -\tfrac{2}{3}$$

The symbols $\sqrt{a}$ and $\sqrt[n]{a}$ are called **radical expressions**.

Note that if a is negative and n is even, then there is no real number b such that $b^n = a$. For this reason, $\sqrt[n]{a}$ is undefined if a is negative and n is even. Later, we will show that by using *complex* numbers, we can determine b such that $b^n = a$ no matter what the value of a.

Let a be a real number. To define a^r, where r is a rational number, we consider the special case where r is a number of the form $1/n$ for a positive integer n. In this case, we define $a^{1/n}$ by the formula

$$a^{1/n} = \begin{cases} \sqrt[n]{a} & \text{if } \sqrt[n]{a} \text{ is defined} \\ \text{undefined} & \text{otherwise} \end{cases}$$

Thus, for example, we have

$$2^{1/2} = \sqrt{2}$$

$$8^{1/3} = \sqrt[3]{8} = 2$$

$$(81)^{1/4} = \sqrt[4]{81} = 3$$

When r is the positive rational number,

$$r = \frac{m}{n}$$

for m and n positive integers and m/n in lowest terms, we define a^r as

$$a^r = a^{m/n} = (a^{1/n})^m = (\sqrt[n]{a})^m = \sqrt[n]{a^m}$$

Here are some examples of numbers raised to rational exponents:

$$4^{3/2} = (\sqrt{4})^3 = 2^3 = 8$$

$$(-2)^{3/2} = (\sqrt{-2}^{\,3}) = \text{undefined}$$

Let a be a nonzero number. We define the **zero power** of a by the formula

$$a^0 = 1, \; a \neq 0$$

Thus, $3^0 = 1$, $(-7)^0 = 1$, $(\frac{2}{3})^0 = 1$, and so forth. Note, however, that 0^0 is undefined.

Suppose that a is nonzero and r is a rational number for which a^r is defined and $r > 0$. Then we define a^{-r} by the formula

$$a^{-r} = \frac{1}{a^r}$$

For example, we have

$$\left(\frac{1}{7}\right)^{-2} = \frac{1}{(\frac{1}{7})^2} = \frac{1}{\frac{1}{49}} = 49$$

$$4^{-3/2} = \frac{1}{4^{3/2}} = \frac{1}{(\sqrt{4})^3} = \frac{1}{2^3} = \frac{1}{8}$$

The following rules are important in manipulating algebraic expressions involving exponents.

LAWS OF EXPONENTS

Let a and b be real numbers and r and s rational numbers. Then the following laws of exponents hold, provided that all of the expressions appearing in a particular equation are defined.

1. $a^r \cdot a^s = a^{r+s}$

2. $(a^r)^s = a^{rs}$

3. $(ab)^r = a^r b^r$

4. $\dfrac{a^r}{a^s} = a^{r-s}, \quad a \neq 0$

5. $\left(\dfrac{a}{b}\right)^r = \dfrac{a^r}{b^r}, \quad b \neq 0$

6. $a^{-r} = \dfrac{1}{a^r}, \quad a \neq 0$

7. $\left(\dfrac{a}{b}\right)^{-r} = \left(\dfrac{b}{a}\right)^r, \quad a, b \neq 0$

➤ **EXAMPLE 1**
Eliminating Negative
and Zero Exponents

Write the following expression in a form that does not involve any negative or zero exponents:

$$\left(\frac{x^{-5}}{x^{-3}}\right)^{-3}$$

Solution

We have

$$\left(\frac{x^{-5}}{x^{-3}}\right)^{-3} = \frac{x^{(-5)\cdot(-3)}}{x^{(-3)\cdot(-3)}} \qquad \textit{(Laws 2 and 5)}$$

$$= \frac{x^{15}}{x^9}$$

$$= x^{15-9} \qquad \textit{(Law 4)}$$

$$= x^6$$

➤ **EXAMPLE 2**
Eliminating Negative
Exponents and Radicals

Write the following expressions in a form that involves no negative exponents or radicals.

$$\sqrt{\frac{x}{x^{-3/2}}}$$

Solution

By definition of a rational exponent, we have

$$\sqrt{\frac{x}{x^{-3/2}}} = \left(\frac{x}{x^{-3/2}}\right)^{1/2}$$

$$= \frac{x^{1/2}}{(x^{-3/2})^{1/2}} \qquad (Law\ 5)$$

$$= \frac{x^{1/2}}{x^{-3/4}} \qquad (Law\ 2)$$

$$= x^{(1/2)-(-3/4)} \qquad (Law\ 4)$$

$$= x^{5/4}$$

Using the laws of exponents, we deduce the following useful properties of radicals:

PROPERTIES OF RADICALS

For a and b, both real and nonnegative:

1. $\sqrt[n]{a}\,\sqrt[n]{b} = \sqrt[n]{ab}$

2. $\sqrt[n]{\dfrac{a}{b}} = \dfrac{\sqrt[n]{a}}{\sqrt[n]{b}}, \quad b \neq 0$

3. $\sqrt[n]{a^n} = |a|, \quad n$ even

4. $\sqrt[n]{a^n} = a, \quad n$ odd

Note that these formulas hold only for values of a, b, and n for which all the radicals appearing are defined.

➢ **EXAMPLE 3**
Simplifying Radical
Expressions

Simplify the following expressions by removing radicals, where possible, and by using as simple as possible an expression in the radical otherwise.

1. $\sqrt[3]{64y^3z^9}$

2. $\dfrac{\sqrt[3]{54x^3}}{\sqrt[3]{16x^2}}$

Solution

1. Applying Property 1 of radicals yields

$$\sqrt[3]{64} \cdot \sqrt[3]{y^3} \cdot \sqrt[3]{z^9} = 4yz^3$$

2. From Property 2, we have

$$\sqrt[3]{\frac{54x^3}{16x^2}} = \sqrt[3]{\frac{27x}{8}} = \frac{3}{2}\sqrt[3]{x}$$

In the discussion above, we have defined the properties of real numbers with rational exponents. It is also possible to define the properties of irrational exponents, such as $2^{\sqrt{2}}$ and 5^{π}. We will provide only a partial definition of irrational exponents in this book (in Chapter 4). However, you should note that all of the laws of exponents stated above hold for irrational exponents as well.

Powers of 10 are commonly used in scientific work, especially in describing numbers that are very large or very small. For example, the distance light travels in a year is approximately 10^{16} meters. The diameter of a DNA molecule is approximately 10^{-8} meters.

Multiplication by 10^{n} is equivalent to shifting the decimal point n places. If n is positive, the shift is to the right, whereas if n is negative, the shift is to the left. Here are some examples:

$$15.354 \times 10^{3} = 15{,}354 \qquad \textit{(shift right 3 places)}$$
$$15.354 \times 10^{-3} = 0.015354 \qquad \textit{(shift left 3 places)}$$

By shifting the decimal point an appropriate number of places, we can write any number as a power of 10 times a decimal with exactly one nonzero digit to the left of the decimal point. For example,

$$15.354 = 1.5354 \times 10^{1}$$
$$0.0058723 = 5.8723 \times 10^{-3}$$

This form of writing a decimal is called **scientific notation**. Its advantage is that very large and very small numbers can be expressed in a very concise format. For instance, 1 followed by 100 zeros can be written in scientific notation as

$$1.0 \times 10^{100}$$

and a decimal point followed by 99 zeros and a 1 can be written in scientific notation as

$$1.0 \times 10^{-100}$$

In scientific notation, the initial decimal is called the **mantissa**, and the power of 10 is called the **exponent**. According to the laws of exponents, to multiply numbers written in scientific notation we multiply mantissas and add the exponents. For example,

$$(2.0 \times 10^{5}) \cdot (3.5 \times 10^{-2}) = \left[(2.0) \cdot (3.5) \cdot (10^{5} \cdot 10^{-2}) \right] = 7.0 \times 10^{3}$$

➤ **EXAMPLE 4**
Gravitation

Newton's law of universal gravitation asserts that the force F of gravitational attraction between two bodies of masses m_1 and m_2 that are at a distance r from one another is given by the formula

$$F = G\frac{m_1 m_2}{r^2}$$

where G is the universal gravitational constant, which in the metric system is equal to

$$G = 6.670 \times 10^{-11}$$

In the metric system, F is measured in newtons, r in meters, and m_1, m_2 in kilograms. The mass of the earth is approximately equal to 5.97×10^{24} kg. Determine the gravitational force exerted by the earth on an asteroid of mass 10^{18} kg at a distance 10^{10} meters from the center of the earth.

Solution

From the given formula, we have

$$F = G\frac{m_1 m_2}{r^2}$$

$$= (6.670 \times 10^{-11})\frac{(5.97 \times 10^{24}) \cdot (10^{18})}{(10^{10})^2}$$

Applying the laws of exponents to multiply the powers of 10, we have

$$F = (6.670) \cdot (5.97) \cdot \frac{10^{-11+24+18}}{10^{20}}$$

$$= 39.82 \times \frac{10^{31}}{10^{20}}$$

$$= 39.82 \times 10^{31-20}$$

$$= 3.982 \times 10^{12} \text{ newtons}$$

In computer or calculator work, the multiplication sign and the 10 are often replaced by the letter E, which stands for *exponent*. For instance, the number

$$1.275 \times 10^{-3}$$

is written for computer or calculator usage as

$$1.275E{-}3$$

Polynomials and Factoring

A **polynomial expression in** x (**polynomial** for short) is an expression of the form

$$a_n x^n + a_{n-1}x^{n-1} + \cdots + a_1 x + a_0 \qquad (1)$$

where n is a nonnegative integer. The numbers $a_n, a_{n-1}, \ldots, a_1$, and a_0 are called the **coefficients** of the polynomial. Here are some examples of polynomials in the variable x:

$$4x^3 - 2x^2 + x - 5, \qquad -5x + 3, \qquad x^2 + 1 \qquad (2)$$

The polynomial with all 0 coefficients is called the **zero polynomial** and is denoted **0**. If a polynomial is nonzero, then it may be written in the form of display (1), with $a_n \neq 0$. In this case, a_n is called the **leading coefficient** and n is called the **degree**. By convention, the zero polynomial does not have a degree. The polynomials in display (2) have leading coefficients 4, -5, and 1, respectively. Their respective degrees are 3, 1, and 2.

Special terms are used to describe polynomials of low degrees. A polynomial of degree 0 has the form a, where a is a nonzero constant, and is called a **constant polynomial**. A polynomial of degree 1 has the form $ax + b$, where a and b are constants ($a \neq 0$). Such a polynomial is called **linear**. A polynomial of degree 2 has the form $ax^2 + bx + c$, where a, b, and c are constants. Such a polynomial is called **quadratic**. Polynomials of degrees 3 and 4 are called **cubic** and **quartic**, respectively.

As a shorthand in notation, we often use minus signs between terms where negative coefficients are present. That is, we write $3x^2 - 2x - 1$ rather than $3x^2 + (-2)x + (-1)$.

Two terms of a polynomial are said to be **similar** if they have the same variables raised to the same powers. For example, the terms $-2xy^2$ and $4xy^2$ are similar, but the terms $4xy^2$ and $4x^2y$ are not similar.

Polynomials may be added and subtracted by combining similar terms. For example, if $P = 2x^2 + 3x - 1$ and $Q = -3x^2 + 4x - 4$, then we have

$$P + Q = [2 + (-3)]\,x^2 + (3 + 4)x + [(-1) + (-4)] = -x^2 + 7x - 5$$

$$P - Q = [2 - (-3)]\,x^2 + (3 - 4)x + [(-1) - (-4)] = 5x^2 - x + 3$$

To multiply two polynomials, we repeatedly apply the laws of exponents, as illustrated in the following example:

$$(x^2 - 1)(2x^2 - x - 3) = x^2(2x^2 - x - 3) - 1(2x^2 - x - 3)$$
$$= 2x^4 - x^3 - 3x^2 - 2x^2 + x + 3$$
$$= 2x^4 - x^3 - 5x^2 + x + 3$$

The discussion above has been concerned exclusively with monomials and polynomials in a single variable, but we may also consider monomials and polynomials in several variables. A **monomial** in several variables is a product of a constant and nonnegative integer powers of the variables. Here are examples of monomials in variables x, y and z:

$$-3xyz, \qquad 14x^2y^3, \qquad 0.5xy^3z^9$$

The **degree of a monomial in several variables** is the sum of the exponents of the variables that appear in the monomial. For instance, the degree of xyz is $1 + 1 + 1 = 3$, and the degree of the monomial x^2yz^3 is $2 + 1 + 3 = 6$.

A **polynomial in several variables** is a sum of monomials in the variables. Here are some examples of polynomials in the variables x, y, and z:

$$-3xyz + 14x^2y^3 - 5xy^3z^9, \qquad x + y + z, \qquad xy + yz + xz^8$$

The **degree of a polynomial in several variables** is the highest degree of the monomials that appear. For instance, the degrees of the three polynomials above are 13, 1, and 9, respectively.

Polynomials with different coefficients are regarded as different. For instance, $x^3 + 1$ is not the same polynomial as x^4, and we write

$$x^3 + 1 \neq x^4$$

An **algebraic expression** is obtained by combining the operations of addition, subtraction, multiplication, and division and by forming radicals applied to real numbers and variables. Here are some examples of algebraic expressions involving the single variable x:

$$3x + 5, \qquad \frac{x^2 - 3}{\sqrt{x}}, \qquad \sqrt[3]{3x^2 - 3x + \frac{2}{\sqrt{x}}}$$

Here are some algebraic expressions involving several variables:

$$xy + y^2, \qquad -13xyz, \qquad \pi r^2 h$$

➤ **EXAMPLE 5**
Products of Algebraic
Expressions

Determine the following product.

$$(2\sqrt{x} - 3x)\left(\frac{2}{x} + x^4\right)$$

Solution

The product is calculated using the laws of exponents. The result is

$$(2\sqrt{x} - 3x)\left(\frac{2}{x} + x^4\right) = 4\frac{\sqrt{x}}{x} + 2x^4 \cdot \sqrt{x} - 6\frac{x}{x} - 3x^5$$

$$= \frac{4x^{1/2}}{x} + 2x^{1/2} \cdot x^4 - 6 - 3x^5$$

$$= \frac{4}{x^{1/2}} + 2x^{9/2} - 6 - 3x^5, \quad x > 0$$

Factoring is the process of writing a polynomial as a product of polynomials. For example, the polynomial $x^2 - x$ can be written as the product $x(x - 1)$. When factoring, always check your results by multiplication.

FACTORIZATION IDENTITIES

Square of Binomial Sum $A^2 + 2AB + B^2 = (A + B)^2$
Square of Binomial Difference $A^2 - 2AB + B^2 = (A - B)^2$
Difference of Squares $A^2 - B^2 = (A + B)(A - B)$
Sum of Cubes $A^3 + B^3 = (A + B)(A^2 - AB + B^2)$
Difference of Cubes $A^3 - B^3 = (A - B)(A^2 + AB + B^2)$

The following example illustrates how these factorization identities may be used to factor a variety of polynomial expressions.

➤ **EXAMPLE 6**
Applying Factorization
Identities

Factor the following polynomials.

1. $4x^2 - 12x + 9$
2. $16x^6 - 25y^4$
3. $8x^4 - 8x$

Solution

1. The expression can be written as

$$4x^2 - 12x + 9$$
$$(2x)^2 - 2 \cdot 2x \cdot 3 + 3^2$$
$$A^2 - 2 \cdot A \cdot B + B^2$$

where $A = 2x$ and $B = 3$, so the Square of a Binomial Difference formula yields

$$4x^2 - 12x + 9 = (2x - 3)^2$$

2. This expression is a difference of two perfect squares, namely

$$16x^6 - 25y^4$$
$$(4x^3)^2 - (5y^2)^2$$

Therefore, by the Difference of Squares formula, with A replaced by $4x^3$ and B replaced by $5y^2$, we have

$$16x^6 - 25y^4 = (4x^3 + 5y^2)(4x^3 - 5y^2)$$

3. Each term has a factor of $8x$. Before we attempt to apply any factorization formula, we factor out this monomial factor

$$8x^4 - 8x$$

$$8x(x^3 - 1)$$

We recognize the second factor as being given by the Difference of Cubes formula, which then yields the factorization

$$8x^4 - 8x = 8x(x - 1)(x^2 + x + 1)$$

➤ **EXAMPLE 7**

Factoring Polynomials

Factor the following polynomials into linear factors with integer coefficients, if possible.

1. $x^2 + 9x + 14$
2. $15x^2 + 2x - 1$

Solution

1. $x^2 + 9x + 14$. Suppose that we have a factorization of the form

$$(ax + b)(cx + d) = acx^2 + (ad + bc)x + bd$$

Matching coefficients, we must have $ac = 1$, $ac + bd = 9$, and $bd = 14$. Use the first and last equations to narrow the choices. A quick check of the various possibilities shows that

$$a = c = 1, \quad b = 7, \quad d = 2$$

So the factorization is

$$(x + 7)(x + 2)$$

2. $15x^2 + 2x - 1$. We are looking for factors of the form $(ax + b)(cx + d)$, where a, b, c, and d are integers. The product ac must equal 15, and the product bd must equal -1. Furthermore, $ad + bc$ must equal 2. Testing all possibilities shows that the only one that works is the factorization $(5x - 1)(3x + 1)$.

Another technique of factoring is to group the terms so that you can apply the distributive law in reverse. The next example illustrates this technique.

➤ **EXAMPLE 8**

Factoring by Grouping

Factor the following polynomial $x^3 - xy^2 + 5x^2 - 5y^2$

Solution

Use parentheses to group the terms

$$(x^3 - xy^2) + (5x^2 - 5y^2) = x(x^2 - y^2) + 5(x^2 - y^2)$$
$$= (x + 5)(x^2 - y^2)$$
$$= (x + 5)(x + y)(x - y)$$

Rational Expressions

A **rational expression in the variable** x is an algebraic expression of the form

$$\frac{P}{Q} \quad \begin{array}{l} \textit{numerator} \\ \textit{denominator} \end{array}$$

where P and Q are polynomials in x, where Q is not the zero polynomial. (As with real numbers, division by the zero polynomial is undefined.)

Suppose that $R = P/Q$ is a rational expression in x. We may evaluate P and Q for all real numbers x. This allows us to calculate R for all values of x for which the denominator Q is nonzero.

For example, consider the rational expression

$$R = \frac{x+1}{(x-2)(x-1)}$$

R is undefined if x equals 1 or 2 because for these values of x, the denominator equals 0.

A rational expression in several variables is a quotient of polynomials in the variables. Here are some examples of rational expressions in the variables x, y, and z:

$$\frac{x+y}{z}, \qquad \frac{x^2+y^2+3xyz}{2x-3y}$$

If A, B, and C are algebraic expressions, then we have the equality of algebraic expressions

$$\frac{AC}{BC} = \frac{A}{B}$$

which is valid for all values of x for which B and C are nonzero. This equation says that any common factors of the numerator and denominator of a rational expression may be eliminated, provided we avoid any values of x for which the eliminated factor is 0.

The process of removing common factors from the numerator and denominator of a rational expression is called **reduction.** If the numerator and denominator of a rational expression have no factors in common, then we say that the rational expression is in the **lowest terms.**

➤ **EXAMPLE 9**
Reducing a Rational
Expression to Lowest Terms

Reduce the following rational expression to lowest terms.

$$\frac{x^5 - x^3}{x^2 + 3x + 2}$$

Solution

Factor the numerator and denominator to obtain the equivalent expression.

$$\frac{x^3(x^2-1)}{(x+1)(x+2)} = \frac{x^3(x-1)(x+1)}{(x+2)(x+1)}$$

The denominator is 0 if x has either of the values -2 or -1, so to avoid division by 0, we must exclude these values. We see that the numerator and denominator have the common factor $x+1$, so the expression is equivalent to

$$\frac{x^3(x-1)}{x+2}, \quad x \neq -1, -2$$

It is clear that the numerator and denominator of this rational expression have no factors in common, so it is in the lowest terms. It is important to note that in the final expression, it seems as if x may equal -1. However, this does not mean that the final expression equals the original expression when $x = -1$ because the original expression is undefined for this value.

WARNING: When you simplify an expression by removing common factors of the numerator and denominator, the results are valid only for values of the variables for which the common factor is nonzero. This is easy to overlook, but may lead to strange results if you later substitute a value for which the common factor is zero.

Let A/Q and B/Q be rational expressions with the same denominator. These expressions may be added or subtracted according to the formula

$$\frac{A}{Q} + \frac{B}{Q} = \frac{A + B}{Q}$$

$$\frac{A}{Q} - \frac{B}{Q} = \frac{A - B}{Q}$$

To add or subtract rational expressions with *different* denominators, it is necessary to write them in equivalent forms with the same denominator. To see how this may be done, consider the rational expressions

$$\frac{x}{x - 1}, \qquad \frac{x + 1}{x + 2}$$

The denominator of the first expression is $x - 1$ and that of the second is $x + 2$. We may replace each expression by one in which numerator and denominator are multiplied by a common factor. Let's multiply the numerator and denominator of the first expression by $x + 2$ and the numerator and denominator of the second by $x - 1$. The two expressions are then replaced, respectively, by the equivalent expressions

$$\frac{x(x + 2)}{(x - 1)(x + 2)}, \qquad \frac{(x + 1)(x - 1)}{(x + 2)(x - 1)}$$

The denominator $(x + 2)(x - 1)$ is the least common multiple of the denominators of the original expressions.

To determine the least common multiple of two polynomials, it is usually simplest to factor them and form the polynomial that consists of the product of the highest power of each factor that appears. For instance, consider the three polynomials

$$(x - 1)^3 (x - 2), \qquad (x - 1)(x - 2)^4, \qquad (x + 1)^2$$

To form the least common multiple, we take the product of the highest power of each factor that appears, namely

$$(x - 1)^3 (x - 2)^4 (x + 1)^2$$

The following example illustrates how rational functions may be added and subtracted by replacing them with rational expressions that have the least common multiple of the given denominators as a common denominator.

➤ **EXAMPLE 10**
Adding Rational Expressions

Determine the sum

$$\frac{1}{x^2 + x + 1} + \frac{1}{x - 1}$$

Solution

In this example, the least common multiple of the denominators is $(x - 1)$ $(x^2 + x + 1)$. Therefore, we have

$$\frac{1}{x^2 + x + 1} + \frac{1}{x - 1} = \frac{1}{x^2 + x + 1} \cdot \frac{x - 1}{x - 1} + \frac{1}{x - 1} \cdot \frac{x^2 + x + 1}{x^2 + x + 1}$$

$$= \frac{(x - 1) + (x^2 + x + 1)}{(x^2 + x + 1)(x - 1)}$$

$$= \frac{x^2 + 2x}{x^3 - 1}, \quad x \neq 1 \qquad \textit{Difference of Cubes}$$

Multiplication of rational expressions is carried out using the following formula:

$$\frac{a}{b} \cdot \frac{c}{d} = \frac{ac}{bd}$$

Thus, for example, we have

$$\frac{x}{x - 1} \cdot \frac{2}{x + 1} = \frac{2x}{(x + 1)(x - 1)}$$

$$= \frac{2x}{x^2 - 1}, \quad x \neq 1, -1$$

Division of rational expressions may be carried out using the formula:

$$\frac{a/b}{c/d} = \frac{a}{b} \cdot \frac{d}{c} = \frac{ad}{bc}$$

For instance, we have

$$\frac{\dfrac{x^2 - 1}{x}}{\dfrac{1}{x - 1}} = \frac{x^2 - 1}{x} \cdot \frac{x - 1}{1}$$

$$= \frac{(x^2 - 1)(x - 1)}{x}$$

$$= \frac{x^3 - x^2 - x + 1}{x}, \qquad x \neq 1, 0$$

The next example combines all of the operations on rational expressions we have discussed to simplify complex rational expressions.

➤ **EXAMPLE 11**
Reducing a Complicated
Rational Expression

Write the following as a rational expression in lowest terms.

$$\frac{a^{-1} + b^{-1}}{a^{-3} + b^{-3}}$$

Solution

The given expression may be written in the form

$$\frac{\dfrac{1}{a} + \dfrac{1}{b}}{\dfrac{1}{a^3} + \dfrac{1}{b^3}}$$

Multiply both the numerator and denominator by $a^3 b^3$, to obtain the equivalent expression

$$\frac{\left(\dfrac{1}{a} + \dfrac{1}{b}\right) \cdot (a^3 b^3)}{\left(\dfrac{1}{a^3} + \dfrac{1}{b^3}\right) \cdot (a^3 b^3)} = \frac{\dfrac{a^3 b^3}{a} + \dfrac{a^3 b^3}{b}}{\dfrac{a^3 b^3}{a^3} + \dfrac{a^3 b^3}{b^3}}$$

$$= \frac{a^2 b^3 + a^3 b^2}{b^3 + a^3}$$

$$= \frac{a^2 b^2 (b + a)}{b^3 + a^3}$$

By the Sum of Cubes factorization formula, we can rewrite the expression in the form

$$\frac{(a^2 b^2)(a + b)}{(a + b)(a^2 - ab + b^2)}$$

Simplifying this last expression to lowest terms gives us the result

$$\frac{a^2 b^2}{a^2 - ab + b^2}, \quad a, b \neq 0, \quad a \neq -b$$

Expressions often involve radicals in the denominator. For example, consider the expression

$$\frac{1}{\sqrt{2}}$$

To remove the radical from the denominator, multiply both numerator and denominator by $\sqrt{2}$ to obtain

$$\frac{1}{\sqrt{2}} = \frac{1 \cdot (\sqrt{2})}{(\sqrt{2}) \cdot (\sqrt{2})} = \frac{\sqrt{2}}{2}$$

The process of removing a radical from a denominator is called **rationalizing the denominator.** This may be accomplished in most situations by multiplying both numerator and denominator by a suitable expression.

➤ **EXAMPLE 12**
Rationalizing Denominators

Rationalize the denominator of

$$\frac{1}{x + \sqrt{x}}$$

Solution

The Difference of Squares factorization identity suggests that we multiply the numerator and denominator by the expression $x - \sqrt{x}$. Doing so gives us the following equivalent expression:

$$\frac{1}{x + \sqrt{x}} = \frac{1}{x + \sqrt{x}} \cdot \frac{x - \sqrt{x}}{x - \sqrt{x}}$$

$$= \frac{x - \sqrt{x}}{x^2 - (\sqrt{x})^2}$$

$$= \frac{x - \sqrt{x}}{x^2 - x}$$

In calculus, it is often necessary to write expressions so that they have no radicals in the *numerator*. This process is called **rationalizing the numerator** and is illustrated in the following example.

> ## ➢ EXAMPLE 13
> Rationalizing a Numerator

Write the expression

$$\frac{\sqrt{x + h} - \sqrt{x}}{h}$$

in a form that includes no radicals in the numerator.

Solution

The Difference of Squares formula suggests that we multiply both numerator and denominator by the expression

$$\sqrt{x + h} + \sqrt{x}$$

This gives us the equivalent expression

$$\frac{(\sqrt{x + h} - \sqrt{x}) \cdot (\sqrt{x + h} + \sqrt{x})}{h \cdot (\sqrt{x + h} + \sqrt{x})}$$

Applying the Sum and Difference formula to the numerator, we have

$$\frac{(\sqrt{x + h})^2 - (\sqrt{x})^2}{h(\sqrt{x + h} + \sqrt{x})}$$

We can simplify the numerator to read

$$\frac{(x + h) - x}{h(\sqrt{x + h} + \sqrt{x})} = \frac{h}{h(\sqrt{x + h} + \sqrt{x})}$$

$$= \frac{1}{\sqrt{x + h} + \sqrt{x}}, \qquad h \neq 0$$

Exercises 1.2

Write the following expressions in a form that does not involve any negative or zero exponents or parentheses and that contains as little multiplication as possible.

1. $a^{-5} \cdot a^2$

2. $t^{-7} \cdot t^7$

3. $q^3 \cdot q^{-6} \cdot q^7$

4. $m^{-3} \cdot m^{-5} \cdot m^{20}$

5. $(3x^5)(-4x^6)$

6. $(6y^{-4})(8y^{-10})$

7. $(4x^3y^3)(9x^{-3}y^{-5})$

8. $(-12p^{-6}t^{11})(-5p^8t^{-7})$

9. $(3ab^4)^2$

10. $(-5x^{-2}y^4)^{-3}$

11. $(2y)^4(3x)^4$

12. $(2ab)^3(3ab)^2$

13. $\dfrac{t^{-12}}{t^{-7}}$

14. $\dfrac{a^{10}}{a^8}$

15. $\dfrac{a^3b^{-3}}{a^{-2}b}$

16. $\dfrac{10x^4y^{-4}}{-2x^{-1}y^3}$

17. $(-3x^2b^{-4})^{-2}$

18. $(-5p^{-3}q^4r)^{-3}$

19. $\left(\dfrac{2a^3b^{-2}}{3a^4b^{-3}}\right)^3$

20. $\left(\dfrac{-4p^5y^{-2}}{5p^{-1}q^6}\right)^{-4}$

21. $\dfrac{(4x^2y^{-1}z^{-3})^{-2}}{(8xy^{-1}z^2)^{-1}}$

22. $\dfrac{(10^{-1}a^{-4}c^5)^{-2}}{(5^{-2}a^3c^6)^{-3}}$

Simplify the following expressions.

23. $\sqrt{32(t+1)^2}$

24. $\sqrt{49x^3}$

25. $\sqrt{3x} \cdot \sqrt{6x}$

26. $\sqrt{8y} \cdot \sqrt{2y}$

27. $\sqrt{4x^2y^4}$

28. $\sqrt[3]{64z^6}$

29. $\sqrt{98a^2b^{-6}}$

30. $\sqrt[5]{-1 \cdot x^{10}y^{-15}}$

31. $\sqrt[3]{9x^2y} \, \sqrt[3]{6xy^3}$

32. $\sqrt[4]{90ab^3} \, \sqrt[4]{16ab}$

33. $\dfrac{\sqrt[3]{54t}}{\sqrt[3]{2t}}$

34. $\dfrac{\sqrt{80a}}{\sqrt{5a}}$

35. $\dfrac{\sqrt{3(2x^2)^3}}{6x}$

36. $\dfrac{\sqrt[3]{9xy}}{\sqrt[3]{3x^{-2}}}$

37. $\dfrac{\sqrt[3]{625a^4b^7}}{\sqrt[3]{5ab^2}}$

38. $\dfrac{\sqrt[3]{40x^5y^2}}{\sqrt[3]{2xy}}$

39. $\sqrt[4]{\dfrac{243a^6b^{-13}c^{15}}{3a^2b^{-9}c^7}}$

40. $\sqrt[5]{\dfrac{160x^9y^{12}}{5xy^2}}$

41. $\sqrt{\sqrt{\sqrt{\sqrt{65,536}}}}$

42. $\dfrac{(-1)^{n+3}}{(-1)^{n+1}}$

43. $\left(\sqrt[12]{\sqrt[6]{a^{80}}}\right)^9 \left(\sqrt[12]{\sqrt[6]{a^{80}}}\right)^9$

44. $\left[\dfrac{(5x^ay^b)^4}{(-5x^ay^b)^3}\right]^5$

Express as a polynomial.

45. $(a-2)(a^2+2a+4)$

46. $(m+n)(m^2-mn+n^2)$

47. $(2a^2b^2-5ab)(3ab^2+2a^2b)$

48. $(2pq^2+10pq)(3p^2q-5pq)$

49. $(5x - 3y)^2$

50. $(a + 7t)^2$

51. $(2ab + 3bc)^2$

52. $(a - b)(a + b)(a^2 + b^2)$

53. $(4x^2 - 7xy)^2$

54. $(3x + t)(3x - t)(9x^2 + t^2)$

Factor the following expressions.

55. $-6x - 3x^2y + 9x$

56. $8a^2 - 2a + 12ab$

57. $x^2 - 4x - 21$

58. $x^2 - 11x + 30$

59. $(5x - 2)(x + 2) + (3x - 8)(5x - 2)$

60. $x(\sqrt{5} + \sqrt{7}) - y(\sqrt{5} + \sqrt{7})$

61. $14a^2b - 12ab^2$

62. $21b^2 + 18xb$

63. $xy - zy - xw + zw$

64. $ab + ac - mb - mc$

65. $2x^3 + x^2 - 8x - 4$

66. $x^3 - 2x^4 + 8y^3 - 16xy^3$

67. $529 - 324x^2$

68. $49 - 9y^2$

69. $8ax - 12x^2 + 4a^2x - 20x$

70. $6x - 3x - 10x^2$

71. $x^3 - 27$

72. $2z^3 - 16$

73. $8ax^2 + 24ax + 18a$

74. $2x^5 + 12x^3 + 18x$

75. $a^{2n} - b^{2n}$

76. $10x^{2y} + x^y - 3$

77. $36t^{2n} - 60t^n + 25$

78. $a^{6n} - b^{3n}$

79. $1 + \dfrac{x^{12}}{1000}$

80. $a^{2x} + 10a^x + 25 - 36y^{2x}$

Perform the indicated operations and write in lowest terms.

81. $\dfrac{x^2 + 5x + 6}{x^2 + 8x + 15} \cdot \dfrac{x^2 + 9x + 20}{x^2 + 6x + 8}$

82. $\dfrac{5x^2 + 30x + 45}{6x^2 - 24} \cdot \dfrac{3x^2 + 12x + 12}{10x^2 - 90}$

83. $\dfrac{t^3 - 1}{t^2 - 1} \div \dfrac{t^2 + t + 1}{t - 1}$

84. $\dfrac{a^3 + c^3}{4a - 4c} \div \dfrac{a^3 + a^2c^2}{2a^2 - 2c^2}$

85. $\dfrac{x^3 + 2x^2 + x + 2}{2x^4 + 6x^2 + 4} + \dfrac{2x^3 + 4x^2 + 6x + 12}{3x^4 - 3x^2 - 18}$

86. $\dfrac{x^2 + 2xy + y^2 + x + y}{x^2 + 6x + 4} + \dfrac{2x^2 + 3xy + y^2 + 2x + y}{x^2 + 3xy + 2y^2}$

87. $\dfrac{3x - 5}{2x + 1} - \dfrac{5 - 3x}{2x + 1}$

88. $\dfrac{7a + 3}{2a + 7} - \dfrac{4a + 7}{2a + 7}$

89. $\dfrac{3x - 7}{x - 3} + \dfrac{2x + 3y}{3 - x}$

90. $\dfrac{3a}{a - b} + \dfrac{2b + 4}{b - a} + \dfrac{6a - b}{a - b}$

91. $\dfrac{2t^2 - 3t + 5}{t^2 + 3t - 5} - \dfrac{t^2 - t + 2}{5 - 3t - t^2}$

92. $\dfrac{x^2 + 6x + 12}{x^2 - 3x - 8} + \dfrac{2x^2 - 6x + 4}{8 + 3x - x^2}$

93. $\dfrac{x - (3x + 2)}{x^2 - 3x - 8} + \dfrac{2x^2 - 6x + 4}{8 + 3x - x^2}$

94. $\dfrac{a - 4 - (5 - a)}{a^3 + 2a^2 + a} - \dfrac{2a - 4 - (a - 3)}{a^2 - 1}$

Write the following as rational expressions in lowest terms.

95. $\dfrac{ab^{-1} + ba^{-1}}{a^{-1} + b^{-1}}$

96. $\dfrac{2a^{-1} + 2y^{-1}}{(x + y)y^{-1}}$

97. $\dfrac{\dfrac{1}{x} + \dfrac{1}{y} + \dfrac{1}{z}}{\dfrac{1}{xy} + \dfrac{1}{yz}}$

98. $\dfrac{\dfrac{x}{y} + \dfrac{y}{x} + 2}{\dfrac{y + 1}{x} + \dfrac{y + 1}{y}}$

99. $\dfrac{\dfrac{1}{x + h} - \dfrac{1}{x}}{h}$

100. $\dfrac{\dfrac{1}{(x + h)^2} - \dfrac{1}{x^2}}{h}$

101. $\dfrac{\dfrac{1}{(x + h)^3} - \dfrac{1}{x^3}}{h}$

102. $\dfrac{a + 1 - 2a^{-1}}{a + 4 + 4a^{-1}}$

Rationalize the denominators of the following expressions.

103. $\dfrac{\sqrt{3x}}{\sqrt{4y}}$

104. $\dfrac{\sqrt[3]{2}}{\sqrt[3]{12}}$

105. $\dfrac{4}{\sqrt{5} + \sqrt{6}}$

106. $\dfrac{6}{\sqrt{3} - \sqrt{8}}$

107. $\dfrac{1}{\sqrt{a} - \sqrt{b}}$

108. $\dfrac{3}{\sqrt{x} + \sqrt{h}}$

109. $\dfrac{\sqrt{a} + \sqrt{2}}{\sqrt{a} - \sqrt{b}}$

110. $\dfrac{2\sqrt{x} + 3\sqrt{3}}{\sqrt{x} - \sqrt{3}}$

Rationalize the numerator. Many of these exercises are like those encountered in calculus.

111. $\dfrac{\sqrt{a + h} - \sqrt{a}}{h}$

112. $\dfrac{2\sqrt{x} + 3\sqrt{y}}{x}$

113. $\dfrac{\sqrt{a} + \sqrt{2}}{\sqrt{a} - \sqrt{2}}$

114. $\dfrac{2\sqrt{x} + 3\sqrt{3}}{\sqrt{x} - \sqrt{3}}$

♦ **Applications**

Convert to scientific notation.

115. 2,000,000,000 lbs (the amount of coffee consumed annually in the United States)

116. 0.0000000000667 (the universal gravitational constant, in N · m²/kg²)

117. 0.0000000000000000016 (1 electron volt, in ft-lb)

118. 5,878,000,000,000 (the distance, in miles, that light travels in one year; also known as 1 light-year)

119. 3.56E12

120. 1.89E − 13

Convert to decimal notation.

121. $\$1.2 \times 10^{10}$ (the amount of money spent on lotteries in 1989)

122. $\$5.8 \times 10^{9}$

123. 3.83×10^{-23} (the number of kilocalories in 1 electron volt)

124. 10^{-6} sec (the amount of time it takes a bullet cap to explode)

125. 5.78 E − 18

126. 7.02 E16

Solve the following problems and give the answers in scientific notation.

127. It takes light 2,200,000 years to travel to earth from the constellation Andromeda. How far is it from earth to Andromeda? (See Exercise 118.)

128. The sun is 93,000,000 miles from the earth. How long does it take for light to travel to the earth from the sun?

129. The moon travels about 1,500,078 miles in one orbit about the earth. About how far is the moon from the center of the earth? (Assume that the orbit of the moon is circular.)

130. The United States imported 6 million barrels of oil each day in a recent year. How many barrels of oil did it import in the entire year?

Solve the following problems.

131. A baseball diamond is a square with 90-ft sides. It is 60.5 ft from home plate to the pitcher's rubber. (See Figure 1.) About how far is it from the pitcher's rubber to second base? (Use the Pythagorean equation for right triangles, $a^2 + b^2 = c^2$ or $c = \sqrt{a^2 + b^2}$, where c is the length of the hypotenuse and a and b are the lengths of the legs.)

132. A circle with area 113.04 m² has a square inscribed as shown in Figure 2. Find the area of the square.

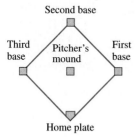

Figure 1

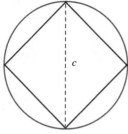

Figure 2

133. The time t, in seconds, it takes an object to fall s feet is given by $t = \frac{1}{4}\sqrt{s}$. TV station KTHI in Fargo, ND, has a transmitting antenna that is 2063 ft high. A tool is dropped from the top by a repairman. (See Figure 3.) How long will it take to fall to the ground?

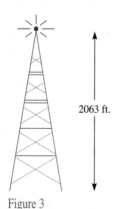

2063 ft.

Figure 3

The *escape velocity* V_0 of a projectile is the initial velocity needed for it to escape the gravitational pull of the planet. Escape velocity (in m/sec) is given by

$$V_0 = \sqrt{\frac{2GM}{R}}$$

where G is the universal gravitational constant, 6.67 $\times 10^{-11}$ newtons $\cdot$ m²/kg², M is the mass of the planet, in kg, and R is its radius, in meters.

134. The mass of the earth is 5.97×10^{24} kg and its radius is 6.38×10^6 m. What velocity is necessary for the rocket to escape the pull of the earth?

135. The mass of Mars is 6.42×10^{23} kg and its radius is 3.45×10^6 m. What is the escape velocity for a rocket leaving Mars?

136. Heron's formula for the area A of a triangle (see Figure 4) is given by

$$A = \sqrt{s(s-a)(s-b)(s-c)}$$

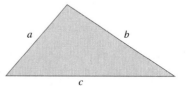

Figure 4

where $s = \frac{1}{2}(a+b+c)$ and a, b, and c are the lengths of the sides of the triangle. What is the area of a triangle with sides of lengths 5.5 ft, 7.2 ft, and 21.1 ft?

137. Suppose an amount of money P is invested in a savings account for time t, in years, at an interest rate of i compounded annually, and grows to an amount A. The interest rate is given by

$$i = \left(\frac{A}{P}\right)^{1/t} - 1$$

Suppose $2000 grows to an amount of $3500 in 10 years. What was the interest rate?

138. The speed of sound V (in m/sec) at temperature T (in degrees Celsius) is given by $V = 331.5(\frac{T+273}{273})^{\frac{1}{2}}$. On a warm day at the beach the temperature is 25°C. What is the speed of sound on such a day?

139. On a cold winter day the temperature is −3°C. What is the speed of sound on such a day?

✍ In Your Own Words

140. Discuss the merits of scientific notation. When would you find scientific notation more convenient that standard decimal notation? Less convenient?

141. Consult your calculator manual to determine how to enter a number in scientific notation. Describe the procedure in words.

142. Consult your calculator manual to determine the upper and lower limits on a number in scientific notation. Test these limits by attempting to enter a number outside them. What happens?

1.3 EQUATIONS

An **equation** is a statement that two algebraic expressions are equal. Examples of equations are

$$3x - 1 = 0, \qquad 2x^2 - 10x + 4 = 0, \qquad \sqrt{x^2 + 1} = -x + 1$$

A number a is called a **solution** of the equation provided that the equation results in a true statement when a is substituted for x. In this case, we say that a **satisfies the equation.** The process of determining the numbers that are solutions of equations is called **solving the equation.**

Various operations transform an equation into an **equivalent equation,** namely one with the same solution.

The following box summarizes the operations that lead to equivalent equations:

OPERATIONS THAT GENERATE EQUIVALENT EQUATIONS

1. Algebraic simplification, such as removing parentheses in applying the associative and distributive rules, combining like terms, performing indicated algebraic operations, applying the laws of exponents.
2. Interchange the two sides of the equation.
3. Add or subtract the same expression from both sides of the equation.
4. Multiply or divide both sides of the equation by the same nonzero number.

For example, the following equations are all equivalent to one another:

$$x^2 + x + 3 = 0$$
$$2x^2 + 2x + 6 = 0$$
$$x^2 = -x - 3$$
$$x^2 + 2x + 3 = x$$

An equation that is satisfied by all values of x is called an **identity.** Some examples of identities are $x + x = 2x$ and $x^3 = x \cdot x^2$ An equation that is satisfied by some but not all values of the variable is called a **conditional equation.**

A **linear equation** is an equation of the form $ax + b = 0$ where a and b are real numbers and $a \neq 0$.

A **quadratic equation** is an equation of the form

$$ax^2 + bx + c = 0$$

where a, b, and c are real numbers and $a \neq 0$. Two methods for solving such equations are factoring and the quadratic formula.

The method of factoring involves factoring the left side of the equation into a product of binomials. For instance, consider the equation

$$2x^2 + 3x + 1 = 0$$
$$(2x + 1)(x + 1) = 0$$
$$2x + 1 = 0 \quad \text{or} \quad x + 1 = 0$$

Solving each of these linear equations, we see that $x = -\frac{1}{2}$ or $x = -1$. Substituting these values of x into the original equation, we see that they are, indeed, solutions.

A second method for solving quadratic equations is based on the **quadratic formula,** which allows us to calculate the solutions of any quadratic equation in terms of the coefficients.

Quadratic Formula

The solutions of the quadratic equation

$$ax^2 + bx + c = 0, \quad a \neq 0$$

are given by the formula

$$x = \frac{-b \pm \sqrt{b^2 - 4ac}}{2a}$$

The use of the $\pm$ in the quadratic formula means that the formula gives one solution corresponding to the positive sign and one to the negative sign.

The expression $b^2 - 4ac$ is called the **discriminant** of the quadratic equation. The sign of the discriminant determines the nature of the solutions to the equation, as follows:

THE SOLUTIONS OF A QUADRATIC EQUATION

Consider the quadratic equation

$$ax^2 + bx + c = 0$$

1. If $b^2 - 4ac > 0$, then the equation has the two different real solutions:

$$x = \frac{-b + \sqrt{b^2 - 4ac}}{2a}, \qquad -x = \frac{-b - \sqrt{b^2 - 4ac}}{2a}$$

2. If $b^2 - 4ac = 0$, the equation has a single real solution:

$$x = -\frac{b}{2a}$$

3. If $b^2 - 4ac < 0$, then the equation has no real solution.

The three possibilities for the sign of the discriminant are illustrated in the following example.

➤ EXAMPLE 1
Quadratic Formula

Solve the following equations for real solutions using the quadratic formula.

1. $2x^2 - 3x - 4 = 0$
2. $9x^2 - 12x + 4 = 0$
3. $x^2 + x + 1 = 0$

Solution

1. From the equation,

$$a = 2, \qquad b = -3, \qquad c = -4$$

So the discriminant has the value

$$b^2 - 4ac = (-3)^2 - 4(2)(-4) = 41$$

Because the discriminant is positive, the equation has two real solutions. They are given by the quadratic formula as

$$x = \frac{-b \pm \sqrt{b^2 - 4ac}}{2a}$$

$$= \frac{-(-3) \pm \sqrt{41}}{2(2)}$$

$$= \frac{3 \pm \sqrt{41}}{4}$$

$$\approx 2.351 \quad \text{or} \quad -0.851$$

2. In this equation, $a = 9$, $b = -12$, $c = 4$, so the discriminant has the value

$$b^2 - 4ac = (-12)^2 - 4(9)(4) = 0$$

The equation has one real solution (two identical real solutions), given by the quadratic formula as

$$x = \frac{-b \pm \sqrt{b^2 - 4ac}}{2a}$$

$$= \frac{-(-12) \pm \sqrt{0}}{2(9)}$$

$$= \frac{12}{18}$$

$$= \frac{2}{3}$$

3. In this case, the discriminant has the value

$$b^2 - 4ac = 1^2 - 4(1)(1) = -3$$

Because the discriminant is negative, the equation has no real solutions.

➤ **EXAMPLE 2**
A Rational Equation

Determine all solutions of the following rational equation:

$$\frac{x}{x - 1} = \frac{6x - 5}{x + 15}$$

Solution

In this example, we may remove the rational expressions by multiplying both sides of the equation by the least common multiple of the denominators, namely by $(x - 1)(x + 15)$

$$x(x + 15) = (6x - 5)(x - 1)$$

$$x^2 + 15x = 6x^2 - 11x + 5$$

$$5x^2 - 26x + 5 = 0$$

We now have a quadratic equation that we can solve using the quadratic formula:

$$x = \frac{-(-26) \pm \sqrt{(-26)^2 - 4(5)(5)}}{2(5)}$$

$$= \frac{26 \pm \sqrt{576}}{10}$$

$$= \frac{26 \pm 24}{10}$$

$$= 5 \text{ or } \tfrac{1}{5}$$

The values $x = 5$ and $x = \frac{1}{5}$ are solutions to the equation. (If 1 or -15 had been a solution to the quadratic equation, we would have had to disqualify that value because it would have made one of the expressions in the equation undefined.)

➤ **EXAMPLE 3**
A Radical Equation

Determine all solutions of the following radical equation:

$$3x + 2\sqrt{x} = 1$$

Solution

Isolate the radical on one side of the equation and square both sides.

$$2\sqrt{x} = 1 - 3x$$

$$(2\sqrt{x})^2 = (1 - 3x)^2 \quad \text{Squaring both sides}$$

$$4x = 1 - 6x + 9x^2$$

$$9x^2 - 10x + 1 = 0$$

We now have a quadratic equation that we can solve with the aid of the quadratic formula.

$$x = \frac{-(-10) \pm \sqrt{(-10)^2 - 4(9)(1)}}{2(9)}$$

$$= \frac{10 \pm \sqrt{64}}{18}$$

$$= \frac{10 \pm 8}{18}$$

$$= 1 \text{ or } \tfrac{1}{9}$$

Check both of the possible solutions to determine whether they are in fact solutions. Substituting $x = \frac{1}{9}$ into the given equation yields

$$3 \cdot \tfrac{1}{9} + 2\sqrt{\tfrac{1}{9}} \stackrel{?}{=} 1$$

$$\tfrac{1}{3} + 2 \cdot \tfrac{1}{3} \stackrel{?}{=} 1$$

$$1 \stackrel{?}{=} 1 \;✔$$

So $\frac{1}{9}$ is a solution. However, if we substitute $x = 1$ into the given equation, we find that

$$3 + 2\sqrt{x} \stackrel{?}{=} 1$$

$$3 \cdot 1 + 2\sqrt{1} \stackrel{?}{=} 1$$

$$3 + 2 \stackrel{?}{=} 1$$

$$5 \stackrel{?}{=} 1 \quad \text{Not true}$$

The two sides are unequal, so $x = 1$ is a solution that arose from the squaring step, but is not a solution of the original equation. The given equation has the single solution $x = \frac{1}{9}$.

➤ **EXAMPLE 4**

Economics

The manufacturer of a very popular model of compact stereo disc system notes that the demand for the product fluctuates with the price. For any price p dollars, the market research department reports that monthly sales s may be predicted by the formula

$$s = 5000 - 10p$$

Suppose that the manufacturer wishes to achieve revenues of $600,000 in a given month. What should be the price of the compact disc players?

Solution

The revenues for the month are given by

$$\text{Revenue} = \text{Sales} \cdot \text{Price}$$

Using the given expression for sales in term of price, we have

$$\text{Revenue} = (5000 - 10p)p$$

To achieve a revenue of $600,000, the following condition must be fulfilled:

$$600,000 = (5000 - 10p)p$$

$$-10p^2 + 5000p - 600,000 = 0$$

$$p^2 - 500p + 60,000 = 0 \qquad \textit{Dividing by } -10$$

We can solve this equation by factoring, but it will take quite a while to work out. However, we can always solve a quadratic equation using the quadratic formula. Let's proceed directly to using the quadratic formula.

$$p = \frac{500 \pm \sqrt{(-500)^2 - 4(1)(60,000)}}{2(1)}$$

$$= \frac{500 \pm \sqrt{10,000}}{2}$$

$$= \frac{500 \pm 100}{2}$$

$$= 300 \quad \text{or} \quad 200$$

In other words, the company can achieve the desired revenues by choosing one of the two prices, $300 or $200, for its compact disc player. Because both prices yield the desired revenue, the company can choose the price on the basis of some criterion other than revenue. The lower price would probably be chosen to generate consumer goodwill.

➤ **EXAMPLE 5**

Architecture

An architect is designing a house featuring Palladian windows, as shown in Figure 1. To use passive solar heating, each window, consisting of four rectangular panes and a semicircular pane, must have 2500 square inches of glass. Determine the dimensions of each window.

Solution

The condition we seek to satisfy is

$$\begin{array}{c}\text{Area of semicircular} \\ \text{region}\end{array} + \begin{array}{c}\text{Area of rectangular} \\ \text{region}\end{array} = 2500$$

Figure 1

The semicircular region has diameter x and radius $\frac{x}{2}$. From the formula for the area of a circle, we have the following formula for a semicircle:

$$\tfrac{1}{2}\pi r^2 = \tfrac{1}{2}\pi \left(\frac{x}{2}\right)^2 = \frac{\pi}{8}x^2$$

The area of the entire rectangular region is

$$2x \cdot x = 2x^2$$

Therefore, x must satisfy the equation

$$\frac{\pi}{8}x^2 + 2x^2 = 2500$$

$$\left(\frac{\pi}{8} + 2\right)x^2 = 2500 \qquad \textit{Factoring out } x^2$$

$$x^2 = \frac{2500}{\left(\dfrac{\pi}{8}\right) + 2}$$

$$x = \pm \sqrt{\frac{2500}{\left(\dfrac{\pi}{8}\right) + 2}} \qquad \textit{Taking square roots}$$

Because x represents a physical length, we can ignore the negative value, so the width of the window is

$$x = \sqrt{\frac{2500}{\left(\dfrac{\pi}{8}\right) + 2}}$$

$$= \sqrt{\frac{20,000}{\pi + 16}} \text{ inches}$$

$$= 32.3241 \text{ inches (calculator)}$$

The height of the rectangular portion of the window is twice this amount, or 64.6482 inches. The circumference of the semicircle is equal to

$$\pi \cdot \text{radius} = \pi\frac{x}{2}$$

$$\approx (3.14159) \cdot \frac{32.3241}{2}$$

$$\approx 50.7745 \text{ inches}$$

Exercises 1.3

Determine the real solutions of the following quadratic equations.

1. $x^2 - 3 = 0$
2. $5x^2 = 0$
3. $x^2 + 4 = 0$
4. $4x^2 + 3 = 0$
5. $3t^2 = -15$
6. $2a^2 = -11$

7. $(x - 2)^2 = 1$
8. $(x + 5)^2 = 20$
9. $(3x + 1)^2 = 9$
10. $(2x - 3)^2 = 5$
11. $x^2 + 5x - 10 = 0$
12. $x^2 + 8x + 7 = 0$
13. $3x^2 + 4x = 1$
14. $8x^2 = 12x + 8$

15. $2x^2 - 4x + 5 = 0$

16. $7x^2 + 10x + 12 = 0$

17. $2x^2 + x = 4$

18. $2x^2 = 5x + 14$

19. $18x^2 = 10x$

20. $13t^2 + 6t = 0$

21. $6x^2 + 5x = 21$

22. $10x^2 - 3x - 1 = 0$

23. $2 + \dfrac{3}{x^2} = \dfrac{1}{x}$

24. $3 - \dfrac{1}{x} + \dfrac{6}{x^2} = 0$

25. $\dfrac{2x - 3}{3x + 5} = \dfrac{x + 7}{3x - 4}$

26. $\dfrac{5 - 2x}{6x + 1} = \dfrac{3 + 2x}{5 - x}$

27. $x^2 + \sqrt{2}x - \sqrt{3} = 0$

28. $\sqrt{5}x^2 + 4x = \sqrt{2}$

29. $(y + 5)(y - 4) = (2y + 1)(3y - 4)$

30. $(2a - 1)(3a + 5) = (a - 3)(4a - 7)$

31. $(-2t)^2 + (4t - 3)^2 + (-1 - t)^2 = 18$

32. $(-4a)^2 + (3a - 1)^2 = (5a - 7)^2$

33. $x - \dfrac{1}{x} = 17$

34. $m + \dfrac{2}{m} = 23$

35. $x^2 - \frac{5}{3}x - \frac{4}{3} = 0$

36. $0.8t^2 - 24t + 90 = 0$

37. $2t^2 + 0.3t = 0.4$

38. $7(p - 0.7) + \dfrac{0.7(3p - 0.5)}{p} = 0$

39. $5800v^2 + 6000v - 10,000 = 0$

40. $2400m^2 = 7700m - 12,000$

41. $x = 7\sqrt{x} + 30$

42. $x - 13\sqrt{x} + 12 = 0$

43. $\sqrt{x + 5} = 3$

44. $\sqrt{x + 10} = 4$

45. $\sqrt{x + 7} + 5 = x$

46. $\sqrt{y + 5} = y - 1$

47. $\sqrt{t - 2} = 3$

48. $11 = \sqrt{4 + m}$

Find all real solutions of the following equations.

49. $2x(x + 3)(5x - 4) = 0$

50. $(5x + 7)(x - 4)(x^2 - 5) = 0$

51. $x^3 - 1 = 0$

52. $8x^3 - 2 = 0$

53. $x^3 + 3x^2 - x - 3 = 0$

54. $3y^3 - 4y^2 - 12y + 16 = 0$

55. $12x^3 - 12x^2 = 0$

56. $4x^3 - 4 = 0$

57. $5x^4 - 5 = 0$

58. $2x - \dfrac{432}{x^2} = 0$

59. $50x - \dfrac{3200}{x^2} = 0$

60. $-(x - x^2)^{-2}(1 - 2x) = 0$

61. $\dfrac{m - \frac{1}{4}}{1 + \frac{1}{4}m} = \dfrac{\frac{3}{2} - m}{1 + \frac{3}{2}m}$

♦ Applications

62. The surface area of a right circular cylinder is 170 m². The radius is one-half the height. Find the dimensions of the cylinder. (Surface area S of a cylinder is given by $S = \pi dh$, where d is the diameter and h is the height.)

63. The bottom of an extension ladder is 8 ft from the wall on which it is leaning. As the ladder is extended to its maximum safe length, the top of the ladder moves up the wall 9 ft. To what length does the ladder extend?

64. The hypotenuse of a right triangle is 17 ft. One leg is 7 ft longer than the other. Find the length of the legs.

65. The manufacturer of a computer printer notes that demand for the product fluctuates with the price. If the price is p dollars, then the number of sales in a year s is given by the formula

$$s = 18,000 - 20p$$

Suppose the manufacturer wishes to achieve revenues of $4,000,000 in a given year. What should be the price of the printers?

66. The formula

$$S = -16t^2 + v_0t + h$$

gives the height S, in feet, after t seconds, of an object thrown upward from a height h, at an initial velocity v_0.

a. A ball is thrown upward at an initial velocity of 96 ft/sec from the top of the Sears Tower, which is 1456 ft tall. How long does it take the ball to reach the ground?

b. After what times will the ball be at a height of 1500 ft?

67. The formula
$$S = -16t^2 + v_0t + h$$
gives the height S, in feet, after t seconds, of an object thrown downward from a height h, at an initial velocity v_0.

a. An object is thrown downward from the roof of a 960 ft high building with an initial velocity of 64 ft/sec. How long does it take to reach the ground?

b. After what amount of time will the object be 500 ft off the ground?

68. The more chairs a factory produces, the less it costs to make each chair. It costs $25 per chair to make 100 chairs. For each additional chair, the average cost per chair for all chairs produced goes down $0.20. How many chairs will be produced when the total cost is $2470?

69. The more fruit trees planted per acre, the less the yield of fruit for each tree. When 30 trees are planted per acre, the yield is 50 bushels per tree. For each additional tree planted, the yield per tree decreases by 2 bushels. How many trees should be planted to yield 900 bushels per acre?

70. One can use the boiling point T of water to estimate the elevation H at which the water is boiling. The quadratic equation that can be used is
$$H = 1000(100 - T) + 580(100 - T)^2$$
where H = elevation, in meters, and T = boiling point, in degrees Celsius.

a. The boiling point is 99.4°C. Find the elevation.

b. What is the boiling point at sea level ($H = 0$)?

c. Denver, Colorado, is 5280 ft (1609 m) above sea level. At what temperature does water boil?

d. Mt. Elbert, outside Denver, is 14,431 ft (4400 m) above sea level and has a road that can be driven to the top. A family drives up Mt. Elbert and boils some water for a tailgate party. How much lower is the boiling point than in Denver?

71. There are many rules for determining the dosage of a medication for a child based on an adult dosage. The following are two such rules:

$$\text{Young's rule: } C = \frac{A}{A + 12}D$$

$$\text{Cowling's rule: } C = \frac{A + 1}{24}D$$

where A = the age of the child, in years, D = adult dosage, and C = child's dosage.
Warning: Do not apply these formulas without consulting a physician. These are estimates that may not apply to a particular medication.

a. An adult dosage of liquid medication is 5 cc (cubic centimeters). Find the child's dosage using each formula for a 3-year-old child.

b. At what ages A are the child's dosages the same in both formulas?

In the following problems, use the compound interest formula
$$A = P\left(1 + \frac{r}{n}\right)^{nt}$$
where P is the principal (original amount), r the annual interest rate, n the frequency of compounding, t the number of years, and A the final amount.

72. What is the interest rate if interest is compounded annually and $1200 grows to $1323 in 2 years?

73. What is the interest rate if interest is compounded annually and $1532 grows to $1737.63 in 2 years?

74. What is the interest rate if interest is compounded semiannually and $2500 grows to $2924.65 in 2 years?

75. $1400 is invested at 10.5 percent for 2 years, interest compounded quarterly. At what interest rate, compounded annually, would this same amount of money have to be invested in order to grow to the same amount in the same time?

76. The length of each side of an equilateral triangle is a. (See Figure 2.)
    a. Find the formula for the height h.
    b. Find the formula for the area A.
    c. Find the formula for the perimeter P.
    d. The area of an equilateral triangle is $25\sqrt{3}$ cm^2. What is the perimeter?

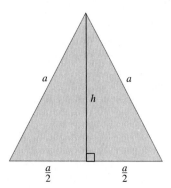

Figure 2

77. An open box is to be folded out of a 15 in. by 25 in. piece of a cardboard by cutting out square corners from the box. (See Figures 3 and 3A.) The area that results at the bottom of the box is 231 in^2. What is the length of each side of the square corners?

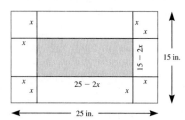

Figure 3

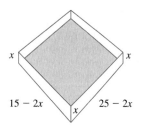

Figure 3A

78. A polygon (see Figure 4) with n sides has D diagonals, where D is given by
$$D = \frac{n(n-3)}{2}$$
    a. How many diagonals does an octagon have?
    b. A polygon has 54 diagonals. How many sides does it have?
    c. Can a polygon with 10 diagonals exist?

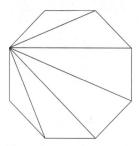

Figure 4

79. There are n teams in a sports league. If each team plays every other team twice, then the total number of games G the league plays is given by $G = n(n-1)$.
    a. There are 10 teams in a certain athletic conference. In their season each team plays every other team twice. What is the total number of games played in a season?
    b. Is it possible for a sports league to have 56 total games in its season when each team plays every other team twice?

✍ **In Your Own Words**

80. Find another formula for solving quadratic equations by rationalizing the numerator of the quadratic formula. What restriction is placed on this formula that does not apply to the quadratic formula?

▌ 1.4 INEQUALITIES

In many applications of algebra, particularly in calculus, inequalities are used extensively. In this section, we consider some of the most common types of inequalities that arise in these applications.

Linear Inequalities

An inequality may involve one or more variables. For example, the inequality
$$3x + 1 > -x - 4$$

involves the variable x. The inequality is valid for some values of x (such as $x = 4$) and is invalid for other values of x (such as $x = -5$). Determining the set of values of the variable for which an inequality holds is called **solving the inequality**.

In many respects, solving an inequality is analogous to solving an equation. The next few examples provide some practice in solving simple inequalities that involve linear expressions.

➤ **EXAMPLE 1**
Solving an Inequality

Solve the inequality $-3x + 1 > -x - 4$.

Solution

We perform allowable operations on the inequality as we would on an equation. First add quantities to both sides to bring all terms involving x to one side of the inequality and all constant terms to the other.

$$-3x + x > -4 - 1$$
$$-2x > -5$$

Now multiply both sides of the inequality by $-\frac{1}{2}$. Because $-\frac{1}{2}$ is negative, we must reverse the inequality sign.

$$(-\tfrac{1}{2})(-2x) < (-\tfrac{1}{2})(-5)$$
$$x < \tfrac{5}{2}$$

At each step, the inequality is replaced by an equivalent one, so the original inequality is satisfied precisely by those values of x satisfying $x < \frac{5}{2}$. A description of all the solutions is given using set builder notation as

$$\{x : x < \tfrac{5}{2}\}$$

This is called the **solution set** of the inequality.

➤ **EXAMPLE 2**
Merchandising

A local sporting goods store earns a profit of $30 per tennis racket. As part of a promotion, it spends $5000 on advertising. How many tennis rackets must be sold if the profit is to be at least $10,000?

Solution

Let q denote the quantity of tennis rackets sold. Then the resulting profit is equal to $30q - 5000$. The condition that the profit must be at least $10,000 may be expressed by the inequality

$$\text{Profit} \geq 10,000$$
$$30q - 5000 \geq 10,000$$

We may solve this last quantity for q:

$$30q \geq 15,000$$
$$q \geq 500$$

In other words, if the profit is to be at least $10,000, the quantity sold must be at least 500.

Many applications involve pairs of inequalities of the form

$$a < b \quad \text{and} \quad b < c,$$

which can also be written as

$$a < b < c$$

For instance, the interest rate I quoted on 30-year fixed-rate mortgages by different banks may be greater than or equal to 9.5 percent and less than 10.75 percent. This corresponds to the double inequality

$$0.095 \le I < 0.1075$$

The solution set of a double inequality consists of the numbers that satisfy both inequalities. That is, the solution set consists of the numbers in common to the two solution sets of the two inequalities.

For example, the solution set of the double inequality $-1 < x < 3$ consists of the numbers in common to the two solution sets $\{x : x > -1\}$ and $\{x : x < 3\}$.

For two sets, A and B, the set that consists of the elements in common is called the **intersection** of A and B and is denoted $A \cap B$. In terms of intersections, the solution set of the double inequality may be written as

$$\{x : x > -1\} \cap \{x : x < 3\}.$$

To solve a double inequality, we must solve two independent inequalities and form the intersection of the respective solution sets. The following example illustrates the procedure.

➤ **EXAMPLE 3**
A Three-Termed Inequality

Solve the inequality

$$-5x < 3x + 1 < 2x + 4$$

Solution

We must determine the solution sets of the inequalities

$$-5x < 3x + 1 \quad \text{and} \quad 3x + 1 < 2x + 4$$

The first inequality may be solved as follows:

$$-5x < 3x + 1$$
$$-5x - 1 < 3x$$
$$-1 < 3x + 5x$$
$$-1 < 8x$$
$$-\tfrac{1}{8} < x$$

The solution set of the first inequality is thus

$$\{x : x > -\tfrac{1}{8}\}$$

The second inequality may be solved as follows:

$$3x + 1 < 2x + 4$$

$$3x - 2x + 1 < 4$$

$$x < 4 - 1$$

$$x < 3$$

The solution set of this inequality is thus

$$\{x : x < 3\}$$

The solution set of the given double inequality consists of the set of numbers in common to the two solution sets just found. That is, the intersection

$$\{x : x > -\tfrac{1}{8}\} \cap \{x : x < 3\}$$

This intersection consists of all numbers x that are greater than $-\frac{1}{8}$ and less than 3, so the solution set is

$$\{x : -\tfrac{1}{8} < x < 3\}$$

In solving inequalities, it is often useful to break the solution into two or more cases. A solution must arise as a solution in one of the cases. If A and B are sets, then the set that consists of the elements in either A or B or both is called the **union** of A and B, and is denoted

$$A \cup B$$

The next example provides an illustration of an inequality whose solution may be expressed as a union.

➤ **EXAMPLE 4**
Polynomial Inequality

Solve the inequality

$$(x + 1)(x + 5) > 0$$

Solution

The inequality requires that the product of two factors be positive. Either both factors must be positive or both factors must be negative. It is simplest to determine this set using a picture like Figure 1. We draw a number line for each factor and draw $+$ where the factor is positive and $-$ where the factor is negative. On a third coordinate axis, we indicate where both factors have the same sign. The solution set is

$$\{x : x > -1\} \cup \{x : x < -5\}$$

This set consists of all values of x that are either greater than -1 or less than -5.

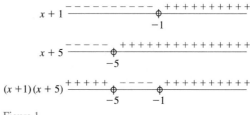

Figure 1

➤ **EXAMPLE 5**
Rational Inequality

Solve the following inequality.

$$\frac{3x + 7}{4x - 5} > 0$$

Solution

In order for the inequality to hold, the numerator and denominator must have the same sign. Reason geometrically as in the previous example (see Figure 2). The solution is the set

$$\{x : x > \tfrac{5}{4}\} \cup \{x : x < -\tfrac{7}{3}\}$$

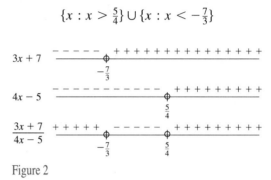

Figure 2

Warning: You might be tempted to solve the inequality by multiplying both sides by $4x - 5$ to clear the denominator. However, this is not a useful way to proceed because the inequality sign may face in either direction after multiplying, depending on the sign of $4x - 5$.

Intervals

The subset of the number line determined by a single or double inequality is called an **interval**. In the preceding examples, we determined solution sets of inequalities that turned out to be intervals, intersections of intervals, or unions of intervals. We classify intervals on the basis of whether their endpoints are included, according to the following definition.

Definition 1
Intervals

Let a and b be real numbers. The **open interval** (a, b) is the set of real numbers

$$(a, b) = \{x : a < x < b\}$$

The **closed interval** $[a, b]$ is the set of real numbers

$$[a, b] = \{x : a \le x \le b\}$$

The **half-open intervals** $[a, b)$ and $(a, b]$ are defined, respectively, as the sets of real numbers

$$[a, b) = \{x : a \le x < b\}$$

$$(a, b] = \{x : a < x \le b\}$$

An interval can be represented as a line segment on the number line. An open interval includes neither endpoint of the segment, a closed interval includes both endpoints, and a half-open interval includes one endpoint. The various types of intervals are illustrated in Figure 3. Note that a bracket at the end of a line segment indicates that the endpoint is included. A parenthesis at the end of a line segment indicates that the endpoint is not included.

Intervals that extend indefinitely to the right or left on the number line are called **infinite intervals.** The solution sets of the inequalities $x > a$ and $x < a$ are the open infinite intervals denoted (a, ∞) and $(-\infty, a)$, respectively. The solution sets of the inequalities $x \geq a$ is $[a, \infty)$ and $x \leq a$ is $(-\infty, a]$. The symbol ∞, or $+\infty$, is read "infinity." The symbol $-\infty$ is read "negative infinity." The infinity symbols indicate intervals extending "infinitely" to the right and left, respectively. The various types of infinite intervals are shown in Figure 4.

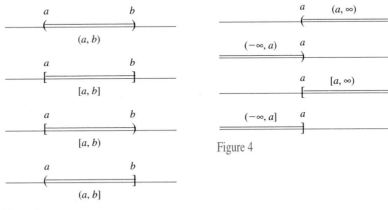

Figure 3

Figure 4

> **EXAMPLE 6**
Sketching Intervals

Sketch the graphs of the following intervals.

1. $(-1, 3)$
2. $(-\infty, 5)$
3. $[0, 1]$
4. $[-2, \infty)$

Solution

1. This is an open interval, so neither endpoint is included. See Figure 5.
2. This is an open infinite interval. The right endpoint is not included and the interval extends indefinitely to the left. See Figure 6.
3. This is a closed interval. Both endpoints are included. See Figure 7.
4. This is a closed infinte interval. The left endpoint is included and the interval extends indefinitely to the right. See Figure 8.

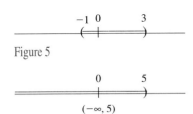

Figure 5

Figure 6

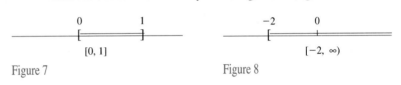

[0, 1]

Figure 7

[-2, ∞)

Figure 8

> **EXAMPLE 7**
Graphing Intersections and Unions

Graph the sets on the number line described as follows:

1. $(-1, 3) \cap (2, 4)$
2. $[0, 5] \cup [5, 7]$
3. $(0, 1) \cup (2, \infty)$

Solution

Figure 9

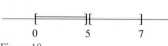

Figure 10

Figure 11

1. The graph consists of the points in common to the intervals $(-1, 3)$ and $(2, 4)$. In Figure 9, we have sketched both intervals. The points in common are those that are shaded twice. These points comprise the interval $(2, 3)$. Note that the endpoints are not included because 2 does not belong to the interval $(2, 4)$ and 3 does not belong to $(-1, 3)$.

2. The graph consists of points that belong to either of the intervals $[0,5]$ and $[5,7]$. In Figure 10, we have shaded each of these intervals. The graph of the union consists of the points that are shaded at least once. It includes the number 5, which is shaded twice, so the indicated union is the interval $[0,7]$.

3. The graph consists of points that belong to either of the intervals $(0, 1)$ and $(2, +\infty)$. Figure 11 shows these intervals as shaded. The union consists of the points that are shaded at least once.

Polynomial Inequalities

An interesting application of intervals arises in solving polynomial inequalities, that is, inequalities of the form

$$p(x) \geq 0 \quad \text{or} \quad p(x) \leq 0$$

where $p(x)$ is a polynomial in the variable x. Let's consider the particular case in which the polynomial may be factored into factors of the form

$$p(x) = A(x - a_1)(x - a_2) \cdots (x - a_n)$$

where the real numbers $a_1, a_2, \ldots, a_n$ are arranged in increasing size

$$a_1 \leq a_2 \leq \cdots \leq a_n$$

Associated with such a polynomial, we have the collection of intervals

$$(-\infty, a_1), (a_1, a_2), \ldots, (a_{n-1}, a_n), (a_n, \infty)$$

These intervals are called **test intervals** for $p(x)$.

Each factor is of constant sign in each test interval. That is, we have the following result.

Constant Sign Theorem

The polynomial $p(x) = A(x - a_1)(x - a_2) \cdots$ is of constant sign in each interval $(-\infty, a_1), (a_1, a_2), \ldots, (a_n \infty)$. That is, if c and d are two numbers in the same interval, then the numbers $p(c)$ and $p(d)$ are either both positive or both negative.

According to the theorem, if one number in a test interval is a solution to the inequality, then so is every other number in the interval. Therefore, the solution set of a polynomial inequality consists of a union of test intervals. To determine which test intervals to include in the union, it suffices to test a single number in each interval. The procedure is illustrated by the following example.

➤ EXAMPLE 8
Solving a Polynomial Inequality

Solve the inequality $2x^3 + x^2 - x < 0$. Graph the solution.

Solution

We begin by factoring the polynomial

$$2x^3 + x^2 - x = x(2x^2 + x - 1) = x(2x - 1)(x + 1)$$

$$= 2(x - (-1))(x - 0)(x - \tfrac{1}{2})$$

In the last factorization above, the factors have been written so that the coefficient of x in each factor is 1 and the constants in the linear factors are in increasing order $(-1 < 0 < \frac{1}{2})$. The test intervals for this polynomial are:

$$(-\infty, -1),\ (-1, 0),\ \left(0, \tfrac{1}{2}\right),\ \text{and} \left(\tfrac{1}{2}, \infty\right)$$

We choose a number in each to act as a test case for that interval: -2, $-\frac{1}{2}$, $\frac{1}{4}$, and 1. We determine the sign of the polynomial at each test number using the factored form $x(2x-1)(x+1)$. For example, at $x = -2$, all three factors are negative. So the product is $(-)(-)(-) = -$. The data is summarized in the following table:

Interval	Test point x	Sign of $2x^3 + x^2 - x$
$(-\infty, -1)$	-2	$-$
$(-1, 0)$	$-\frac{1}{2}$	$+$
$(0, \frac{1}{2})$	$\frac{1}{4}$	$-$
$(\frac{1}{2}, \infty)$	1	$+$

The first and the third intervals have test points that yield negative values for the polynomial, so the solution consists of the union of the first and third intervals:

$$(-\infty, -1) \cup \left(0, \tfrac{1}{2}\right)$$

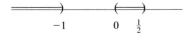

Figure 12

The solution is graphed in Figure 12.

The method used in Example 8 may be used to solve any inequality of the form $p(x) > 0$ where $p(x)$ is a product of linear factors. Note that if we are given an inequality of the form $p(x) > c$ where c is nonzero, then, in order to apply the method, we must transform the inequality into one in which the right side is 0 and the left is a product of linear factors.

Inequalities Involving Absolute Values

Let's now turn to inequalities that involve absolute values. Here are some examples of such inequalities:

$$|2x + 5| < 4$$
$$|x| > 6$$

The key to solving inequalities involving absolute values is the following fundamental result.

FUNDAMENTAL ABSOLUTE VALUE INEQUALITY

Let a be a nonnegative real number. Then x satisfies the inequality $|x| < a$ if and only if $-a < x < a$.

Proof   Recall that $|x|$ is the distance of x from the origin. The equality $|x| < a$ says that the distance of x to the origin is less than that of a. However, the numbers satisfying this property are exactly those that satisfy the inequality

$$-a < x < a$$

The next several examples illustrate how the fundamental absolute value inequality may be applied.

➤ **EXAMPLE 9**
Absolute Value Inequality

Solve the inequality $3|2 - 5x| + 1 < 10$.

Solution

We may simplify the inequality by subtracting 1 from each side and then dividing each side by 3. This gives the inequality

$$|2 - 5x| < 3$$

Use the fundamental absolute value inequality to convert this last inequality to a double inequality

$$-3 < 2 - 5x < 3$$
$$-5 < -5x < 1 \qquad \textit{Subtracting 2 from all three terms}$$
$$1 > x > -\tfrac{1}{5} \qquad \textit{Multiplying by } -\tfrac{1}{5} \textit{ and reversing the inequality}$$
$$-\tfrac{1}{5} < x < 1$$

The solution set of the given inequality is the interval $\left(-\tfrac{1}{5}, 1\right)$.

➤ **EXAMPLE 10**
An Inequality From Calculus

In calculus, one finds inequalities of the form $|x - a| < \delta$, where a is a real number and δ is positive. Solve this inequality and graph its solution.

Solution

The given inequality is equivalent to the double inequality

$$-\delta < x - a < \delta$$
$$-\delta + a < x < \delta + a$$

In other words, the solutions of the inequality consist of all x in the interval $(a - \delta, a + \delta)$. This interval is illustrated in Figure 13.

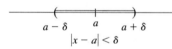

Figure 13

In the preceding examples, we considered inequalities in which an absolute value was less than a constant. In some applications, it is necessary to consider inequalities of the form

$$|x| > a$$

Where a is a positive number. This inequality states that the distance from x to the origin is greater than a. These values of x fall into two categories: Either $x > a$ or $x < -a$. That is, the solution set of the inequality is

$$\{x : x < -a\} \cup \{x : x > a\}$$

The graph of this solution set is shown in Figure 14.

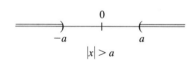

Figure 14

➤ **EXAMPLE 11**
Solution Set a Union

Solve the inequality and graph the solution set.

$$|4x - 1| > 1$$

Solution

The inequality may be replaced by the two inequalities

$$4x - 1 > 1 \quad \text{or} \quad 4x - 1 < -1$$

Solving these inequalities, we have

$$
\begin{array}{ll}
4x - 1 > 1 & 4x - 1 < -1 \\
4x > 1 + 1 & 4x < -1 + 1 \\
4x > 2 & 4x < 0 \\
x > \frac{1}{4} \cdot 2 & x < 0 \\
x > \frac{1}{2} &
\end{array}
$$

The solution set consists of those values of x that satisfy either of the above inequalities.

$$
\{x : x > \tfrac{1}{2}\} \cup \{x : x < 0\} = (-\infty, 0) \cup \left(\tfrac{1}{2}, \infty\right)
$$

The graph of the solution set is sketched in Figure 15.

Figure 15

We may summarize the two types of absolute value inequalities as follows:

INEQUALITIES INVOLVING ABSOLUTE VALUE			
Inequality	Solution	Distance interpretation	Solution in interval
$|x| < a$	$-a < x < a$	The set of x whose distance from the origin is less than a	$(-a, a)$
$|x| > a$	$x > a$ or $x < -a$	The set of x whose distance from the origin is greater than a	$(-\infty, -a) \cup (a, \infty)$

Exercises 1.4

Solve the following inequalities and write the solution set in interval notation. Sketch the solution set.

1. $-2 \le x < 3$
2. $5 < t \le 7$
3. $-4 \le x \le 5$
4. $-5 \le t \le -3$
5. $-2 \le x < 3$
6. $5 \le t < 7$
7. $x \ge -2$
8. $t < 3$

Describe each of the following using set notation.

9. $[-1, 3)$
10. $(-3, -1)$
11. $(-2, \infty)$
12. $(-\infty, 5]$
13. $(-\infty, 2]$
14. $[5, \infty)$

Solve the following inequalities and write the solution sets in interval notation. Do not use a graphing calculator.

15. $-2 \le 5x + 3$
16. $3t + 2 < 17$
17. $-3 - 4x > 0$
18. $6 - 2t < 0$
19. $x + 3 \ge 2x - 1$
20. $5x - 8 < -2(x - 3)$
21. $3 - x \le 4x + 7$
22. $10x - 3 \ge 13x - 8$
23. $3 - (4 - x) \le 3x - (x + 5)$
24. $3 - x - x^2 \ge 2x - x(x + 1)$
25. $-0.3x > 0.2x + 1.3$
26. $-2.6x < 1.2 + 0.04x$
27. $-8 \le 2x \le 10$

28. $-9 < -3t \le -6$

29. $-5 \le 8 - 3x \le 5$

30. $-7 \le 2x - 3 \le 7$

31. $-11 < 2x - 1 \le -5$

32. $3 \le 4y - 3 < 19$

33. $\frac{2}{5}x - 3 \le \frac{4}{5}x - 13$

34. $\frac{3}{5}x \ge \frac{1}{3}x - \frac{1}{10}$

35. $-6 < \dfrac{2 - 3x}{2} \le 13$

36. $1 \le \dfrac{4 + 3t}{4} < 9$

37. $(3x - 2)^2 < 3(x - 2)(3x + 4)$

38. $4(x - 2)(x + 3) \ge (2x + 3)(2x - 3)$

39. $\dfrac{1}{2x + 3} < 0$

40. $\dfrac{-2}{4 - 5x} > 0$

41. $(x + 5)(x - 3) > 0$

42. $(x + 4)(x - 1) < 0$

43. $x^2 - 7x + 12 \ge 0$

44. $x^2 - x \le 20$

45. $t^2 < 25$

46. $12x^2 > 4$

47. $36x^2 - 24x < 0$

48. $3x^2 + 6x + 3 > 0$

49. $(x + 3)(x - 2)(x + 1) > 0$

50. $(x - 4)(x - 5)(x + 2) \le 0$

51. $x^3 + 3x^2 - 10x \le 0$

52. $x^4 - 7x^3 + 12x^2 > 0$

53. $t^3 + 2t^2 - t - 2 < 0$

54. $2x^3 - x^2 - 8x + 4 \ge 0$

55. $\dfrac{x - 3}{x + 1} > 0$

56. $\dfrac{x + 5}{x - 2} < 0$

57. $\dfrac{2x - 3}{x - 4} \le 0$

58. $\dfrac{3 - 2x}{3x + 5} \ge 0$

59. $\dfrac{(x + 1)(x - 2)}{x + 4} < 0$

60. $\dfrac{x + 7}{(x - 3)(x - 2)} > 0$

61. $12x^2 > 0$

62. $x^2 + 1 < 0$

63. $|x| \le 23$

64. $|x| > 23$

65. $|x| > 14$

66. $|x| < 14$

67. $|x - 2| > 4$

68. $|x + 9| \ge 7$

69. $|6 - 3x| < 9$

70. $\left|1 - \dfrac{2}{3}x\right| \le 7$

71. $\left|\dfrac{3x - 2}{4}\right| \le 5$

72. $\left|\dfrac{6 + 7x}{4}\right| < 26$

73. $|3 - 2x| \ge 1 + 2x$

74. $4 \le |2 - x| \le 5$

75. $|x - 1| > |x - 5|$

76. $|2x - 1| < -2$

77. $|m + 3| + |m - 3| < 8$

78. $|y - 3| + |y| \le 4$

79. $\left|\dfrac{x + 1}{2x + 3}\right| < 0$

80. $|x|^2 - 2|x| + 1 \le 9$

81. $|t|^2 + 6|t| + 9 > 25$

82. $\left|\dfrac{x + 3}{x - 2}\right| < 1$

83. $\left|\dfrac{2x - 3}{3x + 4}\right| \ge 2$

84. $|t^2 - 5t - 1| > 5$

85. $|m^2 + m - 4| \le 2$

86. $|x^2 - 1| \le 3$

Determine the values of p at which the quadratic equation will have at least one real number solution.

87. $px^2 + 4x + 8 = 0$

88. $12x^2 + px + 3 = 0$

Solve the inequalities. Write interval notation for the solution sets.

89. $-12x + 12x - x^2 < 0$

90. $\dfrac{(x^2 + 1) - x(2x)}{(x^2 + 1)^2} > 0$

91. $(x - 2)^3 \cdot 2(x + 1) + (x + 1)^2 \cdot 3(x - 2)^2 > 0$

92. $\dfrac{(x - 1)^2(2x - 2) - (x^2 - 2x)(2x - 2)}{(x - 1)^4} > 0$

93. $\dfrac{\dfrac{2x}{3}}{(x^2 - 8)^{2/3}} > 0$

94. Prove or disprove: For any values of a and b, if $a < b$ then $a^2 < b^2$.

♦ **Applications**

95. The base of a triangle is 12 in. For what values of height h is the area 200 in.^{2} or less?

96. A car is traveling at a speed of 65 mph. For what times t can the car travel so that it travels at least 700 miles?

97. A student receives grades of 67, 83, and 78 on three math tests. What scores S can the student score on the fourth test and obtain an average greater than 70 and less than 80?

98. A pitcher's earned run average, ERA, is given by ERA $= 9R/I$, where R is the number of earned runs given up pitching I innings. A pitcher gives up 50 earned runs. What number of innings can be pitched so the pitcher's ERA is 3.00 or less?

99. The following are expressions for total revenue and total cost from the sale of x units of a product. $R(x) = 80x$ and $C(x) = 20x + 10,000$. Total profit $P(x)$ is given by $P(x) = R(x) - C(x)$.

    a. Find the number of units x for which $P(x) > 0$.

    b. Find the number of units x for which $P(x) \geq 0$.

100. The formula $S = -16t^2 + 64t + 960$ gives the height S of an object thrown upward from the roof of a 960-ft building at an initial velocity of 64 ft/sec.

    a. For what times t will the height be greater than 992 ft?

    b. For what times t will the height be less than 960 ft?

101. A polygon with n sides has D diagonals, where D is given by
$$D = \frac{n(n - 3)}{2}$$
For what values of n is $20 \leq D \leq 119$?

102. There are n teams in a sports league. If each team plays every other team twice, then the total number of games G played is given by $G = n(n - 1)$. For what values of n is $G \geq 380$?

103. A company's profits for a certain year are estimated to satisfy the inequality
$$|P - \$3,000,000| \leq \$225,000$$
Determine the interval over which the company's profits will vary.

104. During a strange illness, a patient's temperature T varied according to the inequality
$$|T - 98.6°| \leq 2°$$
Determine the interval over which the patient's temperature varied.

105. A weight is bobbing on a spring (see Figure 16) in such a way that its distance d, in inches, from the top satisfies the inequality
$$|d - 10| < 4$$
Determine the interval over which the distance varies.

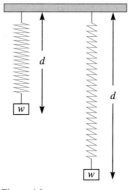

Figure 16

✍ **In Your Own Words**

106. Describe the procedure for solving a polynomial inequality using test intervals.

107. Describe the set of simultaneous solutions to three given inequalities.

1.5 TWO-DIMENSIONAL COORDINATE SYSTEMS

In Section 1.1 we reviewed the notion of a one-dimensional coordinate system. We now introduce two-dimensional coordinate systems, which allow us to assign coordinates to points in a plane. Each point of the plane is identified with an ordered pair (a, b) of real numbers.

Coordinate Systems in a Plane

Start with two perpendicular lines in a plane. On each line, introduce a one-dimensional coordinate system. This pair of one-dimensional coordinate systems is called a **two-dimensional Cartesian coordinate system** in the plane (see Figure 1).

The perpendicular lines are called the **axes**. Usually, the horizontal axis is labeled the x-axis and the vertical axis the y-axis.

The point at which the axes intersect is called the **origin**, and is customarily denoted by the letter O (see Figure 2).

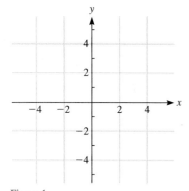

Figure 1

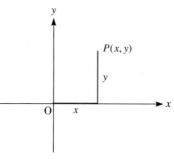

Figure 2

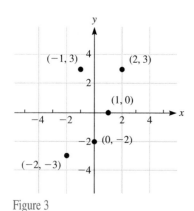

Figure 3

In a Cartesian coordinate system, each point P may be identified with a pair of real numbers (x, y). P is plotted by starting at the origin and moving horizontally a signed distance x and vertically a signed distance y, where signed distance means that movements to the right correspond to positive values of x, and movements to the left correspond to negative values of x; similarly, vertical movements upward correspond to positive values of y, and vertical movements downward correspond to negative values of y. The pair of numbers (x, y) is called the **coordinate representation** of P or the **coordinates** of P. The number x is called the **x-coordinate**, or **abscissa**, of P. The number y is called the **y-coordinate**, or **ordinate**, of P.

NOTE: The coordinate representation (x, y) should not be confused with the open interval from x to y. Although the same notation is used, it will be clear from the context which is meant.

Figure 3 shows the coordinate representations of various points in the plane.

Here are the coordinates of some special points:

1. Origin has both coordinates 0 : (0, 0)
2. Points on the x-axis have y-coordinate 0 : $(x, 0)$
3. Points on the y-axis have x-coordinate 0 : $(0, y)$

The coordinate axes divide the plane into four regions called **quadrants**. The quadrants are numbered counterclockwise starting from the positive x-axis, as shown in Figure 4. In a particular quadrant, the abscissa and ordinate are of constant sign.

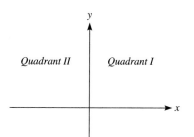

Figure 4

Quadrant	x-coordinate	y-coordinate
I	positive	positive
II	negative	positive
III	negative	negative
IV	positive	negative

The Distance Formula

We may compute the distance between two points in terms of their coordinates using the following formula:

DISTANCE FORMULA

Let $P(x_1, y_1)$ and $Q(x_2, y_2)$ be two points. The distance between them is given by

$$d(P, Q) = \sqrt{(x_2 - x_1)^2 + (y_2 - y_1)^2}$$

See Figure 5.

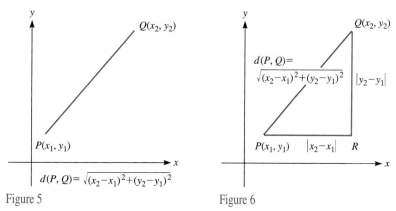

Figure 5   Figure 6

Proof Draw the line segment connecting P and Q and put it in a triangle with sides parallel to the horizontal and vertical axes, as shown in Figure 6. From examining the Figure 6, we see that the horizontal side has length $|x_2 - x_1|$ and that the vertical side has length $|y_2 - y_1|$. Because the triangle PQR is a right triangle, we may apply the Pythagorean theorem to compute the length of the hypotenuse:

$$\begin{aligned}
d(P, Q) &= \sqrt{(RP)^2 + (RQ)^2} \\
&= \sqrt{|x_2 - x_1|^2 + |y_2 - y_1|^2} \\
&= \sqrt{(x_2 - x_1)^2 + (y_2 - y_1)^2}
\end{aligned}$$

♦

> **EXAMPLE 1**
Calculating Distances

Determine the closest distance between $(-1, 4)$ and the x axis.

Solution

The point on the x-axis closest to $(-1, 4)$ is the point on the axis with the same x-coordinate, namely $(-1, 0)$. The desired distance is shown in Figure 7 and by the distance formula equals

$$\sqrt{(-1 - (-1))^2 + (4 - 0)^2} = 4$$

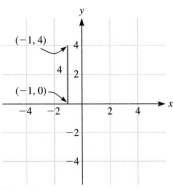

Figure 7

The Midpoint Formula

In many problems, especially those arising in geometry and physics, it is necessary to determine the coordinates of the midpoint of a line from the coordinates of the endpoints.

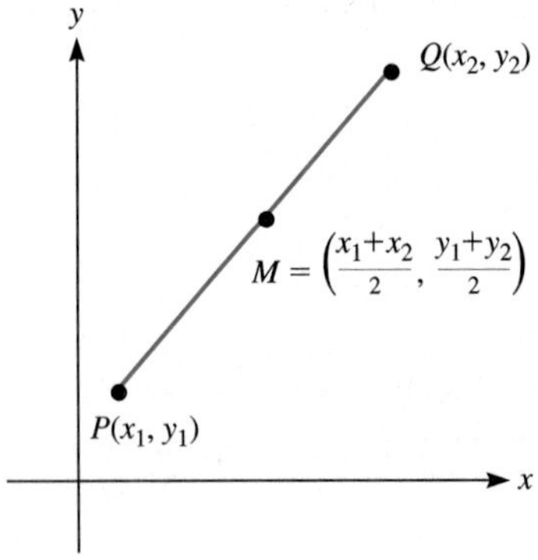

$M = \left(\frac{x_1+x_2}{2}, \frac{y_1+y_2}{2}\right)$

$Q(x_2, y_2)$

$P(x_1, y_1)$

Figure 8

MIDPOINT FORMULA

The midpoint of the line segment with endpoints $P(x_1, y_1)$ and $Q(x_2, y_2)$ is the point

$$M\left(\frac{x_1 + x_2}{2}, \frac{y_1 + y_2}{2}\right)$$

See Figure 8.

Proof Let's prove the result by establishing the equalities:

$$d(M,P) = \tfrac{1}{2}d(P,Q)$$

$$d(M,Q) = \tfrac{1}{2}d(P,Q)$$

These equalities state that M is equidistant from P and Q. Moreover, we can deduce that

$$d(P,M) + d(M,Q) = \tfrac{1}{2}d(P,Q) + \tfrac{1}{2}d(P,Q)$$

$$= d(P,Q)$$

That is, the distance from P to M plus the distance from M to Q equals the distance from P to Q. From an elementary result in geometry, this implies that M lies on the line segment $\overline{PQ}$. We then apply the distance formula:

$$d(M,P) = \sqrt{\left(\frac{x_1 + x_2}{2} - x_1\right)^2 + \left(\frac{y_1 + y_2}{2} - y_1\right)^2}$$

$$= \sqrt{\left(\frac{x_2 - x_1}{2}\right)^2 + \left(\frac{y_2 - y_1}{2}\right)^2}$$

$$= \sqrt{\frac{(x_2 - x_1)^2 + (y_2 - y_1)^2}{4}}$$

$$= \tfrac{1}{2}\sqrt{(x_2 - x_1)^2 + (y_2 - y_1)^2}$$

$$= \tfrac{1}{2}d(Q,P) \qquad \textit{(Distance formula)}$$

$$= \tfrac{1}{2}d(P,Q) \qquad \textit{(the distance from P to Q is equal to the distance from Q to P)}$$

This proves the equality $d(M,P) = \tfrac{1}{2}d(P,Q)$. The proof of the equality $d(M,Q) = \tfrac{1}{2}d(P,Q)$ is similar and will be omitted. ◆

➤ **EXAMPLE 2**
Calculating a Midpoint

Determine the midpoint of the line segment with endpoints $P(4, 1)$ and $Q(8, -5)$.

Solution

By the midpoint formula, the coordinates of the midpoint of the line segment are

$$\left(\frac{x_1 + x_2}{2}, \frac{y_1 + y_2}{2}\right) = \left(\frac{4 + 8}{2}, \frac{1 + (-5)}{2}\right) = (6, -2)$$

Exercises 1.5

Plot the following points on a Cartesian coordinate system.

1.
   a. $(3, 2)$
   b. $(3, -1)$
   c. $(0, 2)$
   d. $(-3, 0)$

2.
   a. $(0, -4)$
   b. $(-2, 3)$
   c. $(5, 0)$
   d. $(3, 4.5)$

3. For each point in Exercise 1, determine the quadrant in which it lies or state the axis it belongs to.

4. For each point in Exercise 2, determine the quadrant in which it lies or state the axis it belongs to.

Determine the distance between the points in Exercises 5–16.

5. $(2, 2)$ and $(-4, -3)$
6. $(6, 10)$ and $(-1, 5)$
7. $(0, -5)$ and $(2, -3)$
8. $(-3, -3)$ and $(3, 3)$
9. $(-2, 5)$ and $(4, 5)$
10. $(-3, 4)$ and $(-3, 8)$
11. $(2a, 3)$ and $(-a, 5)$
12. $(x, 0)$ and $(-2, -3)$
13. (a, b) and $(0, 0)$
14. $(\sqrt{3}, -2)$ and $(0, \sqrt{5})$
15. $(\sqrt{a}, \sqrt{b})$ and $(0, 0)$
16. $(a + b, a - b)$ and $(a - b, a - b)$

17. Determine the distance between $(-1, 4)$ and the y-axis.

18. Determine the distance between $(-2, 3)$ and the x-axis.

19. Determine the distance between the point (a, b) and the origin.

20. Determine the distance between the point (a, b) and the x-axis.

Determine the midpoints of the segments having the following endpoints.

21. $(2, 2)$ and $(-4, -3)$

22. $(6, 10)$ and $(-1, 5)$

23. $(0, -5)$ and $(2, -3)$

24. $(-3, -3)$ and $(3, 3)$

25. $(-2, 5)$ and $(4, 5)$

26. $(-3, 4)$ and $(-3, 8)$

27. $(2a, 3)$ and $(-a, 5)$

28. $(x, 0)$ and $(-2, -3)$

29. (a, b) and $(0, 0)$

30. $(\sqrt{3}, -2)$ and $(0, \sqrt{5})$

31. $(\sqrt{a}, \sqrt{b})$ and $(0, 0)$

32. $(a + b, a - b)$ and $(a - b, a - b)$

♦ Applications

33. Point R on the negative y-axis is twice as far from the point $P(2, 1)$ as from point $Q(0, 0)$. What are the coordinates of point R?

34. Point W on the positive x-axis is three times as far from point $P(-2, 5)$ as from point $Q(-1, 2)$. What are the coordinates of point W?

Consider points $A(0,0)$ and $B(2,5)$ in Exercises 35 and 36.

35. Find the coordinates of point C on the x-axis such that $\angle ABC$ is a right angle.

36. Find the coordinates of point C on the y-axis such that $\angle ABC$ is a right angle.

37. Point $(x, 3)$ is at a distance of 7 units from the point $(-2, 7)$. Find x.

38. Point $(-4, y)$ is at a distance of 5 units from the point $(0, 9)$. Find y.

39. Find the area of the triangle with vertices $A(2, 3)$, $B(-2, -3)$, and $C(5, -3)$.

40. Find the area of the parallelogram with vertices $P(-1, -7)$, $Q(-1, 2)$, $R(3, 5)$, and $S(3, -4)$.

41. Find the point on the y-axis equidistant from points $(-2, 3)$ and $(-5, -7)$.

42. Find the point on the x-axis equidistant from points $(-2, 3)$ and $(-5, -7)$.

Use the distance formula to determine whether each set of three points is on a straight line. (Hint: For three points to be on a straight line, the sum of the distances from the outer points to the middle one equals the distance between the outer points.)

43. $(-3, -4)$, $(2, 3)$, and $(4, 7)$

44. $(-3, -4)$, $(6, 1)$, and $(10, 6)$

Use the distance formula to determine whether the three points are vertices of a right triangle.

45. $(-4, 7)$, $(6, 3)$, and $(-8, -3)$

46. $(-3, 6)$, $(2, 4)$ and $(6, 14)$

47. Find an equation that must be satisfied by any point (x, y) that is equidistant from $(-2, 3)$ and $(5, -4)$.

48. Find an equation that must be satisfied by any point (x, y) that is 5 units from point (h, k).

49. Find the center and radius of a circle that is inscribed in a square with vertices $(7, 5)$, $(2, 10)$, $(7, 15)$, and $(12, 10)$.

Use the distance formula and/or the midpoint formula to prove each of the following. Locate the polygons on a coordinate system. Put one of the points at the origin to ease the proof.

50. The midpoint of the hypotenuse of any right triangle is equidistant from each of the vertices of the triangle.

51. The diagonals of a rectangle have the same length.

52. The line segments joining the midpoints of the sides of any quadrilateral form a parallelogram.

53. The diagonals of any parallelogram bisect each other.

54. If the diagonals of a parallelogram have the same length, then the parallelogram is a rectangle.

55. Find all points $(a, 0)$ whose distance from $(0, 3)$ is greater than $\sqrt{19}$.

✍ In Your Own Words

56. Think of an example of a real-life object that can be thought of as a two-dimensional coordinate system. How would you define distance in this system?

1.6 GRAPHS OF EQUATIONS

In this section, we turn our attention to graphing sets of points and, in particular, to graphing equations.

Graphing Equations

Suppose we are given an equation in two variables x and y, such as

$$3x + 5y = 7$$

$$3x^2 + 4y^3 = 1$$

$$y = 3x^4 + 1$$

A **solution** of such an equation is an ordered pair (c, d) for which the equation is satisfied when $x = c$ and $y = d$. For example, $(0, \frac{7}{5})$ is a solution of the first equation, because if $x = 0$, $y = \frac{7}{5}$, the equation is satisfied:

$$3(0) + 5\left(\tfrac{7}{5}\right) = 7$$

Given an equation in x and y, we may plot all of its solutions as points in the plane. The set of points plotted will generally form some sort of curve, which is called the **graph of the equation**. (See Figure 1.)

The graph of an equation is a way of representing the algebraic properties of the equation geometrically. In the remainder of this section, we will discuss how to sketch the graphs of various sorts of equations and how to read information from graphs.

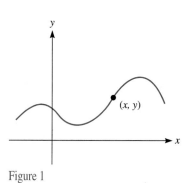

Figure 1

➤ **EXAMPLE 1**
Points with Specific
x- or y-Coordinates

Figure 2 shows the graph of an equation.

1. Determine all points on the graph for which $x = 1$.
2. Determine all points on the graph for which $y = 2$.

Solution

1. Draw a vertical line through the point $x = 1$ on the $x-$axis (see Figure 3). All points on this line have x-coordinate 1. This line intersects the graph of point $(1,0)$, which is the only point on the graph for which $x = 1$.

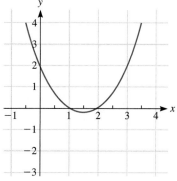

Figure 2

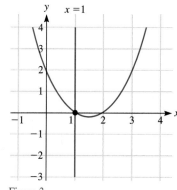

Figure 3

2. Draw a horizontal line through the point $y = 2$ (see Figure 4). All points on this line have y-coordinate 2. This line intersects the graph at points $(0, 2)$, $(3, 2)$, which are the only points on the graph for which $y = 2$.

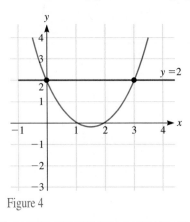

Figure 4

Graphing Linear Equations

The simplest equations in two variables have the form

$$cx + dy + e = 0$$

where c, d, and e are real numbers, with c and d not both zero. Such equations are called **linear equations in two variables**. The graph of such an equation is a straight line and every straight line is the graph of some linear equation in two variables. We won't prove these last statements, but rather will use them as the definition of a straight line.

In order to graph a linear equation in two variables, it suffices to determine two points on the line. The places where the line crosses the coordinate axes are particularly easy to determine, and thus can be used in drawing the graph.

Let us return for a moment to a general equation in x and y. A point at which the graph crosses the x-axis is of the form $(a, 0)$. The number a is called an **x-intercept of the graph**. A point at which a graph crosses the y-axis is of the form $(0, b)$. The number b is called a **y-intercept of the graph** (see Figure 5). Note that the graph of an equation may have many x- and y-intercepts.

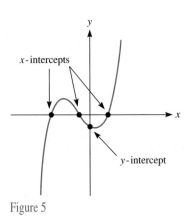

Figure 5

DETERMINING INTERCEPTS OF THE GRAPH OF AN EQUATION

1. x-intercepts: Set $y = 0$ in the equation and solve for x. For each solution $x = a$, the point $(a, 0)$ is an x-intercept.
2. y-intercepts: Set $x = 0$ in the equation and solve for y. For each solution $y = b$, the point $(0, b)$ is a y-intercept of the graph.

A graph's intercepts often provide valuable geometric information. In the case of graphs arising in examples, intercepts often provide information that has meaning in the context of a given application. For example, if we graph the price of a commodity (on the vertical axis) against its demand (on the horizontal axis), the horizontal intercept corresponds to the price at which 0 units of the commodity are demanded (see Figure 6).

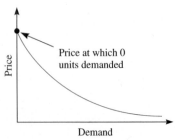

Figure 6

In the following example, we show how to graph a linear equation by determining the intercepts.

➤ EXAMPLE 2
Graphing a Line Using Intercepts

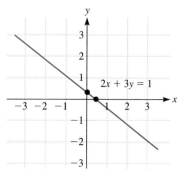

Figure 7

1. Determine the x- and y-intercepts of the graph of the linear equation

$$2x + 3y = 1$$

2. Graph the equation.

Solution

1. To obtain the y-intercept, we substitute $x = 0$ in the given equation.

$$2(0) + 3y = 1$$

$$y = \tfrac{1}{3}$$

so the y-intercept is $\tfrac{1}{3}$. To obtain the x-intercept, we set y equal to 0 in the equation.

$$2x + 3(0) = 1$$

$$x = \tfrac{1}{2}$$

so the x-intercept is $\tfrac{1}{2}$.

2. To obtain the graph, we use the intercepts to plot the points where the graph intercepts the axes, and draw a line through them (see Figure 7).

➤ EXAMPLE 3
Vertical and Horizontal Lines

Graph the following linear equations.

1. $x = 1$
2. $y = 3$

Solution

1. Because y does not appear in the equation, it may assume any value, but the value of x is always 1. Therefore, the graph consists of all points $(1, t)$, where t is any real number. The graph is a vertical line with x-intercept 1 (see Figure 8).

2. Because x does not appear in the equation, it may assume any value, but the value of y is always 3. Therefore, the graph consists of all points $(t, 3)$, where t is any real number. The graph is a horizontal line with y-intercept 3 (see Figure 9).

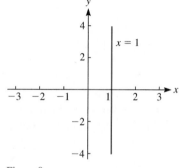

Figure 8

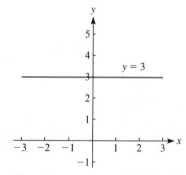

Figure 9

> **EXAMPLE 4**
Demand for a Commodity

A supermarket observes the following relationship between the weekly sales of frozen orange juice q (measured in cans) and the price per can p (measured in dollars):

$$q = -3000p + 12,000$$

1. Graph the linear equation representing the relationship.
2. Give an interpretation for the p-intercept of the graph.

Solution

1. In this example, the horizontal axis is labeled p and the vertical axis q (see Figure 10). The q-intercept is the value of q at which p equals 0. We obtain this value of q by setting p equal to 0 and solving the equation:

$$q = (-3000)0 + 12,000 = 12,000$$

The q-intercept is 12,000. The p-intercept is the value of p for which the value of q is 0:

$$0 = -3000p + 12,000$$
$$p = 4$$

The p-intercept is 4. To graph the equation, we plot the two points at which the graph intersects the axes and draw a line through the plotted points (see Figure 10).

2. The p-intercept is the value of p at which the line crosses the horizontal axis, that is, the point where the value of q is 0, and the value of p is 4. A value of zero for q means that no sales are being made. Because the value of p, price, is \$4 when q is zero, this graph can be interpreted to mean that customers will not buy orange juice when the price is \$4 per can.

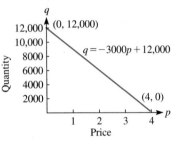

Figure 10

> **EXAMPLE 5**
Graphing an Absolute
Value Equation

Determine the graph of the equation $y = |x + 1|$.

Solution

According to the definition of absolute value, we have

$$|x + 1| = \begin{cases} x + 1 & \text{if } x + 1 \geq 0 \\ -(x + 1) & \text{if } x + 1 < 0 \end{cases}$$

Therefore, the graph consists of two portions, corresponding to values of x satisfying $x \geq -1$ or $x < -1$, namely the graphs of the equations

$$y = x + 1 \quad \text{and} \quad x \geq -1$$
$$y = -x - 1 \quad \text{and} \quad x < -1$$

Each of these equations is linear. However, the values of x are restricted. The graphs are shown in Figure 11.

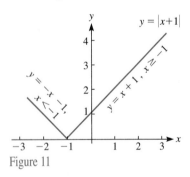

Figure 11

Graphing by Plotting Points

The simplest method for graphing an equation is to plot a number of points that lie on the graph and connect the plotted points with a smooth curve. The next two examples illustrate the procedure for doing this.

> **EXAMPLE 6**
Graphing a Quadratic
Equation

Graph the equation

$$y = x^2$$

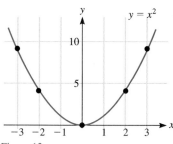

Figure 12

Solution

We determine some points on the graph by using small values of x and determining the corresponding values of y, as shown in the table below. Because the equation expresses y in terms of x, we choose a representative set of values for x and calculate the corresponding values of y. In Figure 12, we plotted the points and drew a smooth curve connecting the points. Note that as $|x|$ increases, so does y. Geometrically, this is reflected in the fact that the graph rises as we proceed either right or left from the origin.

x	y	(x, y)
-3	9	$(-3, 9)$
-2	4	$(-2, 4)$
-1	1	$(-1, 1)$
0	0	$(0, 0)$
1	1	$(1, 1)$
2	4	$(2, 4)$
3	9	$(3, 9)$

➤ EXAMPLE 7
Graphing Where x Is
Expressed in Terms of y

Plot the graph of the following equation

$$y^2 = x$$

Solution

In this example, because x is given in terms of y, it is simpler to choose representative values of y and compute the corresponding values of x. The results of the calculations are contained in the following table. These points are plotted and the resulting graph sketched in Figure 13. Note that as $|y|$ increases, so does the value of x. Geometrically, this means that the graph heads to the right as we move either up or down from the origin.

y	x	(x, y)
-3	9	$(9, -3)$
-2	4	$(4, -2)$
-1	1	$(1, -1)$
0	0	$(0, \ 0)$
1	1	$(1, \ 1)$
2	4	$(4, \ 2)$
3	9	$(9, \ 3)$

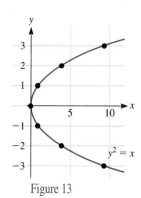

Figure 13

➤ EXAMPLE 8
An Equation Yielding
a Pair of Equations to Graph

Graph the equation

$$x^2 + y^2 = 1$$

Solution

This equation does not allow immediate computation of either variable from the other. To obtain a set of points on the graph, it is easiest to first solve the equation for one variable in terms of the other. Solving for y in terms of x, we obtain

$$y^2 = 1 - x^2$$
$$y = \pm\sqrt{1 - x^2}$$

Using this equation, it is a simple matter to compute a table of points on the graph. The table below gives points corresponding to selected values of x:

$$x = 0, \quad \pm 0.2, \quad \pm 0.4, \quad \pm 0.6, \quad \pm 0.8, \quad \text{and} \pm 1$$

The points are plotted in Figure 14, where we see that the graph appears to be a circle.

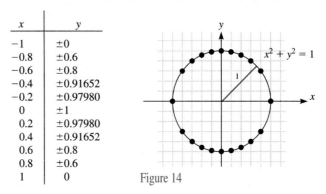

x	y
-1	± 0
-0.8	± 0.6
-0.6	± 0.8
-0.4	± 0.91652
-0.2	± 0.97980
0	± 1
0.2	± 0.97980
0.4	± 0.91652
0.6	± 0.8
0.8	± 0.6
1	0

Figure 14

➤ **EXAMPLE 9**
An Equation with an Undefined x-Value

Sketch the graph of the equation

$$y = \frac{1}{x}, \quad x \neq 0$$

Solution

Whenever you sketch an equation by plotting points, it is necessary to plot a set of points that are representative of the various geometric features the graph possesses. Any algebraic irregularity will likely lead to an interesting geometric feature that should be investigated in detail. In this example, the expression $\frac{1}{x}$ is not defined for x equal to 0. This suggests that the graph of the equation will have some interesting behavior for points (x, y) with x near 0. Therefore, in tabulating a set of points to plot, it is wise to use a number of points with x near 0. The table below includes points for which x is equal to

$$\pm 3, \quad \pm 2, \quad \pm 1, \quad \pm 0.5, \quad \pm 0.2, \quad \text{and} \pm 0.1$$

The points are plotted and the graph sketched in Figure 15.

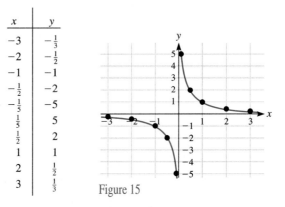

x	y
-3	$-\frac{1}{3}$
-2	$-\frac{1}{2}$
-1	-1
$-\frac{1}{2}$	-2
$-\frac{1}{5}$	-5
$\frac{1}{5}$	5
$\frac{1}{2}$	2
1	1
2	$\frac{1}{2}$
3	$\frac{1}{3}$

Figure 15

Based on the four examples just worked, we can state the following strategy for graphing an equation by plotting points.

GRAPHING AN EQUATION BY PLOTTING POINTS

1. Solve the equation for x in terms of y or for y in terms of x, whichever is easier.
2. If y is expressed in terms of x, choose a representative set of values for x; if x is expressed in terms of y, choose a representative set of values for y.
3. Calculate the other value corresponding to each of the representative values chosen.
4. Plot the points corresponding to the pairs (x, y) calculated.
5. Draw a smooth curve through the points, making sure that the curve has breaks corresponding to undefined values of the variables.

Graphing an equation is a tedious process. In order to obtain an accurate graph, it is often necessary to plot a large number of points. The calculations involved in determining these points can be quite cumbersome.

We should note that plotting points does not always yield, unambiguously, the correct shape of a graph. Given a finite number of points, it is always possible to fit a number of different curves to the data. For instance, in Figure 16, we sketch several different curves that fit the same set of points. The only way to tell which of the graphs is correct is to further examine the algebraic properties of the equation that gave rise to the points.

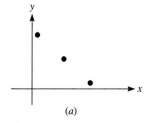 (a)   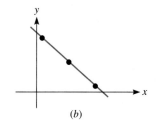 (b)   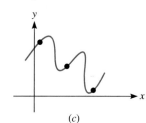 (c)   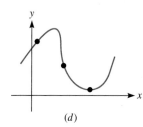 (d)

Figure 16

Let's now consider several pieces of information that allow us to predict geometric properties of the graph and thereby supplement the graphing of points in determining the correct form of a graph.

Circles and Their Equations

There are certain curves that occur so often in applications that it is useful to know their equations in advance. One such family of curves are circles.

A circle with center P and radius r is the set of all points whose distance from P equals r (see Figure 17). Let's determine the equation of such a circle. Let $Q(x, y)$ be a typical point on the circle and let $P(h, k)$ be the center. By the distance formula, the distance from P to Q equals

$$d(P, Q) = \sqrt{(x - h)^2 + (y - k)^2}$$

so that the condition

$$d(P, Q) = r$$

may be rewritten in the form:

$$\sqrt{(x - h)^2 + (y - k)^2} = r$$

$$(x - h)^2 + (y - k)^2 = r^2$$

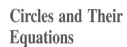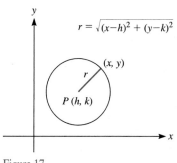

$r = \sqrt{(x-h)^2 + (y-k)^2}$

Figure 17

This proves the following result:

EQUATION OF A CIRCLE

Suppose that a circle has center (h, k) and radius r. Then the circle is the graph of the equation

$$(x - h)^2 + (y - k)^2 = r^2$$

➢ **EXAMPLE 10**
Graphing a Circle

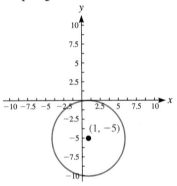

$(x-1)^2 + (y+5)^2 = 5^2$

Figure 18

Describe the graph of the equation

$$(x - 1)^2 + (y + 5)^2 = 25$$

Draw the graph.

Solution

The equation may be written in the form

$$(x - 1)^2 + [y - (-5)]^2 = 5^2$$

This is the equation of a circle with center C at $(1, -5)$ and radius 5. We can draw the graph by plotting the point $(1, -5)$ and then using a compass to draw the circle of the indicated radius with the plotted point as center (refer to Figure 18).

➢ **EXAMPLE 11**
Completing the Square to Graph a Circle

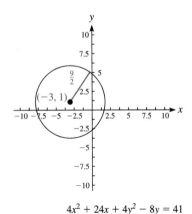

$4x^2 + 24x + 4y^2 - 8y = 41$

Figure 19

Describe the graph of the equation

$$4x^2 + 24x + 4y^2 - 8y = 41$$

Draw the graph.

Solution

The equation has terms x^2 and y^2 with the same coefficient. This suggests that the equation can be transformed into an equivalent equation that is the equation of a circle. We begin by dividing the equation by 4:

$$(x^2 + 6x) + (y^2 - 2y) = \tfrac{41}{4}$$

Make the expressions within the parentheses into perfect squares: Add 9 to the first and 1 to the second and $9 + 1$ to the other side of the equation:

$$(x^2 + 6x + 9) + (y^2 - 2y + 1) = \tfrac{41}{4} + 9 + 1$$

$$(x + 3)^2 + (y - 1)^2 = \tfrac{81}{4}$$

To recognize this as the equation of a circle, we write it in the form

$$(x - (-3))^2 + (y - 1)^2 = \left(\tfrac{9}{2}\right)^2$$

This is the equation of a circle with center at $(-3, 1)$ and radius $\tfrac{9}{2}$ (see Figure 19).

Exercises 1.6

Plot the graph of each equation.

1. $y = 2x + 3$

2. $y = 4 - x$

3. $y = -2$

4. $x = 4$

5. $2x - y = 4$

6. $x + y = 1$

7. $-4x - 3y = 12$

8. $2x + 5y = 10$

9. $y = x^2 + 1$

10. $y = 1 - x^2$

11. $x = y^2 + 3$

12. $x = -y^2$

13. $y = -\dfrac{1}{x}$

14. $y = \dfrac{2}{x}$

15. $y = |x|$

16. $y = |2x - 1|$

17. $y = x^3$

18. $x = y^3$

19. $x = |y + 3|$

20. $x = |y|$

21. $xy = 1$

22. $xy = -0.25$

23. $y = \dfrac{1}{x^2}$

24. $y = \dfrac{1}{|x|}$

25. $x^2 + y^2 = 4$

26. $x^2 + y^2 = 25$

27. $(x - 4)^2 + y^2 = 7$

28. $x^2 + (y + 1)^2 = 5$

29. $(x + 3)^2 + (y - 7)^2 = 13$

30. $(x - 6)^2 + (y + 2)^2 = 11$

31. $x^2 + y^2 + 12x + 19 = -1$

32. $x^2 + 6x + y^2 - 16y + 48 = 0$

33. $x^2 + x + y^2 - y - 5 = 0$

34. $x^2 + y^2 = 4y$

Find an equation of a circle satisfying the following conditions.

35. Center: $(-6, 1)$; radius: 2

36. Center: $(3, -4)$; radius: $\sqrt{3}$

37. Center at the origin; radius: $\sqrt{39}$

38. Center: $(0, a)$; radius: 0.1

Find an equation of a circle satisfying the following conditions.

39. Tangent to the y-axis; center: $(-2, 3)$

40. Tangent to the x-axis; center: $(9, 10)$

41. Tangent to both axes; center in the fourth quadrant; radius 11

42. Tangent to both axes, center in the third quadrant, radius $\frac{2}{3}$

43. Having a diameter whose endpoints are $(-5, 4)$ and $(3, 8)$

44. Having a diameter whose endpoints are $(-3, -5)$ and $(2, -7)$

45. Center: $(-2, -3)$; area: 9π square units

46. Center: $(-2, -3)$; circumference: 22π units

47. Inscribed in a square whose vertices are $(2, 10)$, $(7, 15)$, $(12, 10)$, and $(7, 5)$

Find the center and radius of the circles that have the following equations.

48. $x^2 + y^2 + 3.5x - 4.2y - 10 = 0$

49. $x^2 + y^2 + 10x - 12y - 20 = 0$

50. $144x^2 + 144y^2 + 192x - 216y + 81 = 45$

51. $2x^2 + 2y^2 - 6x + 8y - 100 = 0$

Determine whether each of the following ordered pairs lies on the circle $x^2 + y^2 = 1$.

52. $\left(\dfrac{1}{2}, \dfrac{\sqrt{3}}{2}\right)$

53. $\left(-\dfrac{\sqrt{2}}{2}, \dfrac{\sqrt{2}}{2}\right)$

54. $\left(\sqrt{2}, \sqrt{3}\right)$

55. $(0, 2)$

56. $(1, -1)$

57. $(1, 0)$

58. $(0.9781, 0.2079)$

59. $(-0.2419, -0.9703)$

60. Describe the graph of this equation:
$$(x - 3)^2 + (y + 5)^2 = (x + 7)^2 + (y + 4)^2$$

61. Describe the solution set of this inequality:
$$(x + 3)^2 + (y + 5)^2 \leq 36$$

♦ **Applications**

62. **Spread of an organism**. A certain kind of dangerous organism is released by accident over an area of 3 square miles (see Figure 20). After time t, in years, it spreads over area A, in square miles, where A and t are related by the equation
$$A = 1.8t + 3$$

  a. Graph the linear equation representing the relationship between elapsed time and area.

  b. Give an interpretation of the y-intercept.

63. **Driving costs**. The cost C, in cents per mile, of driving a car is related to the number of years t since 1980 by the equation
$$100C - 7t = 2320$$
where $t = 0$ corresponds to 1980, $t = 1$ to 1981, and so on.

  a. Graph the linear equation representing the relationship between cost and elapsed time. Consider time t on the x-axis and cost C on the y-axis.

  b. Give an interpretation of the y-intercept.

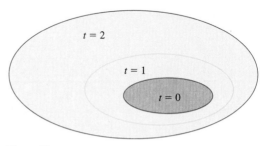

Figure 20

1.7 STRAIGHT LINES

Straight lines are the simplest curves. In many applied situations, experiments or physical laws suggest that certain variables are related to one another via a linear equation. In such circumstances, the relationship between the variables can be depicted using a straight-line graph. In this section, we take a closer look at straight lines and show how to determine equations for them from various sets of data.

Slope of a Straight Line

Let's consider a nonvertical straight line that includes the distinct points (x_1, y_1) and (x_2, y_2). The difference $y_2 - y_1$ measures the vertical change between the two points and is called the **rise**. The difference $x_2 - x_1$ measures the horizontal difference between the two points and is called the **run**. Geometric interpretations of the rise and run are shown in Figure 1. Because the two points are assumed to be distinct and the line nonvertical, the points must have different x-coordinates. That is, $x_2 - x_1 \neq 0$ (the run is nonzero).

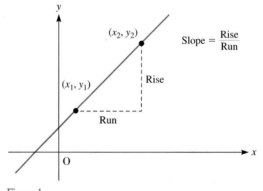

Figure 1

The ratio of the rise to the run measures the steepness of the line and is called the **slope.** Thus, we have the following definition.

Definition 1	Suppose that L is a nonvertical line passing through the points (x_1, y_1) and
Slope of a Line	(x_2, y_2). The **slope** of the line, denoted m, is defined as the ratio

$$m = \frac{\text{rise}}{\text{run}} = \frac{y_2 - y_1}{x_2 - x_1}$$

For a vertical line, the slope is undefined.

➢ **EXAMPLE 1**
Calculating Slope

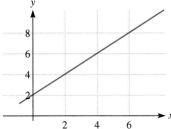

Figure 2

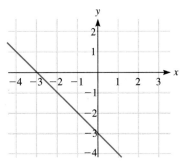

Figure 3

For the straight lines in the following graphs, determine the slope.

1. Figure 2.
2. Figure 3.

Solution

1. To calculate the slope, we need two points on the line. Examining the graph, we see that these points may be taken to be $(2, 4)$ and $(4, 6)$. So the slope of the line is

$$m = \frac{y_2 - y_1}{x_2 - x_1} = \frac{6 - 4}{4 - 2} = 1$$

2. Two points on the graph are $(0, -3)$ and $(2, -5)$. Therefore, the slope is equal to

$$m = \frac{y_2 - y_1}{x_2 - x_1} = \frac{-5 - (-3)}{2 - 0} = -1$$

There's more to the definition of slope than meets the eye! As we shall soon see, the notion of slope is one of the most important measures describing a straight line. However, before we describe any applications of slope, let's make several observations about the definition.

First, note that the value of the slope does not depend on which point is labeled with subscript 1 and which with subscript 2. Indeed, we have

$$\frac{y_2 - y_1}{x_2 - x_1} = \frac{-(y_1 - y_2)}{-(x_1 - x_2)} = \frac{y_1 - y_2}{x_1 - x_2}$$

Next, note that the value of the slope does not depend on which points we choose as (x_1, y_1) and (x_2, y_2), provided, of course, that we choose points on the line. Indeed, suppose that the straight line has the equation

$$Ax + By + C = 0$$

If $B = 0$, the equation involves only x, so the line is vertical and the slope is undefined. So assume that $B \neq 0$. Because (x_1, y_1) and (x_2, y_2) are on the line, both points satisfy the equation of the line. That is,

$$Ax_1 + By_1 + C = 0$$

$$Ax_2 + By_2 + C = 0$$

Subtracting the first equation from the second, we have

$$A(x_2 - x_1) + B(y_2 - y_1) = 0$$

$$A(x_2 - x_1) = -B(y_2 - y_1)$$

$$\frac{y_2 - y_1}{x_2 - x_1} = -\frac{A}{B}, \quad B \neq 0$$

The expression on the left side of the last equation equals the slope. The expression on the right side depends only on the equation of the line and not on the particular points (x_1, y_1) and (x_2, y_2). That is, the value of the slope is independent of the points used to calculate it.

The Greek letter Δ (delta) is often used to denote the change in a variable. For computing the slope, the change in the y variable from one point to the other is $y_2 - y_1$, or the rise. For this reason, the rise is sometimes denoted Δy (read "delta y"). Similarly, the run is denoted Δx (read "delta x"). In this notation, the slope is expressed as

$$m = \frac{\Delta y}{\Delta x}$$

➢ **EXAMPLE 2**
Calculating Slope from Points

A line passes through points $(-1, 5)$ and $(0, 4)$. Determine the slope of the line.

Solution

Using the definition of slope, we have

$$\text{Slope} = \frac{y_2 - y_1}{x_2 - x_1} = \frac{4 - 5}{0 - (-1)} = \frac{-1}{1} = -1$$

Let's now develop an alternative method of computing the slope directly from the equation of a line. Suppose that the equation of a line is

$$Ax + By + C = 0$$

If $B = 0$, then the equation involves only x and the graph is a vertical line, for which the slope is undefined, so assume $B \neq 0$. Then, we may solve the equation for y to obtain

$$y = -\frac{A}{B}x + \left(-\frac{C}{B}\right)$$

As shown above, the expression $-A/B$ equals the slope m of the line. Let's denote the expression $-C/B$ by b. The equation of the line may be written as

$$y = mx + b$$

The number b has a geometric significance. Indeed, if we set x equal to 0, we see that $y = m(0) + b = b$. In other words, the point $(0, b)$ is on the line and b is its y-intercept. Because the equation $y = mx + b$ expresses both the slope and the y-intercept of the line, it is called the **slope–intercept form of the equation of a line**.

We may summarize the above discussion in the following thereom.

EQUATIONS OF STRAIGHT LINES

Let $Ax + By + C = 0$ be the equation of a straight line. If $B \neq 0$, then solving the equation for y in terms of x gives an equivalent equation of the form $y = mx + b$, the slope–intercept form of the equation. Here m is the slope of the line, and b is the y-intercept. If $B = 0$, then the line is vertical. In this case, solving the equation for x yields an equivalent equation of the form $x = a$.

➤ **EXAMPLE 3**
Slope-Intercept Form

Determine the slope–intercept form, the slope, and the y-intercept of the line with equation $5x + 3y - 12 = 0$.

Solution

To determine the slope–intercept form, solve for y in terms of x.

$$5x + 3y - 12 = 0$$
$$3y = -5x + 12$$
$$y = -\tfrac{5}{3}x + 4$$

The last equation is the point–slope form. The slope of the line equals the coefficient of x in the point–slope form. That is, $-\frac{5}{3}$. The y-intercept equals the constant term of the point–slope form, or 4.

The next theorem gives a geometric interpretation of the slope of a line. As we will see, this interpretation is useful in many applied problems.

GEOMETRIC INTERPRETATION OF SLOPE

Suppose that a line has slope m. Starting from any point on the line, if you move h units in the x direction, then it is necessary to move mh units in the y-direction to return to the line (see Figure 4).

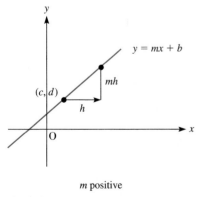

m positive

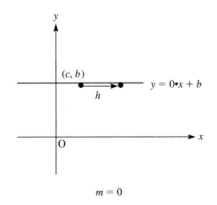

$m = 0$

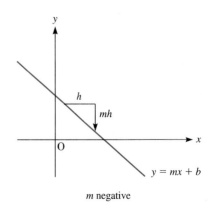

m negative

Figure 4

Proof   As in Figure 4, assume that the initial point on the line is (x_1, y_1). Because the point is on the line, it satisfies the equation

$$y_1 = mx_1 + b$$

If the x-coordinate is changed by h, then the new x-coordinate is $x_1 + h$. The corresponding value of y on the line can be obtained.

$$\begin{aligned} y &= m(x_1 + h) + b \\ &= (mx_1 + b) + mh \\ &= y_1 + mh \end{aligned}$$

That is, the value of y changes from y_1 to $y_1 + mh$. The amount of the change is mh, as claimed in the statement of the theorem.   ◆

➤ **EXAMPLE 4**
Application of Slope

Consider the line with equation

$$y = -2x + 9$$

1. Suppose that x is increased by 5. What is the change in the value of y?
2. Suppose that x is decreased by 100. What is the change in the value of y?

Solution

1. The slope of the line is -2. By the Geometric Interpretation of Slope, a change of h in x results in a change of mh in y, so an increase of 5 in x results in a change of

$$m \cdot 5 = -2 \cdot 5 = -10$$

in y. That is, y decreases by 10.

2. Again, as a consequence of the Geometric Interpretation of Slope, a change of -100 in x results in a change of

$$mh = -2 \cdot (-100) = 200$$

That is, y increases by 200.

➤ **EXAMPLE 5**
Manufacturing

The cost C (in dollars) of manufacturing q television sets in a production run is given by

$$C = 250q + 50,000$$

1. Give an interpretation for the slope of this linear equation.
2. Suppose that production is increased by 100 units. What will be the effect on production cost?

Solution

1. The slope of the line represented by the equation is 250. By the Geometric Interpretation of Slope, if q is increased by 1, that is, if the company produces one more television set, then the value of C, the total cost of manufacturing, changes by

$$250 \cdot 1 = 250$$

That is, the slope equals the cost of producing one additional television set. This additional cost is called **the marginal cost of production** and is used by economists in analyzing the microeconomics of production.

2. If production increases by $h = 100$ units, then the cost C changes by

$$mh = 250 \cdot 100 = 25,000$$

That is, the cost increases by $25,000.

> **EXAMPLE 6**
Distance vs Time

After t hours of driving, a trucker's distance D, in miles, from his delivery point is given by the equation

$$D = -50t + 500$$

Give a physical interpretation of the slope of the corresponding line.

Solution

The relationship connecting D and t is a linear equation expressed in slope–intercept form. The slope is -50, and the vertical-axis intercept (the D-intercept) is 500. According to the preceding discussion, for each increment of time Δt, the truck goes an additional distance ΔD. Moreover, the slope equals the quotient

$$\frac{\Delta D}{\Delta t} = \frac{\text{Change in distance}}{\text{Change in time}}$$

But according to the well-known formula, distance = rate × time. Therefore, the ratio (Change in distance)/(Change in time) is equal to the **rate** at which the distance is changing per unit time. This quantity is called the **velocity** of the truck and is expressed in miles per hour. The statement Velocity $= -50$ means that the distance to the delivery point is decreasing at a rate of 50 miles per hour.

Finding Equations of Straight Lines

In solving problems, it is often necessary to determine equations of straight lines from given data. A number of different forms of equations for straight lines allow us to do just that. Let's discuss these various equations. The first is just the slope–intercept form we have been using in our discussion of slope.

The next formula allows us to determine the equation of a line given its slope and a point on the line.

POINT–SLOPE FORMULA

Suppose that a line has slope m and passes through the point (x_1, y_1). Then an equation for the line is

$$y - y_1 = m(x - x_1)$$

To verify the point–slope formula, first note that point (x_1, y_1) is on the line determined by the equation. Indeed, if we substitute x_1 for x and y_1 for y, we see that the equation is satisfied:

$$y_1 - y_1 = m(x_1 - x_1)$$

$$0 = 0$$

Writing the equation in the form

$$y - y_1 = mx - mx_1$$

$$y = mx + (y_1 - mx_1)$$

$$y = mx + b \qquad \text{where } b = y_1 - mx_1$$

we see that the slope of the line is m.

➤ **EXAMPLE 7**
Point–Slope Form

Determine the equation of a line passing through the point $(3, -2)$ with a slope of 4.

Solution

Using the point–slope formula, we have the equation

$$y - (-2) = 4(x - 3)$$

$$y + 2 = 4x - 12$$

$$y = 4x - 14$$

➤ **EXAMPLE 8**
Line Passing through
Two Points

Determine the equation of a line passing through points $(5, 3)$ and $(2, -2)$.

Solution

The slope of the line is

$$m = \frac{3 - (-2)}{5 - 2} = \frac{5}{3}$$

Therefore, by the point–slope formula, the desired equation is

$$y + 2 = \tfrac{5}{3}(x - 2)$$

$$y = \tfrac{5}{3}x - \tfrac{10}{3} - 2$$

$$y = \tfrac{5}{3}x - \tfrac{16}{3}$$

The following table summarizes the various equations of straight lines we have considered.

EQUATIONS OF LINES

1. General linear equations: $Ax + By + C = 0$ or $Ax + By = D$
2. Slope–intercept form: $y = mx + b$
3. Point–slope form: $y - y_1 = m(x - x_1)$
4. Vertical line: $x = a$
5. Horizontal line: $y = b$

Parallel and Perpendicular Lines

The next two results describe the relationships between slopes of parallel and perpendicular lines.

SLOPES OF PARALLEL LINES

Two distinct nonvertical lines are parallel if and only if they have the same slope and have different y-intercepts. (See Figure 5.)

SLOPES OF PERPENDICULAR LINES

Two lines with slopes m_1 and m_2 are perpendicular if and only if the product of the slopes is -1, that is, if and only if

$$m_1 m_2 = -1 \quad \text{or} \quad m_1 = -\frac{1}{m_2}, \quad \text{where} \quad m_2 \neq 0$$

(See Figure 6.) The proof of this theorem is outlined in the exercises.

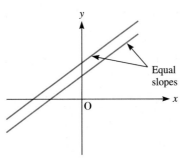

Figure 5

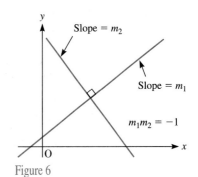

Figure 6

➤ **EXAMPLE 9**
Perpendicular Lines

Determine whether the lines described by the following equations are perpendicular to one another.

$$-4x + 3y = 1, \quad 3x + 4y = 17$$

Solution

Let's apply the preceding result. To do so, we must first determine the slopes of the lines. The slope–intercept forms of the equations are

$$-4x + 3y = 1$$
$$3y = 4x + 1$$
$$y = \tfrac{4}{3}x + \tfrac{1}{3}$$

and

$$3x + 4y = 17$$
$$4y = -3x + 17$$
$$y = -\tfrac{3}{4}x + \tfrac{17}{4}$$

From these equations, we see that the slopes of the lines are $m_1 = \tfrac{4}{3}$ and $m_2 = -\tfrac{3}{4}$, respectively. We then have

$$m_1 m_2 = \tfrac{4}{3} \cdot \left(-\tfrac{3}{4}\right) = -1$$

Therefore, by the preceding result, the two lines are perpendicular.

➤ **EXAMPLE 10**
Parallel Lines

Determine the equation of a line passing through the point $(-1, -2)$ and parallel to the line with equation

$$y = 3x - 18$$

Solution

The line with the given equation has a slope 3. By the Slopes of Parallel Lines theorem, a line parallel to this line has the same slope, 3. Because the desired line must pass through the point $(-1, -2)$, its equation may be derived from the point–slope formula:

$$y - (-2) = 3[x - (-1)]$$
$$y + 2 = 3x + 3$$
$$y = 3x + 1$$

➤ **EXAMPLE 11**
Perpendicular Lines

Determine the equation of a line passing through the point $(7, 4)$ and perpendicular to the line with equation

$$3x - 4y - 12 = 0$$

Solution

The given equation has slope–intercept form

$$y = \tfrac{3}{4}x - 3$$

so the corresponding line has slope $\frac{3}{4}$. By the Slopes of Perpendicular Lines thereom, the slope of m a line perpendicular to it is equal to

$$m = -\frac{1}{\frac{3}{4}} = -\frac{4}{3}$$

Because the perpendicular line must pass through $(7, 4)$, we may derive its equation using the point–slope formula:

$$y - 4 = -\tfrac{4}{3}(x - 7)$$

$$y = -\tfrac{4}{3}x + \tfrac{40}{3}$$

Exercises 1.7

Determine the slopes of the lines having the following equations.

1. $y = 0.3x - 27$

2. $y = \frac{2}{3}x + \frac{4}{5}$

3. $-3y = 18x$

4. $x = 4y$

5. $x - 2y = 6$

6. $6x + 3y = 9$

7. $y + 6 = \frac{2}{3}(x - 3)$

8. $3(x - 3) - 4(y + 5) = 10(x + 7) - 5(y - 8)$

9. Consider the line with equation
   $$y = -12x + 34$$
   a. Suppose that x is increased by 3. What is the change in the value of y?
   b. Suppose that x is decreased by 50. What is the effect on y?

10. Consider the line with equation
    $$y = 3.4x - 15$$
    a. Suppose that x is increased by 200. What is the effect on y?
    b. Suppose that x is decreased by 39. What is the effect on y?

Determine the slope of the line passing through each pair of points.

11. $(0, 0)$ and $(3, 6)$

12. $(4, 0)$ and $(0, 2)$

13. $(1, -1)$ and $(3, -5)$

14. $(-2, -3)$ and $(-1, -6)$

15. $(2, 3)$ and $(-1, 3)$

16. $(-7, 3)$ and $(-7, 2)$

17. $(a, 2a + 3)$ and $(a + h, 2(a + h) + 3)$

18. $(k, -3k)$ and $(k, -3(k + h))$

Determine an equation of the line satisfying the given conditions.

19. Through $(-1, 2)$ with slope -3

20. Through $(4, -2)$ with slope $-\frac{1}{2}$

21. Through $(2, -3)$ with slope undefined

22. Through $(2, -3)$ with slope 0

23. With y-intercept $(0, -1)$ and slope $-\frac{1}{5}$

24. With y-intercept $(0, 0)$ and slope -4

25. Through $(2, 3)$ and $(-1, 5)$

26. Through $(-2, -6)$ and $(4, -10)$

27. Through $(9, -3)$ and $(9, 7)$

28. Through $(4, -6)$ and $(3, -6)$

Find the slope and y-intercept.

29. $y = -\frac{2}{3}x + 1$

30. $2y + 3x = 6$

31. $5x - 3y - 9 = 0$

32. $\frac{2}{3}x - \frac{1}{5}y - \frac{1}{10} = 0$

33. $2y + 3 = 6$

34. $-3x = 6$

Determine whether each pair of equations represents parallel lines, perpendicular lines, or neither.

35. $2x - 3y = 4; -2x + 3y = -8$

36. $y = \frac{4}{5}x - 15; 4y - 5x + 5 = 0$

37. $2x - 3y = 56; x + 4y = 7$

38. $4x - 2y = 86; y = -3x + 7$

39. $x - 5y = 32; x + 11y = 4$

40. $x = -3$; $y = 0$

41. $x = -5y$; $x = 12$

42. $y - 4x = 34$; $y - 4x = -3$

Determine the equation of a line passing through the given point and parallel to the given line.

43. $(-3, -2)$; $2x + 2y = 10$

44. $(-3, 5)$; $9y = 3x + 1$

45. $(3, -1)$; $y = x$

46. $(5, -2)$; $x = 3$

47. $(0, 0)$; $4x - 6y = 2$

48. $(-3, 5)$; $9y = 18$

Determine the equation of a line passing through the given point and perpendicular to the given line.

49. $(0, -2)$; $y = \frac{2}{3}x + 1$

50. $(-1, 3)$; $2x - 4y = 6$

51. $(0, 0)$; $y = x$

52. $(4, 2)$; $-4x = -24$

53. $(-2, 5)$; $4y = 2$

54. $(-3, 5)$; $2x = 9y + 18$

In each case, determine the equation of the line through P that is parallel to the line containing A and B and the equation of the line through P that is perpendicular to the line containing A and B.

55. $A(-2, 3)$; $B(4, 5)$; $P(4, -6)$

56. $A(-1, -8)$; $B(1, 10)$; $P(6, -2)$

57. Prove that the graph of an equation in the form

$$\frac{x}{a} + \frac{y}{b} = 1$$

has x-intercept $(a, 0)$ and y-intercept $(0, b)$.

58. Use the result of Exercise 57 to find the x-intercept and y-intercept of each of the following.

a. $\frac{x}{4} - \frac{y}{2} = 1$

b. $4x - 3y = 12$

59. Find the equation of the perpendicular bisector of the line segment with endpoints $(2, 3)$ and $(6, -7)$.

60. Find the equation of the perpendicular bisector of the line segment with endpoints $(0, 1)$ and $(-4, -5)$.

61. Determine whether points $A(-1, 5)$, $B(-5, 1)$, $C(5, 1)$, and $D(1, 5)$ are vertices of a rectangle. Use slopes.

62. Find k such that the line containing $(-2, k)$ and $(3, 8)$ is parallel to the line containing $(5, 3)$ and $(1, -3)$.

63. Find k such that the line containing $(-2, k)$ and $(3, 8)$ is perpendicular to the line containing $(5, 3)$ and $(1, -3)$.

64. **Slopes of Perpendicular Lines Theorem: Outline of Proof.** To prove a sentence of the type "P if and only if Q," one has two statements to prove: (1) If P, then Q and (2) If Q, then P.

"P if and only if Q" means that P and Q are equivalent statements. Another way to prove "P if and only if Q" is to produce a string of equivalent statements, starting with P and ending with Q. We do so for this proof. Consider the drawing in Figure 7 showing two nonvertical lines L and M with respective equations $y = m_1 x + b_1$ and $y = m_2 x + b_2$. Lines L and M are perpendicular if and only if $\triangle PQR$ is a right triangle and $\triangle PQR$ is a right triangle if and only if $PR^2 + RQ^2 = PQ^2$ from the Pythagorean theorem. Now let (x_1, y_1) and (x_2, y_2) be coordinates of two points on lines L and M, respectively. Use these points and the result of the Pythagorean theorem to complete the proof.

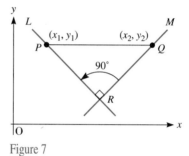

Figure 7

♦ **Applications**

65. **Demand for Orange Juice.** A supermarket observes the following relationship between the weekly sales of frozen orange juice q (measured in cans) and the price p (measured in dollars):

$$q = -3000p + 12,000$$

a. Give an interpretation to the slope of this line.

b. What is the effect on sales of lowering the price $1?

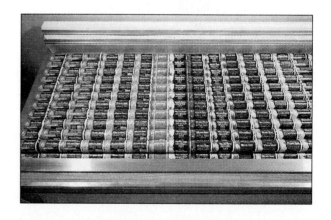

66. In 1983, 2.2 million telephone-answering machines were sold. In 1986, 4.9 million were sold.

    a. Use the two-point equation to find a linear equation relating the number sold N to the year Y.

    b. Use the equation to predict sales in 1996.

67. Highway records show that about 34 percent of drivers aged 20 and 22 percent of those aged 25 will be involved in at least one driving accident within a one-year period.

    a. Use the two-point equation to find a linear equation describing the percentage P of drivers who will have an accident at age A.

    b. Predict the percentage of 30-year-old drivers who will have an accident.

68. **Computer diskettes**. A computer supply catalog recently listed a box of ten $5\frac{1}{4}$-inch diskettes to sell at $1.49 each if 10 are purchased. If 60 are purchased, the price is reduced to $1.29 each.

    a. Use the two-point equation to find a linear equation relating the price P, per diskette, to the number purchased, N.

    b. Predict the price per diskette if 100 diskettes are purchased. The catalog listed the price to be $1.19 if 100 are purchased. How does this compare to the prediction?

69. **Charles's Law.** In 1787, French scientist Jacques Charles noticed that gases expand when heated and contract when cooled according to a linear equation, assuming the pressure remains constant.

Jacques Charles

    a. Suppose a particular gas has a volume $V = 500$ cc when $T = 27°C$ and a volume $V = 605$ cc when $T = 90°C$. Use the two-point equation to find a linear equation relating V and T.

    b. At what temperature does $V = 0$ cc? This is an estimate of what is known as *absolute zero*, the coldest temperature possible.

♦ **Applications**

70. **Straight-line depreciation** is one way a business can compute the value of an item for tax purposes. A company buys an office machine for $9700 on January 1 of a given year. The machine is expected to last for 6 years, at the end of which its trade-in or salvage value will be $1300. If the company figures the decline in value to be the same each year, then the book value or salvage value V after t years, $0 \le t \le 6$, is given by

$$V = \$9700 - \$1400t$$

Give an interpretation of the slope of this line.

71. **Cost**. A clothing manufacturer is planning to market a new type of sweatshirt to college students. The company estimates that fixed costs for setting up the production line are $43,000. Variable costs for producing each sweatshirt are $3.50. Determine an equation that relates cost to the number of sweatshirts produced.

72. A salesperson gets a salary of $24,000 per year plus a commission of 38 percent of sales. Determine an equation that relates salary to sales.

73. **Spread of an Organism**. A certain kind of dangerous organism is released by accident over an area of 3 mi.² After time t, in years, it spreads over area A, in square miles, where A and t are related by the equation

$$A = 1.8t + 3$$

    a. Give an interpretation for the slope of this linear equation.

    b. Suppose that time is increased by 10 years. What will be the effect on the area A affected by the organism?

74. **Driving costs**. The cost C, in cents per mile, of driving a car is related to the number of years t since 1980 by the equation

$$100C - 7t = 2320$$

where $t = 0$ corresponds to 1980, $t = 1$ to 1981, and so on.

    a. Give an interpretation of the slope of this linear equation. Consider time t on the x-axis and cost C on the y-axis.

    b. Suppose that time is increased by 8 years. What will be the effect on the cost C?

75. **Revenue**. The revenue R, in dollars, from the sale of q television sets from a production run is given by

$$R = 409q$$

    a. Give an interpretation of the slope of this linear equation.

    b. Suppose that sales are increased by 2000 units. What will be the effect on revenue?

76. **Temperature Conversion**. Fahrenheit temperatures F and Celsius temperatures C are related by the following linear equation.

$$F = \tfrac{9}{5}C + 32$$

 a. Give an interpretation of the slope of this linear equation.

 b. Suppose that a person's Celsius temperature rises $1°$ C during an illness. What is the change in the person's Fahrenheit temperature?

77. **Tail Length of a Snake**. It has been determined that the total length L and the tail length t, both in millimeters, of females of the snake species *Lampropeltis Polyzona* are related by the linear equation

$$t = 0.143L - 1.18$$

 a. Give an interpretation of the slope of this linear equation.

 b. Suppose a snake is found whose tail is 80 mm longer than another. What will be the difference in total length of the two snakes?

✍ In Your Own Words

78. There is more than one way to define the distance function. For instance, consider the distance function $d(x, y) = |x_1 - x_2| + |y_1 - y_2|$.

 a. How is this definition of distance different from the definition normally used?

 b. Why do you think this definition is called the "Manhattan" distance function?

 c. Describe another model that has another definition for distance.

■ 1.8 CHAPTER REVIEW

Important Concepts, Properties, and Formulas—Chapter 1

Integers	$\{\ldots, -3, -2, -1, 0, 1, 2, 3, \ldots\}$	p. 2
Rational number	Ratio of two integers with nonzero denominator, terminating or infinitely repeating decimal notation.	p. 2
Irrational number	Real number that is not a ratio of two integers, infinite or nonterminating nonrepeating decimal notation.	p. 3
Real numbers	Decimal, terminating or nonterminating.	p. 2
Absolute value	$\|a\| = a$ if $a \geq 0$ $\|a\| = -a$ if $a < 0$ $\|-a\| = \|a\|, \|a\| \geq 0$ $\|ab\| = \|a\| \cdot \|b\|$ $\|a + b\| \leq \|a\| + \|b\|$ $\left\|\dfrac{a}{b}\right\| = \dfrac{\|a\|}{\|b\|}, \quad (b \neq 0)$	p. 5
Distance between points on the number line	$d(A, B) = \|a - b\|$	p. 6

Laws of exponents	$a^r \cdot a^s = a^{r+s}, \quad \dfrac{a^r}{a^s} = a^{r-s}, a \neq 0$ $(a^r)^s = a^{rs}$ $(ab)^r = a^r b^r$ $\left(\dfrac{a}{b}\right)^r = \dfrac{a^r}{b^r}, \quad b \neq 0$ $a^{-r} = \dfrac{1}{a^r}, \quad a \neq 0$ $\left(\dfrac{a}{b}\right)^{-r} = \left(\dfrac{b}{a}\right)^r, \quad a, b \neq 0$	p. 8
Properties of radicals	$\sqrt[n]{ab} = \sqrt[n]{a}\,\sqrt[n]{b}$ $\sqrt[n]{\dfrac{a}{b}} = \dfrac{\sqrt[n]{a}}{\sqrt[n]{b}}, \quad b \neq 0$ $\sqrt[n]{a^n} = \begin{cases} a & \text{if } n \text{ is odd} \\ |a| & \text{if } n \text{ is even} \end{cases}$	p. 8
Formulas for multiplication and factoring	Product of Sum and Difference: $\quad (A + B)(A - B) = A^2 - B^2$	pp. 13–15
	Square of a Binomial Sum: $\quad (A + B)^2 = A^2 + 2AB + B^2$	
	Square of a Binomial Difference: $\quad (A - B)^2 = A^2 - 2AB + B^2$	
	Product of Two Binomials: $(aA + b)(cA + d) = acA^2 + (ad + bc)A + bd$	
	Sum of Cubes: $A^3 + B^3 = (A + B)(A^2 - AB + B^2)$	
	Difference of Cubes: $A^3 - B^3 = (A - B)(A^2 + AB + B^2)$	
Quadratic equation	An equation of the form $\quad ax^2 + bx + c = 0$ where $a \neq 0$	p. 26
Quadratic formula	$x = \dfrac{-b \pm \sqrt{b^2 - 4ac}}{2a}$	p. 27
Distance formula	If $P = (x_1, y_1)$ and $Q = (x_2, y_2)$, then $\quad d(P, Q) = \sqrt{(x_2 - x_1)^2 + (y_2 - y_1)^2}$	p. 47

Midpoint formula	The line joining (x_1, y_1) and (x_2, y_2) has midpoint $$M\left(\frac{x_1 + x_2}{2}, \frac{y_1 + y_2}{2}\right)$$	p. 48
Standard form of a linear equation	$Ax + By = C$	p. 67
Equation of a circle with center (h, k) and radius r	$(x - h)^2 + (y - k)^2 = r^2$	pp. 57–58
Slope of a line	$$m = \frac{y_2 - y_1}{x_2 - x_1}, \quad x_2 - x_1 \neq 0$$ where (x_1, y_1), (x_2, y_2) are any two points on the line with $x_2 - x_1 \neq 0$	pp. 60–61
Slope–intercept formula	Suppose a nonvertical line has slope m and y-intercept $(0, b)$. The slope-intercept equation for the line is $$y = mx + b$$	pp. 63, 67
Point–slope formula	Suppose a line passes through (c, d) and has slope m. The equation of the line is $$y - d = m(x - c)$$	pp. 66–67
Parallel lines	Have the same slope ($m_2 = m_1$)	p. 67
Perpendicular lines	Slope of one is the negative of the multiplicative inverse of the other: $$m_1 = -\frac{1}{m_2}, \quad m_2 \neq 0$$	p. 67

Cumulative Review Exercises—Chapter 1

Simplfy the following expressions using the laws of exponents. Write the answers in a form that does not involve any negative or zero exponents.

1. $(9x^{-2})(-3x^{18})^{-4}$

2. $(8a^3b^5)(-4a^{-5}b^2)$

3. $\left(\dfrac{x^{-5}}{x^{-3}}\right)^{-2}$

4. $(10x^{-2}y^3)(25x^{-3}y^2)^{-1}$

5. $\left(\dfrac{27a^6b^{-5}c^2}{9a^{-4}b^3c^{-4}}\right)^{-3}$

6. $\dfrac{(2x^{-2}y^4z^{-5})^{-4}}{(3x^5y^4z)^{-2}}$

Write the following expressions in a form that involves no negative exponents or radicals.

7. $\sqrt{y^2}$

8. $\sqrt[3]{z^4}$

9. $\sqrt[4]{t^2}$

10. $\sqrt{(a^2 + b^2)^3}$

11. $\sqrt{\dfrac{x^{2/3}}{x^9}}$

12. $\sqrt[4]{a^{-8}b^{12}}$

Simplify the following expressions. Where possible, eliminate fractional or negative exponents and use as few terms as possible.

13. $\dfrac{x^{-3} - y^{-3}}{x^{-2} - y^{-2}}$

14. $\dfrac{\dfrac{1}{a} + \dfrac{1}{b}}{a^2 + 2ab + b^2}$

15. $\dfrac{1}{x} + \dfrac{\dfrac{1}{x^3}}{1 - \dfrac{4}{x^4}}$

16. $\dfrac{ab - 1}{a^3 b^3 - 1}$

17. $\sqrt{12x^2 y z^3}$

18. $\sqrt[3]{\dfrac{27a^3}{64}}$

19. $\dfrac{\sqrt[3]{625a^4 b^7}}{\sqrt[3]{5ab^2}}$

20. $\sqrt{(x + y)^5} \, \sqrt{(x + y)^8}$

21. $\sqrt[4]{(81a^4 b^8)^3}$

22. $4\sqrt{\dfrac{160x^9 y^{12}}{2\sqrt[5]{5xy^2}}}$

23. $\dfrac{\dfrac{2^{n+1} x^{n+1}}{(n + 1)^5}}{\dfrac{2^n x^n}{n^5}}$

24. $\dfrac{\dfrac{\sqrt{(x - 1)^{n/2}}}{(x - 1)^{2n}}}{(x - 1)^{-n/4} x}$

Factor if possible.

25. $8x^2 - 6 - 8x$

26. $49x^2 + 25y^2$

27. $1 + 125a^3$

28. $6(a + 1)^2 + 7(a + 1) - 5$

29. $a^3 - 3a^2 b + 3ab^2 - b^3$

30. $6x^{2a} - 5x^a - 1$

31. $x^{3a} + 27$

32. $7x^2 y - 28y$

33. $4x^3 + 30x^2 + 14x$

34. $4a^6 + 108b^3$

35. $55 - 6x - x^2$

36. $a^3 - \frac{1}{8}$

37. $ax^2 + 2axy + ay^2 + bx^2 + 2bxy + by^2$

38. $x^2 - 10xy + 25y^2$

39. $36t^2 - 16m^2$

40. $a^{2x+4} - b^{6x+10}$

Rationalize the denominators of the following expressions.

41. $\dfrac{5}{\sqrt{6}}$

42. $\dfrac{1 - \sqrt{a}}{1 + \sqrt{a}}$

43. $\dfrac{4}{\sqrt{3} - \sqrt{2}}$

44. $\dfrac{\sqrt[3]{18}}{\sqrt[3]{5}}$

Rationalize the numerators of the following expressions.

45. $\dfrac{\sqrt[4]{t}}{5t}$

46. $\dfrac{\sqrt{3 + h} - \sqrt{3}}{h}$

47. $\sqrt{n^2 - n} - n$

48. $\dfrac{\sqrt{x + 3} - \sqrt{x - 3}}{\sqrt{x + 3} + \sqrt{x - 3}}$

Perform the indicated operations.

49. $(\sqrt{xy} + 2\sqrt{x})(3\sqrt{xy} + 2\sqrt{y})$

50. $\sqrt{1 - a^2} + a^2(1 - a^2)^{\frac{1}{2}}$

51. $\left(x^{\sqrt{3}} + x^{\sqrt{2}} \right)^2$

52. Rationalize the denominator.

$$\dfrac{1}{\sqrt{7} - \sqrt{8} + \sqrt{12}}$$

53. Rationalize the numerator.

$$\dfrac{\sqrt{2(x + h) - 3} - \sqrt{2x - 3}}{h}$$

54. Simplify

$$\left[\left(\sqrt{32} \right)^{(6/5 + \sqrt{6/5})} \right]^{(6/5 - \sqrt{6/5})}$$

Solve these equations and inequalities. Where appropriate, describe your answers using interval notation.

55. $2(2x + 4 - 3x) = 8 - 2x$

56. $4(3x + 5) - 25 - 6(4x - 3) + 8x - 90 = 34 - 3x + 6 - 8(x + 4)$

57. $(2x - 5)(3x + 4) = (x + 2)(6x - 2)$

58. $\dfrac{x}{3} - \dfrac{3x - 5}{2} = 8$

59. $4x^2 + 5x = 1$

60. $4x^2 = 3 + x$

61. $10t^2 - t = 100$

62. $1 - \dfrac{1}{6x} + \dfrac{2}{3x^2} = 0$

63. $\dfrac{5}{2x + 6} = \dfrac{1 - 2x}{4x} + 2$

64. $2x^2 + 10x = 14$

65. $(2a - 3)^2 = (a + 1)^2$

66. $y^2 + \sqrt{3}y - 2 = 0$

67. $\sqrt{x^2 + 7} = x - 1$

68. $15 - 3\sqrt[3]{2x + 1} = 0$

69. $x^2 - 9x^{3/2} + 20x = 0$

70. $\sqrt{\dfrac{x + 1}{4x - 1}} = \dfrac{1}{2}$

71. $\dfrac{2}{x - 2} + \dfrac{2x}{x^2 - 4} = \dfrac{3}{x + 2}$

72. $\dfrac{5}{y^2 + 5y} - \dfrac{1}{y^2 - 5y} = \dfrac{1}{y^2 - 25}$

73. $\dfrac{4x - 20}{x - 9} - \dfrac{16}{x} = \dfrac{144}{x^2 - 9x}$

74. $\dfrac{6t}{t - 5} - \dfrac{300}{t^2 + 5t + 25} = \dfrac{2250}{t^3 - 125}$

75. $m^4 - 7m^2 + 10 = 0$

76. $6t^{-2} = 6t^{-1} + 36$

77. $-3\left(\dfrac{x + 2}{x - 3}\right)^2 - 4\left(\dfrac{x + 2}{x - 3}\right) + 4 = 0$

78. $t^{2/3} = 16$

79. $x^3 + 4x^2 - 9x - 36 = 0$

80. $x - 10\sqrt{x} + 16 = 0$

81. $3x - 4 < 5x + 6$

82. $5 - \dfrac{x - 3}{9} < \dfrac{2 + x}{6}$

83. $3 < 6 - \tfrac{1}{3}x \le 9$

84. $|2x - 5| = 7$

85. Determine the distance between points $(-2, -7)$ and $(3, -4)$.

86. Determine the midpoint of the segment with endpoints $(-2, -7)$ and $(3, -4)$.

87. Determine the slope of the line through points $(-2, -7)$ and $(3, -4)$.

88. Determine the equation of the line through $(-2, -7)$ and $(3, -4)$.

89. Find an equation in standard form of a circle having a diameter with endpoints $(-2, -7)$ and $(3, -4)$.

90. Find the slope and y-intercept of $3x = 6y - 24$.

91. Determine the equation of the line through $(5, -6)$ with slope $-\tfrac{3}{2}$.

Determine whether each set of lines is parallel, perpendicular, or neither.

92. $3x + 6y = 3$, $6y = 10 - 4x$

93. $y = 0.32x + 16$, $1000y = 3125x + 7$

94. $12x - 45y = 17$, $77x - 76y = 87$

95. Find the point on the x-axis equidistant from $(7, 3)$ and $(1, -2)$.

96. Find the distance between $(2\sqrt{2},\ \sqrt{5})$ and $(3\sqrt{2}, -2\sqrt{5})$.

97. Find the midpoint of the line segment with endpoints $(2\sqrt{2},\ \sqrt{5})$ and $(3\sqrt{2}, -2\sqrt{5})$.

Describe and graph each equation in Exercises 98–102.

98. $x^2 + (y - 3)^2 = 9$

99. $(x + 2)^2 + (y + 3)^2 = 5$

100. $x^2 + y^2 - 4x + 2y - 7 = 0$

101. $x^2 + y^2 = 12y$

102. $10x^2 + 10y^2 = 30$

103. Determine the equation of a line through $(-1, -5)$ and perpendicular to the line $3x + 4y = 6$

104. Determine an equation of a line through $(-1, -5)$ and parallel to the line $3x + 4y = 6$

105. Determine an equation of a line through $(-2, 3)$ and perpendicular to the line containing the points $(-3, -4)$ and $(-1, 2)$.

106. Determine an equation of a line through $(-2, 3)$ and parallel to the line containing the points $(-3, -4)$ and $(-1, 2)$

◆ Applications

Convert to scientific notation.

107. 2467 (the boiling point of aluminum, in degrees Celsius)

108. 0.00003 (the probability that Joe DiMaggio would get at least one hit in each of 56 consecutive games)

Convert to decimal notation.

109. 6.48×10^{-2} (the number of grams in one grain)

110. 2.4×10^{13} (the distance of the star Alpha Centauri from Earth)

111. One mole of any element contains 6.02×10^{23} atoms. Given the mass of one mole, as follows, find the mass of one atom of:

   a. Hydrogen, 1.0079 g/mole

   b. Aluminum, 26.98154 g/mole

   c. Lead, 207.2 g/mole

112. A tank contains water H m deep. A hole is pierced in a wall of the tank h m below the water surface (see Figure 1). The stream of water coming from that hole will hit the floor x meters from the wall, where x is given by

$$x = \sqrt{h(H - h)}$$

The water level of an above-ground swimming pool is 3 m above ground. A hole is pierced in the pool 2.5 m above ground. How far from the swimming pool will the water hit the ground?

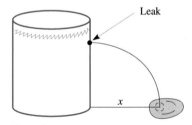

Figure 1

113. The hypotenuse of a triangle is twice as long as the shortest leg of the triangle. Find a formula for the length of the other leg.

114. A circle is inscribed in a square. The area of the circle is 200 cm² (see Figure 2). Find the length of a side of the square. (Use 3.14 for π.)

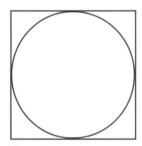

Figure 2

115. A 45°, 45°, 90° triangle has legs of length a. Find the length of the hypotenuse.

116. A cylindrical can with a height twice its diameter has a surface area of 300 square inches. Determine the dimensions of the can.

117. What percent of $225 is $175?

118. A wind is blowing from the west at 50 km/h. A plane flies east 375 km and then back in a total time of 4 hours. How fast does the plane fly in still air?

119. Three pipes can fill a swimming pool in one hour. Alone, pipe A can fill the pool in one third the time pipe B can, and pipe C can fill the pool in one-half the time pipe B can. How long would it take each pipe to fill the pool by itself?

120. A family has budgeted $30 per month for city water. With one month left in the year, the average per month they have spent is $32.50. If water costs $0.85 per hundred cubic feet, how much water can they use in the last month to average $30 for the year?

121. Suppose $50,000 is invested at 8.4 percent. After 5 years, how much is in the account if interest is compounded a. annually, b. semiannually, c. monthly, d. daily (use 360 days per year)?

122. $50,000 is invested for 2 years at an interest rate that is compounded annually. It grows to $59,405. What is the interest rate?

123. To promote business, bank A offers to give a grandfather clock worth $595 to anyone who deposits over $5000 at 9 percent for 10 years, whose interest is compounded annually. Bank B offers no free gift, but compounds interest quarterly, also at 9 percent. Which bank has the better deal for a $5,000 deposit? $10,000?

124. The *period* of a pendulum T, in seconds, is the time it takes to move in one direction and then back, and is given by the formula

$$T = 2\pi \sqrt{\frac{L}{9.8}}$$

where L is the length of the pendulum in meters (see Figure 3).

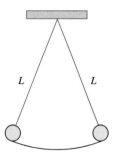

Figure 3

a. How long is a pendulum that has a period of $3\pi/2$ seconds?

b. How long is a pendulum that has a period of 1 second? (Use 3.14 for π.)

c. Solve the formula for L.

125. One leg of a right triangle is 2 cm longer than the other. The hypotenuse is 2 cm longer than the longer leg. What is the perimeter of the triangle?

126. $3500 is deposited at the beginning of the year. At the beginning of the next year, $4400 is deposited in another bank at the same interest rate. At the beginning of the third year, there is a total of $8172.60 in both accounts. If interest is compounded annually, what is the interest rate?

127. A chemist has one solution of sulfuric acid and water that is 20 percent acid and a second that is 75 percent acid. How many liters of each should be mixed together to get 240 liters of a solution that is 40 percent acid?

128. Mercury is a liquid at Celsius temperatures C satisfying the inequality

$$-39° < C < 357°$$

To convert from Celsius to Fahrenheit we use the formula

$$C = \tfrac{5}{9}(F - 32)$$

Find an inequality involving Fahrenheit temperatures F at which mercury is a liquid.

129. A brick is dropped into a well. The sound of the splash is heard 5 seconds later. How deep is the well? (Assume sound travels at 1142 ft/sec.)

130. Two investments are made totaling $120,000. For a certain year, these investments yield $8850 in simple interest. Part of the $120,000 is invested at 9 percent and the rest at 6 percent. Find the amount invested at each rate.

131. The *Highway Code* in Great Britain uses the quadratic equation

$$D = \text{Thinking distance} + \text{Braking distance} = r + \frac{r^2}{20}$$

to estimate, under normal driving conditions, the *stopping distance D* (in feet) of a car traveling r mph.

a. How many feet would it take to stop if you were driving 25 mph? 55 mph? 65 mph?

b. It takes a driver 120 ft to stop a car. How fast was the car moving?

c. For what speeds r is the stopping distance D such that $75 \leq D \leq 175$?

d. Solve the equation for r.

132. The formula

$$S = -16t^2 + v_0 t + h$$

gives the height S, in feet, after t seconds, of an object thrown upward from a height h, at an initial velocity v_0.

a. A ball is thrown upward at an initial velocity of 48 ft/sec from a roof that is 1120 ft high. How long does it take the ball to reach the ground?

b. At what time will the ball be at a height of 1216 ft?

CHAPTER 2

FUNCTIONS AND GRAPHS

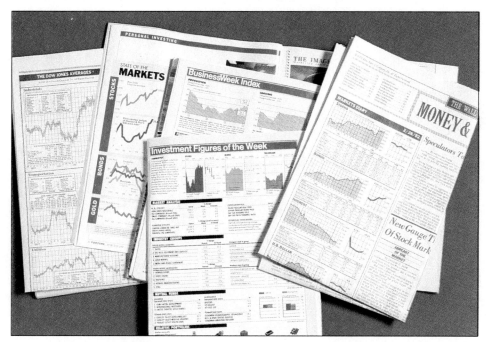

Graphs of stock market prices provide examples of functions that arise in many everyday financial transactions.

T he notion of a function is central in modern mathematics. It arose historically as a generalization of the notion of a "formula," which expresses one variable in terms of another variable. Such relationships arise naturally in applications.

For example, the graph of Figure 1 expresses the relationship between elapsed time and the number of people who have heard a rumor. The graph shows how the rumor is slow to get started and then picks up steam as it spreads. As time passes, most of the population has heard the rumor, and the graph flattens out, indicating the slower spread of the rumor among the few remaining people. The graph is a particular instance of a **logistic curve**, which arises in many mathematical models in such disparate fields as ecology and epidemiology.

Figure 2 shows the relationship between elapsed time and amount remaining of a sample of radioactive carbon 14. The data contained in this graph are used by archaeologists in applying the process of **carbon dating** to determine the ages of ancient artifacts such as mummies and papyrus scrolls.

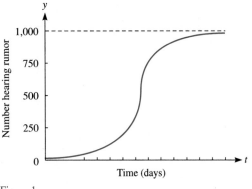

Figure 1

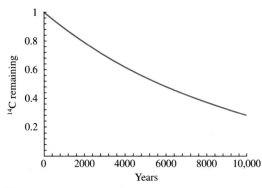

Figure 2

Each of the two graphs depicts a relationship between the variable time, t, and another variable (the number N of people having heard the rumor or the amount A of carbon 14 remaining). For each value of t, the relationship determines the corresponding value of N or A. Such relationships are examples of **functions**, and are the subject of this chapter.

2.1 THE CONCEPT OF A FUNCTION

Definition of a Function

The notion of a function is crucial to all of modern mathematics. All fields of modern mathematics use the terminology of functions to express ideas and results. The notion of a function was not formulated overnight. It took a number of reformulations and generalizations before the modern concept of a function arose. To illustrate the idea of a function, let's start with one of the older formulations. Suppose that a formula expresses the variable y in terms of the variable x in such a way that for each value of x, one value of y is determined. Then we say that **y is a function of x.** For example, consider the formula

$$y = 3x^2$$

For each value of x, this formula determines one corresponding value of y. For instance, if $x = 1$, the corresponding value of y is

$$y = 3 \cdot 1^2 = 3$$

We say that 3 is the **value of the function** at $x = 1$. It is customary to assign letters to denote functions. For instance, the above function can be denoted f, and we may write the function in the form

$$y = f(x)$$

(read "y equals f of x"). This notation indicates that the value of y is determined by the value of x using the function f. We say that x is the **independent variable** and y is the **dependent variable**.

Not every equation that relates x and y defines a function. For example, consider the equation $x = 3y^2$. Here, y is *not* a function of x because for each non-zero value of x, there are *two* values of y satisfying the equation. For instance, the value $x = 3$ corresponds to the two values $y = \pm 1$.

A function may have more than one independent variable. For example, the volume V of a right circular cylinder of radius r and height h is given by the formula

$$V = \pi r^2 h$$

This formula defines a function $C(r, h)$ of the two independent variables r and h. For given values of r and h, the value of the function is given by the formula

$$C(r, h) = \pi r^2 h \quad (r > 0, h > 0)$$

Note that the variables r and h are restricted to be positive because of the physical interpretation of the variables, as the radius and height of a cylinder, respectively.

The previous function f can be viewed as a correspondence that associates the real number x with the real number $3x^2$. In this view, the function is a device for associating elements of one set with elements of a second set. Namely, it associates each x in the set of real numbers $\mathbf{R}$ with the number $3x^2$ in $\mathbf{R}$. The function C associates each point (r, h) in the first quadrant with an element of $\mathbf{R}$.

In many applications of mathematics, it is necessary to consider **correspondences** or **mappings** between sets that have nothing to do with numbers. For example, consider the set of all dots of light on a computer screen. Each dot of light is called a **pixel.** For each image portrayed on the screen, there is a correspondence associating each pixel with its color. We may visualize the correspondence as shown in Figure 3. On the left is the set P of pixels on the screen, and on the right is the set C of possible colors. The correspondence or mapping associates each pixel p with a color c. We give the correspondence the letter S as a name, and we say that S **associates** c **with** p and write

$$c = S(p)$$

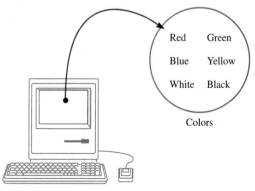

Figure 3

Pictorially, we indicate that c is associated with p by drawing an arrow from p to $S(p)$, as in Figure 4. We may view the correspondence S as a general "formula" that indicates, for each pixel p, the color c. To be sure, this "formula" is not an algebraic one like $y = 3x^2$. However, associating a color with a pixel is no different conceptually from associating the value $3x^2$ with x.

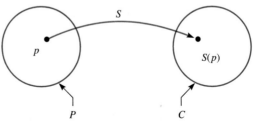

Figure 4

In a similar fashion, we may consider a correspondence F between the set of all people in your math class and the set of months of the year. The correspondence F associates each person with the month in which he or she was born. We might have, for example,

$$F(\text{Jane Smith}) = \text{December}$$

Again, the mapping F can be viewed as a rule that tells you, for each person in your math class, the month of birth.

The preceding examples of correspondences lead to the following definition.

■ **Definition 1** **Function as a** **Correspondence**	Let A and B be sets. A **function** (or **mapping**) f from A to B is a correspondence that associates each element of A with one and only one element of B. If x is an element of A, then the element of B associated with x under the correspondence f is denoted $f(x)$, read "f of x," and is called the **value** of f at x.

If f is a function from A to B, then we write

$$f : A \rightarrow B$$

The set A is called the **domain** of f. That is, the domain of f consists of all x for which the function has a value defined. The subset of B consisting of all values assumed by f is called the **range** of f. If x is in the domain of f, we say that f is **defined at x**. Similarly, if x is not in the domain of f, we say that f **is not defined at x**.

You should view a function as a rule for associating each x in A with one and only one value $f(x)$ in B. You can also think of a function as a machine (say, a computer) into which you feed a value x from A. The machine outputs a unique value $f(x)$ from B. See Figure 5.

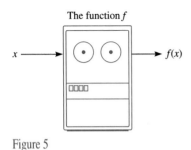

The function f

$$x \longrightarrow f(x)$$

Figure 5

Applications of Functions

Many applied problems give rise to equations involving variables x and y. Many of these equations can be written in an equivalent form in which y is expressed as a function $f(x)$, such as $y = x^2 + 1$.

Any such equation can be regarded as defining a function specifying the value of y in terms of the expression in x. The examples that follow will provide some practice in determining functions that arise in typical applied situations.

➤ **EXAMPLE 1**
Falling Object

A ball is thrown vertically upward from the roof of a skyscraper. Let $h(t)$ denote the height of the ball from the ground t seconds after it is thrown. Then, from the laws of physics, $h(t)$ is given by an expression of the form

$$h(t) = -16t^2 + v_0 t + h_0$$

where v_0 is the initial velocity of the ball and h_0 is the initial height of the ball. Suppose that the building is 1200 feet tall. Furthermore, suppose that an observer sees the ball pass by a window 800 feet from the ground 10 seconds after the ball was thrown. Determine a formula for the function $h(t)$.

Solution

We are given the data $h(0) = 1200$, the height of the ball when it was thrown (at time $t = 0$); and $h(10) = 800$, the height of the ball 10 seconds after it was thrown ($t = 10$). Inserting $t = 0$ into the expression for $h(t)$ gives us

$$h(0) = 1200 = -16(0)^2 + v_0(0) + h_0 = h_0$$

This gives us the value $h_0 = 1200$. Substituting $t = 10$ into the expression for $h(t)$ gives us

$$h(10) = 800 = -16(10)^2 + v_0(10) + 1200$$
$$1200 = 10v_0$$
$$v_0 = 120$$

We can express h as a function of t as follows:

$$h(t) = -16t^2 + 120t + 1200$$

➤ **EXAMPLE 2**
Storage Tank

Large tanks to store industrial chemicals are built in a cylindrical shape 10 feet tall. The paint used for the tanks costs $50 per gallon, and one gallon is sufficient to cover 200 square feet of the tank's surface. Calculate the cost $C(r)$ of painting a tank as a function of its radius r. (Both the top and the sides of the tank must be painted.)

Solution

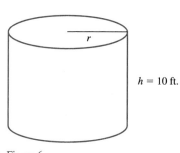

$h = 10$ ft.

Figure 6

Let h denote the height of the tank. (See Figure 6.) From the formulas for the area of a circle and the surface area of a cylinder, the area to be painted equals

$$\text{Area of top} + \text{Area of sides} = \pi r^2 + 2\pi r h$$

Substituting $h = 10$, we have

$$\text{Total area} = \pi r^2 + 2\pi r(10)$$
$$= \pi r^2 + 20\pi r$$

Because one gallon of paint covers 200 square feet, the number of gallons required to paint the entire cylinder is equal to the total area divided by 200:

$$\frac{\pi r^2 + 20\pi r}{200} = \frac{1}{200}\pi r^2 + \frac{1}{10}\pi r$$

We find the cost by multiplying this expression by 50 (the cost of one gallon) and arrive at an expression for the cost C in terms of the radius r:

$$C(r) = \tfrac{1}{4}\pi r^2 + 5\pi r$$

➤ EXAMPLE 3
Distance between Moving Objects

Two cars leave an intersection at the same time, proceeding in perpendicular directions. One car is moving at 45 miles per hour, and the other is moving at 30 miles per hour. Determine an expression for the function $D(t)$ that gives the distance between the two cars as a function of time.

Solution

In t hours, the first car goes $45t$ miles and the second car goes $30t$ miles. The positions of the cars at time t is shown in Figure 7. The distance between the two cars is given by the hypotenuse of a right triangle. So, by the Pythagorean theorem, we have

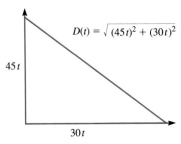

$$D(t) = \sqrt{(45t)^2 + (30t)^2}$$

Figure 7

$$= \sqrt{2025t^2 + 900t^2}$$

$$= \sqrt{(2025 + 900)t^2}$$

$$= \sqrt{2025 + 900} \cdot \sqrt{t^2}$$

$$= \sqrt{2925}\,t$$

(Note that $\sqrt{t^2} = t$ because time t is considered nonnegative.)

Examples of Functions, Domains, and Ranges

Let's explore some of the simplest functions—those that have real values and are specified by formulas involving a single variable.

Suppose that a and b are real constants. We may define the function

$$f(x) = ax + b$$

where x is a real number. Such a function is called a **linear function**.

Similarly, a function of the form

$$f(x) = ax^2 + bx + c, \quad a \neq 0$$

is called a **quadratic function**. A function of the form

$$f(x) = ax^3 + bx^2 + cx + d, \quad a \neq 0$$

is called a **cubic function**. In general, a function in which $f(x)$ is specified as a polynomial expression in x is called a **polynomial function**.

➤ EXAMPLE 4
Calculating Domain and Range

Let f denote the linear function that associates with each real number x the real number $f(x) = 3x + 1$.

1. What is the domain of f?
2. Determine $f(2)$ and $f(-1)$.
3. Determine $f(a)$, $f(a + 1)$, and $f(2t - 1)$, where a and t are real numbers.
4. What is the range of f?

Solution

1. Because x may be any real number, the domain consists of the set of all real numbers **R**. Any value of x is a permissible substitution for x in the formula $3x + 1$.

2. We have

$$f(2) = 3(2) + 1 = 7$$
$$f(-1) = 3(-1) + 1 = -2$$

3. The function may be evaluated at any expression, as long as the value of the expression is contained within the domain. In this case, a and t are presumed to be real numbers. The given expressions are also real numbers and, hence, are in the domain. In each case, we substitute the given expression for x in the definition of the function. In the first case, we replace each occurrence of x by a:

$$f(a) = 3a + 1$$

In the second case, we replace each occurrence of x by $a + 1$:

$$f(a + 1) = 3(a + 1) + 1 = 3a + 4$$

In the third case, we replace each occurrence of x by $2t - 1$:

$$f(2t - 1) = 3(2t - 1) + 1 = 6t - 2$$

4. To determine the range, we must determine those real numbers c that are assumed as values of the function for some x, namely,

$$f(x) = c$$

This equation is equivalent to

$$3x + 1 = c$$
$$x = \tfrac{1}{3}(c - 1)$$

That is, for this value of x, the function f assumes the value c. Because c may be any real number, f assumes all real numbers as values, and its range is **R**, the set of all real numbers.

If we don't specify the domain of a function, we understand the domain to consist of all values of the variable for which the defining expression makes sense. We sometimes refer to this as the **natural domain** of the function. For instance, because the expression $2x^3 - 1$ is defined for all real numbers x, the domain of the function

$$f(x) = 2x^3 - 1$$

consists of all real numbers x. Similarly, because the expression $1/x$ is defined for $x \neq 0$, the domain of the function

$$f(x) = \frac{1}{x}$$

consists of all nonzero real numbers x. Because the expression $\sqrt{x}$ is defined only for nonnegative real numbers, the domain of the function $f(x) = \sqrt{x}$ consists of the nonnegative real numbers x.

The domain is often specified in parentheses after the expression defining the function. For example, the preceding three functions may be defined as follows:

$$f(x) = 2x^3 - 1 \quad \text{(all real } x\text{)}$$

$$f(x) = \frac{1}{x} \quad (x \neq 0)$$

$$f(x) = \sqrt{x} \quad (x \geq 0)$$

In some cases, we wish to specify a domain that is smaller than the natural domain of the defining expression. This may be done by stating the conditions for defining the domain in parentheses after the defining expression. For instance, the domain of the function

$$f(x) = 3x \quad (3 < x \le 5)$$

is the interval $(3, 5]$.

➢ **EXAMPLE 5**
Evaluating a Function

Suppose that we are given the real-valued function

$$f(x) = \sqrt{x - 1}$$

1. Determine the domain of f.
2. Determine the value of $f(10)$ and $f(-1)$.
3. Determine the value of $f(w^2)$.
4. Determine the range of f.

Solution

1. The expression $\sqrt{x - 1}$ produces a real number provided that

$$x - 1 \ge 0$$
$$x \ge 1$$

so the domain is

$$\{x : x \ge 1\} \quad \text{or} \quad [1, \infty)$$

2. For $f(10)$, we have

$$f(10) = \sqrt{10 - 1} = \sqrt{9} = 3$$

Because -1 is not in the domain of f, $f(-1)$ is undefined.

3. We substitute w^2 for x in the definition of the function:

$$f(w^2) = \sqrt{w^2 - 1}$$

The expression on the right gives the value of f at w^2.

4. As x takes values in the domain of f, the expression $x - 1$ assumes as values all nonnegative real numbers, so $\sqrt{x - 1}$ assumes the same set of values as the square root function, namely all nonnegative real numbers.

➢ **EXAMPLE 6**
Determining the Domain of a Quotient

Let h be the function defined by

$$h(x) = \frac{\sqrt{x}}{(x - 1)(x - 2)}$$

1. What is the domain of h?
2. Determine $h(4)$.

Solution

1. The expression on the right requires a number of restrictions on x for it to be defined. First, the radical in the numerator is defined only for nonnegative numbers, that is, for

$$\{x : x \ge 0\}$$

Also, the expression is not defined at values of x for which the denominator is 0. This requires the exclusions $x \neq 1, 2$. Combining these restrictions, we arrive at the domain of h:

$$\{x : x \geq 0 \quad \text{and} \quad x \neq 1, 2\}$$

2. We substitute the given value for x into the expression defining h:

$$h(4) = \frac{\sqrt{4}}{(4-1)(4-2)} = \frac{2}{3 \cdot 2} = \frac{1}{3}$$

In the next example, we evaluate an expression typical of those you will encounter when calculating derivatives in calculus.

➤ **EXAMPLE 7**
Calculus

Let $f(x) = x^2$. Evaluate and simplify the expression
$$\frac{f(a+h) - f(a)}{h}$$
where a and h are real numbers and $h \neq 0$.

Solution

Substituting in the given expression for $f(x)$, we have
$$f(a+h) = (a+h)^2 = a^2 + 2ah + h^2$$

Therefore, we have
$$\frac{f(a+h) - f(a)}{h} = \frac{(a^2 + 2ah + h^2) - a^2}{h}$$
$$= \frac{2ah + h^2}{h}$$
$$= \frac{h(2a + h)}{h}$$
$$= 2a + h$$

Note that we were allowed to cancel h because $h \neq 0$.

➤ **EXAMPLE 8**
Square Root of a Square

Let f be the function defined by
$$f(x) = \sqrt{x^2}$$

1. Determine the domain of f.
2. Find an equivalent expression for f.

Solution

1. For any real number x, the expression x^2 is always nonnegative, so the expression $\sqrt{x^2}$ produces a real number for any real number x.
2. If x is nonnegative, then $\sqrt{x^2} = x$. However, because $(-x)^2 = x^2$, we see that for negative x
$$\sqrt{x^2} = \sqrt{(-x)^2} = -x$$

Thus, we see that
$$\sqrt{x^2} = x \quad (x \geq 0)$$
$$= -x \quad (x < 0)$$

On the right side, we recognize the definition of absolute value, so we have

$$\sqrt{x^2} = |x|$$

as an equivalent expression for f.

$$\sqrt{x^2} = |x|$$

Historical Note

In the 20th century, algebra has evolved in the direction of "abstract algebra," in which the emphasis is on deducing properties of general algebraic structures that are defined by lists of axioms. This point of view has proven to be extremely fertile and has led to the applications of algebra to many disciplines, including physics, chemistry, and engineering. One of the founders of the field of abstract algebra is the German mathematician Emmy Noether. Professor Noether's work was done in the early years of 20th century at Göttingen University, which was the preeminent mathematics department in the world for 75 years. Noether was one of the first women mathematicians who lectured on her work. Notes of her lectures and those of her Göttingen colleague, Emil Artin, were developed into the famous two-volume text *Modern Algebra* by the Dutch mathematician van der Waerden. This text has been in use for almost 75 years and has been the source from which most 20th century algebraists learned their abstract algebra.   ▮▮

Exercises 2.1

For $f(x) = 4x - 5$, find each of the following.

1. $f(0)$ 2. $f(-2)$ 3. $f(-2a)$ 4. $f(a + h)$

For $g(x) = 3x^2 + x - 1$, find each of the following.

5. $g(10)$ 6. $g(-1)$

7. $g(a - h)$ 8. $g(-4t)$

For $f(x) = |5 - x^2|$, find each of the following.

9. $f(0)$ 10. $f(-4)$

11. $f(1.2)$ 12. $f(a + h)$

For $g(x) = \dfrac{1}{x^2} + \dfrac{1}{x} - 1$, find each of the following, if possible.

13. $g(0)$ 14. $g(-2)$ 15. $g\left(\frac{1}{3}\right)$ 16. $g\left(\dfrac{1}{a}\right)$

For $g(x) = 2x^3 - 3x^2$, find each of the following.

17. $g(1)$ 18. $g(-2)$

19. $g(2 - h)$ 20. $g(a + h)$

For $f(x) = (1 + x)^3$, find each of the following.

21. $f(2)$ 22. $f(-1)$

23. $f(a + h)$ 24. $f\left(\dfrac{1}{t}\right)$

For $g(x) = (2x + 1)^2$, find each of the following.

25. $g(0)$ 26. $g\left(-\frac{1}{2}\right)$

27. $g(a + h)$ 28. $g(t + h)$

For $f(x) = 12$, find each of the following.

29. $f\left(-\frac{3}{2}\right)$ 30. $f(0)$

31. $f(a + h)$ 32. $f(t + h)$

Determine the domain and range of each function.

33. $f(x) = 3x + 7$ 34. $g(x) = 5 - 4x$

35. $f(x) = 12$ 36. $g(x) = -4$

37. $f(x) = \dfrac{1}{x - 3}$

38. $f(x) = -\dfrac{2}{3x + 7}$

39. $g(x) = \sqrt{3 - 2x}$

40. $g(x) = \sqrt{1 + 4x}$

41. $g(x) = x^3$

42. $g(x) = 4x^3 + 5$

43. $f(x) = |x|$

44. $g(x) = |5 - 3x|$

45. $g(x) = \sqrt{4 - x^2}$

46. $g(x) = \sqrt{3 - x^2}$

47. $g(x) = \sqrt{x^2 - 5x + 6}$

48. $f(x) = \sqrt{3 + 2x}$

For each function, evaluate and simplify the expressions

$$\frac{f(a + h) - f(a)}{h}, \quad \frac{f(a + h) - f(a - h)}{h}, \quad h \neq 0$$

49. $f(x) = 2x - 3$

50. $f(x) = 4 + 5x$

51. $f(x) = x^2 - 5x + 7$

52. $f(x) = x^3$

53. $f(x) = \sqrt{2x - 3}$

54. $f(x) = x^{-2}$

55. $f(x) = -10$

56. $f(x) = -x$

57. $f(x) = x^4$

58. $f(x) = x^5$

Determine the domain of each function.

59. $f(x) = \dfrac{|x|}{x}$

60. $f(x) = 2x - 4x^{1/2} + 2$

61. $f(x) = \dfrac{1}{x^2 + 3x + 2}$

62. $g(x) = \dfrac{\sqrt{x}}{x^2 - 16}$

63. $g(x) = \sqrt{4x^2}$

64. $f(x) = \sqrt{25x^2}$

65. $f(x) = \sqrt[3]{x}$

66. $f(x) = \dfrac{x^2 - 1}{2x^2 - 3x + 1}$

67. $f(x) = \dfrac{|1 - x|}{x - 1}$

68. $f(x) = x^{3/2}$

69. $f(x) = \dfrac{1}{|4x - 5|}$

70. $f(x) = \dfrac{3}{5x + 9}$

71. $f(x) = \dfrac{1}{\sqrt{x^2 + 5}}$

72. $f(x) = \sqrt{\dfrac{5x - 5}{3x - 3}}$

♦ **Applications**

73. Two cars leave an intersection at the same time, proceeding in perpendicular directions. One car is moving at 65 mph, and the other is moving at 55 mph. Let D denote the distance between the cars. Determine a formula in one variable for $D(t)$, where t is the time the cars travel.

74. From a thin piece of cardboard 20 in. by 20 in., square corners of length x are cut out so the sides can be folded up to make a box. Let V denote the volume of the box. Determine a formula in one variable for $V(x)$.

75. A container company is constructing an open-top, rectangular metal tank with a square base of length x that will have a volume of 300 cubic ft. Let S denote the surface area of the tank. Determine a formula in one variable for $S(x)$.

76. The owner of a 30-unit motel, by checking records of occupancy, knows that when the room rate is $50 a day, all units are occupied. For every increase of x dollars in the daily rate, x units are left vacant. Each unit occupied costs $10 per day to service and maintain. Let R denote the total daily profit. Determine a formula in one variable for $R(x)$.

77. A Palladian window is a rectangle with a semicircle on top, as shown in the figure. Suppose the perimeter of a particular Palladian window is 28 ft. Let A denote the total area of the window. Determine a formula in one variable for $A(x)$.

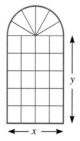

Exercise 77

78. Let A denote the area of a circle with diameter d. Determine a formula in one variable for $A(d)$.

79. A power line is to be constructed from a power station at point A to an island at point Q, which is 1 mile out in the water from a point B on the shore. (See the accompanying figure.) Point B is 3 miles downshore from the power station at A. It costs $6000 per mile to lay the line under the water and $4000 per mile to lay the line underground. Let C denote the total cost of the power line. Determine a formula for $C(x)$.

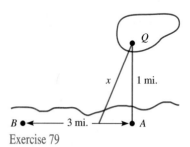

Exercise 79

80. A 28-in. piece of string is to be cut into two pieces. One piece is to be used to form a circle and the other to form a square. Let A denote the total area. Determine a formula for $A(x)$.

81. A ball is thrown vertically downward from the roof of a giant new skyscraper. Let S denote the height of the ball from the ground t seconds after being thrown. Then, $S(t)$ is given by an expression of the form

$$S(t) = -16t^2 + v_0 t + h$$

where v_0 represents the initial velocity and h represents the height of the building. Suppose that the building is 1800 feet tall. Furthermore, suppose that an observer sees the ball pass by a window 700 feet from the ground 8 seconds after the ball was thrown. Determine a formula for $S(t)$.

82. An airplane can travel 500 mph in still air. It travels 2080 miles with a tailwind of w mph and 1440 miles against the same tailwind. Let t denote the total time of travel. Find a function in one variable for $t(w)$.

✍ In Your Own Words

83. Think about the following relation that assigns values to the cells of the first column and first row in a spreadsheet:

$$f(x) = \begin{cases} rownumber & \text{if } x \text{ is in column A} \\ -1 & \text{if } x \text{ is in row 1} \end{cases}$$

Is this a function in the domain for which it is defined? Why or why not?

∎ 2.2 THE GRAPH OF A FUNCTION

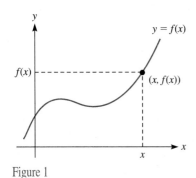

Figure 1

Let $y = f(x)$ be a function with independent variable x and dependent variable y, where x and y both assume real values. The graph of the equation $y = f(x)$ is called the **graph of the function** f. This graph is generally a curve of some sort depicting the relationship between x and y.

A point (x, y) is on the graph precisely when x and y satisfy the equation—that is, provided $y = f(x)$—so the graph consists of the points

$$(x, f(x))$$

where x is in the domain of f. See Figure 1.

From the graph of a function, the function value $f(x)$ may be determined for any value of the independent variable x. Just start at x on the horizontal axis and proceed vertically upward or downward until you get to the graph. The y-coordinate of the corresponding point is the value of $f(x)$. Given the graph of a function, you know the function, in the sense that you can (at least approximately) determine the function value for any x. To put it succinctly, the graph defines the function. For instance, consider the graph of Figure 1 at the beginning of this chapter, which plots the number of people N who have heard a rumor (y-axis) against elapsed time t (t-axis). For each value of t, the graph specifies the number of people who had heard the rumor by time t. That is, the graph specifies N as a function of t.

The domain of a function may be visualized in terms of its graph as a subset of the x-axis. A point a is in the domain provided that there is a point on the graph directly above or below a. For example, in Figure 2, note that a_1 is in the domain but a_2 is not.

The range of a function may be visualized as a subset of the y-axis. A point b on the y-axis is in the range provided that there is a point on the graph at the same height as b. For example, in Figure 3, note that b_1 is in the range but b_2 is not.

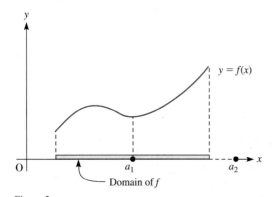

Figure 2

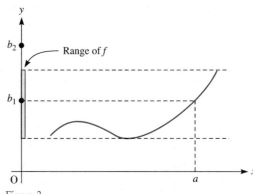

Figure 3

➤ **EXAMPLE 1**
Determining Domain
and Range from the Graph

Each of the following figures shows the graph of a function. From the graph, determine the domain and range of the corresponding function.

1.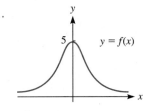

2.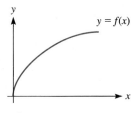

3.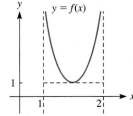

4.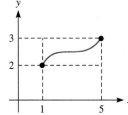

Solution

1. The domain consists of the values of x over which lie points on the graph. In this case, there is a point over every value of x on the x-axis, so the domain is the infinite interval $(-\infty, \infty)$. The range consists of all values y on the y-axis for which there are points on the graph at the same height. In this case, the range is the interval $(0, 5]$.

2. Domain $= [0, \infty)$, range $= [0, \infty)$

3. Domain $= (1, 2)$, range $= [1, \infty)$

4. Domain $= [1, 5]$, range $= [2, 3]$

➤ **EXAMPLE 2**
Graphing a Function

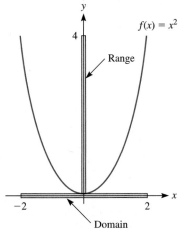

Figure 4

Graph the function

$$f(x) = x^2 \quad (-2 \le x \le 2)$$

From the graph, determine the domain and range of f.

Solution

We graphed the equation

$$y = x^2$$

earlier in the chapter. In Figure 4, we have drawn this graph. The graph of the function consists of the points with x-coordinates satisfying the restriction $-2 \le x \le 2$. The domain and range are indicated on the x- and y-axes, respectively. The domain is the interval $[-2, 2]$, and the range is the interval $[0, 4]$.

➢ **EXAMPLE 3**
Graphing

Sketch the graph of the function

$$f(x) = \sqrt{1 - x^2}$$

Solution

In order for $f(x)$ to be defined, the quantity within the radical must be nonnegative. That is,

$$1 - x^2 \geq 0$$
$$x^2 \leq 1$$
$$\sqrt{x^2} \leq \sqrt{1}$$
$$|x| \leq 1 \qquad \text{(See p. 88.)}$$
$$-1 \leq x \leq 1$$

So the domain is the set

$$\{ x : -1 \leq x \leq 1 \}$$

Next, we must graph the equation

$$y = \sqrt{1 - x^2}$$

We square both sides of this equation to obtain

$$y^2 = 1 - x^2 \qquad \text{and} \qquad y \geq 0$$
$$x^2 + y^2 = 1 \qquad \text{and} \qquad y \geq 0$$

The last equation has as its graph a circle with center $(0, 0)$ and radius 1. The condition $y \geq 0$ restricts the graph to the upper semicircle. See Figure 5.

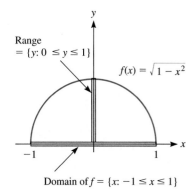

Range $= \{ y: 0 \leq y \leq 1 \}$

$f(x) = \sqrt{1 - x^2}$

Domain of $f = \{ x: -1 \leq x \leq 1 \}$

Figure 5

Let f be a function whose domain and range are subsets of the real numbers. With each x in the domain of f, the function associates a single real number $f(x)$. This means that on a graph of $f(x)$, for a given value of x there is only one point (x, y). To put this another way, a vertical line passing through x on the horizontal axis intersects the graph in exactly one point, namely, the point $(x, f(x))$. See Figure 6. Not *every* vertical line necessarily intersects the graph, of course.

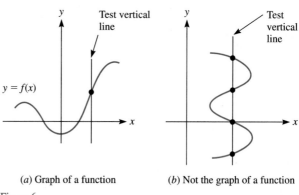

(a) Graph of a function   (b) Not the graph of a function

Figure 6

Vertical lines through various points on the x-axis tell us much about a graph. If x is not in the domain of a function f, then there is no point of the form (x, y)

on the graph, and thus a vertical line through x does not intersect the graph. See Figure 6(a). Furthermore, if a vertical line passing through x intersects the graph in more than one point, the graph does not have a single y associated with x and is not the graph of a function. See Figure 6(b).

We summarize these observations in the following test, which allows us to determine whether a graph in the xy-plane is the graph of a function.

VERTICAL LINE TEST

A set of points in the xy-plane is the graph of a function if and only if each vertical line intersects the graph in at most one point.

> **EXAMPLE 4**
Using the Vertical Line Test

Determine which of the graphs in Figure 7 are the graphs of functions.

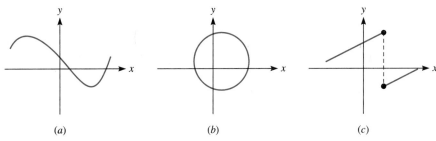

Figure 7

Solution

1. In Figure 8(a), any vertical line either does not intersect the graph or intersects it at a single point. This graph is the graph of a function by the vertical line test.

2. In Figure 8(b), we have drawn a vertical line that intersects the graph in more than one point. According to the vertical line test, the graph is not the graph of a function.

3. In Figure 8(c), we have drawn a vertical line that intersects the graph more than once. According to the vertical line test, the graph is not the graph of a function.

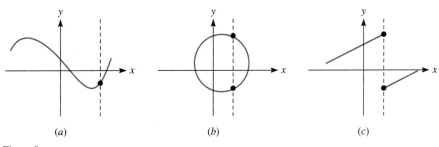

Figure 8

Horizontal and Vertical Translations and Scaling

Up to this point, we have described only one method of sketching the graphs of functions, namely by plotting points. This method is tedious and produces an accurate graph only if you use enough points and take care that these points represent all of the various geometric features of the graph. There are a number of additional methods for sketching graphs of functions. Some of these belong to the province of algebra; others overlap into calculus. In the remainder of this section, we present several methods for sketching graphs using known graphs as a starting point.

The first method we present involves functions whose graphs are obtained by translating known graphs vertically or horizontally.

Suppose that $f(x)$ is a given function and that c is a real number. The graph of the function is the graph of the equation

$$y = f(x)$$

Suppose that we translate this graph vertically a directed distance of c units (upward if $c > 0$ and downward otherwise). Every y-value on the new graph is obtained by adding c to the corresponding y-value on the old graph. If (x, y) is a point on the new graph, $(x, y - c)$ is a point on the old graph, so (x, y) satisfies the equation

$$y - c = f(x)$$

Or, solving for y, we have the equation

$$y = f(x) + c$$

This equation represents a function, which we will call $g(x)$; that is,

$$g(x) = f(x) + c$$

which can be evaluated for each x by adding c to the value of $f(x)$. As we have just seen, the graph of $g(x)$ can be obtained by vertically shifting the graph of $f(x)$. See Figure 9. Rephrasing this geometric fact gives us the following result.

PRINCIPLE OF VERTICAL TRANSLATION

Let $f(x)$ be a function. Then the graph of the function g, defined by

$$g(x) = f(x) + c$$

is obtained by vertically translating the graph of $f(x)$ by c units. If $c > 0$, the shift is upward. If $c < 0$, the shift is downward.

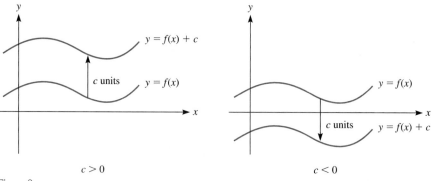

Figure 9

> **EXAMPLE 5**

Applying Translation

Sketch the graph of the function

$$f(x) = \sqrt{1 - x^2} + 4$$

Solution

We have already seen (Example 3) that the graph of the function

$$g(x) = \sqrt{1 - x^2}$$

consists of the upper half of the circle of radius 1 centered at the origin. According to the principle of vertical shifting, the graph of the function $f(x)$ can be obtained by shifting the graph of $g(x)$ upward 4 units. See Figure 10.

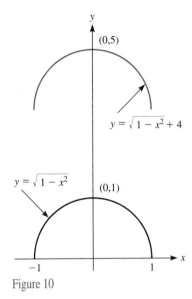

$y = \sqrt{1 - x^2} + 4$

$y = \sqrt{1 - x^2}$

(0,5)

(0,1)

−1   1

Figure 10

We can also sketch graphs by translating known graphs horizontally. Start with a function $f(x)$. Suppose we shift the graph horizontally a directed distance of c units. Then, if (x, y) is a point on the new graph, $(x - c, y)$ is a point on the old graph and satisfies the equation

$$y = f(x - c)$$

That is, we have the following result.

PRINCIPLE OF HORIZONTAL TRANSLATION

Let $g(x)$ be a given function. Then the graph of the function $g(x - c)$ can be obtained by shifting the graph of $g(x)$ by a directed distance of c units horizontally. If $c > 0$, the shift is to the right. If $c < 0$, the shift is to the left. See Figure 11.

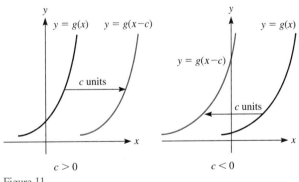

Figure 11

In addition to horizontal and vertical translation, we may apply **scaling** to known graphs. Suppose that $f(x)$ is a given function and that c is a positive number. Consider the graph of the function $cf(x)$. A typical point on this graph is

$$(x, cf(x))$$

where x is in the domain of f. This point can be obtained by replacing the point $(x, f(x))$ with a point with the same x-coordinate but with the y-coordinate scaled to be c times as large. See Figure 12.

Scaling results from multiplying the y-values of a function by a positive number. If the positive number is less than 1, then each y-value shrinks; in this case, the scaling is called **shrinking**. If the positive number is greater than 1, each y-value grows; in this case, the scaling is called **stretching**.

If we multiply a function f by -1, we obtain the function $-f$, whose graph is obtained by reflecting the graph of f in the x-axis. See Figure 13. More generally, if c is *any* number, then the graph of cf can be obtained by first scaling by a factor of $|c|$ and then, if $c < 0$, reflecting the resulting graph in the x-axis.

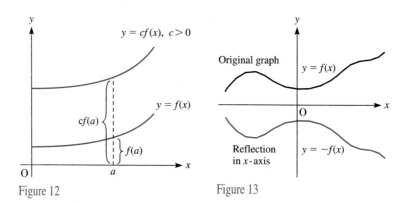

Figure 12   Figure 13

> **EXAMPLE 6**

Applying Translation and Scaling

Sketch the graphs of the following functions.

1. $f(x) = (x - 2)^2$ 2. $f(x) = (x + 3)^2$ 3. $f(x) = 2x^2$
4. $f(x) = -2x^2$ 5. $f(x) = (x - 1)^2 + 3$

Solution

1. By the principle of translation, the graph of $f(x)$ can be obtained by translating the graph of $g(x) = x^2$ to the right 2 units. See Figure 14(*a*).

2. By the principle of translation, the graph of $f(x)$ can be obtained by translating the graph of $g(x) = x^2$ to the left 3 units (because $c = -3$ is less than 0). See Figure 14(*b*).

3. The graph of $f(x) = 2x^2$ can be obtained by scaling the graph of $g(x) = x^2$ by 2. See Figure 14(*c*).

4. The graph of $f(x) = -2x^2$ can be obtained by first scaling the graph of $g(x) = x^2$ by $|-2| = 2$. Because the scaling factor (-2) is less than 0, we must then reflect the graph in the x-axis, yielding the graph in Figure 14(*d*).

5. The graph of $f(x) = (x - 1)^2 + 3$ can be obtained by translating the graph of $g(x) = x^2$ to the right 1 unit and 3 units upward. See Figure 14(*e*).

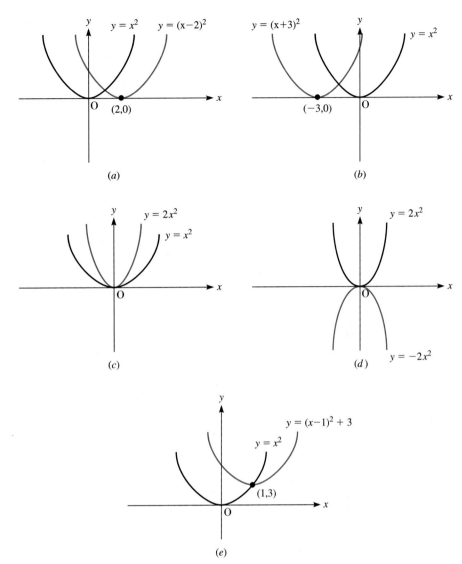

Figure 14

Increasing and Decreasing Functions

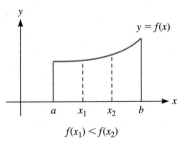

$f(x_1) < f(x_2)$

Figure 15

Up to this point, we have concentrated on developing techniques for sketching graphs of functions. In many applied problems, however, we are given graphs of functions and must describe their features and interpret these features in terms of the applications. One of the most important features is whether a graph is increasing or decreasing for values of x in a particular interval. Specifically, we say that a function $f(x)$ is **increasing on the interval** (a, b) provided that the values of $f(x)$ increase as x moves from left to right through the interval, that is, provided $f(x_1) < f(x_2)$ whenever $a < x_1 < x_2 < b$. Figure 15 shows a graph that is increasing on the interval (a, b). Similarly, we say that $f(x)$ is **decreasing on the interval** (a, b) provided that the values of $f(x)$ decrease as x moves from left to right through the interval, that is, provided $f(x_1) > f(x_2)$ whenever

$a < x_1 < x_2 < b$. Figure 16 shows a graph of a function that is decreasing on the interval (a, b).

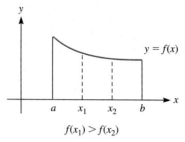

$$f(x_1) > f(x_2)$$

Figure 16

> **EXAMPLE 7**
Income Function

Figure 17 shows the graph of a function that represents the income generated by a young biotechnology firm t years after its founding.

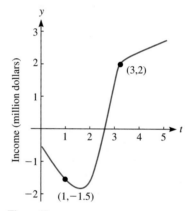

Figure 17

1. Determine the increase in revenue from year 1 to year 3.
2. In what intervals is the graph increasing, and in what intervals is it decreasing?

Solution

1. Reading the graph, we see that the ends of years 1 and 3 correspond, respectively, to the points $(1, -1.5)$ and $(3, 2)$. The change from year 1 to year 3 is given by the difference in y-coordinates, namely, $2 - (-1.5) = 3.5$. The positive sign of the difference indicates that revenue increased from year 1 to year 3, and the amount of increase was $3.5 million.
2. Examining the graph, we see that it is decreasing in the interval $(0, 2)$ and increasing in the interval $(2, 5)$.

Symmetry of Graphs

Let's now see how the geometric notion of symmetry of the graph of an equation with respect to either the x- or the y-axis can be described in terms of functions.

Definition 1 **Even and Odd Functions**	Let f be a function. We say that f is an **even function** provided that $$f(x) = f(-x)$$ for all x in the domain of f. We say that f is an **odd function** provided that $$f(-x) = -f(x)$$ for all x in the domain of f.

➤ **EXAMPLE 8**

Evenness and Oddness

Determine which of the following functions are even, which are odd, and which are neither even nor odd.

1. $f(x) = x^3 - 2x$ 2. $f(x) = \dfrac{1}{x^2 + 1}$

3. $f(x) = x^2 + 3$ 4. $f(x) = (x + 1)^2$

Solution

1. We substitute $-x$ for x in the expression for f:
$$f(-x) = (-x)^3 - 2(-x)$$
$$= -x^3 + 2x$$
$$= -\left(x^3 - 2x\right)$$

We see that this last expression equals $-f(x)$, so f is an odd function.

2. Again we substitute $-x$ for x:
$$f(-x) = \frac{1}{(-x)^2 + 1} = \frac{1}{x^2 + 1}$$

Because the last expression is the same as the expression for $f(x)$, $f(x)$ is an even function.

3. Again we substitute $-x$ for x in the expression for f:
$$f(-x) = (-x)^2 + 3 = x^2 + 3$$

The last expression is just the expression for $f(x)$. In this case, $f(x)$ is an even function.

4. Proceeding as in the previous parts, we have
$$f(-x) = (-x + 1)^2 = x^2 - 2x + 1$$

Because

$$f(x) = (x + 1)^2 = x^2 + 2x + 1$$

we see that $f(-x)$ is equal to neither $f(x)$ nor $-f(x)$ for all x in the domain of f. Thus, f is neither even nor odd.

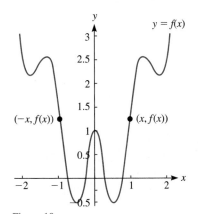

Figure 18

Suppose $f(x)$ is an even function. If $(x, f(x))$ is a point on the graph of f, then so is $(-x, f(-x)) = (-x, f(x))$. These two points are located symmetrically with respect to the y-axis. This is illustrated in Figure 18. In other words, an even function has a graph that is symmetric with respect to the y-axis.

As examples of graphs of even functions, consider the functions in parts 2 and 3 of Example 8. Their graphs are symmetric with respect to the y-axis, as shown in Figures 19 and 20.

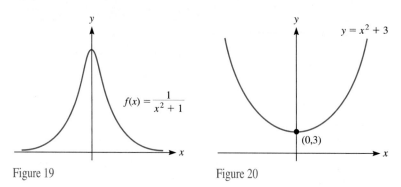

Figure 19   Figure 20

Suppose f is an odd function. Then for each x in the domain of f, the points $(x, f(x))$ and $(-x, -f(x))$ are on the graph. These two points are located symmetrically with respect to the origin. See Figure 21.

As an example of the graph of an odd function, consider the function f defined in part 1 of Example 8. Its graph is illustrated in Figure 22.

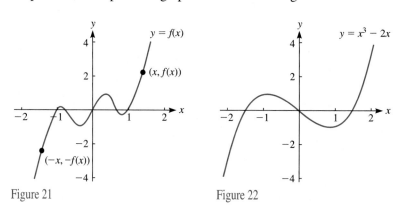

Figure 21   Figure 22

Piecewise-Defined Graphs

All of the functions we have considered so far have been defined using a single formula. However, it is possible to define a function using several formulas, one for each of several parts of the domain. For example, consider the function

$$f(x) = \begin{cases} x & (x < 0) \\ -3x & (x \geq 0) \end{cases}$$

For $x < 0$, we use the first formula to determine $f(x)$; for $x \geq 0$, we use the second formula. For instance, for $x = -3$ and $x = 2$, we have

$$f(-3) = -3, \qquad f(2) = -3(2) = -6$$

A function that is defined using different formulas is said to be **piecewise defined**. The graph of a piecewise-defined function is drawn by separately sketching the graph corresponding to each formula. For example, the graph of the preceding function is shown in Figure 23.

Piecewise-defined graphs are often used in applications, as the next example illustrates.

Figure 23

➤ EXAMPLE 9
Engineering

The voltage on a certain pin of a semiconductor device is equal to either +5 volts or −5 volts. Let $f(t)$ denote the voltage on the pin at time t, where $t = 0$ corresponds to the beginning of a certain circuit test. Graph the function f assuming that the voltage is initially −5 volts for $t < 0$, equal to +5 volts for t between 0 and 2 milliseconds inclusive, and equal to −5 volts thereafter.

1. Write formulas defining $f(t)$.
2. Sketch the graph of $f(t)$.

Solution

1. Using the data given, $f(t)$ is piecewise defined, requiring three defining formulas, as follows:

$$f(t) = \begin{cases} -5 & (t < 0) \\ 5 & (0 \le t \le 2) \\ -5 & (t > 2) \end{cases}$$

2. We sketch the graph of $f(t)$ by separately considering the function over the intervals

$$t < 0, \qquad 0 \le t \le 2, \qquad t > 2$$

In each interval, we use the appropriate definition of $f(t)$ to produce the graph. The result is the graph shown in Figure 24.

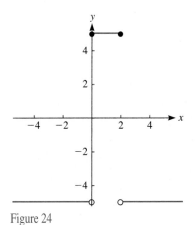

Figure 24

A piecewise-defined function that appears often in applied problems is the **greatest integer function**, denoted by $[x]$. This function is defined for all real numbers x, and its value is given by

$$[x] = \text{greatest integer} \le x$$

Here are the values of the greatest integer function for selected values of x:

$$[5.1] = 5, \qquad [0.17] = 0, \qquad [111] = 111, \qquad [-1.1] = -2$$

Note that if $0 \le x < 1$, the largest integer less than or equal to x is 0, so that

$$[x] = 0 \qquad (0 \le x < 1)$$

For $1 \le x < 2$, the largest integer less than or equal to x is 1, so that

$$[x] = 1 \qquad (1 \le x < 2)$$

Similarly, for any nonnegative integer n, if $n \le x < n + 1$ the largest integer less than or equal to x is n, so that

$$[x] = n \qquad (n \le x < n + 1)$$

The greatest integer function has a separate formula for each interval $n \le x < n + 1$. The graph of the function for x in this interval is a constant. The total graph resembles a series of steps, with a jump occurring at each integer. See Figure 25.

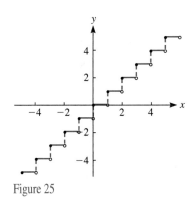

Figure 25

➤ EXAMPLE 10
Cost Function

A cargo service charges by the weight of a package. It charges a flat fee of $2 plus $1 for each whole pound. Determine a formula for the cost $C(x)$ of sending a package weighing x pounds.

Solution

The function $C(x)$ is defined for $x > 0$. The cost of $1 for each whole pound can be expressed in terms of the greatest integer function as $[x]$ dollars. Indeed,

for amounts less than 1 pound, $[x] = 0$; for weights at least 1 pound but less than 2 pounds, $[x] = 2$; and so forth. The function $[x]$ counts 1 for each whole pound. Adding in the flat fee, the desired cost function is given by

$$C(x) = 2 + [x]$$

Exercises 2.2

Determine whether each of the following is the graph of a function.

For each graph of a function, find $f(-2)$, $f(0)$, $f(3)$, and $f(5)$.

1.
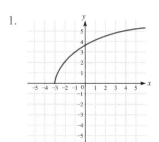

2.
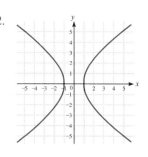

5.
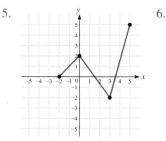

6.
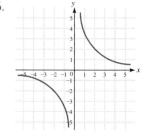

3.
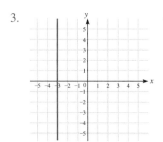

4.
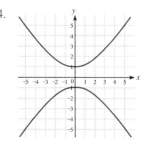

7.
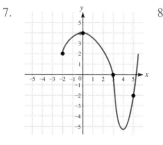

8.
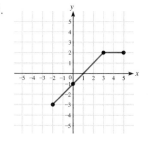

Determine the domain and range of the following functions.

9. See Exercise 5.

10. See Exercise 6.

11. See Exercise 7.

12. See Exercise 8.

Graph each function and from the graph determine the domain and range.

13. $f(x) = 4 - 8x \quad (-1 \le x \le 1)$

14. $f(x) = 2x - 5 \quad (-3 \le x \le 3)$

15. $g(x) = x^2 \quad\quad (-1 \le x \le 1)$

16. $g(x) = x^2 - 3 \quad (-2 \le x \le 2)$

17. $f(x) = x^2 - 3$

18. $g(x) = x^3 - 1$

19. $g(x) = x^3 - 1 \quad (-2 \le x \le 2)$

20. $g(x) = 5 - x^2 \quad (-3 \le x \le 3)$

21. $f(x) = \sqrt{4 - x^2} \quad (-2 \le x \le 2)$

22. $f(x) = \sqrt{-9 + x^2} \quad (|x| \ge 3)$

23. $f(x) = \dfrac{1}{x} \quad (x > 0)$

24. $f(x) = (x - 1)^2 \quad (-1 \le x \le 3)$

Consider the following graph of a function $y = f(x)$ for Exercises 25–34.

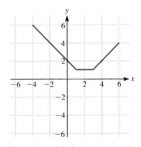

Exercises 25–34

Graph each of the following.

25. $y = f(x) + 1$

26. $y = f(x) - 1$

27. $y = f(x - 1)$

28. $y = f(x + 3)$

29. $y = f(x + 1) - 4$

30. $y = f(x - 2) + 3$

31. $y = -2f(x - 2) + 2$

32. $y = 2f(x) - 5$

33. $y = -\frac{1}{2}f(x) + 1$

34. $2y - 6 = \frac{1}{2}f(x - 4)$

35. a. Graph $f(x) = \sqrt{x}$, and then use that graph to graph parts (b)–(f).

    b. $g(x) = \sqrt{x - 3}$

    c. $g(x) = 2 + \sqrt{x + 1}$

    d. $g(x) = 4\sqrt{x}$

    e. $g(x) = -\sqrt{2x}$

    f. $g(x) = -1 - \frac{1}{2}\sqrt{x - 1}$

36. a. Graph $y = |x|$, and then use that graph to graph parts (b)–(f).

    b. $y = |x + 3|$

    c. $y = |x - 1| - 2$

    d. $y = |-2x|$

    e. $y = -\frac{1}{2}|x| + 2$

    f. $2y - 6 = |x + 1|$

Determine whether each graph is symmetric with respect to the x-axis, the y-axis, or neither.

37. $x^2 + y^2 = 4$

38. $(x - 4)^2 - (y - 2)^2 = 25$

39. $x = y$

40. $y = |x + 2| + |x - 2|$

41. $3x + 3y = 5$

42. $3x^4 - y = 10$

43. $x + 3 = |y|$

44. $|x| + 8 = 2y$

Determine whether each graph is symmetric with respect to the origin.

45. $x^2 - 3y^2 = 4$

46. $(x + 3)^2 + (y - 5)^2 = 16$

47. $2y = \dfrac{5}{3x}$

48. $3x^2 - 7y^2 = 8x$

49. $\dfrac{1}{x} + \dfrac{2}{y} = 3$

50. $|y| = |-2x|$

51. $y = x^7 + x^5 + x^3 + x$

52. $y = x^6 + x^4 + x^2 + 5$

Determine whether each of the following functions is even, odd, or neither.

53. $f(x) = 3x - 2$

54. $f(x) = -5x$

55. $f(x) = 2x^3 - 2x^2 + x$

56. $f(x) = -2x^2$

57. $f(x) = \dfrac{3}{x^2}$

58. $f(x) = \dfrac{5}{x^3 - x}$

59. $f(x) = (x - 3)^2$

60. $f(x) = x^2 + 2x - 4$

61. $f(x) = \left(\dfrac{1}{x}\right)^{-3}$

62. $f(x) = -\left|\frac{2}{3}x\right|$

63. $f(x) = 4x^{2/3}$

64. $f(x) = (x^2 + 8)^2$

Graph each piecewise-defined function.

65. $f(x) = \begin{cases} x & (x \le 1) \\ -x & (x > 1) \end{cases}$

66. $f(x) = \begin{cases} -1 & (x < 0) \\ 0 & (x = 0) \\ 1 & (x > 0) \end{cases}$

67. $f(x) = \begin{cases} x - 1 & \text{if } x < -3 \\ x^2 & \text{if } -3 \le x \le 3 \\ 1 - x & \text{if } x > 3 \end{cases}$

68. $f(x) = \begin{cases} |x| & (x < 2) \\ -x^2 & (2 \le x < 3) \\ 2x - 1 & (x \ge 3) \end{cases}$

69. $f(x) = \begin{cases} \dfrac{x^2 - 9}{x - 3} & (x \ne 3) \\ 2 & (x = 3) \end{cases}$

70. $f(x) = \begin{cases} \dfrac{x^2 - 9}{x + 3} & (x \ne -3) \\ 6 & (x = -3) \end{cases}$

71. $f(x) = \begin{cases} -x - 3 & (x \le -2) \\ 3 - x^2 & (-2 < x < 2) \\ x - 3 & (x \ge 2) \end{cases}$

72. $f(x) = \begin{cases} -3 & (x \text{ not an integer}) \\ 3 & (x \text{ an integer}) \end{cases}$

♦ Applications

73. A parking garage charges $1.00 for the first hour or part of an hour, and $0.50 for each hour or part of an hour thereafter, up to a maximum of $5.00 per day. Graph this function for 0 to 10 hours.

74. It costs $20 to connect to a mainframe computer and $20 per minute or part of a minute of CPU time for the first 30 minutes. After that the cost drops to $10 per minute or part of a minute. Graph this function for the use of 0 to 1 hour of CPU time.

The greatest integer function is given by $f(x) = [x]$, or, in computer language, $f(x) = \text{INT}(x)$. Use the graph of this function to construct the following graphs.

75. $g(x) = [x - 1]$

76. $g(x) = 2[x]$

77. $g(x) = \text{INT}(x)$

78. $f(x) = 3\text{INT}(x + 2) - 1$

79. We define a *decimal place function* as follows: Given a number x, find its decimal notation. Then $f(x)$ is the digit in the fourth decimal place. For example,

$$f\left(\tfrac{2}{5}\right) = f(0.400000) = 0$$
$$f\left(\tfrac{18}{19}\right) = f(0.947368\ldots) = 3$$
$$f\left(-\tfrac{39}{14}\right) = f(-2.785714\ldots) = 7$$

and so on.

   a. Find $f(\tfrac{3}{4})$, $f(\tfrac{14}{23})$, $f(\tfrac{23}{13})$, and $f(\sqrt{2})$.

   b. Determine the domain and range of the function.

   c. Describe the graph of this function.

✍ In Your Own Words

80. For what values of c is the equation $ax + by = c$ symmetric with respect to the origin?

81. For what values of a, b, and c is the equation $ax + by = c$ symmetric with respect to the x-axis? To the y-axis?

■ 2.3 PLOTTING AND GRAPHING USING TECHNOLOGY

Many hand-held calculators have the capability to graph equations. By using such calculators, you can perform many of the graphical tasks of algebra with minimal effort. The following sort discussion is an introduction to graphing calculators using the Texas Instruments TI-81™ calculator to provide specific keystrokes. If you have another brand of graphing calculator, its operation is similar, but you will need to refer to your calculator reference manual for specific keystrokes.

Fundamentals of Graphing Using a Calculator

```
:Y1 =
:Y2 =
:Y3 =
:Y4 =
```

Figure 1

Basic Calculator Techniques   In order to graph an equation using the TI-81, you must first enter the equation. This is done by pressing the key (Y =), which results in a display like the one shown in Figure 1. The screen shown allows you to enter up to four equations to graph simultaneously. For instance, suppose that you wish to graph the equation

$$y = x^2$$

You would move the cursor to the $Y1 =$ line and type

$$\text{X}^2$$

You will find the X on the key (X|T).

You may edit equations using the arrow keys to move the cursor, (Ins) to enable insertion at the cursor, and (Del) to delete the character at the cursor.

Once the equation(s) are entered to your satisfaction, you must set the x-range and y-range you wish to view. For the moment, let's use the default ranges of the calculator, which are

$$-10 \le x \le 10, \qquad -10 \le y \le 10$$

Using the defaults means that we don't need to set the ranges and can view the graph.

To view the graph, press the key [Graph]. The calculator will now display the graph of the specified equation(s) in the specified ranges. You will see a graph as shown in Figure 2.

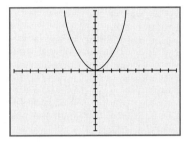

Figure 2

Setting Ranges   To set ranges, press the key [Range]. You will see a display as shown in Figure 3. Use the cursor movement and editing keys to enter the desired minimum and maximum values of x (denoted Xmin and Xmax, respectively) and the desired minimum and maximum values of y (denoted Ymin and Ymax, respectively). For instance, to graph

$$y = x^2$$

in the range

$$-2 \le x \le 2, \qquad -4 \le y \le 4$$

we would set

$$Xmin = -2$$
$$Xmax = 2$$
$$Ymin = -4$$
$$Ymax = 4$$

Pressing [Graph] again, we would see the graph shown in Figure 4.

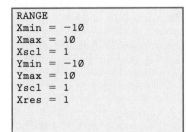

Figure 3

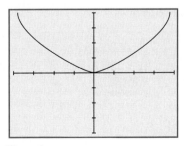

Figure 4

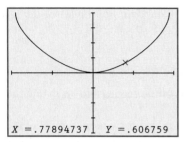

$X = .77894737$ $Y = .606759$

Figure 5

Tracing   You may move a point along a displayed graph and have the calculator display the numerical coordinates of the point. This process is called **tracing**, and is useful in calculating the coordinates of particular points, which you can locate visually (for example, the intersection point of two graphs).

To trace a graph, first display the graph and press the key (Trace). The calculator will display a point on the graph. On the bottom of the display will be the x- and y- coordinates of the point. You may move the point along the curve using the left and right arrow keys. Each keystroke moves the point one dot to the left or right. Each time you move the point, the coordinates of the new point are displayed (see Figure 5).

If several curves are displayed, you may move the point from curve to curve using the up and down arrow keys.

Zooming   You may control the level of detail in a graph by zooming. By zooming out, you step away from the graph and thereby expand the range. By zooming in, you step toward the graph and thereby contract the range. There are a number of methods of zooming, provided from a menu displayed in response to the key (Zoom). Perhaps the most useful method of zooming is by specifying a box on the graph. The portion of the graph in the box specifies the new range.

Here is how to zoom using a box. Press (Zoom). Select the option **Box** from the menu displayed. The calculator will display the graph and a cursor. Using the arrow keys, move the cursor to one corner of the desired box and press the key (Enter). Then move the cursor to the opposite corner of the box. As you move the cursor, you will see the box expand or contract (see Figure 6). When the cursor indicates the desired opposite corner, press the key (Enter). The calculator now redisplays the graph with the designated box as the range. (See Figure 7.)

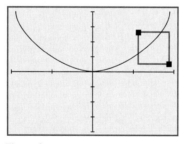

Figure 6

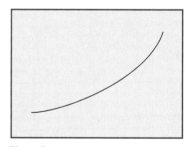

Figure 7

Graphing Calculator Explorations

The following examples illustrate some of the fine points of graphing functions on a calculator. The keystrokes cited are for the TI-81 calculator.

New Technology

Use a graphing calculator to display the graph of

$$y = x^3 - 3x - 2$$

EXAMPLE 1
Graphing a Function

Solution

Enter the function $Y_1 = X^3 - 3X - 2$ in the (Y =) menu (see Figure 8). Use the standard settings in the (Range) menu. Pressing (Graph) yields the graph of the function for $-10 \le x \le 10$. (See Figure 9.)

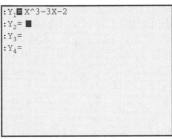

Figure 8

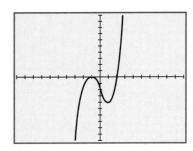

Figure 9

EXAMPLE 2
Graphing a Function with a Limited Domain

Use a graphing calculator to display the graph of
$$y = x^3 - 3x + 2 \quad (-3 \leq x \leq 3)$$

Solution

We may express the limitation $-3 \leq x \leq 3$ in the range by placing the inequalities $(-3 \leq x)$ and $(x \leq 3)$ after the equation defining the function (see Figure 10). Note that the inequality symbols are obtained by pressing [2nd] [Test], then selecting from the menu of symbols provided (using the arrow keys), and finally pressing [ENTER]. Also, note that the minus sign in -3 is obtained by pressing the key [(−)], which is an initial minus sign. When we press [Graph], we see the display in Figure 11. Note that the domain of the displayed graph is $-3 \leq x \leq 3$ as required. Note also that the calculator displays vertical lines correcting the points at the edges of the graph with the x-axis. These line segments should be ignored. They are not part of the function graph as defined mathematically.

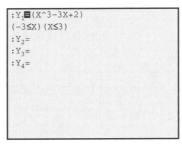

Figure 10

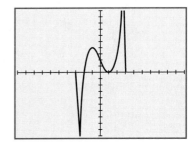

Figure 11

EXAMPLE 3
Graphing a Function Involving a Rational Exponent

Use a graphing calculator to display the graph of
$$y = x^{2/3}$$

Solution

Enter the function as
$$Y_1 = X\char`\^(\tfrac{2}{3})$$

Note that it is necessary to enclose the exponent in parentheses. The calculator exponentiation function $\char`\^$ is defined only for $X \geq 0$. Therefore, even though the function
$$y = x^{2/3} = (\sqrt[3]{x})^2$$

is defined for all x, the graph you see will only include the portion for nonnegative values of x. See Figure 12.

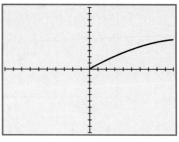

Figure 12   ◄

New Technology

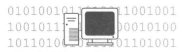

EXAMPLE 4
Graphing a Function with a
Limited Domain

Use a graphing calculator to display the graph of $y = \sqrt{x}$.

Solution

The function has a natural domain $\{x : x \geq 0\}$. Natural domains are built into the calculator and need not be entered separately. We enter the function as
$$Y_1 = \sqrt{X}$$
The graph is shown in Figure 13.

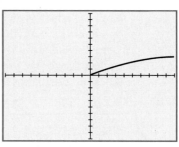

Figure 13   ◄

New Technology

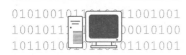

EXAMPLE 5
Graphing a Function

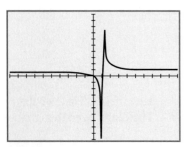

Figure 14

Use a graphing calculator to display the graph of

$$y = \frac{x}{x - 1} \quad (x \neq 1)$$

Solution

The exclusion of $x = 1$ is a natural range restriction because the denominator is 0 for this value of x. This restriction is automatically handled by the calculator. It attempts to evaluate the function at $x = 1$ and finds that it cannot, so it omits this point in drawing the graph. To better see what is going on, narrow the range to $-3 \leq x \leq 3$. The graph is shown in Figure 14. However, in drawing the graph, the calculator attempts to connect consecutive points. This leads to the vertical spike in the picture. Actually, this spike should not be included. The graph consists of two disconnected parts, one to the left of $x = 1$ and one to the right. In studying graphs produced by calculator, you will need to learn to omit consideration of such spikes.

◄

New Technology

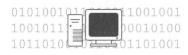

EXAMPLE 6
Determining When Two
Functions Are Equal

Use a graphing calculator to determine the values of x for which $f(x) = g(x)$, where

$$f(x) = x^2 + 2x - 4, \quad g(x) = 2x - 1$$

Solution

Graph $f(x)$ as Y_1 and $g(x)$ as Y_2. With the standard range settings, the display resembles Figure 15. To determine the desired values of x, we must determine the x-coordinates of the points where the graphs intersect. We may do this using the trace feature. Trace along the graph of $f(x)$ until you get to an intersection point and note the x-coordinate shown at the bottom of the screen. From Figure 16 and Figure 17, we see that the desired values of x are approximately given by

$$x \approx -1.789474, \quad 1.7894737$$

To obtain these values more accurately, we may use the box feature to enlarge the portion of the graph near an intersection point and use the trace to locate the intersection point as above. As you enlarge the graph, each horizontal step in the trace is smaller, so the value of x may be determined with greater accuracy.

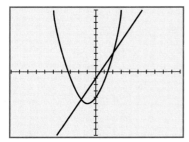

Figure 15

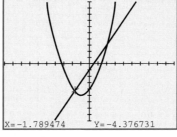

Figure 16

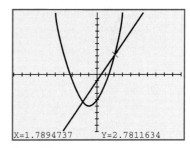

Figure 17

New Technology

EXAMPLE 7
Determining the Domain
and Range of Function
Using a Graphing
Calculator

Use a graphing calculator to determine the domain and range of the function

$$f(x) = 1 + \frac{x}{x^2 + 1} \quad (x \geq 0)$$

Solution

Enter the function as shown in Figure 18. To make the features of the graph more visible use the ranges $-10 \leq x \leq 30$, $0 \leq y \leq 3$. The graph is shown in Figure 19. A point of the graph lies over each point of the x-axis, since the domain of the function consists of all $x \geq 0$. Examining the graph, we see that its points assume all y-values greater than or equal to 1. That is, the domain is the interval $[1, \infty)$.

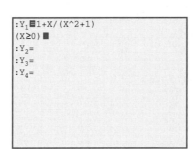

Figure 18

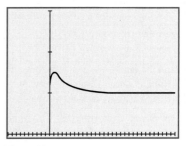

Figure 19

Exercises 2.3

In Exercises 1–8, use a graphing calculator to graph the following equations. Use the default values for the viewing rectangle.

1. $y = 3x - 5$

2. $y = -2x + 3.1$

3. $y = x^2 - x$

4. $y = .3x^2 + .151x$

5. $y = \dfrac{3}{x + 1}$

6. $y = \sqrt{3x - 1}$

7. $y = \dfrac{1}{x^2 + 1}$

8. $y = \dfrac{x - 1}{(x + 1)(x + 2)}$

9. Use the trace feature to determine the y-coordinate on the graph of $y = x^2 - 3x$ when:

   a. $x = 1.2$

   b. $x = 2.1$

   c. $x = -1$

   d. $x = .001$

10. Use the trace feature to determine a point of the graph of $y = \dfrac{x}{x^2 + 1}$ at which $y = 0.1$.

11. Use the zoom feature to enlarge the graph of Exercise 7 in the range $1 \le x \le 2$.

12. Use the zoom feature to enlarge the graph of Exercise 8 in the range $0 \le x \le 1, 0 \le y \le 10$.

13. By using the zoom and trace features, determine the points at which the graph of Exercise 3 crosses the horizontal axis.

14. By using the zoom and trace features, determine to two significant digits the points at which the graph of Exercise 4 crosses the horizontal axis.

15. Graph the equations $y = x^2$ and $y = -2x + 3$ on the same coordinate system. Choose the viewing rectangle so that the two intersection points are visible.

16. Determine the coordinates of the intersection points of the preceding exercise.

17. Graph the equations $y = 1/x$ and $y = 3x^3$ on the same coordinate system. Choose the viewing rectangle so that the intersection points are visible.

18. Determine the coordinates of the intersection points of the preceding exercise.

19. Graph the equations $y = x^2$, $y = 3x^2$ and $y = .6x^2$ on the same coordinate system. What do you observe

about the points where they cross the horizontal axis? Can you make a conclusion about what happens to horizontal axis intersections when a graph is scaled?

20. Graph on the same coordinate system the equation $y = 5x + 3$ and the graph translated 3 units horizontally and 2 units vertically.

21. Graph on the same coordinate system the equation $y = -x^2 + 1$ and the graph translated -1 units horizontally and 3 units vertically.

22. Graph on the same coordinate system the equation $y = -x^2 + x - 1$ and the graph translated -2 units horizontally and -1 unit vertically.

Use a graphing calculator to determine the domains and ranges of the following functions.

23. $f(x) = 3x - 2 \quad (0 \le x \le 2)$

24. $f(x) = 5 - \frac{3}{2}x \quad (x \ge 0)$

25. $f(x) = x^2 - 3x + 1 \quad (1 \le x \le 3)$

26. $f(x) = x^3 - 2x \quad (0 \le x \le 3)$

27. $f(x) = \dfrac{x}{x^2 - 1}$

28. $f(x) = \dfrac{1}{x} + x \quad (0.5 \le x \le 2)$

29. $f(x) = \sqrt{1 - 4x^2}$

30. $f(x) = \dfrac{x}{x^2 + 1}$

31. $f(x) = |x(x + 1)|$

32. $f(x) = \sqrt{x + 9}$

Use a graphing calculator to determine the solutions of the following equations to one decimal place accuracy.

33. $2x^2 = 5x + 14$

34. $18x^2 = 10x$

35. $13t^2 + 6t = 0$

36. $6x^2 + 5x = 21$

37. $10x^2 - 3x - 1 = 0$

38. $2 + \dfrac{3}{x^2} = \dfrac{1}{x}$

39. $3 - \dfrac{1}{x} + \dfrac{6}{x^2} = 0$

40. $\dfrac{2x - 3}{3x + 5} = \dfrac{x + 7}{3x - 4}$

41. $\dfrac{5 - 2x}{6x + 1} = \dfrac{3 + 2x}{5 - x}$

42. $x^2 + \sqrt{2}x - \sqrt{3} = 0$

43. $\sqrt{5}x^2 + 4x = \sqrt{2}$

44. $(y + 5)(y - 4) = (2y + 1)(3y - 4)$

45. $(2a - 1)(3a + 5) = (a - 3)(4a - 7)$

Suppose that $f(x) = 2 - x^2$. Graph the following functions along with f on the same coordinate system.

46. The graph of f translated to the right 4 units.

47. The graph of f translated to the left 6 units.

48. The graph of f translated up by 1 unit.

49. The graph of f translated down by 2 units.

50. The graph of f reflected in the x-axis.

51. The graph of f scaled by a factor of 3.

52. The graph of f scaled by a factor of 2 and then reflected in the x-axis.

53. The graph of f translated up by 1 unit and right by 2 units.

54. The graph of f translated down by 2 units and reflected in the x-axis.

By graphing $f(x)$ and $f(-x)$ on the same coordinate system, determine which of the following functions f are even.

55. $f(x) = x^2 + \dfrac{1}{x^2}$

56. $f(x) = (x^3 + 1)^2 + (x^3 - 1)^2$

57. $f(x) = (2x + 1)^2$

58. $f(x) = (1 - x)^2$

✍ **In Your Own Words**

59. Give two advantages graphing calculators have over manual plotting of functions.

60. Give three pitfalls to watch out for in graphing functions using a graphing calculator.

2.4 OPERATIONS ON FUNCTIONS

It is possible to perform algebraic operations on functions similar to the operations we discussed for algebraic expressions, namely addition, subtraction, multiplication, and division. We can also perform a totally different operation called *composition of functions*. In this section, we introduce these operations and indicate some of their applications.

Sums and Differences of Functions

Let f and g be functions with domains A and B, respectively. We define the sum of f and g, denoted $f + g$, to be the function such that

$$(f + g)(x) = f(x) + g(x)$$

That is, the value of $f + g$ at x is obtained by adding the values of f and g at x. The domain of $f + g$ consists of those x that belong to both A and B. That is, the domain of $f + g$ is the intersection

$$A \cap B$$

➢ **EXAMPLE 1**
Domain of a Sum

Suppose that

$$f(x) = \frac{x}{x - 1}, \quad g(x) = \frac{2}{x - 2}$$

Determine the domain of the sum $f + g$.

Solution

The domains of f and g are given by

$$\text{Domain}(f) = \{x : x \neq 1\}, \quad \text{Domain}(g) = \{x : x \neq 2\}$$

Therefore,

$$\text{Domain}(f + g) = \{x : x \neq 1, 2\}$$

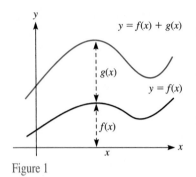

Figure 1

The graph of $f + g$ can be described simply in terms of the separate graphs of f and g. For each x in the domain of $f + g$, the height of the point

$$(x, (f + g)(x))$$

above the x-axis is obtained by adding the corresponding y-coordinates of the graphs of f and g, namely $f(x)$ and $g(x)$. See Figure 1.

Parallel to our definition for sum, we define the difference of f and g by the formula

$$(f - g)(x) = f(x) - g(x)$$

That is, the value of the difference $f - g$ at x equals the value of f at x minus the value of g at x. The domain of $f - g$ consists of those values x that belong to both the domain of f and the domain of g.

Sums and differences of functions appear often in applications. For example, suppose that $R(x)$ and $C(x)$ denote the revenue and cost, respectively, of producing x bushels of corn for a certain agricultural corporation. Then the difference

$$R(x) - C(x)$$

represents the profit from producing x bushels of corn.

Products and Quotients of Functions

The product of f and g, denoted fg, is the function defined by the formula

$$(fg)(x) = f(x)g(x)$$

As with the sum and the difference, the product is defined for values of x that are in the domains of both f and g. That is, the domain of fg is the intersection

$$A \cap B$$

where A and B are the domains of f and g, respectively.

➢ **EXAMPLE 2**
Domain of a Product

Suppose we have the functions

$$f(x) = x + 1, \quad g(x) = x - 1$$

Determine the domain of and an expression for the product fg.

Solution

Each of the functions f and g has as its domain the set **R** of all real numbers, so fg also has as its domain the set **R** of real numbers. Moreover,

$$(fg)(x) = (x + 1)(x - 1) = x^2 - 1$$

➢ **EXAMPLE 3**
Product of Two Functions

Suppose we have the functions

$$f(x) = \sqrt{x}, \quad g(x) = \frac{x - 1}{\sqrt{x}}$$

Determine the domain of and an expression for the product fg.

Solution

We have

$$\text{Domain}(f) = \{ x : x \geq 0 \}, \quad \text{Domain}(g) = \{ x : x > 0 \}$$

The domain of fg is therefore the intersection, or $\{ x : x > 0 \}$. Moreover,

$$(fg)(x) = (\sqrt{x})\left(\frac{x - 1}{\sqrt{x}}\right) = x - 1$$

Note that even though after simplification the expression for the product is defined for all real numbers x, the product fg is defined only for positive values of x (that is, x is in the domain).

A very common mistake is to compute a product, simplify it, and then substitute a value of x that is not in the domain of the product. This sort of error can lead to mathematical nonsense, such as statements like $0 = 1$. To avoid such an error, be careful to note the domains of any functions you compute.

The quotient of f divided by g, denoted f/g, is the function defined by the formula

$$\left(\frac{f}{g}\right)(x) = \frac{f(x)}{g(x)}$$

For the expression on the right to be defined, x must be in both the domain of f and the domain of g, and $g(x)$ must be nonzero. (Otherwise, division by 0 would result.)

➤ **EXAMPLE 4**
Quotient of Two Functions

Suppose

$$f(x) = \frac{x^2 - 1}{x}, \qquad g(x) = \frac{x + 1}{\sqrt{x}}$$

Determine the domain of and an expression for the quotient f/g.

Solution

We have

$$\text{Domain}(f) = \{x : x \neq 0\}, \quad \text{Domain}(g) = \{x : x > 0\}$$

and we have

$$(f/g)(x) = \frac{\dfrac{x^2 - 1}{x}}{\dfrac{x + 1}{\sqrt{x}}}$$

$$= \frac{x^2 - 1}{x} \cdot \frac{\sqrt{x}}{x + 1}$$

$$= \frac{x - 1}{\sqrt{x}}$$

The domain of the quotient f/g is the intersection of the above domains, or

$$\{x : x > 0\}$$

➤ **EXAMPLE 5**
Determining a Quotient
Function

Suppose

$$f(x) = \frac{x}{x^2 + 1}, \qquad g(x) = (x - 1)(x - 2)$$

Determine an expression for the quotient f/g. What is its domain?

Solution

Here

$$\text{Domain}(f) = \mathbf{R}, \quad \text{Domain}(g) = \mathbf{R}$$

The quotient is defined by the formula

$$(f/g)(x) = \frac{x}{(x^2 + 1)(x - 1)(x - 2)}$$

Moreover, $g(x)$ is equal to 0 for $x = 1, 2$. These values must be excluded from the domain of the quotient, which is therefore given by

$$\text{Domain}(f/g) = \{ x : x \neq 1, 2 \}$$

In working with quotients, you must exercise the same care as in working with products with regard to the domain. Even though you may be able to simplify the quotient to an expression with a larger domain, the quotient is defined only for the domain consisting of values common to the domains of the numerator and denominator and for which the denominator is nonzero.

Operations on Functions Using Graphing Technology

If you are using a graphing calculator or graphing software, you will find that most models are able to graph sums, differences, products, and quotients of functions as well as the composition of functions.

For example, suppose you are using the TI-81 and wish to graph the functions $f(x) = 2x$ and $g(x) = x^2$ and their sum $f(x) + g(x)$. One way to do this would be to enter each of these as separate functions, graph them, and then add them together algebraically and graph the result. To carry out the details, begin by setting an appropriate range for the graph. Then using the $\boxed{Y=}$ key, enter $2x$ in Y_1 and x^2 in Y_2. Graph each of the functions separately, as shown in Figure 2. You can see the graphs one at a time as follows: To turn off the graph of one equation, display the list of equations, position the cursor over the equal sign of the equation, and press $\boxed{\text{ENTER}}$. Using this procedure, you may view the graphs one at a time. By repeating the procedure, you may redisplay a graph.

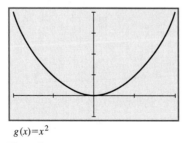

$g(x) = x^2$

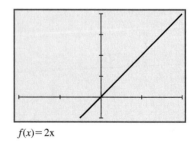

$f(x) = 2x$

Figure 2

Next, let's graph the sum. Return to the $\boxed{Y=}$ menu. Use the $\boxed{\text{2nd}}$ [Y − Var's] key to enter $Y_1 + Y_2$ for Y_3 (see Example 6 for details). Then graph Y_3 without graphing the other two equations. The graph should look like Figure 3.

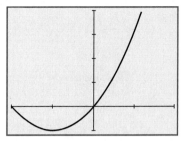

Figure 3

New Technology

EXAMPLE 6
Graphing a Product
of Functions

Let $f(x) = x^3$ and $g(x) = 1/x$ $(x \neq 0)$. Use a graphing calculator to display the graph of $f(x)g(x)$.

Solution

Set Y_1 equal to $f(x)$, Y_2 equal to $g(x)$, and Y_3 equal to $Y_1 * Y_2$. See Figure 4. You may enter Y_1 and Y_2 from the $\boxed{\text{Y-Vars}}$ menu. See Figure 5. To enter Y_1, just place the cursor at the point you wish to enter Y_1 and select $\boxed{\text{Y-Vars}}$ menu. This menu is shown in Figure 6. From this menu, choose Y_1 with the arrow keys and then press $\boxed{\text{Enter}}$. Y_1 is now entered into the equation for Y_3. Press $\boxed{\text{X}}$ and then follow the above procedure to enter Y_2. This completes the entry of the equation for Y_3 as shown in Figure 6. From the $\boxed{\text{Y-Vars}}$ menu, you may turn the graphs of Y_1 and Y_2 off, so you will see only the graph of Y_3. Select the $\boxed{\text{Y-Vars}}$ menu and select $\boxed{\text{Off}}$ from the first line. Then select $\boxed{Y_1}$ from the menu and press $\boxed{\text{ENTER}}$ twice. This turns off the display of Y_1. Follow a similar procedure to turn off the display for Y_2. The final display consists only of the graph for Y_3 as shown in Figure 7. This is the graph of $f(x)g(x)$. Be careful, however; this graph has a hole for $x = 0$ because the product is not defined there. This hole is very hard to see, but you should note its existence from the definition of the functions.

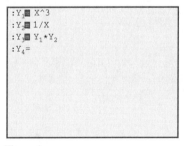

Figure 4

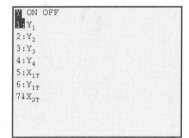

Figure 5

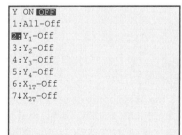

Figure 6

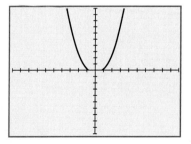

Figure 7

New Technology

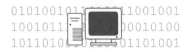

EXAMPLE 7
Graphing a Quotient
of Functions

Let $f(x) = \sqrt{x}$ and $g(x) = x^2 + 1$. Display the graph of $\dfrac{f(x)}{g(x)}$.

Solution

The domain of $f(x)$ is $x \geq 0$ and the domain of $g(x)$ is all real x. Moreover, because $g(x) \neq 0$ for any real x, the denominator is never 0 and the domain of the quotient is $x \geq 0$. Set Y_1 equal to $f(x)$, Y_2 equal to $g(x)$, and Y_3 equal to Y_1/Y_2. (See Figure 8.) To make the features of the graph more evident, choose the range $0 \leq y \leq 2$. The graph of $f(x)/g(x)$ is the graph of Y_3, and is shown in Figure 9.

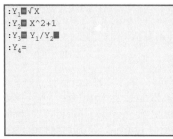

Figure 8

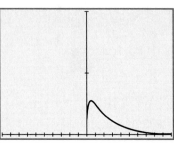

Figure 9

Composition of Functions

In many applications, it is necessary to substitute an expression for the variable of a function $g(x)$. Such substitutions are conveniently described in terms of **composition of functions**, defined as follows:

Definition 1
Composition of Functions

Suppose that f and g are functions. The composite

$$g \circ f$$

is the function defined by the formula

$$(g \circ f)(x) = g(f(x))$$

That is, to compute the value of the composite at x, we first evaluate f at x and then use that value to evaluate g at $f(x)$. We refer to $g \circ f$ as the **composite** of g by f. See Figure 10.

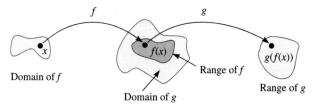

Figure 10

The machine concept of functions gives us a convenient way of viewing a composite. Suppose that f is viewed as one machine and g as a second. The machine corresponding to the composite is a serial combination of the first machine followed by the second.

Let's now determine the domain of a composite. Suppose that x is a value in the domain of $f(x)$. For the value $g(f(x))$ to make sense, the value $f(x)$ must be contained in the domain of g. This is necessary if we are to evaluate $g(z)$ at the value $z = f(x)$. In other words,

$$\text{Domain}\,(g \circ f) = \{\,x : f(x) \text{ belongs to Domain}(g)\}$$

Note that the domain of $g \circ f$ is a subset of the domain of f.

In forming composites, it helps to think in terms of an "inner function" and an "outer function." In forming the expression

$$g(f(x))$$

$f(x)$ is the inner function and $g(x)$ is the outer function.

➤ **EXAMPLE 8**
Calculating a Composite

Suppose

$$f(x) = x + 1, \qquad g(x) = x^2$$

Determine the composite $g \circ f$.

Solution

To form the composite of g by f, we substitute the expression for f wherever x appears in the expression for g. We have $g(x) = x^2$. Therefore, to obtain an expression for the composite, we replace x with $x + 1$ in the expression for g. We have

$$\begin{aligned}
(g \circ f)(x) &= g\,(f(x)) \\
&= g\,(x + 1) \\
&= (x + 1)^2
\end{aligned}$$

➤ **EXAMPLE 9**
Calculating a Composite

Suppose that

$$f(x) = \frac{1}{\sqrt{x}}, \qquad g(x) = x^2 + 1$$

Determine the composite of g by f.

Solution

The expression for the composite of g by f is obtained by substituting $f(x)$ for x in the expression for g:

$$\begin{aligned}
(g \circ f)(x) = g(f(x)) &= g\left(\frac{1}{\sqrt{x}}\right) \\
&= \left(\frac{1}{\sqrt{x}}\right)^2 + 1 \\
&= \frac{1}{x} + 1
\end{aligned}$$

Note that the order of forming composites is important. For instance, if we let f and g be as in the last example, we may form the composite of f by g by substituting $g(x)$ for x in $f(x)$, with the following result:

$$(f \circ g)(x) = f(g(x)) = f(x^2 + 1) = \frac{1}{\sqrt{x^2 + 1}}$$

Thus, in this case, $f \circ g$ is not equal to $g \circ f$.

➢ **EXAMPLE 10**
Domain of a Composite

Suppose that

$$f(x) = x^3 - 1, \qquad g(x) = \sqrt{x} - 1$$

1. Determine the composite

$$g \circ f$$

2. What is the domain of the composite?

Solution

1. To compute the composite, we replace x with $f(x)$ in the expression for $g(x)$:

$$f(x) = x^3 - 1$$
$$(g \circ f)(x) = g(f(x)) = g(x^3 - 1)$$
$$= \sqrt{x^3 - 1} - 1$$

2. To determine the domain of the composite, the value of $f(x)$ must be in the domain of g, which in this case consists of the nonnegative real numbers. The domain of the composite thus consists of all x for which

$$x^3 - 1 \geq 0$$
$$x^3 \geq 1$$
$$x \geq 1$$

That is, the domain of the composite is

$$\{x : x \geq 1\}$$

➢ **EXAMPLE 11**
Writing a Function
as a Composite

Let h be the function defined by

$$h(x) = \sqrt[3]{x^2 + 5}$$

Express h as the composite of two functions.

Solution

Suppose that we define

$$g(x) = \sqrt[3]{x}, \qquad f(x) = x^2 + 5$$

Then we have

$$g(f(x)) = h(x)$$

In other words, h is the composite of g by f. Note that there is no unique way to express a function as a composite. For example, there are other functions we can use instead of f and g as we defined them here, such as $g(x) = \sqrt[3]{(x + 5)}$ and $f(x) = x^2$. However, the functions we chose were the obvious ones that do the job.

Graphing Technology and Composites

The method for graphing a composite function on the TI-81 is similar to the method for graphing products and quotients. Consider the functions

$$f(x) = \frac{1}{\sqrt{x}} \quad \text{and} \quad g(x) = x^2 + 1$$

as given in Example 9. The composite function we are looking for is $g(f(x))$. To graph this, first enter $f(x)$ in Y_1 in the $\boxed{Y=}$ menu. Now move to Y_2 and enter $Y_1{}^\wedge 2 + 1$. Then graph Y_2. The results should resemble Figure 11.

The power of a graphing utility is the ease with which we can graph many functions. For instance, by entering $g(x)$ in Y_1 and $1/\sqrt{Y_1}$ in Y_2 and graphing Y_2, we have the graph of $f(g(x))$ as shown in Figure 11(a). The statement in Example 9 that $f(g(x)) \neq g(f(x))$ is clearly true by looking at the two graphs.

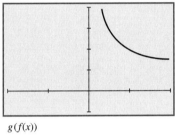

$g(f(x))$

Figure 11

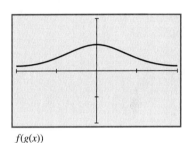

$f(g(x))$

Figure 11a

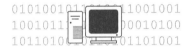
EXAMPLE 12
Graphing a Composite of Functions

Let $f(x) = x - 1/x$ and $g(x) = 1/x$. Display the graph of $g(f(x))$.

Solution

We have:

$$g(f(x)) = \frac{1}{f(x)}$$

So set Y_1 equal to $f(x)$ and Y_2 equal to $1/Y_1$ (see Figure 12). The graph of Y_2 is the graph of the composite $g(f(x))$ (see Figure 13). To make the features of the graph more visible, we have chosen the range $-3 \le x \le 3$. Moreover, we have turned off the graph of Y_1.

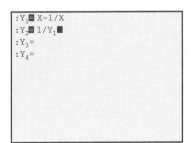

Figure 12

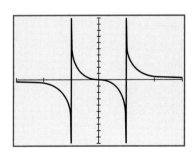

Figure 13

Exercises 2.4

For each of the following functions find the domain of

f, g, $f + g$, $f - g$, fg, ff, f/g, g/f, $f \circ g$, and $g \circ f$.

Then find

$(f + g)(x)$, $(f - g)(x)$, $(fg)(x)$, $(ff)(x)$, $(f/g)(x)$, $(g/f)(x)$, $(f \circ g)(x)$, and $(g \circ f)(x)$.

1. $f(x) = x - 4$, $g(x) = x + 5$

2. $f(x) = x^2 - 4$, $g(x) = 2x + 3$

3. $f(x) = 2x^2 + x - 3$, $g(x) = x^3$

4. $f(x) = \sqrt{x}$, $g(x) = x^2$

Let $f(x) = x^2 - 1$ and $g(x) = 2x + 3$. Find each of the following.

5. $(f - g)(3)$

6. $(f + g)(-1)$

7. $(f - g)(x)$

8. $(f + g)(x)$

9. $(fg)(3)$

10. $\left(\dfrac{f}{g}\right)(-1)$

11. $\left(\dfrac{g}{f}\right)(-1)$

12. $\left(\dfrac{f}{g}\right)(-1.5)$

13. $(fg)(x)$

14. $\left(\dfrac{f}{g}\right)(x)$

15. $\left(\dfrac{g}{f}\right)(x)$

16. $(f \circ f)(x)$

17. $(f \circ g)(x)$

18. $\left(f \cdot (f \circ f)\right)(x)$

19. $(g \circ g)(x)$

20. $(g \circ f)(x)$

Find $(f \circ g)(x)$ and $(g \circ f)(x)$.

21. $f(x) = \frac{2}{3}x$, $g(x) = \frac{3}{2}x$

22. $f(x) = x + 5$, $g(x) = x - 5$

23. $f(x) = 2x + 1$, $g(x) = \dfrac{x - 1}{2}$

24. $f(x) = \frac{4}{3}x + \frac{16}{3}$, $g(x) = \frac{3}{4}x - 4$

25. $f(x) = x^3 - 5$, $g(x) = \sqrt[3]{x + 5}$

26. $f(x) = \sqrt[7]{x + 1}$, $g(x) = x^7 - 1$

27. $f(x) = \sqrt{x - 2}$, $g(x) = x^2 + 2$

28. $f(x) = \sqrt[4]{x}$, $g(x) = x^4$

29. $f(x) = \dfrac{1}{x + 1}$, $g(x) = \dfrac{1 - x}{x}$

30. $f(x) = \dfrac{x^2 + 1}{x^2 - 1}$, $g(x) = \dfrac{2x - 5}{3x + 4}$

31. $f(x) = 3$, $g(x) = -5$

32. $f(x) = -4$, $g(x) = 3x + 1$

In each of the following, find $f(x)$ and $g(x)$ such that $h(x) = (f \circ g)(x)$. Answers may vary, but try to choose the most obvious answer.

33. $h(x) = (4x^3 - 1)^5$

34. $h(x) = \sqrt[3]{x^2 + 1}$

35. $h(x) = \dfrac{1}{(x + 5)^4}$

36. $h(x) = \dfrac{1}{\sqrt{7x + 2}}$

37. $h(x) = \dfrac{x^3 + 1}{x^3 - 1}$

38. $h(x) = \left|4x^2 - 3\right|$

39. $h(x) = \left(\dfrac{1 + x^3}{1 - x^3}\right)^4$

40. $h(x) = \left(\sqrt{x} + 3\right)^4$

41. $h(x) = \sqrt{\dfrac{x + 2}{x - 2}}$

42. $h(x) = \sqrt{2 + \sqrt{2 + x}}$

43. $h(x) = (x^5 + x^4 + x^3 - x^2 + x - 2)^{63}$

44. $h(x) = (x - 7)^{2/3}$

In Exercises 45–49, graph each equation. Then graph $y = (f + g)(x)$ by adding y-coordinates.

45. $f(x) = x^2$, $g(x) = 2x - 3$

46. $f(x) = x$, $g(x) = \dfrac{1}{x}$

47. $f(x) = \sqrt{x}$, $g(x) = 3 - x^2$

48. $f(x) = 2 - x^2$, $g(x) = x^2 - 2$

Exercises 49–60 refer to the graph of the function $f(x)$ shown in Figure 14. In each exercise, draw the graph of the indicated function:

49. $f(x) + 1$

50. $f(x) + 3$

51. $f(x + 2)$

52. $f(x + 1)$

53. $f(x - 1)$

54. $f(x - 5)$

55. $-f(x)$

56. $-f(x - 1)$

57. $f(-x)$

58. $-f(-x + 1)$

59. $f(-x + 2)$

60. $1 - f(x)$

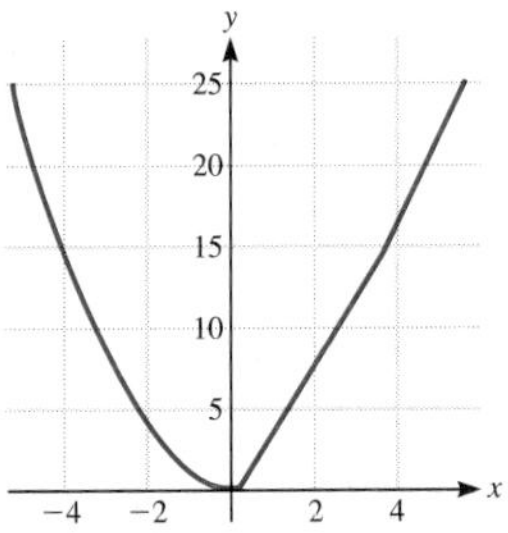

Figure 14

♦ **Applications**

61. An airplane is 200 ft from the control tower at the end of the runway. It takes off at a speed of 125 mph (see Figure 15).

    a. Let a be the distance the plane travels down the runway. Find a formula for a in terms of the time t the plane travels. That is, find an expression for $a(t)$.

    b. Let P be the distance of the plane from the control tower. Find a formula for P in terms of the distance a. That is, find an expression for $P(a)$.

    c. Find $(P \circ a)(t)$. Explain the meaning of this function.

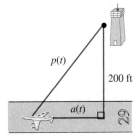

Figure 15

62. A tanker has an oil spill. Oil is moving away from the tanker in a circular pattern such that the radius is increasing at the rate of 10 ft per hour (see Figure 16).

    a. Find a function $r(t)$ for the radius in terms of time t.

    b. Find a function $A(r)$ for the area of the oil spill in terms of the radius r.

    c. Find $(A \circ r)(t)$. Explain the meaning of this function.

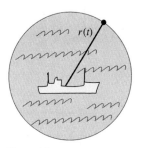

Figure 16

63. A baseball diamond is a square 90 ft on each side. A batter runs to first base at the rate of 25 ft/sec (see Figure 17).

    a. Find a function $b(t)$ for the distance b of the runner from first base in terms of time t.

    b. Find a function $h(b)$ for the distance of the runner from second base. This distance is to be expressed in terms of the distance b. Note that, as shown in the drawing, this is the straight-line distance to second base.

    c. Find $(h \circ b)(t)$. Explain the meaning of this function.

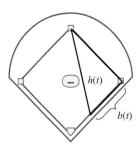

Figure 17

In Exercises 64 and 65 compute the total profit function $P(x) = R(x) - C(x)$.

64. $R(x) = 50x$, $C(x) = 2x^3 - 12x^2 + 40x + 10$

65. $R(x) = 50x - 0.5x^2$, $C(x) = 10x + 3$

66. Consider $f(x) = 3x - 2$ and $g(x) = 5x + b$. Find b such that $(f \circ g)(x) = (g \circ f)(x)$ for all real numbers x.

67. Consider $f(x) = 3x - 2$ and $g(x) = mx + b$. Find m and b such that $(f \circ g)(x) = (g \circ f)(x) = x$ for all real numbers x.

➡ **Technology**

Use a graphing calculator to graph on a single coordinate system $f(x)$, $g(x)$, and $f(x)/g(x)$. What do you observe about the behavior of the graph of the quotient at zeros of the denominator?

68. $f(x) = x + 1$, $\quad g(x) = x$

69. $f(x) = x^2 - x$, $\quad g(x) = (x + 1)^2$

70. $f(x) = x^2 - 4x + 4$, $\quad g(x) = x - 2$

71. $f(x) = \sqrt{x + 1}$, $\quad g(x) = x^2$

Graph $f(g(x))$, where f and g are given by the following:

72. $f(x) = x - 4$, $\quad g(x) = x + 5$

73. $f(x) = x^2 - 4$, $\quad g(x) = 2x + 3$

74. $f(x) = 2x^2 + x - 3$, $\quad g(x) = x^3$

75. $f(x) = \sqrt{x}$, $\quad g(x) = x^2$

✍ **In Your Own Words**

76. Explain why the composition of two even functions is even.

77. Explain why the sum of two even functions is even.

78. Explain why the sum of two odd functions is odd.

79. Explain why the composition of two odd functions is odd.

80. Explain why the product of an even function and an odd function is odd.

81. Form a conjecture and prove a result about the product of two odd functions.

82. Form a conjecture and prove a result about the product of two even functions.

2.5 INVERSE FUNCTIONS

One of the important tasks of algebra is defining functions that are suitable for solving the widest variety of real-world problems. Throughout the book, we will introduce new functions required to describe various events and processes. One general method for constructing new functions from known ones is by using inverse functions. In this section, we define inverse functions and develop an initial facility in dealing with them. Many of the functions we will encounter later in the book (such as the inverse trigonometric functions and logarithmic functions) are constructed as inverses of known functions. Connected with our task is the notion of a one-to-one function, which we now discuss.

One-to-One Functions

Suppose that f is a function with domain A and range B. For every element c in B, there is some x in the domain A such that $f(x) = c$. However, there may be more than one x for which f has the value c. For instance, consider the function $f(x) = x^2$. In this case, there are two values of x for which $f(x) = 4$, namely $x = 2$ and $x = -2$. On the other hand, for some functions, each value c in the range corresponds to exactly one value x for which $f(x) = c$. Let's assign such functions a name.

Definition 1 **One-to-One Function**	We say that f is a **one-to-one function** provided that for each c in the range, there is precisely one x in the domain such that $f(x) = c$. To put it another way, f is one-to-one provided that f has distinct (different) values at distinct numbers in the range.

When the domain and range are both subsets of the set of real numbers, we can use a simple geometric test to determine whether a function is one-to-one. On the graph of $f(x)$, the values of x for which $f(x) = c$ are the y-coordinates of the points at which the graph intersects the horizontal line $y = c$. Generally, there may be several such values of x, that is, many such points of intersection (see Figure 1(a)). However, if $f(x)$ is one-to-one, then **every** line $y = c$ intersects the graph of $f(x)$, at no more than one point (see Figure 1(b)). Thus, we have the following geometric criterion for one-to-one functions:

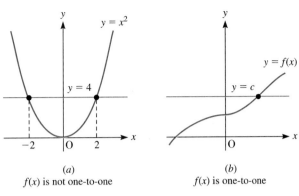

(a)
$f(x)$ is not one-to-one

(b)
$f(x)$ is one-to-one

Figure 1

HORIZONTAL LINE TEST FOR ONE-TO-ONE FUNCTIONS

A function f is one-to-one if and only if each horizontal line intersects the graph of f at no more than one point.

As an example of a one-to-one function, consider the function $f(x) = \sqrt{x}$ ($x \geq 0$), whose graph is shown in Figure 2. Note that if a horizontal line intersects the graph, it intersects it at precisely one point. (Of course, not every horizontal line intersects the graph.) The function f is one-to-one.

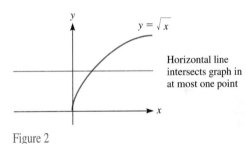

Horizontal line intersects graph in at most one point

Figure 2

➢ **EXAMPLE 1**

Determining if a Function Is One-to-One

Determine which of the following functions are one-to-one.

1. $f(x) = -2x + 5$
2. $f(x) = \sqrt{1 - x^2}$ $(0 \leq x \leq 1)$
3. $f(x) = x^2$

Solution

1. Suppose that we are given a value y in the range of f (which is the set of all real numbers). A particular value of x satisfies $f(x) = y$ precisely if
$$-2x + 5 = y$$
Solving this equation for x in terms of y, we obtain
$$x = -\tfrac{1}{2}(y - 5)$$
This equation shows that for a particular value of y, there is a single value of x such that $f(x) = y$. Thus, f is one-to-one. The graph of f is shown in Figure 3(a). From the horizontal line test, we obtain a second proof that the function is one-to-one.

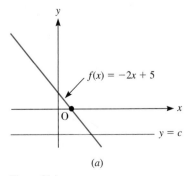

(a)

Figure 3(a)

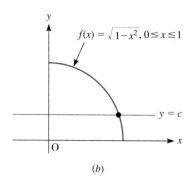

$f(x) = \sqrt{1-x^2}, 0 \le x \le 1$

(b)

Figure 3(b)

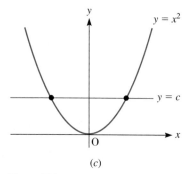

(c)

Figure 3(c)

2. Again suppose that y belongs to the range of f. The condition $f(x) = y$ is equivalent to

$$\sqrt{1 - x^2} = y$$
$$1 - x^2 = y^2 \quad \text{and} \quad x \text{ in } [0, 1]$$
$$x^2 = 1 - y^2$$
$$x = \pm \sqrt{1 - y^2}$$

At first glance, it appears that for a given value of y, there are two possible values of x, one for each choice of sign. However, because x is in [0, 1], the negative sign cannot hold and there is a single choice for x, namely

$$x = \sqrt{1 - y^2}$$

Because a single value of x corresponds to a given value of y, the function f is one-to-one. In Figure 3(b), we have sketched the graph of f. It is easy to verify that the horizontal line test holds in this case.

3. We have sketched the graph of f in Figure 3(c). It is clear that the horizontal line test fails. For instance, the figure shows a horizontal line that intersects the graph in two places. This is reflected in the algebra as follows: If we solve the equation $f(x) = y$ for x in terms of y, we have

$$y = x^2$$
$$x = \pm \sqrt{y}$$

For all $y \ne 0$, the numbers $\sqrt{y}$ and $-\sqrt{y}$ are different (one is positive and the other negative). Therefore, for $y \ne 0$ there are two distinct values of x corresponding to y. This is confirmed by examining the graph.

Inverse of a Function

Let $y = f(x)$ be a one-to-one function. Each value of y is assumed exactly once, so we may create a new function that associates a value of y with the corresponding value of x. As an example of how this is done, let's consider a concrete example, using the function

$$y = f(x) = -2x + 5$$

which is defined for any real value of x. We previously showed that this function is one-to-one and assumes as values all real values of y. Suppose we are given a value of y. What value of x corresponds? To find out, we solve the above equation for x in terms of y:

$$-2x = y - 5$$
$$x = -\tfrac{1}{2}(y - 5)$$

That is, if y is any real number, then the given function f has value y at the x-value $-\tfrac{1}{2}(y-5)$. The correspondence that associates y with its x-value $-\tfrac{1}{2}(y-5)$ is called the **inverse function** of f and is denoted f^{-1} (read: f *inverse*). That is, we have

$$f^{-1}(y) = -\tfrac{1}{2}(y - 5) \qquad y = \text{any real number}$$

More generally, suppose that f is a one-to-one function with domain A and range B. For each y in B, there is exactly one x in A for which $f(x) = y$. The correspondence that associates y with its corresponding x-value is a function with domain B and range A. This function is called the **inverse function** of f and is denoted f^{-1}. (Note that -1 here is not an exponent, but stands for the inverse function.)

We can represent the inverse function graphically as follows. Recall that we can represent a function from A to B as an arrow from an element x of A to the element $y = f(x)$ in B. The inverse function is represented by the arrow in the opposite direction, from y to x. See Figure 4.

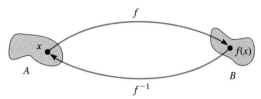

Figure 4

Examples of Inverse Functions

Consider the function

$$f(x) = \sqrt{1 - x^2} \qquad (0 \le x \le 1)$$

In Example 1, we determined that the function f is one-to-one and that the equation $f(x) = y$ is equivalent to the equation

$$x = \sqrt{1 - y^2} \qquad (0 \le y \le 1)$$

This means that the inverse function of f is given by

$$f^{-1}(y) = \sqrt{1 - y^2} \qquad (0 \le y \le 1)$$

Or, replacing the variable y with the customary variable x, we have

$$f^{-1}(x) = \sqrt{1 - x^2} \qquad (0 \le x \le 1)$$

Note that in this particular example, the inverse of the function is the function itself.

We can condense the preceding computations into the following method for determining the inverse of a one-to-one function f.

DETERMINING THE INVERSE FUNCTION OF f

1. Determine that $f(x)$ has an inverse by determining that it is one-to-one.
2. Write the equation $f(x) = y$.
3. Solve the equation for x in terms of y:

$$x = \text{expression in } y$$

4. Write

$$f^{-1}(y) = \text{expression in } y$$

5. Replace y throughout with x to obtain the inverse function

$$f^{-1}(x) = \text{expression in } x$$

➤ **EXAMPLE 2**
Calculating Inverses

Determine the inverse functions of the following functions. Determine the domain and range of the inverse.

$$1. \ f(x) = 3x + 5 \qquad 2. \ f(x) = \sqrt{x}, \qquad x \geq 0$$

Solution

1. The graph of f is a nonvertical straight line. By the horizontal line test, f is one-to-one and therefore has an inverse function. The domain and range of f both consist of the set **R** of all real numbers. Therefore, the domain and range of f^{-1} are also the set **R**. To obtain a formula for f^{-1}, we write

$$y = 3x + 5$$

We now solve this equation for x in terms of y:

$$y - 5 = 3x$$

$$x = \tfrac{1}{3}(y - 5)$$

We now interchange x and y to obtain

$$y = \tfrac{1}{3}(x - 5)$$

$$f^{-1}(x) = \tfrac{1}{3}(x - 5)$$

2. Figure 5 shows the graph of $y = \sqrt{x}$. By inspecting the graph, we see that it passes the horizontal line test, so the function is one-to-one. Therefore, f has an inverse function. Both the domain and range of f are the set of all nonnegative real numbers. Therefore, both the domain and range of the inverse function are the set of all nonnegative real numbers. We now obtain a formula for the inverse function by first solving the equation for x in terms of y:

$$y = \sqrt{x}$$

$$x = y^2 \quad \text{and} \quad y \geq 0$$

Interchanging x and y, we have

$$y = x^2 \quad \text{and} \quad x \geq 0$$

$$f^{-1}(x) = x^2$$

That is, the inverse function is $f^{-1}(x) = x^2 \qquad (x \geq 0)$.

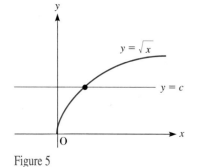

Figure 5

The following facts about inverse functions follow directly from their definition.

INVERSE FUNCTIONS

1. The inverse function f^{-1} is defined if and only if the function f is one-to-one.
2. The domain of the inverse function is the range of f.
3. The range of the inverse function is the domain of f.
4. The inverse function satisfies the property

$$f^{-1}(y) = x \text{ if and only if } f(x) = y$$

Suppose that we start with a number x in the domain of the function f. The function value $f(x)$ lies in the range of f (the domain of the inverse function). The definition of the inverse function says

$$f^{-1}(f(x)) = x \qquad (1)$$

Similarly, suppose that y is a number in the domain of f^{-1} (i.e., in the range of f). Then the definition of the inverse function says that

$$f(f^{-1}(x)) = x \qquad (2)$$

The last two formulas can be phrased in terms of composition of functions:

$$(f^{-1} \circ f)(x) = x \qquad x \text{ in domain } (f) \qquad (3)$$
$$(f \circ f^{-1})(x) = x \qquad x \text{ in domain } (f^{-1}) \qquad (4)$$

These formulas state important relationships between a function and its inverse. We will be using these formulas in many special cases to perform algebraic manipulations in expressions that involve both a function and its inverse.

The Graph of the Inverse Function

From the definition of the inverse function, we can see that the point (a, b) is on the graph of f if and only if (b, a) is on the graph of the inverse of f. Indeed, we have

$$b = f(a) \quad \text{if and only if} \quad a = f^{-1}(b)$$

The points (a, b) and (b, a) are located symmetrically with respect to the line $y = x$, which bisects the first and third quadrants. (See Figure 6.)

Here is a proof of this geometric fact. The line through (a, b) and (b, a) has slope

$$\frac{a - b}{b - a} = -1$$

and so is perpendicular to line $y = x$, (which has slope 1). Moreover, the equation of this line is

$$y - b = -(x - a)$$
$$y = -x + b + a$$

and thus intersects line $y = x$ at the point where

$$x = -x + b + a$$

$$x = \frac{b + a}{2}, \qquad y = \frac{b + a}{2}$$

Therefore, the distance from (a, b) to line $y = x$ is

$$\sqrt{\left(a - \frac{b + a}{2}\right)^2 + \left(b - \frac{b + a}{2}\right)^2} = \sqrt{\left(\frac{a - b}{2}\right)^2 + \left(\frac{b - a}{2}\right)^2}$$

Similarly, the distance from (b, a) to line $y = x$ is

$$\sqrt{\left(b - \frac{b + a}{2}\right)^2 + \left(a - \frac{b + a}{2}\right)^2} = \sqrt{\left(\frac{b - a}{2}\right)^2 + \left(\frac{a - b}{2}\right)^2}$$

That is, points (a, b) and (b, a) have equal perpendicular distances to line $y = x$.

To better understand the relationship of points (a, b) and (b, a), think of line $y = x$ as a mirror that reflects images on one side of it to the other side,

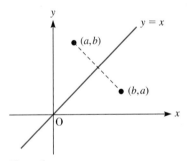

Figure 6

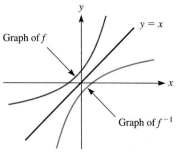

Graph of f

$y = x$

Graph of f^{-1}

Figure 7

preserving all distances. Then point (a, b) is the reflection of point (b, a) and vice versa. This fact provides us with a simple method for sketching the graph of the inverse function.

GRAPH OF AN INVERSE FUNCTION

The graph of the inverse function $f^{-1}(x)$ is the reflection of the graph of f in line $y = x$. See Figure 7.

> **EXAMPLE 3**
Graphing Inverses

Sketch the graph of the inverse function of $f(x)$.

1. $f(x) = -2x + 5$
2. $f(x) = \sqrt{1 - x^2}$ $(0 \le x \le 1)$

Solution

1. We first sketch the graph of $f(x)$, as in Figure 8. To obtain the graph of the inverse, we reflect this graph in the line $y = x$.
2. We proceed as in part 1. We first graph f and then reflect that graph in the line $y = x$. (See Figure 9.) Note that in this case, the reflected graph is the same as the original. This is a geometric confirmation of the fact we discovered earlier in this section by computation, namely, that the inverse of the function

$$f(x) = \sqrt{1 - x^2} \quad (0 \le x \le 1)$$

is the function $f(x)$ itself.

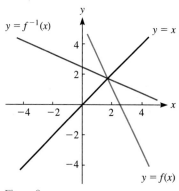

Figure 8

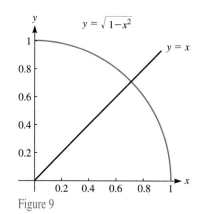

Figure 9

New Technology

EXAMPLE 4
Checking Graphically that a
Function Is One-To-One

Determine if the following functions are one-to-one using a graphing calculator.

1. $f(x) = x^2 + 2x$ $(-1 \le x \le 1)$
2. $f(x) = x + \dfrac{1}{x}$ $(x \ne 0)$

Solution

1. The graph of $f(x)$ is shown in Figure 10. From the horizontal line test, we see that the function is not one-to-one.

2. The graph of $f(x)$ is shown in Figure 11. From the horizontal line test, we see that the function is not one-to-one.

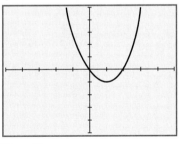

Figure 10

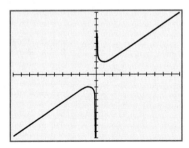

Figure 11

New Technology

EXAMPLE 5
Graphing an Inverse Function

Graph the function $f(x) = 3x + 1$ and its inverse function $g(x) = (x - 1)/3$ on the same coordinate system, along with the graph of $y = x$.

Solution

Graph $f(x)$ as Y_1 and $g(x)$ as Y_2 and $Y_3 = X$. The three graphs are shown in Figure 12. Note that the graphs of Y_1 and Y_2 are reflections of each other in line Y_3. This is a graphical demonstration that they are inverses of one another.

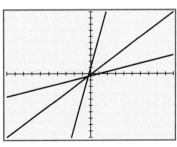

Figure 12

New Technology

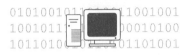

EXAMPLE 6
Proving Graphically that
Functions Are Inverses of
One Another

The inverse of

$$f(x) = x^2 - 4 \qquad (x \geq 0)$$

is

$$g(x) = \sqrt{x + 4} \qquad (x \geq -4)$$

Prove graphically that

$$g(f(x)) = x \qquad (x \geq 0)$$

Solution

To prove that f and g are inverses of one another, set Y_1 equal to $f(x)$ and Y_2 equal to $\sqrt{Y_1 + 4} = g(f(x))$. Finally, set

$$Y_3 = X \qquad (X \geq 0)$$

(See Figure 13.) Graphing Y_2 and Y_3, we see that the two graphs are the same. (See Figure 14.) This provides a graphical demonstration that

$$g(f(x)) = x \qquad (x \geq 0)$$

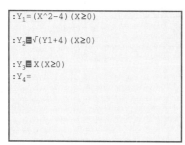

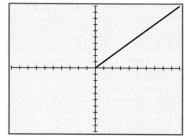

Figure 13

Figure 14 ◄

New Technology

EXAMPLE 7
Domain and Range of an
Inverse Function

Use a graphing calculator to determine the domain and range of the inverse function of

$$f(x) = \frac{x - 1}{x + 1} \quad (x > -1)$$

Solution

The graph of f is shown in Figure 15, using the range $-1 \leq x \leq 5$. From the graph, we see that f is one-to-one and therefore has an inverse function. The range of the inverse is the domain of f, that is, $\{x : x > -1\}$. The domain of the inverse is the range of f, which, from the figure, is the set of all real numbers.

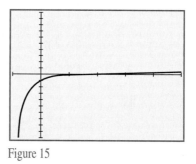

Figure 15 ◄

New Technology

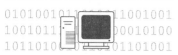

EXAMPLE 8
Evaluating an Inverse
Function

Suppose that

$$f(x) = \frac{1}{x^2 + 1} \quad (x \geq 0)$$

Use a graphing calculator to determine $f^{-1}(0.25)$.

Solution

The graph of f is shown in Figure 16. If $y = f(x)$, then $x = f^{-1}(y)$. To determine $f^{-1}(0.25)$, we need to determine the value of x for which $f(x) = 0.25$. To do this, we trace along the curve until we arrive at a point at which the y-

coordinate (= the value of f) is equal to 0.25. The desired y value is approximately 1.736.

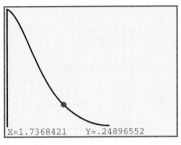

X=1.7368421   Y=.24896552

Figure 16   ◀

Exercises 2.5

The following are graphs of functions. Determine which functions have inverse functions. For those that do, draw the graph of the inverse function.

1.
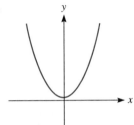

2.
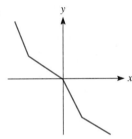

3.
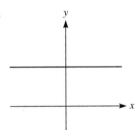

4.
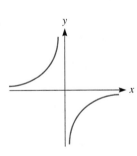

5.
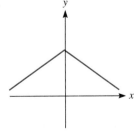

6.
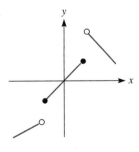

Graph each of the following functions. Determine which functions have an inverse. For those that do, find a formula for $f^{-1}(x)$ and draw its graph.

7. $f(x) = 5 - 8x$

8. $f(x) = 4$

9. $f(x) = -8$

10. $f(x) = \dfrac{x - 2}{4}$

11. $f(x) = x^2 - 3$

12. $f(x) = \sqrt{x}$ $(x \geq 0)$

13. $f(x) = \sqrt{4 - x^2}$ $(0 \leq x \leq 2)$

14. $f(x) = \dfrac{1}{x}$ $(x \neq 0)$

15. $f(x) = \sqrt{4 - x^2}$ $(-2 \leq x \leq 2)$

16. $f(x) = \sqrt{4 - x^2}$ $(-2 \leq x \leq 0)$

In each of the following exercises, determine whether the functions are inverses of each other by calculating $f \circ g(x)$ and $g \circ f(x)$.

17. $f(x) = \dfrac{2x}{3}, \quad g(x) = \dfrac{3x}{2}$

18. $f(x) = x + 5, \quad g(x) = x - 5$

19. $f(x) = 2x + 1, \quad g(x) = \dfrac{x - 1}{2}$

20. $f(x) = \dfrac{4x}{3} + \dfrac{16}{3}, \quad g(x) = \dfrac{3x}{4} - 4$

21. $f(x) = x^3 - 5, \quad g(x) = \sqrt[3]{x + 5}$

22. $f(x) = \sqrt[7]{x + 1}, \quad g(x) = x^7 - 1$

23. $f(x) = \sqrt{x - 2}, \quad g(x) = x^2 + 2$

24. $f(x) = \sqrt[4]{x}$ $(x \geq 0), \quad g(x) = x^4$

25. $f(x) = \dfrac{1}{x+1}$, $g(x) = \dfrac{1-x}{x}$

26. $f(x) = \dfrac{x^2+1}{x^2-1}$, $g(x) = \dfrac{2x-5}{3x+4}$

27. $f(x) = 3$, $g(x) = \frac{1}{3}$

28. $f(x) = -4$, $g(x) = 3x - 1$

For each of the following functions, find a formula for $f^{-1}(x)$.

29. $f(x) = 2x - 3$

30. $f(x) = 7x + 2$

31. $f(x) = \dfrac{2x}{3} + \dfrac{3}{5}$

32. $f(x) = 1.6x - 6.4$

33. $f(x) = x^2$ $(x \geq 0)$ 34. $f(x) = 4x^2$ $(x \leq 0)$

35. $f(x) = \sqrt{x+2}$ $(x \geq -2)$

36. $f(x) = \sqrt{7x-2}$ $(x \geq \frac{2}{7})$

37. $f(x) = x^5$

38. $f(x) = x^3 - 2$

39. $f(x) = \dfrac{2}{x}$

40. $f(x) = -\frac{4}{7}x$

41. $f(x) = \dfrac{2x-3}{3x+1}$

42. $f(x) = \dfrac{7x+2}{8x-3}$

43. $f(x) = \sqrt{25 - x^2}$ $(-5 \leq x \leq 0)$

44. $f(x) = \sqrt{49 - x^2}$ $(0 \leq x \leq 7)$

45. $f(x) = \sqrt[3]{x+8}$

46. $f(x) = \sqrt[5]{6-3x}$

47. For $f(x) = 23{,}457x - 3456$, find $(f \circ f^{-1})(739)$ and $(f^{-1} \circ f)(5.00023)$.

48. For $f(x) = 677x^3$, find $(f \circ f^{-1})(958.34)$ and $(f^{-1} \circ f)(\frac{765}{577})$.

49. The following formulas for conversion between Fahrenheit and Celsius temperatures have been considered several times in the text:

$$F = \tfrac{9}{5}C + 32, \qquad C = \tfrac{5}{9}(F - 32)$$

Show that these functions are inverses of one another.

50. Find a formula for the area A of a circle in terms of its radius r. Then find a formula for the radius r in terms of the area A. Show that the functions defined by these formulas are inverses of one another.

51. Let $f(x) = mx + b$, $m \neq 0$. Find a formula for $f^{-1}(x)$.

52. For what integers n does $f(x) = x^n$ have an inverse?

53. Determine whether or not an even function has an inverse.

54. Find examples of odd functions that are inverses of one another.

55. Suppose that f and g both have inverses. Show that $(f \circ g)^{-1}(x) = (g^{-1} \circ f^{-1})(x)$.

56. The function f is increasing over an interval provided that $a < b$ for a and b in the interval always implies that $f(a) < f(b)$. Prove that if f is increasing over an interval, then it has an inverse.

➡ **Technology**

Use a graphing calculator to determine whether or not the following functions are one-to-one.

57. $f(x) = x^2 + 2x$ $(0 \leq x \leq 2)$

58. $f(x) = 1 - x^2$ $(0 \leq x \leq 5)$

59. $f(x) = \dfrac{2}{x}$ $(x > 0)$

60. $f(x) = \dfrac{x}{1-x}$

For each of the following functions $f(x)$, use a graphing calculator to determine whether or not it has an inverse function. If the inverse exists, calculate $f^{-1}(2)$.

61. $f(x) = x^2 - 3.5x + 1$ 62. $f(x) = -3.9x^3 + 9.6$

63. $f(x) = \dfrac{x}{4x+3}$, $x > 0$

64. $f(x) = \dfrac{1}{x^2 - 17}$ 65. $f(x) = |3x - 11|$

66. $f(x) = \sqrt{x^2 - x}$, $x \geq 1$

✎ **In Your Own Words**

67. Write a paragraph defining the concept of a one-to-one function and providing examples of functions that are and functions that are not one-to-one. Use complete sentences.

68. Explain the procedure for determining the inverse of a one-to-one function.

69. What happens when you attempt to apply the procedure of Exercise 68 to a function that is not one-to-one?

70. Does an inverse function always exist for a decreasing function? Explain your reasoning.

▌ 2.6 VARIATION AND ITS APPLICATIONS

Many applications involve equations in which one variable is expressed in terms of other variables. Such an equation expresses a **functional relationship** between the variables. In this section, we will explore some of the most common types of functional relationships.

Direct Variation

Let's begin with the simplest type of functional relationship between two variables:

Definition 1
Direct Variation

We say that y **varies directly as** x (or that y is **proportional to** x) provided that

$$y = kx$$

for some constant k. The constant k is called the **proportionality factor**.

When y varies directly as x, any increase in x results in a proportional increase in y, and any decrease in x results in a proportional decrease in y. Moreover, if y varies directly as x, y is a linear function of x with a constant term of zero.

➤ EXAMPLE 1
Direct Variation

Suppose that y varies directly as x. Moreover, suppose that the value of y is 100 when $x = 4$.

1. Determine the functional relationship between y and x.
2. What is the value of y when $x = 5$?

Solution

1. Because y varies directly as x, we have

$$y = kx$$

for some positive constant k. We may determine the constant k using the given data. Substitute the values $y = 100$ and $x = 4$ into this equation to obtain

$$100 = k(4)$$
$$k = 25$$

The functional relationship between x and y is given by

$$y = 25x$$

2. Using the functional relationship just derived, we see that when $x = 5$, we have

$$y = 25x = 25(5) = 125$$

➤ EXAMPLE 2
Hooke's Law

Hooke's law from physics states that the distance a spring stretches from its rest position is directly proportional to the amount of force applied. Suppose that a spring is stretched 8 in. by a force of 50 lb. Suppose that a force of 60 lb is applied. How far will the spring be stretched? (See Figure 1.)

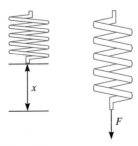

Figure 1

Solution

Denote the force acting on the spring by F and the distance the spring stretches by x. Then Hooke's law states that F varies directly as x. Therefore, we have

$$F = kx$$

for some constant k. Inserting the given data into this equation, we find that

$$F = kx$$
$$50 = k(8)$$
$$k = \tfrac{25}{4}$$
$$F = \tfrac{25}{4}x$$

Therefore, when F is equal to 60 pounds, we have

$$60 = \tfrac{25}{4}x$$
$$\frac{60 \cdot 4}{25} = x$$
$$x = \tfrac{48}{5} \text{ in.}$$

When a force of 60 lb is applied, the spring is stretched $\tfrac{48}{5}$ or 9.6 inches.

In many applications, functional relationships of the form

$$y = kx^m$$

are required, where m and k are constants. This is a generalization of the formula for direct variation. In the case of a functional relationship of this form, we say that **y varies directly as x^m**.

Inverse Variation

Definition 2 **Inverse Variation**	We say that **y varies inversely as x^m** (or that **y is inversely proportional to x^m**), provided that $$y = \frac{k}{x^m}$$ for some constant k.

Here are some examples of functional relationships in which y varies inversely as some power of x:

$$y = \frac{5}{x^2}$$

$$y = \frac{0.3}{x}$$

$$y = -\frac{5}{x^4}$$

➤ **EXAMPLE 3**
Inverse Variation

Suppose that y varies inversely as x. Furthermore, suppose that when $x = 0.1$, the value of y is 20. Determine y as a function of x.

Solution

Because y varies inversely as x, we have

$$y = \frac{k}{x}$$

for some constant k. We may determine the value of k from the given data by substituting $x = 0.1$ and $y = 20$ into the equation to obtain

$$20 = \frac{k}{0.1}$$

$$k = 2$$

Replacing k with this value, 2, we have

$$y = \frac{2}{x}$$

➤ **EXAMPLE 4**
Electrical Resistance

The electrical resistance of an electrical wire of a certain length varies inversely with the square of the radius of the wire. Suppose that a certain wire has a radius of 0.05 in., and the resistance is measured to be 1000 ohms. What will be the resistance for a wire of radius 0.1 in.? (See Figure 2.)

Solution

Let R denote the resistance and r the radius of the wire. Then we have

$$R = \frac{k}{r^2}$$

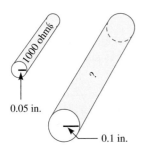

Figure 2

for some constant k. To determine k, we substitute $R = 1000$ and $r = 0.05$ into the equation, obtaining

$$1000 = \frac{k}{(0.05)^2}$$

$$k = 2.5$$

Therefore, we have the functional relationship

$$R = \frac{2.5}{r^2}$$

In particular, if $r = 0.1$ in., then

$$R = \frac{2.5}{(0.1)^2} = 250 \text{ ohms}$$

➤ **EXAMPLE 5**
Newton's Law of Gravitation

The force of gravitational attraction between two bodies varies inversely as the square of the distance between them. Suppose that when two space rocks are 100 miles apart, the force of gravitational attraction between them is 50 lb. What

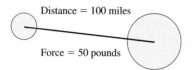

Distance = 100 miles

Force = 50 pounds

Figure 3

is the gravitational force between the rocks when they are 30 miles apart? (See Figure 3.)

Solution

Let F denote the force of gravitational attraction and d the distance between the rocks. Then we are given that

$$F = \frac{k}{d^2}$$

for some constant k. Substituting $F = 50$ and $d = 100$ into the equation, we have

$$50 = \frac{k}{100^2}$$
$$k = 500,000$$

This means that the gravitational force between the rocks is given by the functional relationship

$$F = \frac{500,000}{d^2}$$

To determine the amount of the force when the rocks are 30 miles apart, we substitute $d = 30$ into this equation to obtain

$$F = \frac{500,000}{30^2} = 555.56 \text{ lb}$$

Other Forms of Variation

Other types of variation correspond to other types of relationships that occur in applications. A commonly encountered form of variation is joint variation, which is defined as follows.

Definition 3 **Joint Variation**	We say that y **varies jointly as** x **and** t (or y **is proportional to** x **and** t) provided that $$y = ktx$$ for some constant k.

➤ **EXAMPLE 6**
Joint Variation

Suppose that y varies jointly as x and t. What is the effect if both x and t are increased by 50 percent?

Solution

We are given that

$$y = kxt$$

If x and t are each increased by 50 percent, then x is replaced by $1.5x$ and t is replaced by $1.5t$. This means that y is replaced by

$$k(1.5x)(1.5t) = 2.25(kxt)$$

But kxt is the original value of y, so the original value of y is multiplied by 2.25.

Other types of variation are obtained by combining the types of variation already introduced. For example, to say that y varies jointly as a and b and inversely as the square root of c means that y is defined by a relationship of the form

$$y = k\frac{ab}{\sqrt{c}}$$

for some constant k. Similarly, to say that y varies jointly with the square of a and the fourth power of b means that y satisfies a relationship of the form

$$y = ka^2b^4$$

for some constant k. Other types of variation are defined similarly.

Exercises 2.6

1. y varies directly as x, and the value of y is 0.04 when $x = 2.8$.

    a. Determine y as a function of x.

    b. What is the value of y when $x = 350$?

2. y varies directly as x, and the value of y is 5.6 when $x = 3.2$.

    a. Determine y as a function of x.

    b. What is the value of y when $x = 0.64$?

3. y varies directly as the square of x, and $y = 0.8$ when $x = 0.2$.

    a. Determine y as a function of x.

    b. What is the value of y when $x = 3.7$?

4. y varies inversely as x, and $y = 2.6$ when $x = 1.2$.

    a. Determine y as a function of x.

    b. What is the value of y when $x = 0.64$?

5. y varies inversely as the square of x, and $y = 2$ when $x = 0.7$.

    a. Determine the relationship between y and x, also known as an *equation of variation*.

    b. What is the value of y when $x = 0.1$?

6. y varies jointly as x and z, and $y = 3.4$ when $x = 4$ and $z = 5$.

    a. Determine the equation of variation between y, x, and z.

    b. What is the value of y when $x = 8.4$ and $z = 505$?

7. y varies jointly as x and z and inversely as w. It is known that $y = 52$ when $x = 2$, $z = 4$, and $w = \frac{1}{2}$.

    a. Determine the relationship between y, x, z, and w.

    b. What is the value of y when $x = 12$, $z = 24$, and $w = 15$?

8. y varies jointly as x and the square of z, and $y = 152.1$ when $x = 1.8$ and $z = 6.5$.

    a. Determine the relationship between y, x, and z.

    b. What is the value of y when $x = 0.0034$ and $z = 23.4$?

9. y varies jointly as x and z and inversely as the square of w. It is known that $y = 22$ when $x = 14$, $z = 10$, and $w = 12$.

    a. Determine the relationship between y, x, z, and w.

    b. What is the value of y when $x = 12.4$, $z = 1400$, and $w = 5$?

10. y varies jointly as x and z and inversely as the product of w and p. It is known that $y = 16$ when $x = 23$, $z = 31$, $w = 42$, and $p = 19$.

    a. Determine the relationship between y, x, z, w, and p.

    b. What is the value of y when $x = 12.4$, $z = 1400$, $w = 35.8$, and $p = 200$?

11. y varies inversely as the square of x and directly as the cube of z. It is known that $y = 3$ when $x = 4$ and $z = 6$.

    a. Determine the relationship between y, x, and z.

    b. What is the value of y when $x = 6$ and $z = 3$?

12. y varies directly as the square root of x and inversely as the product of the cube of z and the square of w. It is known that $y = 25$ when $x = 16$, $z = 2$, and $w = 40$.

    a. Determine the functional relationship between y, x, z, and w.

    b. What is the value of y when $x = 25$, $z = 4$, and $w = 20$?

Exercises 13–16 each present a situation involving variation. Determine a functional relationship, list the variation constant, and describe the variation.

13. The area A of a circle and its radius r.

14. The area A of a triangle and its base b and height h.

15. The distance d a car travels at a constant speed of 65 mph in t hours.

16. The simple interest I on a principal of P dollars at interest rate r for t years.

♦ Applications

17. **Water in a carrot**. The amount of water A in a raw carrot varies directly as the carrot's weight w. Suppose that a 25-g carrot contains 22 g of water. How much water does a 32-g carrot contain?

18. **Radiation energy**. The amount of energy E emitted by radiation varies directly as the wavelength L of the radiation. Suppose that one type of x-ray has a wavelength of 10^{-6} cm and emits 2×10^{-10} erg. How much energy is emitted from infrared light, which has a wavelength of 7×10^{-4} cm?

19. **Temperature of a gas**. The temperature of a gas varies jointly as its pressure P and volume V. A tank contains 100 l of oxygen under 15 atm of pressure at 20°C. If the volume of the gas is decreased to 75 l and the pressure is increased to 22 atm, what is the temperature of the gas?

20. **Period of a pendulum**. The period of a pendulum varies directly as the square root of the length of the pendulum. A 3.15-m pendulum has a period of 3.56 sec. What is the period of a 10-m pendulum?

21. **Stopping distance**. The stopping distance of a car on a certain surface varies directly as the square of the velocity before the brakes are applied. A car traveling 55 ft/sec needs 79 ft to stop. What is the stopping distance for a car on the same road surface at a velocity of 95 ft/sec?

22. **Illumination**. The illumination I from a light source varies inversely as the square of the distance d from the source. A flashlight is shining at a painting on a wall 8 ft away. At what distance should the flashlight be placed so that the amount of light is doubled? (See figure.)

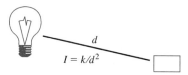

$$I = k/d^2$$

23. **Speed from skid marks**. Police have used the formula $s = \sqrt{30 f d}$ to estimate the speed s (in mph) at which a car was traveling if it skidded d feet. The variable f is the coefficient of friction, determined by the kind of road (concrete, asphalt, gravel, tar) and whether the road was wet or dry. Following are some values of f.

	Concrete	Tar
Wet	0.4	0.5
Dry	0.8	1.0

a. Determine a functional relationship between s and d on a dry tar road. At 40 mph, about how many feet will a car skid? A car leaves a skid mark of about 141 ft. How fast was it going?

b. Determine a functional relationship between s and d on a wet concrete road. At 55 mph, about how many feet will a car skid? A car leaves a skid mark of about 208 ft. How fast was it going?

24. **Force of attraction**. In Newton's law of gravitation, the force F with which two masses attract each other varies directly as the product of the masses M and m and inversely as the square of the distance d between the masses. What happens to the force of attraction when the distance between the two masses is increased from 3 ft to 12 ft? (See figure.)

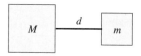

25. **Gas pressure**. The pressure P of a given quantity of gas varies directly as the absolute temperature T and inversely as the volume V. At what temperature T, with $V = 200$ in.^{3}, is the pressure three times as great as when $T = 300°$ and $V = 500$ in.^{3}?

26. **Kepler's law** states that the time t for a planet to make a circuit of the sun is directly proportional to the planet's mean distance d from the sun raised to the $\frac{3}{2}$ power. The mean distances from the sun are 93 million miles for the earth and 141 million miles for Mars. In days, how long will it take Mars to make one orbit of the sun? (Use 365 days for the time it takes the earth to orbit the sun.)

27. **Accuracy of a speedometer**. Even if a car's speedometer is not accurate, its reading varies directly with the speed of the car. A speedometer reads 38 mph when the car is actually going 42 mph. How fast is the car going when the speedometer reads 60 mph?

28. **Power of a windmill**. Within certain limits, the power P, in watts, generated by a windmill is proportional to the cube of the velocity V, in miles per hour, of the wind, as given by the equation

$$P = 0.015V^3$$

The constant 0.015 is a typical value that does not apply in all situations. (See figure.)

a. How much power would be generated by a continuous 6-mph wind?

b. How much power would be generated by a continuous 3-mph wind?

c. By what fraction is the power changed by cutting the wind speed in half?

d. How fast would the wind speed need to be to produce 120 W of power?

(*Note:* The fact that the wind generates such little power relative to its speed is one of the frustrations of recent attempts to use windmills as an alternative energy source. Also, the functional relationship does not apply for higher and higher wind speeds. That is, the design of a windmill may not have a tripling effect on power; consider the effect of a tornado as an example.)

29. **Safe length of a beam**. The safe length S of a wooden beam varies directly as the width w and the square of the height h and inversely as the length L. An old house has beams that are 3 in. wide by 10 in. high by 18 ft long. A remodeler wants to replace these with beams that are 2 in. wide and the same length. What height would the new beams have to be to have twice the safe length as the old beams? (See figure.)

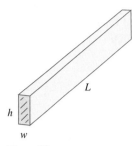

Figure 29

30. **Wavelength of an electronic particle.** The wavelength w of a particle is inversely proportional to its momentum M, which is its mass m times its velocity v. The mass of an electron is 9.19×10^{-28} g. Traveling at one-tenth the speed of light (3×10^3 m/sec), its wavelength is 2.5×10^{-11} m. What is the wavelength of a 150-g baseball traveling at 40 m/sec? Give your answer in scientific notation.

31. **Electrical attraction or repulsion**. The force E of electrical attraction or repulsion between two charged particles varies jointly as the magnitudes C_1 and C_2 of the two charges and inversely as the square of the distance d between them. If the force of repulsion between two electrons, each of charge -1.6×10^{-19} coulombs and 1×10^{-8} cm apart, is 2.3×10^{-13} dyn, what is the force of attraction between an electron and a proton if a proton's charge is 1.6×10^{-19} coulombs and the particles are 8×10^{-9} cm apart? Give your answer in scientific notation.

32. If a varies jointly as c and d and inversely as the cube of b, what can be said of d in relation to a, b, and c?

33. Suppose x varies directly as the square of y.

a. How does y vary with respect to x? (Assume that all values of x and y are positive.)

b. If k is the variation constant for x varying with respect to y, what is the variation constant for y varying with respect to x?

34. Suppose y varies jointly as x and the square of z and inversely as w.

a. How does x vary with respect to w, y, and z?

b. How does w vary with respect to x, y, and z?

c. If k is the variation constant for y varying with respect to x, z, and w, what is the variation constant for part (*a*)? For part (*b*)?

2.7 CHAPTER REVIEW

Important Concepts, Properties, and Formulas—Chapter 2

Function	Let A and B be sets. A function f from set A to set B is a correspondence that associates each element of set A with one and only one element of set B.	p. 82
Vertical line test	A set of points in the x-y plane is the graph of a function if and only if each vertical line intersects the set at no more than one point.	p. 93
Vertical translation	To obtain the graph of $f(x) = g(x) + c$, start with the graph of $g(x)$ and: $$\begin{cases} \text{shift upward } |c| \text{ units} & \text{if } c > 0 \\ \text{shift downward } |c| \text{ units} & \text{if } c < 0 \end{cases}$$	p. 94
Horizontal translation	To obtain the graph of $f(x) = g(x - c)$, start with the graph of $g(x)$ and: $$\begin{cases} \text{shift right } |c| \text{ units} & \text{if } c > 0 \\ \text{shift left } |c| \text{ units} & \text{if } c < 0 \end{cases}$$	p. 95
Vertical scaling	To obtain the graph of $f(x) = cg(x)$, start with the graph of $g(x)$ and: $$\begin{cases} \text{stretch vertically } |c| \text{ units} & \text{if } c > 1 \\ \text{shrink vertically } |c| \text{ units} & \text{if } 0 < |c| < 1 \end{cases}$$ Then reflect the result across the x-axis if $c < 0$.	p. 95
Even function	f is even provided $f(x) = f(-x)$ for all x in the domain of f.	p. 99
Odd function	f is odd provided $f(-x) = -f(x)$ for all x in the domain of f.	p. 99
Operations on functions	Let f and g be functions.	p. 111ff
	The sum of f and g is $(f + g)(x) = f(x) + g(x)$.	
	The difference of f and g is $(f - g)(x) = f(x) - g(x)$.	
	The product of f and g is $(fg)(x) = f(x)g(x)$.	
	The quotient of f and g is $\left(\dfrac{f}{g}\right)(x) = \dfrac{f(x)}{g(x)}, \quad g(x) \neq 0$	
	The composition of f and g is $(f \circ g)(x) = f(g(x))$.	

Inverse of a function	Exists provided that f is one-to-one. Exists if the graph passes the horizontal line test.	p. 125
	If f^{-1} is the inverse function, then $$f(f^{-1}(y)) = y, \quad f^{-1}(f(x)) = x$$ for all x in the domain and y in the range of f.	
Variation	y varies directly with x: $y = k \cdot x$	p. 133
	y varies inversely with x: $y = \dfrac{k}{x}$	p. 134
	y varies jointly with x and z: $y = kxz$	p. 136

Cumulative Review Exercises—Chapter 2

Determine whether each of the following is the graph of a function.

1. Figure 1(a).
2. Figure 1(b).
3. Figure 1(c).
4. Figure 1(d).

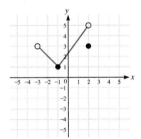

Figure 1(a)

Figure 1(b)

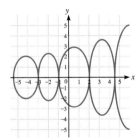

Figure 1(c)   Figure 1(d)

For $f(x) = x^2 + 3x - 4$, find each of the following.

5. $f(-1)$ 6. $f(5)$ 7. $f(0)$
8. $f(a + h)$ 9. $f(a - 1)$ 10. $f(0.75)$

For each function, evaluate and simplify the expression
$$\frac{f(a + h) - f(a)}{h}, \quad h \neq 0$$

11. $f(x) = x^2 + 3x - 4$ 12. $f(x) = 3\sqrt{x}$

13. $f(x) = x$ 14. $f(x) = 1 - \dfrac{1}{x}$

Find the domain and range of each function.

15. $f(x) = -5x + 8$ 16. $f(x) = x^3 - 1$

17. $f(x) = \sqrt{6 - x}$ 18. $g(x) = \dfrac{x^2 - 1}{x + 1}$

19. $g(x) = 56.7$

20. $g(x) = -\dfrac{3}{x^2 + x - 30}$

21. $f(x) = \sqrt{64 - x^2}$

Find the domain of each function.

22. $f(x) = \sqrt{24 - 5x - x^2}$

23. $f(x) = 1 - \dfrac{1}{x}$

24. $g(x) = 3x^{45} - 2x^{34} + 6$

25. $f(x) = \sqrt{|x|(x + 1)}$ 26. $g(x) = \dfrac{3}{|2x - 6|}$

27. $f(x) = \sqrt{49x^2}$ 28. $f(x) = \dfrac{1}{\sqrt{x - 1}}$

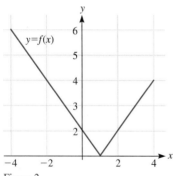

Figure 2

Consider the graph of $y = f(x)$ in Figure 2. Graph each of the functions described in Exercises 29–32.

29. $y = f(x) - 3$

30. $y = f\left(\frac{1}{2}x\right) + 3$

31. $y = -f(x)$

32. $y = 2f(x - 1)$

Determine whether the graphs of the functions in Exercises 33–38 are symmetric with respect to either of the coordinate axes or the origin.

33. $y = x^2 + 4$

34. $y = -x^{1/3}$

35. $y = \sqrt{-x^2 + 3}$

36. $y = -\frac{1}{2}x + 3$

37. $y = -x^2 + 4$

38. $x = y^{5/3}$

Determine whether each function is even, odd, or neither.

39. $f(x) = |4 - x|$

40. $f(x) = (x + 3)^2$

41. $f(x) = |x + 1| + |x - 1|$

42. $f(x) = -x^2 - 2$

43. $f(x) = (x^4 + 4)(x^3 - x)$

44. $f(x) = \dfrac{x}{x + 1}$

45. $f(x) = \sqrt{x^2 + 1}$

46. $f(x) = \sqrt[3]{x}$

47. $f(x) = \sqrt{x^4 + 1}$

For $f(x) = x^2 - 1$ and $g(x) = 2x + 3$, find each of the following.

48. The domain of f

49. The domain of g

50. *a.* The domain of f/g

51. *a.* The domain of g/f

b. $\left(\dfrac{f}{g}\right)(x)$

b. $\left(\dfrac{g}{f}\right)(x)$

52. *a.* The domain of $f + g$

53. *a.* The domain of $f \circ g$

b. $(f + g)(x)$

b. $(f \circ g)(x)$

54. *a.* The domain of $g \circ f$

b. $(g \circ f)(x)$

Find $(f \circ g)(x)$ and $(g \circ f)(x)$.

55. $f(x) = x^2 - 2x + 1,\quad g(x) = \dfrac{3}{x^2}$

56. $f(x) = x^3 + 2,\quad g(x) = \sqrt[3]{x - 2}$

57. $f(x) = x^3,\quad g(x) = x^{11}$

58. $f(x) = \dfrac{1}{2 - x},\quad g(x) = \dfrac{2x - 1}{x}$

59. $f(x) = \dfrac{2x + 7}{3x - 4},\quad g(x) = \dfrac{5}{x^2}$

In each of the following, find $f(x)$ and $g(x)$ such that $h(x) = (f \circ g)(x)$. Answers may vary, but try to choose the most obvious answer.

60. $h(x) = (3x^2 - 5x + 1)^{11}$

61. $h(x) = \sqrt[5]{\dfrac{x^3 + 1}{x^3 - 1}}$

62. $h(x) = (x^3 + 2)^5 + (x^3 + 2)^3 - 2(x^3 + 2)^2 - 7$

63. $h(x) = \left|\dfrac{x^2 - 5}{x^2 + 5}\right|$

64. Determine a function whose graph is the equation of a line through $(-1, -5)$ and perpendicular to line $3x + 4y = 6$.

65. Determine a function whose graph is an equation of a line through $(-1, -5)$ and parallel to line $3x + 4y = 6$.

66. Determine a function whose graph is an equation of a line through $(-2, 3)$ and perpendicular line containing points $(-3, -4)$ and $(-1, 2)$.

67. Determine a function whose graph is an equation of a line through $(-2, 3)$ and parallel to the line containing the points $(-3, -4)$ and $(-1, 2)$.

68. y varies jointly as x and w and inversely as the cube of z, and $y = 24$ when $x = 20$, $w = 15$, and $z = 5$.

a. Determine the relationship between y, x, w, and z.

b. What is the value of y when $x = 10$, $w = 2.4$, and $z = 20$?

69. Give a piecewise definition for the function g whose graph is shown in Figure 3.

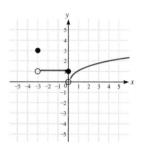

Figure 3

70. For

$$f(x) = \frac{1}{x - 4} \quad \text{and} \quad g(x) = 4$$

find the domain of $f \circ g$ and $g \circ f$.

71. For the following functions, find $(f \circ g)(x)$ and $(g \circ f)(x)$.

$$f(x) = \begin{cases} x^2 & (-3 \le x \le 0) \\ 1 - x & (0 < x \le 2) \\ \dfrac{1}{x - 2} & (2 < x) \end{cases}$$

$$g(x) = \begin{cases} 2x - 7 & (-3 \le x \le 0) \\ x^3 & (0 < x \le 2) \\ \dfrac{x + 3}{x} & (2 < x) \end{cases}$$

Determine whether these functions are inverses of each other by calculating $(f \circ g)(x)$ and $(g \circ f)(x)$.

72. $f(x) = -\frac{4}{5}x, \quad g(x) = -\frac{5}{4}x$

73. $f(x) = 2x - 5, \quad g(x) = \frac{5}{2} + \frac{1}{2}x$

74. $f(x) = x^2, \quad g(x) = \sqrt{x}$

75. $f(x) = x^5, \quad g(x) = \sqrt[5]{x}$

For each of the following, find a formula for $f^{-1}(x)$.

76. $f(x) = 9x - 14$

77. $f(x) = \dfrac{3x - 4}{4 - 2x}$

78. $f(x) = \sqrt{16 - x^2} \quad (0 \le x \le 4)$

79. $f(x) = x^3 - 1$

80. For $f(x) = 45{,}677x^5 + 0.0457$, find
$$(f \circ f^{-1})(78.8999)$$
and
$$(f^{-1} \circ f)(-2{,}344{,}789)$$

♦ **Applications**

81. **Loudness of Sound.** Suppose you are sitting at a distance d from a stereo speaker. You and the speaker are outside. The loudness L of the sound is inversely pro-

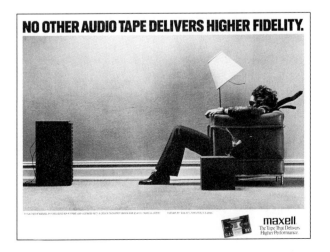

portional to the square of d. What happens to the sound if you move three times the distance d from the speaker?

82. **Safe Load.** The safe load S for a rectangular beam of fixed length varies jointly as the width w and the square of the height h.

    a. Determine a relationship between S, w, and h.

    b. Which of the following choices will give the strongest single beam that can be cut from a cylindrical log of diameter 25 cm? (See Figure 4.)

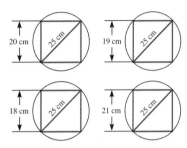

Figure 4

83. **Growth Rate of Yeast Cells.** The rate of growth G of yeast cells under a certain laboratory condition varies jointly as the number of cells N and $700 - N$ that are present. Determine a relationship between G, N, and $700 - N$.

84. **Temperature conversion.** Celsius temperatures C and Fahrenheit temperatures F are related by the linear equation

$$C = \tfrac{5}{9}(F - 32)$$

    a. Give an interpretation of the slope of this equation.

    b. Suppose a person's Fahrenheit temperature dropped 1° F during an illness. What is the effect on the person's Celsius temperature?

85. **Sales Commissions.** A salesperson gets a base salary of $25,000 plus a commission of 24 percent of sales that exceed $400,000.

    a. Determine a function that relates salary and sales.

    b. Graph this function.

86. **Boyle's Law.** Boyle's law asserts that the volume V of a gas at a constant temperature is inversely proportional to the pressure P.

    a. Determine a functional relationship between V and P.

    b. A tank contains 16 cubic feet of oxygen at a pressure of 50 pounds per square inch. What volume of oxygen will be occupied if the pressure is changed to 15 pounds per square inch?

87. **Force on a Batted Ball.** The force F exerted by a softball bat of mass m being swung at speed v varies jointly

as m and the square of v according to the equation

$$F = \tfrac{1}{2}mv^2$$

By what factor would a batter have to decrease velocity to exert the same force with a 38-oz softball bat as with a 35-oz softball bat?

88. **Kelvin temperature** K is related to Celsius temperature C by the linear equation

$$K = C + 273.15$$

   a. What is the effect of a 1° change in Celsius temperature on the Kelvin temperature?

   b. **Absolute zero** is the coldest temperature possible and occurs when $K = 0$. To what Celsius temperature does 0° K correspond? To what Fahrenheit temperature does it correspond?

89. A rectangular box with a volume of 560 cubic feet is to be constructed with a square base and top, each with sides of length x. The cost per square foot for the bottom is \$0.35, for the top is \$0.20, and for the sides is \$0.11. Let C denote the total cost of the box. Determine a formula in one variable for $C(x)$.

90. A boat travels 50 miles upstream and 50 miles back downstream. The speed of the stream is 2 mph. Let t denote the total time of the trip. Find a formula in one variable for $t(b)$, where b is the speed of the boat in still water.

91. A tank in the shape of an inverted cone has a cross section that is an equilateral triangle with dimensions as shown in Figure 5. The volume of a cone is given by

$$V = \tfrac{1}{3}\pi r^2 h$$

Find a function $V(h)$ in one variable for the volume in terms of the height h.

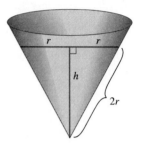

Figure 5

92. A carpet warehouse offers discounts to buyers of large quantities. If x is the number of square yards of carpet, then $C(x)$ is the cost, where

$$C(x) = \begin{cases} 12x & (0 \le x \le 50) \\ 10x & (50 < x \le 200) \\ 9x & (200 < x) \end{cases}$$

Graph this function for $0 \le x \le 300$.

POLYNOMIAL AND RATIONAL FUNCTIONS

The symphonic shell shown has a cross-section that is a curve called a **parabola**, *which is the graph of a quadratic function. The orchestra is seated in the center of the shell, and the sound it produces is reflected off the shell. Due to a geometric property of the parabola and the way sound is reflected, the sound waves are projected forward.*

W e begin this chapter with a study of quadratic functions. We will introduce optimization problems, which call for maximizing or minimizing a quantity. Optimization problems are one of the main topics in a calculus course. However, by studying the properties of quadratic functions, we will be able to handle some special optimization problems. In this chapter, we will also learn to sketch the graphs of polynomial and rational functions and will develop the theory of polynomial equations.

3.1 QUADRATIC FUNCTIONS AND OPTIMIZATION PROBLEMS

Polynomial functions provide a rich source of mathematical models. Let's begin by defining these functions.

Definition 1
Polynomial Function

A **polynomial function** of the variable x is a function of the form
$$f(x) = a_n x^n + a_{n-1} x^{n-1} + \cdots + a_1 x + a_0$$
where n is a nonnegative integer and $a_n, a_{n-1}, \ldots, a_1, a_0$ are real numbers.

The domain of a polynomial function consists of all real numbers. The range is a subset of the real numbers and may or may not consist of all real numbers. The range of a polynomial function will be discussed later in the chapter.

Here are some examples of polynomial functions.

$$f(x) = 5$$
$$f(x) = -0.1x + 0.001$$
$$f(x) = 3x^2 + 100x - 310$$
$$f(x) = -x^3 + x^2 - x + 1$$

As the first two examples illustrate, constant and linear functions are special cases of polynomial functions. Because we have already explored the properties of these functions and their graphs, let's turn to the next most complicated polynomial functions, namely those for which n equals 2.

Definition 2
Quadratic Function

A **quadratic function** is a function defined by an expression of the form
$$f(x) = ax^2 + bx + c$$
where a, b, and c are real numbers and a is nonzero.

Here are some examples of quadratic functions.

$$f(x) = x^2$$
$$f(x) = \tfrac{1}{2}x^2 + 3x - 1$$
$$f(x) = 0.001x^2 + 5.07$$

Graphs of Quadratic Functions

We have already graphed the function $y = x^2$ and shown that its graph is bowl-shaped, opening upward, with the bottom of the bowl at the origin (see Figure 1). The graphs of the quadratic functions

$$f(x) = ax^2, \qquad a > 0$$

have the same general appearance as that of $f(x) = x^2$. The coefficient a controls the width of the opening. The smaller the value of a, the wider the opening. Figure 2 shows the graphs of such quadratic functions for various positive numbers a.

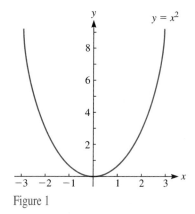

Figure 1

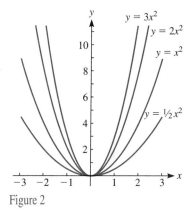

Figure 2

The graph of

$$f(x) = ax^2, \qquad a < 0$$

is obtained by reflecting the graph of $f(x) = ax^2$ in the x-axis; thus, it is a bowl-shaped curve that opens downward (see Figure 3).

The graph of

$$f(x) = a(x - h)^2 + k \tag{1}$$

is obtained by translating the graph of $f(x) = ax^2$ and is therefore a bowl-shaped graph with bottom or top at (h, k). The graph opens upward if $a > 0$ and downward if $a < 0$ (see Figure 4).

By completing the square, every quadratic function may be written in form (1), which is called the **standard form** of the quadratic function. The bowl-shaped curve arising as the graph of a quadratic function is called a **parabola**.

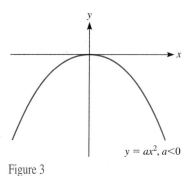

Figure 3

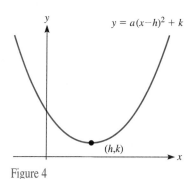

Figure 4

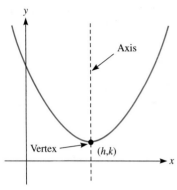

Figure 5

The bottom or top of the bowl of a parabola is called its **vertex**. In the graph of (1), the vertex is at (h, k). The vertical line through the vertex is called the **axis** of the parabola (see Figure 5).

Let's summarize this information for easy reference:

STANDARD FORM OF A QUADRATIC FUNCTION

The quadratic function $f(x) = ax^2 + bx + c$ may be written in the form
$$f(x) = a(x - h)^2 + k$$
for suitable real numbers h and k. The graph of the function is a parabola with vertex (h, k). The parabola opens upward if $a > 0$ and downward if $a < 0$.

➤ **EXAMPLE 1**
Standard Form
for a Quadratic Function

Consider the quadratic function $f(x) = 2x^2 + 6x - 10$.

1. Write $f(x)$ in the form (1).
2. Determine the vertex and the axis of the graph of $f(x)$.

Solution

1. Complete the square:
$$\begin{aligned} f(x) &= 2x^2 + 6x - 10 \\ &= 2(x^2 + 3x) - 10 \\ &= 2\left(x^2 + 3x + \left(\tfrac{3}{2}\right)^2\right) - 10 - 2\left(\tfrac{3}{2}\right)^2 \\ &= 2\left(x + \tfrac{3}{2}\right)^2 - \tfrac{29}{2} \end{aligned}$$

This is of the form (1) with
$$a = 2, \qquad h = -\tfrac{3}{2}, \qquad k = -\tfrac{29}{2}$$

2. The vertex is at $(h, k) = \left(-\tfrac{3}{2}, -\tfrac{29}{2}\right)$. The axis is the vertical line through the vertex and thus has equation $x = -\tfrac{3}{2}$.

➤ **EXAMPLE 2**
Determining a Quadratic
Function from Geometric
Information

Determine the equation of a quadratic function whose graph passes through the point $(1, -2)$ and has vertex $(4, 5)$.

Solution

By equation (1), the quadratic function has the form
$$f(x) = a(x - 4)^2 + 5$$
for some value of a. Because the graph must pass through $(1, -2)$, we must have
$$\begin{aligned} f(1) &= -2 \\ a(1 - 4)^2 + 5 &= -2 \\ 9a + 5 &= -2 \\ a &= -\tfrac{7}{9} \\ f(x) &= -\tfrac{7}{9}(x - 4)^2 + 5 \end{aligned}$$

Because the function $y = ax^2$ is even, its graph is symmetric about the y-axis. Thus, the graph of (1), which is obtained by translating the graph of $y = ax^2$, is symmetric about its axis. To sketch the graph of a quadratic function, we may proceed as follows:

GRAPHING A QUADRATIC FUNCTION $f(x)$

1. Write the quadratic function in standard form: $f(x) = a(x - h)^2 + k$.
2. Plot the vertex (h, k).
3. Determine a point on the graph. The simplest such point to determine is usually the one corresponding to $x = h + 1$. Plot this point.
4. Draw a smooth curve from the vertex through the plotted point.
5. Complete the graph by drawing a symmetrically placed curve on the other side of the vertex.

➤ **EXAMPLE 3**
Graphing Quadratic Functions

Graph the following quadratic functions.

1. $f(x) = 3x^2 - 12x - 10$
2. $f(x) = -2x^2 + 12x$

Solution

1. As a first step, we write the function in standard form:

$$3x^2 - 12x - 10 = 3(x^2 - 4x) - 10$$
$$= 3(x^2 - 4x + 4) - 10 - 3 \cdot 4$$
$$= 3(x - 2)^2 - 22$$

The vertex is $(2, -22)$. When $x = 3$, the value of $f(x)$ is

$$f(3) = 3 \cdot 1^2 - 22 = -19$$

so $(3, -19)$ is on the graph. Draw a smooth curve from $(2, -22)$ through $(3, -19)$. Now draw a symmetric curve on the left of the vertex. The completed graph is shown in Figure 6.

2. We put the function into standard form:

$$f(x) = -2x^2 + 12x$$
$$= -2(x^2 - 6x)$$
$$= -2(x^2 - 6x + 9) - (-2) \cdot 9$$
$$= -2(x - 3)^2 + 18$$

The vertex of the parabola in this case is $(3, 18)$. Another point on the graph is $(4, 16)$. Draw a smooth curve from the vertex through $(4, 16)$ and complete the graph via symmetry. The completed graph is shown in Figure 7.

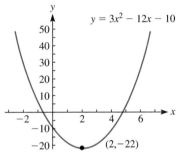

Figure 6

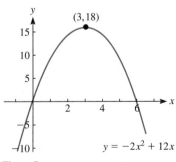

Figure 7

Applications of Quadratic Functions

The graph of a quadratic function is a parabola that opens either upward or downward. In the case of a graph that opens upward, the bottom point on the graph is called the **minimum point** of the function. Its y-value is the least value that

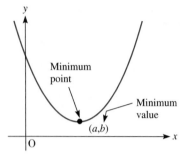

Figure 8

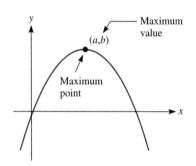

Figure 9

the function assumes and is called the **minimum value** of the function (see Figure 8).

Analogously, in the case of a graph that opens downward, the graph has a top point, called the **maximum point** of the function. The y-value of the maximum point is the largest value that the function assumes and is called the **maximum value** of the function (see Figure 9).

The maximum or minimum point of the graph of a quadratic function is called the **vertex** or **extreme point** of the graph. The maximum or minimum value of the function is called the **extreme value**.

The extreme point of a quadratic function can easily be determined. Consider the quadratic function

$$f(x) = ax^2 + bx + c$$

Assume that $a > 0$, so that the graph of f opens upward. The function then has a minimum point. Write the function in the form

$$f(x) = a\left(x^2 + \frac{b}{a}x\right) + c$$

$$= a\left(x^2 + \frac{b}{a}x + \frac{b^2}{4a^2}\right) + \left(c - \frac{ab^2}{4a^2}\right)$$

$$= a\left(x + \frac{b}{2a}\right)^2 + \frac{4ac - b^2}{4a}$$

$$= a(x - h)^2 + k$$

where

$$(h, k) = \left(-\frac{b}{2a}, \frac{4ac - b^2}{4a}\right)$$

Recall that a perfect square is nonnegative. Because we are assuming that a is positive, the expression

$$a\left(x + \frac{b}{2a}\right)^2$$

is nonnegative—it is the product of a positive number and a nonnegative number. Therefore, for any value of x, we have

$$f(x) \geq k$$

In other words, the value of $f(x)$ is always at least k. Moreover, for

$$x = -\frac{b}{2a}$$

the value of f is exactly k because

$$f\left(-\frac{b}{2a}\right) = a\left(-\frac{b}{2a} + \frac{b}{2a}\right)^2 + k$$

$$= 0 + k$$

$$= k$$

We have proved that $f(x)$ is always at least k, so we have located the minimum point of f, which is $[-b/2a, f(-b/2a)]$.

If $a < 0$, the graph opens downward, and f has a maximum point whose coordinates are given by the same formulas just derived for the minimum point. The proof is similar to the one just given. Thus, we have proved the following important result.

EXTREME POINT OF A QUADRATIC FUNCTION

Let $f(x) = ax^2 + bx + c$ be a quadratic function. The extreme point of its graph is

$$(h, k) = \left(-\frac{b}{2a}, f\left(-\frac{b}{2a} \right) \right)$$

If $a > 0$, then the extreme point is a minimum point. If $a < 0$, then the extreme point of f is a maximum point.

Now that we have determined the extreme point of a quadratic function, we may determine its range. Consider a quadratic function

$$f(x) = ax^2 + bx + c, \qquad a \neq 0$$

In case $a > 0$, the graph of the function is a parabola opening upward. The range of the function consists of the possible y-coordinates of points on the graph. These consist of the values $[m, \infty)$, where m is the minimum value of the function. That is,

---

range of f is the interval $[f\left(-\dfrac{b}{2a} \right), \infty)$ if $a > 0$

Similarly, if $a < 0$, the graph is a parabola opening downward and we have

---

range of f is the interval $(-\infty, f\left(-\dfrac{b}{2a} \right)]$ if $a < 0$

We have shown that the extreme point of a parabola is at the vertex. Many applied problems can be reduced to determining the coordinates of the extreme point. Such problems are called **optimization problems** because in the applied context, the maximum or minimum point represents an optimum condition of some sort (maximum profit, minimum cost, for example). In the next three examples, we illustrate a number of practical uses of quadratic functions.

➤ **EXAMPLE 4**
Projectile Motion

A ball is thrown vertically upward with an initial velocity of 100 feet per second.

1. Determine a function $h(t)$ that gives the height of the ball at time t seconds after it is thrown.

2. How long does it take the ball to reach its maximum height?

3. What is the maximum height the ball reaches?

Solution

1. In Chapter 2 we used the result from physics that gives the height function $h(t)$ as

$$h(t) = -16t^2 + v_0 t + h_0$$

where v_0 is the initial velocity of the ball (in feet per second) and h_0 is the initial height of the ball (in feet). In this case, we have the formula

$$h(t) = -16t^2 + 100t$$

(because the initial height h_0 equals 0).

2. The height function is a quadratic function. In Figure 10 we have drawn the path of the ball $t \geq 0$ and labeled the extreme point, which is a maximum. (This is consistent with the fact that the coefficient of the squared term is -16, which is negative.) According to the preceding proof, the maximum point occurs at

$$t = -\frac{b}{2a} = -\frac{100}{2 \cdot (-16)} = \frac{25}{8}$$

That is, the ball reaches its maximum height after $\frac{25}{8} = 3.125$ seconds. You should note that the curve shown in Figure 10 is the physical path of the ball but does not represent the graph of the function $h(t)$.

3. The maximum height of the ball is given by the maximum value of the function. This maximum value equals

$$h\left(-\frac{b}{2a}\right) = h\left(\frac{25}{8}\right)$$

$$= -16\left(\frac{25}{8}\right)^2 + 100\left(\frac{25}{8}\right)$$

$$= 156.25 \text{ ft}$$

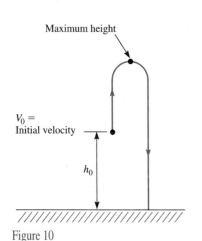

Maximum height

$V_0 =$ Initial velocity

h_0

Figure 10

> ## EXAMPLE 5
Maximizing Revenue

A company manufactures upholstered chairs. If it manufactures x chairs, then the revenue $R(x)$ per chair is given by the function

$$R(x) = 200 - 0.05x$$

How many chairs should the company manufacture to maximize total revenue?

Solution

We first find a function $T(x)$ that expresses the total revenue T received in terms of the number of chairs manufactured x. We have

$$T(x) = \text{Number of chairs} \cdot \text{Revenue per chair}$$

Therefore,

$$T(x) = x \cdot R(x)$$
$$= x(200 - 0.05x)$$
$$= -0.05x^2 + 200x$$

Thus, $T(x)$ is a quadratic function. Figure 11 shows the graph of this function. Note that the graph is a parabola opening downward. In economic terms, the graph says that initially as the number of chairs x is increased, the total rev-

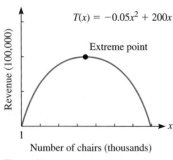

$T(x) = -0.05x^2 + 200x$

Extreme point

Revenue (100,000)

Number of chairs (thousands)

Figure 11

enue $T(x)$ increases. At a certain level of production, the total revenue is maximized. Increasing the number of chairs further actually decreases total revenue. The maximum occurs for

$$x = -\frac{b}{2a} = \frac{200}{2(-0.05)} = 2000$$

That is, the revenue is at a maximum when 2000 chairs are manufactured.

➤ EXAMPLE 6
Architecture

Suppose that an architect wishes to design a house with a fenced back yard. To save fencing cost, he wishes to use one side of the house to border the yard and to use cyclone fence for the other three sides. The specifications call for using 100 feet of fence. What is the largest area the yard can contain?

Solution

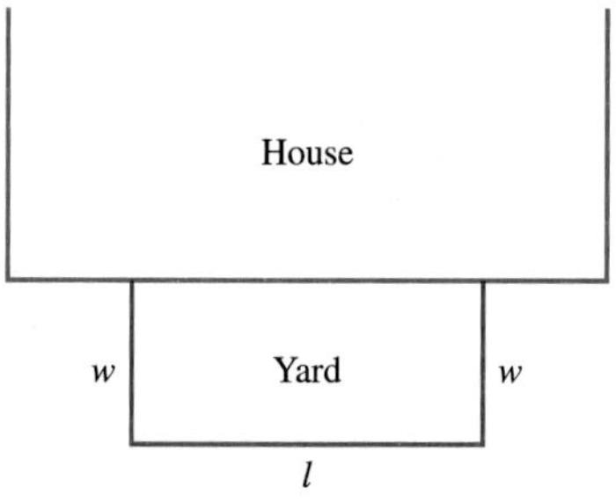

Let l denote the length of the yard and w the width (see Figure 12). Because the builder wishes to use 100 feet of fence, the diagram in the figure shows that

$$l + 2w = 100$$

The area of the yard, which is to be maximized, is given by the expression

$$\text{Area} = lw$$

We may express the area as a function $A(w)$, as follows. Solve the first equation for l in terms of w:

$$l = 100 - 2w$$

Now substitute this expression for l into the expression for the area:

$$\text{Area} = lw = (100 - 2w)w = -2w^2 + 100w$$

That is, the area function $A(w)$ is given by

$$A(w) = -2w^2 + 100w$$

This is a quadratic function. The coefficient of the square term is -2, which is negative, so the extreme point is a maximum point. The value of w for which the maximum value of $A(w)$ occurs is given by

$$w = -\frac{b}{2a} = -\frac{100}{2(-2)} = 25 \text{ ft}$$

For this value of w, the corresponding value of l is

$$l = 100 - 2w = 100 - 2(25) = 50 \text{ ft}$$

In other words, if the area of the back yard is to be maximized, the yard should be 25 feet wide and 50 feet long. The maximum area of the yard equals

$$A(25) = -2(25)^2 + 100(25) = 1250 \text{ sq ft}$$

House

w Yard w

l

Figure 12

■ New Technology

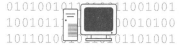

EXAMPLE 7
Translating a Parabola Using a Calculator

Translate the graph of $f(x) = 2x^2 - 3x + 1$ two units to the left and 5 units down. Graph both the original and translated graphs on a single coordinate system.

Solution

The translated graph is the graph of

$$f(x - (-2)) + (-5) = f(x + 2) - 5$$
$$= 2(x + 2)^2 - 3(x + 2) - 4$$

Enter $f(x)$ as Y_1 and the above expression for Y_2 (see Figure 13). In order to view the resulting graphs, it is helpful to zoom out several times. See Figure 14.

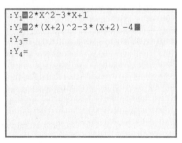

Figure 13

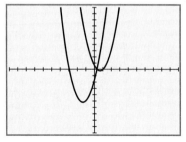

Figure 14

Exercises 3.1

Match each of the following functions with its graph among the graphs (a)–(f).

1. $f(x) = 2x^2$

2. $f(x) = -x^2$

3. $f(x) = 2x^2 - x$

4. $f(x) = -(x + 1)^2 + 3$

5. $f(x) = (x - 3)^2 + 1$

6. $f(x) = x(x - 1)$

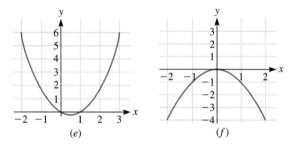

(e) (f)

Determine the equations of each of the parabolas shown in the following graphs.

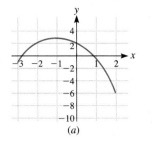

(a)

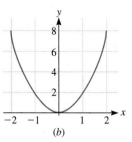

(b)

7.

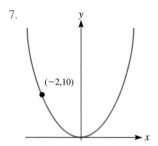

$(-2,10)$

8.

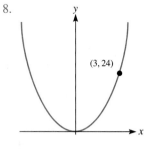

$(3, 24)$

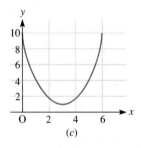

(c)

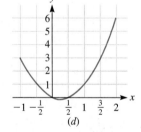

(d)

9.

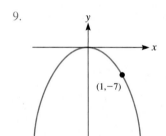

$(1, -7)$

10.

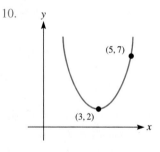

$(5, 7)$

$(3, 2)$

11.

12.

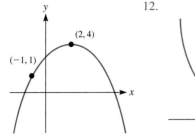

For each of the following quadratic functions:

a. Find the vertex.

b. Determine whether the graph opens upward or downward.

c. Graph the function.

13. $f(x) = -2x^2$

14. $f(x) = 2\frac{1}{4}x^2$

15. $f(x) = (x - 2)^2$

16. $f(x) = -2(x + 1)^2$

17. $f(x) = 2(x - 3)^2$

18. $f(x) = -1.5(x + 2)^2$

19. $f(x) = -(x + 3)^2 - 2$

20. $f(x) = \frac{1}{2}(x - 4)^2 - 3$

For each of the following quadratic functions:

a. Determine the standard quadratic form.

b. Find the vertex.

c. Determine whether the graph opens upward or downward.

d. Find the maximum or minimum value.

e. Graph the function.

21. $f(x) = x^2 + 6x + 5$

22. $f(x) = -13 - 8x - x^2$

23. $f(x) = -\frac{1}{4}x^2 - 3x - 1$

24. $f(x) = 2x^2 + x - 1$

Find the maximum or minimum value.

25. $f(x) = -3(x + 1)^2$ 26. $f(x) = 4(x - 3)^2$

27. $f(x) = 5(x - 1)^2 + 3$ 28. $f(x) = 12(x - 5)^2 + 2$

29. $f(x) = -2x^2 - x + 15$ 30. $f(x) = x^2 - 6x + 9$

31. $f(x) = 3x^2 - x - 14$ 32. $f(x) = -x^2 + 1$

33. $f(x) = 21.34x^2 - 456x + 2000$

34. $f(x) = \frac{2}{3}x^2 + \frac{5}{6}x - \frac{7}{12}$

35. $f(x) = -\sqrt{2}x^2 + 4\sqrt{3}x - \sqrt{7}$

36. $f(x) = \$240,000x^2 - \$560,000x + \$4,500,000$

Solve. (Functions for many of these exercises were developed in Exercise Set 2.1.)

37. Of all the numbers whose sum is 30, find the two that have the maximum product. That is, find x and y that yield the maximum value of $P = xy$, where $x + y = 30$.

38. Of all the numbers whose difference is 13, find the two that have the minimum product.

39. The sum of the height and the base of triangle is 166 cm. Find the dimensions of such a triangle that has maximum area.

40. The sum of the base and height of a parallelogram is 145 yd. Find the dimensions of such a parallelogram that has maximum area.

♦ **Applications**

41. A rancher wants to fence a rectangular area next to a river, using 150 yd of fencing, as shown in the accompanying figure.

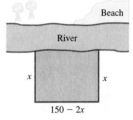

$150 - 2x$

Exercise 41

a. Determine a formula in one variable for the area $A(x)$.

b. At what value of x will the area be a maximum?

c. What is the largest area that can be enclosed?

42. A rancher wants to enclose two rectangular areas with a fence, as shown in the accompanying figure, next to a river, using 380 yd of fencing.

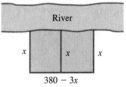

$380 - 3x$

Exercise 42

a. Determine a formula in one variable for the enclosed area $A(x)$.

b. At what value of x will the area be a maximum?

c. What is the largest area that can be enclosed?

43. A carpenter is building a rectangular room with a fixed perimeter of 74 ft. What are the dimensions of the largest room that can be built? What is its area?

44. Of all the rectangles that have a perimeter of 98 ft, find the dimensions of the one with the largest area. What is its area?

45. A clothing firm is coming out with a new line of suits. If it manufactures x suits, then the revenue $R(x)$ per suit is given by the function

$$R(x) = 400 - 0.16x$$

How many suits should the company manufacture in order to maximize total revenue?

46. An electronics firm is marketing a new low-price stereo. It determines that its total revenue from the sale of x stereos is given by the function

$$R(x) = 280x - 0.4x^2$$

The firm also determines that its total cost of producing x stereos is given by

$$C(x) = 5000 + 0.6x^2$$

Total profit is given by

$$P(x) = R(x) - C(x)$$

   a. Find the total profit.

   b. How many stereos must the company produce and sell to maximize profit?

   c. What is the maximum profit?

47. The owner of a 30-unit motel, by checking records of occupancy, knows that when the room rate is $50 a day, all units are occupied. For every increase of x dollars in the daily rate, x units are left vacant. Each unit occupied costs $10 per day to service and maintain. Let R denote the total daily profit.

   a. Determine a formula in one variable for $R(x)$.

   b. Find a value of x for which $x \geq 0$ and R is a maximum.

   c. What is the maximum profit?

48. A Palladian window is a rectangle with a semicircle on top (see figure). Suppose the perimeter of a particular Palladian window is to be 28 ft. Let A denote the total area of the window.

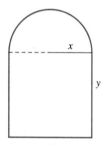

Exercise 48

   a. Determine a formula in one variable for $A(x)$.

   b. For what value of x will A be a maximum?

   c. What is the maximum possible area?

49. A 28-in. piece of string is to be cut into two pieces (see figure). One piece is to be used to form a circle and the other to form a square. Let A denote the total area.

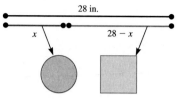

Exercise 49

   a. Determine a formula for $A(x)$.

   b. Find a value of x for which A is a minimum.

   c. What is the minimum possible area?

50. A ball is thrown vertically upward from the roof of a skyscraper. Let S denote the height of the ball from the ground t seconds after being thrown. $S(t)$ is given by an expression of the form

$$S(t) = -16t^2 + v_0 t + h$$

where v_0 represents the initial velocity and h represents the height of the building.

   Suppose that the building is 1800 ft tall. Furthermore, suppose that an observer sees the ball pass by a window 900 ft from the ground 8 seconds after the ball was thrown.

   a. Determine a formula for $S(t)$.

   b. How long does it take the ball to reach its maximum height?

   c. What is the maximum height the ball reaches?

51. A soybean farmer has 15 tons of beans on hand. If he sold them now, his profit would be $300 per ton. If he waits to sell, he can add 3 tons of beans each week, but he will lose $20 per ton for each week he delays in selling his beans.

   a. In how many weeks should he sell to maximize his profits?

   b. What would his maximum profit be?

52. The power P, in watts, delivered in an electric circuit in which a current I is flowing is found by the formula

$$P = EI - RI^2$$

where E is the voltage and R is the resistance. Find the maximum power that can be delivered by a 200-volt circuit with a resistance of 5.5 ohms.

53. Find a such that

$$f(x) = ax^2 + 4x - 5$$

has a maximum value at $x = 7$.

54. Find b such that the function

$$f(x) = 2x^2 + bx - 3$$

has a minimum value of 100.

55. Find c such that the function

$$f(x) = 0.1x^2 + 7x + c$$

has a minimum value of -127.

Find the maximum or minimum value.

56. $f(x) = (1 - c)x^2 - 3cx + 12$

57. $f(x) = -(t + 2)x^2 + (t^2 - 4)x - 16$

58. Find a quadratic function that has $(-2, 3)$ as a vertex and contains the point $(-1, 8)$.

59. Find the vertex of the quadratic function

$$f(x) = px^2 - 3x + p$$

60. Graph $f(x) = |x^2 - 3|$.

61. Graph $f(x) = |6x - 5 - x^2|$.

62. Graph $x = |(y + 3)^2 - 4|$.

⇒ **Technology**

Use a graphing calculator to determine the maximum or minimum value of each of the following functions. Determine the value to two significant digits.

63. $y = 5.78x^2 + 3.11x$

64. $y = -12x^2 + 13x - 11$

65. $y = \dfrac{x^2}{21} + \dfrac{11x}{2} + \dfrac{5}{23}$

66. $y = 11(x - 3.4)(x + 7.9)$

67. $y = |x^2 - 3x + 4|$

68. $y = |1.1 - x^2|$

■ 3.2 GRAPHS OF POLYNOMIAL FUNCTIONS OF DEGREE *n*

As we have already seen, the graph of a linear function is a straight line, and the graph of a quadratic function is a parabola. Many applied problems, especially those in science and engineering, require us to deal with polynomial functions of degree 3 or higher. In this section, we develop some elementary methods for sketching the graphs of such functions.

Graphs of the Functions $y = ax^n$

Among the most elementary polynomial functions are those of the form

$$y = ax^n$$

where n is a positive integer and a is a real constant. We begin our study of the graphs of polynomial functions by examining the graphs of these functions.

Consider the function

$$f(x) = x^3$$

To sketch its graph, we first tabulate points corresponding to a set of representative positive values of x. Using these points, we plot the portion of the graph for $x > 0$.

We see that as x increases, so does $f(x)$. In fact, as x increases without bound, the value of $f(x)$ does the same, so we draw the graph heading steadily upward without bound as x increases.

Next, we note that the function is odd because

$$f(-x) = (-x)^3 = -x^3 = -f(x)$$

Therefore, its graph is symmetric with respect to the origin, so the portion of the graph for $x < 0$ is obtained by reflecting the portion for $x > 0$ in the origin. The sketch of the graph is shown in Figure 1.

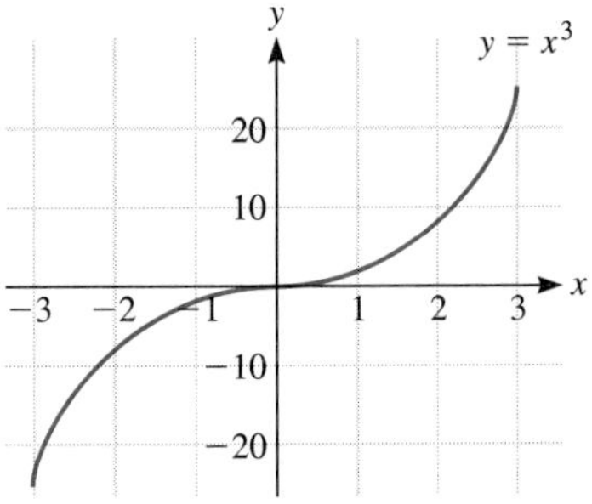

Starting from the graph of Figure 1, we can sketch the graphs of many other functions using the geometric transformations of scaling, reflection, horizontal translation, and vertical translation. The next example illustrates some of the possibilities.

x	$f(x)$
0	0
0.2	0.008
0.5	0.125
1	1
2	8
3	27

Figure 1

➤ EXAMPLE 1
Graphing Cubics

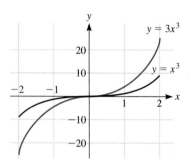

Figure 2

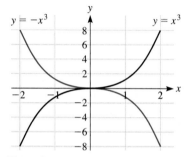

Figure 3

Sketch the graphs of the following functions.

1. $f(x) = 3x^3$
2. $f(x) = -x^3$
3. $f(x) = (x - 1)^3$
4. $f(x) = (x + 2)^3 + 3$
5. $f(x) = -2(x - 1)^3 - 4$

Solution

1. We scale the graph of Figure 1 using a factor of 3. See Figure 2.
2. We reflect the graph of Figure 1 in the x-axis because the scaling factor is negative. See Figure 3.
3. We translate the graph of Figure 1 one unit to the right. See Figure 4.
4. We translate the graph of Figure 1 two units to the left and 3 units upward. See Figure 5.

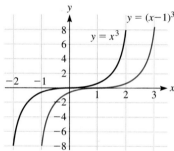

Figure 4

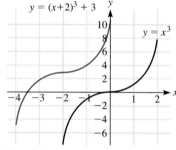

Figure 5

5. We perform the following geometric transformations to Figure 1: First, we scale the graph using a factor of 2 . Then, we reflect the graph in the x-axis. Finally, we translate the graph 1 unit to the right and 4 units downward. See Figure 6.

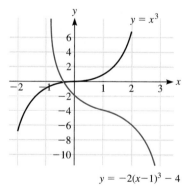

Figure 6

Let's consider the graphs of the functions

$$y = x^n, \qquad n \geq 2$$

When *n* is odd, the function is odd, just as we observed above for the case $n = 3$. When *n* is odd, the graph has the same general shape as the graph of Figure 1. In Figure 7, we show the graphs corresponding to several small odd values of *n*.

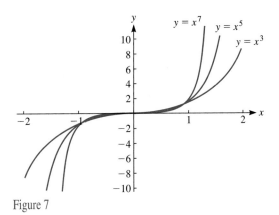

Figure 7

If *n* is even, the function

$$y = x^n$$

is even, like the function $y = x^2$. In this case, the graph is symmetric with respect to the *y*-axis and has the general shape of a parabola. In Figure 8, we have sketched the graphs corresponding to several small even values of *n*. Note that as *n* increases, the graph tends to be flatter for *x* between -1 and 1 and to rise more steeply for $x < -1$ and $x > 1$.

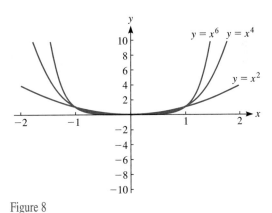

Figure 8

By using scaling, reflection, and translation with the basic graphs in Figures 7 and 8, we can sketch the graphs of all functions of the form

$$y = ax^n, \qquad n \geq 2$$

Graphs of Other Polynomial Functions

Let's turn our attention to graphing more general polynomial functions. A full discussion of this subject really belongs in calculus, where the tools necessary to describe the various specific geometric features of these graphs are developed. However, using a few simple ideas, we can make some inroads into the problem of sketching graphs of polynomial equations and come up with rough sketches of such graphs.

Definition 1 **Zero of a Polynomial**	Let $f(x)$ be a polynomial function. A number x for which $$f(x) = 0$$ is called a **zero** of f.

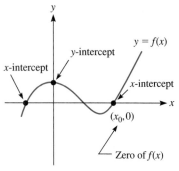

Figure 9

Suppose that x_0 is a real zero of f. The point $(x_0, 0)$ is an x-intercept of the graph of f (see Figure 9). In general, a polynomial function $f(x)$ has a number of zeros, which can be found as solutions to the polynomial equation $f(x) = 0$. Later in the chapter, we will discuss these zeros more fully and develop techniques for determining them.

Figure 10 shows the graph of a polynomial function of degree greater than 2. Notice that the graph consists of a number of "peaks" and "valleys." A point on the graph that is either the top of a peak or the bottom of a valley is called a **turning point** or an **extreme point**. Using calculus, it is possible to prove that a polynomial of degree n has at most $n - 1$ extreme points.

Note that in the interval between two consecutive extreme points, the graph is either increasing or decreasing. (If the graph changes from increasing to decreasing or vice versa, then there must be an extreme point within the interval. This fact can be proven using calculus.)

The following is another result that is proven in more general form in calculus and that is useful in sketching graphs of polynomials.

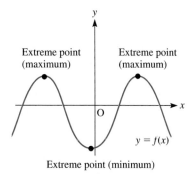

Figure 10

INTERMEDIATE VALUE THEOREM FOR POLYNOMIALS

Suppose that $f(x)$ is a polynomial with real coefficients and that a and b are real numbers with $a < b$. Suppose that one of $f(a)$ and $f(b)$ is positive and the other negative. Then $f(x)$ has a real zero in the interval (a, b).

As a consequence of the Intermediate Value theorem, we conclude that if a and b are two consecutive real zeros of f, then for x between a and b, the values of f are all of one sign, either all positive or all negative. (If not, then by the Intermediate Value theorem, there would be a zero between the supposedly consecutive zeros.) We can determine which sign prevails by evaluating f for a **test value** lying between a and b. Using the signs of f in each of the intervals determined by consecutive real zeros allows us to produce a rough sketch of the graph, as shown in the following two examples.

➤ **EXAMPLE 2**
Graphing a Polynomial Function

Sketch the graph of the function
$$f(x) = 3(x + 2)(x - 1)(x - 4)$$

Solution

The zeros of f are determined as the solutions of the equation
$$3(x + 2)(x - 1)(x - 4) = 0$$

A product can be zero only if one of the factors is 0. In this case, that means that either $x + 2 = 0$, $x - 1 = 0$, or $x - 4 = 0$. That is, the zeros of x are -2, 1, and 4. These three real zeros divide the x-axis into four intervals, namely

$$(-\infty, -2), (-2, 1), (1, 4), (4, +\infty)$$

In each of these intervals, we choose a test value and determine the sign of f at the test value. The results are summarized in the following table.

Interval	Test Value x	$f(x)$	Sign of f
$(-\infty, -2)$	-3	$3(-1)(-4)(-7)$	$-$
$(-2, 1)$	0	$3(2)(-1)(-4)$	$+$
$(1, 4)$	2	$3(4)(1)(-2)$	$-$
$(4, +\infty)$	5	$3(7)(4)(1)$	$+$

We now may sketch the graph. Begin by plotting the points corresponding to the x-intercepts, namely the points $(-2, 0)$, $(1, 0)$, and $(4, 0)$. Next, we consult the table of test intervals. The sign of f in a test interval determines whether the graph is above or below the x-axis in that interval. For the first interval listed, the value of f is negative, so the graph lies below the x-axis throughout the interval. As x becomes large, either positively or negatively, the values of f become arbitrarily large in absolute value. (This is a general property of polynomial functions that we will assume without proof.) Because f is negative in the first interval, the values must get arbitrarily large in the negative direction as x becomes large in the negative direction. This allows us to sketch the graph of f for x in the first interval.

In the second interval, f is positive. At the left and right endpoints of the interval, the value of $f(x)$ is 0. This means that the graph must have a turning point somewhere in the interval. Knowing this allows us to make a sketch of the graph for x in the second interval. (We have drawn the graph with one turning point in this interval. It might have more than one, but without calculus we have no way of knowing.)

In the third interval, f is negative. Again, because both endpoints give the value 0 for $f(x)$, we can use the same reasoning as in the second interval to sketch the graph corresponding to x in this interval. Finally, the last interval includes values of x that are arbitrarily large. For these values, $f(x)$ grows arbitrarily large. Because the sign of $f(x)$ is positive, the growth is in the positive direction. The final sketch of the graph is shown in Figure 11.

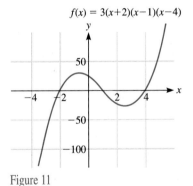

$f(x) = 3(x+2)(x-1)(x-4)$

Figure 11

➤ EXAMPLE 3
Graphing a Fourth-Degree Polynomial

Sketch the graph of the function

$$f(x) = x^4 - 4x^2$$

Solution

Factor the expression on the right to obtain

$$f(x) = x^2(x^2 - 4) = x^2(x + 2)(x - 2)$$

Examining the factors on the right, we see that the zeros of f are -2, 0, and 2. These divide the x-axis into four intervals, as shown in the following table. As in the preceding example, we choose a test value in each interval and determine the sign of f at the test value. The results are summarized in the following table.

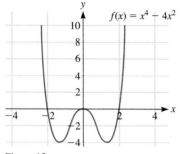

Figure 12

Interval	Test Value x	$f(x)$	Sign of f
$(-\infty, -2)$	-3	$(-3)^2(-1)(-5)$	$+$
$(-2, 0)$	-1	$(-1)^2(1)(-3)$	$-$
$(0, 2)$	1	$(1)^2(3)(-1)$	$-$
$(2, +\infty)$	3	$(3)^2(5)(1)$	$+$

In Figure 12, we have sketched the graph of f by first plotting the zeros and then drawing a section of the graph corresponding to each of the intervals in the table. The accuracy of the graph is enhanced by noting that the function $f(x)$ is even, so the graph is symmetric with respect to the y-axis.

Polynomial Functions and Technology

Let's illustrate how graphing calculators may be applied to solve problems involving polynomials.

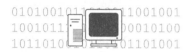

New Technology

EXAMPLE 4
Graphing a Fourth-Degree Polynomial Function

Graph the following fourth-degree polynomial function and determine the number of maximum and minimum points.

$$f(x) = x^4 + x^3 - 2x^2 + x - 1$$

Solution

The graph is shown in Figure 13. There is a single minimum point and no maximum point. Note that by using the graphing calculator to produce the graph, we do not need to factor the polynomial, as we did in the preceding example. This can often be a difficult task. ◀

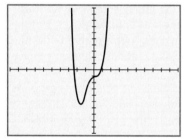

Figure 13

In the preceding section, we learned to solve optimization problems that involve determining the maximum or minimum point of a quadratic function. Many applied problems require you to determine maximum or minimum points of other functions. In calculus, you will learn techniques for solving such problems. However, using a graphing calculator, we may approximately locate maximum or minimum points of functions by tracing their graphs. The following example illustrates the procedure.

New Technology

EXAMPLE 5
Solving an Optimization Problem Using a Calculator

Determine the maximum and minimum points of the function $f(x) = x^3 - 3x + 5$.

Solution

Graph the function to obtain the display in Figure 14. The graph has one maximum point and one minimum point. To determine the coordinates of these points, we use the trace function. To obtain greater accuracy, we use the Box command to enlarge the areas of the graph around the maximum and minimum points. In Figure 15 we trace the curve near the minimum point, whose coordinates are approximately $(1.0094183, 3.0002669)$. In Figure 16, we trace the curve near the maximum point, whose coordinates are approximately $(-.9894737, 6.9996688)$. In calculus, you will learn to show that the respective minimum and maximum points are exactly $(1, 3)$ and $(-1, 7)$. The graphing calculator is limited to providing only approximations to these exact values.

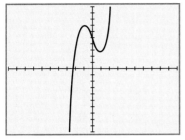

Figure 14

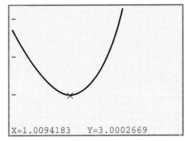

X=1.0094183   Y=3.0002669

Figure 15

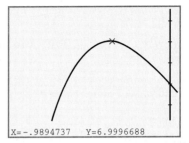

X=-.9894737   Y=6.9996688

Figure 16 ◄

Exercises 3.2

Sketch the graph of each of the following functions.

1. $f(x) = -x^2$
2. $f(x) = 0.25x^2$
3. $f(x) = \frac{1}{2}x^3$
4. $f(x) = 4x^3$
5. $f(x) = (x - 2)^3 + 1$
6. $f(x) = -(x + 1)^3 - 4$
7. $f(x) = x(x - 1)(x + 1)$
8. $f(x) = (x + 3)(x - 1)(x + 2)$
9. $f(x) = 0.25x^4$
10. $f(x) = -x^4$
11. $f(x) = -x^5$
12. $f(x) = -x^6$
13. $f(x) = -x^2 + 2$
14. $f(x) = -.25x^2$
15. $f(x) = x^3 - 3x + 22$
16. $f(x) = x^3 - x^2 - 2x$
17. $f(x) = x^4 - 2x^2$
18. $f(x) = x^4 - 6x^2$
19. $f(x) = x^3 - 2x^2 + x - 1$
20. $f(x) = x^3 + 2x^2 - 5x - 6$
21. $f(x) = x^4 - 4x^3 - 4x^2 + 16x$
22. $f(x) = x^4 - 2x^3 + 3x - 5$
23. $f(x) = \frac{1}{3}x^3 - \frac{1}{2}x^2 - 2x + 1$
24. $f(x) = \frac{8}{3}x^3 - 2x + \frac{1}{3}$
25. $f(x) = (x - 2)^2(x + 3)$
26. $f(x) = x^2(x + 2)(x^2 + 1)$

♦ Applications

27. **Maximizing volume**. From a thin piece of cardboard 8 in. by 8 in., square corners are cut out so that the sides can be folded up to make a box (see figure). Let *V* denote the volume of the box.

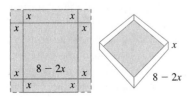

Exercise 27

a. Determine a polynomial function in one variable for $V(x)$.

b. Graph the function on the interval [0, 4].

c. Use the graph to estimate the maximum value of the function and the value of *x* at which it occurs.

28. **Deflection of a beam**. A beam rests at two points, *A* and *B*, and has a concentrated load applied to the center of the beam, as shown in the accompanying figure. Let *y* denote the deflection of the beam at a distance of *x* units measured from the beam to the left of the weight. The deflection depends on the elasticity of the board, the load, and other physical characteristics. Suppose under certain conditions that *y* is given by

$$y = \frac{1}{12}x^3 - \frac{1}{16}x$$

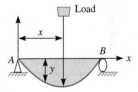

Exercise 28

a. Graph the function on the interval [0, 10].

b. Use the graph to estimate the minimum value of the function and the value of *x* at which it occurs.

➡ Technology

Graph each of the following functions using a graphing calculator. Use the resulting graph to estimate the *x*-intercepts.

29. $f(x) = x^3 - 3x^2 - 144x - 140$
30. $f(x) = x^4 - 2x^3$
31. $f(x) = 6x^5 - 24x^3$
32. $f(x) = x^2(x + 5)(x^2 + 3x + 5)^2$

33. You are given the following total revenue and total cost functions.

$$R(x) = 100x - x^2, \quad C(x) = \frac{1}{3}x^3 - 6x^2 + 89x + 100$$

a. Total profit is given by $P(x) = R(x) - C(x)$. Find $P(x)$.

b. Graph $R(x)$, $C(x)$, and $P(x)$ using the same set of axes.

c. Estimate the value or values of x at which $P(x)$ has a maximum. Estimate the maximum value.

♦ **Applications**

34. **Medical dosage.** The function

$$N(t) = -0.045t^3 + 2.063t + 2$$

gives the blood levels, in parts per million, of a certain dosage of medication after time t, in hours.

a. Graph the function.

b. Estimate the maximum value of the function and the time t at which it occurs.

c. The *minimum effective dosage* of the medication occurs for values of t for which $2 \leq N(t)$. Use the graph to estimate the times for which the medication has its minimum effective dosage.

35. **Temperature during an illness.** A patient's temperature during an illness was given by

$$T(t) = 98.6 - t^2(t - 4)$$

a. Graph the function over the interval [0, 4].

b. Estimate the maximum value of the function and the time t at which it occurs.

Match the following functions with the corresponding graph among graphs (a)–(f).

36. $f(x) = x^3$ 37. $f(x) = x^4 + 3$

38. $f(x) = x(x + 1)(x - 2)$

39. $f(x) = x^3 - 3x + 1$ 40. $f(x) = -x^4$

41. $f(x) = (x^2 - 4)(x^2 - 9)$

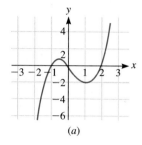

(a)

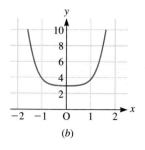

(b)

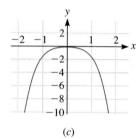

(c)

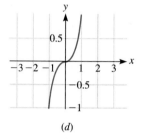

(d)

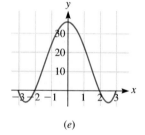

(e)

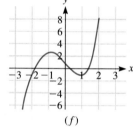

(f)

3.3 COMPLEX NUMBERS

Definition of the Complex Numbers

The need to solve equations has motivated the introduction of larger and more comprehensive number systems. The first number system used by humans consisted of the counting numbers 1, 2, 3, ... However, to perform certain subtraction operations among such numbers, say, $5 - 5$ or $12 - 28$, mathematicians created zero and the negative numbers, which, along with the counting numbers, form the set of integers **I**.

The integers are closed under addition, subtraction, and multiplication. However, certain division operations, such as $\frac{5}{3}$ or $\frac{128}{9}$, result in numbers that are not integers. The desire to perform division among all integers motivated mathematicians to extend the set of integers to the set of rational numbers **Q**. The rational numbers are closed under division in the sense that if a and b are rational numbers with b nonzero, then a/b is also a rational number.

As we pointed out in Chapter 1, the decimal expressions of all rational numbers are terminating or repeating. To deal with numbers with nonterminating and nonrepeating decimal expressions, mathematicians defined the set of real numbers **R**, which contains rational and irrational numbers. The real numbers are closed under some arithmetic operations with respect to which the rational numbers are not. For example, if a is a rational number, $\sqrt{a}$ is not always a rational number, as in the case of $\sqrt{2}$, but if a is a positive real number, then $\sqrt{a}$ is a real number.

As comprehensive as the set of real numbers appears to be, certain elementary arithmetic operations among real numbers are not possible using real numbers. For example, there is no real number whose square is -1. To put this another way, it is impossible to compute $\sqrt{-1}$ within the real numbers because the square of any real number is nonnegative and cannot be -1.

Just as the real numbers do not contain a square root of -1, they do not contain the square root of any negative number. This is a serious shortcoming of the real numbers and has motivated mathematicians to create a larger number system, called the **complex numbers,** in which taking square roots is always possible. We define the complex numbers as follows.

Definition 1 **Complex Numbers**	The **imaginary unit**, denoted i, is a number whose square is -1. That is, $$i^2 = -1, \qquad i = \sqrt{-1}$$ A **complex number** is a number of the form $$a + bi,$$ where a and b are real numbers, and $i^2 = -1$. The set of complex numbers is denoted **C.**

Here are some examples of complex numbers:

$$2 + 3i, \qquad \tfrac{1}{2} - 4i, \qquad 5 + 0i, \qquad \sqrt{2}i$$

If $a + bi$ is a complex number, then a is called its **real part** and bi is called its **imaginary part**.

Definition 2 **Equality of Complex Numbers**	Two complex numbers $a + bi$ and $c + di$ are **equal** if and only if $a = c$ and $b = d$.

The definition of equality of complex numbers is useful in solving equations involving complex numbers, as we will see later in this section.

It is customary to write a complex number $a + 0i$ as a and to consider it the same as the real number a. In this way, the complex numbers contain all the real numbers as a subset.

Figure 1 shows the relationships between the complex numbers and the various number systems we introduced earlier.

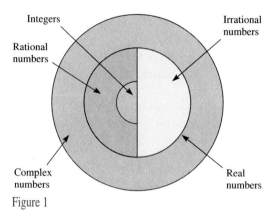

Figure 1

Arithmetic Involving Complex Numbers

Definition 3 **Addition and Multiplication of Complex Numbers**	Let $\alpha = a + bi$ and $\beta = c + di$ be complex numbers. Their sum and product are the numbers defined by the formulas $$\alpha + \beta = (a + c) + (b + d)i$$ $$\alpha \cdot \beta = (ac - bd) + (ad + bc)i$$

For example, we have

$$(5 - 3i) + (4 + 2i) = (5 + 4) + (-3 + 2)i = 9 - i$$
$$(2 + 4i)(3 - 7i) = 2 \cdot 3 + 2 \cdot (-7i) + (4i)(3) + (4i)(-7i)$$
$$= 6 - 14i + 12i - 28i^2$$
$$= 6 - 2i - 28(-1)$$
$$= 6 - 2i + 28$$
$$= 34 - 2i$$

The definition of addition is about what you might expect, but the definition of multiplication of complex numbers may seem strange. However, this is just the definition we get assuming that the distributive law holds, and using the fact that $i^2 = -1$. Indeed, we have the following motivation for the definition of the product of two complex numbers:

$$\alpha \cdot \beta = (a + bi)(c + di)$$
$$= ac + (ad)i + (bc)i + (bd)i^2$$
$$= ac + (-bd) + (ad + bc)i \qquad \textit{Because } i^2 = -1$$
$$= (ac - bd) + (ad + bc)i$$

Addition and multiplication of complex numbers have the same fundamental properties as addition and multiplication of real numbers. In other words, they obey the commutative, associative, and distributive laws: If α, β, γ are complex numbers, then

$$\alpha + \beta = \beta + \alpha \qquad \textit{Commutative law of addition}$$
$$(\alpha + \beta) + \gamma = \alpha + (\beta + \gamma) \qquad \textit{Associative law of addition}$$

$$\alpha\beta = \beta\alpha \qquad \text{\textit{Commutative law of multiplication}}$$
$$\alpha(\beta\gamma) = (\alpha\beta)\gamma \qquad \text{\textit{Associative law of multiplication}}$$
$$\alpha(\beta + \gamma) = (\alpha \cdot \beta) + (\alpha \cdot \gamma) \qquad \text{\textit{Distributive law}}$$

Complex numbers have an **additive identity,** namely 0 (or $0 + 0i$), with the property

$$\alpha + 0 = \alpha$$
$$a + bi + 0 = a + bi$$

A complex number $\alpha = a + bi$ has an **additive inverse,** namely, $-\alpha = -a - bi$, with the property

$$\alpha + (-\alpha) = 0$$
$$(a + bi) + (-a - bi) = 0$$

Subtraction of complex numbers is defined, as for real numbers, in terms of the additive inverse:

$$\alpha - \beta = \alpha + (-\beta)$$

Thus, for instance,

$$(5 - 4i) - (3 - 3i) = (5 - 4i) + (-3 + 3i) = 2 - i$$

Addition and subtraction of complex numbers is consistent with the corresponding operations for real numbers. Indeed, if we consider two real numbers a and c, then

$$(a + 0i) + (c + 0i) = (a + c) + 0i + 0i = a + c$$
$$(a + 0i) - (c + 0i) = (a + 0i) + (-c + 0i) = (a - c) + 0i + 0i = a - c$$

The number 1 (or $1 + 0i$) is the **multiplicative identity**:

$$\alpha \cdot 1 = \alpha$$
$$(a + bi) \cdot 1 = a + bi$$

➤ **EXAMPLE 1**
Arithmetic of Complex Numbers

Perform the following calculations involving complex numbers:

1. $2(3 + i) - i(2 - 4i)$
2. $(1 + i)^2$

Solution

1. $2(3 + i) - i(2 - 4i) = (6 + 2i) + (-2i + 4i^2)$
   $$= (6 + 2i) + (-2i - 4) \qquad \text{\textit{Because } } i^2 = -1$$
   $$= 2 + 0i$$

2. $(1 + i)^2 = (1 + i)(1 + i)$
   $$= (1 - 1) + (1 + 1)i \qquad \text{\textit{Definition of multiplication}}$$
   $$= 2i$$

Just as with the real number system, the only way for a product of complex numbers to be zero is for at least one of the factors to be zero. That is,

$$\alpha\beta = 0 \quad \text{implies} \quad \alpha = 0 + 0i \quad \text{or} \quad \beta = 0 + 0i$$

We will find this fact of great use in determining solutions of polynomial equations in the complex number system.

Powers of i

The complex number i has the property that its square is -1. Using this fact and the laws of exponents, we can easily calculate higher powers of i. Here are the first few:

$$i^3 = i^2 \cdot i = (-1) \cdot i = -i$$
$$i^4 = i^2 \cdot i^2 = (-1)(-1) = 1$$

Using the values of the first four powers of i, we may compute higher powers, as follows:

$$i^5 = i^4 \cdot i = 1 \cdot i = i$$
$$i^6 = i^4 \cdot i^2 = 1 \cdot (-1) = -1$$
$$i^7 = i^4 \cdot i^3 = 1 \cdot (-i) = -i$$
$$i^8 = i^4 \cdot i^4 = 1 \cdot 1 = 1$$

You should note that the powers of i repeat each time the exponent is increased by 4. Moreover, if the exponent is divisible by 4, then the power of i is 1. We can use these facts to rapidly calculate any power of i. For example, let's calculate

$$i^{57}$$

First, divide the exponent by 4 and write it in the form

$$57 = 14 \cdot 4 + 1$$

Substitute this expression for the exponent of i:

$$
\begin{aligned}
i^{57} &= i^{14 \cdot 4 + 1} \\
&= i^{14 \cdot 4} \cdot i^1 \\
&= \left(i^4\right)^{14} \cdot i \\
&= (1)^{14} i \\
&= i
\end{aligned}
$$

Square Roots of Negative Real Numbers

Mathematicians introduced the complex numbers to create a number system that contains square roots of negative numbers. Suppose that p is a positive real number. Then, as we have observed, there is no real number whose square is $-p$. However, there is a complex number with this property:

$$\sqrt{p}\,i$$

For we have

$$
\begin{aligned}
\left(\sqrt{p}\,i\right)^2 &= \left(\sqrt{p}\right)^2 i^2 \\
&= p(-1) \\
&= -p
\end{aligned}
$$

The quantity $\sqrt{p}\,i$ is called the **principal square root** of $-p$.

Historical Note: Origin of Complex Numbers

Square roots of negative numbers were used for centuries in calculating solutions to equations of degrees 2, 3, and 4. However, mathematicians regarded them with great suspicion. They realized, of course, that they were not real numbers, but they were not quite sure of the legitimacy of working with them. Accordingly, square roots of negative numbers were called **imaginary numbers** or just **imaginaries**. The German mathematician Karl Friedrich Gauss in 1799 gave a formal definition of the complex numbers that provided them with a logical foundation and the basis for legitimate proofs of theorems regarding them. ∎

➤ **EXAMPLE 2**
Square Roots
of a Negative Number

Determine two square roots of -4.

Solution

One square root of -4 is given by the principal square root, $\sqrt{4}i = 2i$. A second square root is given by the negative of this quantity, $2i$, since

$$(-2i)^2 = (-2)^2 i^2$$
$$= 4 \cdot (-1) = -4$$

➤ **EXAMPLE 3**
Complex Products

Compute the product
$$\left(x - \sqrt{-1}\right)\left(x + \sqrt{-1}\right)$$

Solution

Apply the distributive law:

$$\left(x - \sqrt{-1}\right)\left(x + \sqrt{-1}\right) = x^2 + \sqrt{-1}x - \sqrt{-1}x - \left(\sqrt{-1}\right)^2$$
$$= x^2 - (-1) \qquad \text{\small\textit{Using }}\left(\sqrt{-1}\right)^2 = -1$$
$$= x^2 + 1$$

The preceding example can also be read in reverse—not as a multiplication example, but as a factorization. Starting with our solution, a sum of squares,

$$x^2 + 1$$

we can arrive at the factorization

$$(x - i)(x + i)$$

In a similar fashion, we can factor the sum of squares

$$x^2 + y^2$$

into the linear factors

$$(x - yi)(x + yi)$$

To verify this fact, we just multiply the linear factors as in Example 3. Let's record this fact for future reference.

FACTORING A SUM OF SQUARES
$x^2 + y^2 = (x + yi)(x - yi)$

Note that the sum of two squares is not factorable using real numbers, yet we can factor it using complex numbers.

**Quadratic Equations
Revisited**

The quadratic equation
$$ax^2 + bx + c = 0$$

has solutions given by the quadratic formula

$$x = \frac{-b \pm \sqrt{b^2 - 4ac}}{2a}$$

As we have seen, in case the discriminant $b^2 - 4ac$ is negative, the equation has no real solutions. However, in this case, the formula gives two distinct complex solutions.

➤ **EXAMPLE 4**
Quadratic with Complex Zeros

Determine all solutions of the equation $x^2 + x + 1 = 0$.

Solution

By the quadratic formula, the solutions are given by

$$x = \frac{-b \pm \sqrt{b^2 - 4ac}}{2a}$$

$$= \frac{-1 \pm \sqrt{1^2 - 4 \cdot 1 \cdot 1}}{2 \cdot 1}$$

$$= \frac{-1 \pm \sqrt{-3}}{2}$$

$$= \frac{-1 \pm \sqrt{3}i}{2}$$

Conjugation

Definition 4	
Conjugate of a Complex Number	Let $a + bi$ be a complex number. Its conjugate is the number $a - bi$ and is denoted $\overline{a + bi}$.

Here are some examples of conjugates:

$$\overline{2 + 3i} = 2 - 3i, \qquad \overline{5} = 5, \qquad \overline{\tfrac{1}{2}i} = -\tfrac{1}{2}i, \qquad \overline{-8 - 7i} = -8 + 7i$$

The following formulas are very useful.

PROPERTIES OF CONJUGATES

Let $\alpha = a + bi$ be a complex number.

Conjugate of a conjugate: $\overline{\overline{\alpha}} = \alpha$
Sum of conjugates: $\alpha + \overline{\alpha} = 2a$
Difference of conjugates: $\alpha - \overline{\alpha} = 2bi$
Product of conjugates: $\alpha\overline{\alpha} = a^2 + b^2$

Each nonzero complex number $\alpha = a + bi$ has a multiplicative inverse α^{-1} such that

$$\alpha\alpha^{-1} = 1$$

In fact, we can use conjugates to explicitly calculate the multiplicative inverse, as given in the following result.

MULTIPLICATIVE INVERSE OF A COMPLEX NUMBER

Let $\alpha = a + bi$ be a nonzero complex number. Then

$$\alpha^{-1} = \frac{\overline{\alpha}}{a^2 + b^2}$$

Proof   We have

$$\alpha^{-1} = \frac{1}{\alpha} = \frac{1 \cdot \overline{\alpha}}{\alpha \cdot \overline{\alpha}} = \frac{\overline{\alpha}}{a^2 + b^2}$$

◆

Division of complex numbers is defined in terms of the multiplicative inverse, just as it is for real numbers. Assume that α and β are complex numbers with $\beta \neq 0$. The quotient α/β is then defined as

$$\frac{\alpha}{\beta} = \alpha\beta^{-1}$$

➢ **EXAMPLE 5**
Inverse and Division

Express the following complex numbers in the form $a + bi$.

1. $(5 - 4i)^{-1}$

2. $\dfrac{3 + i}{1 + 2i}$

Solution

1. We may write the complex number as $1/(5 - 4i)$. Multiply the numerator and denominator by the conjugate $5 + 4i$:

$$\frac{1}{5 - 4i} = \frac{1}{5 - 4i} \cdot \frac{5 + 4i}{5 + 4i} = \frac{5 + 4i}{5^2 + 4^2} = \frac{5}{41} + \frac{4}{41}i$$

2. We may calculate the quotient by multiplying numerator and denominator by the conjugate of the denominator:

$$\frac{3 + i}{1 + 2i} = \frac{3 + i}{1 + 2i} \cdot \frac{1 - 2i}{1 - 2i}$$

$$= \frac{3 \cdot 1 + (-2)i^2 + (1 - 3 \cdot 2)i}{1^2 + 2^2}$$

$$= \frac{5 - 5i}{5}$$

$$= 1 - i$$

More Examples Involving Complex Numbers

In many problems, it is necessary to deal with polynomials with complex coefficients. The next two examples provide some practice in calculating with such polynomials.

➤ **EXAMPLE 6**
Polynomial With Complex
Coefficients

Calculate the product of the following binomials with complex coefficients.

$$\left(x - \frac{1 + \sqrt{3}i}{2}\right)\left(x - \frac{1 - \sqrt{3}i}{2}\right)$$

Solution

The product equals

$$\left[\left(x - \tfrac{1}{2}\right) - \frac{\sqrt{3}}{2}i\right]\left[\left(x - \tfrac{1}{2}\right) + \frac{\sqrt{3}}{2}i\right]$$

$$= \left(x - \tfrac{1}{2}\right)^2 + \left(\frac{\sqrt{3}}{2}\right)^2$$

$$= x^2 - x + \tfrac{1}{4} + \tfrac{3}{4}$$

$$= x^2 - x + 1$$

➤ **EXAMPLE 7**
Solving an Equation with
Complex Coefficients

Find all solutions (x, y) of the following equation:

$$(2x + 1) + (-y + 3)i = 6 + \sqrt{2}i$$

Solution

By the definition of equality for complex numbers, in order for two complex numbers to be equal, their corresponding real and imaginary parts must be equal. In this case,

$$2x + 1 = 6 \quad \text{and} \quad -y + 3 = \sqrt{2}$$

$$x = \tfrac{5}{2} \quad \text{and} \quad y = 3 - \sqrt{2}$$

Exercises 3.3

Express in terms of i.

1. $\sqrt{-3}$
2. $\sqrt{-4}$
3. $\sqrt{-81}$
4. $\sqrt{-27}$
5. $\sqrt{-98}$
6. $-\sqrt{-18}$
7. $-\sqrt{-49}$
8. $-\sqrt{-125}$
9. $4 - \sqrt{-60}$
10. $6 - \sqrt{-84}$
11. $\sqrt{-4} + \sqrt{-12}$
12. $-\sqrt{-76} + \sqrt{-125}$

Simplify.

13. i^7
14. i^{11}
15. i^{24}
16. i^{35}
17. i^{42}
18. i^{64}
19. i^9
20. $(-i)^{71}$

Simplify to the form $a + bi$.

21. $7 + i^4$
22. $-18 + i^3$
23. $i^4 - 26i$
24. $i^5 + 37i$

25. $i^2 + i^4$
26. $5i^5 + 4i^3$
27. $i^5 + i^7$
28. $i^{84} - i^{100}$
29. $1 + i + i^2 + i^3 + i^4$
30. $i - i^2 + i^3 - i^4 + i^5$
31. $5 - \sqrt{-64}$
32. $\sqrt{-12} + 36i$
33. $\dfrac{8 - \sqrt{-24}}{4}$
34. $\dfrac{9 + \sqrt{-9}}{3}$
35. $\dfrac{\sqrt{-16}}{\sqrt{-25}}$
36. $\dfrac{\sqrt{-9}}{\sqrt{-36}}$
37. $(3 + 2i) + (2 + 4i)$
38. $(2 - 5i) + (3 + 6i)$
39. $(3 + 4i) + (3 - 4i)$
40. $(2 + 5i) + (-2 - 5i)$
41. $(9 + 12i) - (7 + 8i)$
42. $(10 - 4i) - (6 + 2i)$
43. $6i - (7 - 4i)$
44. $45 - (23 + 5i)$
45. $(1 - 3i)(2 + 4i)$
46. $(-2 + 3i)(6 - 7i)$
47. $(2 - 5i)(2 + 5i)$
48. $(-5 + 7i)(-5 - 7i)$
49. $2i(4 - 3i)$
50. $5i(-8 + 6i)$

51. $(5 - 2i)^2$

52. $(3 + i)^2$

53. $\dfrac{1 + 2i}{3 - i}$

54. $\dfrac{2 + i}{2 - i}$

55. $\dfrac{\sqrt{3} - i}{\sqrt{3} + i}$

56. $\dfrac{\sqrt{2} + i}{\sqrt{2} - i}$

57. $\dfrac{5 - 3i}{i}$

58. $\dfrac{\sqrt{2} - i}{i}$

59. $\dfrac{i}{1 - i}$

60. $\dfrac{4}{3 + 10i}$

61. $\dfrac{2 + i}{(2 - i)^2}$

62. $\dfrac{5 - i}{(5 + i)^2}$

63. $(1 - 2i)^{-1}$

64. $(1 + i)^{-1}$

65. $\dfrac{(1 + i)(2 - i)}{(4 - 2i)(5 - 3i)}$

66. $\dfrac{(2 - 3i)(5 - 6i)}{(9 + 2i)(4 + 3i)}$

67. $\dfrac{1 + i}{1 - i} + \dfrac{2 - i}{2 + i}$

68. $\dfrac{5 - 2i}{3 + 2i} - \dfrac{7 - i}{4 + i}$

69. $(1 + i)^{-2}$

70. $\left(\sqrt{3} - 2i\right)^{-2}$

Determine two square roots of each of the following.

71. -5

72. -9

73. -64

74. -17

Calculate each product.

75. $(x - 2i)(x + 2i)$

76. $(x + 5i)(x - 5i)$

77. $\left(x - \dfrac{1 - \sqrt{2}i}{3}\right)\left(x - \dfrac{1 + \sqrt{2}i}{3}\right)$

78. $\left(x - \dfrac{\sqrt{3} + 2i}{4}\right)\left(x - \dfrac{\sqrt{3} - 2i}{4}\right)$

Factor.

79. $x^2 + 4$

80. $x^2 + 25$

81. $x^2 + 3$

82. $x^2 + 5$

83. $a^2 + b^2$

84. $x^2 + 16y^2$

Find all solutions (x, y) of the following equations, where x and y are real numbers.

85. $5x + 8i = -9 + yi$

86. $(-2x + y)i - 8 = 3x - 4y + 9i$

87. $x^2 + 4yi = 6i + 9$

88. $y^3 + 5xi = 40i - 8$

Let $\alpha = a + bi$ be any complex number. Prove each of the following.

89. $\alpha + \overline{\alpha} = 2a$

90. $\alpha - \overline{\alpha} = 2bi$

91. $\alpha\overline{\alpha} = a^2 + b^2$

Suppose α and β are any complex numbers. Prove the following.

92. $\overline{\alpha + \beta} = \overline{\alpha} + \overline{\beta}$ (The conjugate of a sum is the sum of the conjugates.)

93. $\overline{\alpha \cdot \beta} = \overline{\alpha} \cdot \overline{\beta}$ (The conjugate of a product is the product of the conjugates.)

94. $\overline{\alpha^n} = (\overline{\alpha})^n$, where n is a positive integer. (The conjugate of a power is the power of the conjugate.)

95. $\overline{\alpha - \beta} = \overline{\alpha} - \overline{\beta}$ (The conjugate of a difference is the difference of the conjugates.)

96. $\overline{\alpha} = \alpha$ if α is a real number. (The conjugate of a real number is the same number.)

97. Prove that the sum of a complex number and its conjugate is a real number.

Simplify each of the following conjugates of polynomials in the complex number α.

98. $\overline{\alpha^3 + \alpha^2 + \alpha + 1}$

99. $\overline{4\alpha^2 - 25}$

100. $\overline{5\alpha^4 - 6\alpha^3 + 2\alpha^2 + 3\alpha - 23}$

✍ **In Your Own Words**

101. Describe in words a procedure for determining if a quadratic equation has nonreal solutions.

102. What is the conjugate of the conjugate of a complex number?

■ 3.4 THE DIVISION ALGORITHM

In the preceding section, we saw how knowing the zeros of a polynomial function provided a first piece of information for sketching the graph. In this section and the next, we will discuss a number of theorems that provide information about the zeros of polynomial functions. In this section, we relate the zeros to the linear factors of the polynomial via a fundamental result called the division algorithm.

Long Division of Polynomials

We can divide two polynomials to produce a quotient and a remainder. As an illustration, consider the quotient

$$\frac{x^2 + 3x + 1}{x}$$

We can divide a polynomial by a monomial by dividing each term of the polynomial by the monomial:

$$\frac{x^2 + 3x + 1}{x} = \frac{x^2}{x} + \frac{3x}{x} + \frac{1}{x} = x + 3 + \frac{1}{x}$$

We may express this result purely in terms of polynomials by multiplying each term by the denominator x to obtain

$$x^2 + 3x + 1 = x(x + 3) + 1$$

The polynomial $x + 3$ is the quotient and 1 is the remainder.

You may calculate the quotient and remainder using the process of long division, which is included in first courses in algebra. The next example recalls the procedure.

➤ **EXAMPLE 1**

Long Division

Calculate the quotient and remainder when $f(x) = 2x^3 + 5x^2 - 10x - 7$ is divided by $g(x) = 2x - 1$.

Solution

Because we are seeking the quotient and remainder for the division $f(x)/g(x)$, we organize the calculation as a long division problem of $f(x)$ divided by $g(x)$.

$$
\begin{array}{r}
x^2 \quad +3x \quad -\frac{7}{2} \quad \textit{quotient} \\
2x - 1 \overline{\smash{\big)}\, 2x^3 \quad +5x^2 \quad -10x \quad -7} \\
\underline{2x^3 \quad -x^2} \\
6x^2 \quad -10x \\
\underline{6x^2 \quad -3x} \\
-7x \quad -7 \\
\underline{-7x \quad +\frac{7}{2}} \\
-\frac{21}{2} \quad \textit{remainder}
\end{array}
$$

Therefore, in this example, the quotient $q(x)$ and remainder $r(x)$ are

$$q(x) = x^2 + 3x - \tfrac{7}{2}, \qquad r(x) = -\tfrac{21}{2}$$

That is, we have

$$\frac{f(x)}{g(x)} = q(x) + \frac{r(x)}{g(x)}$$

or, multiplying by $g(x)$,

$$f(x) = q(x)g(x) + r(x)$$

Note that the degree of the denominator $g(x)$ is 1 and the degree of the remainder is 0, which is less than the degree of $g(x)$. Note also that because both $f(x)$ and $g(x)$ have all rational coefficients, so do the quotient and remainder. However, observe that even though both $f(x)$ and $g(x)$ have all integer coefficients, the quotient and remainder have some noninteger coefficients.

The division algorithm states that a computation like the one just carried out can always be carried out for computing the quotient of a polynomial $f(x)/g(x)$.

DIVISION ALGORITHM FOR POLYNOMIALS

Suppose that $f(x)$ and $g(x)$ are polynomials with complex coefficients, with $g(x)$ not the zero polynomial. Then unique polynomials $q(x)$ and $r(x)$ exist such that

1. $f(x) = g(x)q(x) + r(x)$
2. Either $r(x)$ is the zero polynomial, or the degree of $r(x)$ is less than the degree of $g(x)$.

We will omit the proof of this result.

The polynomial $q(x)$ is called the **quotient** and $r(x)$ the **remainder** when $f(x)$ is divided by $g(x)$. If $f(x)$ and $g(x)$ have real (respectively rational) coefficients, then $q(x)$ and $r(x)$ have real (respectively rational) coefficients.

Factoring and Zeros

Let's consider the special case when a polynomial $f(x)$ is divided by a linear polynomial of the form $x - a$. By the division algorithm, we may express the division in the form

$$f(x) = (x - a)q(x) + r(x)$$

where $q(x)$ is the quotient and $r(x)$ is the remainder. From the statement of the division algorithm, either $r(x)$ is the zero polynomial or the degree of $r(x)$ is less than the degree of $x - a$, which is 1. That is, in either case, $r(x)$ is a constant polynomial.

We can determine the value of this constant by replacing x in the preceding equation with a. We then obtain

$$\begin{aligned} f(a) &= (a - a)q(a) + r(a) \\ &= 0 + r(a) \\ &= r(a) \end{aligned}$$

Thus, the value $r(a)$ is equal to $f(a)$. Because $r(x)$ is a constant polynomial, however, this means that $r(x)$ is equal to $f(a)$. This proves the following useful result.

REMAINDER THEOREM

Let $f(x)$ be a polynomial. Then the remainder on dividing $f(x)$ by $x - a$ is the constant polynomial $f(a)$. That is,

$$f(x) = (x - a)q(x) + f(a)$$

where $q(x)$ is the quotient when $f(x)$ is divided by $x - a$.

➤ **EXAMPLE 2**
Calculating the Remainder

Suppose that

$$f(x) = -x^4 + 5x^2 - 2$$

What is the remainder when $f(x)$ is divided by $x - 3$?

Solution

By the Remainder theorem, the remainder is the constant polynomial $f(3)$, which equals

$$-(3)^4 + 5(3)^2 - 2 = -81 + 45 - 2 = -38$$

That is, the remainder is the constant polynomial $r(x) = -38$.

We can draw the following simple, but important, consequence of the Remainder theorem.

FACTOR THEOREM

Let $f(x)$ be a polynomial and a a complex number. Then $x - a$ is a factor of $f(x)$ if and only if a is a zero of $f(x)$.

Proof   Suppose that a is a zero of the polynomial $f(x)$. Then $f(a)$ is equal to 0 and the equation in the Remainder theorem reads

$$f(x) = (x - a)q(x)$$

That is, $x - a$ is a factor of $f(x)$. Conversely, suppose that $x - a$ is a factor of $f(x)$. Then there is a polynomial $g(x)$ such that

$$f(x) = (x - a)g(x)$$

Substituting a for x in this equation, we see that

$$f(a) = (a - a)g(a) = 0$$

That is, $f(a)$ is equal to 0 and a is a zero of $f(x)$.   ◆

Combining the factor theorem with what we already know about zeros of polynomials, we now have a number of equivalent ways of recognizing real zeros of a polynomial.

REAL ZEROS OF POLYNOMIAL FUNCTIONS

Suppose that $f(x)$ is a polynomial with real coefficients and a is a real number. Then the following are all equivalent ways of stating that a is a zero of $f(x)$.

1. a is a solution of the equation $f(x) = 0$.
2. $x - a$ is a factor of $f(x)$.
3. The graph of $f(x)$ has $(a, 0)$ as an x-intercept.

➢ **EXAMPLE 3**
Applying the Factor Theorem

Suppose that $f(x)$ is a cubic polynomial with zeros $\frac{1}{2}$, 1, and 3. Further, suppose that $f(0)$ is equal to 4. Determine $f(x)$.

Solution

Because $\frac{1}{2}$, 1, and 3 are all zeros of $f(x)$, the Factor theorem asserts that $f(x)$ has as factors each of the polynomials $x - \frac{1}{2}$, $x - 1$, and $x - 3$. Therefore, $f(x)$ has as a factor the polynomial

$$\left(x - \tfrac{1}{2}\right)(x - 1)(x - 3)$$

Because $f(x)$ is a cubic polynomial,

$$f(x) = c\left(x - \tfrac{1}{2}\right)(x - 1)(x - 3)$$

for some constant c. We are also given that $f(0) = 4$. To use this fact, we substitute 0 for x into the above equation for $f(x)$ to obtain

$$4 = f(0) = c\left(0 - \tfrac{1}{2}\right)(0 - 1)(0 - 3)$$

$$4 = -\tfrac{3}{2}c$$

Solving this equation for c gives us

$$c = -\tfrac{8}{3}$$

Finally, substituting this back into the formula for $f(x)$ gives us

$$f(x) = -\tfrac{8}{3}\left(x - \tfrac{1}{2}\right)(x - 1)(x - 3)$$

This answer is acceptable as it stands. Or, if you wish, you may multiply out the right-hand side to obtain the cubic polynomial expression for $f(x)$. However, this expression is preferable because it is clear just by looking at it that it satisfies the conditions specified in the problem.

In our above discussion of the Factor and Remainder theorems, we assumed that the polynomial $f(x)$ has real coefficients and that the zero a is real. However, as we have seen in the preceding section, quadratic polynomials, even if their coefficients are real numbers, can have zeros that are complex. In order to accomodate such zeros, we must extend our notion of polynomial to include those in which the coefficients are complex numbers. If we do so, then the Division Algorithm continues to hold, with the provision that the quotient and remainder now possibly have complex coefficients. Moreover, the Factor and Remainder theorems remain true word for word.

> **EXAMPLE 4**
Application
of the Factor Theorem

Suppose that $f(x)$ is a polynomial with zeros $1 + i$ and $1 - i$. What can be said about $f(x)$?

Solution

By the Factor theorem (for polynomials with complex coefficients), we conclude that $f(x)$ is divisible by the polynomials $x - (1 + i)$ and $x - (1 - i)$. Thus, $f(x)$ is divisible by

$$[x - (1 + i)][x - (1 - i)] = x^2 - 2x + 2$$

Here is an interesting consequence of the Factor theorem.

NUMBER OF ZEROS THEOREM

A polynomial of degree n has at most n distinct zeros.

Proof   If $f(x)$ has distinct zeros

$$a_1, a_2, \ldots, a_k$$

then $f(x)$ has as factors the polynomials

$$x - a_1, x - a_2, \ldots, x - a_k$$

Because all the given zeros are assumed distinct, these factors are all different, so that $f(x)$ has as a factor the polynomial

$$(x - a_1)(x - a_2) \cdots (x - a_k)$$

Because this is a polynomial of degree k (a product of k linear factors) and is a factor of $f(x)$, k is at most equal to n. This means that the number of distinct zeros of f is at most equal to n. ◆

Synthetic Division

In the preceding discussion, we showed the significance of dividing a polynomial $f(x)$ by polynomials of the form $x - a$. The procedure we have given for performing this division is often tedious and time-consuming. There is a simple algorithm, called **synthetic division**, for carrying out such calculations in a much simpler and faster manner.

Let's begin by working out an example and showing how the efficient organization of long division calculations leads naturally to synthetic division. Consider the problem of dividing

$$3x^5 - 2x^4 - 5x^3 + x^2 - x + 4$$

by $x - 2$. Here is the traditional method of performing the division:

$$
\begin{array}{r}
3x^4 + 4x^3 + 3x^2 + 7x + 13 \\
x - 2 \,\overline{\big)\, 3x^5 - 2x^4 - 5x^3 + x^2 - x + 4} \\
\underline{3x^5 - 6x^4} \\
4x^4 - 5x^3 \\
\underline{4x^4 - 8x^3} \\
3x^3 + x^2 \\
\underline{3x^3 - 6x^2} \\
7x^2 - x \\
\underline{7x^2 - 14x} \\
13x + 4 \\
\underline{13x - 26} \\
30
\end{array}
$$

To simplify this computation, the first thing to note is that the calculations are completely contained in the various coefficients. It is not really necessary to include the variables at all. Moreover, because we will always be dividing by a polynomial of the form $x - a$, we may omit the coefficient 1 corresponding to x. This leaves a computation of the following form:

$$
\begin{array}{r|rrrrrr}
 & 3 & 4 & 3 & 7 & 13 \\
-2 & 3 & -2 & -5 & 1 & -1 & 4 \\
 & \underline{3} & \underline{-6} & & & & \\
 & & 4 & -5 & & & \\
 & & \underline{4} & \underline{-8} & & & \\
 & & & 3 & 1 & & \\
 & & & \underline{3} & \underline{-6} & & \\
 & & & & 7 & -1 & \\
 & & & & \underline{7} & \underline{-14} & \\
 & & & & & 13 & 4 \\
 & & & & & \underline{13} & \underline{-26} \\
 & & & & & & 30
\end{array}
$$

Note that this array of numbers still contains a lot of duplicate information. Consider the first step of the division, to obtain the first 3 in the quotient (top row). This is just a duplicate of the first 3 in the dividend (second row, inside the division sign). Moreover, the third row duplicates the 3 again. Let's simplify the computation by writing the 3 only once, in the dividend. The third row then consists of the single number −6, which is computed as 3 (from the quotient) times the −2 in the divisor (second row, outside the division sign). The next entry in the quotient, 4, is calculated by subtracting −2 minus −6. But we only write 4 once, in the quotient. The fourth row is determined by bringing down the −5 from the dividend, and the fifth row as the product 4 (quotient) times −2 (dividend), and so forth. This is how the abbreviated computation is shown:

$$
\begin{array}{r|rrrrrr}
 & 3 & 4 & 3 & 7 & 13 & \\
-2 & 3 & -2 & -5 & 1 & -1 & 4 \\
\hline
 & -6 & & & & & \\
\hline
 & & -5 & & & & \\
 & & -8 & & & & \\
\hline
 & & & 1 & & & \\
 & & & -6 & & & \\
\hline
 & & & & -1 & & \\
 & & & & -14 & & \\
\hline
 & & & & & 4 & \\
 & & & & & -26 & \\
\hline
 & & & & & 30 & \\
\end{array}
$$

Rather than subtracting numbers at each stage, let's perform addition. This can be done by changing the sign of the −2 in the divisor. The multiplications will then result in the negatives of the previous results, so we may add the entries at each step. The computation now looks like this:

$$
\begin{array}{r|rrrrrr}
 & 3 & 4 & 3 & 7 & 13 & \\
2 & 3 & -2 & -5 & 1 & -1 & 4 \\
\hline
 & 6 & & & & & \\
\hline
 & & -5 & & & & \\
 & & 8 & & & & \\
\hline
 & & & 1 & & & \\
 & & & 6 & & & \\
\hline
 & & & & -1 & & \\
 & & & & 14 & & \\
\hline
 & & & & & 4 & \\
 & & & & & 26 & \\
\hline
 & & & & & 30 & \\
\end{array}
$$

Let's compress the entire display into three lines as follows:

$$
\begin{array}{r|rrrrrr}
 & 3 & 4 & 3 & 7 & 13 & \\
\hline
2 & 3 & -2 & -5 & 1 & -1 & 4 \\
 & & 6 & 8 & 6 & 14 & 26 & 30 \\
\end{array}
$$

Finally, let's combine the dividend row at the top and the remainder 30 at the bottom into a single row and place it at the bottom. Moreover, tradition dictates turning the division sign inside out. (This surely doesn't affect the calculation!)

$$
\begin{array}{r|rrrrrr}
2 & 3 & -2 & -5 & 1 & -1 & 4 \\
  &   & 6 & 8 & 6 & 14 & 26 \\
\hline
  & 3 & 4 & 3 & 7 & 13 & 30
\end{array}
$$

Note how simply the bottom row may be calculated: The initial 3 in the quotient is a copy of the 3 in the dividend in the first row. Then compute 6 as 2 (divisor) times 3 (quotient). Compute 4 by adding -2 and 6. Now compute 8 as 4 (quotient) times 2 (divisor), and so forth. The quotient corresponds to the terms of the last row, except for the last, which is the remainder.

A similar procedure may be used to calculate the quotient of any polynomial $f(x)$ divided by $x - a$.

SYNTHETIC DIVISION

To divide $f(x)$ by $x - a$ using synthetic division:

1. Write the coefficients of f across the top row.
2. Start with the leftmost column. Bring down the leading coefficient into the third row.
3. Multiply the entry in the third row by a.
4. Move one column to the right and put the result of step 3 in the second row.
5. Add rows 1 and 2 of the current column and put the result in the third row.
6. Repeat steps 3–5 for each of the columns in turn.
7. The numbers in the third row are the coefficients of the quotient and remainder.

The following examples illustrate the mechanics of synthetic division and show how it may be applied to get information about a polynomial $f(x)$.

➤ **EXAMPLE 5**
Synthetic Division

Use synthetic division to determine the quotient and remainder when $f(x) = 4x^5 - 2x^4 + x^3 - 7x^2 + 3$ is divided by $x - 2$.

Solution

We use the algorithm for synthetic division. Note that we have included a zero to represent the zero coefficient of the x term. This is because there must be one column corresponding to each power of x in the divisor. If a power of x is missing, we must use 0 as a placeholder in the corresponding column. Be careful of this point. Leaving out a 0 placeholder is an easy mistake to make.

$$
\begin{array}{r|rrrrrr}
2 & 4 & -2 & 1 & -7 & 0 & 3 \\
  &   & 8 & 12 & 26 & 38 & 76 \\
\hline
  & 4 & 6 & 13 & 19 & 38 & 79
\end{array}
$$

The coefficients of the quotient are given by the first five entries in the third row of the table, making the quotient

$$q(x) = 4x^4 + 6x^3 + 13x^2 + 19x + 38$$

The last entry in the table, 79, gives the remainder.

➤ EXAMPLE 6
Application of Synthetic Division

Use synthetic division to determine $f(-3)$, where

$$f(x) = -x^4 - 5x^3 + 4x^2 - 9x + 10$$

Solution

By the Remainder theorem, the value of $f(-3)$ is the remainder when $f(x)$ is divided by $x - (-3)$. We use synthetic division to determine the value of this remainder.

$$
\begin{array}{r|rrrrr}
-3 & -1 & -5 & 4 & -9 & 10 \\
   &   & 3 & 6 & -30 & 117 \\
\hline
   & -1 & -2 & 10 & -39 & 127
\end{array}
$$

The last entry in the table, the remainder, provides the value of $f(-3)$. That is,

$$f(-3) = 127$$

➤ EXAMPLE 7
Another Application of Synthetic Division

Use synthetic division to prove that -2 is a zero of the polynomial

$$f(x) = x^6 - 4x^4 + 16x^2 - 64$$

Solution

According to the Remainder theorem, -2 is a zero of $f(x)$ if and only if the remainder is zero on dividing $f(x)$ by $x - (-2)$. We can determine this remainder using synthetic division.

$$
\begin{array}{r|rrrrrrr}
-2 & 1 & 0 & -4 & 0 & 16 & 0 & -64 \\
   &   & -2 & 4 & 0 & 0 & -32 & 64 \\
\hline
   & 1 & -2 & 0 & 0 & 16 & -32 & 0
\end{array}
$$

The final entry in the synthetic division table, the remainder, is 0. Therefore, $f(-2) = 0$, so -2 is a zero of $f(x)$.

Exercises 3.4

For Exercises 1–16, a polynomial $f(x)$ and a divisor $g(x)$ are given. Calculate the quotient $q(x)$ and the remainder $r(x)$.

1. $f(x) = x^5 - 2x^4 + x^3 - 5$
   $g(x) = x - 2$

2. $f(x) = 2x^5 - 3x^3 + 2x^2 - x + 3$
   $g(x) = x + 1$

3. $f(x) = 2x^3 + x^2 - 13x + 6$
   $g(x) = x + 2$

4. $f(x) = 6x^4 - 5x^3 + 9x^2 + 2x - 10$
   $g(x) = 2x - 1$

5. $f(x) = 3x^4 - x^3 + 2x - 6$
   $g(x) = 3x - 2$

6. $f(x) = 2x^5 - 3x^3 + 2x^2 - x + 3$
   $g(x) = 4x + 5$

7. $f(x) = x^5 - 2x^4 + x^3 - 5$
   $g(x) = x^2 - 2$

8. $f(x) = 2x^5 - 3x^3 + 2x^2 - x + 3$
   $g(x) = x^2 + 1$

9. $f(x) = x^3 + 125$
   $g(x) = x + 5$

10. $f(x) = x^3 - 1$
    $g(x) = x - 1$

11. $f(x) = x^4 - 3x^2 + 8$
    $g(x) = x^2 - 1$

12. $f(x) = 2x^5 - 3x^3 + 2x^2 - x + 3$
    $g(x) = x^3 + 1$

13. $f(x) = x^3 + x^2$
    $g(x) = x^2 + x - 2$

14. $f(x) = x^5 - x^4$
    $g(x) = x^4 - x^3 + 1$

15. $f(x) = x^6 - y^6$
    $g(x) = x - y$

16. $f(x) = x^5 + b^5$
    $g(x) = x + b$

17. Suppose that $f(x)$ is a quadratic polynomial with zeros -1 and 2, and $f(3) = -10$. Determine f.

18. Suppose that $f(x)$ is a quadratic polynomial with zeros 5 and -2, and $f(1) = 7$. Determine f.

19. Suppose that $f(x)$ is a quadratic polynomial with zeros i and $-i$, and $f(-2) = 6$. Determine f.

20. Suppose that $f(x)$ is a quadratic polynomial with zeros $2 - i$ and $2 + i$, and $f(1) = 7$. Determine f.

21. Suppose that $f(x)$ is a cubic polynomial with zeros -1, 0, and 1, and $f(2) = 24$. Determine f.

22. Suppose that $f(x)$ is a cubic polynomial with zeros -1, 0, and 2, and $f(1) = -24$. Determine f.

23. Suppose that $f(x)$ is a fifth-degree polynomial with zeros 0, -2, -1, i, and $-i$; and $f(1) = -24$. Determine f.

24. Suppose that $f(x)$ is a fifth-degree polynomial with zeros 1, -2, -5, i, and $-i$; and $f(1) = -24$. Determine f.

25. Suppose that $f(x)$ is a fourth-degree polynomial with zeros $1 - \sqrt{2}$, $1 + \sqrt{2}$, $4 \pm 3i$; and $f(0) = -2$. Determine f.

26. Suppose that $f(x)$ is a fourth-degree polynomial with zeros

$$\frac{2 + \sqrt{3}}{5}, \quad \frac{2 - \sqrt{3}}{5}, \quad \frac{1}{2} - \frac{3}{2}i, \quad \frac{1}{2} + \frac{3}{2}i$$

and $f(0) = 1$. Determine f.

27. Suppose that $f(x)$ is a polynomial with zeros $2 - i$ and $2 + i$. What can be said about $f(x)$?

28. Suppose that $f(x)$ is a polynomial with zeros $1 + \sqrt{5}$ and $1 - \sqrt{5}$. What can be said about $f(x)$?

29. For $f(x) = 2x^3 - 3x^2 + x - 1$,

    a. Find $f(-2)$.

    b. Find the remainder when $f(x)$ is divided by $x + 2$.

    c. Find $f(1)$.

    d. Find the remainder when $f(x)$ is divided by $x - 1$.

30. For $f(x) = 22x^4 - x^3 + x^2 - x + 1$,

    a. Find $f(2)$.

    b. Find the remainder when $f(x)$ is divided by $x - 2$.

    c. Find $f(-1)$.

    d. Find the remainder when $f(x)$ is divided by $x + 1$.

31. For $f(x) = 2x^2 + ix + 1$,

    a. Find $f(i)$.

    b. Find the remainder when $f(x)$ is divided by $x - i$.

32. For $f(x) = 3x^2 - ix - 1$,

    a. Find $f(-i)$.

    b. Find the remainder when $f(x)$ is divided by $x + i$.

Use synthetic division to find the quotient and remainder when $f(x)$ is divided by $g(x)$.

33. $f(x) = 4x^5 + 2x^3 - x + 5$

    $g(x) = x + 2$

34. $f(x) = 2x^5 - 3x^3 + 2x^2 - x + 3$

    $g(x) = x + 1$

35. $f(x) = 2x^3 + x^2 - 13x + 6$

    $g(x) = x - 2$

36. $f(x) = 6x^4 - 5x^3 + 9x^2 + 2x - 10$

    $g(x) = x - 1$

37. $f(x) = x^3 + 125$ 38. $f(x) = x^3 - 1$

    $g(x) = x + 5$ $g(x) = x - 1$

39. $f(x) = 3x^4 - 4x^3 - 19x^2 + 8x + 12$

    $g(x) = x - \frac{2}{3}$

40. $f(x) = 4x^5 + 2x^3 - x + 5$

    $g(x) = x + \frac{3}{4}$

41. $f(x) = x^6 - y^6$ 42. $f(x) = x^5 + b^5$

    $g(x) = x - y$ $g(x) = x + b$

Use synthetic division to determine the given function values.

43. $f(x) = 2x^4 - x^3 - 3x^2 + 2$; find $f(1)$, $f(-2)$, $f(4)$.

44. $f(x) = x^4 - 3x^3 + 5x - 2$; find $f(-2)$, $f(0)$, $f(2)$.

45. $f(x) = x^3 - 4x^2 + 9$; find $f(-1)$, $f(3)$, $f(5)$.

46. $f(x) = 2x^3 - 5x^2 + x - 3$; find $f(-2)$, $f(1)$, $f(4)$.

47. $f(x) = 2x^3 - 3x^2 + 5x - 1$; find $f(3)$, $f(-5)$, $f(11)$, $f(\frac{2}{3})$.

48. $f(x) = x^5 - 3x^4 + 2x^2 - x - 5$; find $f(-3)$, $f(20)$, $f(-\frac{1}{2})$.

Use synthetic division to determine whether each number is a zero of the given polynomial.

49. -1, 1; $f(x) = x^4 + 2x^3 + 3x^2 - 2$

50. -1, 2; $f(x) = x^4 - 2x^3 + x^2 - x - 2$

51. i, $2i$; $f(x) = 2x^4 - x^3 + x^2 - x - 1$

52. i, 0; $f(x) = 3x^4 - 2x^2 + 5x$

53. $\frac{1}{3}$, -5; $f(x) = 6x^3 + 31x^2 + 4x - 5$

54. 3, $-\frac{1}{2}$; $f(x) = x^3 + \frac{17}{2}x^2 + 19x + \frac{15}{2}$

Factor each polynomial $f(x)$. Then use the result of the factoring to solve the equation $f(x) = 0$.

55. $f(x) = x^4 - x^3 - 19x^2 - 11x + 30$

56. $f(x) = x^3 + 5x^2 - 12x - 36$

57. $f(x) = x^4 + 2x^3 - 13x^2 - 14x + 24$

58. $f(x) = x^3 - 4x^2 + x + 6$

59. Solve $x^4 + 2x^3 - 13x^2 - 14x + 24 < 0$.

60. Solve $x^3 - 4x^2 + x + 6 > 0$.

61. Find k such that when $2x^3 + x^2 - 5x + 2k$ is divided by $x + 1$ the remainder is 6.

62. Find those values of k such that $x - 3$ is a factor of $kx^3 - 5x^2 - 2kx + 3$.

For Exercises 63–66, a polynomial $f(x)$ and a divisor $g(x)$ are given. Calculate the quotient $q(x)$ and the remainder $r(x)$.

63. $f(x) = (x + 1)^2 - 2(x + 1) + 2$
    $g(x) = x + 1$

64. $f(x) = (x - 2)^3 + 7(x - 2) + 4$
    $g(x) = x - 2$

65. $f(x) = (x + 1)^3 4(x - 5)^3 + (x - 5)^4 3(x + 1)^2$
    $g(x) = x + 1$

66. $f(x) = (2x^2 - 5x - 3)^3$
    $g(x) = 2x + 1$

67. *a.* Find a way to use synthetic division to divide a polynomial by a divisor of the form $ax + b$.

    b. Use the result of (*a*) to find the quotient and remainder when $f(x)$ is divided by $g(x)$, where
    $$f(x) = 5x^3 + 4x^2 - 3x - 7$$
    $$g(x) = 2x + 3$$

68. Prove that $x - a$ is a factor of $x^n - a^n$ for all natural numbers n.

69. Prove that $x + a$ is a factor of $x^n + a^n$ for all odd positive natural numbers n.

70. Prove that $x + a$ is a factor of $x^n - a^n$ for all even natural numbers n.

71. Show that $x^n - a^n = (x - a)(x^{n-1} + x^{n-2}a + \cdots + a^{n-1})$ for all natural numbers n.

72. Show that $x^n - a^n = (x + a)(x^{n-1} - x^{n-2}a + x^{n-3}a^2 - \cdots - a^{n-1})$ for all even natural numbers n.

73. Show that $x^n + a^n = (x + a)(x^{n-1} - x^{n-2}a + x^{n-2}a^2 - \cdots + a^{n-1})$ for all odd natural numbers n.

➡ **Technology**

74. Let $f(x) = x^3 - 5x + 1$, $g(x) = x^2 - 2x$.

    a. Calculate $q(x)$, $r(x)$ using the division algorithm.

    b. Use a graphing calculator to graph $f(x)/g(x)$ and $q(x)$ on the same coordinate system.

    c. Zoom out the graph of (*b*) several times. Describe what you see.

75. Repeat the preceding exercise with $f(x) = x^4 - 2x^3 + 10$, $g(x) = 3x^2 - x$.

76. Can you generalize the graphical phenomenon exhibited in the preceding two exercises? Test your generalization using four more sets of polynomials.

3.5 THE ZEROS OF A POLYNOMIAL

In this section, we introduce additional results that provide information about the zeros of a polynomial with complex coefficients.

The Fundamental Theorem of Algebra

A polynomial with real coefficients may have complex numbers as zeros. Therefore, it might be the case that some polynomials with complex coefficients have zeros belonging to some number system larger than the complex numbers. However, this is not the case, as asserted by the following theorem.

FUNDAMENTAL THEOREM OF ALGEBRA

Let $f(x)$ be a nonconstant polynomial with complex coefficients. Then $f(x)$ has at least one complex zero.

Any proof of the Fundamental Theorem of Algebra requires advanced mathematical techniques well beyond the scope of this text. However, in what follows we will assume this result without proof.

An immediate consequence of the Fundamental Theorem of Algebra is the following result.

FACTORIZATION THEOREM

Let $f(x)$ be a polynomial with complex coefficients and leading coefficient a and positive degree n. Then $f(x)$ can be written in the form

$$f(x) = a(x - \alpha_1)(x - \alpha_2)\cdots(x - \alpha_n)$$

where

$$\alpha_1, \alpha_2, \ldots, \alpha_n$$

are complex numbers.

Proof   By the Fundamental Theorem of Algebra, $f(x)$ has a complex zero α_1. By the Factor theorem, this implies that $f(x)$ has $x - \alpha_1$ as a factor.

$$f(x) = (x - \alpha_1)f_1(x)$$

If $f_1(x)$ is a constant polynomial, then $f_1(x) = a$. If not, then $f_1(x)$ has a degree of at least 1, and the Fundamental Theorem of Algebra implies that $f_1(x)$ has a complex zero α_2. Again applying the Factor theorem, we see that

$$f_1(x) = (x - \alpha_2)f_2(x)$$
$$f(x) = (x - \alpha_1)(x - \alpha_2)f_2(x)$$

If $f_2(x)$ is a constant polynomial, then $f_2(x) = a$. Otherwise, repeat the same reasoning again. After n repetitions of the reasoning, we will arrive at the desired factorization of f.   ♦

➢ **EXAMPLE 1**
Factoring into Linear Factors

Factor the following polynomial into linear factors with complex coefficients:
$$f(x) = 6x^4 + 6x^3 + 18x^2$$

Solution

We begin by noting that we may factor out $6x^2$ from each term to obtain
$$6x^2(x^2 + x + 3)$$

To factor the second quadratic polynomial, we use the quadratic formula to determine its zeros.

$$\frac{-1 \pm \sqrt{1^2 - 4(1)(3)}}{2} = \frac{-1 \pm \sqrt{-11}}{2} = \frac{-1 \pm \sqrt{11}i}{2}$$

Therefore, the factorization of the given polynomial is

$$6x^4 + 6x^3 + 18x^2 = 6x^2\left(x - \frac{-1 + \sqrt{11}i}{2}\right)\left(x - \frac{-1 - \sqrt{11}i}{2}\right)$$

Historical Note:
Formulas for the Solution
of Polynomial Equations

The quadratic formula gives us a way of calculating the zeros of a quadratic polynomial using the operations of addition, subtraction, multiplication, division, and extraction of roots. It settles, once and for all, the problem of determining the zeros of quadratic polynomials, and in a very effective manner. It is natural to expect that such formulas exist for polynomials of other degrees. In the 16th century, the Renaissance mathematician Tartaglia developed a formula for calculating the zeros of a cubic equation. This formula was plagiarized by his rival Cardano, and history erroneously calls the result **Cardano's formula**. This formula is complex, but it does allow us to calculate the zeros from the coefficients of the equation and resembles, in broad shape, the quadratic formula. Farrari, a contemporary of Cardano, developed a similar formula for equations of the fourth degree. However, four centuries passed and, in spite of many attempts, mathematicians were unable to find an equation for the zeros of equations of the fifth degree. The reason for this failure was explained by the 19th-century mathematician Abel, who proved that there exist no formulas for equations of the fifth or higher degrees. Attempts to explain the reason for this phenomenon and to explain the general nature of the zeros of equations of higher degree led to the incredibly fertile field of mathematics called **the theory of equations**. In its turn, the theory of equations led to the field of mathematics called **group theory**, which is now a fundamental tool in the fields of crystallography, astrophysics, elementary particle theory, and quantum mechanics. | ∎

Zeros and Their Multiplicities

It can be proved that the factorization given by the Fundamental Theorem of Algebra is unique. That is, any two such factorizations of $f(x)$ differ only in the order of their linear factors. The zeros

$$\alpha_1, \alpha_2, \ldots, \alpha_n$$

may not all be different. The number of times a particular zero appears among the factors is called the **multiplicity** of the zero. For many purposes, it is convenient to count a zero of multiplicity m as m zeros. In this case, we say that *zeros are counted according to their multiplicities*. For example, consider the polynomial $f(x) = (x - 1)^3$. This polynomial has the single zero 1 of multiplicity 3, which is counted as three zeros.

Suppose that the polynomial $f(x)$ has degree n. When zeros are counted according to their multiplicities, the factors listed above consist of n zeros. That is, we have the following result.

NUMBER OF ZEROS OF A POLYNOMIAL

Suppose that $f(x)$ is a polynomial with complex coefficients and positive degree n. If zeros are counted according to their multiplicities, then $f(x)$ has exactly n zeros.

➤ **EXAMPLE 2**
Determining Multiplicities of Zeros

Determine the multiplicities of the zeros of the polynomial
$$f(x) = x^5 (x - 1)^6 (x + 7)^2 (x + i)^2 (x - i)^2$$

Solution

The zeros can be read off from the distinct factors appearing in the factorization of $f(x)$. The multiplicities can be read off from the exponents, which indicate the number of times the particular factor is present. Here are the results:

$$\alpha_1 = 0, \text{ multiplicity} = 5$$
$$\alpha_2 = 1, \text{ multiplicity} = 6$$

$$\alpha_3 = -7, \text{ multiplicity } = 2$$
$$\alpha_4 = -i, \text{ multiplicity } = 2$$
$$\alpha_5 = i, \text{ multiplicity } = 2$$

Polynomials with Real Coefficients

Polynomials with real coefficients form a very important class of polynomials because these polynomials occur most frequently in applied problems. Let's develop some information about the zeros of these polynomials.

CONJUGATE ROOT THEOREM

Suppose that $f(x)$ is a polynomial with real coefficients. If α is a zero of $f(x)$, then so is its complex conjugate $\overline{\alpha}$.

The proof of the Conjugate Root theorem requires several important properties of complex conjugates, which we establish first. Let α and β be complex numbers. Then we have the following.

PROPERTIES OF COMPLEX CONJUGATES

1. $\overline{\alpha + \beta} = \overline{\alpha} + \overline{\beta}$
2. $\overline{\alpha\beta} = \overline{\alpha} \cdot \overline{\beta}$
3. α is real if and only if $\alpha = \overline{\alpha}$

For the proofs of these three facts, suppose that

$$\alpha = a + bi, \qquad \beta = c + di$$

for real numbers a, b, c, and d. Then we have

$$\overline{\alpha + \beta} = \overline{(a + bi) + (c + di)}$$
$$= \overline{(a + c) + (b + d)i}$$
$$= (a + c) - (b + d)i$$
$$= (a - bi) + (c - di)$$
$$= \overline{\alpha} + \overline{\beta}$$

Thus, we have proved Property 1.

The proof of Property 2 is similar and is left to the exercises.

For the proof of Property 3, note that if α is real, then $b = 0$ and

$$\overline{\alpha} = \overline{a + 0i} = a - 0i = a = \alpha$$

Conversely, if

$$\alpha = \overline{\alpha}$$

then

$$a + bi = \overline{a + bi}$$
$$a + bi = a - bi$$
$$2bi = 0$$
$$b = 0$$

That is,

$$\alpha = a + bi = a + 0i$$

is real. This proves Property 3.

Proof of the Conjugate Root Theorem   Let

$$f(x) = a_n x^n + a_{n-1} x^{n-1} + \cdots + a_0$$

Because all the coefficients of $f(x)$ are real, by Property 3 of complex conjugates we have

$$\overline{a_j} = a_j \quad (j = 0, 1, \ldots, n) \tag{1}$$

Because α is a zero of $f(x)$, we have

$$0 = f(\alpha) = a_n \alpha^n + a_{n-1} \alpha^{n-1} + \cdots + a_0$$

Taking the complex conjugates of the left and right sides of this equation, we have

$$\overline{0} = \overline{a_n \alpha^n + a_{n-1} \alpha^{n-1} + \cdots + a_0}$$

However, because 0 is real, it is its own conjugate. Moreover, applying Properties 1 and 2, we see that

$$0 = \overline{a_n \alpha^n + a_{n-1} \alpha^{n-1} + \cdots + a_0}$$

$$= \overline{a_n \alpha^n} + \overline{a_{n-1} \alpha^{n-1}} + \cdots + \overline{a_0}$$

$$= \overline{a_n} \cdot \overline{(\alpha)}^n + \overline{a_{n-1}} \cdot \overline{(\alpha)}^{n-1} + \cdots + \overline{a_0}$$

However, using Equation (1), the preceding equation may be written

$$a_n (\overline{\alpha})^n + a_{n-1} (\overline{\alpha})^{n-1} + \cdots + a_0 = 0$$

The left side of this last equation is just $f(x)$ with x replaced by $\overline{\alpha}$. That is, we have

$$f(\overline{\alpha}) = 0$$

This completes the proof of the theorem.   ◆

An immediate consequence of Conjugate Root theorem is the following.

CONJUGATE ROOT THEOREM (ALTERNATE FORM)

The nonreal zeros of a polynomial with real coefficients occur in complex-conjugate pairs.

> **EXAMPLE 3**
Zeros of a Fifth-Degree Polynomial

How many real zeros can a polynomial of degree 5 with real coefficients have?

Solution

By the alternate form of the Conjugate Root theorem, the nonreal zeros occur in pairs. The number of pairs can be 0, 1, or 2. (Any more pairs would result in more than five zeros.) These three possibilities correspond to the following numbers of real zeros:

$$0 \text{ pairs}: \quad 5 - 0 = 5 \text{ real zeros}$$
$$1 \text{ pair}: \quad 5 - 2 = 3 \text{ real zeros}$$
$$2 \text{ pairs}: \quad 5 - 4 = 1 \text{ real zero}$$

> **EXAMPLE 4**
Determining Complex
Zeros

Determine the zeros of the polynomial

$$x^6 - 1$$

Solution

We proceed by factoring the polynomial using the various factorization formulas from Chapter 1.

$$x^6 - 1 = (x^3)^2 - 1$$
$$= (x^3 - 1)(x^3 + 1)$$
$$= (x - 1)(x^2 + x + 1)(x + 1)(x^2 - x + 1)$$

From this factorization, we see that the zeros are solutions of the equations

$$x - 1 = 0, \qquad x + 1 = 0$$
$$x^2 + x + 1 = 0, \qquad x^2 - x + 1 = 0$$

The first two equations give the zeros 1, -1. To solve the last two equations, we use the quadratic formula:

$$x^2 + x + 1 = 0$$

$$x = \frac{-1 \pm \sqrt{(1)^2 - 4(1)(1)}}{2(1)}$$

$$= \frac{-1 \pm \sqrt{3}i}{2}$$

In a similar fashion, the last equation yields the zeros

$$x = \frac{1 \pm \sqrt{3}i}{2}$$

The six zeros of the given polynomial are

$$\pm 1, \frac{\pm 1 \pm \sqrt{3}i}{2}$$

where all possible combinations of signs are allowed in the final expression.

The Fundamental Theorem of Algebra says that a polynomial with real co-efficients can be factored in the form

$$f(x) = a(x - \alpha_1)(x - \alpha_2) \cdots (x - \alpha_n)$$

where the complex numbers

$$\alpha_1, \alpha_2, \ldots, \alpha_n$$

are the zeros of $f(x)$. According to the alternate form of the Conjugate Root theorem, the nonreal zeros occur in complex–conjugate pairs. If such a pair of zeros is

$$\beta, \overline{\beta}$$

then the product

$$(x - \beta)(x - \overline{\beta})$$

is contained in the factorization of $f(x)$ and is equal to

$$x^2 - (\beta + \overline{\beta})x + \beta\overline{\beta}$$

The coefficients of this polynomial are real because if we let $\beta = a + bi$, then
$$\beta + \overline{\beta} = (a + bi) + (a - bi) = 2a$$
and
$$\beta\overline{\beta} = (a + bi)(a - bi) = a^2 + b^2$$
and both of the expressions on the right of these two equations are real numbers. This shows that the factor corresponding to a complex-conjugate pair of zeros, namely
$$(x - \beta)(x - \overline{\beta})$$
is a real polynomial. On the other hand, if a zero α is real, the factor $x - \alpha$ has real coefficients. This proves the following important result about the factorization of polynomials with real coefficients.

POLYNOMIAL FACTORIZATION OVER THE REALS

Let $f(x)$ be a polynomial of positive degree with real coefficients. Then $f(x)$ can be factored into a product of linear and quadratic polynomials having real coefficients, where the quadratic factors have no real zeros.

➢ **EXAMPLE 5**
Factoring into Real Factors

Factor the following polynomial into polynomials with real coefficients.
$$x^4 + x^3 + 5x^2 + 4x + 4$$

Solution

Note that the coefficient 4 appears twice. This suggests that we can group terms together so that we can factor out a 4. In addition to grouping the last two terms together, we can rewrite the middle term as
$$x^2 + 4x^2$$
Now we can factor two groups of terms:
$$(x^4 + x^3 + x^2) + (4x^2 + 4x + 4) = x^2(x^2 + x + 1) + 4(x^2 + x + 1)$$
$$= (x^2 + 4)(x^2 + x + 1)$$

Notice that the zeros of each of the two factors are nonreal complex numbers because their respective discriminants are negative (-16 for the first and -3 for the second). Therefore, the polynomials cannot be factored further into real polynomials.

Note, however, that by using the quadratic formula to determine the zeros of each of the factors, we see that the given polynomial may be factored into complex polynomials
$$(x + 2i)(x - 2i)\left(x - \frac{-1 + \sqrt{3}i}{2}\right)\left(x - \frac{-1 - \sqrt{3}i}{2}\right)$$

Descartes' Rule of Signs

It is possible to derive information about the zeros of a polynomial $f(x)$ with real coefficients by examining the number of changes of sign in the sequence of coefficients. Consider the polynomial
$$5x^6 - 3x^4 + x^3 + 0.5x^2 - 10x + 4$$

The sequence of signs in this polynomial is

$$+\ -\ +\ +\ -+$$

The number of changes in this sequence is four: the change from the first to the second, from the second to the third, from the fourth to the fifth, and from the fifth to the sixth. Note that *no change in signs is recorded for terms with zero coefficients*. The number of changes in the sequence of signs is related to the number of real zeros of the polynomial by the following famous result of Descartes.

DESCARTES' RULE OF SIGNS

Let $f(x)$ be a polynomial with real coefficients and with nonzero constant term.

1. The number of positive real zeros is at most equal to the number of changes m in the sequence of signs of $f(x)$ and differs from m by an even integer.
2. The number of negative real zeros is at most equal to the number of changes n in the sequence of signs for $f(-x)$ and differs from n by an even integer.

➤ **EXAMPLE 6**
Number of Real Zeros

Apply Descartes' Rule of Signs to the polynomial
$$5x^6 - 3x^4 + x^3 + 0.5x^2 - 10x + 4$$

Solution

The polynomial has a nonzero constant term, so we may apply Descartes' Rule of Signs. As we have already seen, the sequence of signs for this polynomial is $+\ -\ +\ +\ -+$, and there are four changes in this sequence, so $m = 4$. By Descartes' Rule of Signs, the polynomial has at most four positive real zeros, and the actual number differs from 4 by an even integer. Thus, the number of positive real zeros is either 4, 2, or 0.

To analyze the possibilities for negative real zeros, we form $f(-x)$:
$$5(-x)^6 - 3(-x)^4 + (-x)^3 + 0.5(-x)^2 - 10(-x) + 4$$
$$= 5x^6 - 3x^4 - x^3 + 0.5x^2 + 10x + 4$$

The sequence of signs for $f(-x)$ is

$$+\ -\ -\ +\ +\ +$$

There are two changes in sign, so $n = 2$. Descartes' Rule of Signs then states that the polynomial has at most two negative real zeros, and the number of negative real zeros is either 2 or 0.

Thus, Descartes' Rule of Signs shows that there are six possibilities, as shown in the following table:

Positive Real Zeros	Negative Real Zeros	Nonreal Zeros	Total Zeros
4	2	0	6
4	0	2	6
2	2	2	6
0	0	6	6
0	2	4	6
2	0	4	6

> **EXAMPLE 7**
Study of Real Zeros

Use Descartes' Rule of Signs to analyze the zeros of the equation

$$x^3 + 2x^2 - x + 1$$

Solution

There are two changes in the sequence of signs of $f(x)$. By Descartes' Rule of Signs, this means that there are at most two positive zeros. Moreover, the number of positive zeros differs from 2 by an even integer. This means that the number of positive zeros is either 2 or 0.

To analyze the negative zeros, we form $f(-x)$:

$$f(-x) = (-x)^3 + 2(-x)^2 - (-x) + 1 = -x^3 + 2x^2 + x + 1$$

There is a single change in sign. Again from Descartes' Rule of Signs, this means that there is at most one negative zero. Because the actual number of negative zeros differs from 1 by an even integer, there is only one possibility: a single negative zero.

Combining the various possibilities, we see that there are either two positive zeros and one negative zero or no positive zeros, one negative zero and two nonreal zeros.

Descartes' Rule of Signs assumes that the constant term of the polynomial is nonzero. However, this is not much of a restriction when it comes to analyzing zeros, for a polynomial with a zero constant term has a power of x as a factor. By factoring out this power of x, we can arrive at a polynomial with a nonzero constant term to which Descartes' Rule can be applied. For example, consider the polynomial

$$f(x) = x^5 - 3x^3 + 12x^2$$

It has zero constant term but can be written in the form

$$f(x) = x^2(x^3 - 3x + 12)$$

The first factor on the right corresponds to a zero with multiplicity 2. We may investigate the remaining zeros by applying Descartes' Rule of Signs to the polynomial

$$x^3 - 3x + 12$$

Exercises 3.5

Factor each of the following polynomials into linear factors with complex coefficients.

1. $f(x) = 6x^4 + 5x^3 + 5x^2$

2. $f(x) = 3x^6 + 6x^5 - 3x^4$

3. $f(x) = (x^2 + 1)(x^2 + 2x + 3)$

4. $f(x) = (x^2 + 4)(x^2 - 4x - 5)$

5. $f(x) = x^3 - 1$

6. $f(x) = x^3 + 8$

7. $f(x) = x^3 + 3x^2 + 5x + 15$ (*Hint:* Consider factoring by grouping.)

8. $f(x) = x^3 - 5x^2 + 7x - 35$ (*Hint:* Consider factoring by grouping.)

9. $f(x) = (x^2 + x + 3)^2$

10. $f(x) = (x^2 - x - 5)^2$

Determine the zeros and their multiplicities of the following polynomials.

11. $f(x) = 3x(x - 2)^2(2x - 5)^3$

12. $f(x) = (2x - 3)(x - 4)^2(x + 1)^2$

13. $f(x) = (x^2 - 4)^2(x + 2)^3$

14. $f(x) = (2x^2 - 5x - 3)^3$

15. $f(x) = (x - 2)^3(x + 1)$

16. $f(x) = (x^2 - x - 6)^2(x^2 + 2x - 3)^2$

17. $f(x) = x(x - 3)^2(x^2 + 1)$

18. $f(x) = x^2(x + 5)(x^2 + 3x + 5)^2$

19. $f(x) = x^5(x^2 + x - 1)^3$

20. $f(x) = x^4(x^2 + 5)^3(x^2 - x)$

21. How many real zeros can a polynomial of degree 3 with real coefficients have?

22. How many real zeros can a polynomial of degree 4 with real coefficients have?

23. How many real zeros can a polynomial of degree 2 with real coefficients have?

24. How many real zeros can a polynomial of degree 7 with real coefficients have?

Factor each polynomial into polynomials with real coefficients and then into polynomials with complex coefficients. Then find the zeros.

25. $x^6 - 64$

26. $t^6 - 1$

27. $x^3 - 1$

28. $x^3 + 1$

29. $m^3 + 8$

30. $x^3 - 8$

31. $y^4 - 4$

32. $y^4 - 9$

33. $r^7 + 2r^5 + r^3$

34. $p^6 + 2p^4 + p^2$

35. $x^4 - x^3 + 3x^2 - 4x - 4$

36. $x^4 - 2x^3 + 7x^2 - 18x - 18$

37. $x^4 - 2x^2 - 3$

38. $x^4 - 2x^2 - 35$

39. $r^3 - r^2 - 8r + 12$

40. $t^3 - 4t^2 - 12t + 48$

Use Descartes' Rule of Signs to analyze the zeros of each polynomial.

41. $5x^4 + x^3 - 2x^2 - 3x + 5$

42. $8x^6 - 3x^5 - 2x^3 + x^2 - 3x + 8$

43. $11x^{10} + 8x^3 + x^2 - x + 2$

44. $x^8 - x^7 + 3x^6 + 2x^2 - 1$

45. $5x^4 + x^3 - 2x^2 - 3x + 5$

46. $8x^6 - 3x^5 - 2x^3 + x^2 - 3x + 8$

47. $x^5 - x^4 - 3x^3 + 2x^2 - x - 1$

48. $10x^4 - 3x^3 + 2x^2 + x + 4$

49. $11x^{10} + 8x^3 + x^2 - x + 2$

50. $x^8 - x^7 + 3x^6 + 2x^2 - 1$

51. $10x^6 - 3x^5 + 2x^4 + x^3 + 4x^2$

52. $x^8 - x^7 - 3x^6 + 2x^5 - x^4 - x^3$

For each polynomial, certain zeros are given. Find the remaining zeros.

53. $x^3 + 5x^2 + 9x + 5; -2 - i$

54. $3x^3 + 2x^2 + 3x + 2; i$

55. $x^4 - 3x^3 - 2x^2 + 6x; 3$

56. $x^4 - 4x^3 + 8x^2 - 8x - 5; 1 + 2i$

57. $x^4 - 6x^2 - 8x - 3; -1$ is a zero of multiplicity 2

58. $x^4 - 5x^3 + 6x^2 + 4x - 8; 2$ is a zero of multiplicity 3

59. $x^5 - 12x^3 + 46x^2 - 85x + 50; -5, 1 + 2i$

60. $12x^5 + 4x^4 + 7x^3 + 14x^2 - 34x + 12; -\sqrt{2}i, -\dfrac{3}{2}$

61. Prove: $\overline{\alpha\beta} = \overline{\alpha} \cdot \overline{\beta}$

3.6 CALCULATING ZEROS OF REAL POLYNOMIALS

In the preceding section, we obtained a great deal of information about the zeros of polynomials. We discussed the Fundamental Theorem of Algebra and related theorems, which state that a polynomial of degree n with complex coefficients has n complex zeros, provided that zeros are counted according to their multiplicities. We then discussed the special features of real polynomials and their zeros. In this section, we continue the discussion by considering the rational zeros of polynomials with integer coefficients. Then we discuss the approximation of real zeros of real polynomials.

Polynomials with Rational Coefficients

Suppose that $f(x)$ is a polynomial with rational coefficients and that we wish to determine the zeros of $f(x)$. The first point to notice is that by multiplying $f(x)$ by a common denominator of the coefficients, we can obtain a polynomial with the same zeros but with integer coefficients. For the sake of our discussion, we can assume that the coefficients are all integers.

The simplest zeros of $f(x)$ are the rational zeros, that is, the zeros of the form c/d, where c and d are integers and d is nonzero. We can determine all such zeros using the following result.

RATIONAL ZERO THEOREM

Let

$$f(x) = a_n x^n + a_{n-1} x^{n-1} + \cdots + a_1 x + a_0$$

be a polynomial with integer coefficients. Suppose that c/d is a rational zero of $f(x)$, where c and d are integers, with d nonzero and the fraction c/d in lowest terms. Then c is a factor of the constant coefficient a_0, and d is a factor of the leading coefficient a_n.

The proof of this theorem belongs to either a course in abstract algebra or a course in the theory of numbers, and will be omitted here.

Using the Rational Zero theorem, we reduce the problem of determining the rational zeros of $f(x)$ to testing a finite number of possibilities obtained by factoring the leading and constant coefficients of $f(x)$ and forming all possible fractions c/d. The next two examples illustrate how to organize the calculations.

> **EXAMPLE 1**
Rational Zeros

Find all rational zeros of the polynomial
$$x^4 - 5x^3 + 13x^2 - 35x + 42$$

Solution

In this case, the leading and constant coefficients are given respectively by
$$a_n = 1, \quad a_0 = 42$$
By the Rational Zero theorem, if c/d is a rational zero, then c is a factor of 42 and d is a factor of 1. If necessary, we may multiply both c and d by -1, so we may assume that d is positive. This means that $d = 1$. The choices for c are as follows:
$$\pm 1, \ \pm 2, \ \pm 3, \ \pm 6, \ \pm 7, \ \pm 14, \ \pm 21, \ \pm 42$$
This leads to 16 possibilities for rational zeros. We now test each of them in turn to determine which ones work. One possible method of testing them is to use synthetic division to evaluate the remainder when $f(x)$ is divided by $x - c$. After tackling the computations, taking each value of c in turn, we find zero remainders for $c = 2, 3$ and for no other values of c. This means that the only rational zeros of $f(x)$ are $c/d = \frac{2}{1} = 2$ and $\frac{3}{1} = 3$.

> **EXAMPLE 2**
Determine Rational Zeros

Determine all rational zeros of the polynomial
$$2x^3 - 3x^2 - 10x + 15$$

Solution

In this case, the leading and constant coefficients are given respectively by
$$a_n = 2, \quad a_0 = 15$$
By the Rational Zero theorem, if c/d is a rational zero, then c is a factor of 15 and d is a factor of 2. If necessary, we may multiply both c and d by -1, so we may assume that d is positive. This means that $d = 1$ or 2. The choices for c are $\pm 1, \ \pm 3, \ \pm 5, \ \pm 15$. This leads to 16 possibilities for rational zeros. We now test each of them in turn to determine which ones work. The only possibility that is a zero is $c/d = \frac{3}{2}$, so the only rational zero of the given polynomial is $\frac{3}{2}$.

➤ **EXAMPLE 3**
Factoring from Zeros

Factor the polynomial

$$f(x) = x^3 - 5x^2 - 3x - 18$$

into linear factors.

Solution

We first look for rational zeros to obtain any simple linear factors of $f(x)$. According to the Rational Zero theorem, any rational zeros can be written in the form c/d, where c is a factor of -18 and d is a factor of 1. This leads to the following choices for c/d:

$$\pm 1, \pm 2, \pm 3, \pm 6, \pm 9, \pm 18$$

As in the preceding example, we determine which of these are zeros using synthetic division. We get nonzero remainders for c/d equal to the first six choices. However, for the choice $+6$, synthetic division yields

$$
\begin{array}{r|rrrr}
6 & 1 & -5 & -3 & -18 \\
  &   & 6 & 6 & 18 \\
\hline
  & 1 & 1 & 3 & 0
\end{array}
$$

This shows that division by $x - 6$ yields a zero remainder, so $x - 6$ is a factor of $f(x)$. Moreover, from the final row of the synthetic division, we read off the factorization

$$f(x) = (x - 6)(x^2 + x + 3)$$

The quadratic polynomial $x^2 + x + 3$ has a negative discriminant and hence has two complex zeros, which may be determined by the quadratic formula. They are

$$\frac{-1 \pm \sqrt{-11}}{2} = \frac{-1 \pm \sqrt{11}i}{2}$$

This leads to the desired factorization

$$f(x) = (x - 6)\left(x - \frac{-1 + \sqrt{11}i}{2}\right)\left(x - \frac{-1 - \sqrt{11}i}{2}\right)$$

Historical Note:
Evariste Galois

The theory of polynomial equations is a vast branch of mathematics called **Galois theory**, named after its founder, Evariste Galois, who made his ingenious discoveries in the early 19th century. Galois was an unrecognized prodigy who failed the entrance examination to college because his examiners didn't understand what he was talking about. He submitted a paper for a prize given by the French Academy, only to have it lost by Augustin Cauchy, one of the most important mathematicians of the age. Totally ignored, Galois delved into radical politics and was challenged to a duel by a monarchist (the French monarchy had recently been restored after Napoleon had been exiled to Elba). Anticipating his death in the duel, Galois spent the night before hastily recording his mathematical discoveries. The next day he was killed, just short of his 21st birthday. Mathematicians have spent more than a century building on Galois' fertile ideas.   ▐ ▆

➤ **EXAMPLE 4**
Architecture

The cost of building an office building of n floors is equal to

$$100n^3 + 3000n^2 + 50{,}000$$

Suppose that the total cost of construction is \$450,000. How many stories is the building?

Solution

We must solve the following equation for n:

$$100n^3 + 3000n^2 + 50,000 = 450,000$$
$$100n^3 + 3000n^2 - 400,000 = 0$$
$$n^3 + 30n^2 - 4000 = 0$$

Let's apply the Rational Zero theorem to the polynomial

$$f(x) = x^3 + 30x^2 - 4000$$

We are looking for a positive zero. Because the leading coefficient is 1, the Rational Zero theorem asserts that the zero must be a factor of 4000. If we factor 4000, we see that

$$4000 = 2^5 5^3$$

The factors of 4000 are obtained by multiplying a certain number of factors of 2 and a certain number of factors of 5. These are 1, 2, 4, 5, 8, 10, 16, 20, 32, ... Using synthetic division, we may test these factors one by one to determine which is a zero of $f(x)$. After a bit of calculation, we see that the only one that works is 10, so 10 is a solution of the equation, and the building has 10 stories.

Bounds on the Zeros of Real Polynomials

As we have just seen, the rational zeros of a polynomial with integer coefficients may be determined by examining a finite list of possibilities. Often, however, this list can be long, resulting in a large number of possibilities to check. The number of possibilities can be narrowed by using the following result.

UPPER AND LOWER BOUNDS FOR ZEROS

Let $f(x)$ be a polynomial with real coefficients and positive leading coefficient.

1. Let $a > 0$ be chosen so that the third row in synthetic division of $f(x)$ by $x - a$ has all positive or zero entries. Then all real zeros of $f(x)$ are less than or equal to a. The number a is called an **upper bound** for the zeros of f.
2. Let $b < 0$ be chosen so that the third row in the synthetic division of $f(x)$ by $x - b$ has alternating signs. Then all real zeros of $f(x)$ are greater than or equal to b. The number b is called a **lower bound** for the zeros of f.

Proof

1. Because the third row of synthetic division of $f(x)$ by $x - a$ has all nonnegative entries, we see that

$$f(x) = (x - a)q(x) + c$$

where c and the coefficients of $q(x)$ are all positive or zero. Assume that $t > a$. Then

$$f(t) = (t - a)q(t) + c > 0$$

so that t is not a zero of f. That is, any zero of f must be less than or equal to a.

2. The proof is similar to that of part 1 and is left as an exercise.   ◆

The next example will give you an idea of how this result may be applied to obtain information about the zeros of a real polynomial.

➤ EXAMPLE 5
Calculating Upper and Lower Bounds for Zeros

Let
$$f(x) = x^5 - x^3 + 2x^2 - 2x + 3$$
Determine upper and lower bounds for the real zeros of $f(x)$.

Solution

In a search for an upper bound for the zeros, we may try the various positive integers in turn and perform synthetic division. For the integer $a = 1$, we have

$$
\begin{array}{r|rrrrrr}
1 & 1 & 0 & -1 & 2 & -2 & 3 \\
  &   & 1 & 1 & 0 & 2 & 0 \\
\hline
  & 1 & 1 & 0 & 2 & 0 & 3
\end{array}
$$

Because all entries in the last row are nonnegative, we see that 1 is an upper bound for the zeros of f. To obtain a lower bound, we test negative integers and look for third rows with alternating signs. For $a = -1$, we have

$$
\begin{array}{r|rrrrrr}
-1 & 1 & 0 & -1 & 2 & -2 & 3 \\
   &   & -1 & 1 & 0 & -2 & 4 \\
\hline
   & 1 & -1 & 0 & 2 & -4 & 7
\end{array}
$$

The signs don't alternate. Next we try $a = -2$:

$$
\begin{array}{r|rrrrrr}
-2 & 1 & 0 & -1 & 2 & -2 & 3 \\
   &   & -2 & 4 & -6 & 8 & -12 \\
\hline
   & 1 & -2 & 3 & -4 & 6 & -9
\end{array}
$$

The signs alternate, so -2 is a lower bound for the zeros of f.

Approximating Real Zeros

For many applications, it is sufficient to determine a real zero with an accuracy of a certain number of decimal places. We will present a method for determining such approximations based on the Intermediate Value theorem, stated in Section 3.2. Recall that this result says that if there are two points at which the graph of a polynomial lies on opposite sides of the x-axis, then the graph crosses the x-axis at some point between these points (see Figure 1).

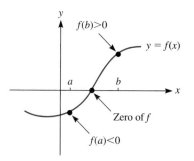

Figure 1

➤ **EXAMPLE 6**
Isolating a Zero

Show that the polynomial

$$f(x) = 3x^3 - x + 1$$

has a zero in the interval $(-1, 0)$.

Solution

We note that

$$f(-1) = 3(-1)^3 - (-1) + 1 = -1 < 0$$
$$f(0) = 3(0)^3 - 0 + 1 = 1 > 0$$

Therefore, $f(-1)$ and $f(0)$ have opposite signs. The Intermediate Value theorem states that the given polynomial has a zero in the interval $(-1, 0)$.

We can place a real zero in an interval as small as we wish by applying the Intermediate Value theorem to successively smaller intervals. Suppose that we determine from using the Intermediate Value theorem that $f(x)$ has a zero within a certain interval. We may divide this interval into subintervals of equal length. For one of these subintervals, the values of $f(x)$ at the endpoints are of opposite signs. This means that a real zero lies within that subinterval. We may now break this subinterval into equal parts and apply the same procedure again. In this way we can approximate the zero to any desired degree of accuracy. This method of approximating real zeros is called the **method of bisection**. The next example provides the details of a generalization of the method of bisection, in which each interval is subdivided into 10 equal subintervals (rather than 2). Using 10 subintervals allows determination of one additional decimal place accuracy with each iteration.

➤ **EXAMPLE 7**
Approximating a Zero

Determine the real zero of the function

$$f(x) = 3x^3 - x + 1$$

from the preceding example with an accuracy of one tenth.

Solution

In the preceding example, we showed that for this function

$$f(-1) < 0, \qquad f(0) > 0$$

so a zero lies in the interval $(-1, 0)$. We can divide this interval into the 10 subintervals:

$$(-1, -0.9), (-0.9, -0.8), \ldots, (-0.1, 0)$$

We calculate the values of $f(x)$ at the endpoints of each of these intervals. As a result of these calculations, we determine that

$$f(-0.9) = -0.287 < 0, \qquad f(-0.8) = 0.264 > 0$$

Thus, the zero lies between -0.9 and -0.8. To pin down the zero to the nearest tenth, evaluate the function at the midpoint of the interval $(-0.9, -0.8)$, to obtain

$$f(-0.85) = 0.00763 > 0$$

Therefore, the zero lies in the interval $(-0.9, -0.85)$. That is, to the nearest tenth the zero equals -0.9.

> **EXAMPLE 8**
Approximation of Zero
to Specified Accuracy

Determine the real zero of $f(x) = x^4 - 2x^2 + 3x - 1$, which lies between 0 and 1. Calculate the zero to within 0.001.

Solution

We begin by tabulating the values of $f(x)$ for x between 0 and 1, at intervals of 0.1.

We have used the computer program *Mathematica*™ to prepare the following table:

x	$f(x)$
0	-1
0.1	-0.7199
0.2	-0.4784
0.3	-0.2719
0.4	-0.0944
0.5	0.0625
0.6	0.2096
0.7	0.3601
0.8	0.5296
0.9	0.7361
1.0	1.0

You may verify the entries using your calculator. According to the shaded rows, the zero lies between 0.4 and 0.5, because the value of $f(x)$ changes sign in this interval. We now tabulate the values of the function for x between 0.4 and 0.5 at intervals of 0.01. We arrive at the following table:

x	$f(x)$
0.4	-0.0944
0.41	-0.0779424
0.42	-0.061683
0.43	-0.045612
0.44	-0.029719
0.45	-0.0139938
0.46	0.00157456
0.47	0.0169968
0.48	0.0322842
0.49	0.047448
0.5	0.0625

The shaded rows show that the zero lies between 0.45 and 0.46. Finally, we tabulate the function values for x between 0.45 and 0.46 at intervals of 0.001. The following table shows that the zero lies between 0.458 and 0.459:

x	$f(x)$
0.45	-0.0139938
0.451	-0.01243
0.452	-0.0108679
0.453	-0.00930727
0.454	-0.00774819
0.455	-0.00619065
0.456	-0.00463462
0.457	-0.0030801
0.458	-0.00152706
0.459	0.0000244838
0.46	0.00157456

To three significant digits, the zero is therefore 0.459.

Approximating Real Zeros Using a Graphing Calculator

The next example illustrates another method for approximating the real zeros of a polynomial (or of any function, for that matter) using a graphing calculator.

New Technology

EXAMPLE 9
Approximating Real Zeros Graphically

Approximate the zeros of the polynomial function

$$f(x) = x^4 - x^3 + 3x^2 - 4x - 4$$

Solution

Begin by graphing the function using the standard range settings $-10 \le x \le 10$. See Figure 2. It appears that the only possible zeros lie in the interval $-2 \le x \le 2$. Change the range to this interval and regraph. See Figure 3. We now see that there is a zero in the interval $-2 \le x \le -1$ and another in $1 \le x \le 2$. Using a box to zoom in on each of these ranges to each of these intervals and using the trace, we find that these zeros are approximately: -1.418504 and 1.418974. See Figures 4 and 5.

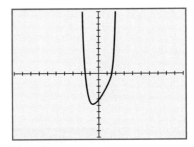

Figure 2

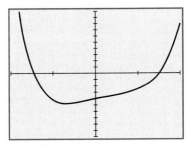

Figure 3

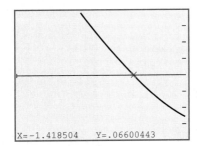

Figure 4

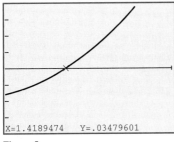

Figure 5

Exercises 3.6

Determine all rational zeros of the following polynomials.

1. $2x^3 - 11x^2 + 17x - 6$ 2. $6x^4 - 5x^2 + 1$

3. $3x^3 - 2x^2 - 3x - 12$ 4. $x^3 - 3x^2 - 3x - 4$

5. $5x^5 - 4x^4 - 50x^3 + 40x^2 + 125x - 100$

6. $x^5 - 2x^3 + 5x^2 - 10$

7. $x^4 - 25$ 8. $z^6 - 1$

Determine upper and lower bounds for the real zeros of $f(x)$.

9. $f(x) = 2x^4 - x^3 + 3x^2 - 5$

10. $f(x) = 3x^3 + 4x^2 + 12$

11. $f(x) = x^5 + 3x^4 - x^2 + 14$

12. $f(x) = 2x^4 - 3x^2 - 9x + 1$

13. $f(x) = 16x^{12} - 11x^{10} + 2x^9 - 3x^8 + 4x - 6$

14. $f(x) = 4x^4 + x^3 - 6x^2 + 10$

15. $f(x) = x^4 - 25$ 16. $f(x) = x^6 - 1$

Find the rational zeros of the following polynomials.

17. $9x^3 - 18x^2 + 11x - 2$ 18. $3x^3 + 4x^2 + 12$

19. $4x^3 - x^2 - 100x + 25$ 20. $2x^4 - 3x^2 - 9x + 1$

21. $4x^4 - 21x^2 - 25$ 22. $81x^4 - 16$

23. $x^4 - 25$ 24. $x^6 - 1$

25. $\frac{3}{2}x^3 + \frac{11}{4}x^2 - \frac{3}{4}x - \frac{1}{2}$ 26. $\frac{3}{4}x^3 - \frac{2}{3}x^2 - \frac{5}{6}x - \frac{1}{2}$

Show that each of the following polynomials has a zero in the given interval.

27. $f(x) = x^3 + x + 1; (-1, 0)$

28. $f(x) = x^3 - x + 1; (-2, -1)$

29. $f(x) = x^3 - 3x^2 + 4x - 5; (2, 3)$

30. $f(x) = x^3 + x - 3; (1, 2)$

31. $f(x) = x^3 + 2x^2 + 8x - 2; (0, 1)$

32. $f(x) = x^3 - 3x^2 - 5; (3, 4)$

33. $f(x) = x^4 - x^2 - 3; (-1, -2)$ and $(1, 2)$

34. $f(x) = x^4 - x^3 - 1; (-1, 0)$ and $(1, 2)$

Approximate the real zeros of each of the following polynomial functions with an accuracy of 0.0001 using the method of bisection.

35. $f(x) = x^3 + x + 1$

36. $f(x) = x^3 - x + 1$

37. $f(x) = x^3 - 3x^2 + 4x - 5$

38. $f(x) = x^3 + x - 3$

39. $f(x) = x^3 + 2x^2 + 8x - 2$

40. $f(x) = x^4 - x^2 - 3$

41. $f(x) = x^4 - x^3 - 1$

42. $f(x) = x^4 - 3x^3 - 2x - 3$

43. $f(x) = x^5 - 3x^4 + 5x^3 - 7x^2 + 6$

44. $f(x) = -x^5 + 4x^4 - 2x^3 + x^2 - 4x + 1$

♦ Applications

45. **Cost of a building.** The cost of building an office building of n floors is equal to

$$100n^3 + 3000n^2 + 50,000$$

a. Suppose that the total cost of construction is $3,487,500. How many stories is the building?

b. Suppose that the total cost of construction is about $776,000. Use the method of bisection to estimate how many stories are in the building. Give your answer correct to the nearest one.

46. **Volume of a box.** An open box of volume 108 in.³ can be made from a piece of cardboard that is 10 in. by 15 in. by cutting a square from each corner and folding up the sides (see figure). What is the length of a side of the squares? Use the method of bisection to find an approximate answer, if appropriate.

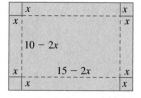

47. **Deflection of a beam.** A beam rests at two points A and B and has a concentrated load applied to the center of the beam, as shown in the figure. Let y denote the deflection of the beam at a distance of x units from the left end of the beam to the left edge of the weight. The deflection depends on the elasticity of the board, the load, and other physical characteristics. Suppose under certain conditions that y is given by

$$y = \tfrac{1}{12}x^3 - \tfrac{1}{16}x$$

Use the method of bisection to estimate the number of units x at which the deflection will be 40 units.

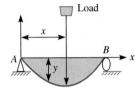

Load

48. **Break-even points.** Given the following total revenue and total cost functions:

$$R(x) = 100x - x^2, \qquad C(x) = \tfrac{1}{3}x^3 - 6x^2 + 89x + 100$$

a **break-even** point occurs at a nonnegative value of x for which $R(x) = C(x)$. Find all the break-even points for the functions. Use the method of bisection if appropriate.

49. **Medical dosage.** The function

$$N(t) = 0.045t^3 + 2.063t + 2$$

gives the blood level in parts per million of a certain dosage of medication after time t, in hours. Find all times in the interval [0, 10] for which the concentration is 4 parts per million. Use the method of bisection if appropriate.

50. **Temperature during an illness.** A patient's temperature during an illness was given by

$$T(t) = 98.6 - t^2(t - 4)$$

Find the times t at which the patient's temperature was 100°. Use the method of bisection where appropriate.

Find all the zeros. Use any procedures studied in this chapter.

51. $x^3 - 3x - 2$

52. $x^3 + 2x^2 + 9$

53. $x^3 - x^2 - 3x + 2$

54. $x^3 + 4x^2 + 7x + 6$

55. $x^4 - x^3 - 5x^2 + 7x - 2$

56. $x^4 + x^3 + x^2 - x - 2$

57. $6x^4 + x^3 + 4x^2 + x - 2$

58. $4x^4 - 4x^3 - 5x^2 + x + 1$

59. $4x^5 - 24x^4 + 25x^3 + 39x^2 - 38x - 24$

60. $x^5 + x^4 - 9x^3 - 5x^2 + 16x + 12$

61. $3x^4 - 3x^2 + 18$ 62. $x^4 - 10x^2 + 23$

Solve. Use any procedures studied in this chapter.

63. $x^3 - 4x^2 + x + 6 = 0$

64. $24x^3 + 10x^2 - 13x - 6 = 0$

65. $x^4 + 3x^2 = 20$

66. $2x^3 + 15x^2 = -20x - 3$

67. $4x^5 + 4x = 17x^3$

68. $x^5 + 60x + 72 = 15x^3 + 10x^2$

69. Prove that $\sqrt{2}$ is irrational by considering the zeros of the polynomial function $f(x) = x^2 - 2$.

70. Prove that $\sqrt{5}$ is irrational by considering the zeros of the polynomial function $f(x) = x^2 - 5$.

➡ Technology

71. Use a graphing calculator to approximate the zeros in Exercises 51–62. Do this by zooming.

72. Use a graphing calculator to approximate all real zeros of the function $f(x) = 3x^4 + x - 1$.

73. **Cardano's formula.** Any cubic polynomial equation can be reduced to one of the form

$$y^3 + by^2 + cy + d = 0$$

by dividing on both sides by the leading coefficient. Furthermore, we can obtain a reduced form by making the substitution

$$y = x - \tfrac{1}{3}b$$

This will yield an equation of the type

$$x^3 + px + q = 0$$

where

$$p = c - \frac{b^2}{3} \qquad \text{and} \qquad q = d - \frac{bc}{3} + \frac{2b^3}{27}$$

To solve the latter equation we first make the following computations:

$$a = \sqrt[3]{-\frac{q}{2} + \sqrt{\frac{q^2}{4} + \frac{p^3}{27}}}$$

$$b = \sqrt[3]{-\frac{q}{2} - \sqrt{\frac{q^2}{4} + \frac{p^3}{27}}}$$

Then the three cube roots of the reduced cubic equation are

$$x_1 = a + b$$

$$x_2 = \frac{-a + b}{2} + \frac{a - b}{2}\sqrt{3}i$$

$$x_3 = \frac{-a + b}{2} - \frac{a - b}{2}\sqrt{3}i$$

Use the preceding results to find the roots of the following cubics for which you approximated the real zeros in the preceding exercises. Then find approximations using your calculator or computer and compare.

a. $f(x) = x^3 + x + 1$ b. $f(x) = x^3 - x + 1$

c. $f(x) = x^3 - 3x^2 + 4x - 5$

d. $f(x) = x^3 + x - 3$

e. $f(x) = x^3 + 2x^2 + 8x - 2$

74. Use the method of bisection to approximate all the real zeros of the function

$$f(x) = 1 + \sqrt{x - 2} - \sqrt[3]{x + 6}$$

on the interval $(2, 6)$.

3.7 RATIONAL FUNCTIONS

Recall that a rational function is a function given by an expression of the form

$$f(x) = \frac{p(x)}{q(x)}$$

where $p(x)$ and $q(x)$ are polynomials with real coefficients. The domain of such a function consists of all real numbers x for which $q(x)$ is nonzero. In this section, we will learn to sketch the graphs of rational functions. As we will see, the significant new feature is the behavior of the functions for x near a zero of the denominator $q(x)$.

To simplify our discussion, we will initially assume that $p(x)$ and $q(x)$ have no factors in common.

Vertical Asymptotes

We begin with a graph of a typical rational function.

➤ EXAMPLE 1
Graph of a Rational Function

Sketch the graph of the rational function

$$f(x) = \frac{1}{x^2}$$

Solution

First note that $f(x)$ is not defined for $x = 0$, the value at which the denominator is 0. Values of $f(x)$ for x near 0 therefore display special characteristics. Note that

$$f(-x) = \frac{1}{(-x)^2} = \frac{1}{x^2}$$

so $f(x)$ is even, and its graph is symmetric with respect to the y-axis. Because the graph is symmetric with respect to an axis, we can restrict ourselves at first to the positive values of x. Here is a table of the values of $f(x)$ at some representative values that decrease through positive numbers and get increasingly close to 0. We say that these values of x **approach zero from the right.**

x	$f(x)$
1	1
0.1	100
0.01	10,000
0.001	1,000,000

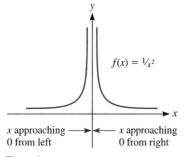

$f(x) = 1/x^2$

x approaching 0 from left ——►|◄—— x approaching 0 from right

Figure 1

The data in the table suggest that as x approaches zero from the right, the values of $f(x)$ increase without bound. Geometrically, this means that, as x approaches 0 from the right, the graph of $f(x)$ rises without bound and approaches arbitrarily close to the y-axis without ever touching it. We say that the y-axis is a **vertical asymptote** of the graph. Using the data in the table, we can make a sketch of the graph for $x > 0$. By symmetry, the portion of the graph corresponding to $x < 0$ can be obtained by reflection in the y-axis (see Figure 1).

Suppose that the rational function $f(x)$ is undefined for $x = a$ (that is, the value a would make the denominator of the function equal to zero). As we did in Example 1 for $a = 0$, we can examine the behavior of $f(x)$ as x approaches any value a from the right. If the values of $f(x)$ increase without bound as x approaches a from the right, we say that $f(x)$ *approaches positive infinity from the right* and write $f(x) \to +\infty$ as $x \to a^+$. [See Figure 2(a).] If, as x approaches a from the right, the values of $f(x)$ decrease without bound (say, -1, -100, $-10,000$, etc.), then we say that $f(x)$ *approaches negative infinity from the right* and write $f(x) \to -\infty$ as $x \to a^+$. [See Figure 2(b).]

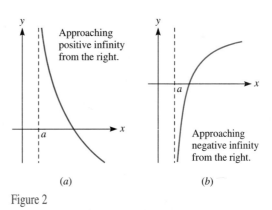

Approaching positive infinity from the right.

Approaching negative infinity from the right.

(a)   (b)

Figure 2

In a similar fashion, we can consider x approaching a from the left, that is, through values smaller than a. If the values of $f(x)$ increase without bound when x approaches a from the left, we say that $f(x)$ *approaches positive infinity from the left* and write $f(x) \to +\infty$ as $x \to a^-$. [See Figure 3(a).] If the values of $f(x)$ decrease without bound when x approaches a from the left, then we say that $f(x)$ *approaches negative infinity from the left* and write $f(x) \to -\infty$ as $x \to a^-$. [See Figure 3(b).]

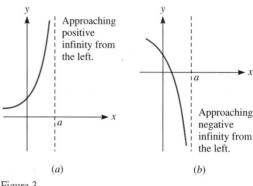

Figure 3

<table>
<tr><td>**Definition 1**
Vertical Asymptote</td><td>If, as x approaches a, the graph of a rational function $f(x)$ approaches positive infinity or negative infinity from either the right or left, then we say that the line $x = a$ is a **vertical asymptote** of the graph.</td></tr>
</table>

Graphs of rational functions (assumed to be written in lowest terms) have a vertical asymptote corresponding to each zero of the denominator. It is necessary to allow x to approach such values from both the right and the left to determine the behavior of the graph in the vicinity of the asymptote. The next two examples illustrate how to do this.

➤ **EXAMPLE 2**
Graph with Vertical Asymptote

Sketch the graph of the rational function

$$y = \frac{1}{x + 1}$$

Solution

The function is undefined for $x = -1$, the value at which the denominator equals 0. This means that the line $x = -1$ is a vertical asymptote of the graph. To determine the behavior of the graph in the vicinity of this asymptote, we allow x to approach -1 from both the right and the left.

As x approaches -1 from the right, $x > -1$, so the value of $f(x)$ is positive. Moreover, as x approaches -1 from the right, the values of $f(x)$ increase without bound. That is, they approach positive infinity. This information allows us to sketch the graph for $x > -1$. We first sketch the asymptote $x = -1$. Then we draw in the portion of the graph to the right of the asymptote (see Figure 4).

Next, we consider values of x when x approaches -1 from the left. These values for x are less than -1, so $x + 1$ is negative, as are the values of $f(x)$. Moreover, as x approaches -1 from the left, the values of $f(x)$ decrease without bound. That is, $f(x)$ approaches negative infinity (see Figure 4).

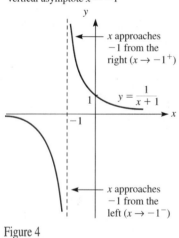

Vertical asymptote $x = -1$

Figure 4

Horizontal Asymptotes

In drawing an accurate sketch of a graph, it is helpful to indicate how the function behaves for values of x that are very large in either the positive or the negative direction. These values lie far to the right or left on the number line.

If x increases without bound through positive numbers, then we say that x *approaches positive infinity*. We may visualize this concept by imagining the value of x as represented by a point on the number line moving indefinitely far to the right. If x increases without bound through negative numbers, then we say that x *approaches negative infinity.* We may visualize this concept by imagining the value of x as represented by a point on the number line moving indefinitely far to the left.

As x approaches either positive or negative infinity, the values of a rational function $f(x)$ may exhibit several different behaviors. On one hand, the value of $f(x)$ may approach positive infinity or negative infinity. These possibilities are shown in Figure 5(*a*) and (*b*), respectively.

On the other hand, the values of $f(x)$ may approach a real number c. Two ways in which this may occur are illustrated in Figure 5(*c*).

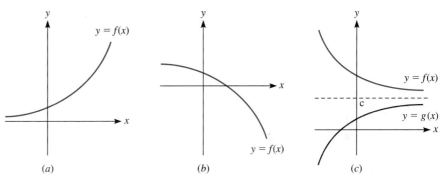

| (a) | (b) | (c) |

$f(x)$ approaches positive infinity as x approaches positive infinity.

$f(x)$ approaches negative infinity as x approaches positive infinity.

$f(x)$ and $g(x)$ approach the asymptote $y = c$ as x approaches positive infinity.

Figure 5

| **Definition 2** **Horizontal Asymptote** | Suppose that the values of the rational function $f(x)$ approach a real number c as x approaches either positive or negative infinity. We then say that the line $y = c$ is a **horizontal asymptote** of the graph of $f(x)$. |

As x approaches positive infinity, so do all positive powers of x. However, negative powers of x approach 0 (because the reciprocal of a large number is small).

As x approaches negative infinity, negative powers of x still approach 0. However, a positive even power of x approaches positive infinity, and a positive odd power of x approaches negative infinity. (A positive even power of a large negative number is a large positive number; a positive odd power of a large negative number is a large negative number.)

Using these facts, we can determine the behavior of a rational function as x approaches positive infinity or negative infinity. The next example illustrates how this may be done.

➤ **EXAMPLE 3**
Graph with Horizontal Asymptotes

Determine the horizontal asymptotes, if any, of the rational function

$$f(x) = \frac{x^2}{3x^2 - 4x - 1}$$

Solution

Divide the numerator and denominator by the largest power of x in the denominator, namely x^2, to obtain the following expression for $f(x)$:

$$f(x) = \frac{\dfrac{x^2}{x^2}}{\dfrac{3x^2 - 4x - 1}{x^2}} = \frac{1}{3 - \dfrac{4}{x} - \dfrac{1}{x^2}}$$

As x approaches either positive infinity or negative infinity, the terms

$$\frac{4}{x} \quad \text{and} \quad \frac{1}{x^2}$$

approach 0, and the value of $f(x)$ approaches

$$\frac{1}{3 + 0 + 0} = \frac{1}{3}$$

That is, the graph has as horizontal asymptote the line $y = \frac{1}{3}$. This asymptote is shown along with a sketch of the graph in Figure 6.

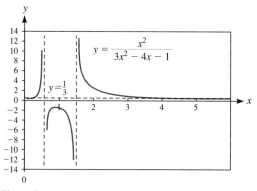

Figure 6

Note that whereas a graph can never cross a vertical asymptote, it is perfectly possible for a graph to cross a horizontal asymptote. In fact, there are graphs that oscillate an infinite number of times around a horizontal asymptote, approaching closer with each oscillation.

Oblique Asymptotes

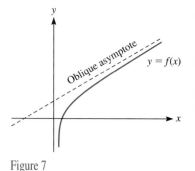

Figure 7

A line is said to be **oblique** if it is nonhorizontal and nonvertical. In our discussion above, we introduced both horizontal and vertical asymptotes. Analogously, we have the concept of an **oblique asymptote**, which is a line that a graph approaches as x approaches either positive infinity or negative infinity (see Figure 7).

Here is an example to illustrate how oblique asymptotes arise. Consider the rational function $f(x) = (x^2 + 3x + 1)/x$. By performing the indicated division, we arrive at the following expression for $f(x)$:

$$f(x) = x + 3 + \frac{1}{x}$$

It is clear that as x approaches either positive infinity or negative infinity, the term $1/x$ approaches 0, so the function value approaches the value of the linear function $x + 3$. Geometrically, this means that the graph of $f(x)$ approaches the graph of $x + 3$ as x approaches either positive or negative infinity. That is, the line $y = x + 3$ is an oblique asymptote of the graph (see Figure 8).

In a similar fashion, suppose that we are given a rational function $f(x) = p(x)/q(x)$, where the degree of $p(x)$ exceeds the degree of $q(x)$ by 1. Then we may perform division to write the rational function in the form

$$f(x) = ax + b + \frac{r(x)}{q(x)}$$

where the degree of $r(x)$ is less than the degree of $q(x)$. This ensures that $r(x)/q(x)$ approaches 0 as x approaches $+\infty$ or $-\infty$. In this case, the line $y = ax + b$ is an oblique asymptote of the graph of $f(x)$ (see Figure 9).

We can summarize our discussion of horizontal and oblique asymptotes with the following result.

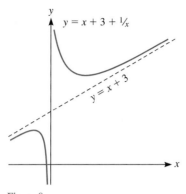

Figure 8

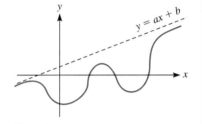

Figure 9

ASYMPTOTES OF A RATIONAL FUNCTION

Let $f(x) = p(x)/q(x)$ be a rational function, where $p(x)$ is a polynomial of degree m and leading coefficient c, and $q(x)$ is a polynomial of degree n and leading coefficient d.

1. If $m < n$, then the x-axis is a horizontal asymptote of the graph of $f(x)$, and $f(x)$ approaches 0 as x approaches either positive infinity or negative infinity.

2. If $m = n$, then the line $y = c/d$ is a horizontal asymptote of the graph of $f(x)$, and $f(x)$ approaches c/d as x approaches either positive infinity or negative infinity.

3. If $m = n + 1$, then we may write $f(x)$ in the form $f(x) = ax + b + r(x)/q(x)$, where $r(x)$ has degree less than n. In this case, $y = ax + b$ is an oblique asymptote of the graph of $f(x)$.

4. If $m > n$, then the graph of $f(x)$ has no horizontal asymptotes. Moreover, as x approaches positive infinity or negative infinity, $f(x)$ approaches either positive infinity or negative infinity.

Some Detailed Examples

Let's work out some examples that use everything we have learned about sketching graphs of rational functions.

> **EXAMPLE 4**
Asymptotes of Rational Functions

Describe the behaviors of the following rational functions as x approaches positive infinity and negative infinity.

1. $f(x) = \dfrac{1}{2x - 1}$

2. $f(x) = \dfrac{2x^3 - 1}{(x - 1)(x + 1)}$

3. $f(x) = \dfrac{(x - 1)(3x - 5)}{(2x + 7)(x + 1)}$

Solution

1. In this example, the degree of the numerator is less than the degree of the denominator. By Statement 1 on page 207, the graph has the x-axis

as a horizontal asymptote. That is, as x approaches positive or negative infinity, the values of $f(x)$ approach 0.

2. In this case, the degree of the numerator exceeds the degree of the denominator by 1, so the graph has no horizontal asymptotes but has an oblique asymptote. To determine the oblique asymptote, divide the numerator by the denominator:

$$\frac{2x^3 - 1}{x^2 - 1} = 2x + \frac{2x}{x^2 - 1}$$

The equation shows that $y = 2x$ is an oblique asymptote of $f(x)$.

3. In this case, the degree of the numerator equals the degree of the denominator, so by Statement 2 on page 207, the graph has a horizontal asymptote. The leading coefficient of the numerator is 3, and the leading coefficient of the denominator is 2, so the horizontal asymptote is the line $y = \frac{3}{2}$. As x approaches either positive infinity or negative infinity, the value $f(x)$ approaches $\frac{3}{2}$.

Sketching Graphs of Rational Functions

Here is a general, organized approach to graphing rational functions.

GRAPHING RATIONAL FUNCTIONS

Suppose that $f(x) = p(x)/q(x)$ is a rational function. To sketch the graph of $f(x)$:

1. Determine any common factor of the numerator and denominator. The zeros of this common factor are points at which the rational function is undefined. Replace the given expression of the rational function by one in lowest terms.

2. Determine the horizontal and oblique asymptotes using the procedure on page 207. If there are none, determine the behavior of $f(x)$ as x approaches both positive infinity and negative infinity.

3. Determine the vertical asymptotes by calculating the real zeros of the denominator. For each zero a, the line $x = a$ is a vertical asymptote.

4. Determine the zeros of $f(x)$ by determining the zeros of the numerator. These are the x-intercepts of the function.

5. Divide the x-axis into intervals determined by the real zeros of $f(x)$ and by the vertical asymptotes. For each interval, choose a test value and determine the sign of $f(x)$ in that interval. Then use the signs to determine whether the graph is above or below the x-axis in each interval.

6. Draw the horizontal and vertical asymptotes and plot the zeros.

7. Draw the portion of the graph corresponding to each interval using the information developed in steps 1–6.

The next example illustrates how to apply this general approach to sketching a graph with vertical asymptotes.

➤ **EXAMPLE 5**
Graph with Horizontal and Vertical Asymptotes

Determine all asymptotes of the rational function

$$f(x) = \frac{(x - 2)(x - 4)}{(x - 1)(x - 3)}$$

Sketch the graph of the function.

Solution

There is a vertical asymptote corresponding to each zero of the denominator. There are two such zeros, namely 1 and 3, so the vertical asymptotes are the lines $x = 1$ and $x = 3$. Because the numerator and denominator have the same degree, there is a horizontal asymptote. In finding the horizontal asymptote, use the expanded form of $f(x)$, that is,

$$f(x) = \frac{x^2 - 6x + 8}{x^2 - 4x + 3}$$

The leading coefficients of the numerator and denominator are each 1, so the horizontal asymptote is the line $y = \frac{1}{1} = 1$.

To sketch the graph, we must first determine the zeros of $f(x)$. These are just the zeros of the numerator. In this case, the numerator is factored for us, so we may read off the zeros directly, namely 2 and 4. The zeros and the vertical asymptotes determine a division of the x-axis into intervals:

$$(-\infty, 1), (1, 2), (2, 3), (3, 4), (4, +\infty)$$

In each of these intervals, the value of $f(x)$ has a constant sign. Let's choose a test point in each interval and determine the corresponding sign. The results can be summarized in the accompanying table.

We now turn to the graph itself. First we draw asymptotes and plot the zeros. We now plot the section of the graph lying in each interval.

For the leftmost interval, $(-\infty, 1)$, the graph approaches the horizontal asymptote $y = 1$ from above as x approaches negative infinity. On the right side of the interval, the graph must approach the vertical asymptote $x = 1$. The graph must head upward as it does so; otherwise, it would cross the x-axis within the interval and there are no zeros of $f(x)$ within the interval.

In the second interval, $(1, 2)$, the function value is negative. This means that the graph must head downward to the asymptote on the left. Moreover, the graph crosses the x-axis at $x = 2$. In the third interval, $(2, 3)$ the graph is positive. So the graph must increase in its approach to the asymptote $x = 3$ on the right.

In the fourth interval, $(3, 4)$, the function value is negative. This means that the graph must head downward to the asymptote on the left. Moreover, the graph crosses the x-axis at $x = 4$.

In the fifth interval, $(4, +\infty)$, the graph rises from the x-axis on the left and must approach the horizontal asymptote $y = 1$ from below as x approaches positive infinity.

The completed graph is sketched in Figure 10.

Interval	Test Point	Sign of $f(x)$
$(-\infty, 1)$	0	$+$
$(1, 2)$	$\frac{3}{2}$	$-$
$(2, 3)$	$\frac{5}{2}$	$+$
$(3, 4)$	$\frac{7}{2}$	$-$
$(4, +\infty)$	5	$+$

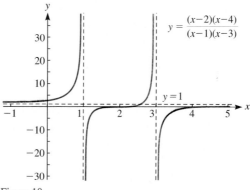

Figure 10

> **EXAMPLE 6**
Inventory Management

The cost $C(x)$ of managing an inventory of x cases of a certain brand of canned beans is determined by a team of financial analysts to be given by the function

$$C(x) = \frac{10,000}{x} + 3x$$

Sketch the graph of $C(x)$.

Solution

Because x represents a number of cases of canned goods, x must be positive, so the domain of the function $C(x)$ is the set of all positive integers. If we write the expression for $C(x)$ in the form

$$\frac{3x^2 + 10,000}{x}$$

we see that $C(x)$ is a rational function. The only zero of the denominator is 0, so there is a single vertical asymptote, namely, $x = 0$. Moreover, because the degree of the numerator is one more than the degree of the denominator, there is an oblique asymptote and no horizontal asymptotes. If we write

$$C(x) = 3x + \frac{10,000}{x}$$

then we see that the line $y = 3x$ is an oblique asymptote.

We now sketch the graph beginning with the single vertical asymptote $x = 0$. Because $C(x)$ is positive throughout the domain, the graph must rise in approaching the asymptote from the right. We have drawn the oblique asymptote, and show the graph approaching this asymptote as x approaches positive infinity. The completed sketch is shown in Figure 11.

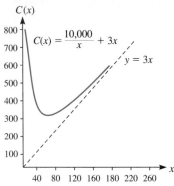

Figure 11

In case the numerator and denominator of a rational function have a common factor, the rational function will not be defined at the zeros of the common factor, and the graph will have "holes" in it corresponding to the values of x at which the function is undefined. The next example illustrates this phenomenon.

> **EXAMPLE 7**
Rational Function
with Cancellation

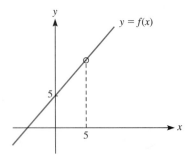

Figure 12

Sketch the graph of the function

$$f(x) = \frac{x^2 - 25}{x - 5}$$

Solution
Because

$$\frac{x^2 - 25}{x - 5} = \frac{(x + 5)(x - 5)}{x - 5}$$

we see that $f(x)$ has the equivalent expression $f(x) = x + 5$ as long as x is not equal to a zero of the common factor $x - 5$. That is, we have

$$f(x) = x + 5, \qquad x \neq 5$$

The graph of this function is shown in Figure 12.

Graphing Rational Functions Using Graphing Technology

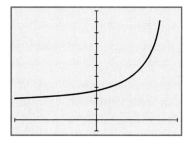

Figure 13

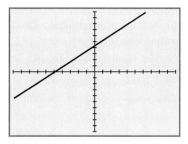

Figure 14

If you are using a graphing utility, you have probably seen that even very complicated rational functions can be graphed easily. You may then wonder why it is important to be able to sketch these graphs by hand or to understand the algebra if the technology will do all of the graphing for you. The answer to this is that sometimes a graph on a calculator may not look anything like the graph of the function.

For instance, look at Figure 13. This is a graph of

$$f(x) = \frac{(x-2)(x-4)}{(x-1)(x-3)}$$

which was sketched in Figure 10. The two graphs do not look similar at all. Why? Figure 13 was graphed only in the range $-1 \le x \le 1$. Without a knowledge of asymptotes and rational functions, you might be fooled into thinking that this is what the graph looks like. This is one danger involved in using a graphing utility.

Rational functions provide another example of how you could get a misleading graph from a graphing utility. Figure 14 was produced on a TI-81 graphing calculator. It is the graph of

$$f(x) = \frac{x^2 - 25}{x - 5}$$

Again notice that this is not the same graph that was shown in Figure 12 in Example 7. The reason is that the graphing calculator can only evaluate the function at a certain number of points, and it uses these points to plot the graph. It happened that $x = 5$ was not one of the points evaluated, so the graph does not show the hole that we know is there. Even using a graphing utility's zoom feature does not guarantee that it will show these holes. Figure 15 shows the same function graph in the range $4.95 \le x \le 5.05$. Even this close-up view does not show the hole. Without some understanding of the concepts involved, the technology could mislead us.

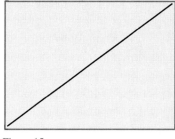

Figure 15   ◄

EXAMPLE 8
Graphing a Rational Function Using a Calculator

Determine the asymptotes of the function $f(x) = \dfrac{x+1}{x^2 - x - 4}$.

Solution

Graph the function using the standard range $-10 \le x \le 10$. See Figure 16. we see that there is a single horizontal asymptote, namely $y = 0$, and two vertical asymptotes. Using the trace and observing the value of the x-coordinate, we determine that the vertical asymptotes are $x = a$ and $x = b$, where $a \approx -1.55$, $b \approx 2.68$.

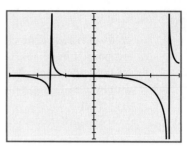

Figure 16   ◀

New Technology

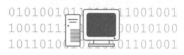

EXAMPLE 9
An Optimization Problem Involving a Rational Function

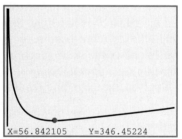

X=56.842105   Y=346.45224

Figure 17

In Example 6, we graphed the function

$$C(x) = \frac{10,000}{x} + 3x$$

which gives the cost of managing an inventory of x cases of canned beans. We saw that this function had a single minimum point. Approximate the minimum value of this function.

Solution

First, use a graphing calculator to graph the function within the range $0 \le x \le 200$, $0 \le y \le 3000$. See Figure 17. Use the trace to locate this point. Move from left to right and locate the point at which the y-coordinate stops decreasing. This point is approximately $(56.84, 346.45)$. Therefore, the minimum value of the function is approximately 346.45.

                                                                                              ◀

Exercises 3.7

For each function:

a. Determine all the asymptotes.

b. Sketch the graph of the function.

1. $f(x) = \dfrac{2}{x^2}$

2. $f(x) = \dfrac{1}{x^3}$

3. $f(x) = -\dfrac{1}{x^3}$

4. $f(x) = -\dfrac{3}{x^2}$

5. $f(x) = \dfrac{1}{x + 1}$

6. $f(x) = \dfrac{1}{x - 1}$

7. $f(x) = \dfrac{-2}{x + 2}$

8. $f(x) = \dfrac{-2}{(x - 3)^2}$

9. $f(x) = \dfrac{1}{(x + 3)^2}$

10. $f(x) = \dfrac{x + 1}{2 - x}$

11. $f(x) = \dfrac{3 - x}{x + 2}$

12. $f(x) = \dfrac{2x + 3}{16 - 5x}$

13. $f(x) = \dfrac{8 - 2x}{10 + 3x}$

14. $f(x) = \dfrac{1}{x^2 + 1}$

15. $f(x) = -\dfrac{1}{x^2 + 3}$

16. $f(x) = \dfrac{1}{x^2 - 4}$

17. $f(x) = -\dfrac{1}{x^2 - 9}$

18. $f(x) = \dfrac{2x}{x^2 - x - 6}$

19. $f(x) = \dfrac{x - 3}{2x^2 + 5x - 3}$

20. $f(x) = \dfrac{x^2 - 9}{x + 2}$

21. $f(x) = \dfrac{x^2 - 4}{x - 3}$

22. $f(x) = \dfrac{x^2 - 1}{x^2 + x - 6}$

23. $f(x) = \dfrac{x^2 - 9}{x^2 - x - 2}$

24. $f(x) = \dfrac{x^2 - x - 2}{x + 2}$

25. $f(x) = \dfrac{x^2 - 2x - 8}{x + 4}$

26. $f(x) = x + \dfrac{4}{x}$

27. $f(x) = x + \dfrac{1}{x}$

28. $f(x) = \dfrac{2x^2 - x - 3}{3x^2 - 2x - 8}$

29. $f(x) = \dfrac{x^2 + 2x - 3}{3x^2 - 1}$

30. $f(x) = \dfrac{x^3 + 1}{x}$

31. $f(x) = \dfrac{x^3 - 2}{4x}$

32. $f(x) = \dfrac{x^2 - 1}{x - 1}$

33. $f(x) = \dfrac{x^2 - 9}{x + 3}$

34. $f(x) = \dfrac{(x + 2)(x^2 - 9)}{x + 2}$

35. $f(x) = \dfrac{(x - 1)(3 - x)^2}{x - 1}$

36. $f(x) = \dfrac{x^2 + 2x - 15}{x + 5}$

37. $f(x) = \dfrac{x^2 + x - 2}{x - 1}$

38. $f(x) = \dfrac{x^2 - 4x + 3}{x^2}$

39. $f(x) = \dfrac{4x^2}{x^2 - 1}$

40. $f(x) = \dfrac{x^3 - 2x^2 - 8x}{x^2 - 9}$

41. $f(x) = \dfrac{x^3 + 2x^2 - 3x}{x^2 - 25}$

42. $f(x) = \dfrac{x^3 - 4x^2 + x + 6}{x^2 + x - 2}$

♦ **Applications**

43. **Earned run average**. The earned run average, ERA, of a pitcher is given by ERA $= 9R/I$, where $R =$ the number of earned runs allowed and $I =$ the number of innings pitched.

   a. Suppose $R = 1$. Graph the function over the interval $(0, 20]$.

   b. Suppose $R = 2$. Graph the function over the interval $(0, 20]$.

   c. Suppose $R = 5$. Graph the function over the interval $(0, 20]$.

44. **Distance as a function of time.** The distance from Dallas, Texas, to El Paso, Texas, is 617 miles. A car makes the trip in time t, in hours, at various speeds r, in miles per hour.

   a. Find a rational function for t in terms of r.

   b. Graph the function.

45. **Loudness of sound.** A person is sitting at a distance d from a stereo speaker. (See figure.) The loudness L of

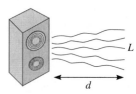

Exercise 45

the sound varies inversely as the square of the distance d according to the equation

$$L = \frac{k}{d^2}$$

where k depends on the original intensity and the units used to measure d and L. Graph this function assuming $k = 0.1$.

46. **Inventory management.** The cost $C(x)$ of managing an inventory of x television sets is determined by a team of financial analysts to be given by the function

$$C(x) = 5x + \frac{50,000}{x} + 22,500$$

Sketch the graph of this function over the interval $(0, +\infty)$.

47. **Weight above the earth.** A person's weight W_h at a height h above sea level satisfies the equation

$$W_h = \left(\frac{r}{r + h}\right)^2 W_0$$

where W_0 is the person's weight at sea level; h is the height, in miles, of the person above sea level; and r is the radius of the earth, in miles. Assume that $r \approx 4000$ miles.

   a. Suppose a person weighs 200 lb at sea level. Find a rational function for the person's weight h miles above the earth.

   b. Graph this function on the interval $[0, 16,000]$.

   c. At what height would the person weigh 100 lb?

48. **Optics.** Consider the accompanying figure. F is the focal length of the lens in mm, where an object u meters from the lens of a camera has an image v millimeters focused on the film from the lens. A law of optics asserts that $1/F = 1/u + 1/v$. Suppose the focal length is 50 mm.

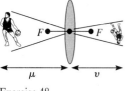

Exercise 48

   a. How far is the lens from the film if the camera is focused on an object that is 6 meters away from the camera?

   b. Let v be a dependent variable. Find a rational function for v in terms of u.

   c. Graph v.

Graph.

49. $f(x) = \left| \dfrac{1}{x - 1} \right|$

50. $f(x) = \dfrac{2x}{|x^2 - 2x + 15|}$

51. $f(x) = \dfrac{x^4 - 3x^3 - 21x^2 + 43x + 60}{x^4 - 6x^3 + x^2 + 24x - 20}$

Graph using the same set of axes. Try to discover a pattern for graphing a function and its reciprocal.

52. $f(x) = x^2$ and $g(x) = \dfrac{1}{x^2}$

53. $f(x) = x^2 - 4$ and $g(x) = \dfrac{1}{x^2 - 4}$

54. $f(x) = \dfrac{x^2}{x^2 - 1}$ and $g(x) = \dfrac{x^2 - 1}{x^2}$

55. Sketch the general shape of the graph of $y = 1/x^n$ for any even natural number.

56. Sketch the general shape of the graph of $y = 1/x^n$ for any odd natural number.

➡ **Technology**

Sketch a graph of each of the following. Use computer graphing software, a calculator that sketches graphs, or just a computer or calculator to find function values.

57. $f(x) = \dfrac{1}{x^{26} + 4} - 3$

58. $f(x) = (x + 1)^{2/3}$

59. $f(x) = (x^2 - 1)^{2/3}$

60. $f(x) = \dfrac{x^{28} - x^2 + 2}{x^{28} - x + 11}$

Match the following functions with graphs (a)–(f).

61. $f(x) = \dfrac{x}{(x - 1)(x + 2)}$

62. $f(x) = \dfrac{1}{x + 3}$

63. $f(x) = \dfrac{x^2 + 1}{x}$

64. $f(x) = \dfrac{1}{x(x + 1)(x + 2)}$

65. $f(x) = \dfrac{x^2}{x^2 + 1}$

66. $f(x) = \dfrac{1}{x^2 + 1}$

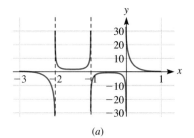

(a)

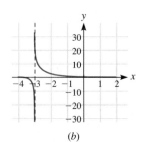

(b)

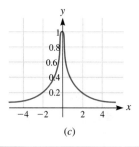

(c)

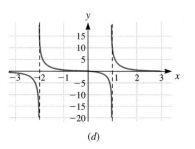

(d)

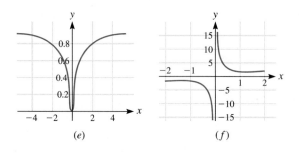

(e)   (f)

The graphs of rational functions $f(x)$ and $g(x)$ are shown in graphs (a) and (b). Sketch the graphs of the following functions.

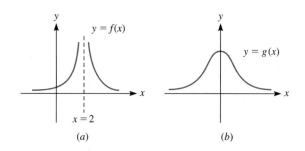

(a)   (b)

67. $f(x + 1)$

68. $f(x) + 3$

69. $f(x - 2) + 3$

70. $g(x + 1) + 1$

71. $g(x - 4)$

72. $g(x) - 2$

✐ **In Your Own Words**

73. Write a detailed procedure for determining the vertical asymptotes of a rational function.

74. Write a detailed procedure for determining the horizontal asymptotes of a rational function.

75. Give an applied example of a rational function with a horizontal asymptote. Explain the applied significance of the horizontal asymptote.

76. Give an applied example of a rational function with a vertical asymptote. Explain the applied significance of the vertical asymptote.

3.8 CHAPTER REVIEW

Important Concepts, Properties, and Formulas—Chapter 3

Polynomial function	$f(x) = a_n x^n + a_{n-1} x^{n-1} + \cdots + a_0$	p. 146
Quadratic function	$f(x) = ax^2 + bx + c, \quad a \neq 0$	p. 146
Standard form for a quadratic function	$f(x) = a(x - h)^2 + k$, where $$h = -\frac{b}{2a}, \qquad k = \frac{4ac - b^2}{4a}$$ and where the extreme point (h, k) is maximum if $a < 0$ and minimum if $a > 0$.	p. 148
Descartes' rule of signs	Number of positive real zeros is at most the number of sign changes m of $f(x)$ and differs from m by a multiple of two. The number of negative real zeros is at most the number of sign changes n of $f(-x)$ and differs from n by a multiple of 2.	p. 190
Asymptotes of a rational function	If $$f(x) = \frac{p(x)}{q(x)}$$ where $p(x)$ is of degree m and leading coefficient c, and $q(x)$ is a polynomial of degree n and leading coefficient d, then: 1. If $m < n$, then the x-axis is a horizontal asymptote of the graph of $f(x)$, and $f(x)$ approaches 0 as x approaches either positive infinity or negative infinity. 2. If $m = n$, then the line $y = c/d$ is a horizontal asymptote of the graph of $f(x)$, and $f(x)$ approaches c/d as x approaches either positive infinity or negative infinity. 3. If $m = n + 1$, then we may write $f(x)$ in the form $$f(x) = ax + b + \frac{r(x)}{q(x)}$$ where $r(x)$ has degree less than n. In this case, $y = ax + b$ is an oblique asymptote of the graph of $f(x)$. 4. If $m > n$, then the graph of $f(x)$ has no horizontal asymptotes. Moreover, as x approaches positive infinity or negative infinity, $f(x)$ approaches either positive infinity or negative infinity. (It is simplest to separately determine which it approaches in each instance.)	p. 207

Graphing rational functions	1. Determine any common factor of the numerator and denominator. The zeros of this common factor are points at which the rational function is undefined. Replace the given expression of the rational function by one in lowest terms.	p. 208
	2. Determine the horizontal and oblique asymptotes. If there are none, determine the behavior of $f(x)$ as x approaches both positive infinity and negative infinity.	
	3. Determine the vertical asymptotes by calculating the real zeros of the denominator. For each zero a, the line $x = a$ is a vertical asymptote.	
	4. Determine the zeros of $f(x)$ by determining the zeros of the numerator. These are the x-intercepts of the function.	
	5. Divide the x-axis into intervals determined by the real zeros of $f(x)$ and by the vertical asymptotes. For each interval, choose a test value and determine the sign of $f(x)$ in that interval. Then use the signs to determine whether the graph is above or below the x-axis in each interval.	
	6. Draw the horizontal and vertical asymptotes and plot the zeros.	
	7. Draw the portion of the graph corresponding to each interval using the information developed in steps 1–6.	

Cumulative Review Exercises—Chapter 3

For each of the following quadratic functions:

a. Determine the standard form.

b. Find the vertex.

c. Determine whether the graph opens upward or downward.

d. Find the maximum or minimum value.

e. Graph the function.

1. $f(x) = x^2 - 3x$
2. $f(x) = 2(x - 1)^2 + 4$
3. $f(x) = 2x^2 + 12x + 2$
4. $f(x) = \frac{1}{2}x^2 + 2x + 5$
5. $f(x) = -\frac{1}{3}x^2 + \frac{2}{5}x - \frac{4}{5}$

Find the maximum or minimum value.

6. $f(x) = 2x^2 + 17x - 35$
7. $f(x) = \frac{1}{4}x^2 - 20x + \frac{2}{3}$
8. $f(x) = 23.45x^2 - 34.67x - 1$
9. $f(x) = -\sqrt{2}x^2 + \sqrt{6}x + \sqrt{8}$

10. Find two numbers whose sum is 31 and whose product is a maximum.

11. The Metro Bus Company charges a person 20 cents to ride the bus. They now carry 8000 passengers per day.

They want to raise the fare, but for every 5-cent raise in the fare they will lose 800 passengers. What fare should they charge to maximize passenger income?

12. **Total profit.** A small business is marketing a new tape recorder. It determines that its total revenue from the sale of x tape recorders is given by the function

$$R(x) = 44x - x^2$$

The firm also determines that its total cost of producing x tape recorders is given by

$$C(x) = 700 - 36x$$

Total profit is given by

$$P(x) = R(x) - C(x)$$

a. Find the total profit function $P(x)$.

b. How many tape recorders must the company produce and sell to maximize profit?

c. What is the maximum profit?

Sketch graphs of each of the following functions.

13. $f(x) = (x^2 - 1)(x^2 - 4)$

14. $f(x) = 4x^2 - x^3$

15. $f(x) = (x + 4)(x - 1)(x - 3)$

16. $f(x) = -(x - 2)^3 - 3$ 17. $f(x) = (x + 3)^2(x - 2)$

18. $f(x) = 2x^4 + 8$ 19. $f(x) = -\dfrac{2}{x^2}$

20. $f(x) = \dfrac{1}{x + 1}$ 21. $f(x) = \dfrac{8}{x^2 - 4}$

22. $f(x) = \dfrac{x^2 + 2x - 3}{x^2 - x - 2}$ 23. $f(x) = 2x + \dfrac{2}{x}$

24. $f(x) = \dfrac{1}{x^2 + 2}$

25. $f(x) = \dfrac{x^3 + 12x^2 + 4x + 48}{x^3 + 2x^2 - 3x}$

26. $f(x) = \dfrac{x^2 - 5}{x - 4}$

27. $f(x) = \dfrac{(x - 2)(x^2 - 1)}{x - 2}$

28. $f(x) = \dfrac{x^2 - 4}{x - 2}$

29. $f(x) = \dfrac{x^3 + 2}{x}$

30. $f(x) = \dfrac{x^3 + x^2 - 6x}{x^2 + x - 2}$

In Exercises 31 and 32, a polynomial $f(x)$ and a divisor $g(x)$ are given. Calculate the quotient $q(x)$ and the remainder $r(x)$.

31. $f(x) = x^5 + 3x^4 - 2x^2 + 6x - 3$

   $g(x) = x - 1$

32. $f(x) = x^5 + 3x^4 - 2x^2 + 6x - 3$

   $g(x) = x^3 - x^2 + 4$

33. Suppose that $f(x)$ is a quadratic polynomial with zeros $\frac{1}{2}$ and $-\frac{2}{3}$, and $f(-1) = 5$. Determine f.

34. Suppose that $f(x)$ is a quadratic polynomial with zeros $-2 + \sqrt{3}$ and $-2 - \sqrt{3}$, and $f(-1) = 5$. Determine f.

35. Suppose that $f(x)$ is a quadratic polynomial with zeros

$$\frac{1 + 3i}{2} \quad \text{and} \quad \frac{1 - 3i}{2}$$

and $f(-1) = 5$. Determine f.

36. Suppose that $f(x)$ is a fourth-degree polynomial with zeros $1 + \sqrt{2}, 1 - \sqrt{2}, 3$, and -2; and $f(0) = 6$. Determine f.

37. Suppose that $f(x)$ is a fifth-degree polynomial with zeros $2i, -2i, 3$, and -1 with muliplicity 2; and $f(1) = -6$. Determine f.

Use synthetic division to find the quotient and remainder when $f(x)$ is divided by $g(x)$.

38. $f(x) = 4x^5 + 6x + 1$

   $g(x) = x + 2$

39. $f(x) = x^4 - 2x^3 - 5x^2 - x + 1$

   $g(x) = x + 1$

40. $f(x) = \dfrac{1}{3}x^2 + \dfrac{5}{6}x - 3$

   $g(x) = 2x + 9$

41. $f(x) = x^4 - 2x^3 - 2x - 1$

   $g(x) = x + i$

Use synthetic division to determine the given function values.

42. $f(x) = x^4 - 2x^3 - 2x - 1$; find $f(0)$, $f(-1)$, and $f(i)$.

43. $f(x) = x^5 + x^4 - 2x^3 - 2x^2 + x - 1$; find $f(-1)$, $f(2)$, and $f(-i)$.

Use synthetic division to determine whether each number is a zero of the given polynomial.

44. $2, 3, -2; f(x) = x^3 - 9x^2 + 23x - 15$

45. $-1, i, 2; f(x) = x^4 - x^3 - 7x^2 - 7x - 2$

Factor each polynomial $f(x)$. Then use the result of the factoring to solve the equation $f(x) = 0$.

46. $f(x) = x^3 - 100$

47. $f(x) = x^3 - 6x^2 + 3x + 10$

For each polynomial, determine the zeros and their multiplicities.

48. $f(x) = 5x^3(x - 4)^2(2x + 3)$

49. $f(x) = (3x^3 - 6x^2 + 3x)^2$

50. $f(x) = (x^2 + 1)^2(x^2 - 1)^3$

51. How many real zeros can a polynomial of degree 5 have?

52. How many real zeros can a polynomial of degree 6 have?

Use Descartes' Rule of Signs to analyze the zeros of each polynomial.

53. $6x^4 - 30x^3 - 32x^2 - 21$

54. $3x^6 - 5x^4 + 2x^3 - x - 4$

55. $8x^5 + 3x^4 - x^3 + 5x^2 - 3$

56. Find k so that the first polynomial is a factor of the second.

$$x - 2; \qquad 2x^3 + 3x^2 + kx + 10$$

For each polynomial, certain zeros are given. Find the remaining zeros.

57. $f(x) = 2x^4 + 4x^3 + 3x^2 - 6x - 9; -1 - i\sqrt{2}$

58. $f(x) = x^3 - x^2 - 7x + 3; \ 3$

59. $f(x) = x^4 + 2x^3 + x^2 + 60x + 144; -3$ is a root of multiplicity 2.

60. $f(x) = x^4 - 4x^3 + 18x^2 + 8x - 40; 2 + 4i$

61. Find all possible rational zeros of

$$4x^6 - 5x^5 + 6x^4 - 20x^3 - 10x^2 + 6$$

Determine upper and lower bounds for the real zeros of $f(x)$.

62. $f(x) = 6x^4 - 30x^3 - 32x^2 - 23x + 2$

63. $f(x) = 3x^6 - 5x^4 + 2x^3 - x - 4$

64. $f(x) = 8x^5 + 3x^4 - x^3 + 5x^2 - 3$

Find only the rational zeros.

65. $6x^4 - 30x^3 - 32x^2 - 23x + 2$

66. $3x^6 - 5x^4 + 2x^3 - x - 4$

67. $8x^5 + 3x^4 - x^3 + 5x^2 - 3$

68. $x^5 - 2x^4 + x^3 - 1$

Solve. Use any procedures studied in this chapter.

69. $(x - 1)(x^2 + x + 1) = 0$

70. $x^3 + 31x = 10x^2 + 30$

71. $5x^2 = x^3 + 5x + 3$

72. $12x^6 - 52x^5 + 91x^4 - 82x^3 + 40x^2 - 10x + 1 = 0$

73. $x^3 + 2 = 2x^2 + 7x + 2$

74. $x^4 + x^2 + 1 = 0$

75. $2x^4 - 3x^3 + x^2 + 4x - 2 = 0$

Show that each of the following polynomials has a zero in the given interval.

76. $f(x) = x^3 - 2x - 5; (2, 3)$

77. $f(x) = x^4 - x^3 - 2x^2 - 6x - 4; (-1, 0)$

Approximate the real zeros of each of the following polynomial functions with an accuracy of 0.0001 using the method of bisection.

78. $f(x) = x^3 - 2x - 5; (2, 3)$

79. $f(x) = x^4 - x^3 - 2x^2 - 6x - 4; (-1, 0)$

80. $f(x) = x^5 - x + 1; (-2, 0)$

◆ Applications

81. An open box of volume 132 cubic inches can be made from a piece of cardboard that is 10 in. by 15 in. by cutting a square from each corner and folding up the sides. What is the length of a side of the squares? Use the method of bisection to find an approximate answer, if appropriate.

82. **Electrical resistance.** Suppose three resistors are connected in parallel, as shown in the figure.

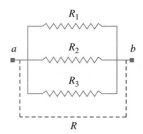

Exercise 82

The result of the three resistances is an equivalent resistance R, given by

$$\frac{1}{R} = \frac{1}{R_1} + \frac{1}{R_2} + \frac{1}{R_3}$$

Suppose $R = 1$, the second resistance is two more than the first, and the third resistance is three more than the first (see figure).

a. Find a polynomial equation in R_1 that can be used to find the resistances.

b. Find each of the three resistances. (Resistance is always a nonnegative unit.) Use the method of bisection if appropriate.

83. **Inventory management.** The cost $C(x)$ of managing an inventory of x pool tables is determined by a team of financial analysts to be given by the function

$$C(x) = 10x + \frac{4000}{x} + 1600$$

Sketch the graph of this function over the interval $(0, +\infty)$.

84. There are many rules for determining the dosage of a medication for a child based on an adult dosage. The following are two such rules:

$$\text{Young's rule:} \quad C = \frac{A}{A + 12}D$$

$$\text{Cowling's rule:} \quad C = \frac{A + 1}{24}D$$

where A = the age of the child, in years; D = adult dosage; and C = child's dosage. (*Warning:* Do not apply these formulas without consulting a physician. These are estimates and may not apply to a particular medication.) An adult dosage of a liquid medication is 5 cc (cubic centimeters).

a. Sketch a graph of Young's rule for the value of A in the interval [0, 18].

b. Sketch a graph of Cowling's rule for values of A in the interval [0, 18].

c. For what ages A are the child's dosages the same?

Sketch a graph of each function and find the x-intercepts. Approximate where appropriate.

85. $f(x) = 3x^3 - 16x^2 + 12x + 16$

86. $f(x) = x^2 + \dfrac{1}{x^2}$

87. $f(x) = \dfrac{1}{x^2 - 1}$

88. $f(x) = \dfrac{2x^2}{x^2 - 16}$

89. $f(x) = \dfrac{x}{x - 3}$

90. Find a such that

$$f(x) = ax^2 - \tfrac{2}{3}x + 7$$

has a maximum value of 85.

91. Find c such that $f(x) = \tfrac{8}{7}x^2 - \tfrac{3}{4}x + c$ has a minimum value of 226.

92. Prove that $\sqrt{3}$ is irrational by considering the zeros of the polynomial function $f(x) = x^2 - 3$.

➡ **Technology**

93. Use computer software or a graphing calculator to sketch a graph of the following function. Then approximate the real zeros.

$$f(x) = 3x - (x - 1)^{3/2}, \qquad (x \geq 1)$$

CHAPTER 4

EXPONENTIAL AND LOGARITHMIC FUNCTIONS

*The spiral exhibited in this cross section of a nautilus is a curve called a **logarithmic spiral** and is an illustration of the natural logarithmic function appearing in nature.*

A pplied mathematics uses various functions that describe common situations. In this chapter, we add to our repertoire of functions the exponential and logarithmic functions, which are necessary for describing many situations you may encounter in fields such as radioactive decay, bacterial growth, hearing, and the transmission of light to ocean depths.

4.1 EXPONENTIAL FUNCTIONS WITH BASE *a*

Suppose that a is a positive number. We have already defined the power a^x where x is a rational number. This suffices to define a^x where x is a finite decimal (because such a decimal represents a rational number).

In many applications, it is necessary to consider a^x where x is an irrational number, such as $x = \pi$ or $x = \sqrt{2}$. Such powers can be defined using a sequence of rational approximations to x. For example, to define the power $3^{\sqrt{2}}$, we approximate $\sqrt{2}$ by the sequence of decimal values

$$1.4, \ 1.41, \ 1.414, \ 1.4142, \ \ldots$$

Corresponding to this sequence of approximations, we have the following approximations of $3^{\sqrt{2}}$:

$$3^{1.4} = 4.655536\ldots$$
$$3^{1.41} = 4.706965\ldots$$
$$3^{1.414} = 4.727695\ldots$$
$$3^{1.4142} = 4.728733\ldots$$

On the basis of these approximations, we see that $3^{\sqrt{2}}$ is approximately equal to 4.73. In fact, using more precise approximations of the value of the exponent, we get ever closer approximations of the value of

$$3^{\sqrt{2}} = 4.728804\ldots$$

Using similar approximations, we can define the value of a^x for any real number x.

Calculating a^x

To calculate a^x, it is most convenient to use a scientific calculator. Such calculators have a key, usually labeled $\boxed{x^y}$, for calculating powers. To calculate a^x, first enter the value of a. Then press the $\boxed{x^y}$ key. Next enter the value of x. Finally, press the $\boxed{=}$ key. These instructions will work for a calculator using algebraic logic. For example, to calculate

$$2.57^{3.4875}$$

first key in the number 2.57. Then press the $\boxed{x^y}$ key. Next, key in the exponent 3.4875. Finally, press the $\boxed{=}$ key.

Some calculators, most notably Hewlett-Packard calculators, use "reverse Polish logic." For such calculators, a different order of data entry must be used: First enter the value of a and press the $\boxed{\text{ENTER}}$ key. Next enter the value of x. Finally, press the $\boxed{x^y}$ key. No matter what type of calculator you have, the approximate value you get is 26.893.

Laws of Exponents

It is possible to prove that the laws of exponents, introduced in Chapter 1 for rational exponents, continue to hold for real exponents. In fact, we can state the following result.

LAWS OF REAL EXPONENTS

Let a and b be positive real numbers, and let x and y be any real numbers. Then the following laws of exponents hold:

1. $a^x a^y = a^{x+y}$
2. $(a^x)^y = a^{xy}$
3. $(ab)^x = a^x b^x$
4. $\dfrac{a^x}{a^y} = a^{x-y}$

Definition 1
Exponential Function
with Base a

Let a be a positive real number. Then the **exponential function with base a** is the function defined by the formula

$$f(x) = a^x, \quad a \neq 1$$

Note that we excluded the possibility $a = 1$ so that the constant function $f(x) = 1$ would not be considered as an exponential function.

Here are some examples of exponential functions with various bases:

$$f(x) = 2^x$$
$$g(x) = 3^x$$
$$h(x) = \sqrt{5}^{-x}$$
$$k(x) = \left(\tfrac{1}{2}\right)^x$$

Because a^x is defined for every real number x, the domain of an exponential function is the set **R** of real numbers. Moreover, a^x is positive, so the range of an exponential function is contained in the set

$$\{y : y \text{ is positive}\}$$

Actually, it can be proved that the range of an exponential function is precisely the set of all positive numbers. We will assume this fact throughout the text.

➤ **EXAMPLE 1**
Calculating Exponentials

Let $f(x) = 2^x$ be the exponential function with base 2. Calculate the following:

1. $f(1)$ 2. $f(2)$ 3. $f(3)$ 4. $f(-1)$
5. $f(-2)$ 6. $f(0)$ 7. $f\left(\tfrac{1}{2}\right)$

Solution

1. $f(1) = 2^1 = 2$
2. $f(2) = 2^2 = 4$
3. $f(3) = 2^3 = 8$
4. $f(-1) = 2^{-1} = \dfrac{1}{2^1} = \dfrac{1}{2}$
5. $f(-2) = 2^{-2} = \dfrac{1}{2^2} = \dfrac{1}{4}$

6. $f(0) = 2^0 = 1$

7. $f\left(\frac{1}{2}\right) = 2^{1/2} = \sqrt{2}$

Graphs of Exponential Functions

As we have seen in the case of polynomial and rational functions, the graph of a function provides valuable information about the function (such as where it increases or decreases, its zeros, and its asymptotes). With this in mind, let's learn to sketch the graphs of exponential functions. The following example discusses the graph of a particular exponential function.

➤ **EXAMPLE 2**
Graphing an Exponential

Sketch the graph of the function $f(x) = 2^x$.

Solution

We may record the results of Example 1 in a table.

x	$f(x)$
-2	$\frac{1}{4}$
-1	$\frac{1}{2}$
0	1
$\frac{1}{2}$	$\sqrt{2}$
1	2
2	4
3	8

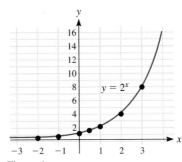

Figure 1

In Figure 1, we plot the points listed in the table and connect them with a smooth curve. Note that as x increases, the graph rises very sharply. As x approaches negative infinity, the graph has the negative x-axis as an asymptote.

For $a > 1$, the graph of $f(x) = a^x$ has the same general shape as the graph in Figure 1. As x increases, so does the value of $f(x)$. Moreover, as x approaches negative infinity, the graph approaches the negative x-axis as an asymptote. Furthermore, the larger the value of a, the steeper the rate at which the graph rises. In Figure 2, we have drawn the graphs of several exponential functions corresponding to assorted values of a that are larger than 1.

In the next example, we will consider a particular exponential function for which $0 < a < 1$.

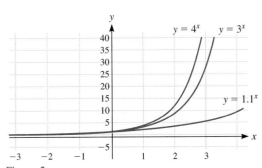

Figure 2

➢ **EXAMPLE 3**
Graphing a Decreasing
Exponential

x	$f(x)$
-3	8
-2	4
-1	2
0	1
1	$\frac{1}{2}$
2	$\frac{1}{4}$
3	$\frac{1}{8}$

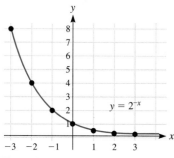

Figure 3

Sketch the graph of the exponential function
$$f(x) = \left(\tfrac{1}{2}\right)^x$$

Solution

To get a sense of the shape of the graph, we plot points corresponding to some representative values of x. The calculations require the use of the laws of real exponents. For instance, we have
$$f(-1) = \left(\frac{1}{2}\right)^{-1} = \frac{1}{\frac{1}{2}} = 2$$

The accompanying table summarizes various values of the function. You should verify these values as an exercise.

In Figure 3, we have plotted the points corresponding to the table entries and drawn a smooth curve through the points. Note that as x increases, the graph of $f(x)$ decreases. Moreover, as x approaches positive infinity, the graph approaches the positive x-axis as an asymptote. As x decreases and approaches negative infinity, the function values increase without bound and approach infinity.

For $0 < a < 1$, the graph of $f(x) = a^x$ has the same general shape as the graph shown in Figure 3. In Figure 4, we have sketched the graphs of several exponential functions with base values a that are less than 1 and greater than 0. Note that the smaller the value of a, the steeper the graph.

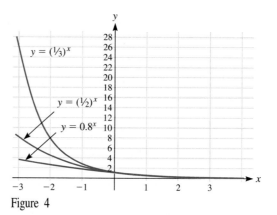

Figure 4

**Applications of the
Exponential Functions**

Exponential functions abound in applications such as radioactive decay, bacteria growth, and the spread of epidemics. Such functions typically arise, multiplied by a constant, in the form
$$f(x) = Aa^x$$
where A is the constant. The following examples illustrate some of the instances in which such functions arise.

➢ **EXAMPLE 4**
Compound Interest

Suppose that P dollars are deposited in a bank account paying annual interest at a rate r and compounded n times per year. After a length of time t, in years, the amount $A(t)$ in the account is given by the formula
$$A(t) = P\left(1 + \frac{r}{n}\right)^{nt}$$

Suppose that the interest rate is 10 percent, the amount is compounded annually, and the initial deposit is $10,000.

1. Write the function $A(t)$ in terms of an exponential function.
2. Determine the amount in the account after 1, 2, and 10 years.
3. Graph the function $A(t)$.

Solution

1. In this example, we have

$$P = 10,000, \qquad r = 0.1, \qquad n = 1$$

In this case, the compound interest formula reads

$$A(t) = 10,000 \cdot \left(1 + \frac{0.1}{1}\right)^{1 \cdot t}$$

$$A(t) = 10,000 \cdot (1.1)^t$$

2. The desired amounts are given, respectively, by the function values $A(1)$, $A(2)$, and $A(10)$. Computing these values, we have

$$A(1) = 10,000 \cdot (1.1)^1 = \$11,000.00$$
$$A(2) = 10,000 \cdot (1.1)^2 = \$12,100.00$$
$$A(10) = 10,000 \cdot (1.1)^{10} = \$25,937.43$$

To calculate these values, we used a calculator.

3. We know the general shape of the graph of the exponential function $f(t) = 1.1^t$ from Figure 2. The graph of $A(t)$ is obtained by scaling the graph of $f(t)$ by a factor of 10,000. From part 2, we have several points that help calibrate the graph. The graph must pass through these points and increase rapidly with increasing values of t. From the context of the application, only nonnegative values of t are relevant, so we limit the domain of the graph to $t > 0$. A sketch of the graph is given in Figure 5.

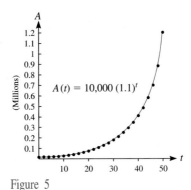

$A(t) = 10,000 (1.1)^t$

Figure 5

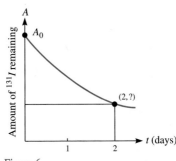

Figure 6

➤ EXAMPLE 5
Radioactive Decay

The decay of radioactive materials is described by an exponential function. Suppose that A_0 denotes the amount of a radioactive substance present at time 0, and $Q(t)$ represents the quantity present at time t. As time passes, the amount of radioactive material decays, with the amount approaching 0 as time increases. Suppose H denotes the time it takes for a quantity of the radioactive material to decay to half its original amount. The number H depends on the particular radioactive material and is called the **half-life** of the material. The function $Q(t)$ is then given by the expression

$$Q(t) = A_0 2^{-t/H}$$

Radioactive iodine (^{131}I) is used in various medical diagnostic tests. Its half-life is 8 days.

1. Write the formula for $Q(t)$ for radioactive iodine in terms of an exponential function.
2. What proportion of the original radioactive iodine is present 48 hours after it is swallowed? (See Figure 6.)

Solution

1. We are given that $H = 8$ days. Because we are not given the original amount of radioactive iodine, we leave it in the formula as A_0. We then have

$$Q(t) = A_0 2^{-t/8}$$
$$= A_0 2^{-(1/8)\cdot t}$$
$$= A_0 (2^{-1/8})^t$$
$$= A_0 0.917^t$$

The function $Q(t)$ is a constant, A_0, times an exponential function, a^t, where a is equal to 0.917.

2. When t equals 48 hours $= 2$ days, the amount of radioactive iodine remaining is $Q(2)$, which is equal to

$$Q(2) = A_0 2^{-(1/8)\cdot 2} = A_0 2^{-1/4} = 0.840896 A_0$$

That is, after 48 hours, about 84 percent of the radioactive iodine remains.

➤ EXAMPLE 6
Exponential Growth

Colonies of certain bacteria exhibit exponential growth. At time 0 hours, suppose that there are N_0 bacteria. Then $N(t)$, the number of bacteria at time t, in hours, is given by

$$N(t) = N_0 2^{t/D}$$

where D denotes the time, in hours, that it takes the colony to double in size.

1. Write $N(t)$ in terms of an exponential function.
2. Suppose that the colony initially contains 5000 bacteria and the doubling time is 4 hours. How many bacteria are there after 10 hours?
3. Consider again the bacterial colony of part 2. How long will it be before the colony contains 160,000 bacteria?

Solution

1. By the laws of exponents, we may write

$$N(t) = N_0 2^{t/D}$$
$$= N_0 2^{(1/D)\cdot t}$$
$$= N_0 (2^{1/D})^t$$

If we let $a = 2^{1/D}$, then the formula takes the form of a constant, N_0, times an exponential function, a^t:

$$N(t) = N_0 a^t$$

2. We are given that the doubling time $D = 4$ hours and $N_0 = 5000$, so the formula for $N(t)$ is given by

$$N(t) = 5000 \cdot 2^{t/4}$$

The number of bacteria after 10 hours is given by the value $N(10)$, which equals

$$5000 \cdot 2^{10/4} = 5000 \cdot 2^{5/2} = 5000 \cdot \sqrt{2^5} = 28,284$$

Note that the last number was rounded to the nearest integer. The formula for the number of bacteria, like most mathematical descriptions of physical phenomena, is only approximate, and answers must be interpreted to conform to physical reality (in this case, the fact that there are no fractional bacteria and that any bacterial count is an approximation.).

3. We must determine the value of *t* for which the following equation holds:

$$N(t) = 160,000$$

Using the equation for $N(t)$ derived in part 2, we have

$$5000 \cdot 2^{t/4} = 160,000$$
$$2^{t/4} = 32$$
$$2^{t/4} = 2^5$$

In the last equation, we have two powers of 2 that are equal. For this equation to hold, the exponents must be equal. (Look at the graph of 2^x. It is constantly increasing, so that no two values of *x* have the same value for *y*.) Thus, we have the equation

$$\frac{t}{4} = 5$$
$$t = 20$$

That is, after 20 hours, the bacteria colony has 160,000 bacteria.

New Technology

EXAMPLE 7
Graphing an Exponential Function

Consider the function

$$f(x) = 3 \cdot 1.1^{-3x} + 1$$

1. Display the graph of $f(x)$.
2. Determine the asymptotes of the graph.

Solution

1. To get a reasonable idea of the shape of the graph, let's use the range $0 \le x \le 10, 0 \le y \le 5$. See Figure 7.
2. The line $y = 1$ is a horizontal asymptote.

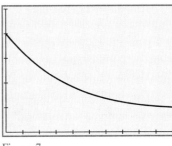

Figure 7   ◄

New Technology

Exploring the Laws of Real Exponents with a Graphing Utility

Two advantages of graphing utilities are that they allow you to graph many functions quickly and that they easily tabulate and graph functions that would not be convenient to work with otherwise. In this discussion, we will look at how the power to graph several functions quickly can aid in understanding the Laws of Real Exponents. It is important to recognize that graphing these functions, no matter how many we graph, does not prove the truth of these laws, but it will visually demonstrate their truth for the specific examples we use.

The first law states that $a^x a^y = a^{x+y}$ where *a* is a positive real number and *x, y* are any real numbers. We can visually demonstrate this on a graphing utility by first graphing the functions $f(x) = a^x$ and $g(x) = a^x$. We need a specific example for the purpose of demonstration, so we'll look at $f(x) = 2^x$ and $g(x) = 2^{x/2}$. Let $h(x) = f(x)g(x)$. On a TI-81, these two functions would be entered as Y_1 and Y_2. Next, in Y_3, enter $Y_1 * Y_2$. If the range is set to $-3 \leq x \leq 3$, $-1 \leq y \leq 10$, the resulting graphs can then be seen on the same system as in Figure 8(a). As we move from the bottom to the top of the right side of the figure, the functions plotted are $g(x)$, $f(x)$, and $h(x)$.

To demonstrate the law $a^x a^y = a^{x+y}$, we must now graph the function $k(x) = 2^{x+x/2} = 2^{3x/2}$. Clear the other functions from the $\boxed{Y=}$ entry screen and enter $k(x)$ in Y_1. The resulting graph is shown in Figure 8(b). A visual comparison shows that this is the same as the graph of $h(x)$ in Figure 8(a).

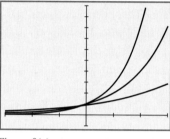

Figure 8(a)

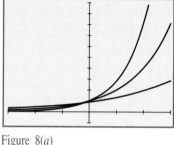

Figure 8(b)

Exercises 4.1

1. Estimate 4^π by making the following calculations:

   a. 4^3 b. $4^{3.1}$ c. $4^{3.14}$ d. $4^{3.141}$

   e. $4^{3.1415}$ f. $4^{3.14159}$ g. $4^{3.141592}$

   h. What seems to be the value of 4^π to two decimal places?

2. Estimate $5^{\sqrt{3}}$ by making the following calculations:

   a. 5^1 b. $5^{1.7}$ c. $5^{1.73}$ d. $5^{1.732}$

   e. $5^{1.73205}$ f. $5^{1.7320508}$ g. $5^{1.732050808}$

   h. What seems to be the value of $5^{\sqrt{3}}$ to five decimal places?

3. a. For $f(x) = 3^x$, find $f(1)$, $f(2)$, $f(3)$, $f(0)$, $f(-1)$, $f(-2)$, $f(-3)$, and $f\left(\frac{1}{2}\right)$.

   b. Sketch a graph of $f(x) = 3^x$.

4. a. For $f(x) = \left(\frac{1}{3}\right)^x$, find $f(1)$, $f(2)$, $f(3)$, $f(0)$, $f(-1)$, $f(-2)$, $f(-3)$, and $f\left(\frac{1}{2}\right)$.

   b. Sketch a graph of $f(x) = \left(\frac{1}{3}\right)^x$.

Sketch a graph of each of the following:

5. $f(x) = 4^x$ 6. $f(x) = 6^x$

7. $f(x) = \left(\frac{1}{4}\right)^x$ 8. $f(x) = \left(\frac{1}{5}\right)^x$

9. $f(x) = \left(\frac{3}{4}\right)^x$ 10. $f(x) = \left(\frac{2}{3}\right)^x$

11. $f(x) = (2.4)^x$ 12. $f(x) = (3.6)^x$

13. $f(x) = (0.38)^x$ 14. $f(x) = (0.76)^x$

15. $y = 5^x$

16. $y = 3^x$

17. $y = 10^x$

18. $y = 2.7^x$

19. $f(x) = 2^{-x}$

20. $f(x) = 2^{x-1}$

21. $f(x) = 2^{-(x-3)}$

22. $f(x) = 2^x - 4$

23. $f(x) = 2^{|x|}$

24. $f(x) = 2^x + 2^{-x}$

25. $f(x) = 2^{-x^2}$

26. $f(x) = \left|2^x - 4\right|$

27. $f(x) = 2^x - 2^{-x}$

28. $f(x) = 2^{-(x+3)^2}$

29. $f(x) = 1 - 2^{-x^2}$

30. $f(x) = x + 2^{-x^2}$

♦ Applications

31. **Compound interest**. Suppose that P dollars are deposited in a bank account paying annual interest at a rate i and compounded n times per year. After a length of time t, the amount $A(t)$ in the account is given by the formula

$$A(t) = P\left(1 + \frac{i}{n}\right)^{nt}$$

Suppose that the interest rate is 8 percent, that the amount is compounded annually, and the initial deposit is $10,000.

   a. Write the function $A(t)$ in terms of an exponential function.

   b. Determine the amount in the account after 1, 2, 3, and 10 years.

   c. Graph the function $A(t)$.

32. **Radioactive decay.** A radioactive decay function is given by

$$Q(t) = A_0 2^{-t/H}$$

where A_0 denotes the amount of a radioactive substance present at time $t = 0$, and H is the half-life. The radioactive isotype cobalt 60 (^{60}Co) has a half-life of 5.3 years. It is used for the radiotherapy treatment of cancer tumors.

 a. Write the formula for $Q(t)$ for radioactive cobalt in terms of an exponential function.

 b. Suppose a radiotherapy device contains 4 grams of ^{60}Co. That is, A_0 is 4 grams. How much will be present after 1 day ($\frac{1}{365}$ yr)? 1 year? 2 years? 10.6 years? 20 years?

33. **Radioactive decay.** Plutonium, a common product and ingredient of nuclear reactors, is a great concern to those opposed to the building of such reactors. The half-life of plutonium is 23,105 years.

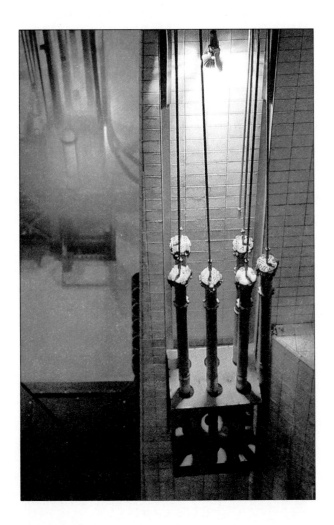

 a. Write the formula for $Q(t)$ for plutonium in terms of an exponential function.

 b. A nuclear reactor contains 400 kilograms of plutonium. What proportion of this original amount of plutonium is present after 1 year? 2 years? 200 years?

34. **Exponential growth.** The population of the world was about 5 billion on January 1, 1987, and is doubling every 43 years. Suppose we let $t = 0$ correspond to 1987 and consider the world population N_0 to be 5 billion at $t = 0$. The population after t years is $N(t)$, where

$$N(t) = N_0 2^{t/D}$$

and D is the doubling time.

 a. Write $N(t)$ in terms of an exponential function.

 b. What will the world population be in 43 years? 86 years? In 1997? In 2000?

35. **Exponential growth.** The population of Texas was 16,370,000 on January 1, 1985, and is doubling every 23 years. Suppose we let $t = 0$ correspond to 1985 and consider the population of Texas N_0 to be 16,370,000 at $t = 0$. The population after t years is $N(t)$, where

$$N(t) = N_0 2^{t/D}$$

and D is the doubling time.

 a. Write $N(t)$ in terms of an exponential function.

 b. What will the population of Texas be in 23 years? In 46 years? In 1995? In 2000?

36. **Exponential growth.** An investment grows exponentially in such a way that it doubles every 10 years. An initial amount of N_0 dollars at time $t = 0$ will grow to an amount $N(t)$ in t years, where

$$N(t) = N_0 2^{t/D}$$

and D is the doubling time.

 a. Write $N(t)$ in terms of an exponential function assuming the initial amount invested is $100,000.

 b. What will be the value or amount of the investment after 5 years? After 10 years? After 20 years? After 32 years?

 c. How many years will it take for the amount of the investment to reach $1 million?

37. **Light filtration.** A certain type of tinted automobile glass 1 cm thick allows only 75 percent of the light to pass through.

 a. How much light will pass through such glass when its thickness is 2 cm? 3 cm? 10 cm? x cm?

 b. Let $P(x)$ represent the percentage of light that passes through glass x cm thick. Find a formula for $P(x)$.

 c. Sketch a graph of $P(x)$ for $x \geq 0$.

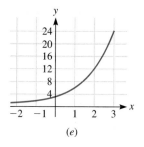

(e)

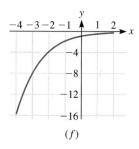

(f)

Match the following functions with their corresponding graphs (a)–(j).

38. $f(x) = 2^x$

39. $f(x) = 2^{-x}$

40. $f(x) = 2^{x+1}$

41. $f(x) = 2^x + 1$

42. $f(x) = 2^{-(x+1)}$

43. $f(x) = 2^{x-3}$

44. $f(x) = 2^x - 3$

45. $f(x) = 3 \cdot 2^x$

46. $f(x) = -2^x$

47. $f(x) = -2^{-x}$

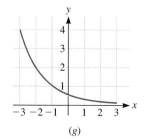

(g)

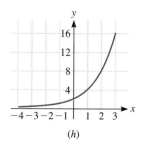

(h)

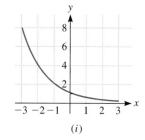

(i)

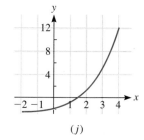

(j)

➡ Technology

Use a graphing calculator to graph the following functions for $0 \le x \le 5$. Experiment with the y range to include points corresponding to all values of x.

48. $f(x) = 3^x$

49. $f(x) = 2.5 \cdot 2^x$

50. $f(x) = 8.71 \cdot 2^{5x}$

51. $f(x) = -3.1 \cdot 2^{-1.1x}$

52. $f(x) = 2^x - 2^{-x}$

53. $f(x) = 2^{2^x}$

54. Determine all points at which the graphs of $f(x) = 2^x$ and $f(x) = 3^x - 2.5^x$ intersect. Approximate coordinates to two significant digits.

55. Graph the function $f(x) = 1000(1 - 2^{-x})$. By zooming out, determine any asymptotes of the graph.

✍ In Your Own Words

56. Describe the sort of physical situation that would best be modeled by an exponential function of the form $f(x) = a^x$, $a > 1$.

57. Describe the sort of physical situation that would best be modeled by an exponential function of the form $f(x) = a^x$, $0 < a < 1$.

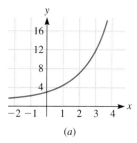

(a)

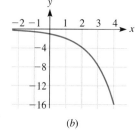

(b)

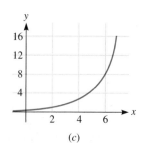

(c)

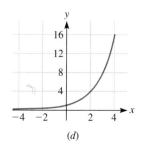

(d)

58. Visually demonstrate the laws of real exponents using a graphing utility. Look at functions with various integer, irrational, and rational bases and exponents.

59. In the Laws of Real Exponents, explain why the base must be a positive number. Why is this restriction not necessary for the exponents?

4.2 THE NATURAL EXPONENTIAL FUNCTION

The Natural Number *e*

In the preceding section, we introduced the exponential function with base a. In this section, we will choose the "standard" function from this infinite family of exponential functions. This so-called natural exponential function is the most convenient one to use in many modeling problems and in calculus.

Any positive real number $a \neq 1$ can serve as the base for an exponential function. The most important value for a is a particular real number designated e. The number e is an irrational number whose value is approximately $2.718281. \ldots$ Although this number may seem to come out of nowhere, it arises quite naturally in many applications of mathematics in the biological and physical sciences as well as in business. To introduce the number e, let's consider a problem in compound interest.

Suppose that P dollars are deposited for t years in a bank account yielding interest at an annual rate i and compounded n times per year. As we stated in the preceding section, the value of the account at time t, namely, $A(t)$, is given by the formula

$$A(t) = P \left(1 + \frac{i}{n} \right)^{nt}$$

Suppose that the initial deposit is \$1, the length of time is 1 year, and the rate of interest is 100 percent. In this case, the compound interest reads

$$A(1) = 1 \cdot \left(1 + \frac{1}{n} \right)^{n \cdot 1} = \left(1 + \frac{1}{n} \right)^{n}$$

Let's consider the effect of different periods of compounding on the amount after 1 year. This amounts to computing the value of $A(1)$ for various values of n. We have computed assorted values on a calculator and listed them in the accompanying table.

n	Amount from Compounding n Times per Year
1	2
2	2.25
3	2.370370
4	2.441406
10	2.5937425
50	2.691588
100	2.7048138
1000	2.7169239
10,000	2.7181459
100,000	2.718268
1,000,000	2.7182805
10,000,000	2.7182817

Note that as the frequency of compounding increases without bound, the amount in the account at the end of the year approaches an amount approximately equal to $2.718\ldots$. This is rather surprising. Your intuition might suggest that as the money is compounded more and more frequently (every day, every hour, every second, every millisecond), the amount in the account would grow without limit. However, this is not the case. As the frequency of compounding is increased without bound, the amount in the account approaches a specific real number, denoted e, whose decimal expansion is given by $2.718281\ldots$. Let's record these results for future reference.

$$\left(1 + \frac{1}{n}\right)^n \text{ approaches } e \text{ as } n \text{ becomes arbitrarily large}$$
$$e = 2.718281$$

Historical Note: Euler and e

The number e was first defined by the 18th century mathematician Euler (pronounced "Oiler"). Although born in Switzerland, Euler spent most of his life in Russia as court mathematician to Empress Catherine the Great. One of the greatest mathematicians of the century, Euler made contributions to pure mathematics, fluid dynamics, mechanics, and astronomy. Among his many accomplishments, Euler wrote the first textbook on calculus, *Introductio in Analysin Infinitorum*, in 1740. Euler, who was amazingly prolific, wrote more than 60 large volumes. Euler became blind toward the end of his life, but he continued his mathematical research by dictating calculations to a servant he trained specially for the purpose. | ∎

Calculations Involving e

You can calculate with e just as you would with any other decimal number. In fact, expressions involving e may be evaluated using the e^x key found on most scientific calculators. Some calculators don't have an e^x key but use other techniques for calculating powers of e. For such calculators, consult your manual for details.

Algebraic expressions involving e^x can be manipulated using the fact that e^x is just an ordinary power to which the laws of exponents apply. The next example provides some practice in carrying out such algebraic manipulations.

➤ EXAMPLE 1
Calculating with e

Simplify the following expressions involving e.

1. $e^3 e^{-1} e^{-5}$ 2. $(e + 1)^2$ 3. $(e^x)^5$
4. $(e^x + e^{-x})^2$ 5. $(e^x + 1)(3e^x - 4)$ 6. $\sqrt{e^x}$

Solution

1. By the laws of exponents, we have
$$e^3 e^{-1} e^{-5} = e^{3+(-1)+(-5)} = e^{-3}$$

2. Use the formula for the square of a binomial to obtain
$$(e + 1)^2 = e^2 + 2 \cdot e \cdot 1 + 1^2$$
$$= e^2 + 2e + 1$$

3. By the laws of exponents, we have
$$(e^x)^5 = e^{x \cdot 5} = e^{5x}$$

4. Multiply out the binomials and apply the laws of exponents to obtain
$$(e^x + e^{-x})^2 = (e^x)^2 + 2e^x e^{-x} + (e^{-x})^2$$
$$= e^{2x} + 2e^0 + e^{-2x}$$
$$= e^{2x} + e^{-2x} + 2$$

5. Multiply out and apply the laws of exponents to show

$$(e^x + 1)(3e^x - 4) = e^x \cdot 3e^x - 4e^x + 3e^x - 4$$
$$= 3e^{2x} - e^x - 4$$

6. Recalling that taking the square root is equivalent to raising to the power $\frac{1}{2}$, we see that

$$\sqrt{e^x} = (e^x)^{1/2} = e^{x \cdot 1/2} = e^{x/2}$$

The Natural Exponential Function

The exponential function with base e is the most important exponential function. To indicate its importance, let's assign it a name as follows:

Definition 1
Natural Exponential Function

The **natural exponential function** is the function $f(x)$ defined by

$$f(x) = e^x$$

The natural exponential function is a special case of the more general exponential function a^x. In particular, the domain of the natural exponential function is the set **R** of all real numbers, and the range is the set of all positive real numbers.

In the preceding section, we considered applications that involved exponential functions of the form

$$f(x) = a^x$$

where a is a positive constant. Later in this chapter, we will show that every such function can be written in the alternative form

$$f(x) = e^{Ax}$$

where A is a constant. For this reason, functions of the alternative form are called **exponential functions.**

The case $A = 0$ is uninteresting since it corresponds to $f(x)$ = the constant function 1. The nonzero values of A give rise to two sorts of functions:

$$f(x) = e^{kx}$$
$$f(x) = e^{-kx}$$

where k is a positive constant.

The first sort of function increases very rapidly as x increases. The rapid rate of increase of such functions is known as **exponential growth**. In Figure 1, we show the graph of a typical function $f(x) = e^{kx}$ for $k > 0$.

For $k > 0$, the function $f(x) = e^{-kx}$ shows exponential decay and rapidly approaches 0 as x approaches positive infinity. In Figure 2, we show the graph of this sort of function.

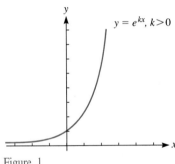

$y = e^{kx}, k > 0$

Figure 1

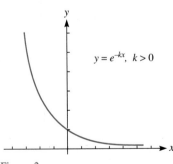

$y = e^{-kx}, k > 0$

Figure 2

Applications of the Natural Exponential Function

The natural exponential function is used in countless applications in fields ranging from biology and medicine to physics and engineering. The following examples provide illustrations of three of these applications.

➤ **EXAMPLE 2**
Bacteria Growth

Suppose that at time t (in hours), the number $N(t)$ of *Escherichia coli* bacteria in a culture is given by the formula $N(t) = 5000e^{0.1t}$. How many bacteria are in the culture at time 5 hours?

Solution

At time $t = 5$, the number of bacteria equals $N(5)$, which by the given formula equals

$$5000e^{0.1(5)} = 5000e^{0.5} = 8244$$

We used a calculator to determine the value of the exponential function. Furthermore, we rounded the result to the nearest integer because the answer represents a number of bacteria.

➤ **EXAMPLE 3**
Skydiving

A skydiver jumps out of an airplane and executes a free-fall for a period of time. The diver's velocity does not increase indefinitely but approaches a limiting velocity, called the **terminal velocity** and denoted v_T. The value of the terminal velocity depends on the size and position of the particular skydiver. Let $v(t)$ denote the velocity of a skydiver in free-fall t seconds after jumping from an airplane. The laws of physics can be used to show that $v(t)$ can be described in terms of exponential functions, as follows:

$$v(t) = v_T(1 - e^{-kt})$$

where k is a positive constant. Suppose that for a certain skydiver the terminal velocity is 150 feet per second and k is equal to 0.1. What is the velocity of the skydiver 10 seconds after jumpoff? The graph of the function in this case is shown in Figure 3.

Solution

From the given data, $v_T = 150$ and $k = 0.1$. Inserting these values in the equation for the function, we derive the following formula for $v(t)$:

$$v(t) = 150(1 - e^{-0.1t})$$

The velocity 10 seconds after jumpoff is equal to $v(10)$, which has the value

$$150(1 - e^{-0.1 \cdot 10}) = 150(1 - e^{-1})$$

We used a calculator to evaluate the above expression, which is equal to 94.818. That is, after 10 seconds of free-fall, the skydiver is falling with a velocity of 94.818 feet per second.

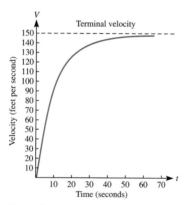

Figure 3

➤ **EXAMPLE 4**
Spread of an Epidemic

Suppose that a long-lasting epidemic, against which there is no immunity, spreads through a town with P inhabitants. Let $N(t)$ denote the number of people infected t days after the epidemic breaks out. One model of the spread of the epidemic gives the following formula for $N(t)$:

$$N(t) = \frac{P}{1 + Ae^{-kt}}$$

where A and k are constants. The graph of the function $N(t)$ is shown in Figure 4. The graph is called a **logistic curve.** Suppose that the town has 100,000 inhabitants and that statistics show that A has the value 5000 and k the value

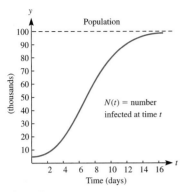

Figure 4

1. Note that the line $y = P$ is a horizontal asymptote of the graph. This indicates that after a long period of time, almost everyone is infected with the epidemic. How many people are sick after 5 days? After 10 days?

Solution

According to the given data, the value of P is 100,000, the value of A is 5000, and the value of k is 1. Inserting these values into the formula for $N(t)$, we have

$$N(t) = \frac{100,000}{1 + 5000e^{-t}}$$

The number of people sick after 5 days is

$$N(5) = \frac{100,000}{1 + 5000e^{-5}} = \frac{100,000}{1 + 5000 \cdot 0.006738} = 2883$$

We used a calculator to determine both the value of e^{-5} and the value of the entire expression.

In a similar fashion, we calculate the number of people sick after 10 days as

$$N(10) = \frac{100,000}{1 + 5000e^{-10}} = 81,500$$

Note the virulence of the epidemic. After only 10 days, more than 81 percent of the population is sick!

The Natural Exponential Function and Graphing Calculators

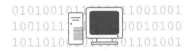

New Technology

EXAMPLE 5

Graphing Exponential Functions

On a single coordinate system, graph the functions

$$f_1(x) = e^{-x} + 2$$
$$f_2(x) = e^{-2x} + 2$$
$$f_3(x) = e^{-3x} + 2$$

Determine the asymptotes of the graphs.

Solution

Enter the three functions using the $\boxed{e^x}$ key to indicate the exponential function. For example, to enter the function e^{-x}, use the keystrokes $\boxed{e^x}\boxed{(-)}\boxed{X}$.

To see all three graphs clearly, we should restrict the standard range to something like $-2 \le x \le 2$. See Figure 5.

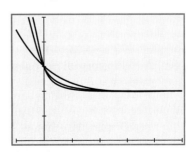

Figure 5

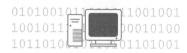

New Technology

EXAMPLE 6
Graphing an Exponential Function

Use a graphing calculator to display the graph of the function

$$v(t) = 150\left(1 - e^{-0.1t}\right)$$

describing the skydiver of Example 3. Describe the behavior of this function.

Solution

Even though the variable in the function is t, on the calculator you must use the variable X. A suitable range (determined by trial and error) is $0 \leq X \leq 50$, $0 \leq Y \leq 160$. See Figure 6. The graph increases as x increases, approaching the asymptote $y = 150$.

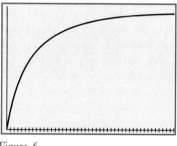

Figure 6   ◄

New Technology

EXAMPLE 7
A Logistic Curve

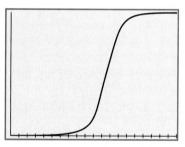

Figure 7

Sketch the graph of the function

$$N(t) = \frac{100,000}{1 + 5000e^{-t}}$$

describing the spread of the epidemic of Example 4. Describe the behavior of this function.

Solution

Trial and error shows that a suitable range is $0 \leq x \leq 20$, $0 \leq y \leq 100,000$. See Figure 7. Initially, the number of people sick is small and the epidemic spreads quickly, indicated by the steepness of the curve. The curve flattens out, indicating a slower spread of the epidemic. The approach to the asymptote indicates that almost everyone is eventually infected.

◄

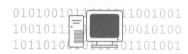

New Technology

EXAMPLE 8
The Normal Curve

The **normal probability curve** is the graph of the function

$$f(x) = \frac{1}{\sqrt{2\pi}}e^{-x^2/2}$$

Display the graph of this function. What symmetry is exhibited by this graph?

Solution

We enter the function $f(x)$ as shown in Figure 8. Note that the number π may be entered using the π key. Because the exponential function decreases very rapidly, we use the range $-2 \le x \le 2$. The normal probability curve is shown in Figure 9. Visually, we see that the curve is symmetric with respect to the y-axis. This graphically reflects the fact that the function is even.

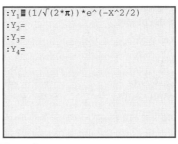

Figure 8

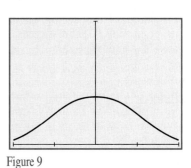

Figure 9   ◀

Exercises 4.2

Simplify the following expressions involving e. Where appropriate, approximate values using a calculator with an e^x key.

1. $e^5 e e^{-3}$

2. $e^{-4} e^6 e^0$

3. $e^5 e^{-2} + 2e^3 e^4$

4. $5e^{-2} - 4e^6 e^{-8}$

5. $\dfrac{e^5 - e^{-5}}{e^5 + e^{-5}}$

6. $\dfrac{e^2 + 4e^5}{e^3}$

7. $(e - 1)^2$

8. $(2 + e)^2$

9. $(e^4)^x$

10. $(2e^{-x})^4$

11. $(e^x - e^{-x})^2$

12. $(1 + 2e^x)^2$

13. $(e^x - 1)^3$

14. $(e^x + e^{-x})^3$

15. $(e^{x-2})(4e^x + 5)$

16. $(e^x + e^{-x})(e^x - e^{-x})$

17. $(e^x - 3)(e^x + 3)$

18. $(e^x - 1)(e^{2x} + e^x + 1)$

19. $\sqrt{e^{2x}}$

20. $\sqrt[3]{e^{6x}}$

21. $e^{0.1654} + e^{-2.34}$

22. $2e^{-4.6} - e^{5.71}$

◆ Applications

23. **Skydiving.** The velocity of a particular skydiver $v(t)$ can be described in terms of exponential functions as follows:

$$v(t) = v_T(1 - e^{-kt})$$

where k is a positive constant. Suppose that for a certain skydiver the terminal velocity v_T is 160 miles per hour and k is equal to 0.2, and t is in seconds.

a. What is the velocity of the skydiver 1 second after jumpoff? After 2 sec? After 5 sec? After 10 sec? After 20 sec?

b. Sketch a graph of the function for $t \ge 0$.

24. **Advertising.** A company introduces a new product on a trial run in a city. It advertises the product on TV and gathers data concerning the percentage $P(t)$ of the people in the city who bought the product after it was advertised a certain number of times t. It determines that the **limiting effect** of the advertising is 75 percent. This means that no matter how many times the company advertises its product, the percentage of people who buy the product will never reach or exceed 75 percent, although the percentage will get closer and closer to 75 percent. The company determines that $P(t)$ can be described in terms of exponential functions as follows:

$$P(t) = 0.75(1 - e^{-kt})$$

where $k = 0.05$.

a. What percentage of the city buys the product after 1 advertisement? After 2 ads? After 10 ads? After 20 ads? After 50 ads?

b. Sketch a graph of $P(t)$.

25. **Advertising.** Repeat Exercise 24 assuming that the limiting effect is 50 percent and $k = 0.08$.

26. **Advertising.** Repeat Exercise 24 that assuming that the limiting effect is 100 percent and $k = 0.06$.

27. **Medicine.** Even after a new medication has been extensively tested and approved by the FDA, it takes time for it to be accepted for usage by doctors. They first have to become aware of it and then try it on their patients. The usage approaches a limiting value of 100 percent, or 1, as a function of t, in months. The percentage $P(t)$ of doctors who use the medication can be described in terms of exponential functions as follows:

$$P(t) = 100\%(1 - e^{-kt})$$

where $k = 0.3$.

a. What percentage of the doctors have accepted the medication after 1 month? After 2 months? After 5 months? After 10 months? After 1 year?

b. Sketch a graph of $P(t)$ for $t \geq 0$.

28. **Psychology.** The Hullian model of learning asserts that the probability $P(t)$ of mastering a certain concept after t learning trials is given by

$$P(t) = 1 - e^{-kt}$$

where k is a constant that depends on the difficulty of the learning. An educational psychologist analyzes the amount of time it takes to learn a concept in mathematics and finds that $k = 0.18$.

a. What is the probability of learning the concept after 1 trial? After 2 trials? After 3 trials? After 5 trials? After 13 trials?

b. Sketch a graph of the function $P(t)$.

29. **Agriculture.** A farmer is growing soybeans in a field. The more fertilizer the farmer spreads, the greater his or her yield up to some **limiting value** L. The yield per acre $Y(n)$ using n pounds of fertilizer can be described in terms of exponential functions as

$$Y(n) = L(1 - e^{-kn})$$

where k is a constant that depends on the crop and the fertilizer. (This mathematical model makes an assumption that may not always work in practice because most fertilizers will kill a crop if applied excessively, but past some point it is not cost-effective to keep adding more fertilizer to get just a slight change in yield.) Suppose for a certain crop $k = 0.04$ and the limiting value is 30 bushels per acre.

a. Find $Y(n)$ when $n = 1, 2, 5, 6, 10, 20, 30,$ and 40.

b. Sketch a graph of $Y(n)$ for $n \geq 0$.

30. **Spread of an epidemic.** Suppose that a long-lasting epidemic against which there is no immunity spreads through a town with P inhabitants. Let $N(t)$ denote the number of people infected t days after the epidemic breaks out. One model of the spread of the epidemic gives the following formula for $N(t)$:

$$N(t) = \frac{P}{1 + Ae^{-kt}}$$

where A and k are constants. Suppose that the town has 4000 inhabitants and that statistics show that $A = 200$ and $k = 2$.

a. How many people are sick after 5 days? After 10 days? After 20 days? After 30 days?

b. Sketch a graph of $N(t)$ for $t \geq 0$.

31. **Limited population growth.** A ship carrying 100 passengers is shipwrecked on a small island and never rescued. The amount of food and fresh water on the island limits the growth of the population to a limiting value of 540, which the population approaches but never reaches. The number of people $N(t)$ on the island after time t is given by

$$N(t) = \frac{P}{1 + Ae^{-kt}}$$

where A and k are constants and t is time in years. For the people on the island, $P = 540, A = 4.4,$ and $k = 3$.

a. What is the population after 10 years? After 20 years? After 30 years? After 100 years?

b. Sketch a graph of the logistic curve $N(t)$ for $t \geq 0$.

32. **Spread of a rumor.** In a college with a population of 800, a group of students spread the rumor "Study algebra and trigonometry from this book and you will get an A in calculus." The number of people $N(t)$ who have heard the rumor after t minutes is given by

$$N(t) = \frac{P}{1 + Ae^{-kt}}$$

where A and k are constants. In this situation, $P = 800,$ $A = 132,$ and $k = 0.4$.

a. How many people have heard the rumor after 1 minute? After 2 minutes? After 6 minutes? After 12 minutes? After 30 minutes? After 1 hour?

b. Sketch a graph of the logistic curve $N(t)$ for $t \geq 0$.

33. **Electricity.** (See figure.) If an electromotive force is suddenly connected to a circuit containing a resistor and a capacitor, the charge does not immediately reach its equilibrium value but approaches it exponentially. For a circuit with resistance R ohms, capacitance C farads, and electromotive force E volts, the charge q (in coulombs) after time t seconds is given by

$$q = CE(1 - e^{-t/RC})$$

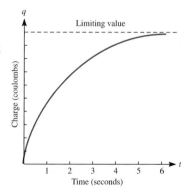

Exercise 33

a. What is the limiting value of the charge?

b. A car battery (12 V) is connected to a circuit containing a 10,000-ohm resistor and a 10-microfarad capacitor. What will the charge be after 0.05 seconds?

Simplify each of the following expressions.

34. $\dfrac{(1 + e^x)(1 + e^{-x}) - (1 + e^x)(1 - e^{-x})}{(1 + e^x)^2}$

35. $\dfrac{(e^x + e^{-x})(e^x + e^{-x}) - (e^x - e^{-x})(e^x - e^{-x})}{(e^x + e^{-x})^2}$

For each of the following functions, find and simplify

$$\frac{f(x + h) - f(x)}{h}, \qquad h \neq 0$$

36. $f(x) = e^x$

37. $f(x) = \dfrac{e^x - e^{-x}}{2}$

38. $f(x) = \dfrac{e^x + e^{-x}}{2}$

39. $f(x) = e^{-3x}$

➡ **Technology**

Use a graphing calculator to graph the following functions in the range $-4 \leq x \leq 4$. Experiment with the y range so that a point is included for each given x-value.

40. $f(x) = e^x$

41. $f(x) = 5e^{3x}$

42. $f(x) = e^{-0.1x}$

43. $f(x) = \dfrac{e^x - e^{-x}}{e^x + e^{-x}}$

44. $f(x) = \dfrac{1}{e^x + 1}$

45. $f(x) = 5 - 5e^{-x}$

Use a graphing calculator to solve the following exercises:

46. Exercise 23.

47. Exercise 28.

48. Exercise 29.

49. Exercise 30.

◼ 4.3 LOGARITHMIC FUNCTIONS

Inverse functions provide us with a useful way of creating new functions from given ones. In this section, we introduce the logarithmic functions by taking the inverse of the exponential function.

Definition of Logarithmic Functions

Suppose that a is a positive real number. We have seen that the exponential function $f(x) = a^x$ for $a \neq 1$ is one-to-one with domain the set **R** of all real numbers, and range the set of all positive real numbers. Because the exponential function with base a is one-to-one, it has an inverse function, which depends on the value of a. The inverse function is denoted $\log_a x$ and is called the **logarithm with base a**.

The domain of $\log_a x$ is the range of $f(x)$, that is, the set of all positive real numbers. The range of $\log_a x$ is the domain of a^x, that is, the set **R** of all real numbers.

Definition 1
Logarithmic Function with Base a

Let x be a positive real number. Then

$$\log_a x = y \quad \text{if and only if} \quad a^y = x$$

We may put this definition into words as follows:

$\log_a x$ is the exponent to which a must be raised to obtain x.

For example, because $3^2 = 9$, we have

$$\log_3 9 = 2$$

That is, 2 (or $\log_3 9$) is the exponent to which 3 must be raised to obtain 9. Similarly, because

$$2^{10} = 1024$$

we have

$$\log_2 1024 = 10$$

That is, 10 (or $\log_2 1024$) is the exponent to which 2 must be raised to obtain 1024.

A word about notation: In writing logarithmic functions, it is common to omit the parentheses if no confusion will result. For example, it is common to write $\log_a x$ instead of $\log_a (x)$. However, in cases where you are taking the logarithm of a complicated expression, it is best to use the parenthetic notation.

Properties of Logarithms

For each fact about raising a number to a power, there is a corresponding fact about logarithms. For example, the property

$$a^1 = a$$

gives us the following property of logarithms:

$$\log_a a = 1$$

That is, 1 is the power to which a must be raised to obtain a.

Similarly, the property

$$a^0 = 1$$

yields the property of logarithms

$$\log_a 1 = 0$$

That is, 0 is the power to which a must be raised to obtain 1.

Because $y = \log_a x$ implies that

$$a^y = x,$$

we see that

$$a^{\log_a x} = x, \qquad x > 0 \tag{1}$$

Similarly, if

$$y = \log_a x$$

then

$$y = \log_a(a^y)$$

so that

$$\log_a(a^x) = x, \qquad x = \text{any real number} \tag{2}$$

Let's record (1) and (2) for future reference.

FUNDAMENTAL PROPERTIES OF LOGARITHMS

1. $a^{\log_a x} = x, \quad x > 0$
2. $\log_a a^x = x, \quad x$ any real number

Fundamental Property 1 merely restates the fact that $\log_a x$ is the exponent to which a must be raised to obtain x. (In the first property, the logarithm is in the exponent.) The value on the left side of the second property is the exponent to which a must be raised to obtain a^x, and this exponent is clearly x.

Note that these two fundamental properties are just special cases of the two formulas for inverse functions expressed as composites of functions, which we stated in Chapter 2. Indeed, if we set $f(x) = a^x$ and $g(x) = \log_a x$, then because f and g are inverses of one another, we have

$$f(g(x)) = f(\log_a x) = a^{\log_a x} = x$$
$$g(f(x)) = g(a^x) = \log_a(a^x) = x$$

The two fundamental properties of logarithms are very useful in simplifying expressions involving logarithmic functions, as the following example illustrates.

➤ **EXAMPLE 1**
Calculations Involving Logarithms

Calculate the values of the following expressions:

1. $\log_{10} 10^{5.781}$ 2. $\log_{10} 1000$
3. $\log_2 \frac{1}{8}$ 4. $4^{\log_4 11.1}$
5. $\log_a a^{3x}$ 6. $a^{\log_a x^3}$

Solution

1. By Fundamental Property 2, $\log_a a^x = x$, so the value of the expression is $x = 5.781$.

2. Because $1000 = 10^3$, we see that $\log_{10} 1000 = \log_{10} 10^3 = 3$ (Fundamental Property 2).

3. We have
$$\log_2 \frac{1}{8} = \log_2 2^{-3}$$
$$= -3 \qquad \text{(Fundamental Property 2)}$$

4. $4^{\log_4 11.1} = 11.1$

5. $\log_a a^{3x} = 3x$

6. $a^{\log_a x^3} = x^3$

The Graph of $\log_a x$

The function $\log_a x$ is the inverse of the function a^x. As we have seen, this relationship implies that the graph of $\log_a x$ can be obtained by reflecting the graph of a^x in the line $y = x$. In Figure 1, we have sketched the graphs of the two functions together with the line $y = x$, where $a > 1$.

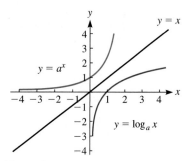

Figure 1

> **EXAMPLE 2**
Graphing Logarithmic
Functions

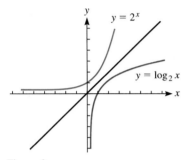

Figure 2

Sketch the graphs of the following functions:

1. $y = \log_2 x$ 2. $y = \log_2 (x - 1)$ 3. $y = \log_2 x + 3$

Solution

1. The function $y = \log_2 x$ is the inverse of the function $y = 2^x$. Figure 2 shows the graph of $y = 2^x$ in blue. To obtain the graph of the inverse function, we reflect this graph in the line $y = x$. This graph is shown in red in Figure 2. The y-axis is a vertical asymptote of the latter graph, and the graph increases without bound as x approaches positive infinity.

2. According to our discussion on translating graphs of functions, the graph of $y = \log_2(x - 1)$ can be obtained by translating the colored graph of Figure 2 one unit to the right. The sketch of the graph appears in Figure 3. Note that the vertical asymptote is now the line $x = 1$.

3. According to our discussion on translating graphs of functions, the graph of $y = \log_2 x + 3$ can be obtained by translating the graph $\log_2 x$ three units upward. The sketch is shown in Figure 4.

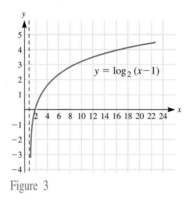

Figure 3

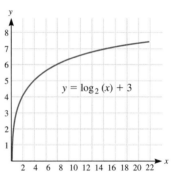

Figure 4

Properties of $\log_a x$

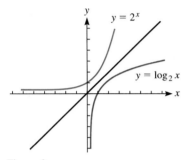

Figure 5

By examining the graph in Figure 5, we can verify the following properties of the function $\log_a x$. (We have already observed and stated several of these, but we record them here for convenient reference.)

PROPERTIES OF LOGARITHMIC FUNCTIONS

Suppose that $a > 1$. Then we have the following properties of $\log_a x$:

1. The domain of $\log_a x$ is the set of all positive real numbers. The range is the set **R** of all real numbers.
2. As x increases, the value of $\log_a x$ increases.
3. The y-axis is a vertical asymptote of the graph of $\log_a x$.
4. As x approaches positive infinity, the value of $\log_a x$ approaches positive infinity.
5. The function $\log_a x$ is one-to-one.

The logarithmic function $\log_a x$ has a number of important algebraic properties. The following result is of great help in simplifying expressions involving logarithmic functions.

LAWS OF LOGARITHMS

The function $\log_a x$ has the following properties for $x, y > 0$:

1. The logarithm of a product is the sum of the logarithms.

$$\log_a(xy) = \log_a x + \log_a y$$

2. To compute the logarithm of x raised to a power, multiply the logarithm of x by the exponent.

$$\log_a x^y = y \log_a x$$

3. The logarithm of a quotient is the difference of the logarithms.

$$\log_a\left(\frac{x}{y}\right) = \log_a x - \log_a y$$

Proof

1. By Fundamental Property 1, we have

$$x = a^{\log_a x}$$
$$y = a^{\log_a y}$$

Multiplying these expressions for x and y gives

$$xy = a^{\log_a x} \cdot a^{\log_a y}$$

By the laws of exponents, the expression can be rewritten as

$$xy = a^{\log_a x + \log_a y}$$

The last equation states that $\log_a x + \log_a y$ is the exponent to which a must be raised to obtain the value xy. By the definition of the logarithmic function, this means that

$$\log_a x + \log_a y = \log_a xy$$

2. Again we use Fundamental Property 1 to obtain $x = a^{\log_a x}$, and we raise both sides of the equation to the power y:

$$x^y = (a^{\log_a x})^y = a^{y \log_a x}$$

This equation states that $y \log_a x$ is the exponent to which a must be raised to obtain x^y. From the definition of the logarithmic function, this means that

$$y \log_a x = \log_a x^y$$

3. Write the expression x/y in the form

$$xy^{-1}$$

and apply what we just proved in parts 1 and 2:

$$
\begin{aligned}
\log_a\left(\frac{x}{y}\right) &= \log_a(xy^{-1}) && \textit{Laws of exponents}\\
&= \log_a x + \log_a y^{-1} && \textit{(part 1)}\\
&= \log_a x + (-1)\log_a y && \textit{(part 2)}\\
&= \log_a x - \log_a y
\end{aligned}
$$

♦

Algebraic Expressions Involving Logarithms

The laws of logarithms may be used to simplify expressions involving logarithmic functions, as the following two examples illustrate.

> ## EXAMPLE 3
Simplifying Logarithmic Expressions

Write the following expressions as logarithms of a single number.

1. $\log_5 [(x - 1)(x - 2)] - \log_5 (x - 2)$
2. $\log_{10} \sqrt{x} + \log_{10} x^{3/2}$

Solution

1. By Law of Logarithms 1, we have

$$\log_5 [(x - 1)(x - 2)] - \log_5 (x - 2)$$
$$= \left[\log_5 (x - 1) + \log_5 (x - 2)\right] - \log_5 (x - 2)$$
$$= \log_5 (x - 1)$$

2. By Law of Logarithms 2, we have

$$\log_{10} \sqrt{x} + \log_{10} x^{3/2} = \log_{10} x^{1/2} + \log_{10} x^{3/2}$$
$$= \tfrac{1}{2} \log_{10} x + \tfrac{3}{2} \log_{10} x$$
$$= 2 \log_{10} x$$

> ## EXAMPLE 4
Further Practice with the Laws of Logarithms

Write the following expressions using a single logarithm.

1. $3 \log_a x + 4 \log_a y + \tfrac{1}{2} \log_a z$
2. $\log_2 \dfrac{x}{y} - \log_2 \dfrac{y^2}{z}$

Solution

1. Applying Law of Logarithms 2, we may write the given expression in the form

$$\log_a x^3 + \log_a y^4 + \log_a z^{1/2}$$

We may now apply Law 1 to combine the three terms to obtain the result
$$\log_a (x^3 y^4 z^{1/2})$$

2. By Law of Logarithms 3, we have

$$\log_2 \frac{x}{y} - \log_2 \frac{y^2}{z} = \log_2 \left[\frac{x}{y} \div \frac{y^2}{z}\right]$$
$$= \log_2 \left[\frac{x}{y} \cdot \frac{z}{y^2}\right]$$
$$= \log_2 \left(\frac{xz}{y^3}\right)$$

Equations Involving Logarithms

We may solve equations involving logarithms and exponentials of a variable by using the following approach.

> ### SOLVING EXPONENTIAL AND LOGARITHMIC EQUATIONS
>
> 1. Solving an equation involving an exponential of a variable
>   a. Transform the equation so that it has a single exponential on one side of the equation and a number on the other side.
>   b. Take logarithms of both sides of the equation.
>   c. Apply the laws of logarithms to solve for the variable.
> 2. Solving an equation involving a logarithm of a variable
>   a. Transform the equation so that it has a single logarithm on one side of the equation and a number on the other side.
>   b. Apply the appropriate exponential fuction to both sides of the equation.
>   c. Apply the laws of logarithms to solve for the variable.

The next example illustrates how this approach is carried out in practice.

➤ **EXAMPLE 5**
Solving Logarithmic Equations

Solve the following equations for x:

1. $\log_5 (3x - 1) = 2$
2. $7^{2x} = 4$

Solution

1. For an equation involving logarithms, the best strategy is to use the properties of logarithms to eliminate them from the equation. In this case, we can raise 5 to the quantity on both sides of the equation to obtain

$$5^{\log_5(3x-1)} = 5^2$$

By Fundamental Property 1, the left side of the equation equals $3x - 1$ and the equation reads

$$3x - 1 = 5^2$$
$$3x - 1 = 25$$
$$x = \tfrac{26}{3}$$

2. Apply the function $\log_7$ to both sides of the equation:

$$\log_7 7^{2x} = \log_7 4$$
$$2x \log_7 7 = \log_7 4$$
$$2x = \log_7 4$$
$$x = \frac{\log_7 4}{2}$$

➤ **EXAMPLE 6**
More Solving Logarithmic Equations

Solve the following equation:

$$\log_2 (x + 1) + \log_2 (x - 1) = 3$$

Solution

We may combine the two terms on the left using the first Law of Logarithms:

$$\log_2[(x + 1)(x - 1)] = 3$$
$$\log_2 (x^2 - 1) = 3$$
$$2^{\log_2 (x^2-1)} = 2^3$$
$$x^2 - 1 = 8$$

$$x^2 = 9$$
$$x = \pm 3$$

We now test these potential solutions to the equation to determine whether they satisfy the equation. For $x = 3$,

$$\begin{aligned}
\log_2(x + 1) + \log_2(x - 1) &= \log_2(3 + 1) + \log_2(3 - 1) \\
&= \log_2 4 + \log_2 2 \\
&= 2 + 1 \\
&= 3
\end{aligned}$$

3

so $x = 3$ is a solution to the equation. Note, however, that for $x = -3$, the expression $\log_2(x + 1)$ is not even defined, so $x = -3$ is not a solution.

➤ **EXAMPLE 7**

Binary Representation of Numbers

A **binary number** is a number composed of a sequence of digits 0 and 1, such as 1001 or 111101000111. It is possible to represent any decimal integer as a binary number. For example, 189 can be written as the binary number 10111101. (For the moment, it is of no concern to us how to determine this binary representation.) The electronic circuits of a computer work with numbers in binary form, which accounts for the great significance of binary representation of integers. Suppose that x is a positive integer. It can be shown that the number of digits $B(x)$ in its binary representation is given by the formula

$$B(x) = \text{INT}(\log_2 x) + 1$$

where INT denotes the greatest integer function, that is, $\text{INT}(x)$ is equal to the greatest integer less than or equal to x. How many binary digits are required to represent the integer 32,767? The integer 32,768?

Solution

To solve this problem, we must first obtain some feeling for the values assumed by the function $\log_2 x$. We begin by tabulating the various values of the powers of 2 and the corresponding values of $\log_2$:

x	$\log_2 x$	x	$\log_2 x$
1	0	1024	10
2	1	2048	11
4	2	4096	12
8	3	8192	13
16	4	16,384	14
32	5	32,768	15
64	6	65,536	16
128	7	131,072	17
256	8	262,144	18
512	9	524,288	19

According to the resulting table, the value of $\log_2 32,767$ is greater than 14 and less than 15. Therefore, the value of $\text{INT}(\log_2 32,767)$ is 14, and the value of $B(32,767)$ is equal to $14 + 1 = 15$. That is, it takes 15 binary digits to represent the decimal number 32,767.

Also according to the table, the value of $\log_2 32,768$ is exactly 15, so the value of $\text{INT}(\log_2 32,767)$ is 15, and the value of $B(32,768)$ is $15 + 1 = 16$. That is, it takes 16 binary digits to represent the decimal number 32,768.

The next example shows how the properties of logarithms may be applied in dealing with the exponential functions that arise in describing radioactive decay.

➤ **EXAMPLE 8**
Radioactive Decay

A physicist measures the decay of a radioactive isotope. She finds that after 20 days, only 40 percent of the original amount of the isotope remains. Determine the half-life of the isotope.

Solution

Let A_0 denote the original quantity of the radioactive isotope, let $Q(t)$ denote the quantity left after time t, and let H denote the half-life of the isotope. Earlier in the chapter, we stated the result from physics that

$$Q(t) = A_0 2^{-t/H}$$

In this example, we are given that

$$Q(20) = 0.4A_0$$

Inserting the value 20 for t in the first formula and substituting the value $0.4A_0$ for $Q(20)$, we have

$$0.4A_0 = A_0 2^{-20/H}$$
$$0.4 = 2^{-20/H}$$

To solve this equation, we take logarithms with base 2 of both sides to obtain

$$\log_2 0.4 = \log_2(2^{-20/H})$$
$$\log_2 0.4 = -\frac{20}{H}$$
$$(\log_2 0.4)H = -20$$
$$H = -\frac{20}{\log_2 0.4}$$

This gives an expression from which a numerical value for the half-life H may be computed. We will show how to compute the numerical value for H in the next section.

Exercises 4.3

Determine the equivalent exponential equation.

1. $\log_2 8 = 3$
2. $\log_5 3125 = 5$
3. $\log_3 243 = 5$
4. $\log_9 729 = 3$
5. $\log_2 \frac{1}{8} = -3$
6. $\log_5 \frac{1}{25} = -2$
7. $\log_x 3 = -12$
8. $\log_4 Q = t$

Determine the equivalent logarithmic equation.

9. $4^2 = x$
10. $20^0 = 1$
11. $10^{-3} = 0.001$
12. $27^{2/3} = 9$
13. $\sqrt{4} = 2$
14. $\sqrt[3]{27} = 3$
15. $x^{-3} = 0.01$
16. $a^b = c$

Simplify the following expressions.

17. $\log_x x$
18. $\log_e e$
19. $\log_{10} 10$
20. $\log_{23} 23$
21. $\log_m 1$
22. $\log_{10} 1$

23. $\log_e 1$
24. $\log_k 1$
25. $25^{\log_{25} 4}$
26. $9^{\log_9 3x}$
27. $\log_{10} 10^{-3}$
28. $\log_8 8^k$
29. $e^{\log_e 0.8241}$
30. $10^{\log_{10} 25}$
31. $\log_2 8$
32. $\log_3 9^2$
33. $y^{\log_7 (2x-1)}$
34. $\log_5 5^{(3x+5)}$
35. $\log_m m^{x^4}$
36. $t^{\log_t (x^2+3)}$

Sketch graphs of the following functions.

37. a. $f(x) = \log_3(x)$
    b. $f(x) = \log_3(x+1)$
    c. $f(x) = \log_3(x) - 2$
38. a. $f(x) = \log_4(x)$
    b. $f(x) = \log_4(x-1)$
    c. $f(x) = \log_4(x) + 3$
39. $f(x) = \log_5(x)$
40. $f(x) = \log_{10}(x)$
41. $f(x) = \log_2(x-3)$
42. $f(x) = \log_2(x+3)$
43. $f(x) = \log_2 |x|$
44. $f(x) = |\log_2(x)|$

45. $f(x) = \log_2 (x^2)$

46. $f(x) = \log_2 (\sqrt{x})$

47. $f(x) = \log_2 \left(\frac{1}{x}\right)$

48. $f(x) = \frac{1}{\log_2 x}$

Write the following expressions using a single logarithm.

49. $\frac{1}{3}(2\log_{10} 8 - 6\log_{10} 3) - 2\log_{10} 2 + \log_{10} 3$

50. $\frac{1}{2}(\log_{10} 6 + \log_{10} 4 - \log_{10} 12 + \log_{10} 2) + 2\log_{10} 5$

51. $\frac{1}{2}\log_a 36 - \frac{1}{3}\log_a 8$

52. $2\log_a 25x - \log_a 125x$ 53. $2\log_b \sqrt{bx} - \log_b x$

54. $\frac{1}{2}\log_a 8x^2 - \frac{1}{6}\log_a 8x^3$

55. $\log_a (x^2 - 9) - \log_a (x - 3)$

56. $\log_b (x^2 - 1) - [\log_b (x + 1) + \log_b (x - 1)]$

57. $\log_3 (2x - 3) - \log_3 (2x^2 - x - 3) + \log_3 (3x + 3)$

58. $\log_a (3x^2 - 5x - 2) - \log_a (x^2 - 4) - \log_a (3x + 1)$

59. $\log_a (x^3 - y^3) - \log_a (x^2 + xy + y^2)$

60. $\log_a (m + n) + \log_a (m^2 - mn + n^2)$

In Exercises 61–70, express the given logarithm in terms of $\log x$, $\log y$, and $\log z$.

61. $\log_b \left(\frac{x\sqrt{y}}{z^2}\right)$

62. $\log_c \left(x^3 \sqrt[4]{\frac{y}{z}}\right)$

63. $\log_2 (x^2 y^2 z^4)$

64. $\log_5 x^3 (yz)^4$

65. $\log_3 \frac{x^4 y}{z^2}$

66. $\log_2 \frac{1}{x^2 y z^3}$

67. $\log_{10} \frac{x^2 y^{2/3}}{z^4}$

68. $\log_a \sqrt{\frac{x}{y^2 z^3}}$

69. $\log_{10} \sqrt{x^3 \sqrt{yz^5}}$

70. $\log_6 \frac{x^3 y^2 z^{-1}}{x y^4 z^{-2}}$

In Exercises 71–76, write the given expression in terms of logarithms of polynomials of degree no higher than 2.

71. $\log_a (m^3 - n^3)$

72. $\log_t (P^3 - 8Q^3)$

73. $\log_a \sqrt[3]{x^2 - 6x + 5}$

74. $\log_a \sqrt{\frac{x^2 - y^2}{x + y}}$

75. $\log_b (x^3 + 8)^2$

76. $\log_b (6x^2 + 7xy + 2y^2)$

77. Graph $f(x) = 2^x$ and $f^{-1}(x) = \log_2 x$ using the same set of axes.

78. Graph $f(x) = \log_3 x$ and $f^{-1}(x) = 3^x$ using the same set of axes.

Given that

$\log_7 2 = 0.3562$, $\log_7 3 = 0.5646$, $\log_7 5 = 0.8271$

find each of the following by applying laws of logarithms.

79. $\log_7 4$

80. $\log_7 6$

81. $\log_7 \frac{2}{3}$

82. $\log_7 \frac{3}{5}$

83. $\log_7 \frac{7}{3}$

84. $\log_7 \frac{3}{7}$

85. $\log_7 \frac{1}{7}$

86. $\log_7 \frac{1}{49}$

87. $\log_7 50$ (*Hint:* $50 = 2 \cdot 5^2$)

88. $\log_7 90$

89. $\log_7 \sqrt{5}$

90. $\log_7 \sqrt[3]{2}$

91. $\log_7 \sqrt[5]{24}$

92. $\log_7 \sqrt{\frac{3}{5}}$

93. $\log_7 0.9^{2/3}$

94. $\log_7 \left(\frac{8}{9}\right)^{2.3}$

Binary representation of numbers. Use the function

$$B(x) = \text{INT}[\log_2 (x) + 1]$$

to determine how many digits are required to represent the following integers in binary representation.

95. 78,723

96. 123,876

97. 235,669

98. 380,000

♦ **Applications**

99. A physicist measures the decay of a radioisotope. She finds that after 33 days, only 40 percent of the original amount remains. Determine the half-life of the isotope.

100. Xenon 133 is a radioactive chemical. It is known that 1000 grams of this chemical will decay to 246.6 grams in 10 days. What is its half-life?

101. The population of Mexico City was 14.5 million in 1980 and 21.5 million in 1989. Find the doubling time of the population. Assume exponential growth.

102. *Escherichia coli* bacteria is a common cause of urinary tract infections. A population of 10,000,000 will grow to 10,352,650 in 1 minute. Find the doubling time of the population. Assume exponential growth.

Solve each of the following equations.

103. $\log_x 125 = 3$

104. $\log_x 3 = 2$

105. $3 \cdot 4^x = 1$

106. $56^x = 0$

107. $\log_3 81 = x$

108. $\log_7 x = -2$

109. $\log_{13} x = 3$

110. $\log_x 256 = 4$

111. $\log_x \frac{1}{36} = -2$

112. $\log_5 \frac{1}{25} = x$

113. $\log_x 5 = 3$

114. $\log_5 1 = x$

115. $5^{x^2 - 5} = 14$

116. $2^{5x+7} = 16$

117. $5^{3x-1} = 125^{2x}$

118. $4^{x^2} = 8^{2/3} \cdot 2^{-3x}$

119. $b^x = 5$

120. $\frac{1}{8} = 4^{x-3}$

121. $5^x = 7$

122. $4^{3x+5} = 13$

123. $16^{-x} = 1024$

124. $8^{2x-5} = 2^{x+3}$

125. $4^{2x-1} + 8 = 40$

126. $2^{2-3x} + 2 = 18$

127. $5^{2x}(25^{x^2}) = 125$

128. $\left(\frac{1}{25}\right)^{x+3} = 0.2^{x+9}$

129. $24^{\log_{24} (x^2 - 8)} = 1$

130. $5^{-2 \cdot \log_5 x} = 4$

True or false.

131. $\log_a \frac{xy}{wz} = \log_a x + \log_a y - \log_a w + \log_a z$

132. $\log_2 12 - (\log_2 2 - \log_2 5) = \log_2 30$

133. $\log_a (P + Q) = \log_a P + \log_a Q$

134. $(\log_a P)(\log_a Q) = \log_a PQ$

135. $\dfrac{\log_a P}{\log_a Q} = \log_a \dfrac{P}{Q}$

136. $\log_a (ax)^n = n + n \log_a x$

137. $\log_a x^2 = 2 \log_a x$

138. $\log_a 5x = 5 \log_a x$

Solve each of the following inequalities by examining graphs of functions.

139. $\log_3 x \geq 0$ 140. $\log_3 x < 0$

141. $\log_2 (x + 3) < 0$ 142. $\log_2 (x - 4) \geq 0$

143. $2^x < 1$ 144. $3^x \geq 1$

145. $2^{x-2} < 1$ 146. $3^{x+2} \geq 1$

Simplify.

147. $\log_a (x^3 + 3x^2 y + 3xy^2 + y^3)$

148. $\log_b (a^4 - 4a^3 b + 6a^2 b^2 - 4ab^3 + b^4)$

Prove each of the following.

149. $\log_b \left(\dfrac{1}{P}\right) = -\log_b P$

150. $\log_a \left(\dfrac{x + \sqrt{x^2 - 1}}{x - \sqrt{x^2 - 1}}\right) = 2 \log_a (x + \sqrt{x^2 - 1})$

151. Suppose that $f(x) = 3^x$. Determine $f^{-1}(x)$.

152. Suppose that $f(x) = \log_2 x$. Determine $f^{-1}(x)$.

153. Suppose that $f(x) = 2^x + 2^{-x}$. Prove that $f(x)$ is one-to-one. Determine $f^{-1}(x)$.

154. Suppose that $f(x) = 2^{x^2}$, $x \geq 0$. Prove that $f(x)$ is one-to-one. Determine $f^{-1}(x)$.

155. Visually demonstrate the laws of logarithms using a graphing utility. Look at functions with various integer, irrational, and rational bases.

4.4 COMMON LOGARITHMS AND NATURAL LOGARITHMS

Rather than a single function, $\log_a x$ is really an infinite family of functions, one for each value of a. It is very inconvenient to have calculations stated in terms of many different logarithmic functions. For instance, calculations involving the strength of earthquakes and intensity of light are most conveniently phrased in terms of the function $\log_{10} x$, whereas calculations involving information theory are most conveniently phrased in terms of the function $\log_e x$. In this section, we examine more closely the two logarithmic functions that are most commonly used, namely, $\log_{10} x$ and $\log_e x$.

Common Logarithms

The decimal number system is based on the number 10. Place values are determined by powers of 10. Therefore, it should come as no surprise that logarithms to base 10 play a special role in applications.

Definition 1 **Common Logarithms**	Logarithms to base 10 are called **common (or Briggs) logarithms,** and the function $\log_{10} x$ (also written $\log x$), is called the **common logarithmic** function.

Here are some special values of $\log_{10} x$:

$$\log_{10} 10 = 1$$
$$\log_{10} 100 = \log_{10} 10^2 = 2$$
$$\log_{10} 1000 = \log_{10} 10^3 = 3$$
$$\log_{10} 0.01 = \log_{10} 10^{-2} = -2$$

Before the common availability of scientific calculators, values of common logarithms were determined from tables. However, there is little need for such tables today. Scientific calculators have a key for calculating values of the common logarithm function. This key is usually labeled log or log₁₀ . On most calculators, to compute the value of $\log_{10} x$, press the log key, and then enter the value of x and press ▢.

➤ **EXAMPLE 1**
Calculators and Logarithms

Calculate the following with the use of a calculator.

1. $\log_{10} 50.39$
2. $\log_{10} \frac{5000}{4.1}$

Solution

We assume the use of a TI-81 calculator.

1. Here are the keystrokes needed: $\boxed{\log_{10}}$ 50.39 $\boxed{\text{ENTER}}$. The answer is 1.7023.
2. Here are the keystrokes needed: $\boxed{\log_{10}}$ (5000 ÷ 4.1) $\boxed{\text{ENTER}}$. The answer is 3.08619.

**Historical Note:
Common Logarithms
and Calculations**

Before the invention of computers, common logarithms were used to perform arithmetic calculations. By the properties of logarithms, the logarithm of a product is the sum of the logarithms. To multiply two numbers, one could first determine their logarithms, add them, and then determine the number whose logarithm is the sum. Values of common logarithms were determined from tables, and it was commonplace for most textbooks in the sciences to include a table of common logarithms to aid the computations in the text. Book-length tables of logarithms, carried out to many decimal places, were considered a standard tool for anyone who needed to perform lengthy scientific calculations, especially those involving powers.

Common logarithms are named for the 17th century English mathematician Henry Briggs. In some old books, common logarithms were called Briggs' logarithms. Although common logarithms are an anachronism for computational purposes today, it is impossible to overemphasize the great advance in calculation that they afforded to scientists of the 17th through the 19th centuries. Important calculations in astronomy, physics, and chemistry became possible only after tables of logarithms became available.

So important were tables of logarithms to calculation that when the Works Progress Administration (WPA) was looking for jobs for unemployed scientists and mathematicians during the Great Depression, it commissioned a new set of logarithm tables, carried out to 14 decimal places.   ❙ ∎

➤ **EXAMPLE 2**
Loudness of Sound

The loudness D of a sound, as measured in decibels, may be calculated in terms of the sound's intensity I using the formula

$$D(I) = 10 \log\left(\frac{I}{I_0}\right)$$

where I is measured in watts per square centimeter and I_0 is the minimum perceptible intensity, set by international standard at 10^{-16} watts per square centimeter.

1. Determine the number of decibels in a human conversation if the intensity of the sound of a human voice is 10^{-10} watts per square centimeter.
2. An ordinance bans noise of more than 80 decibels after midnight. What is the intensity of an 80-decibel sound?

Solution

1. The number of decibels in a human conversation is equal to
$$D(10^{-10})$$
By the given formula, this quantity is equal to
$$10\log\left(\frac{10^{-10}}{10^{-16}}\right) = 10\log(10^{-10-(-16)})$$
$$= 10\log(10^6)$$
$$= 10 \cdot 6$$
$$= 60 \text{ decibels}$$

2. The intensity I of an 80-decibel sound satisfies the equation
$$80 = 10\log\left(\frac{I}{I_0}\right)$$
$$8 = \log\left(\frac{I}{I_0}\right)$$
$$10^8 = 10^{\log(I/I_0)}$$
$$10^8 = \frac{I}{I_0}$$
$$I = 10^8 I_0$$
$$I = 10^8 10^{-16}$$
$$= 10^{-8} \text{ watts per square centimeter}$$

Natural Logarithms

The number e assumes its importance because mathematical formulas (particularly in calculus) involving exponential and logarithmic functions assume an especially simple form when expressed in terms of base e. We have already explored such exponential functions. Let's discuss the corresponding logarithmic functions.

Definition 2
Natural Logarithmic Function

Logarithms to base e are called **natural logarithms**. The function $f(x) = \log_e x$, $x > 0$ is called the **natural logarithm function** and is denoted $\ln x$.

Here, the notation "ln" is shorthand for the Latin expression for "natural logarithm." This notation has been in use for several hundred years and has become traditional.

As with the common logarithmic function, it has become traditional to omit the parentheses, where possible, in writing the natural logarithm function. That is, we write $\ln x$ rather than $\ln(x)$, and $\ln e^x$ rather than $\ln(e^x)$. We will use this notation as long as it does not result in confusion. However, in taking the natural logarithm of complicated expressions, we will always use the parentheses for clarity.

To calculate values of the natural logarithmic function, we can use a scientific calculator, which has a key labeled [ln] for this purpose. Calculate $\ln x$ by pressing the [ln] key, keying in the value of x, and then pressing the [ENTER] key.

Applying to the natural logarithmic function the properties previously established for logarithmic functions in general allows us to state the following facts:

PROPERTIES OF NATURAL LOGARITHMS

1. $\ln 1 = 0$
2. $\ln e = 1$
3. $\ln e^x = x$
4. $e^{\ln x} = x, \qquad x > 0$

➢ **EXAMPLE 3**
Using Laws of Logarithms

Determine the values of the following expressions.

1. $\ln e^5$
2. $(\ln e)^5$
3. $e^{5 \ln(2e)}$

Solution

1. By Property 3 of natural logarithms, $\ln e^x = x$, so the value of the expression is $x = 5$.

2. The value of $\ln e$ is 1, so the value of the given expression equals
$$(1)^5 = 1$$

3. By Law of Logarithms 2 in Section 4.3, we have
$$e^{5 \ln(2e)} = e^{\ln(2e)^5}$$

By Property 4 of natural logarithms, $e^{\ln 2e} = 2e$, so
$$e^{\ln(2e)^5} = (2e)^5 = 32e^5$$

The laws of logarithms proved in Section 4.3 are also valid in the special case of the natural logarithmic function.

LAWS OF NATURAL LOGARITHMS

Let $x, y > 0$. Then the following properties of the natural logarithmic function hold.

1. $\ln xy = \ln x + \ln y$
2. $\ln x^a = a \ln x$
3. $\ln \dfrac{x}{y} = \ln x - \ln y$

➢ **EXAMPLE 4**
Ecology

A reservoir has become polluted from an industrial waste spill. The pollution has caused a buildup of algae. The number of algae $N(t)$ present per 1000 gallons of water t days after the spill is given by the formula

$$N(t) = 100e^{2.1t}$$

How long will it take before the algae count reaches 10,000 per 1000 gallons?

Solution

We are asked to determine the value of t satisfying the equation

$$N(t) = 10,000$$

Inserting this value for $N(t)$ into the given formula, we obtain

$$10,000 = 100e^{2.1t}$$
$$e^{2.1t} = 100$$

To solve the last equation, we take natural logarithms of each side and apply the properties of natural logarithms to simplify the resulting equation:

$$\ln(e^{2.1t}) = \ln 100$$
$$2.1t = \ln 100$$
$$t = \frac{\ln 100}{2.1} = \frac{4.6052}{2.1} = 2.1929$$

To compute the final numerical value, we used a scientific calculator. Thus, the number of algae per 1000 gallons of water will reach 10,000 after 2.1929 days. That is, the number of algae will equal 10,000 per 1000 gallons in a little more than 2 days.

➢ EXAMPLE 5
Archaeology

If $Q(t)$ denotes the amount of radioactive carbon 14 present t years after an organism dies, then $Q(t)$ is given by the formula

$$Q(t) = A_0 e^{-0.00012t}$$

where A_0 denotes the amount of carbon 14 present at the time the organism died. Suppose that archaeologists discover a papyrus scroll in which they measure carbon 14 in an amount equal to 30 percent of that present in freshly manufactured papyrus. How old is the papyrus?

Solution

We are asked to determine the value of t for which

$$Q(t) = 0.3A_0$$

Using this value for $Q(t)$, we derive the equation

$$0.3A_0 = A_0 e^{-0.00012t}$$

Dividing both sides by A_0, we remove any dependence of the equation on that number. We obtain

$$e^{-0.00012t} = 0.3$$

To solve this equation, we take natural logarithms of both sides and apply the properties of natural logarithms:

$$\ln(e^{-0.00012t}) = \ln 0.3$$
$$-0.00012t = \ln 0.3$$
$$t = -\frac{\ln 0.3}{0.00012}$$

To obtain a numerical value for t, we use a scientific calculator to evaluate the quantity on the right side:

$$-\frac{\ln 0.3}{0.00012} = -\frac{-1.204}{0.00012} = 10,033$$

That is, the piece of papyrus is 10,033 years old.

In the preceding discussion, we saw that the values of common and natural logarithms can be calculated using scientific calculators. However, these are the only special cases of the function $\log_a x$ that can be calculated directly in this way. For any a, however, we can compute $\log_a x$ using the following formula.

CHANGE OF BASE FORMULA

Let a and b be positive numbers. Then

$$\log_a x = \frac{\log_b x}{\log_b a}$$

Proof   Begin with Fundamental Property 1 of logarithms:

$$x = a^{\log_a x}$$

Then

$$\log_b x = \log_b a^{\log_a x} \qquad \text{\textit{Taking} } \log_b \text{ \textit{of both sides}}$$
$$= \log_a x \cdot \log_b a \qquad \text{\textit{Law of Logarithms 2}}$$

Therefore,

$$\log_a x = \frac{\log_b x}{\log_b a} \qquad \qquad \blacklozenge$$

The Change of Base formula asserts that any logarithmic function is a constant multiple of any other given logarithmic function. In particular, we can calculate logarithms with base a in terms of either natural logarithms or common logarithms using the formulas

$$\log_a x = \frac{\ln x}{\ln a}$$
$$\log_a x = \frac{\log x}{\log a}$$

By setting a equal to 10 in the first formula, we have the following relationship between common and natural logarithms:

$$\log x = \frac{\ln x}{\ln 10}$$

➤ **EXAMPLE 6**
Using Change
of Base Formula

Calculate $\log_{100} 10$ without using a calculator.

Solution

By the Change of Base formula, we have

$$\log_{100} 10 = \frac{\log_{10} 10}{\log_{10} 100}$$
$$= \tfrac{1}{2}$$

because $\log_{10} 10 = 1$ and $\log_{10} 100 = 2$.

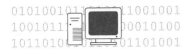

New Technology

EXAMPLE 7
Graphing a Function
Involving a Natural
Logarithm

Sketch the graph of the function

$$f(x) = x \ln x, \qquad x > 0$$

Solution

Use the range $0 \le x \le 10$, $-1 \le y \le 20$. The graph is shown in Figure 1. Note that the calculator does not plot a point for the function $\ln x$ for $x = 0$ because this value is not defined.

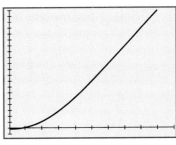

Figure 1 ◀

Exercises 4.4

Find each common logarithm without using a calculator.

1. $\log_{10} 10$
2. $\log_{10} 100$
3. $\log_{10} 1000$
4. $\log_{10} 10,000$
5. $\log_{10} 0.1$
6. $\log_{10} 0.01$
7. $\log_{10} 0.001$
8. $\log_{10} 0.0001$

Find each of the following using a calculator.

9. $\log_{10} 83.47$
10. $\log_{10} 124.7$
11. $\log_{10} 3.456$
12. $\log_{10} 9.87$
13. $\log_{10} 0.458$
14. $\log_{10} 0.00778$
15. $\log_{10} (6.732 \times 10^6)$
16. $\log_{10} (4.012 \times 10^8)$
17. $\log_{10} (9.152 \times 10^{49})$
18. $\log_{10} (1.302 \times 10^{28})$
19. $\log_{10} (8.6022 \times 10^{-4})$
20. $\log_{10} (5.554432 \times 10^{-7})$
21. $\log_{10} (1.21223 \times 10^{-23})$
22. $\log_{10} (9.902 \times 10^{-19})$

♦ Applications

23. **Loudness of sound.** The loudness D of a sound, as measured in decibels, may be calculated in terms of the sound's intensity I using the formula

$$D(I) = 10 \log \left(\frac{I}{I_0} \right)$$

where I is measured in watts per square centimeter and I_0 is the minimum perceptible intensity, set by international standard at 10^{-16} watts per square centimeter. Determine the number of decibels from the sound of an elevated train if the intensity is 10^{-3} watts per square meter.

24. Determine the number of decibels of the sound of a riveter if the intensity is 3.2×10^{-3} watts per square meter.

25. The noise of rustling leaves is 10 decibels. What is the intensity of a 10-decibel sound?

26. Sound actually hurts the ear at levels of 120 decibels or higher. What is the intensity of a 120-decibel sound?

27. Two sounds have intensities I_1 and I_2, respectively. Show that the difference in their sound levels can be expressed as

$$D(I_1) - D(I_2) = 10 \log \frac{I_1}{I_2}$$

28. Find a formula for the sum of the sound levels of two sounds with intensities I_1 and I_2.

29. Solve the sound level formula for the intensity I.

30. **Earthquake magnitude.** The magnitude R of an earthquake, as measured on what is called the Richter scale, may be calculated in terms of an intensity I, where

$$R(I) = \log \frac{I}{I_0}$$

and I_0 is a minimum perceptible intensity. What is the value of $R(I_0)$?

31. Find the magnitudes of earthquakes whose intensities are

a. $10I_0$ b. $100I_0$ c. $1000I_0$ d. $10,000I_0$
e. $100,000I_0$ f. $100,000,000I_0$

32. One earthquake has a magnitude 10 times as intense as another. How much higher is its magnitude on the Richter scale?

33. One earthquake has a magnitude 100 times as intense as another. How much higher is its magnitude on the Richter scale?

34. One of the worst earthquakes ever recorded was near Lebu, Chile, on May 21–23, 1960. It had an intensity of $10^{8.9} \cdot I_0$. What was its magnitude on the Richter scale?

35. Two earthquakes have intensities I_1 and I_2, respectively. Show that the difference in their magnitudes can be expressed as

$$R(I_1) - R(I_2) = \log \frac{I_1}{I_2}$$

36. Find a formula for the sum of the magnitudes of two earthquakes with intensities I_1 and I_2.

37. Solve the Richter scale magnitude formula for I.

38. **pH in chemistry.** In chemistry, the pH of a substance is defined by

$$pH = -\log_{10} H^+$$

where H^+ represents the concentration of hydrogen ions, in moles per liter. Find the pH of each of the following:

   a. Normal blood, $H^+ = 3.4 \times 10^{-8}$

   b. Ammonia, $H^+ = 1.6 \times 10^{-11}$

   c. Wine, $H^+ = 4.0 \times 10^{-4}$

39. Given the following pH values, find the corresponding hydrogen ion concentrations, H^+.

   a. Eggs, pH $= 7.8$

   b. Tomatoes, pH $= 4.2$

40. Solve the pH formula for H^+.

41. A substance is said to be an **acid** if its pH is less than 7. For what hydrogen ion concentrations is a substance an acid?

42. A substance is said to be a **base** if its pH is greater than 7. For what hydrogen ion concentrations is a substance a base?

Determine the values of the following expressions.

43. $\ln 1$ 44. $\ln e^2$ 45. $\ln e^{-7}$ 46. $(\ln e)^{-7}$

47. $(\ln e)^2$ 48. $e^{\ln t}$ 49. $e^{\ln 7k}$ 50. $e^{3\ln(4e)}$

51. $e^{-4\ln(5e^2)}$ 52. $\ln \sqrt{e^8}$ 53. $\ln \sqrt{e^6}$

54. $\ln e^2 + 4^3$ 55. $\ln e^3 + 5^3$

Simplify the following expressions.

56. $\dfrac{1}{2\ln 10}(10)^{2x}$ 57. $\dfrac{\ln x^2}{x}, x > 0$ 58. $\dfrac{\ln n}{\ln n^2}$

59. $\dfrac{1}{n \ln(n^3)}$ 60. $e^{2\ln x}$ 61. $e^{\ln x - 2\ln y}$

62. $\frac{1}{2}(\ln 4 - \ln 1)$ 63. $\ln 8 - \ln 2$

64. $\frac{1}{2}e^{2(\ln 6)+1} - \frac{1}{2}e^{2(\ln 2)+1}$

65. $2^x(\ln 3)3^{2x} \cdot 2 + 3^{2x}(\ln 2)2^x$

66. $\ln(4^2 + 4 - 1) - \ln(1^2 + 1 - 1)$

67. $-\frac{1}{3}\ln \frac{8}{4} + \frac{1}{3}\ln(3 + \sqrt{10})$

68. $\dfrac{x(\ln 4)4^{1/x}\left(-\dfrac{1}{x^2}\right) - 4^{1/x}}{x^2}$

♦ Applications

69. **Ecology.** A reservoir has become polluted due to an industrial waste spill. The pollution has caused a buildup of algae. The number of algae $N(t)$ present per 1000 gallons of water t days after the spill is given by the formula

$$N(t) = 100e^{2.1t}$$

How long will it take before the algae count reaches 20,000 per 1000 gallons?

70. **Archaeology.** Use the formula of Example 5. Suppose archaeologists find a Chinese artifact that has lost 40 percent of its carbon 14. How old is the artifact?

71. **The population of the United States.** (See figure.) The population of the United States was 240 million on January 1, 1986. Think of that date as $t = 0$. The population $P(t)$ t years later is given by

$$P(t) = P_0 e^{kt}$$

where $P_0 = 240$ million and $k = 0.009$.

   a. Estimate the population of the United States in 1995 $(t = 9)$. Estimate the U.S. population in 2001.

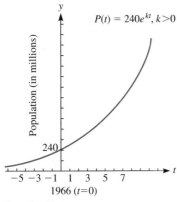

$P(t) = 240e^{kt}, k>0$

Population (in millions)

240

-5 -3 -1 1 3 5 7

1966 $(t=0)$

Exercise 71

b. After what period of time will the population be double that in 1986?

72. Exponential population growth is often modeled by the function

$$P(t) = P_0 e^{kt}$$

where P_0 is the initial population at time $t = 0$ and k is a positive constant, called the *exponential growth rate*.

a. Show that k and the doubling time T are related by the formula

$$k = \frac{\ln 2}{T}$$

b. The growth rate of the U.S. population is 0.9 percent. Find the doubling time.

c. The doubling time of the population of Texas is 23 years. Find the exponential growth rate.

73. **Population growth of California.** The population of California was 23,669,000 in 1980 and was estimated to be 27,526,000 in 1990. Assuming this was exponential growth:

a. Find the exponential growth rate k.

b. Find the equation of exponential growth.

c. Estimate the population in 2000.

74. **Interest compounded continuously**. A method of compounding interest that used to be quite prevalent was *compounding continuously*. It is used less now because the public did not understand it. A "population" of money, P_0, invested at interest rate k, compounded continuously, grows to an amount $P(t)$ in t years, where

$$P(t) = P_0 e^{kt}$$

Suppose $10,000 is invested in a bank account at 8 percent interest, compounded continuously.

a. How much is in the account after 1 year? After 2 years? After 5 years? After 10 years?

b. How much would be in the account after the time periods in part (a) if interest had been compounded annually?

c. What is the doubling time of this population of money?

d. Suppose the money doubles in value every 9.9 years. What is the interest rate?

75. A new father deposits $250 in a bank account paying $10\frac{1}{4}$ percent, compounded continuously. How much will be in the account in 18 years?

76. If the father in Exercise 75 wanted to have $30,000 in the bank account after 18 years, how much should he now invest?

77. What would the interest rate have to be if the father wanted the $250 to grow to $40,000 in 18 years?

78. **Atmospheric pressure**. At altitude a, under standard conditions of temperature, the atmospheric pressure $P(a)$ is given by

$$P(a) = P_0 e^{-0.0000385a}$$

At sea level $P_0 = 1013$ millibars.

a. What is the pressure at 1000 ft? At 2000 ft? At 10,000 ft? At 30,000 ft?

b. A device for measuring atmospheric pressure reads 1013 millibars on the ground floor of the Empire State Building and 965.4038 millibars at the top of the building. How tall is the Empire State Building?

79. **Magnitude of a star**. In astronomy, the **magnitude** M of a star is a measure of the brightness of the star. (Stars with the greatest magnitude are the faintest.) The magnitude of stars made visible through a telescope depends on the aperture a of the telescope, as given by the formula

$$M(a) = 9 + \log_{10} a$$

Calculate the magnitude of the faintest star made visible by a telescope with these apertures:

a. 3 inches.

b. 30 inches.

80. **Beer-Lambert law**. A beam of light enters a medium, such as water or smoky air, with initial intensity I_0. Its intensity is decreased depending on the thickness (or concentration) of the medium. The intensity I at a depth (or concentration) of x units is given by

$$I = I_0 e^{-\mu x}$$

The constant μ (pronounced "mu"), called the **coefficient of absorption**, varies with the medium. Light through smog has $\mu = 1.4$.

a. What percentage of I_0 remains at a depth of sea water that is 1 m? 2 m? 3 m?

b. Plant life cannot exist below 10 meters. What percentage of I_0 remains at 10 meters?

81. **Salvage value**. A business estimates that the salvage value V of a piece of machinery after t years is given by $V(t) = \$80,000 e^{-t}$.

a. What did the machinery cost initially?

b. What is the salvage value after 4 years?

82. **Length of a parabolic arc**. (See figure.) When an object is thrown into the air at an angle, it follows a parabolic path. If the height H to which the object rises and the horizontal distance D are known, the actual distance traveled L, or the length of the arc, is given by

$$L = 2D\left\{\sqrt{N^2 + \tfrac{1}{16}} + \tfrac{1}{16}N\left[\ln 4\left(N + \sqrt{N^2 + \tfrac{1}{16}}\right)\right]\right\}$$

where $N = H/D$.

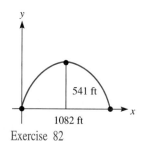

Exercise 82

A cannonball shot at an angle of 45° with an initial velocity of 180 mph will rise to a height of 541 ft and will travel a horizontal distance of 1082 ft. What is the actual distance (the length of the parabola) the ball travels?

83. A soccer ball kicked at an angle of 37° with an initial speed of 50 ft/sec will rise to a height of 14 ft and will travel a horizontal distance of 75 ft. What is the actual distance the ball travels?

84. **Newton's law of cooling**. An object hotter than its surroundings will cool down to its surroundings and a cool object will warm up to its surroundings exponentially. If a body of temperature T_1 is placed in surroundings of temperature T_0, it will cool or warm to temperature $T(t)$ after time t, in minutes, where

$$T(t) = T_0 + |T_1 - T_0|e^{-kt}$$

a. An object of temperature 98.4°C is placed in a 26°C room. After 12 seconds ($\tfrac{1}{5}$ min), the temperature is 66.4°C. What will its temperature be 5 minutes after placement in a room?

b. Another object, at water's freezing temperature (0°C), is placed in the same room. After 30 seconds, it has warmed up to 10°C. What will its temperature be in 10 minutes?

Use the Change of Base formula to find the following logarithms.

85. $\log_7 50$

86. $\log_{12} 265$

87. $\log_{23} 0.0089$

88. $\log_8 2.347$

Convert each of the following to natural logarithms.

89. $\log 4578$

90. $\log 2.347$

91. $\log x^2$

92. $\log \sqrt{x}$

Convert each of the following to common logarithms.

93. $\ln 0.987$

94. $\ln 78.56$

95. $\ln \sqrt{y}$

96. $\ln t^4$

Simplify.

97. $(e^{e^x})^2$

98. $\dfrac{x^2}{2}\ln\left[x(x^2 + 1)\right] + \tfrac{1}{2}\ln(x^2 + 1) - \tfrac{3}{4}x^2$

Simplify by rationalizing the numerator of the fractional expression involving z.

99. $\dfrac{1}{\sqrt{3}}\ln\left|\dfrac{z - \sqrt{3}}{z + \sqrt{3}}\right|$

➡ **Technology**

100. **Population growth of Hawaii**. The population of Hawaii in 1980 was 964,691. The total area of Hawaii is 6425 square miles. The growth rate was 1.8 percent per year. Assuming this was exponential growth, after how many years will the population be such that there will be one person for every square yard of land?

Use a graphing calculator to approximate the solutions to the following equations to two significant digits.

101. $3^x - 2^x = 40$

102. $5^x = 5^{-x} - 4$

103. $3^x + x^2 = 50$

104. $2^x - 3^x + e^x = x^2 + x^3$

Match the following functions with the graphs (a)–(d).

105. $f(x) = \ln x$ 106. $f(x) = \ln x + 1$

107. $f(x) = \ln x - 1$ 108. $f(x) = -\ln x$

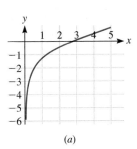
(a)

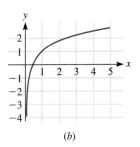
(b)

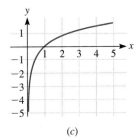
(c)

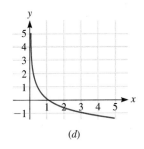
(d)

4.5 EXPONENTIAL AND LOGARITHMIC EQUATIONS

In Section 4.4, we saw how equations involving logarithms and exponential functions arise in applications. Actually, the discussion of that section was only the tip of the iceberg. Equations involving logarithms and exponential functions are common throughout mathematics, and it is important to become proficient in solving them. In this section, we will solve a number of typical equations of this sort as a first step in developing that proficiency.

➤ **EXAMPLE 1**
Exponential Equation

Solve the following equations:

1. $5^{x^2} = 3$
2. $e^{8x+2} = 3^x$

Solution

1. The presence of the variable in the exponent suggests that we take logarithms of both sides. But logarithms to what base? Actually, it doesn't matter because logarithms to all bases are related to one another by the Change of Base formula. It is fairly standard to perform all calculations in terms of the natural logarithmic function. Taking natural logarithms of both sides yields

$$\ln 5^{x^2} = \ln 3$$

Therefore, we obtain

$$x^2 \cdot \ln 5 = \ln 3$$

$$x^2 = \frac{\ln 3}{\ln 5}$$

$$x = \pm\sqrt{\frac{\ln 3}{\ln 5}} \approx \pm 0.8262$$

2. Again we take natural logarithms of the expressions on both sides to obtain

$$\ln e^{8x+2} = \ln 3^x$$

Therefore, we have

$$8x + 2 = x \ln 3$$

In spite of the complex appearance of this equation, it is just a linear equation in x because $\ln 3$ is a constant. Collecting terms involving x on the left side of the equation yields

$$(8 - \ln 3)x = -2$$

$$x = -\frac{2}{8 - \ln 3}$$

$$\approx -0.2898$$

➤ **EXAMPLE 2**
Logarithmic Equation

Solve the equation

$$\ln \frac{x - 1}{2x} = 4$$

Solution

To eliminate the natural logarithmic function on the left side of the equation, we express both sides as a logarithm with base e:

$$e^{\ln[(x-1)/2x]} = e^4$$

$$\frac{x - 1}{2x} = e^4$$

$$x - 1 = 2e^4 x$$

$$x - 2e^4 x = 1$$

$$(1 - 2e^4)x = 1$$

$$x = \frac{1}{1 - 2e^4}$$

$$\approx -0.0092425$$

➤ **EXAMPLE 3**
Hyperbolic Function

The function sinh (x) (pronounced "sinch x"), called the **hyperbolic sine** of x, is defined by the equation

$$\sinh(x) = \frac{e^x - e^{-x}}{2}$$

Determine all values of x for which sinh (x) is equal to $\frac{1}{4}$.

Solution

The equation sinh $(x) = \frac{1}{4}$, is equivalent to

$$\frac{e^x - e^{-x}}{2} = \frac{1}{4}$$

$$2e^x - 2e^{-x} = 1$$

To eliminate the negative exponent, we multiply both sides by e^x to obtain

$$2e^x e^x - 2e^{-x} e^x = e^x$$
$$2e^{2x} - 2e^0 - e^x = 0$$
$$2e^{2x} - e^x - 2 = 0$$

To solve this last equation, we write it in the form

$$2(e^x)^2 - e^x - 2 = 0$$

This equation is a quadratic equation in the variable e^x. Applying the quadratic formula gives us the solutions

$$e^x = \frac{1 \pm \sqrt{(-1)^2 - 4(2)(-2)}}{4} = \frac{1 \pm \sqrt{17}}{4}$$

To determine the value of x, we take natural logarithms of both sides. Recall, however, that the domain of the natural logarithmic function consists of the positive real numbers. In particular, the natural logarithmic function is undefined at the value of the right side corresponding to the minus sign because this gives a negative value for the right side. The only possible value of x corresponds to

$$\ln e^x = x = \ln\left(\frac{1 + \sqrt{17}}{4}\right) \approx 0.2475$$

> **EXAMPLE 4**
A More Complex
Logarithmic Equation

Solve the equation

$$\ln(x^2 + 2x) = 0$$

Solution

Raise e to the power indicated by both sides of the equation:

$$e^{\ln(x^2 + 2x)} = e^0$$
$$x^2 + 2x = 1$$
$$x^2 + 2x - 1 = 0$$
$$x = \frac{-2 \pm \sqrt{2^2 - 4(1)(-1)}}{2} = \frac{-2 \pm \sqrt{8}}{2}$$
$$= \frac{-2 \pm 2\sqrt{2}}{2} = -1 \pm \sqrt{2}$$

New Technology

EXAMPLE 5
Solving a Logarithmic
Equation

Determine the approximate values of the solutions of the equation

$$8 \ln x = 2x$$

Solution

Plot the two functions

$$Y_1 = 8 \ln X$$
$$Y_2 = 2X$$

on the same coordinate system. See Figure 1. We use the trace to locate the point of intersection, which is approximately $(1.3684211, 2.5092605)$. The *x*-coordinate of this point is the solution to the equation. That is, the solution of the equation is approximately 1.3684211. An equation involving both polynomial and logarithmic expressions cannot usually be solved exactly. However, graphing calculators can provide numerical approximations to the solutions.

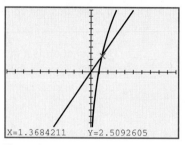

X=1.3684211   Y=2.5092605

Figure 1   ◀

Exercises 4.5

Solve the following equations.

1. $7^{x^2} = 4$

2. $8^{x^2} = 5$

3. $e^{5x-2} = 4^x$

4. $e^{7x+4} = 5^{2x}$

5. $\ln\left(\dfrac{x+1}{3x}\right) = 6$

6. $\ln\left(\dfrac{2x-3}{5x+2}\right) = 7$

7. $e^t = 100$

8. $e^t = 60$

9. $e^{-t} = 0.01$

10. $e^{-0.02t} = 0.06$

11. $184.50 = 100e^{10k}$

12. $52 = 25e^{9k}$

13. $90 = 25e^{-10k} + 75$

14. $25e^{-0.05t} + 75 = 80$

15. $\ln(x^2 + 3x) = 0$

16. $\ln(x^2 - 5x) = 0$

17. $\log_3(6x + 5) = 2$

18. $\log_4(8 - 3x) = 3$

19. $\log x + \log(x + 9) = 1$

20. $\log(x - 9) + \log x = 1$

21. $\log(x + 3) + \log x = 1$

22. $\log(x - 3) + \log x = 1$

23. $\log_4(x + 5) + \log_4(x - 5) = 2$

24. $\log_5(x - 5) + \log_5(x + 2) = 2$

25. $\log_2 x - \log_2(x - 3) = 5$

26. $\log_3(2x - 5) - \log_3(3x + 1) = 4$

27. $(\log_2 x)^2 + \log_2 x - 6 = 0$

28. $(\log x)^2 - \log x = 6$

29. $\dfrac{\log(x + 1)}{\log x} = 2$

30. $\dfrac{\log(x + 2)}{\log x} = 2$

31. $2 = 5\log x + 12(\log x)^2$

32. $\log(x - 1) + \log(x + 2) = \log_7 7$

33. $\log_5 x^3 + \log_5 x = 4$

34. $3\log x = \log 125$

35. $\log_3(\log_2 x) = 1$

36. $\log_5(\log_6 x) = 0$

37. $e^{2x} + 5e^x - 14 = 0$

38. $e^{4x} - e^{2x} = 12$

39. $e^x + e^{-x} - 1 = 0$

40. $5e^x - 3e^{-x} = 8$

41. The function sinh (x), called the hyperbolic sine of x, is defined by

$$\sinh(x) = \frac{e^x - e^{-x}}{2}$$

Determine all values of x for which sinh (x) equals 8.

42. The function cosh (x), called the **hyperbolic cosine** of x, is defined by

$$\cosh(x) = \frac{e^x + e^{-x}}{2}$$

Determine all values of x for which cosh (x) equals 3.

43. The function tanh (x), called the **hyperbolic tangent** of x, is defined by

$$\tanh(x) = \frac{e^x - e^{-x}}{e^x + e^{-x}}$$

Determine all values of x for which tanh (x) equals $\frac{1}{2}$.

44. Determine all values of x for which tanh (x) equals $-\frac{1}{4}$.

Solve the following equations.

45. $|\log_3 x| = 2$

46. $\log_2 |x| = 5$

47. $\log_2 \sqrt[3]{x - 12} = 2 - \log_2 \sqrt[3]{x}$

48. $\log \sqrt{\dfrac{x + 5}{x - 4}} = 2$

49. $\log x^{\log x} = 9$

50. $x^{\log x} = 1000x$

51. $3^{\log_2 x} = 3x$

52. $\sqrt{\dfrac{(e^{3x} \cdot e^{-7x})^{-6}}{e^{-x}/e^x}} = e^{10}$

Solve each of the following equations for t.

53. $P(t) = P_0 e^{kt}$

54. $N(t) = N_0 e^{-kt}$

55. $N = \dfrac{P}{1 + Ae^{-kt}}$

56. $T = T_0 + |T_1 - T_0|e^{-kt}$

57. $\dfrac{e^t + e^{-t}}{2} = x$

58. $\dfrac{e^t - e^{-t}}{2} = x$

59. $\dfrac{e^t - e^{-t}}{e^t + e^{-t}} = x$

60. $v = v_T(1 - e^{-kt})$

61. $I = \dfrac{E}{R}(1 - e^{-Rt/L})$

62. $N = GB^t$

Prove each of the following.

63. $\tanh(x) = \dfrac{\sinh(x)}{\cosh(x)}$

64. $\left[\cosh(x)\right]^2 - \left[\sinh(x)\right]^2 = 1$

65. $\sinh(a + b) = \sinh(a)\cosh(b) + \cosh(a)\sinh(b)$

66. $\sinh(x)$ is an odd function.

67. $\sinh(2x) = 2\sinh(x)\cosh(x)$

Determine a function of the form $f(x) = Ce^{kx}$ that has the properties shown in the graphs.

68.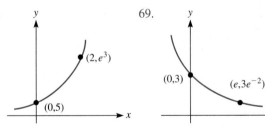

69.

70.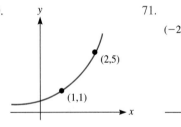

71.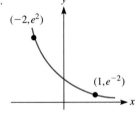

72.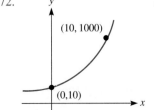

73.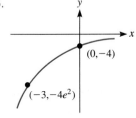

→ **Technology**

Use a graphing calculator to approximate numerically the solutions to the following equations.

74. $\log(x + 1) + \log(x - 3) = 4.1$

75. $(\log x)^2 - 3x = 1$

76. $\log \log x = 0.1x$

77. $e^x = 3\log x$

78. $\log x + 10\sqrt{x} = 4$

79. $e^{3.7x} - e^{1.5x} = 8$

■ 4.6 CHAPTER REVIEW

Important Concepts, Properties, and Formulas—Chapter 4

Laws of exponents	$a^x a^y = a^{x+y}$ $(a^x)^y = a^{xy}$ $(ab)^x = a^x b^x$	p. 222
Compound interest formula	$A(t) = P\left(1 + \dfrac{i}{n}\right)^{nt}$	p. 224
Radioactive decay	$Q(t) = A_0 2^{-t/H} \quad Q(t) = Q_0 e^{-kt}$	p. 225
Exponential growth	$N(t) = N_0 2^{t/D} \quad P(t) = P_0 e^{kt}$	p. 226
Other exponential formulas	$v(t) = v_T(1 - e^{-kt})$ $N(t) = \dfrac{P}{1 + Ae^{-kt}}$	p. 234

Properties and laws of logarithms	$\log_a x = y$ equivalent to $a^y = x$	pp. 239, 240, 242, 243, 253, 254
	$\log_a a = 1; \log_a 1 = 0$	
	$a^{\log_a y} = y; \log_a a^x = x$	
	$\log_a(xy) = \log_a x + \log_a y$	
	$\log_a\left(\frac{x}{y}\right) = \log_a x - \log_a y$	
	$\log_a x^y = y \log_a x$	
	$\log x = \log_{10} x, \quad \ln x = \log_e x$	
	$\ln xy = \ln x + \ln y, \quad \ln\frac{x}{y} = \ln x - \ln y$	
	$\log_a x = \dfrac{\log_b x}{\log_b a}, \quad \ln x^k = k \ln x$	

Cumulative Review Exercises—Chapter 4

Determine the equivalent exponential equation.

1. $\log_e 2 = 0.6931$
2. $\log_{10} 2 = 0.3010$
3. $\log_{0.5} 0.25 = 2$
4. $\log_{0.25} 0.5 = \frac{1}{2}$
5. $\ln p = k$
6. $\log_3 \sqrt{27} = \frac{3}{2}$

Determine the equivalent logarithmic equation.

7. $2^1 = 2$
8. $2^{-1} = \frac{1}{2}$
9. $e^2 = 7.3891$
10. $16^x = 4.015$

♦ **Applications**

11. **Exponential growth.** The population of Kenya was 20 million on January 1, 1985, and is doubling every 23 years. Suppose we let $t = 0$ correspond to 1985. The population after t years is $N(t)$ where

    $$N(t) = N_0 2^{t/D}$$

    and D is the doubling time.

    a. Write $N(t)$ in terms of an exponential function.

    b. What will the population of Kenya be in 23 years? In 46 years? In 1995? In 2000?

    c. How many years will it take for the population of Kenya to reach 25 million?

12. **Radioisotope power supply.** A space satellite has a radioisotope power supply. The power output $W(t)$ is given by the equation

    $$W(t) = W_0 e^{-t/250}$$

    where t is the time in days, $W(t)$ is measured in watts, and $W_0 = 50$ watts.

    a. How much power is available at the beginning of the flight?

    b. How much power will be available after 30 days? After 1 year?

    c. What is the half-life of the power supply?

    d. The equipment aboard the satellite requires 10 watts of power to operate properly. How long can the satellite operate properly?

Sketch a graph of each of the following functions.

13. $f(x) = 2^x$
14. $f(x) = \log_2 x$
15. $f(x) = 2^{-(x-1)^2}$
16. $f(x) = |3^x - 4|$
17. $f(x) = \log_2(x + 3) - 5$
18. $f(x) = \log_3 |x|$
19. $f(x) = |\log_3 x|$
20. $f(x) = \log_3\left(\frac{3}{x}\right)$

Solve each of the following equations.

21. $\log_9 81 = x$
22. $\log_x \frac{9}{16} = 2$
23. $\log_2 x = -\frac{1}{2}$
24. $\ln x = 4$
25. $\log_8 x = -\frac{2}{3}$
26. $\log x = -3$
27. $\log_2(x^2 - 1) - \log_2(x + 1) + \log_2(x - 1) = 0$
28. $2\log_6 \sqrt{6x} - \log_6 x = 1$
29. $a^{3\log_a x} x = 16$
30. $\log_3 x = -2$
31. $27^{2x+3} - 9 = 72$
32. $\log_4(x - 2) + \log_4(x - 8) = 2$
33. $e^{-2x} = 0.01$
34. $\log_2[\log_4(\log_{10} x)] = -1$
35. $\log_{10}[\log_2(\log_x 25)] = 0$
36. $3e^x + 2 = 7e^{-x}$
37. $3(\log_8 x)^2 + \log_8 x - 2 = 0$

38. $\dfrac{e^x + e^{-x}}{2} = \dfrac{1}{3}$

39. $\log x^{\log x} = 16$

40. $21 = 60(1 - e^{-10k})$

41. $|\log_4 x| = 2$

42. $10^x = 3.6308$

Simplify the following expressions involving e. Where appropriate, approximate values using a calculator with an e^x key.

43. $\sqrt{e^{4x}}$

44. $2700(e^{0.008 \cdot 10} - e^{0.008 \cdot 0})$

45. $(e^x + e^y)(e^x - e^y)$

46. $(e^x + 2)(e^{2x} - 2e^x + 4)$

◆ Applications

47. **Sales of a firm.** A firm's sales S increase toward a limiting value of \$20 million and satisfy the following function:

$$S(t) = \$20(1 - e^{-kt})$$

where $k = 0.11$ and time t is in years.

   a. Find $S(t)$ for $t = 0, 1, 2, 5,$ and 10 years.

   b. Sketch a graph of $S(t)$ for $t \geq 0$.

   c. After what time will the sales be \$10 million?

48. **Spread of a rumor**. In a college with a population of 6400, a group of students spread the rumor "People who study mathematics are people you can count on!" The number of people $N(t)$ who have heard the rumor after t minutes is given by

$$N(t) = \dfrac{P}{1 + Ae^{-kt}}$$

where A and k are constants. In this situation, $P = 6400$, $A = 8$, and $k = 0.23$.

   a. How many people start the rumor at $t = 0$?

   b. How many people have heard the rumor after 1 minute? After 2 min? After 6 min? After 12 min? After 30 min? After 1 hr?

   c. Sketch a graph of the **logistic curve** $N(t)$ for $t \geq 0$.

   d. How much time passes before half the students have heard the rumor?

Simplify each of the following expressions.

49. $6^{\log_6(5x-2)}$

50. $\log_5 25^{(3x-1)}$

51. $\log_b b$

52. $\ln 1$

53. $\log 10^3$

54. $\log_4 4x$

55. $\log_x x^4$

56. $4^{2\log_4 x + (1/2)\log_4 9}$

57. $2\log_a x - 3\log_a y + \log_a(x + y)$

58. $\log_a x^3 - \log_a \sqrt{x}$

59. $\frac{1}{3}\left[\log_a x - \frac{1}{2}\log_a w\right] + 3\log_a v$

60. $\log(1 - t^3) - \log(1 + t + t^2)$

Express each of the following as a combination of logarithms.

61. $\log_3 \dfrac{x^3(y + z)}{x^2}$

62. $\log_a\left(\dfrac{x^2 y}{\sqrt{xz}}\right)$

Given that

$$\log_b 3 = 1.099 \quad \text{and} \quad \log_b 5 = 1.609$$

find each of the following by applying laws of logarithms.

63. $\log_b b^2$

64. $\log_b 3b$

65. $\log_b b^b$

66. $\log_b \dfrac{b}{b + 1}$

67. $\log_b \sqrt[3]{b}$

68. $\log_b \frac{1}{5}$

69. $\log_b \sqrt{b}$

70. $\log_b (b^{4.7\sqrt{b}})$

Binary representation of numbers. Use the function

$$B(x) = \text{INT}[\log_2(x) + 1]$$

to determine how many digits are required to represent the following integers in binary representation.

71. 178,723

72. 443,876

◆ Applications

73. A physicist measures the decay of a radioisotope. She finds that after 13 days, 80 percent of the original amount remains. Determine the half-life of the isotope.

74. The population of Maryland was 4,216,000 in 1980 and is estimated to be 4,491,000 in 1990. Find the doubling time of the population.

75. Calculate the pH of a substance whose hydrogen ion concentration is 8.2×10^{-12} mol/L.

76. Determine the number of decibels in busy street traffic if the intensity of the sound is 10^{-5} watts per square meter.

77. **Medicine.** A patient takes a dosage A_0 of a medication. The medication leaves the circulatory system through the liver, the urinary system, and the sweat glands, and by being consumed by those organs that use it. The amount of the medication present after time t, in hours, is given by

$$A(t) = A_0 e^{-kt}$$

where k is a constant that depends on the metabolism of the patient and the medication. A patient takes a 2 mg tablet for which $k = 0.3$.

   a. What amount of the medication will be in the patient's system after 1 hr? 2 hr? 4 hr? 6 hr?

   b. After what amount of time will the amount of medication in the patient's system be 1 mg?

78. For a certain medication and patient, an initial dosage of 5 mg will decrease to 3 mg after 1 hr.

    a. Find the decay constant k.

    b. Find an equation of exponential decay.

    c. After what amount of time will the amount of medication in the patient's system be 2 mg?

79. The population growth rate of Kuwait is 5.9 percent per year. What is the doubling time?

80. The growth rate of a population of rabbits is 12 percent per day. What is the doubling time?

81. **Population growth of Arizona.** The population of Arizona was 2,286,000 in 1980 and was estimated to be 2,580,000 in 1990. Assuming this was exponential growth:

    a. Find the exponential growth rate k.

    b. Find the equation of exponential growth.

    c. Estimate the population in 2010.

82. Refer to the previous exercise. When will the population of Arizona increase by 50 percent over its 1990 level?

Solve each of the following equations for t.

83. $\dfrac{e^t + e^{-t}}{e^t - e^{-t}} = y$

84. $N = e^{e^t}$

85. $\log_a y = 1 + n \log_a t$

86. $y = an^t$

87. $A = P\left(1 + \dfrac{i}{n}\right)^{nt}$

88. $0.49 = 0.5(1 - e^{-0.07/t})$

89. $400 = \dfrac{4800}{6 + 794e^{-800(12.2)t}}$

For each of the following functions, find and simplify

$$\frac{f(x + h) - f(x)}{h}, \qquad h \neq 0$$

90. $f(x) = e^{-x}$

91. $f(x) = \ln x$

Prove each of the following.

92. $\left(\log_x a\right)\left(\log_a b\right) = \log_x b$

93. $\log_a\left(\dfrac{x - \sqrt{x^2 - h}}{h}\right) = -\log_a\left(x + \sqrt{x^2 - h}\right)$

94. $\log_x b = \dfrac{1}{\log_b x}$

95. If $\log_a y = a + \frac{1}{2}\log_a x$, then $y = a^a \cdot \sqrt{x}$.

THE TRIGONOMETRIC FUNCTIONS

A sextant is a device used in navigation. By finding the angles of elevation of particular stars or planets, sailors can accurately determine their positions.

T rigonometry originated in the work of the ancient Babylonian astronomers and the Greek geometers who developed the mathematics of angles and triangles. The **trigonometric functions,** originally introduced for solving problems about angles and triangles, are important aids in constructing mathematical models of various physical phenomena, including electrical currents, sound waves, and predator-prey interaction.

In Chapters 5 through 7, we will study trigonometry from the perspective of both geometry (angles and triangles) and functions (mathematical modeling).

5.1 ANGLES AND RADIAN MEASURE

Angles

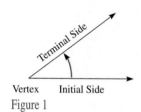

Vertex　　Initial Side

Figure 1

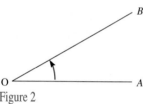

Figure 2

An **angle** is formed by rotating a ray or line segment about its endpoint. The point around which the ray or line segment is rotated is called the **vertex** of the angle. (See Figure 1.) The starting position of the ray or line segment is called the **initial side**, and the ending position is called the **terminal side**.

The angle in Figure 2 is formed by rotating the line segment OA about the point O. The initial side is OA and the terminal side is OB. We denote this angle as $\angle AOB$. (The vertex is the middle letter.) This notation is especially useful for identifying angles contained in geometric figures such as triangles, rectangles, and pentagons.

Counterclockwise rotation is considered positive, whereas clockwise rotation is considered negative. In many instances, we will specify the sign of an angle by indicating the direction of rotation with an arrow pointing from one side to the other. (See Figure 3.)

An angle can consist of more than one complete rotation. For example, Figure 4(a) shows an angle that consists of $1\frac{1}{2}$ rotations in the positive direction, Figure 4(b) shows an angle that consists of $\frac{5}{6}$ of a rotation in the positive direction, and Figure 4(c) shows an angle that consists of $2\frac{1}{4}$ rotations in the negative direction.

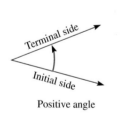

Positive angle

Figure 3

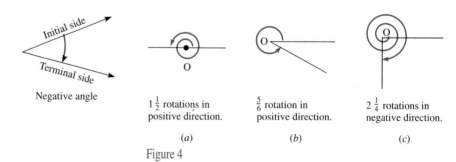

$1\frac{1}{2}$ rotations in positive direction.

(a)

$\frac{5}{6}$ rotation in positive direction.

(b)

$2\frac{1}{4}$ rotations in negative direction.

(c)

Figure 4

We will use letters, such as x and t, to denote angles. We will also bow to tradition and denote angles using lowercase Greek letters, such as α (alpha), β (beta), γ (gamma), and θ (theta).

Angles are often studied by placing them in a two-dimensional coordinate system, with the vertex at the origin and the initial side on the positive x-axis.

Such angles are said to be in **standard position**. Figure 5 shows several angles in standard position.

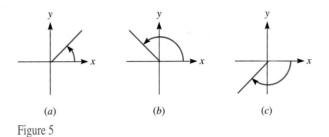

(a)   (b)   (c)

Figure 5

Degree Measure

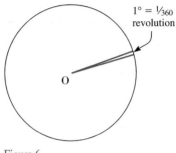

$1° = \frac{1}{360}$ revolution

Figure 6

The most familiar method for measuring angles is the system of degrees, minutes, and seconds. In this system, a complete rotation consists of 360 equal parts, called **degrees**. Degrees are denoted using the symbol °. An angle of 1° consists of $\frac{1}{360}$ of a complete rotation. (See Figure 6.)

Here are some important angles and their measures in degrees:

Quarter revolution	90°
Half revolution	180°
Three-quarter revolution	270°
One revolution	360°

Figure 7 shows a circle with various angles in degrees. These angles are in standard position.

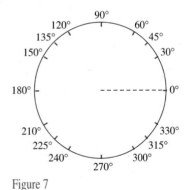

Figure 7

Each degree consists of 60 equal parts, called **minutes.** Each minute consists of 60 equal parts, called **seconds**. Minutes are denoted by the symbol ′ and seconds by the symbol ″. Using this notation, we have the following relationships:

$$1' = 60'', \qquad 1° = 60' = 3600''$$

➤ **EXAMPLE 1**
Drawing Angles

Draw the following angles θ in standard position.

1. $\theta = 30°$ 2. $\theta = 45°$ 3. $\theta = -90°$ 4. $\theta = 750°$

Solution

The specified angles are drawn in Figure 8(a)–(d).

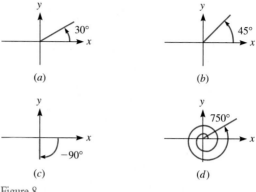

Figure 8

The most common types of angles are given special names. A 90° angle is called a **right angle**. A 180° angle is called a **straight angle**. A positive angle less than 90° is said to be **acute,** whereas an angle greater than 90° but less than 180° is said to be **obtuse**. Two acute angles whose measures add up to 90° are said to be **complementary**. Two angles whose measures add up to 180° are said to be **supplementary**. Figure 9 shows angles illustrating all of these terms.

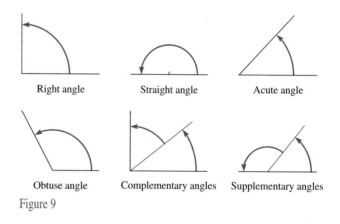

Figure 9

> **EXAMPLE 2**
Complementary and
Supplementary Angles

Suppose that $\theta = 41°28'$. Determine the measure of an angle that is

1. Complementary to θ.   2. Supplementary to θ.

Solution

1. The measure of an angle complementary to θ equals
$$90° - \theta = 90° - 41°28'$$
To perform the indicated subtraction, we write 90° as 89° + 1° or 89°60′. The difference then equals
$$89°60' - 41°28' = 48°32'$$

2. The measure of an angle supplementary to θ equals
$$180° - \theta = 180° - 41°28' = 179°60' - 41°28' = 138°32'$$

Angles that have the initial and terminal same sides are called **coterminal**. (See Figure 10.) For two angles to be coterminal, they must differ from one another by an integral number of complete rotations; that is, they must differ by an integral multiple of 360°.

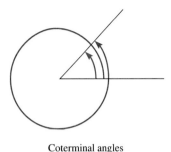

Coterminal angles
Figure 10

> **EXAMPLE 3**
Determining Coterminal Angles

Determine a coterminal angle that lies between 0° and 360° for the following angles:

    1. 530°   2. −90°   3. −400°

Solution

1. We must add or subtract a multiple of 360° so that the remainder lies between 0° and 360°. In this case, we subtract 360° from 530° to obtain the coterminal angle 170°.
2. We add 360° to −90° to obtain the coterminal angle 270°.
3. We add 2 · 360° to −400° to obtain the coterminal angle 320°.

In many applications, such as navigation and astronomy, angles are most commonly expressed in degrees, minutes, and seconds. However, for performing computations involving angles, it is easiest to express angles in terms of decimals (e.g., 42.5° rather than 42°30′). This way of expressing angles is called **decimal degrees**. The following two examples show how to convert between degrees, minutes, and seconds and decimal degrees.

> **EXAMPLE 4**
Converting to Decimal Degrees

Approximate the angle 13°32′50″ as a decimal to the nearest ten-thousandth of a degree.

Solution

Each minute is $\frac{1}{60}$ of a degree, and each second is $\frac{1}{3600}$ of a degree. Therefore, we have

$$13°32'50'' = 13 + \left(\tfrac{32}{60}\right)° + \left(\tfrac{50}{3600}\right)°$$

Because we wish to approximate the angle to the nearest ten-thousandth of a degree, we convert these fractions to decimals to the nearest ten-thousandth:

$$13° + 0.5333° + 0.0139° = 13.5472°$$

➢ **EXAMPLE 5**
Converting to
Degrees, Minutes,
and Seconds

Approximate the angle $32.519°$ in terms of degrees, minutes, and seconds.

Solution

Because a minute is $\frac{1}{60}$ of a degree, express the fractional part as 60ths of a degree:

$$32.519° = 32° + \left(\frac{60 \cdot 0.519}{60}\right)° = 32° + \left(\frac{31.14}{60}\right)°$$

The last term is $31'$ plus a fractional number of minutes. Because one second is $\frac{1}{60}$ of a minute, write this fraction in the form

$$32°31.14' = 32°31' + \left(\frac{60 \cdot 0.14}{60}\right)''$$

$$= 32°31'8.4''$$

If we wish the answer rounded to the nearest second, we have $32°31'8''$. The preceding calculation may be carried out automatically on scientific calculators that implement single keystrokes to convert decimal degrees to or from degrees, minutes, and seconds.

Radian Measure

The degree system of angle measurement dates back to the ancient Babylonians. However, the number 360 is rather artificial. The system of **radian measure** is a much more natural measure of angle measurement, not based on such an artificial number. To understand this unit of measurement, consider a circle of radius 1, with its center at the origin. Such a circle is called a **unit circle**. Place an angle in standard position in the circle. (See Figure 11.) As a measure of the size of the angle, let us take the distance traversed along the circle in moving from the initial side to the terminal side of the angle. In Figure 12, the angle θ is measured by the length of the shaded arc along the circle. The units for this system of measurement are called **radians**.

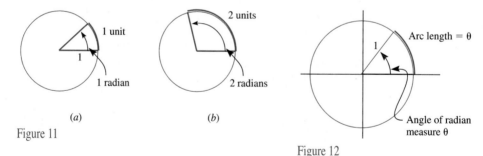

(a)
Figure 11

(b)

Figure 12

Definition 1 **Radian Measure**	For an angle inscribed in the unit circle, with vertex at the origin, the **radian measure** of the angle is the distance traveled along the unit circle from the end of the initial side to the end of the terminal side.

One radian is the measure of an angle corresponding to a distance of 1 unit along the unit circle. (See Figure 11(a).) Two radians is the measure of an angle corresponding to a distance of 2 units along the circle. (See Figure 11(b).)

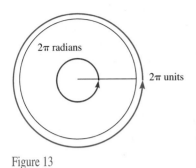

Figure 13

Because a unit circle has radius 1, its circumference is equal to

$$C = 2\pi \cdot \text{radius} = 2\pi$$

That is, one complete rotation contains 2π radians. (See Figure 13.)

We can state the connection between the radian and degree systems of angle measurement as follows.

CONVERSION BETWEEN DEGREES AND RADIANS

$$360° = 2\pi \text{ radians}, \qquad 1° = \frac{\pi}{180} \text{ radians} \tag{1}$$

Using formula (1), we deduce the following simple procedure for converting angles from degrees to radians and vice versa:

---

1. To convert an angle from degrees to radians, multiply by $\pi/180$.
2. To convert an angle from radians to degrees, multiply by $180/\pi$.

Here are some commonly used angles measured in degrees and their equivalent measurements in radians:

$$180° = \pi \text{ radians}$$

$$90° = \frac{\pi}{2} \text{ radians}$$

$$60° = \frac{\pi}{3} \text{ radians}$$

$$45° = \frac{\pi}{4} \text{ radians}$$

$$30° = \frac{\pi}{6} \text{ radians}$$

These angles are shown in Figure 14.

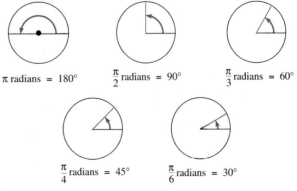

Figure 14

➤ **EXAMPLE 6**
Converting to Radians

Convert the following angles to radian measure.

   1. 72°   2. −300°   3. 270°

Solution

1. To convert to radians, we multiply degree measure by $\pi/180$:

$$72° = 72° \cdot \frac{\pi}{180} \text{ radians} = \frac{2\pi}{5} \text{ radians}$$

2. Proceeding as in part 1, we have

$$-300° = -300° \cdot \frac{\pi}{180} \text{ radians} = -\frac{5\pi}{3} \text{ radians}$$

3. We have

$$270° = 270° \cdot \frac{\pi}{180} \text{ radians} = \frac{3\pi}{2} \text{ radians}$$

➤ **EXAMPLE 7**
Converting to Degrees

Convert the following angles to degree measure (decimal degrees):

   1. $\frac{\pi}{5}$ radians   2. $\frac{7\pi}{2}$ radians   3. 1 radian

Solution

To convert from radian measure to degree measure, multiply by $180/\pi$:

1. $\dfrac{\pi}{5}$ radians $= \dfrac{\pi}{5} \cdot \dfrac{180}{\pi} = 36°$

2. $\dfrac{7\pi}{2}$ radians $= \dfrac{7\pi}{2} \cdot \dfrac{180}{\pi} = 630°$

3. 1 radian $= 1 \cdot \dfrac{180}{\pi} = \dfrac{180}{3.14159\ldots} = 57.2958\ldots°$

Radian measure is the official unit of angle measurement in the international metric system. Moreover, radian measure is the preferred unit of angle measurement from a mathematical point of view, because in terms of radians, the main formulas of calculus assume their simplest form. Throughout this book, *all angles will be in radians unless explicitly stated otherwise.* That is, when we speak of an angle of measure 3, we will mean an angle of measure 3 radians.

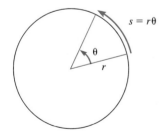

Figure 15

LENGTH OF A CIRCULAR ARC

A positive angle θ with vertex at the center of a circle of radius r determines an arc whose length is given by the formula

$$\text{Length of arc} = s = r\theta$$

(See Figure 15.)

Proof   To prove this formula, draw the circle in which the angle θ is contained. Suppose the arc is of length s. Draw a unit circle with the same center. By definition of radian measure, the length of the arc determined by θ on the unit circle is

θ. From elementary geometry, the ratio of the arc lengths is the same as the ratio of the corresponding radii. That is, we have

$$\frac{r}{1} = \frac{s}{\theta}$$

$$s = r\theta$$

This proves the formula.   ◆

> **EXAMPLE 8**
Length of a Circular Arc

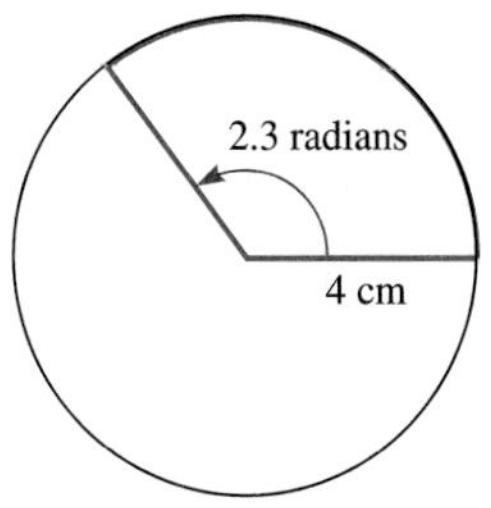

Suppose that an angle inscribed within a circle of radius 4 centimeters has measure 2.3 radians. Determine the length of the arc defined by the angle. (See Figure 16.)

Solution

We are given the angle 2.3 radians. By the above formula, the length of the arc defined by the angle is $r\theta = 4 \cdot 2.3 = 9.2$ centimeters.

Figure 16

> **EXAMPLE 9**
Angle Defined
by a Circular Arc

Suppose that a central angle of a circle of radius 12 meters intercepts an arc of length 14 meters. Find the radian measure of the angle.

Solution

Let θ denote the measure of the angle in radians. By the above formula, we see that

$$\text{Arc length} = \text{Radius} \cdot \text{Angle measure}$$

$$14 = 12\theta$$

$$\theta = \tfrac{14}{12} = \tfrac{7}{6} \text{ radians}$$

That is, the angle has measure $\tfrac{7}{6}$ radians.

Suppose that a wheel is rotating about an axis perpendicular to the wheel and passing through its center. Further, suppose that a line is drawn from the center of the wheel to a point P on its rim. The **angular speed** ω of the wheel is the angle generated by P in 1 unit of time. The **linear speed** V of the wheel is the distance P travels in 1 unit of time. (See Figure 17.) To obtain the relationship between ω and V, start from the familiar formula

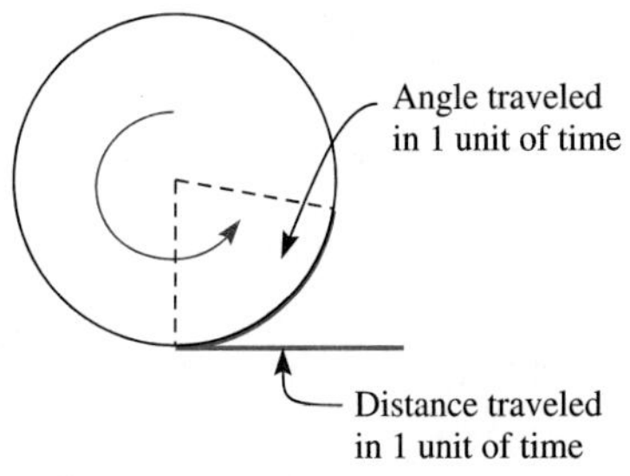

$$\text{speed} = \frac{\text{distance}}{\text{time}}$$

$$V = \frac{\text{distance } P \text{ travels in 1 unit of time}}{1 \text{ unit of time}}$$

Each second, P travels through an angle ω and covers an arc of length $r\omega$. Therefore we have the formula

$$V = r\omega$$

Angle traveled
in 1 unit of time

Distance traveled
in 1 unit of time

Figure 17

> **EXAMPLE 10**
Relating Linear Speed
and Angular Speed

Suppose that the wheels on a tractor have a radius of 3 feet and that the angular speed of the tires is 20 radians per second. What is the linear speed of the tractor in miles per hour?

Solution

Here $r = 3$ft, $\omega = 20$ radians/sec, so that the linear speed equals

$$V = 3 \cdot 20 = 60 \text{ ft/sec}$$

Therefore, the tractor is moving at 60 ft/sec. We now convert this velocity to miles per hour:

$$60 \text{ ft/sec} = (60 \text{ ft/sec}) \cdot (3600 \text{ sec/hr}) \cdot \left(\tfrac{1}{5280} \text{ mi/ft}\right) = 40.91 \text{ mi/hr}$$

Many scientific calculators provide for entry of angles in either degree or radian measure. Furthermore, many scientific calculators also allow for entry of angles in degrees, minutes, and seconds, in addition to entry in decimal degrees. To set the method to be used for angle entry, just press the appropriate key on the calculator. The radian measure key is usually labeled RAD, the decimal degree key DEG, and the degree-minute-second key DMS. Radians and decimal degrees are entered as decimal numbers. However, degrees, minutes, and seconds are entered as a sequence of digits. For instance, on some calculators $36°2'15''$ is entered in DMS mode as 360215. (Note that the minutes and seconds always are entered as two digits, using a leading zero where necessary.) Refer to your calculator manual for specific instructions about using degree and radian measure and about converting from one to the other.

Exercises 5.1

Draw the following angles in standard position.

1. $30°$ 2. $150°$ 3. $-60°$ 4. $-390°$
5. $-45°$ 6. $-30°$ 7. $480°$ 8. $-90°$

Given the angle θ, determine the measure of the angle that is (a) complementary to θ, and (b) supplementary to θ.

9. $\theta = 36°24'$ 10. $\theta = 78°11'$
11. $\theta = 56°34'53''$ 12. $\theta = 89°42'23''$

Approximate each angle in terms of decimal degrees to the nearest ten-thousandth of a degree.

13. $\theta = 36°24'$ 14. $\theta = 78°11'$
15. $\theta = 56°34'53''$ 16. $\theta = 89°42'23''$
17. $\theta = -35°48'$ 18. $\theta = -78°13'$
19. $\theta = 142°34'$ 20. $\theta = 244°46'10''$

Approximate each angle in terms of degrees, minutes, and seconds.

21. $46.327°$ 22. $78.807°$
23. $-72.25°34'53''$ 24. $-189.62°$

25. $364.045°$ 26. $780.0042°$
27. $-1.6556°$ 28. $-222.22222°$

Convert the following angles to radian measure.

29. $36°$ 30. $108°$ 31. $135°$
32. $-156.25°$ 33. $-240°$ 34. $325°$
35. $244.48°$ 36. $-90°$

Convert the following angles to degree measure (decimal degrees).

37. $\dfrac{5\pi}{6}$ 38. $\dfrac{17\pi}{5}$ 39. $-\dfrac{7\pi}{4}$ 40. $-\dfrac{3\pi}{2}$

41. 3.521 42. -1.873 43. $-\dfrac{13\pi}{6}$ 44. $\dfrac{14\pi}{3}$

♦ **Applications**

45. In 35 minutes, the minute hand of a clock rotates $210°$. What is this in radians? (See figure.)

46. The measure of one of the base angles of an isosceles triangle is $40°$. (See figure.) What is the measure of each angle of the triangle in radians?

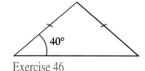

Exercise 45   Exercise 46

47. One angle has a measure of $3\pi/4$ radians. What is the degree measure of it and its supplement?

48. One angle has a measure of 72°. What is the radian measure of it and its complement?

49. What is the angle measure, in degrees and radians, of each angle of an isosceles right triangle?

50. A clock's pendulum is 1 m long. From one side to the other it moves through an angle of 0.2 radians. Through what distance does the tip of the pendulum move?

51. A baseball player swings a bat, completely level, through an angle of 2.79 radians. If the bat is 39 inches long and the batter's arm is 30 inches, through what distance does the tip of the bat move?

52. For a circle with a 7-cm radius, how long is an arc associated with a central angle of 65°?

53. What is the radius of a circle for which a central angle of 135° sweeps out an arc of length 9.42 ft?

54. In five hours, the tip of the needle of a circular thermometer moves 3 in. The needle is 5 in. long. What is the measure of the angle of rotation?

55. A door built in the corner of a house can rotate through 2.09 radians before it is stopped by a wall. If the door is 3 ft wide, through what distance can the edge of the door move?

56. A clock in Copenhagen, Denmark, makes one complete revolution every 25,753 years. The clock is 20 ft in diameter. How far does a point on the outside of the clock move every decade?

57. Find the linear speed of an exercise bicycle if the wheel has a radius of 10 in. and the angular speed is 0.256 radian/sec.

58. A building in Milwaukee has a clock with a minute hand 20 ft long. What is the linear speed of the tip of the minute hand in in./sec?

59. One end of a 6-inch rod is in the center of a centrifuge. The other end moves at 4500 mph. What is the angular speed in rev/sec?

60. A 150-ft diameter ferris wheel makes one revolution in 65 seconds. What is the linear speed of each cab on the wheel, in mph?

61. A person is seated on the end of a see-saw whose total length is 5 m. The see-saw moves up and down through a 28° angle every 3 seconds. Through what distance does the person move in a minute?

62. The Tour de France is a bicycle race held in France. The average speed of one race was 15 mph. The diameter of a wheel is 26 in. What is the average angular speed in radians/hr?

✍ **In Your Own Words**

63. Explain the definition of a radian.

64. Describe a procedure for converting degrees, minutes, and seconds to radian measurement.

➡ **Technology**

Convert the following angles to radians using your calculator. If your calculator supports direct entry of degrees, minutes, and seconds, use this feature.

65. 59°41′38.11″

66. 139°10″

67. −58°49′

68. 759°10′14″

Convert the following angles to degrees, minutes, and seconds.

69. 1.90317 radians

70. 2.07312 radians

71. −2 radians

72. 0.05732 radians

5.2 TRIGONOMETRIC FUNCTIONS

In the preceding section, we introduced angle measurement using both the degree and radian systems. In this section, we define six functions that describe an angle.

Definition of Sine and Cosine

We define the sine and cosine of an angle t in geometric terms as follows:

> **Definition 1**
> **Sine and Cosine of an Angle**
>
> Draw a unit circle, and let t be the measure of an angle in standard position. (Recall that this means that the vertex of the angle is at the origin and the initial side is aligned with the positive x-axis.) (See Figure 1.) The terminal side of the angle intersects the circle at a point that we label P. The **cosine of the angle** t, denoted $\cos t$, is the x-coordinate of P. The **sine of the angle** t, denoted $\sin t$, is the y-coordinate of P.

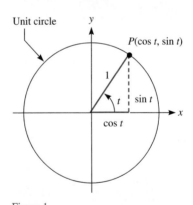

Figure 1

For instance, consider the angle t shown in Figure 2. The horizontal coordinate gives the value of the cosine, and the vertical value of the coordinate gives the value of the sine. Thus, we have

$$\cos t = 0.6, \qquad \sin t = 0.8$$

For some angles t (measured in either radians or degrees), we may determine the value of the corresponding sine and cosine by embedding the angle within a unit circle and using the geometry to determine the coordinates of the point P.

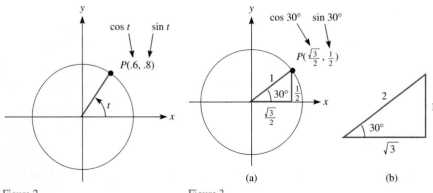

Figure 2

Figure 3

> **EXAMPLE 1**
Determine Sine
and Cosine of 30°

Determine $\sin 30°$ and $\cos 30°$.

Solution

Figure 3(*a*) shows an angle of measure 30° in standard position. The triangle shown is a 30°–60°–90° triangle. Recall from geometry that such a triangle has sides proportional to 1, $\sqrt{3}$, and 2, as shown in Figure 3(*b*). In Figure 3(*a*), the hypotenuse is a radius of the unit circle and is therefore equal to 1. So this triangle corresponds to the triangle in Figure 3(*b*) with similarity factor $\frac{1}{2}$. The coordinates of the point P are therefore $\left(\sqrt{3}/2, 1/2\right)$. According to Definition 1,

$$\cos 30° = \frac{\sqrt{3}}{2}, \qquad \sin 30° = \tfrac{1}{2}$$

> **EXAMPLE 2**
Determining Sine
and Cosine of 45°

Determine $\sin 45°$ and $\cos 45°$.

Solution

Figure 4(*a*) shows the angle 45° in standard position. A 45°–45°–90° triangle has sides proportional to 1, 1, and $\sqrt{2}$, as shown in Figure 4(*b*). The hypotenuse of the triangle in Figure 4(*a*) is 1 because it is a radius of the unit circle. Therefore, this triangle corresponds to the triangle in Figure 4(*b*) with similarity factor $1/\sqrt{2} = \sqrt{2}/2$, so point P has coordinates $\left(\sqrt{2}/2, \sqrt{2}/2\right)$, and by Definition 1, we have

$$\sin 45° = \frac{\sqrt{2}}{2}, \qquad \cos 45° = \frac{\sqrt{2}}{2}$$

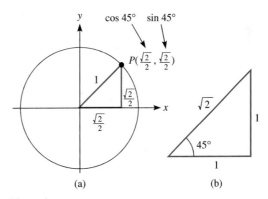

(a)   (b)

Figure 4

> **EXAMPLE 3**
Determining Sine
and Cosine of 0°

Determine $\sin 0°$ and $\cos 0°$.

Solution

Figure 5 shows the angle 0° in standard position. We see that P has coordinates $(1, 0)$, so that $\sin 0° = 0$, $\cos 0° = 1$.

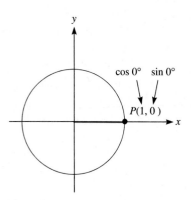

Figure 5

There are two important points to remember about Definition 1:

1. The circle used to define the sine and cosine is a **unit** circle (radius 1). If you use a circle of some other radius, the coordinates of the point P will not give the value of the sine and cosine of the angle t.
2. The order of the coordinates is important. The horizontal coordinate is the cosine, and the vertical coordinate is the sine. Be careful not to interchange the coordinates.

The angle in a trigonometric function may be given in either degrees (as in the previous examples) or radians.

➤ **EXAMPLE 4**
Calculating Sine and Cosine of Angles in Radians

Determine the values of $\sin t$ and $\cos t$ for the following angles t:

1. $\dfrac{\pi}{2}$　　2. π　　3. 5π　　4. $-\dfrac{3\pi}{2}$

Solution

1. The angle of radian measure $\pi/2$ (or 90°) is drawn in Figure 6. The coordinates of P are (0, 1). Therefore, by Definition 1,

$$\cos \frac{\pi}{2} = 0, \qquad \sin \frac{\pi}{2} = 1$$

2. The angle of measure π radians is a straight angle. (See Figure 7.) From the coordinates of P, we see that

$$\cos \pi = -1, \qquad \sin \pi = 0$$

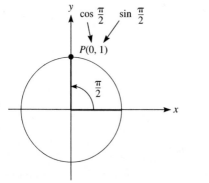

Figure 6

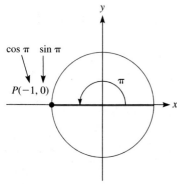

Figure 7

3. The angle 5π can be drawn by going two complete revolutions and then another half revolution in the positive direction. This results in the point P shown in Figure 8. From the coordinates of P, we see that the trigonometric function values for 5π are the same as those for π, namely,

$$\cos 5\pi = -1, \qquad \sin 5\pi = 0$$

4. The angle is obtained by making three quarter-revolutions in the negative direction. Refer to Figure 9. From the coordinates of P, which are $(0, 1)$, we see that

$$\cos\left(-\frac{3\pi}{2}\right) = 0, \qquad \sin\left(-\frac{3\pi}{2}\right) = 1$$

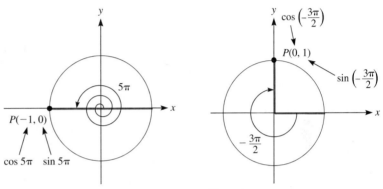

Figure 8   Figure 9

In Examples 1–3, we used the geometry of known triangles to determine the values of sine and cosine for various angles. To do such calculations, it is helpful to use a reference angle as follows.

Definition 2 **Reference Angle**	Let t be an angle in standard position. The corresponding **reference angle** t' is the acute angle between the terminal side of t and the horizontal axis. (See Figure 10.)

Figure 11 shows the reference angles corresponding to three given angles. Note that the reference angle is always acute, no matter what the angle t. If the angle is embedded in a unit circle, then the reference angle defines a triangle, as shown in Figure 11. We may use this triangle to determine the values of sine and cosine.

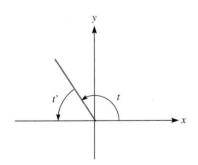

Figure 10

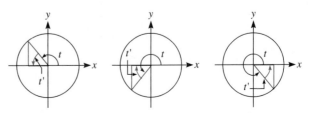

Figure 11

> **EXAMPLE 5**
Calculating
Reference Angles

Determine the reference angles for the following angles:

1. $t = -70°$ 2. $t = 255°$ 3. $t = \dfrac{5\pi}{3}$ radians

Solution

1. See Figure 12(*a*).
2. See Figure 12(*b*).
3. See Figure 12(*c*).

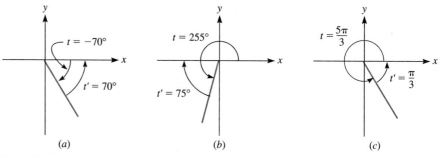

Figure 12

> **EXAMPLE 6**
Calculating Sine and Cosine
Using Reference Angles

Calculate $\sin t$ and $\cos t$ for $t = 5\pi/6$.

Solution

Draw the angle $5\pi/6$ in standard position, as shown in Figure 13. Now draw the corresponding reference angle, which is $\pi/6$. Draw the associated triangle defined by the reference angle and the unit circle. This is just a 30°–60°–90° triangle with hypotenuse 1 (because the unit circle has radius 1). Therefore, the legs are of length 1/2 and $\sqrt{3}/2$, as shown in the figure. Because the angle $5\pi/6$ is in the second quadrant, point P has a positive y-coordinate and a negative x-coordinate. Therefore, we see that $P = (-\sqrt{3}/2, 1/2)$, so that

$$\sin \frac{5\pi}{6} = \tfrac{1}{2}, \qquad \cos \frac{5\pi}{6} = -\frac{\sqrt{3}}{2}$$

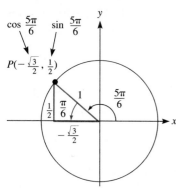

Figure 13

The Other Trigonometric Functions

In addition to the trigonometric functions $\sin t$ and $\cos t$, there are four other trigonometric functions associated with the angle t. These functions are called the **tangent**, denoted $\tan t$, the **secant**, denoted $\sec t$, the **cotangent**, denoted $\cot t$, and the **cosecant**, denoted $\csc t$. These trigonometric functions are defined in terms of $\sin t$ and $\cos t$ using the following formulas:

$$\tan t = \frac{\sin t}{\cos t}, \qquad \cos t \neq 0$$

$$\sec t = \frac{1}{\cos t}, \qquad \cos t \neq 0$$

$$\csc t = \frac{1}{\sin t}, \qquad \sin t \neq 0$$

$$\cot t = \frac{1}{\tan t} = \frac{\cos t}{\sin t}, \qquad \sin t \neq 0$$

Note that each of the four additional trigonometric functions are defined only for certain angles t. We will explore the domains of these functions in more detail later in the chapter.

➤ **EXAMPLE 7**
Calculating Trigonometric Functions

Determine the values of $\tan t$, $\sec t$, $\csc t$, and $\cot t$ for $t = 5\pi/6$.

Solution

From the previous example, we have

$$\sin \frac{5\pi}{6} = \tfrac{1}{2}, \qquad \cos \frac{5\pi}{6} = -\frac{\sqrt{3}}{2}$$

From these values, we apply the definitions of the tangent and secant:

$$\tan \frac{5\pi}{6} = \frac{\sin \dfrac{5\pi}{6}}{\cos \dfrac{5\pi}{6}} = \frac{\tfrac{1}{2}}{-\dfrac{\sqrt{3}}{2}} = -\frac{\sqrt{3}}{3}$$

$$\sec \frac{5\pi}{6} = \frac{1}{\cos \dfrac{5\pi}{6}} = \frac{1}{-\dfrac{\sqrt{3}}{2}} = -\frac{2\sqrt{3}}{3}$$

$$\csc \frac{5\pi}{6} = \frac{1}{\sin \dfrac{5\pi}{6}} = \frac{1}{\tfrac{1}{2}} = 2$$

$$\cot \frac{5\pi}{6} = \frac{1}{\tan \dfrac{5\pi}{6}} = \frac{1}{-\dfrac{\sqrt{3}}{3}} = -\sqrt{3}$$

Signs of the Trigonometric Functions

The signs of the trigonometric functions do not change if the angle remains within a single quadrant. Let's determine these signs. Recall that the sine of an angle is given by the y-coordinate of the point determined by the angle on the unit circle.

Thus, $\sin t$ is positive for angles in quadrants I and II and negative for angles in quadrants III and IV. The cosine of an angle is determined by the x-coordinate of the point determined by the angle on the unit circle. Thus, $\cos t$ is positive for t in quadrants I and IV and negative in quadrants II and III.

The signs of the other trigonometric functions can be determined directly from their respective definitions. For instance, consider the tangent function, given by the formula

$$\tan t = \frac{\sin t}{\cos t}$$

If t is in quadrant I, the numerator and denominator are both positive, so the value of $\tan t$ is positive. If t is in quadrant II, then the numerator is positive and the denominator is negative, so the value of $\tan t$ is negative. We use similar reasoning for the other quadrants. Using such arguments, we derive the scheme in Figure 14 for determining the signs of the various trigonometric functions by the quadrant of the angle.

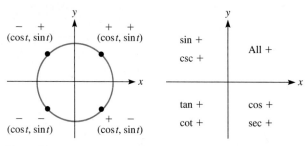

Figure 14

Numerical Approximations

The values of the trigonometric functions for particular angles t are (except for special angles) determined by referring to a table or by using a calculator. At the end of the book, you will find a short table giving the values of $\sin t$, $\cos t$, and $\tan t$ for angles t given in radians. To use these tables, look up the desired value of t in the left column and read off the corresponding values of the trigonometric functions. Tables are limited in the accuracy of the angle allowed and the number of significant digits of the trigonometric functions. In the tables provided, the angle t is given in hundredths of a radian, and the trigonometric functions are given to five significant digits.

Scientific calculators have keys for computing the values of the trigonometric functions $\sin t$, $\cos t$, and $\tan t$, usually to an accuracy of 10 decimal places. The following example illustrates how a typical algebraic-logic calculator may be used to calculate the values of various trigonometric functions. In this example, and throughout the remainder of the book, we will assume five significant digits for trigonometric functions. (This will give consistency with the results derived using tables.)

➤ EXAMPLE 8
Numerical Approximations for Sine and Cosine

Use a scientific calculator to determine the approximate values of the following trigonometric functions.

1. $\sin 42°31'$ 2. $\cos 72°15'27''$ 3. $\tan 38.523°$ 4. $\sec 63.581$

Solution

The following solution assumes the use of a calculator with algebraic logic and a DMS (Degrees-Minutes-Seconds) mode of operation. (The steps for your calculator may be different.)

1. Set the calculator on DMS angle measurement. Then enter the following key sequence:

$$\boxed{\text{SIN}}\, 4231 =$$

The answer is 0.67580.

2. Check that the angle measurement mode is on DMS. (There is no need to set the angle measurement mode for each new calculation.) Then enter the following key sequence:

$$\boxed{\text{COS}}\, 721527 =$$

The answer is 0.30474.

3. Set the angle measurement mode to DEG. Then enter the following key sequence:

$$\boxed{\text{TAN}}\ 38.523\ =$$

The answer is 0.79609.

4. Set the angle measurement mode to RAD. Recall that the secant is the reciprocal of the cosine. To compute the given value, enter the following key sequence:

$$\boxed{\text{COS}}\ 63.581\ \text{X}^{-1}\ =$$

The answer is 1.3656.

Historical Note: Numerical Calculations and Tables of Functions

Before the advent of scientific calculators, the values of trigonometric functions were determined from tables developed by mathematicians. To ensure accurate calculations, it was necessary to use tables that recorded the values of trigonometric functions to precise fractions of a degree and to a large number of significant figures. Some tables recorded the trigonometric function values for angle intervals of one second and to 14 decimal places. So important were such tables that they were a necessary purchase for every student in any field of science and engineering. During the first third of the 20th century, the world's center for science was Germany, and there were no tables of trigonometric function values produced in English. Rather, it was necessary to use tables imported from Germany! In order to read the explanation of the tables' operation, it was therefore necessary for students to learn scientific German, a standard requirement of all engineering and science students of that era. ❚❚

Solving Trigonometric Equations

In many applications, it is necessary to calculate the measure of an angle given the value of a particular trigonometric function. Such calculations can be performed using the **inverse trigonometric functions.** The inverse of a particular trigonometric function is denoted by placing a superscript -1 after the name of the function. For instance, the inverses of the functions sin, cos, and tan are denoted, respectively,

$$\sin^{-1}, \quad \cos^{-1}, \quad \tan^{-1}$$

Note that the superscripts -1 are not exponents but rather are used to denote inverse functions. We will discuss the inverse trigonometric functions in detail in Chapter 6. For now, however, we just need the following definitions:

$$\sin^{-1} x = \text{the angle in } [-90°, 90°] \text{ whose sine is } x$$

$$\cos^{-1} x = \text{the angle in } [0°, 180°] \text{ whose cosine is } x$$

$$\tan^{-1} x = \text{the angle in } [-90°, 90°] \text{ whose tangent is } x$$

For example, because $\sin 45° = \sqrt{2}/2$, we have

$$\sin^{-1}\left(\frac{\sqrt{2}}{2}\right) = 45°$$

Similarly, because $\tan 45° = 1$, we have

$$\tan^{-1} 1 = 45°$$

Scientific calculators allow you to determine values of the inverse trigonometric functions. To do this, most calculators require you to enter two keystrokes to indicate an inverse trigonometric function. The first is the inverse key, usually labeled [2nd] (or [INV]), and the second is the name of the trigonometric function. For example, if we wish to know the angle in radian measure whose sine is 0.37821, we must calculate

$$\sin^{-1}.37821$$

Here's how to do so using the TI-81 calculator. Be sure the calculator is set on radian measure, and then enter the following sequence of keystrokes:

$$\boxed{\text{2nd}}\;\boxed{\text{SIN}^{-1}}.37821\;\boxed{\text{ENTER}}$$

The calculator displays the value 0.38786, indicating that the value of the inverse trigonometric function is 0.38786 radians.

➤ **EXAMPLE 9**
Numerical Trigonometric
Equations

Approximate the acute angle t (measured in radians) that satisfies the following equations:

1. $\sin t = 0.71382$ 2. $\cos t = 0.6$
3. $\tan t = 5.11278$ 4. $\sec t = 4.90245$

Solution

In each case, we seek an angle for which the value of a trigonometric function is given. The angle is the value of the appropriate inverse trigonometric function. In each case, we use a scientific calculator set to radians to determine the values of the inverse trigonometric functions.

1. We have
$$t = \sin^{-1}0.71382 \approx 0.79494 \text{ radians}$$

2. We have
$$t = \cos^{-1}0.6 = 0.92730 \text{ radians}$$

3. We have
$$t = \tan^{-1}5.11278 = 1.3776 \text{ radians}$$

4. Most calculators do not have a key for the inverse secant function. To solve the given equation, we first express the secant in terms of the cosine
$$\sec t = 4.90245$$
$$\frac{1}{\cos t} = 4.90245$$
$$\cos t = \frac{1}{4.90245} = 0.20398$$
$$t = \cos^{-1}0.20398 = 1.3654 \text{ radians}$$

Exercises 5.2

Determine the values of the six trigonometric functions, if they exist, for each value of t. Do not use a calculator.

1. $t = \dfrac{5\pi}{4}$ 2. $t = \dfrac{3\pi}{4}$ 3. $t = \dfrac{7\pi}{6}$

4. $t = \dfrac{11\pi}{6}$ 5. $t = \dfrac{4\pi}{3}$ 6. $t = \dfrac{8\pi}{3}$

7. $t = -\dfrac{\pi}{4}$ 8. $t = -\dfrac{\pi}{3}$ 9. $t = -\dfrac{3\pi}{4}$

10. $t = -\dfrac{7\pi}{3}$ 11. $t = -\dfrac{14\pi}{3}$ 12. $t = -6\pi$

13. $t = 360°$ 14. $t = 180°$ 15. $t = 240°$

16. $t = 300°$ 17. $t = 315°$ 18. $t = -90°$

19. $t = -270°$ 20. $t = 150°$ 21. $t = -150°$

22. $t = 570°$ 23. $t = -570°$ 24. $t = -480°$

25. $t = 675°$ 26. $t = -\dfrac{25\pi}{4}$ 27. $t = \dfrac{35\pi}{6}$

28. $t = 32\pi$ 29. $t = 43\pi$ 30. $t = 50\pi$

Determine the values of the six trigonometric functions for the angle t defined by the point P on the unit circle.

31. $P = \left(\frac{3}{5}, \frac{4}{5}\right)$ 32. $P = \left(\frac{4}{5}, -\frac{3}{5}\right)$

33. $P = \left(\frac{5}{13}, -\frac{12}{13}\right)$ 34. $P = \left(-\frac{12}{13}, -\frac{5}{13}\right)$

➡ Technology

Use a scientific calculator to determine the approximate values of the following trigonometric functions. Round to four decimal places.

35. $\tan 89°$ 36. $\cos 71°$ 37. $\sin 34°$

38. $\tan 11°34'$ 39. $\sec 23°18'$ 40. $\cot 1°57'$

41. $\csc 88°59'$ 42. $\cos 23°34'18''$ 43. $\sin 17°51'2''$

44. $\tan 21°48'32''$ 45. $\cot 75°55'25''$ 46. $\csc 56.678°$

47. $\cos 76.1213°$ 48. $\sin 4.34°$ 49. $\tan 59.9°$

50. $\sin 0.763$ 51. $\cos 1.4$ 52. $\tan 1.2113$

53. $\sec 0.87$ 54. $\csc 3\pi$ 55. $\cot 0.8\pi$

56. $\sin 1$ 57. $\cos 0.6$

Approximate the solution θ for the following equations for an acute angle θ. Give your answer in radians and in decimal degrees.

58. $\sin \theta = 0.8654$ 59. $\cos \theta = 0.1121$

60. $\tan \theta = 4.31126$ 61. $\cot \theta = 16.34$

62. $\sec \theta = 6.83012$ 63. $\csc \theta = 1.3457$

64. $\tan \theta = 10.7811$ 65. $\sin \theta = 0.9876$

66. $\cos \theta = 0.2$ 67. $\sec \theta = 8$

68. $\csc \theta = 24$ 69. $\cos \theta = 0.32$

5.3 TRIGONOMETRIC FUNCTIONS OF ACUTE ANGLES

In the preceding section, we introduced the trigonometric functions of an angle t. In case t is acute, we can give an alternate definition of the trigonometric functions in terms of right triangles. This alternate definition is very useful in applications of trigonometry.

The Trigonometric Functions—Geometric Definition

Suppose that we are given a right triangle with an acute angle θ. Associated with θ are three lengths, the hypotenuse, the opposite side, and the adjacent side of the triangle. (See Figure 1.) We may calculate the values of the trigonometric functions directly from the geometry of the triangle as follows:

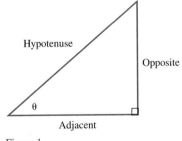

Figure 1

TRIGONOMETRIC FUNCTIONS OF AN ACUTE ANGLE	
$\sin \theta = \dfrac{\text{Opposite}}{\text{Hypotenuse}}$	(1)
$\cos \theta = \dfrac{\text{Adjacent}}{\text{Hypotenuse}}$	(2)
$\tan \theta = \dfrac{\text{Opposite}}{\text{Adjacent}}$	(3)
$\cot \theta = \dfrac{\text{Adjacent}}{\text{Opposite}}$	(4)
$\sec \theta = \dfrac{\text{Hypotenuse}}{\text{Adjacent}}$	(5)
$\csc \theta = \dfrac{\text{Hypotenuse}}{\text{Opposite}}$	(6)

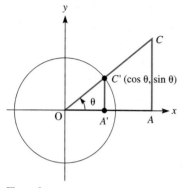

Figure 2

To prove these formulas, draw a unit circle and place the triangle so that the adjacent side lies along the positive x-axis. (See Figure 2.) Label the triangle AOC as in the figure. Let C' be the point at which the line $\overline{OC}$ intersects the unit circle, and draw the right triangle $A'OC'$ as shown in the figure. Note that triangles AOC and $A'OC'$ are similar because they have two angles in common. Let $d(a, b)$ denote the distance between points a and b in the plane. Because the triangles are similar, there is a positive fraction k such that:

$$d(A', C') = kd(A, C), \qquad d(O, A') = kd(O, A), \qquad d(O, C') = kd(O, C)$$

Moreover, because $\overline{OC'}$ is a radius of a unit circle, we have

$$d(O, C') = 1$$

From the definition of the sine and cosine, we have:

$$\cos\theta = d(O, A')$$
$$= \frac{d(O, A')}{1}$$
$$= \frac{d(O, A')}{d(O, C')}$$
$$= \frac{kd(O, A)}{kd(O, C)}$$
$$= \frac{d(O, A)}{d(O, C)}$$
$$= \frac{\text{Adjacent}}{\text{Hypotenuse}}$$

This proves formula (2). The proof of (1) is similar. To prove (3), note that $\tan\theta = \sin\theta/\cos\theta$, so that by formulas (1) and (2)

$$\tan\theta = \frac{\dfrac{\text{Opposite}}{\text{Hypotenuse}}}{\dfrac{\text{Adjacent}}{\text{Hypotenuse}}}$$
$$= \frac{\text{Opposite}}{\text{Hypotenuse}} \cdot \frac{\text{Hypotenuse}}{\text{Adjacent}}$$
$$= \frac{\text{Opposite}}{\text{Adjacent}}$$

This proves formula (3). The remainder of the formulas are proved similarly.

➤ **EXAMPLE 1**
Determining Values
of Trigonometric Functions

Suppose that the right triangle ABC has an acute angle θ as shown in Figure 3. Determine the values of $\sin\theta$, $\cos\theta$, and $\tan\theta$.

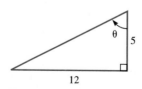

Figure 3

Solution

From the figure, we determine that

$$\text{Opposite} = 12, \qquad \text{Adjacent} = 5$$

Moreover, by the Pythagorean theorem,

$$\text{Hypotenuse} = \sqrt{\text{Opposite}^2 + \text{Adjacent}^2}$$

$$= \sqrt{5^2 + 12^2}$$

$$= \sqrt{169}$$

$$= 13$$

From formulas (1)–(3), we then have

$$\sin \theta = \frac{\text{Opposite}}{\text{Hypotentuse}} = \frac{12}{13}$$

$$\cos \theta = \frac{\text{Adjacent}}{\text{Hypotenuse}} = \frac{5}{13}$$

$$\tan \theta = \frac{\text{Opposite}}{\text{Adjacent}} = \frac{12}{5}$$

NOTATION IN LABELING TRIANGLES

It is customary to use a labeling convention that simplifies the description of a triangle. We may label the sides a, b, c, the vertices A, B, C, and the angles α, β, γ, with each vertex and angle assumed to be opposite the side with the corresponding letter. That is, vertex A and angle α are opposite side a, vertex B and angle β are opposite side b, and vertex C and angle γ are opposite side c. This labeling convention will be used throughout the remainder of this book.

➤ **EXAMPLE 2**
Computing Trigonometric
Functions from a Triangle

Consider the triangle shown in Figure 4. Determine the values of the following trigonometric functions.

1. $\sin \alpha$ 2. $\sin \beta$ 3. $\tan \alpha$ 4. $\csc \beta$

Solution

1. The sine of an angle equals the ratio Opposite/Hypotenuse. In the triangle shown, side $\overline{AB}$ is the hypotenuse, and $\overline{BC}$ is the opposite side. Therefore,

$$\sin \alpha = \frac{d(B, C)}{d(A, B)} = \frac{8}{10} = 0.8$$

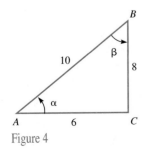

Figure 4

2. For the angle β, the hypotenuse remains unchanged, but the opposite side is $\overline{AC}$. Therefore,

$$\sin\beta = \frac{d(A, C)}{d(A, B)} = \tfrac{6}{10} = 0.6$$

3. We have

$$\tan\alpha = \frac{\text{Opposite}}{\text{Adjacent}}$$

$$= \frac{d(B, C)}{d(A, C)}$$

$$= \tfrac{8}{6}$$

$$= \tfrac{4}{3}$$

4. We have

$$\csc\beta = \frac{\text{Hypotenuse}}{\text{Opposite}}$$

$$= \frac{d(A, B)}{d(A, C)}$$

$$= \tfrac{10}{6}$$

$$= \tfrac{5}{3}$$

➤ **EXAMPLE 3**
Computing Trigonometric
Functions from
a Given One

Suppose that θ is an acute angle for which $\cos\theta = \tfrac{5}{7}$. Determine the values of the other five trigonometric functions.

Solution

Draw a triangle having the angle θ, as in Figure 5. Because the cosine is the ratio of the adjacent side divided by the hypotenuse, we have labeled the adjacent side 5 and the hypotenuse 7. By the Pythagorean theorem,

$$\text{Opposite} = \sqrt{7^2 - 5^2} = \sqrt{24} = 2\sqrt{6}$$

Therefore, by formulas (1)–(6), we have

$$\sin\theta = \frac{\text{Opposite}}{\text{Hypotenuse}} = \frac{2\sqrt{6}}{7}$$

$$\tan\theta = \frac{\text{Opposite}}{\text{Adjacent}} = \frac{2\sqrt{6}}{5}$$

$$\sec\theta = \frac{\text{Hypotenuse}}{\text{Adjacent}} = \tfrac{7}{5}$$

$$\csc\theta = \frac{\text{Hypotenuse}}{\text{Opposite}} = \frac{7}{2\sqrt{6}} = \frac{7\sqrt{6}}{12}$$

$$\cot\theta = \frac{\text{Adjacent}}{\text{Opposite}} = \frac{5}{2\sqrt{6}} = \frac{5\sqrt{6}}{12}$$

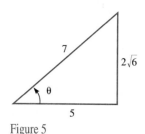

Figure 5

Exercises 5.3

Given each triangle in Exercises 1–12 , find the six trigonometric function values of the angle θ.

1.

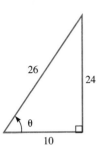

2.

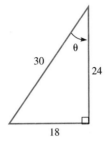

3.

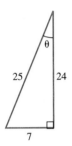

4.

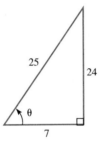

5.

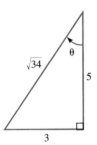

6.

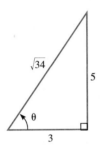

7.

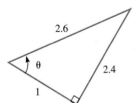

8.

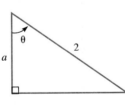

9.

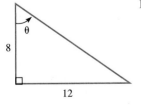

10.

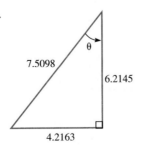

11.

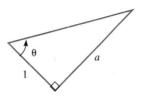

12.

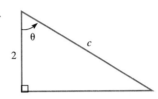

In Exercises 13–20, a value of a trigonometric function is given. Find the five other trigonometric values of the angle θ. Assume θ is acute.

13. $\sin \theta = \frac{2}{3}$

14. $\cos \theta = \dfrac{1}{\sqrt{7}}$

15. $\tan \theta = \sqrt{2}$

16. $\cot \theta = 3$

17. $\sec \theta = \frac{25}{7}$

18. $\csc \theta = \dfrac{4}{\sqrt{5}}$

19. $\cos \theta = a, \quad 0 < a < 1$

20. $\tan \theta = \dfrac{1}{v}$

➡ Technology

In Exercises 21–24, a value of a trigonometric function is given. Find the five other trigonometric values of the angle θ. Use your calculator and round to four decimal places.

21. $\cos \theta = 0.1762$

22. $\sin \theta = 0.9974$

23. $\tan \theta = 6.7899$

24. $\sec \theta = 8$

◆ Applications

25. The top of a 13-ft ladder leaning against a wall touches the wall 12 ft above the ground. Find the six trigonometric function values of the angle the ladder makes with the ground. (See figure.)

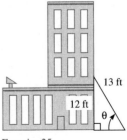

Exercise 25

26. A guy wire reaches from the top of a 20-ft pole to a point on the ground 15 ft from the bottom of the pole. (See figure.) What are the six trigonometric function values of the angle the wire makes with the ground?

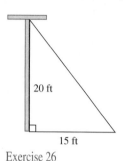

20 ft

15 ft

Exercise 26

27. The formula

$$\cos \theta \approx 1 - \frac{\theta^2}{2} + \frac{\theta^4}{24} - \frac{\theta^6}{720}$$

can be used to approximate values of the cosine function when θ is radian measure. Use the formula to approximate cos 0.2 and compare it with the result found directly on a calculator.

28. The formula

$$\tan \theta \approx \theta + \frac{\theta^3}{3} + \frac{2\theta^5}{15} + \frac{17\theta^7}{315}$$

can be used to approximate values of the tangent function when θ is radian measure. Use the formula to approximate tan 1.4 and compare it with the result found directly on a calculator.

✍ In Your Own Words

29. How large must an angle be before it differs from its tangent in the third decimal place? Use your calculator.

30. How large must an angle be before its sine differs from its tangent in the third decimal place? Use your calculator.

▌ 5.4 TRIGONOMETRY APPLICATIONS

Using the definitions of the various trigonometric functions along with their numerical values, it is possible to determine unknown angles and sides of a right triangle from known angles and sides. The process of determining all unknown sides and angles is called **solving the triangle**. In the next examples, we illustrate how the trigonometric functions can be used to solve triangles, starting from various sets of known information.

➤ EXAMPLE 1
Solving a Triangle

The right triangle of Figure 1 has one acute angle and one side specified. Determine the other acute angle and the lengths of the other two sides.

Solution

Let's determine the various components one at a time. Let's first determine the angle α. The sum of the angles of a triangle is 180°. Therefore, because we know one acute angle and because the other angle is a right angle, we have

$$\alpha = 180° - 90° - 38°41' = 51°19'$$

Next, let's determine the length of side a. This side is the side opposite the angle given as 38°41'. The adjacent side for this angle has length 55. Because we know the adjacent side and want to find the opposite side, we use a trigonometric function that involves both:

$$\tan = \frac{\text{Opposite}}{\text{Adjacent}}$$

$$\tan 38°41' = \frac{a}{55}$$

Using a calculator to determine the value of the tangent, we derive the equation

$$\frac{a}{55} \approx 0.800673589$$

$$a = 44.037$$

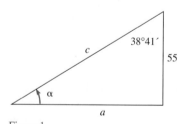

Figure 1

Now that we know the lengths of two sides, we can determine the length of the third side, c, by applying the Pythagorean theorem

$$c^2 = a^2 + 55^2$$
$$= (44.037)^2 + 55^2$$
$$= 4964.3$$
$$c = 70.458$$

➤ **EXAMPLE 2**
Solving a Right Triangle

The right triangle shown in Figure 2 has two sides specified. Determine the length of the unknown side and the measures of the two acute angles.

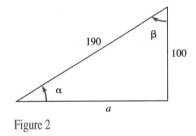

Figure 2

Solution

In the right triangle shown, the hypotenuse has length 190 and one side has length 100. Therefore, by the Pythagorean theorem, we have

$$a^2 = 190^2 - 100^2$$
$$= 26,100$$
$$a = \sqrt{26,100} = 161.555$$

To find the measure of α, we can use the tangent function, which relates the value of a to the measure of the angle α:

$$\tan \alpha = \frac{\text{Opposite}}{\text{Adjacent}}$$
$$= \frac{100}{a}$$
$$= \frac{100}{161.555} = 0.618984246$$
$$\alpha = \tan^{-1} 0.618984246$$
$$\approx 31.757°$$

Because the acute angles of a right triangle are complementary, we can determine the value of β from the value of α:

$$\beta = 90° - \alpha$$
$$= 58.243°$$

The process of solving a triangle can be used in many applied problems, as we illustrate in the next few examples.

➢ EXAMPLE 3
Surveying

A surveyor wishes to measure the distance from a point A on one bank of a river to a point B on the opposite side. To do so, he stakes out a point C 100 yards from point A on the same side of the river, as shown in Figure 3. He then uses a transit (a device for measuring angles) and determines that the angle formed by the lines AC and BC is 37°12′. What is the distance between the two points A and B across the river from one another?

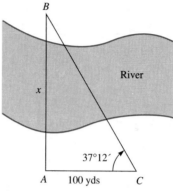

Figure 3

Solution

We have denoted the distance we wish to determine by x in Figure 3. We can relate x to the known angle and side through the tangent function:

$$\tan 37°12' = \frac{\text{Opposite}}{\text{Adjacent}} = \frac{x}{100}$$

$$0.75904 \approx \frac{x}{100}$$

$$x \approx 75.90$$

That is, the distance across the river is ≈ 75.9 yards.

➢ EXAMPLE 4
Architecture

A housing developer wishes to create a triangular lot at the corner of two streets. The sides forming the corner are 125 feet and 160 feet long. Assuming the angle at the corner is 90°, determine the other two angles. (These angles are usually recorded in the land records of the local jurisdiction.) (See Figure 4.)

Solution

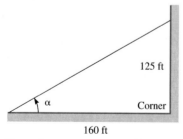

Figure 4

The angle α can be related to the two known legs of the triangle by the equation

$$\tan \alpha = \frac{\text{Opposite}}{\text{Adjacent}}$$

$$= \frac{125}{160}$$

$$\alpha = \tan^{-1} 0.78125$$

$$\approx 37.999°$$

The angle β is the complement of α. Therefore, we have

$$\beta = 90° - \alpha$$

$$\approx 52.001°$$

➤ EXAMPLE 5
Surveying

A surveyor wishes to determine the height of a building. She measures a distance of 1500 feet from the center of the base of the building and uses a transit to determine the angle of inclination to the top of the building. This angle turns out to be 24°57′. How high is the building? (See Figure 5.)

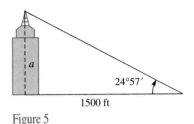

Figure 5

Solution

Let the height of the building be denoted a. Then a is related to the given data of the problem via the equation

$$\tan 24°57' = \frac{\text{Opposite}}{\text{Adjacent}}$$

$$= \frac{a}{1500}$$

$$0.46524567 = \frac{a}{1500}$$

$$a = 697.87 \text{ ft}$$

That is, the building is 697.87 feet tall.

Historical Note:
The Mathematics
of Surveying

The field of surveying as currently practiced was founded by the great German mathematician Karl Friedrich Gauss in the early 19th century. At that time, academic appointments paid very meager wages. Gauss invented the modern field of surveying while working summers for the geodetic summer of Hannover, determining the boundaries of local land holdings. The mathematical basis for modern surveying is exactly as Gauss set it out almost one and one-half centuries ago.   ❙■

Exercises 5.4

Solve these triangles.

1.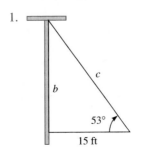

2.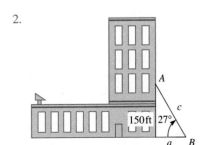

3.

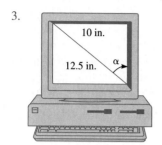

4.

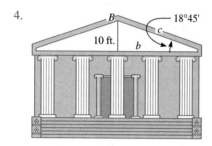

Solve these right triangles. Standard position and lettering is used.

5. $a = 3.5$, $b = 6.8$ 6. $a = 11.7$, $b = 3.1$

7. $b = 23$, $c = 38$ 8. $b = 2.3$, $c = 4.1$

9. $A = 23°40'$, $b = 1.42$ 10. $B = 74°10'$, $b = 21.3$

11. $B = 56°8'$, $a = 340$

12. $A = 48°50'$, $a = 340.4$

13. $a = 32.1$, $b = 40.3$ 14. $c = 103.5$, $a = 39.2$

15. $A = 37°43'$, $a = 1250$

16. $B = 58°22'$, $a = 24,000$

17. $a = 32.5$, $c = 63.4$ 18. $b = 8.0$, $c = 14.0$

◆ Applications

19. **Surveying.** A surveyor wishes to measure the distance from a point on one side of a river to a point on the opposite side. She stakes out two points 100 ft apart on the same side of the river. She then uses a transit (a device for measuring angles) and determines that the angle between the line through the two stakes and a line through the second stake and the point across the river is 43°39'. What is the distance between the two opposite points across the river from each other?

20. **Architecture.** A housing developer wishes to create a triangular lot at the corner of a street. The sides bordering on the two sides of the corner are 135 ft and 210 ft, respectively. Determine the angles formed by the vertices of the lot.

21. **Forestry.** The largest tree in Sequoia National Park in California is called General Sherman. (See figure.) To find its height, park rangers marked off a distance of 185

feet from its base. At that point they found the angle of elevation to the top of the tree to be 55°50'. How tall is General Sherman?

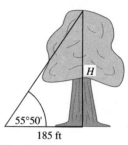

Exercise 21

22. At a point 15 m from the base of a tree, a fireman looks up and sees a cat stuck in the tree 35 m above ground level. The fireman places the bottom of a ladder at his feet and extends it so it reaches the cat in the tree. How long is the extended ladder? What angle does the ladder make with the ground? With the tree?

23. A photographer attaches her camera to a tripod. The camera must be 1.5 m from the floor to get the best shot, and the tripod legs are each 1.6 m long. What is the diameter of the circle on the floor that includes the three tripod legs?

24. A picnic table bench was made with legs 40 cm long. The bench needs a support attached from the middle of the bench to a point $\frac{3}{4}$ of the way down the leg. The distance from the leg to the middle of the bench is 75 cm. How long must the support be? What angle does the support make with the bench? With the leg?

25. The wood frame for the side of a house must be propped up on its concrete slab with a wood plank. (See figure.) The wood frame is 15 ft tall, the slab is 1 ft higher than ground level, and the plank is 20 ft long. How far is the bottom end of the plank from the slab? What angle does the plank make with the ground?

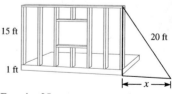

Exercise 25

26. A diver 2 m tall on a 3 m-high board sees a spectator 2 m tall standing at the other end of an Olympic-sized pool, 100 m long. How far is the diver from the spectator? What is the angle of elevation from the spectator's head to the diver?

27. An airplane takes off from a runway at an angle of elevation of 4°20'. Two minutes later the plane is at an altitude of 2000 ft. How far is the airplane from the point of takeoff? (See figure.)

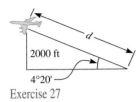

2000 ft

4°20'

Exercise 27

28. A 747 jet airliner takes off at an angle of elevation of 35° and travels a distance in the air of 2500 ft. What is its altitude?

29. An artillery battery spotter locates a target that is 6480 yards west and 5720 yards north of her position. What angle along the ground should she use to fire on the target? What angle would she yet have to determine?

30. You have the task of putting up a new TV antenna. It will be 50 ft off the ground, and you will need three guy wires to hold it up. Each guy wire will be at an angle of 48° from level ground and connected to the middle of the antenna. How much guy wire will you need in all?

31. In order for paratroopers to land on a target when jumping from a C-47 traveling at 145 mph at an altitude of 1200 ft, they jump when the angle of depression to the target is 24°10'. (See figure.) How far from the target horizontally will they be when they jump? How far will they travel in the air?

32. **Measuring cloud height.** To measure the height of clouds one can use a light such as those used at shopping centers to draw attention to a sale. (See figure.) The light is pointed straight up at the clouds. From a point 150 ft away, an angle of elevation is measured and found to be 72°50'. How high are the clouds?

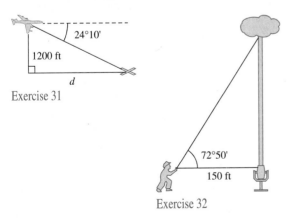

24°10'

1200 ft

d

Exercise 31

72°50'

150 ft

Exercise 32

33. Find c and b. (See figure.)

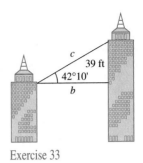

c   39 ft

42°10'

b

Exercise 33

34. Find h. (See figure.)

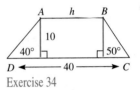

A   h   B

10

40°   50°

D ◄──── 40 ────► C

Exercise 34

35. Engineers wish to build a hanging bridge between two mountain peaks. The angle of elevation from the top of one peak to the top of the other is 18°40'. The taller peak is 2400 m higher than the shorter one. How long will the bridge have to be?

36. **Civil engineering.** For safety reasons, civil engineers are taught when building a road down a mountain that the angle of depression can be no more than 60°. (See figure.) The distance from point A on a mountain to point B on a mountain is 4 miles, and point A is 2.5 miles above point B. Can the engineers safely build a road straight down from A to B? What is the angle of depression?

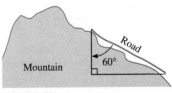

Road

Mountain   60°

Exercise 36

37. A regular octagon has sides 40.3 ft long and is inscribed in a circle. Find its radius.

38. A regular hexagon is inscribed in a circle whose radius is 5.6 m. Find the perimeter of the hexagon.

39. A regular pentagon has a perimeter of 70 yd and is inscribed in a circle. Find the radius of the circle.

40. The altimeter of a Navy reconnaissance airplane records 5000 ft as it passes over its carrier. At that moment it sights a submarine just under the surface. The angle of

depression from the plane to the submarine is 25°40'. How far is it from the sub to the carrier?

41. In reference to Exercise 40, the plane spots a second submarine behind the first. The angle of depression to that submarine is 20°55'. (See figure.) How far apart are the two submarines?

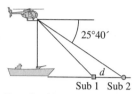

Exercise 41

42. **Aerial navigation.** Trigonometry has extensive application to aerial navigation. (See figure.) Pilots learn that directions are measured in degrees, clockwise from north. North is 0°, east is 90°, south is 180°, and west is 270°. Suppose an airplane leaves an airfield and flies 200 miles in a direction of 240°. How far south of the airfield is the plane? How far west?

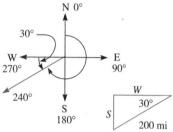

Exercise 42

43. The numbers on the end of a runway indicate the direction a plane would be taking off or landing. For exam-

ple, in the movie *Airport* the pilot was most concerned that he be able to land on runway 29. This meant that he would be landing at a direction of 290°, to the nearest 10°. What would the direction be if the pilot were landing on the other end of this runway?

44. From a hot air balloon 4000 ft high, the angles of depression to two houses in line with the balloon are 46°30' and 32°40'. How far apart are the houses?

45. A **degree** may seem like a very small unit, but at large distances an error of 1 degree can be very significant. Suppose a laser beam is directed toward the visible center of the moon and that it misses its assigned target by 30 seconds. Approximate how far it is, in miles, from its assigned target. Assume that the distance from the earth to the moon is 234,000 miles and that, because of the extreme distance, the surface of the moon is flat.

46. In the accompanying figure, $\angle BA_1C_1 = \angle BA_2C_2 = 25.6°$, the distance from A_1 to A_2 is 425 ft, from A_1 to C_1 is 52 ft, and from D to C_2 is 79 ft. Find the distance from A_2 to D.

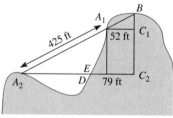

Exercise 46

47. Two tracking stations s miles apart measure the angles of elevation of a weather balloon to be α and β. (See figure.) Derive a formula for the altitude h of the weather balloon in terms of s, α, and β.

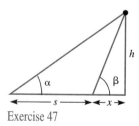

Exercise 47

48. An aircraft carrier is due east of a destroyer. A lighthouse is 10 miles south of the destroyer. From the car-

rier, the bearing to the lighthouse is S 54°28′ W. How far is the carrier from the destroyer? How far is the carrier from the lighthouse?

49. **Carpentry**. A table is designed so that the leg is inclined at an angle of 75° to the top. (See figure.) The width of the wood is to be 3 inches, and it is to be 18 inches from the floor. How long a piece of wood is needed for a leg? Find $x + y$.

50. Find a formula for the area A of a right triangle in terms of b, c, and $\sin \alpha$.

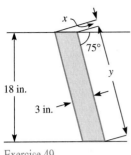

Exercise 49

5.5 THE GRAPHS OF SINE AND COSINE

In this section, we will study $\sin t$ and $\cos t$ with emphasis on their properties as functions of the variable t.

The Functions Sine and Cosine

For each angle of radian measure t, we have defined the number $\sin t$. The correspondence

$$f(t) = \sin t$$

associates to the real number t the value $\sin t$ and also is therefore a function of the real variable t. Similarly, the correspondence

$$g(t) = \cos t$$

associates to the real number t the value $\cos t$, and also is a function of the real variable t.

Recall that the cosine function is defined as the x-coordinate of a certain point on the unit circle. Because x-coordinates of points on the unit circle lie between -1 and 1, we have the inequality

$$-1 \le \cos t \le 1 \tag{1}$$

Similarly, because the sine is defined as the y-coordinate of a certain point on the unit circle, we have the inequality

$$-1 \le \sin t \le 1 \tag{2}$$

Inequalities (1) and (2) say that the ranges of the sine and cosine functions are subsets of the interval $[-1, 1]$. In fact, because any real number in this interval can be the x-coordinate (or y-coordinate) of a point on the unit circle, every number in the interval $[-1, 1]$ is the value of $\cos t$ (or $\sin t$) for some value of t. That is, the range of both the sine and cosine functions is precisely the interval $[-1, 1]$.

Sketching the Graphs of sin t and cos t

Let's first sketch the graph of the function $\sin t$. To do this, we draw a unit circle and the coordinate system for the graph side by side. Imagine an angle t whose terminal side moves counterclockwise, beginning at the positive x-axis. For each position of the terminal side, we plot the vertical height of the point P shown in Figure 1. As the angle increases, the vertical height starts at 0 and increases to 1 when t equals $\pi/2$. The height then decreases from 1 to -1 as the angle goes from $\pi/2$ to $3\pi/2$. Finally, the height increases from -1 to 0 as the angle goes from $3\pi/2$ to 2π. The height of P gives the value of $\sin t$. The graph of the height, as shown in Figure 2, is the graph of $\sin t$ for t in the interval $[0, 2\pi]$.

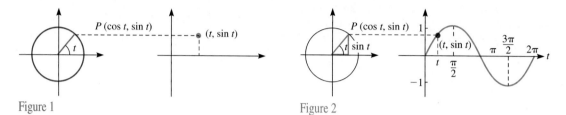

Figure 1   Figure 2

For angles t larger than 2π, we allow the terminal side of the angle in Figure 2 to continue rotating in the counterclockwise direction. We plot the height of the endpoint P of the terminal side of the angle. This height equals the y-coordinate of P, which equals $\sin t$. In this way, we arrive at the graph of $\sin t$ for the remaining positive values of t. For negative values of t, we start the angle t at 0, rotate it in the clockwise (negative) direction, and plot the height of P. This results in the portion of the graph in Figure 3 that lies to the left of the y-axis.

The horizontal position of a point on the unit circle equals the value of the cosine of the corresponding angle. We may rotate the terminal side of the angle t and plot the horizontal position of the point P to sketch the graph of $\cos t$. The graph is shown in Figure 4.

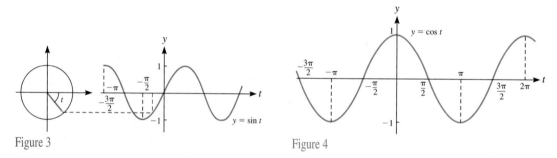

Figure 3   Figure 4

By examining the graphs of $\sin t$ and $\cos t$, we see that they repeat horizontally every 2π units. This is because the vertical and horizontal positions of P repeat after each revolution of 2π. This repetitive character of the sine and cosine functions can be expressed by the equations

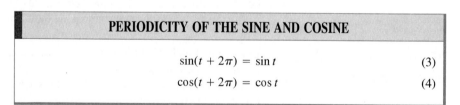

PERIODICITY OF THE SINE AND COSINE

$$\sin(t + 2\pi) = \sin t \qquad (3)$$

$$\cos(t + 2\pi) = \cos t \qquad (4)$$

These equations say that the values of the sine and cosine functions repeat every 2π units. Mathematicians say that the sine and cosine functions are **periodic** and have the **period** 2π.

> **EXAMPLE 1**
Reading the Sine Graph

Determine all values of t for which $\sin t = 0.5$.

Solution

Because the sine function is periodic, it suffices to determine the solutions in the interval $[0, 2\pi]$. By Definition 1 in Section 5.2, we must determine the points P on the circle whose y-coordinate is 0.5. Examining Figure 5, we see that by the definition of the sine function the possible values of t are $\pi/6$ and $5\pi/6$.

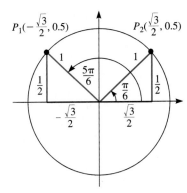

Figure 5

Moreover, these are the only solutions to the equation in the interval $[0, 2\pi]$. By periodicity, the set of all solutions is

$$\left\{ \ldots, -\frac{11\pi}{6}, \frac{\pi}{6}, \frac{13\pi}{6}, \ldots \right\} \cup \left\{ \ldots, -\frac{7\pi}{6}, \frac{5\pi}{6}, \frac{17\pi}{6}, \ldots \right\}$$

Zeros, Maximum Values, Minimum Values

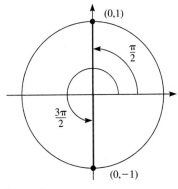

Figure 6

The zeros of $\cos t$ correspond to points on the unit circle with x-coordinate 0, namely, $(0, 1)$ and $(0, -1)$. (See Figure 6.) The corresponding angles t are $\pi/2$ and $3\pi/2$. Of course, we may add to these angles any number of complete revolutions, that is, any integer multiple of 2π. So the complete set of zeros of $\cos t$ is

$$\left\{ \ldots, -\frac{5\pi}{2}, -\frac{3\pi}{2}, -\frac{\pi}{2}, \frac{\pi}{2}, \frac{3\pi}{2}, \frac{5\pi}{2}, \ldots \right\}$$

Using similar reasoning, we find that the zeros of $\sin t$ are the values of t corresponding to the points with y-coordinate 0, as shown in Figure 7. The corresponding values of t are

$$\{ \ldots, -2\pi, -\pi, 0, \pi, 2\pi, \ldots \}$$

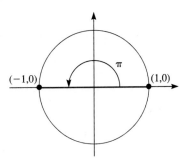

Figure 7

We can summarize these results in the following statements.

ZEROS OF THE SINE AND COSINE FUNCTIONS

1. The function $\sin t$ equals 0 for precisely the following values of t :

$$\{\ldots, -2\pi, -\pi, 0, \pi, 2\pi, \ldots\}$$

2. The function $\cos t$ equals 0 for precisely the following values of t:

$$\left\{\ldots, -\frac{5\pi}{2}, -\frac{3\pi}{2}, -\frac{\pi}{2}, \frac{\pi}{2}, \frac{3\pi}{2}, \frac{5\pi}{2}, \ldots\right\}$$

The maximum value of $\sin t$ (the largest value assumed by the function) is 1 and occurs for angles corresponding to the points $(0, 1)$ on the unit circle. (See Figure 8.) These angles are

$$t = \frac{\pi}{2} + 2k\pi \qquad (k = 0, \pm1, \pm2, \ldots)$$

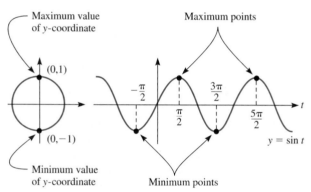

Figure 8

The minimum value of $\sin t$ (the smallest value assumed by the function) is -1 and occurs for angles corresponding to the point $(0, -1)$. (See Figure 8.) These angles are

$$t = \frac{3\pi}{2} + 2k\pi \qquad (k = 0, \pm1, \pm2, \ldots)$$

The minimum value of $\cos t$ is -1 and occurs for angles corresponding to the point $(-1, 0)$ on the unit circle. (See Figure 9.) These angles are

$$t = \pi + 2k\pi = (2k + 1)\pi \qquad (k = 0, \pm1, \pm2, \ldots)$$

The maximum value of $\cos t$ is 1 and occurs for angles corresponding to the point $(1, 0)$ on the unit circle. (See Figure 9.) These angles are

$$t = 2k\pi \qquad (k = 0, \pm1, \pm2, \ldots)$$

The accompanying table summarizes the facts we have determined about the sine and cosine functions.

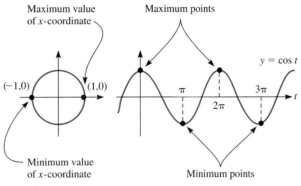

Figure 9

	$\sin t$	$\cos t$
Domain	All real numbers t	All real numbers t
Range	$[-1, 1]$	$[-1, 1]$
Zeros	$t = k\pi$ $(k = 0, \pm1, \pm2, \ldots)$	$t = \pi/2 + k\pi$ $(k = 0, \pm1, \pm2, \ldots)$
Maximum value	1 at $t = \pi/2 + 2k\pi$ $(k = 0, \pm1, \pm2, \ldots)$	1 at $t = 2k\pi$ $(k = 0, \pm1, \pm2, \ldots)$
Minimum value	-1 at $t = 3\pi/2 + 2k\pi$ $(k = 0, \pm1, \pm2, \ldots)$	-1 at $t = \pi + 2k\pi$ $(k = 0, \pm1, \pm2, \ldots)$

Periodic Phenomena

Nature provides us with many examples of repetitive phenomena. For example, the length of the day in a given location varies with the time of year, but those lengths repeat themselves each year. Sound of a particular frequency consists of repeated vibrations a certain number of times per second. The heights of the tides vary on a 28-day cycle that depends on the position of the moon relative to the earth. Blood pressure varies in a cycle determined by respiration.

From these examples, you can see the importance of repetitive phenomena. We must be able to describe such phenomena mathematically and to solve problems concerning them. Many repetitive phenomena can be described using functions of the form

$$f(x) = a \sin(bx + c), \qquad g(x) = a \cos(bx + c)$$

where a, b, and c are real constants.

In this section, we will study such functions, their graphs, and how they may be used to describe repetitive phenomena.

Throughout this section, we will consider the trigonometric functions as functions of a real variable. To emphasize that the variable may or may not arise as the measure of an angle, we will denote it by x rather than t.

Periodicity

Let's begin by providing a mathematical description of a repetitive phenomenon.

Definition 1
Period of a Function

We say that the function $f(x)$ is **periodic** provided that there is a real number c such that

$$f(x + c) = f(x)$$

for all real numbers x. We say that c is a **period** of f.

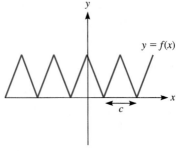

y

$y = f(x)$

x

c

Figure 10

As an example of a periodic function, consider

$$f(x) = \sin x$$

Earlier in this section, we saw that

$$\sin(x + 2\pi) = \sin x$$

for all real numbers x. In other words, $\sin x$ is periodic with period 2π.

Suppose that $f(x)$ has period c. Then the values of $f(x)$ repeat if x is increased by c units. In particular, the graph of $f(x)$ repeats every c units along the x-axis. Figure 10 shows the repetitive nature of the graph of a periodic function.

From our knowledge of the periodicity of the sine and cosine functions, we can deduce the following result.

PERIODS OF GENERAL SINE AND COSINE FUNCTIONS

Let b be a nonzero real number and a and c any real numbers. The functions

$$f(x) = a \sin(bx + c), \qquad g(x) = a \cos(bx + c)$$

have period $2\pi/b$.

Proof   We will prove the result for the function $f(x) = a \sin(bx + c)$. If x is any real number, then

$$f\left(x + \frac{2\pi}{b}\right) = a \sin\left[b\left(x + \frac{2\pi}{b}\right) + c\right]$$

$$= a \sin\left(bx + b \cdot \frac{2\pi}{b} + c\right)$$

$$= a \sin(bx + c + 2\pi)$$

$$= a \sin(bx + c) \qquad \textit{(because } \sin x \textit{ has period } 2\pi)$$

$$= f(x)$$

That is, whenever x is increased by $2\pi/b$, the value of $f(x)$ is unchanged, so $f(x)$ has period $2\pi/b$. The proof of the result for the function $g(x) = a \cos(bx + c)$ is similar and will be omitted.   ♦

> **EXAMPLE 2**
Calculating Periods

Determine the periods of the following functions.

1. $\sin(3x)$ 2. $\sin(2\pi x)$

3. $\sin\left(\dfrac{x}{3}\right)$ 4. $\cos\left(\dfrac{5\pi x}{3}\right)$

Solution

In each example, the period is equal to $2\pi/b$, where b is the coefficient of x.

1. $\dfrac{2\pi}{b} = \dfrac{2\pi}{3}$ 2. $\dfrac{2\pi}{b} = \dfrac{2\pi}{2\pi} = 1$

3. $\dfrac{2\pi}{b} = \dfrac{2\pi}{(1/3)} = 6\pi$ 4. $\dfrac{2\pi}{b} = \dfrac{2\pi}{5\pi/3} = \frac{6}{5}$

The value of the function $f(x) = \sin bx$ oscillates between -1 and 1 with period $2\pi/b$. The next example illustrates the procedure for graphing such functions.

➤ **EXAMPLE 3**
Graphing General Sine Functions

Sketch the graphs of the following functions.

1. $\sin(2x)$ 2. $\sin(2\pi x)$ 3. $\sin\left(\dfrac{x}{3}\right)$

Solution

1. The period of $\sin(2x)$ is $2\pi/b = 2\pi/2 = \pi$, which is half the period of $\sin x$. This means that every cycle on the graph of $\sin x$ corresponds to two cycles on the graph of $\sin(2x)$. The graphs of both functions are sketched in Figure 11. This may be easily checked by graphing both functions simultaneously using a graphing calculator.

2. The period of $\sin(2\pi x)$ is $2\pi/b = 2\pi/2\pi = 1$. Its graph is sketched in Figure 12.

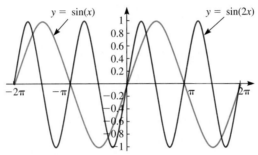

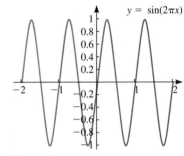

Figure 11   Figure 12

3. The period of $\sin x/3$ equals $2\pi/b = 2\pi/1/3 = 6\pi$. This period is three times the period of $\sin x$. Figure 13 shows the graphs of both functions.

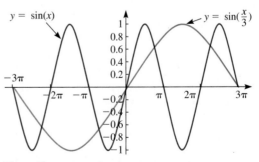

Figure 13

Amplitude

The graphs of the functions $a \sin(bx)$ and $a \cos(bx)$ may be obtained by scaling the graphs of $\sin(bx)$ and $\cos(bx)$, respectively, by a factor $|a|$ and, in case a is negative, reflecting the graphs in the x-axis.

Definition 2	The number $|a|$ is called the **amplitude** of the graph. Geometrically, the ampli-
Amplitude of a General Sine	tude equals one-half the height of the graph. Figure 14 illustrates these interpre-
or Cosine Graph	tations of the amplitude.

The next few examples illustrate the use of the amplitude in graphing.

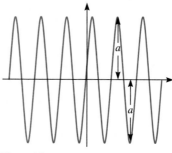

Figure 14

➢ **EXAMPLE 4**
Graphing a General
Sine Function

Sketch the graph of the function

$$y = 2\sin(3x)$$

Solution

The period of the function is equal to $2\pi/b = 2\pi/3$. Moreover, the amplitude equals 2, so the function values range from -2 to 2. We sketch the graph by first drawing one cycle, in the shape of a sine curve, for x in the interval $[0, 2\pi/3]$, which corresponds to the first period. We then sketch the remainder of the graph by repeating the portion just sketched. (See Figure 15.)

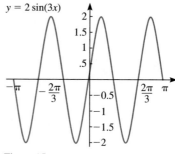

Figure 15

➢ **EXAMPLE 5**
Graphing Another General
Sine Function

Sketch the graph of the function

$$y = -2\sin(6x)$$

Solution

The period of the function equals $2\pi/b = 2\pi/6 = \pi/3$. The amplitude is 2, so the function values range between -2 and 2. Because the value of a is negative, the graph is the graph of a sine curve reflected in the x-axis. We first sketch one period of the graph for x in the interval $\left[0, \frac{\pi}{3}\right]$. Then we sketch the remainder of the graph by repeating the portion already sketched. (See Figure 16.)

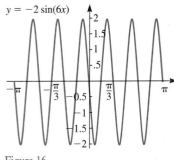

Figure 16

➤ EXAMPLE 6
Graphing a General Cosine Function

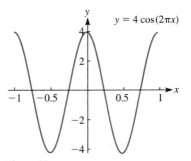

Figure 17

Sketch the graph of the function

$$y = 4\cos(2\pi x)$$

Solution

The period in this example is $2\pi/2\pi = 1$. The amplitude equals 4. The graph resembles the graph of $\cos x$ except that one cycle corresponds to a distance 1 along the x-axis instead of 2π. We sketch one cycle of the graph corresponding to the interval $[0, 1]$. Then we complete the sketch by repeating the portion already sketched to the right and the left. (See Figure 17.)

Many instances of motion are periodic in nature. Such motion involves a pattern that is repeated at fixed time intervals and is called **harmonic motion.** Examples of harmonic motion are pendulums and objects suspended from a spring. Harmonic motion may be described using trigonometric functions, as the following example illustrates.

➤ EXAMPLE 7
Harmonic Motion

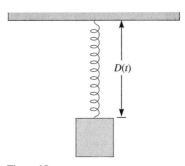

Figure 18

A block is attached to a hanging spring and set in motion by stretching the spring from its rest position and releasing it. At time t seconds after the block is set in motion, the distance of the block $D(t)$ from its rest position is given by formula

$$D(t) = 7\cos(6\pi t)$$

where $D(t)$ is measured in inches. (See Figure 18.)

1. How many bounces per second does the block execute?
2. How far does the block move between extreme ends of a bounce?
3. Sketch the graph of $D(t)$.

Solution

1. The period of the function $D(t)$ is equal to $2\pi/b = 2\pi/6\pi = 1/3$. This means that each bounce of the block takes 1/3 second, or, equivalently, there are 3 bounces per second.

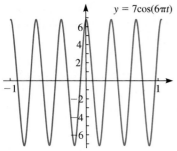

Figure 19

2. The extreme ends of a bounce are located at a distance $|a|$ in both the positive and the negative direction from the rest position of the block, where $|a|$ is the amplitude of $D(t)$. From the given formula, the amplitude equals 7, so the distance traveled in one bounce equals the distance down to 0 followed by the distance down to -7. That is, the distance equals $2 \cdot 7$, or 14 inches.

3. The graph of $D(t)$ is sketched in Figure 19.

Phase Shift

By writing the function in the form

$$f(x) = a \sin(bx + c)$$

$$= a \sin b\left[x - \left(-\frac{c}{b}\right)\right]$$

we see that the function $f(x)$ can be obtained by replacing x with $x - (-c/b)$ in the function $h(x) = \sin(bx)$. Recall our general discussion in Chapter 2, in which we showed that the graph of $f(x - h)$ can be obtained by translating the graph of $f(x)$ horizontally by $|h|$ units. If h is positive, the translation is to the right, whereas if h is negative, the translation is to the left. The number $-c/b$ is called the **phase shift** of the function $a \sin(bx + c)$.

To obtain the graph of the function $a \sin(bx + c)$, we first determine the graph of $a \sin(bx)$, determine the phase shift, and then translate the initial graph horizontally by an amount equal to the phase shift. This general graphing procedure is illustrated in the next two examples.

➤ **EXAMPLE 8**
A Function
with Phase Shift

Consider the function

$$f(x) = 3 \sin\left(2x - \frac{\pi}{3}\right)$$

1. Determine the amplitude, period, and phase shift.
2. Sketch the graph of the function.

Solution

1. Write the function in the form

$$f(x) = 3 \sin 2\left(x - \frac{\pi}{6}\right)$$

The amplitude is the absolute value of the coefficient of sine, or 3. To obtain the period, we divide 2π by the coefficient of x, or 2, so the period is $2\pi/2 = \pi$. The phase shift is the constant term subtracted from x, $\pi/6$.

2. We start from the graph of $\sin x$. Then we scale vertically by the amplitude, which is 3, to obtain the graph of $3 \sin x$. The scaled graph varies vertically from -3 to 3. Next, we scale horizontally by a factor of 1/2 to obtain the function $3 \sin 2x$. This compresses the graph horizontally, so that a period fits in half of its original horizontal space. Finally, we translate the graph to the right by the phase shift $\pi/6$ to obtain the graph of $3 \sin 2(x - \pi/6)$. The completed sketch is shown in Figure 20.

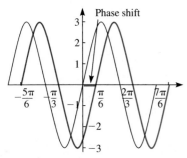

Figure 20

Another Application

Sine and cosine functions may be shifted and scaled to arrive at a variety of new functions, as we have seen. In the next example, we show how a function that arises by vertically shifting a sine function can be used in an applied situation.

➤ EXAMPLE 9
Blood Pressure

A person's blood pressure changes in a rhythmic fashion, with a periodicity set by the beating of the heart. Suppose that a person's blood pressure at time t minutes is given by the formula

$$P(t) = 100 + 25 \sin(160\pi t)$$

1. Determine the rate at which the person's heart is beating.
2. Sketch the graph of $P(t)$.
3. What are the maximum and minimum blood pressure readings?

Solution

1. The function $P(t)$ is periodic with period equal to $2\pi/b = 2\pi/160\pi = \frac{1}{80}$. That is, each period of the function lasts for $\frac{1}{80}$ second. Because each period corresponds to one heartbeat (that is, one complete blood pressure cycle), the rate at which the heart is beating is 80 beats per minute.

2. The function $P(t)$ is obtained by adding the constant 100 to the trigonometric function $25 \sin (160\pi t)$. From our general principles of graphing developed in Chapter 2, the graph of $P(t)$ may be obtained by first graphing the trigonometric function and then translating the graph upward by 100 units. The graph is shown in Figure 21.

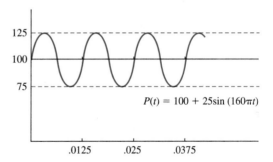

Figure 21

3. The amplitude of the trigonometric function $25 \sin(160\pi t)$ is $|a| = 25$. By examining the graph in Figure 21, we see that the value of $P(t)$ oscillates between $100 - 25 = 75$ and $100 + 25 = 125$. That is, the minimum blood pressure is 75 and the maximum blood pressure is 125.

Exercises 5.5

For each of the following functions, (a) find the period, (b) find the amplitude, and (c) sketch the graph.

1. $f(x) = \cos x$
2. $f(x) = \sin x$
3. $f(x) = \sin\left(\frac{1}{2}x\right)$
4. $f(x) = \cos(2x)$
5. $f(x) = \cos(2\pi x)$
6. $f(x) = \sin(\pi x)$
7. $f(x) = \cos(3x)$
8. $f(x) = \cos\left(\frac{1}{4}x\right)$
9. $f(x) = 2 \sin x$
10. $f(x) = 3 \cos x$
11. $f(x) = -\frac{1}{2} \cos x$
12. $f(x) = -2 \sin x$
13. $f(x) = 2\cos(2x)$
14. $f(x) = 3 \sin(3x)$
15. $f(x) = 4 \sin(2\pi x)$
16. $f(x) = 2\cos(\pi x)$

17. $f(x) = -3\cos(\pi x)$

18. $f(x) = -0.25\sin(6\pi x)$

19. $f(x) = -\sin\left(\dfrac{2\pi x}{3}\right)$ 20. $f(x) = 6\sin\left(\dfrac{\pi x}{4}\right)$

For each of the following functions, (*a*) find the period, (*b*) find the amplitude, (*c*) find the phase shift, and (*d*) sketch the graph.

21. $f(x) = \sin\left(x - \dfrac{\pi}{2}\right)$ 22. $f(x) = \sin\left(x + \dfrac{\pi}{4}\right)$

23. $f(x) = 3\cos(x + \pi)$ 24. $f(x) = -2\sin(x - \pi)$

25. $f(x) = -2\sin(4x + \pi)$

26. $f(x) = -2\sin\left(x + \dfrac{3\pi}{2}\right)$

27. $f(x) = 3\cos\left(2x - \dfrac{\pi}{3}\right)$ 28. $f(x) = 4\cos(\pi x - 2)$

29. $f(x) = \frac{3}{2}\sin(2\pi x + 1)$

30. $f(x) = \frac{5}{2}\cos(2\pi x + 2)$

31. $f(x) = -2 + \sin x$ 32. $f(x) = 3 - 2\cos x$

33. $f(x) = 1 + 2\sin(\pi x)$

34. $f(x) = -3 + \frac{1}{2}\cos(2x - \pi)$

For each of the following functions, find (*a*) the period, (*b*) the amplitude, and (*c*) the phase shift.

35. $f(x) = -6\cos(2x - \pi)$

36. $f(x) = \frac{1}{3}\sin\left(-3x - \dfrac{3\pi}{4}\right)$

37. $f(x) = 5\cos(4\pi x + 2)$

38. $f(x) = -\frac{1}{6}\sin\left(2x + \dfrac{2\pi}{3}\right)$

39. $f(x) = -\frac{2}{3}\sin\left(3x + \dfrac{\pi}{4}\right)$

40. $f(x) = 4\sin\left(3\pi x - \dfrac{5\pi}{4}\right)$

♦ Applications

41. **Physics—vibrations.** A weight is attached to a spring and set in motion by stretching the spring from its rest position and releasing it. At time *t* seconds after the block is set in motion, the distance of the block $D(t)$ from its rest position is given by

$$D(t) = 5\sin(4\pi t)$$

where $D(t)$ is measured in inches.

a. How many bounces per second does the weight execute?

b. How far does the block move between extreme ends of a bounce?

c. What are the maximum and minimum extremes of the bounces?

d. Sketch the graph of $D(t)$.

42. **Blood pressure.** A person's blood pressure changes in a rhythmic fashion, with a periodicity set by the beating of the heart. Suppose a person's blood pressure at time *t* minutes is given by the formula

$$P(t) = 101 + 24\cos(160\pi t)$$

a. Sketch the graph of $P(t)$.

b. Determine the rate at which the person's heart is beating.

c. What are the maximum and minimum blood pressure readings?

43. **Temperature during an illness.** The temperature of a patient during a 12-day illness is given by

$$T(t) = 101.6° + 3\sin\left(\dfrac{\pi}{8}t\right)$$

a. What are the maximum and minimum temperatures of the patient during the illness?

b. Sketch a graph of the function over the interval [0,12].

44. **Satellite location.** (See figure.) A satellite circles the earth in such a way that it is *y* miles from the equator (north or south, height above the earth not considered), *t* minutes after its launch, where

$$y(t) = 5000\left[\cos\dfrac{\pi}{45}(t - 10)\right]$$

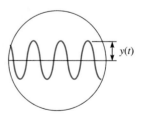

Exercise 44

a. What is the period of *y*?

b. What is the furthest the satellite gets from the equator?

c. Sketch a graph of $y(t)$.

45. **Business: periodic sales.** A company in a northern climate has sales of skis as given by

$$S(t) = 7\left[1 - \cos\left(\dfrac{\pi}{6}t\right)\right]$$

where t is time, in months, $t = 0$ corresponds to July 1, and $S(t)$ is in thousands of dollars.

a. Sketch a graph of $S(t)$ over a two-year interval [0, 24].

b. What is the minimum sales and when does it occur?

c. What is the maximum sales and when does it occur?

d. What is the period of the function? Does this seem reasonable? Explain.

46. **Roller coaster height**. (See figure.) A roller coaster is constructed in such a way that its height H (in meters) x meters from the starting point is given by

$$H(x) = 20 \sin\left(\frac{\pi}{60}x - 2\pi\right)$$

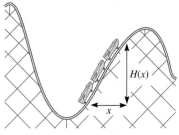

Exercise 46

a. Sketch a graph of $H(x)$ over the interval [0, 480].

b. What are the maximum and minimum heights of the function?

c. What is the period of the function?

d. Describe the roller coaster if $H(x)$ is given by

$$H(x) = 20 \sin\left(\frac{\pi}{60}x - 2\pi\right) + 20$$

47. **Height of a ferris wheel.** (See figure.) You are riding on a ferris wheel. At time t, in seconds, you are at height y, in feet, where

$$y(t) = 53 - 50 \sin\left[\frac{\pi}{5}\left(t - \frac{5}{2}\right)\right]$$

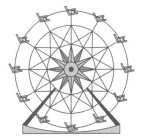

Exercise 47

a. Find the height of the ferris wheel when $t = 0, 2, 3,$ 6, 8, 9, 10 seconds.

b. What is the lowest height of the wheel?

c. What is the maximum height of the wheel?

d. How long does it take for the wheel to go around once?

e. Sketch a graph of the function.

48. **Number of hours of daylight.** The number of hours H of daylight for a city depends on its latitude (how far north) and on the day D of the year. Such a function is given by

$$H(D) = 12 + A \sin\left[\frac{2\pi}{365}(D - 80)\right]$$

where leap year is ignored and 80 corresponds to the 80th day of the year, March 21, on which the vernal equinox occurs, when there are 12 hours of daylight and the daylight begins to increase. The constant A depends on the latitude. New Orleans is 30° north. For New Orleans, the constant A is about 2.3.

a. Suppose your birthday is on August 9. How many hours of daylight are there on August 9?

b. How many hours of daylight are there on December 1?

c. Sketch a graph of the function.

d. What is the period of the function?

e. Which day has the most hours of daylight? The least?

49. **Biorhythmic cycles.** Some believe that a person's life has cycles of good and bad days that begin at birth. There are supposedly three cycles, or **biorhythms:**

Physical: A cycle represented by a sine function with a period of 23 days:

$$P(t) = \sin\left(\frac{2\pi}{23}t\right)$$

Emotional (sensitivity): A cycle represented by a sine function with a period of 28 days:

$$E(t) = \sin\left(\frac{2\pi}{28}t\right)$$

Mental (intellectual): A cycle represented by a sine function with a period of 33 days:

$$M(t) = \sin\left(\frac{2\pi}{33}t\right)$$

Time t is measured in days, with $t = 0$ at birth. Let us consider 50 days in your life following your 19th birthday.

a. Using the same set of axes, sketch graphs of the three biorhythm functions over the interval [6940, 6990].

b. Suppose one could use these curves to gauge when to do important activities. Then one plan might be to carry out such activities when all the curves are high. Over what interval do you think you would do best on a test over this course? The worst?

50. **Predator–prey models.** In the Hudson Bay area, the population of lynx, a member of the cat family, and the snowshoe hare, a member of the rabbit family, interact. When the population of prey (rabbits), is high, there is a lot of food for the predator (lynx), and this causes the population of lynx to increase. As the population of lynx increases, they eat more and more rabbits, and this causes the population of rabbits to decrease. Then the lynx population decreases because there are fewer rabbits to eat. Suppose over a certain area during a four-year period of time t, the populations of rabbits and lynx, respectively, are given by the following trigonometric functions:

$$R(t) = 80 + 30 \sin\left(\frac{\pi}{2}t\right)$$

and

$$L(t) = 11 + 2 \cos\left(\frac{\pi}{2}t\right)$$

where the populations are in animals per square mile and time t is in years.

a. Sketch graphs of both functions using the same set of axes over the four-year interval [0, 4].

b. Find the maximum and minimum values of the rabbit population.

c. Find the maximum and minimum values of the lynx population.

d. Carefully examine the two curves in terms of the discussion of the interactive increase and decrease of their populations.

51. **Wave functions.** Water waves, sound waves, motion of a vibrating guitar string—all of these are examples of wave motion. The general equation of wave motion is given by

$$y = A \sin\left(\frac{t}{T} - \frac{x}{\lambda}\right)$$

where y = vertical height of wave, x = distance from the origin of the wave, t = time the wave has traveled, and T = time required for the wave to travel one wavelength λ (the Greek letter lambda).

a. Assuming y is a function of time t, find the amplitude, period, and phase shift.

b. Assuming y is a function of x, find the amplitude, period, and phase shift.

52. **Water wave.** The cross-section of a water wave is given by

$$y = 2 \sin\left(\frac{\pi}{4}x - \frac{\pi}{4}\right)$$

Sketch a graph of the function.

53. **Electrical current.** The current i of a wire at time t passing through a magnetic field is given by

$$i(t) = I \sin(\omega t + \alpha)$$

a. Find the amplitude, period, and phase shift.

b. Sketch a graph of the function when $I = 150$, $\omega = 150\pi$, and $\alpha = -\pi/6$.

Find a function of the type $f(x) = a \sin(bx + c)$ or $f(x) = a \cos(bx + c)$, depending on which is requested, whose graph is the given function or satisfies the given condition.

54. sin

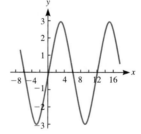

55. cos

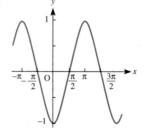

56. cos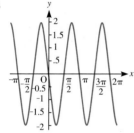

57. cos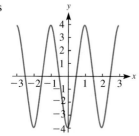

58. sin: amplitude $= 4.3$, period $= 7$, phase shift $= -10$.

59. cos: amplitude $= 2.8$, period $= 1$, phase shift $= 6$.

➡ **Technology**

Use a computer software package or graphing calculator to graph each of the following functions and approximate the zeros on the interval $[-\pi, \pi]$.

60. $f(x) = -\sqrt{43}\cos(0.38x + 0.42) + 4.67$

61. $f(x) = \sqrt{23}\sin(1.42x + 3.43)$

62. $f(x) = \sqrt{23}\sin(1.42x + 0.43) - 8.2$

63. $f(x) = -0.5\ln|\sin(0.89x - 10.67)|$

Use a graphing calculator to graph the following functions. Use the graph to approximate the period, amplitude, and phase shift.

64. $y = 4\sin\dfrac{3\pi}{4}x$

65. $y = -\cos 3x$

66. $y = 1.6\cos 3\pi x$

67. $y = -2.2\sin(x + 1)$

68. Use a graphing calculator to determine the zeros of the function $y = x + \sin x$.

69. Use a graphing calculator to determine the zeros of the function $y = e^x - \sin x$.

Match each of the functions with its graph.

70. $f(t) = 2\sin t$

71. $f(t) = 3\cos t$

72. $f(t) = \sin\dfrac{t}{2}$

73. $f(t) = 2\cos 3t$

74. $f(t) = 3\cos 2t$

75. $f(t) = \sin\left(t + \dfrac{\pi}{4}\right)$

76. $f(t) = 2\cos t + 3$

77. $f(t) = -\sin 2t$

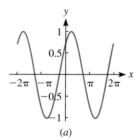

(a)

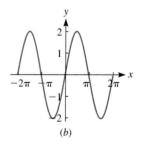

(b)

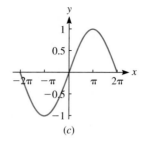

(c)

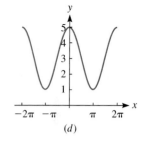

(d)

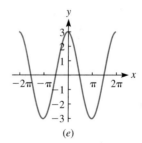

(e)

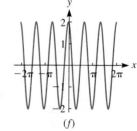

(f)

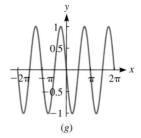

(g)

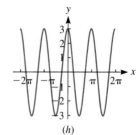

(h)

5.6 GRAPHS OF OTHER TRIGONOMETRIC FUNCTIONS

In the preceding section, we sketched the graphs of functions of the form $a \sin(bx + c)$. We will now determine the domains and sketch the graphs of the other trigonometric functions.

The Tangent Function

Recall that the tangent function is defined as

$$\tan x = \frac{\sin x}{\cos x}$$

It is defined for all values of x for which the denominator, $\cos x$, is nonzero. That is, the domain of $\tan x$ is

$$\left\{ x : x \neq \ldots - \frac{3\pi}{2}, -\frac{\pi}{2}, \frac{\pi}{2}, \frac{3\pi}{2}, \ldots \right\}$$

Because both $\sin x$ and $\cos x$ have period 2π, the same is clearly true of $\tan x$. However, as we shall prove in the next chapter, $\tan x$ actually has period π. For purposes of sketching the graph of the tangent function, let us assume this result.

To sketch the graph of the tangent function, we need only sketch the portion corresponding to x in the interval $[0, \pi]$, and extend the sketch to other values of x using the periodicity of the function. To sketch the graph within this interval, we will plot a set of representative points (or use a graphing calculator). The function is undefined at the midpoint. To obtain the behavior of the graph near the midpoint, we plot points corresponding to several nearby values of x to the left and to the right. The accompanying table lists a set of representative points on the graph, including points nearby and on both sides of $\pi/2 = 1.57080.\ldots$ As x approaches the midpoint from the left, the function values approach ∞. As x approaches the midpoint from the right, the function values approach $-\infty$.

x	$\tan x$
0	0
$\pi/6$	$\sqrt{3}/3 = 0.57735$
$\pi/3$	$\sqrt{3} = 1.73205$
1.5	14.1014
1.55	48.0785
1.6	-34.2325
1.65	-12.5993
$2\pi/3$	$-\sqrt{3} = -1.73205$
$5\pi/6$	$-\sqrt{3}/3 = -0.57735$
π	0

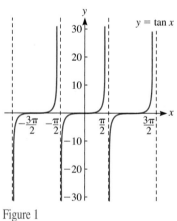

Figure 1

Figure 1 shows a sketch of the resulting graph. Note that the graph has as asymptotes the vertical lines

$$x = \frac{k\pi}{2}, \quad k = \pm 1, \pm 3, \pm 5, \ldots$$

➤ EXAMPLE 1
Sketching a Scaled Tangent Function

Sketch the graph of the function $f(x) = \tan(3x)$.

Solution

The graph has the same general shape as the graph of $\tan x$ except that the scale on the x-axis is changed. The asymptotes occur for

$$3x = \frac{k\pi}{2}, \quad k = \pm 1, \pm 3, \pm 5, \ldots$$

$$x = \frac{k\pi}{6}, \qquad k = \pm1, \pm3, \pm5, \dots$$

In particular, the first asymptotes on either side of the origin are $x = \pm\pi/6$. The distance between the asymptotes is $\pi/3$, which is the period. Figure 2 shows several periods of the graph.

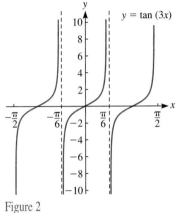

Figure 2

> **EXAMPLE 2**

Sketching a Scaled and Translated Tangent Function

Sketch the graph of the function $f(x) = \tan(3x - \pi/4)$.

Solution

Write the function in the form $f(x) = \tan 3\,(x - \pi/12)$. We may obtain the function from the tangent function in two steps. First we scale horizontally by a factor of 1/3. This compresses the periods of tangent function to intervals of width $\pi/3$. We then translate to the right by the phase shift $\pi/12$. The resulting graph is shown in Figure 3.

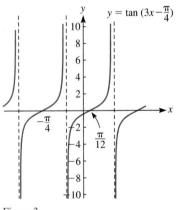

Figure 3

The Secant Function

The function $\sec x$ is defined as

$$\sec x = \frac{1}{\cos x}$$

The domain consists of all x for which the denominator is nonzero, that is,

$$\left\{ x : x \neq \dots, -\frac{3\pi}{2}, -\frac{\pi}{2}, \frac{\pi}{2}, \frac{3\pi}{2}, \dots \right\}$$

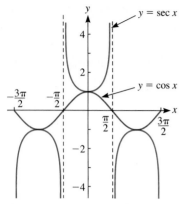

Figure 4

This is the same domain as for the function tan x. To graph sec x, we first sketch the graph of cos x. On the same coordinate system, we plot, for each value of x, a point with height equal to the reciprocal of cos x. The accompanying table gives some points to plot. The resulting graph is shown in Figure 4. Note that the vertical asymptotes occur at the zeros of cos x, namely, at

$$x = \pm\frac{\pi}{2}, \pm\frac{3\pi}{2}, \pm\frac{5\pi}{2}, \ldots$$

x	$\cos x$	$\sec x$
0	1	1
$\pi/4$	0.70711	1.41421
$\pi/2$	0	Undefined
$3\pi/4$	-0.70711	-1.41421
π	-1	-1
$5\pi/4$	-0.70711	-1.41421
$3\pi/2$	0	Undefined
$7\pi/4$	0.70711	1.41421
2π	1	1

➢ **EXAMPLE 3**
Sketching a Scaled
Secant Function

Sketch the graph of the function $f(x) = \sec x/2$.

Solution

The graph has the same shape as the graph of sec x, but the x-axis is rescaled. The first pair of vertical asymptotes occur at

$$\frac{x}{2} = \pm\frac{\pi}{2}$$

$$x = \pm\pi$$

Because sec $x/2 = 1/\cos(x/2)$, the period of sec $x/2$ is the same as that of cos $x/2$, which is $2\pi/1/2 = 4\pi$, so sec $x/2$ repeats every 4π units. Figure 5 shows its graph.

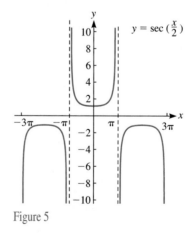

Figure 5

**The Cosecant and
Cotangent Functions**

The function cot x is defined by

$$\cot x = \frac{\cos x}{\sin x}$$

and may be graphed in a manner similar to tan x. The resulting graph is shown in Figure 6. The period of cot x is π. The asymptotes are

$$x = k\pi, \qquad k = 0, \pm1, \pm2, \ldots.$$

The function csc x is defined by

$$\csc x = \frac{1}{\sin x}$$

and may be graphed in a manner similar to sec x. The resulting graph is shown in Figure 7. The period of csc x is 2π and the asymptotes are

$$x = \frac{k\pi}{2}, \qquad k = 0, \pm1, \pm2, \ldots.$$

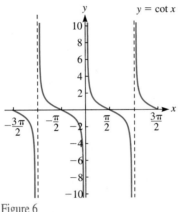

Figure 6

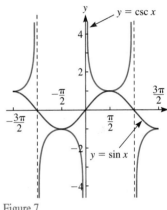

Figure 7

Graphs of Sums

In many applications, it is necessary to consider functions that are sums, with one or more of the summands being a trigonometric function. The next two examples illustrate how such functions may be graphed.

> **EXAMPLE 4**
Graph of a Sum
of Trigonometric Functions

Sketch the graph of $\sin x + 3\cos 2x$.

Solution

In Figure 8, we have sketched on separate axes the graphs of the summands $\sin x$ and $3\cos 2x$. Because the first summand has period 2π and the second has period π, we see that the sum has period 2π. To graph the sum for x between

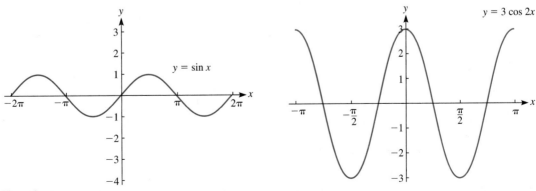

Figure 8

0 and 2π, we tabulate the function for various special values of x and plot the corresponding points. We then extend the graph to all x using periodicity. Figure 9 shows the plotted points and the graph of the sum.

x	$\sin x$	$3\cos 2x$	$\sin x + 3\cos 2x$
0	0	3	3
$\pi/4$	$\sqrt{2}/2$	0	$\sqrt{2}/2$
$\pi/2$	1	-3	-2
$3\pi/4$	$\sqrt{2}/2$	0	$\sqrt{2}/2$
π	0	3	3
$5\pi/4$	$-\sqrt{2}/2$	0	$-\sqrt{2}/2$
$3\pi/2$	-1	-3	-4
$7\pi/4$	$-\sqrt{2}/2$	0	$-\sqrt{2}/2$
2π	0	3	3

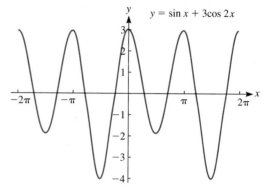

Figure 9

➢ EXAMPLE 5
Graphing by Addition of Ordinates

Sketch the graph of the function $x + \sin x$.

Solution

In Figure 10, we have plotted the graphs of the functions x and $\sin x$ on separate coordinate systems. The graph of the sum is obtained by forming, for each value of x, the sum of the corresponding heights on the two graphs. We can reason out the shape of the sum as follows. On each successive interval of length 2π along the x-axis, the graph of $\sin x$ oscillates between -1 and 1. The graph of x is a straight line. On each successive interval of length 2π along the x-axis, the graph of the sum oscillates around the line: when $\sin x$ is negative, the graph is below the line; when $\sin x$ is positive, the graph is above the line. This results in the third graph shown in Figure 10.

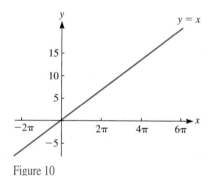
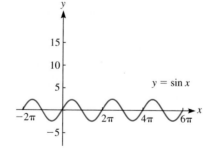
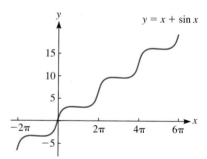

Figure 10

More Complex Graphs

➢ EXAMPLE 6
Damped Oscillation

Sketch the graph of the function $e^{-0.3x}\sin x$, $x > 0$.

Solution

The function $\sin x$ is periodic with period 2π. As x increases, the value of $e^{-0.3x}$ decreases and approaches 0 as x approaches positive infinity. On each successive interval of length 2π along the x-axis, the value of $e^{-0.3x}\sin x$ oscillates like the graph of $\sin x$. However, the amplitude of the oscillations steadily decreases, approaching 0 as x approaches infinity. The graph, obtained from a graphing calculator, is shown in Figure 11. A graph of this sort is called a **damped sine wave**. Such graphs depict vibrations that dissipate over time, such as sound vibrations traveling in water.

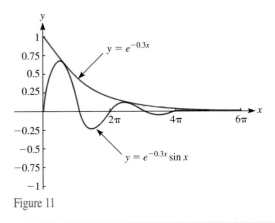

Figure 11

Trigonometric Functions and Graphing Calculators

Graphing functions formed from trigonometric functions can be a tedious affair involving much calculation, as some of the above examples show. Graphing calculators, properly used, can be very helpful in graphing such functions. The next few examples illustrate some of the possibilities.

New Technology

010100101100011001001
100101110010010100
101101000101101001

EXAMPLE 7
The Graph of a Sum

Use a graphing calculator to display at least two periods of the graph of

$$f(x) = \sin x + 2\sin 2x + 3\sin 3x$$

Solution

The first term has period 2π, the second has period π, and the third has period $2\pi/3$. The period of the sum is the least common multiple of the period of the terms, namely 2π. (This number is a common multiple of all the periods and is the least such number.) Two periods corresponds to an interval on the x-axis of length $4\pi \approx 12.56$, so we may choose the standard range $-10 \le x \le 10$, which will show $20/2.09 =$ more than three periods. The graph is shown in Figure 12.

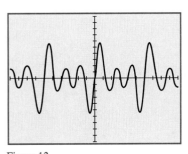

Figure 12

◄

EXAMPLE 8
A Graph with an Infinite Number of Oscillations Near $x = 0$

Use a graphing calculator to display the graph of

$$f(x) = \sin\frac{1}{x}, \qquad x \neq 0$$

Solution

As x approaches 0 through positive values, the quotient $1/x$ increases without bound and goes through an unlimited number of periods of the sine function. The graph of $f(x)$ oscillates infinitely often between -1 and 1 as x approaches 0 through positive values. The graph is shown in Figure 13. Note that the poor resolution of the calculator is not able to truly capture the wild behavior of the graph near $x = 0$. For instance, the oscillations should, in each instance, run from $y = -1$ to $y = +1$. However, the graph makes it appear as if the heights decrease near 0. This is due to the limited resolution of the calculator screen. If you zoom in, you will get a more accurate picture of the graph near $x = 0$.

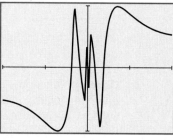

Figure 13 ◄

EXAMPLE 9
Another Oscillating Graph

Use a graphing calculator to display the graph of

$$f(x) = x \sin x$$

Solution

The factor $\sin x$ causes the graph to oscillate between 0 and 1, whereas the factor x^2 grows as x increases in absolute value. The combination of the two factors is shown in Figure 14. Note that because of the limited resolution of the screen, the graph appears flat near $x = 0$. It is actually curved, as you may see by zooming in.

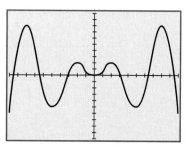

Figure 14 ◄

Exercises 5.6

Sketch the graph of each function.

1. $f(x) = \tan x$

2. $f(x) = \cot x$

3. $f(x) = \csc x$

4. $f(x) = \sec x$

5. $f(x) = \cot\left(x + \dfrac{\pi}{2}\right)$

6. $f(x) = \tan\left(x - \dfrac{\pi}{4}\right)$

7. $f(x) = 2\sec(2x)$

8. $f(x) = -2\sec(x + \pi)$

9. $f(x) = \sin x - \cos x$

10. $f(x) = \cos x - \cos\left(\dfrac{x}{2}\right)$

11. $f(x) = 2\cos x + \sin(2x)$

12. $f(x) = \cos x - \sin(2x)$

13. $f(x) = \sin x - \sin(2x)$

14. $f(x) = 2\sin x + \sin(2x)$

15. $f(x) = 3\sin x + \cos\left(\dfrac{x}{3}\right)$

16. $f(x) = 2\cos\left(\dfrac{x}{2}\right) + \sin x$

17. $f(x) = 3\cos x + \sin(2x)$

18. $f(x) = 3\sin x + \cos(2x)$

19. $f(x) = x + \cos x$

20. $f(x) = \dfrac{x}{2} + \sin x$

21. $f(x) = x - \sin x$

22. $f(x) = \dfrac{x}{2} - \cos x$

23. $f(x) = 2^{-x}\sin x, \quad x \geq 0$

24. $f(x) = 2^{-x}\cos x, \quad x \geq 0$

25. $f(x) = e^{-0.4x}\cos x, \quad x \geq 0$

26. $f(x) = e^{-0.4x}\sin x, \quad x \geq 0$

27. $f(x) = e^x\sin x, \quad x \geq 0$

28. $f(x) = 2^x\cos x, \quad x \geq 0$

29. $f(x) = |\sin x|$

30. $f(x) = |\cos x| - 2$

31. $f(x) = \sin|x|$

32. $f(x) = \cos|x - \pi|$

33. $f(x) = |\tan x|$

34. $f(x) = |\sec x|$

35. $f(x) = x\sin x$

36. $f(x) = x^2\cos x$

37. $f(x) = e^{\sin x}$

38. $f(x) = \cos^2 x$

♦ Applications

39. **Damped oscillations.** A spring will not oscillate up and down forever unless it is reset in motion. Suppose the motion of a spring is given by

$$D(t) = 7e^{-0.1t}\cos(6\pi t)$$

where $D(t)$ is measured in inches and time t is in seconds.

a. Find $D(t)$ when $t = 0, 1, 2, 3, 4, 5, 6, 7, 8, 9, 10$.

b. Sketch the graph of $D(t)$.

c. What is the maximum value of $D(t)$?

d. What is the minimum value of $D(t)$?

40. **Temperature during an illness.** A patient's temperature fluctuates during an illness according to the formula $T(t) = 98.6° + 2e^{-t}\cos 2\pi t$, where t is measured in days. Sketch a graph of the function on the interval $[0, \infty)$.

41. **Rotating spotlight.** A spotlight 180 ft from a prison wall rotates in a circle once every 4 seconds. At $t = 0$, the spotlight is pointing directly at point Q. The spotlight is pointing at a distance d from Q after t seconds. (See figure.)

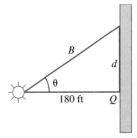

Exercise 41

a. Express θ in terms of time t.

b. Find a function describing d in terms of time t.

c. Sketch a graph of the function on the interval $[0, \infty]$.

d. Explain the meaning of the undefined values of d.

e. Explain the meaning of negative values of d.

42. **Business profit.** The profits of a company fluctuate during a 12-month period according to the formula

$$P(t) = \$50{,}000(\sin t + \cos t)$$

where t is time in months.

a. Sketch a graph of t over the interval $[0, 12]$.

b. Explain the meaning of the negative values of the function.

➡ Technology

Use a computer software package or graphing calculator to graph each of the following and approximate the zeros.

43. $f(x) = \dfrac{\sin x}{x}, \quad -10 \leq x \leq 10$

44. $f(x) = 10\dfrac{\sin x^2}{x}, \quad -4 \leq x \leq 4$

45. $f(x) = 2\dfrac{\cos x^2}{x}, \quad -4 \leq x \leq 4$

46. $f(x) = \dfrac{\cos x - 1}{x}, \quad -8 \leq x \leq 8$

Match each of the following functions with its graph.

47. $f(t) = \tan 2t$

48. $f(t) = \sec 3t$

49. $f(t) = \tan t + 5$

50. $f(t) = \tan\left(t + \dfrac{\pi}{4}\right)$

51. $f(t) = \sec\left(t + \dfrac{\pi}{2}\right)$

52. $f(t) = 2\csc t$

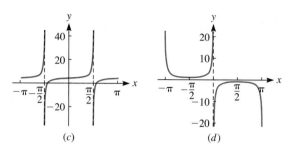

(c)   (d)

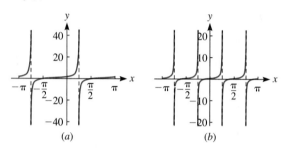

(a)   (b)

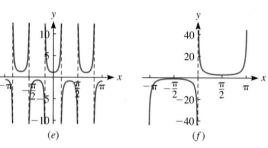

(e)   (f)

5.7 FUNDAMENTAL IDENTITIES FOR TRIGONOMETRIC FUNCTIONS

Many of the most significant properties of the trigonometric functions are expressed as identities that the functions satisfy. In this section, we state and prove the simplest of these identities.

Powers of Trigonometric Functions

A special notation for powers of the trigonometric functions has become standard through centuries of use. For powers of $\sin t$, for example, we write $\sin^n t$ rather than $(\sin t)^n$, where n is an integer. The one exception to this notation is the case $n = -1$. We reserve the notation $\sin^{-1} t$ to mean the inverse function of $\sin t$, and $(\sin t)^{-1}$ to mean the reciprocal $1/\sin t$.

Similar notation is used for powers of other trigonometric functions. For instance, we write $\cos^4 t$ instead of $(\cos t)^4$, $\tan^{-2} t$ instead of $(\tan t)^{-2}$, and so forth.

The Pythagorean Identities

In performing calculations and simplifying expressions involving the trigonometric functions, you will find useful the various identities expressing relationships among the trigonometric functions. In this section, we present the fundamental identities. The first and most basic of the trigonometric identities is the following trigonometric version of the Pythagorean theorem.

PYTHAGOREAN IDENTITIES, I
Let t be any real number. Then
$$\sin^2 t + \cos^2 t = 1 \qquad (1)$$

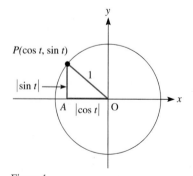

Figure 1

Proof   Draw the angle t inscribed in a unit circle, as shown in Figure 1. From the definition of $\sin t$ and $\cos t$, the point P has the coordinates $(\cos t, \sin t)$. In Figure 1, the right triangle POA has hypotenuse of length 1 and legs of length

$|\cos t|$ and $|\sin t|$, respectively. Therefore, by the Pythagorean theorem, the sum of the squares of the legs equals the square of the hypotenuse. That is,

$$|\sin t|^2 + |\cos t|^2 = 1^2$$
$$\sin^2 t + \cos^2 t = 1 \qquad \blacklozenge$$

> **EXAMPLE 1**

Applying Pythagorean Identities

Suppose that t is an angle in the third quadrant for which $\sin t$ equals $-\frac{3}{4}$. What is the value of $\cos t$?

Solution

By Pythagorean identity (1), we have

$$\sin^2 t + \cos^2 t = 1$$
$$\left(-\tfrac{3}{4}\right)^2 + \cos^2 t = 1$$
$$\cos^2 t = 1 - \tfrac{9}{16} = \tfrac{7}{16}$$
$$\cos t = \pm\sqrt{\frac{7}{16}} = \pm\frac{\sqrt{7}}{4}$$

Because the angle lies in the third quadrant, its cosine (which is the x-coordinate of P) is negative. Therefore, in the above equation, the negative sign prevails and

$$\cos t = -\frac{\sqrt{7}}{4}$$

We may write the Pythagorean identity in the form
$$\sin^2 t = 1 - \cos^2 t$$
In this form, the identity may be used to replace an even power of $\sin t$ by an expression involving only powers of $\cos t$. In a similar fashion, we may write the Pythagorean identity in the form
$$\cos^2 t = 1 - \sin^2 t$$
In this form, the identity may be used to replace an even power of $\cos t$ by an expression involving only powers of $\sin t$. The next example illustrates how to apply these alternate forms of the Pythagorean identity.

> **EXAMPLE 2**

Reducing to Sines

Write the following expression in terms of $\sin t$:
$$\sin^4 t + \cos^4 t$$

Solution

We can write the given expression in the form
$$\sin^4 t + (\cos^2 t)^2$$

Now we use Pythagorean identity (1) to replace $\cos^2 t$ in the above expression with the equivalent expression $1 - \sin^2 t$:

$$\sin^4 t + (1 - \sin^2 t)^2 = \sin^4 t + 1 - 2\sin^2 t + \sin^4 t$$
$$= 2\sin^4 t - 2\sin^2 t + 1$$

> **EXAMPLE 3**
Reducing to a Single
Trigonometric Function

Write the following expression in terms of integer powers of a single trigonometric function.

$$\sin^4 t \cos t$$

Solution

The expression involves both the sine and cosine functions. Note, however, that the sine appears to an even power whereas the cosine appears to an odd power. Pythagorean identity (1) allows us to make a substitution for an even power of either the sine or cosine. This suggests that we make a substitution to replace the power of the sine and write the given expression in terms of powers of $\cos t$:

$$\sin^4 t \cos t = (\sin^2 t)^2 \cos t$$
$$= (1 - \cos^2 t)^2 \cos t$$
$$= (1 - 2\cos^2 t + \cos^4 t)\cos t$$
$$= \cos t - 2\cos^3 t + \cos^5 t$$

From Pythagorean identity (1), we may deduce the following Pythagorean identities for the tangent and secant functions.

PYTHAGOREAN IDENTITIES, II
$\tan^2 t + 1 = \sec^2 t$ (2)
$\cot^2 t + 1 = \csc^2 t$ (3)

Proof   To prove identity (2), we start with the fundamental identity

$$\sin^2 t + \cos^2 t = 1$$

We divide both sides by $\cos^2 t$ to obtain

$$\frac{\sin^2 t}{\cos^2 t} + \frac{\cos^2 t}{\cos^2 t} = \frac{1}{\cos^2 t}$$

$$\left(\frac{\sin t}{\cos t}\right)^2 + 1 = \left(\frac{1}{\cos t}\right)^2$$

Using the definitions of $\tan t$ and $\sec t$, this last formula may be written in the form

$$\tan^2 t + 1 = \sec^2 t$$

To prove identity (3), we start from the fundamental identity

$$\sin^2 t + \cos^2 t = 1$$

and divide both sides by $\sin^2 t$ to obtain

$$\frac{\sin^2 t}{\sin^2 t} + \frac{\cos^2 t}{\sin^2 t} = \frac{1}{\sin^2 t}$$

$$1 + \left(\frac{\cos t}{\sin t}\right)^2 = \left(\frac{1}{\sin t}\right)^2$$

Using the definitions of $\csc t$ and $\cot t$, the last identity can be written in the equivalent form

$$1 + \cot^2 t = \csc^2 t \qquad \blacklozenge$$

The following example shows how the identities just proved can be used to simplify trigonometric expressions.

> **EXAMPLE 4**

Applying Pythagorean Identities

Express each of the following in terms of a single trigonometric function:

1. $\sec^4 t - (2 + \tan t)^2$
2. $\sec^3 t \cot^2 t$
3. $(\sec t + \tan t)(\sec t - \tan t)$

Solution

1. First expand the second term to obtain the expression

$$\sec^4 t - (4 + 4\tan t + \tan^2 t) = \sec^4 t - 4 - 4\tan t - \tan^2 t$$

Identity (2) connects $\sec^2 t$ and $\tan^2 t$. Because $\tan t$ appears to an odd power, to use the identity, we express $\sec^4 t$ in terms of powers of $\tan t$:

$$\left(\sec^2 t\right)^2 - 4 - 4\tan t - \tan^2 t$$
$$= \left(\tan^2 t + 1\right)^2 - 4 - 4\tan t - \tan^2 t$$
$$= \left(\tan^4 t + 2\tan^2 t + 1\right) - 4 - 4\tan t - \tan^2 t$$
$$= \tan^4 t + \tan^2 t - 4\tan t - 3$$

2. None of the identities directly relates the secant function to the cotangent function. For our initial simplification, we express both functions in terms of $\sin t$ and $\cos t$:

$$\sec^3 t \cot^2 t = \left(\frac{1}{\cos t}\right)^3 \left(\frac{\cos t}{\sin t}\right)^2$$

$$= \frac{1}{\cos^3 t} \cdot \frac{\cos^2 t}{\sin^2 t}$$

$$= \frac{1}{\sin^2 t \cos t}$$

Let's now use identity (1) to express everything in terms of $\cos t$:

$$\frac{1}{\sin^2 t \cos t} = \frac{1}{(1 - \cos^2 t)\cos t}$$

$$= \frac{1}{\cos t - \cos^3 t}$$

3. Multiply out binomial expressions to obtain the product

$$(\sec t + \tan t)(\sec t - \tan t)$$
$$= \sec^2 t - \tan^2 t$$
$$= (1 + \tan^2 t) - \tan^2 t \qquad \textit{(Identity 2)}$$
$$= 1$$

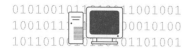

New Technology

Verifying Trigonometric Identities with a Graphing Utility

Graphing utilities may be used to provide convincing evidence for the truth or falsity of trigonometric identities. Consider the identity of Example 2 in this section:

$$\sin^4 x + \cos^4 x = 2 \sin^4 x - 2 \sin^2 x + 1$$

(We have changed the variable from t to x to facilitate the use of a graphing utility.) Graph the function

$$f(x) = \sin^4 x + \cos^4 x$$

corresponding to the left-hand side using a graphing utility. Figure 2(a) shows this graph on a TI-81 graphing calculator. Now graph the function

$$g(x) = 2 \sin^4 x - 2 \sin^2 x + 1$$

corresponding to the right-hand side. (See Figure 2(b).) By comparing the two graphs on the same coordinate system, we see that they are the same. That is, $f(x) = g(x)$ for x in the range graphed. This provides visual evidence that the equation is an identity.

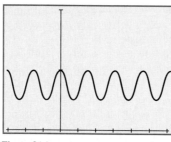

Figure 2(a)

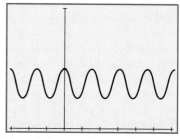

Figure 2(b)

Using a graphing calculator provides a useful method of checking the verifications of the more complicated identities that will be studied in the next chapter. It is important to remember that you need to graph the functions in a range at least as large as the largest period in order to view the complete behavior of the graph.   ◄

The Parity of the Trigonometric Functions

Let's determine the parity (evenness or oddness) of the various trigonometric functions. As with the Pythagorean identities, the basic facts are those for the functions $\sin t$ and $\cos t$, from which all else is derived. These basic facts can be summarized as follows.

PARITY IDENTITIES

1. The function $\sin t$ is odd. That is, we have the identity

$$\sin(-t) = -\sin t \qquad (4)$$

2. The function $\cos t$ is even. That is, we have the identity

$$\cos(-t) = \cos t \qquad (5)$$

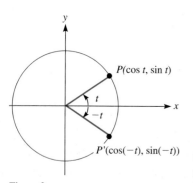

Figure 3

Proof   In Figure 3, we show a unit circle with the angles t and $-t$ drawn. Note that these two angles are reflections of one another in the x-axis. This is because the two angles are obtained by rotating their terminal sides in opposite directions, by the same absolute amounts. Let P and P' denote the points at which these two angles' terminal sides intersect the unit circle, as shown in the figure. From the unit circle definition of $\sin t$ and $\cos t$, the coordinates of P and P' are, respectively, $P(\cos t, \sin t)$ and $P'(\cos(-t), \sin(-t))$. Because P and P' are symmetric with respect to the x-axis, however, these points have the same x-coordinates, and their y-coordinates are negatives of one another. That is, we have

$$\cos(-t) = \cos t, \qquad \sin(-t) = -\sin t$$

From the parity identities above, we can deduce the following parity identities for the other trigonometric functions.

$$\tan(-t) = -\tan(t) \tag{6}$$
$$\sec(-t) = \sec t \tag{7}$$
$$\csc(-t) = -\csc t \tag{8}$$
$$\cot(-t) = -\cot t \tag{9}$$

We will leave the proofs of these formulas for the exercises.   ◆

Exercises 5.7

In each case a trigonometric value and a quadrant are given. Find the values of the other five trigonometric functions.

1. $\sin t = -\frac{2}{5}$; third quadrant.

2. $\sin t = \frac{1}{3}$; second quadrant.

3. $\cos t = -\frac{2}{3}$; second quadrant.

4. $\cos t = \frac{3}{8}$; fourth quadrant.

5. $\sin t = -\dfrac{\sqrt{5}}{4}$; fourth quadrant.

6. $\cos t = -\dfrac{\sqrt{3}}{5}$; third quadrant.

7. $\tan t = -\frac{2}{3}$; second quadrant.

8. $\tan t = 2$; third quadrant.

Write the following expressions in terms of $\sin t$.

9. $\sin^4 t - \cos^4 t$

10. $\cos t(\tan t - \sec t)$

11. $4 + 5\cos^2 t$

12. $\sin^3 t + \sin t \cos^2 t$

Write the following expressions in terms of $\cos t$.

13. $\sin^4 t + \cos^4 t$

14. $\sin t(\cot t + \csc t)$

15. $(\sin t - \cos t)(\sin t + \cos t)$

16. $\sin^6 t - \cos^6 t$

Write the following expressions in terms of integer powers of a single trigonometric function and simplify where possible.

17. $\cos^4 t \sin t$

18. $\csc^3 t \cot^2 t$

19. $(\cot t - \tan t)(\cot t + \tan t)$

20. $(\cos t + \sin t)(\cos t - \sin t)$

21. $\csc^4 t - (2 + \cot t)^2$

22. $\sin^6 t - \cos^6 t$

23. $\cot^3 t + \tan^3 t$

24. $\cos^3 t - \sec^3 t$

25. $\cos t + \sin^2 t$

26. $\sin^2 t - \cos^2 t$

27. $(\sin t - \csc t)^2$

28. $(\cos t + \sec t)^2$

29. $\dfrac{\sin^2 t \cos t}{\cos^2 t \sin t}$

30. $\dfrac{\cos^2 t \sin t}{\sin^2 t \cos t}$

31. $\dfrac{42 \sin^3 \theta}{6 \cos^2 \theta}$

32. $\dfrac{4 \sin \theta \cos^3 \theta}{22 \sin^2 \theta \cos \theta}$

33. $\dfrac{\sin^2 x - 2\sin x + 1}{\sin x - 1}$

34. $\dfrac{\cos y + 3}{\cos^2 y + 6\cos y + 9}$

35. $\dfrac{\cos \alpha + 5}{\cos^2 \alpha + 10\cos \alpha + 25}$

36. $\dfrac{\sin^2 \beta - 36}{\sin \beta + 6}$

37. $\dfrac{\sec^2 t + \tan^2 t}{\sec^4 t - \tan^4 t}$

38. $\dfrac{\sin^4 t + \cos^4 t}{\sin^2 t - \cos^2 t}$

39. $\dfrac{5\cos^3 y}{\sin^2 y} \cdot \left(\dfrac{\sin y}{5\cos y}\right)^2$

40. $\left(\dfrac{3\sin p}{\cos p}\right)^2 \cdot \dfrac{\cos^3 p}{3\sin^2 p}$

41. $\dfrac{1}{\csc^2 t} + \left(\dfrac{\cot t}{\csc t}\right)^2$

42. $\dfrac{1}{\cos^2 t} - \left(\dfrac{\sin t}{\cos t}\right)^2$

43. $\dfrac{9\sin^4 x - 16}{2\cos^2 x - 10} \cdot \dfrac{\cos^4 x - 25}{6\sin^2 x - 8}$

44. $\dfrac{\sin^2 x - 49}{2\cos x + 1} \cdot \dfrac{6\cos x + 3}{7\sin x + 49}$

45. $(\tan^4 x + 2\tan^2 x + 1)^2$

46. $\cot^4 t + 2\cot^2 t + 1$

47. $\dfrac{\sin^6 t - \cos^6 t}{\sin^2 t - \cos^2 t}$

48. $\dfrac{\sin \beta + \cos \beta}{\sin^3 \beta + \cos^3 \beta}$

Prove each of the following.

⟶ Technology

49. $\tan(-t) = -\tan t$

50. $\csc(-t) = -\csc t$

51. $\sec(-t) = \sec t$

52. $\cot(-t) = -\cot t$

53.–84. Visually verify your results for Exercises 17–48 using a graphing utility.

▌ 5.8 CHAPTER REVIEW

Important Concepts, Properties, and Formulas—Chapter 5

Conversion between degrees and radians	To convert from degrees to radians, multiply by $\pi/180$.	p. 273
	To convert from radians to degrees, multiply by $180/\pi$.	p. 273
Length of a subtended arc	$s = r\theta$	p. 274
Important identities	$\tan \theta = \dfrac{\sin \theta}{\cos \theta}$, $\cot \theta = \dfrac{\cos \theta}{\sin \theta}$, $\sec \theta = \dfrac{1}{\cos \theta}$, $\csc \theta = \dfrac{1}{\sin \theta}$, $\sin(t + 2\pi) = \sin t$, $\cos(t + 2\pi) = \cos t$, $\tan(t + \pi) = \tan t$, $\cot(t + \pi) = \cot t$, $\sin(-t) = -\sin t$, $\cos(-t) = \cos t$ $\tan(-t) = -\tan t$, $\cot(-t) = -\cot t$, $\sec(-t) = \sec t$, $\csc(-t) = -\csc t$	pp. 282, 283, 300, 326, 327
Pythagorean identities	$\sin^2 t + \cos^2 t = 1$ $\tan^2 t + 1 = \sec^2 t$ $\csc^2 t = 1 + \cot^2 t$	pp. 322, 324
Graphs of the trigonometric functions $f(x) = a \sin(bx + c)$ **and** $g(x) = a \cos(bx + c)$	Period $= \dfrac{2\pi}{b}$ Phase shift $= -c/b$ Amplitude $= |a|$	pp. 304, 306, 308

▌ Cumulative Review Exercises—Chapter 5

Given the angle θ, determine the measure of the angle that is (a) complementary to θ and (b) supplementary to θ.

1. $\theta = 47°51'$

2. $\theta = 74°11'38''$

Approximate each angle in terms of decimal degrees to the nearest ten-thousandth of a degree.

3. $47°51'$

4. $74°11'38''$

5. $-124°34'$

6. $-22°22'22''$

Approximate each angle in terms of degrees, minutes, and seconds.

7. $56.789°$

8. $-13.002°$

9. $117.0406°$

10. $56.23°$

Convert the following angles to radian measure.

11. $75°$

12. $1°$

13. $-150°$

14. $234.46°$

Convert the following angles to degree measure (decimal degrees).

15. 1

16. $\dfrac{11\pi}{3}$

17. -13π

18. 11.26

♦ Applications

19. Find the measure in degrees and radians of each angle of an equilateral triangle.

20. Find the radius of a circle with a central angle of 60° that sweeps out an arc length of 5.0 cm.

21. For a clock with a 24-in. diameter, the minute hand travels 4.71 in. What is the central angle formed by this movement?

22. In a clock with a 12-in. diameter, the minute hand travels 4.71 in. What is the central angle formed by this movement?

23. In one complete stroke, the tip of a bird's wing travels 0.78 m. If the wing is 0.3 m long, through what angle does the wing move?

24. A car travels at a linear speed of 55 mph. The radius of a tire on the car is 14 inches. What is the angular speed of the tire in radians/hr?

25. A $33\frac{1}{3}$ rpm record has a radius of 15.2 cm. What is the linear speed, in cm/sec, of a point on the outside of the record?

Given each triangle in Exercises 26 and 27, find the six trigonometric function values of the angle θ.

26.

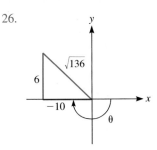

27.

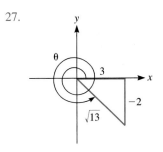

Use a scientific calculator to determine the values of the following trigonometric functions. Round to four decimal places.

28. $\sin 545°$

29. $\tan(-56.78°)$

30. $\cos(3.42 \text{ radians})$

31. $\sec(-2.337 \text{ radians})$

32. $\cot 29°31'12''$

33. $\sin 122°23'$

34. $\csc 6\pi$

35. $\cos\left(\dfrac{-23\pi}{5}\right)$

In each case a trigonometric value and a quadrant are given. Find the values of the other five trigonometric functions.

36. $\sin\theta = \frac{2}{7}$, first quadrant

37. $\cos\theta = -\dfrac{3}{\sqrt{13}}$, second quadrant

38. $\tan\theta = 5$, first quadrant

39. $\cot\theta = -\dfrac{1}{k}$, fourth quadrant

Solve the following equations for an acute angle θ. Give your answer in radians and in decimal degrees.

40. $\sin\theta = 0.5624$

41. $\tan\theta = 12.345$

42. $\sec\theta = 10$

43. $\cos\theta = 0.9923$

Find the values of the six trigonometric functions, if they exist, for each value of t. Give exact answers.

44. $t = 0°$

45. $t = \dfrac{2\pi}{3}$

46. $t = \dfrac{-5\pi}{6}$

47. $t = \dfrac{7\pi}{6}$

48. $t = -300°$

49. $t = 9\pi$

50. $t = -\dfrac{5\pi}{4}$

51. $t = \dfrac{43\pi}{2}$

Determine the solution set of each equation.

52. $\cos t = -1$

53. $\sin t = -\dfrac{\sqrt{3}}{2}$

54. $\sin t = \frac{1}{2}$

55. $\tan t = \sqrt{3}$

Write the following expressions in terms of integer powers of a single trigonometric function, and simplify where possible.

56. $(5 - 4\cos t)(4 + 5\cos t)$

57. $\tan t - \sec t \csc t$

58. $\sin^2 t \sec^2 t - \sin^2 t$

59. $\csc^3 t - \cot^2 t - \csc t + 1$

60. $\dfrac{1}{\sec t + 1} - \dfrac{1}{\sec t}$

61. $\dfrac{\tan \theta}{\tan \theta + \sin \theta} - \dfrac{1 - \cos \theta}{\sin^2 \theta}$

62. $(1 + \sin x)(1 - \sin x)$

63. $1 + \tan^2 x$

64. $1 - \sin^2 \theta$

65. $\dfrac{\sec x \cot^2 x \cos x}{\cot^2 x \sec^2 x - 1}$

66. $\dfrac{1 + \sec x}{\sec x - 1} + \dfrac{1 + \cos x}{\cos x - 1}$

67. $\dfrac{\tan^3 t + \cot^3 t}{\tan t + \cot t}$

68. $\sqrt{\dfrac{\cot^2 x}{\tan^2 x}}$

69. $\cos^2 \alpha + \sin^2 \alpha$

Solve each equation for the indicated trigonometric value.

70. $\sin t : \sin^2 t - \sin t = 0$

71. $\tan t : \sqrt{\tan t} = 12$

72. $\cos \theta : \cos^2 \theta - 1 = 0$

73. $\sec x : \sec^2 x - 2 \sec x - 2 = 0$

74. $\sin \theta : \ln \sin \theta = 0$

Rationalize the denominator.

75. $\dfrac{8}{\sqrt{\sin \beta}}$

76. $\sqrt{\dfrac{5 - \cos \theta}{5 + \cos \theta}}$

77. Give the period of the six trigonometric functions.

78. Give the domain of the six trigonometric functions.

79. Give the range of the six trigonometric functions.

80. Give the amplitude of the cosine function.

For each of the following functions, (a) find the period, (b) find the phase shift, (c) find the amplitude, and (d) sketch the graph.

81. $f(x) = -\sin x$

82. $f(x) = -3\sin\left(x + \dfrac{\pi}{2}\right)$

83. $f(x) = 2\sin\left(\tfrac{1}{2}x\right)$

84. $f(x) = 3\cos\left(2x - \dfrac{\pi}{2}\right)$

Sketch the graph.

85. $f(x) = \csc x$

86. $f(x) = |\cos x| + 1$

87. $f(x) = 2\sin x + \cos x$

88. $f(x) = \sin(\pi x) - \cos(\pi x)$

89. $f(x) = e^{-0.2x} \sin x$

90. $f(x) = e^{\ln \sin x}$

Find a function of the type $f(x) = a \sin(bx + c)$ or $f(x) = a \cos(bx + c)$, depending on which is requested, whose graph is the given function or that satisfies the given condition.

91. sin :

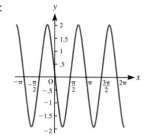

92. cos :

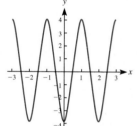

93. cos : amplitude $= 7.2$, period $= \pi$, phase shift $= \pi$

94. sin : amplitude $= 0.3$, period $= 12$, phase shift $= -4$

Simplify. Do not use a calculator.

95. $\dfrac{\pi}{2}\left(\sin^2 \dfrac{\pi}{4} + \sin^2 \dfrac{3\pi}{4} + \sin^2 \dfrac{5\pi}{4} + \sin^2 \dfrac{7\pi}{4}\right)$

96. $\dfrac{\pi}{5}\left(\sin^2 \dfrac{\pi}{12} + \sin^2 \dfrac{5\pi}{12} + \sin^2 \dfrac{7\pi}{12} + \cos^2 \dfrac{5\pi}{12} + \cos^2 \dfrac{\pi}{12}\right)$

Use a scientific calculator to approximate to four decimal places.

97. $\dfrac{1}{4}\left(\sin \dfrac{\pi}{8} + \sin \dfrac{3\pi}{8} + \sin \dfrac{5\pi}{8} + \sin \dfrac{7\pi}{8}\right)$

◆ Applications

98. The angle of elevation to the top of a pole is $35°$. How high is the pole if it casts a 40-ft shadow on the ground?

99. A chord of a circle is 30 in. long and subtends an angle $20°$ at the center. Find the radius of the circle. (See figure.)

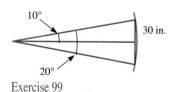

Exercise 99

100. A building 283 ft high casts a shadow 100 ft long. What is the angle of elevation? (See figure.)

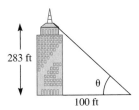

283 ft

θ

100 ft

Exercise 100

101. An airplane takes off from a runway at a constant angle of elevation of 5°40′. Two minutes later it is at an altitude of 1,800 ft. How far is the airplane from the point of takeoff?

102. **Aerial navigation.** An airplane leaves an airfield and flies 150 miles in a direction of 330°. How far west of the airfield is the airplane? How far north?

103. **Area of a conical tent.** A tent in the shape of a cone has a center pole that is 12 ft high. The vertical angle of the tent is 56°. Find the surface area of the tent. (See figure.)

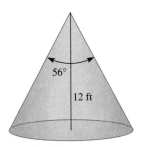

56°

12 ft

Exercise 103

104. From a hot air balloon one mile high, two cities are spotted on the same side of the balloon. The angle of depression to one city is 36°40′, and the angle to the other city is 22°11′. Find the distance between the cities.

105. **Surveying: measuring the height of a mountain.** (See figure.) Standard right triangle methods for finding the height of a tree or a flagpole can break down when one tries to find the height of a mountain, because foothills and the base of the mountain itself interfere with measuring the horizontal distance from the observer to the point B directly under the top of the mountain. However, the problem may often be solved by making two observations. Suppose P and Q are two points at altitude h meters, as shown in the figure. Suppose also that the angles at which the top of the mountain is seen from P and Q are 32° and 20.5°, respectively. How high is the mountain? (*Hint:* Find two expressions for y in terms of x.)

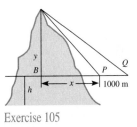

y

B P Q

x 1000 m

h

Exercise 105

106. **Height of an air valve.** (See figure.) The air valve on a tire is 8 in. from the center of the tire, and the radius of the tire is 14 in. The car travels at about 55 mph. The height h of the valve from the ground after time t, in seconds, is given by

$$h(t) = 14 - 8\cos(39\pi t)$$

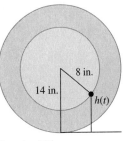

8 in.

14 in.

$h(t)$

Exercise 106

a. Find the height of the valve at $t = 0, 0.1, 0.2, 0.3, 0.4, 0.5, 2, 3, 4, 10$.

b. Sketch a graph of the function on the interval [0, 6].

c. The car travels 500 miles. At what height is the valve at the end of the trip?

ANALYTIC TRIGONOMETRY

Models involving the trigonometric functions are used to describe many real-world phenomena, such as the cyclical growth and decline of populations of predators and their prey.

I n the previous chapter, we introduced the trigonometric functions and proved some of their elementary properties. Let's explore some further properties of these functions, mostly in the form of identities, which prove useful in applying trigonometry both to other fields and to various mathematical disciplines, such as analytic geometry and calculus.

6.1 TRIGONOMETRIC IDENTITIES

In this section, we will discuss identities that involve the trigonometric functions. On one hand, we will use trigonometric identities to solve various trigonometric equations. On the other, we will use trigonometric identities to simplify various expressions, a process that is especially useful in calculus.

Elementary Trigonometric Identities

Let's begin by reviewing the elementary trigonometric identities we have proven thus far. First are the identities that are used to define the trigonometric functions in terms of the sine and cosine or are immediate consequences of these identities. Second are the Pythagorean identities. Third are the parity identities. All of these identities are summarized in the following table.

ELEMENTARY TRIGONOMETRIC IDENTITIES

Definition Identities

$$\sin x = \frac{1}{\csc x} \qquad \csc x = \frac{1}{\sin x}$$

$$\cos x = \frac{1}{\sec x} \qquad \sec x = \frac{1}{\cos x}$$

$$\tan x = \frac{\sin x}{\cos x} = \frac{1}{\cot x} \qquad \cot x = \frac{\cos x}{\sin x} = \frac{1}{\tan x}$$

Pythagorean Identities

$$\sin^2 x + \cos^2 x = 1$$

$$\tan^2 x + 1 = \sec^2 x$$

$$\cot^2 x + 1 = \csc^2 x$$

Parity Identities

$$\sin(-x) = -\sin x \qquad \csc(-x) = -\csc x$$

$$\cos(-x) = \cos x \qquad \sec(-x) = \sec x$$

$$\tan(-x) = -\tan x \qquad \cot(-x) = -\cot x$$

In the following examples, we will discuss a number of applications of the elementary identities.

➤ **EXAMPLE 1**
Evaluating a Trigonometric Function

Suppose that $\cot x = -2$ and that x is in the second quadrant. Determine the value of $\sin x$.

Solution

By the Pythagorean identity relating $\cot x$ and $\csc x$, we have

$$\cot^2 x + 1 = \csc^2 x$$

$$\csc x = \pm \sqrt{\cot^2 x + 1}$$

$$= \pm \sqrt{(-2)^2 + 1}$$

$$= \pm \sqrt{5}$$

Because $\sin x = 1/\csc x$, we see that

$$\sin x = \frac{1}{\pm \sqrt{5}}$$

$$= \pm \frac{\sqrt{5}}{5} \qquad \text{(Rationalizing denominator)}$$

Because x is in the second quadrant, the sine is positive, so that

$$\sin x = \frac{\sqrt{5}}{5}$$

➤ **EXAMPLE 2**
Proving a Trigonometric Identity

Prove the identity

$$\cos(-x)\sin(-x) + \tan x = \frac{\sin^3 x}{\cos x}$$

Solution

Looking at the given expression, we see that the parity identities for sine and cosine may be applied. Moreover, by replacing the tangent by its definition in terms of the sine and cosine, we can hope to simplify the expression further:

$$\cos(-x)\sin(-x) + \tan x = \cos x \cdot (-\sin x) + \tan x \qquad \text{(Parity identities)}$$

$$= -\sin x \cos x + \frac{\sin x}{\cos x} \qquad \text{(Definition of } \tan x)$$

$$= \frac{-\sin x \cos^2 x + \sin x}{\cos x} \qquad \text{(Adding fractions)}$$

$$= \frac{\sin x(-\cos^2 x + 1)}{\cos x} \qquad \text{(Factoring)}$$

$$= \frac{\sin x \cdot \sin^2 x}{\cos x} \qquad \text{(Pythagorean identity)}$$

$$= \frac{\sin^3 x}{\cos x}$$

➤ **EXAMPLE 3**
Reducing an Expression to a
Single Trigonometric Function

Write the expression $\tan^3 x \cos x - \tan^2 x$ in terms of $\sin x$.

Solution

Start by factoring $\tan^2 x$ from both terms, with the hope of using the Pythagorean identity for $\tan^2 x$:

$$\tan^3 x \cos x - \tan^2 x = \tan^2 x(\tan x \cos x - 1)$$

$$= \tan^2 x \left(\frac{\sin x}{\cos x} \cdot \cos x - 1 \right) \qquad \textit{(Definition of} \tan x)$$

$$= \tan^2 x(\sin x - 1)$$

$$= (\sec^2 x - 1)(\sin x - 1) \qquad \textit{(Pythagorean identity)}$$

$$= \left(\frac{1}{\cos^2 x} - 1 \right)(\sin x - 1) \qquad \textit{(Definition of} \sec x)$$

$$= \frac{1 - \cos^2 x}{\cos^2 x} \cdot (\sin x - 1)$$

$$= \frac{\sin^2 x}{1 - \sin^2 x}(\sin x - 1) \qquad \textit{(Pythagorean identity)}$$

$$= \frac{\sin^2 x}{(1 - \sin x)(1 + \sin x)}(\sin x - 1) \qquad \textit{(Factoring)}$$

$$= -\frac{\sin^2 x}{1 + \sin x}$$

➤ **EXAMPLE 4**
Solving a Trigonometric
Equation

Solve the equation

$$\sec^2 x - \tan x = 1, \qquad -\frac{\pi}{2} < x < \frac{\pi}{2}$$

Solution

In order to solve an equation such as this, it is usually helpful to transform it using identities so that the transformed equation involves a single trigonometric function. In this case, by applying the Pythagorean identity involving $\sec^2 x$, we have

$$(\tan^2 x + 1) - \tan x = 1$$

$$\tan^2 x - \tan x = 0$$

$$\tan x(\tan x - 1) = 0$$

$$\tan x = 0 \quad \text{or} \quad \tan x = 1$$

$$x = 0 \quad \text{or} \quad x = \frac{\pi}{4}$$

➤ **EXAMPLE 5**
Clearing Trigonometric
Fractions

Write the following expression in nonfractional form:

$$\frac{1}{\sec x - 1}$$

Calculations such as this are useful in calculus.

Solution

Multiply both numerator and denominator by $\sec x + 1$ so that we may apply the Pythagorean identity for the tangent:

$$\frac{1}{\sec x - 1} \cdot \frac{\sec x + 1}{\sec x + 1} = \frac{\sec x + 1}{\sec^2 x - 1}$$

$$= \frac{\sec x + 1}{\tan^2 x}$$

$$= \frac{1}{\tan^2 x} \cdot (\sec x + 1)$$

$$= \cot^2 x (\sec x + 1)$$

➤ **EXAMPLE 6**
Calculating the Sum
of Rational Trigonometric
Expressions

Calculate the following sum and simplify:

$$\frac{1}{\sin y - 1} + \frac{1}{\sin y + 1}$$

Solution

Add the two fractions to obtain

$$\frac{1}{\sin y - 1} + \frac{1}{\sin y + 1} = \frac{1}{\sin y - 1} \cdot \frac{\sin y + 1}{\sin y + 1} + \frac{1}{\sin y + 1} \cdot \frac{\sin y - 1}{\sin y - 1}$$

(Put fractions over common denominator)

$$= \frac{\sin y + 1}{\sin^2 y - 1} + \frac{\sin y - 1}{\sin^2 y - 1} \qquad \text{(Multiply out denominators)}$$

$$= \frac{(\sin y - 1) + (\sin y + 1)}{\sin^2 y - 1} \qquad \text{(Add fractions)}$$

$$= \frac{2 \sin y}{- \cos^2 y} \qquad \text{(Because } \sin^2 y + \cos^2 y = 1\text{)}$$

$$= -2 \sin y \cdot \frac{1}{\cos^2 y}$$

$$= -2 \sin y \sec^2 y \qquad \text{(Definition of secant)}$$

There are many different ways to simplify expressions, leading to equivalent, but different, answers. In simplifying trigonometric expressions, a good rule to follow is to eliminate fractions and radicals and to use as few different trigonometric functions as possible.

Verifying Trigonometric Identities

Trigonometric identities are useful for writing trigonometric expressions in equivalent forms, a process that is needed over and over in more advanced mathematics.

The identities stated at the beginning of the section are the basic identities, on which many of the other identities are based. You should memorize these identities. (Actually, you will be using them all so frequently that you probably won't need to make a special effort.)

In learning to manipulate trigonometric expressions, it is helpful to prove the validity of trigonometric identities. Given an alleged identity, here is a useful approach for doing this:

TIPS FOR VERIFYING TRIGONOMETRIC IDENTITIES

1. Replace the more complex side of the equation with a series of equivalent expressions until you arrive at the simpler side.
2. In making the replacements of part 1, use the various identities you already know.

The next two examples provide concrete illustrations of how to prove identities.

> **EXAMPLE 7**
Proving an Identity

Verify the following trigonometric identity.

$$\tan x + 1 = \sec x(\sin x + \cos x)$$

Solution

The right side involves three different trigonometric functions. Let's write it in terms of sines and cosines:

$$\sec x(\sin x + \cos x) = \frac{1}{\cos x}(\sin x + \cos x) \qquad \textit{(Definition of } \sec x\textit{)}$$

$$= \frac{\sin x}{\cos x} + 1 \qquad \textit{(Distributive property)}$$

$$= \tan x + 1 \qquad \textit{(Definition of } \tan x\textit{)}$$

We have transformed the right side into the left side using allowable algebraic operations and known identities. This proves that the identity is valid.

> **EXAMPLE 8**
Using One Identity
to Prove Another

Verify the following trigonometric identity:

$$\frac{\sin^3 x - \cos^3 x}{\sin x - \cos x} = 1 + \sin x \cos x$$

Solution

This identity is more complicated. However, we can apply the same principles as we did in the preceding example. The more complex side is the left, so let's transform the left side into the right. The numerator of the left is a difference of cubes. This suggests that we factor it using the Difference of Cubes factorization formula to obtain

$$\frac{(\sin x - \cos x)(\sin^2 x + \sin x \cos x + \cos^2 x)}{\sin x - \cos x} = \sin^2 x + \sin x \cos x + \cos^2 x$$

$$\textit{(Cancel common factor)}$$

$$= 1 + \sin x \cos x$$

(Because $\sin^2 x + \cos^2 x = 1$)

We have transformed the left side into the right, so the identity is valid.

In the preceding examples, we proved the validity of several identities. Note, however, that to prove that an identity is *not valid*, it is necessary to exhibit only a single value of the variables for which the right side does not equal the left. For example, consider the alleged identity

$$\sin x + \cos x = \tan x$$

This identity is not valid. Indeed, if we substitute x equals 0, we have

$$\sin 0 + \cos 0 = \tan 0$$
$$0 + 1 = 0$$

which is an incorrect statement.

An identity is a mathematical statement that is valid for all values of the variable, not just for a limited set of values. In particular, this implies that if we graph both sides of an identity, we will obtain the same graph. However, if an equation is not an identity, the two sides will have different graphs. This observation points the way to an application of graphing calculators. You may check the plausibility of an identity by graphing both sides. If you get two different graphs, the equation is not an identity. If you get the same graph, then the equation is an identity. Of course, because the calculator graphs only a finite number of points, such a comparison of graphs cannot yield a proof of an identity. However, to prove that an identity is false, you need only exhibit a single value of x for which the two sides differ, and this can be done using a calculator.

➢ **EXAMPLE 9**
Trigonometric Substitution

Given the expression $\sqrt{a^2 - t^2}$, make the substitution $t = a \sin x, 0 < x < \pi/2$, and write the expression in terms of the variable x. This sort of substitution is called a **trigonometric substitution**.

Solution
Substitute $a \sin x$ for t in the given expression to obtain

$$\sqrt{a^2 - t^2} = \sqrt{a^2 - (a \sin x)^2}$$
$$= \sqrt{a^2 - a^2 \sin^2 x}$$
$$= \sqrt{a^2(1 - \sin^2 x)}$$

Applying the Pythagorean identity, we can replace $1 - \sin^2 x$ with $\cos^2 x$ to obtain

$$\sqrt{a^2 \cos^2 x} = \sqrt{a^2} \sqrt{\cos^2 x}$$

Because x lies in the interval $(0, \pi/2)$, the value of $\cos x$ must be positive. Therefore, the above expression is equivalent to

$$|a| \cos x$$

This is the desired expression of $\sqrt{a^2 - t^2}$ in terms of x.

New Technology

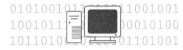

Using a Graphing
Utility to Visually
Demonstrate an Identity

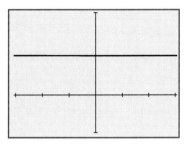

Figure 1

In the last chapter, we used a graphing utility to visually reinforce some simple trigonometric identities. This technique is very useful, so we will now spend a little time examining it further. For instance, consider the graph in Figure 1. This is the graph of

$$f(x) = \sin^2 x + \cos^2 x$$

for x in the interval $-\pi \le x \le \pi$. Note that because the periods of $\sin x$ and $\cos x$ are 2π, this interval contains a full period for each function. (In fact, it can be shown that the periods of $\sin^2 x$ and $\cos^2 x$ are both π, so that the interval contains two full periods.) The graph shown is the straight line $y = 1$. By periodicity, the graph of $f(x)$ coincides with this line for all values of x (and not just in the interval $-\pi \le x \le \pi$). Thus, we have visual demonstration of the identity

$$\sin^2 x + \cos^2 x = 1$$

This technique of visually demonstrating identities becomes most useful when the trigonometric identities are complicated. Although we will not give explicit examples of this technique in each section of the chapter, you should attempt to use it to further examine the examples and exercises (if you have access to a graphing utility).   ◄

Exercises 6.1

Write the following trigonometric expressions in terms of $\sin x$.

1. $\csc^2 x - \cot^2 x$

2. $\cot x (\cos x - 1)$

3. $\dfrac{\tan^2 x}{1 + \tan^2 x}$

4. $\dfrac{\cot^2 x + 1}{\cot^2 x}$

5. $\cos x \cot x$

6. $\sec x \tan x$

7. $\csc x \sec x \cot x$

8. $\cos x \cot x + \sin x$

9. $(1 - \sin^2 x)(\sec^2 x - 1)$

10. $\sec x [(\sin x + \cos x)^2 - 1]$

Write the following in terms of $\cos x$.

11. $\tan x \sin x$

12. $\tan x \csc x$

13. $\sec^2 x$

14. $\cot^2 x$

15. $\dfrac{1 + \tan^2 x}{\tan^2 x}$

16. $\dfrac{\cot^2 x}{1 + \cos^2 x}$

17. $\dfrac{1 - \sec^2 x}{1 - \cos^2 x}$

18. $1 + \cot^2 x$

Write the following in terms of a single trigonometric function.

19. $\tan^2 x \sec x \cos x$

20. $\dfrac{\cos x}{\csc x - \sin x}$

21. $\csc x \sec x - \tan x$

22. $\dfrac{\sec^2 x}{\cos^2 x} - \dfrac{\sin^2 x}{\cos^4 x} - 1$

23. $\dfrac{\tan x (1 - \sin^2 x)}{\sin x}$

24. $\dfrac{\cos x}{1 - \sin x} - \tan x$

Make the given substitution for t, with $0 < x < \pi/2$, and write the given expressions in terms of the variable x.

25. $t = a \tan x$; for $\sqrt{a^2 + t^2}$

26. $t = a \tan x$; for $\sqrt{\dfrac{1}{t^2 + a^2}}$

27. $t = 5 \sin x$; for $\dfrac{t^2}{\sqrt{25 - t^2}}$

28. $t = 5 \sin x$; for $\sqrt{25 - t^2}$

29. $t = a \sec x$; for $\sqrt{t^2 - a^2}$

30. $t = a \sin x$; for $t \sqrt{(a \cos x)^2 + t^2}$

31. $t = \frac{4}{3} \tan x$; for $\sqrt{16 + 9t^2}$

32. $t = 4 \tan x$; for $t^2 \sqrt{16 + t^2}$

Prove each of the following trigonometric identities.

33. $1 = \cos^2 a + \cos^2 a \tan^2 a$

34. $\cos^2 \beta + 1 = 2 \cos^2 \beta + \sin^2 \beta$

35. $1 - 2 \sin^2 y = \cos^2 y - \sin^2 y$

36. $\tan x \sin x + \cos x = \sec x$

37. $\dfrac{\tan a - 1}{\tan a + 1} = \dfrac{1 - \cot a}{1 + \cot a}$ 38. $\dfrac{1 + \sin \beta}{\cos \beta} = \dfrac{\cos \beta}{1 - \sin \beta}$

39. $\dfrac{\cos \alpha}{\cos \alpha - \sin \alpha} = \dfrac{1}{1 - \tan \alpha}$

40. $\dfrac{1 - \tan^2 a}{1 - \sec^2 a} = 1 - \cot^2 a$

41. $\dfrac{\cos \alpha + \sin \alpha}{\cos \alpha} = 1 + \tan \alpha$

42. $\dfrac{\cos a}{\sec a - \tan a} = 1 + \sin a$

43. $\dfrac{1 + \csc x}{\cot x + \cos x} = \sec x$

44. $1 + \sec^2 a \tan^2 a = \tan^4 a + \sec^2 a$

Prove each of the following identities.

45. $1 + \sin x = \dfrac{\cos x}{\sec x - \tan x}$

46. $\dfrac{\tan^2 a - 1}{\tan a - \cot a} = \tan a$

47. $-4 \sec \alpha \tan \alpha = \dfrac{1 - \sin \alpha}{1 + \sin \alpha} - \dfrac{1 + \sin \alpha}{1 - \sin \alpha}$

48. $\sin^3 \beta \cos \beta - \sin^5 \beta \cos \beta = \sin^3 \beta \cos^3 \beta$

49. $\dfrac{\csc^2 \theta - \cot^2 \theta \csc^2 \theta}{\cot^2 \theta} = \sec^2 \theta - \csc^2 \theta$

50. $\dfrac{\sin x}{\csc x} = 1 - \dfrac{\cos x}{\sec x}$

51. $\dfrac{1 + \cos \theta}{\sin \theta} + \dfrac{\sin \theta}{1 + \cos \theta} = 2 \csc \theta$

52. $2 \sin \alpha (1 + \cos \alpha) = \dfrac{2 \sin^3 \alpha}{1 - \cos \alpha}$

53. $\dfrac{1}{\cos x} = \dfrac{\csc(-x)}{\cot(-x)}$

54. $-\cot a = \cos(-a) \csc(-a)$

55. $(\tan \beta \sin \beta)^2 = \tan^2 \beta - \sin^2 \beta$

56. $\dfrac{\sin^3 \theta + \cos^3 \theta}{\sin \theta + \cos \theta} = 1 - \sin \theta \cos \theta$

57. $\dfrac{\sin^2 x}{1 - \cos x} = 1 + \cos x$

58. $\dfrac{\sec x}{\sec x - 1} - \dfrac{\sec x + 1}{\tan^2 x} = 1$

59. $\sec^2 x - 2 \sec x \cos x + \cos^2 x = \tan^2 x - \sin^2 x$

60. $\left(\dfrac{1 + \cos x}{\sin x}\right)^2 = \dfrac{\tan x + \sin x}{\tan x - \sin x}$

61. $\dfrac{\sin x}{1 + \cos x} = \csc x - \cot x$

62. $\dfrac{1 + \cot x}{1 - \cot x} + \dfrac{1 + \tan x}{1 - \tan x} = 0$

➡ **Technology**

Use a graphing calculator to check whether the following equations are identities. For those that appear to be identities, prove them. For those that are not, determine a value of the variable for which the two sides are unequal.

63. $\sin x = \sqrt{1 - \cos^2 x}$ 64. $(\sin x + \cos x)^2 = 1$

65. $\sqrt{\tan^2 \theta + \cot^2 \theta} = \tan \theta + \cot \theta$

66. $(\tan \theta)^2 = (\sin \theta \csc \theta)^2$

67. $\dfrac{\cos x + 1}{\sin x \cos x} - \dfrac{\sin x}{\cos x} = \dfrac{1 + \cos x}{\sin x}$

68. $\dfrac{\sec x}{\sin x} = \dfrac{\sec x + \csc x}{\sin x + \cos x}$

69. $\dfrac{\sin x}{\sin x \cot x + \cos x} = \tfrac{1}{2} \tan x$

70. $\sqrt{\tan^2 x + 1} = \sec x$

71. $\cos x = \csc x \tan x$

72. $\dfrac{1 - \sin x}{\cos x} - \dfrac{\cos x}{1 + \sin x} = 0$

Use a graphing utility to graph the following expressions in an appropriate range. Then use the graph to predict an identity for the expression.

73. $2 \sin x \cos x$ 74. $\dfrac{1 + \cos 2x}{2}$

75. $-4 \sin^3 x - 3 \sin x$

6.2 THE ADDITION AND SUBTRACTION IDENTITIES

Among the most important identities satisfied by the trigonometric functions are the addition and subtraction formulas, which allow you to calculate the trigonometric functions for the sum $x + y$ in terms of the trigonometric functions for x and y. In this section, we will prove these identities and a number of others that follow from them. For future reference, here is a summary of the identities we will prove in this section.

SUMMARY OF IDENTITIES FOR SECTION 6.2

Addition Identities

$$\sin(x + y) = \sin x \cos y + \cos x \sin y$$
$$\cos(x + y) = \cos x \cos y - \sin x \sin y$$
$$\tan(x + y) = \frac{\tan x + \tan y}{1 - \tan x \tan y}$$

Subtraction Identities

$$\sin(x - y) = \sin x \cos y - \cos x \sin y$$
$$\cos(x - y) = \cos x \cos y + \sin x \sin y$$
$$\tan(x - y) = \frac{\tan x - \tan y}{1 + \tan x \tan y}$$

Cofunction Identities

$$\cos\left(\frac{\pi}{2} - x\right) = \sin x \qquad \sin\left(\frac{\pi}{2} - x\right) = \cos x \qquad \tan\left(\frac{\pi}{2} - x\right) = \cot x$$

Symmetry Identities

$$\sin x = \sin(\pi - x) \qquad \cos x = -\cos(\pi - x)$$

Let's begin with the formulas for the cosine function.

ADDITION AND SUBTRACTION IDENTITIES FOR COSINE

Let x and y be any real numbers. Then we have the following identities:
$$\cos(x + y) = \cos x \cos y - \sin x \sin y \tag{1}$$
$$\cos(x - y) = \cos x \cos y + \sin x \sin y \tag{2}$$

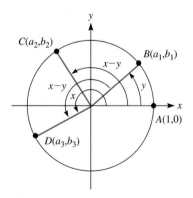

Figure 1

Proof   Interpret x and y as measures of angles. Draw the angles x, y, and $x - y$ in a unit circle, as shown in Figure 1. The angles AOC and BOD are both equal to $x - y$, so the arcs AC and BD are equal. By elementary geometry, this implies that the chords BD and AC are equal.

Set $A = (1, 0)$, $B = (a_1, b_1)$, $C = (a_2, b_2)$, and $D = (a_3, b_3)$. By the distance formula, we have the equation

$$\sqrt{(a_2 - 1)^2 + (b_2 - 0)^2} = \sqrt{(a_3 - a_1)^2 + (b_3 - b_1)^2}$$

Remove the radicals by squaring both sides. Then expand the expressions using the Product of Binomials formula:

$$a_2^2 - 2a_2 + 1 + b_2^2 = a_3^2 - 2a_3 a_1 + a_1^2 + b_3^2 - 2b_3 b_1 + b_1^2$$

Because the points A, B, C, and D lie on a unit circle, we have $a_1^2 + b_1^2 = 1$, $a_2^2 + b_2^2 = 1$, $a_3^2 + b_3^2 = 1$. Substituting these equations into the previous equation, we have

$$2 - 2a_2 = 2 - 2a_3 a_1 - 2b_3 b_1$$
$$a_2 = a_3 a_1 + b_3 b_1$$

However, from the unit circle definition of the trigonometric functions, we have

$$(a_1, b_1) = (\cos y, \sin y)$$
$$(a_2, b_2) = (\cos(x - y), \sin(x - y))$$
$$(a_3, b_3) = (\cos x, \sin x)$$

Substituting these values into the last displayed equation, we have

$$a_2 = a_3 a_1 + b_3 b_1$$
$$\cos(x - y) = \cos x \cos y + \sin x \sin y$$

This completes the proof of identity (2).

To prove identity (1), replace x with $-x$ in identity (2):

$$\cos(-x - y) = \cos(-x)\cos y + \sin(-x)\sin y$$
$$\cos[-(x + y)] = \cos x \cos y + (-\sin x)\sin y \qquad \textit{(Parity formulas)}$$

Again applying the parity formula for $\cos(-(x + y))$, we have

$$\cos(x + y) = \cos x \cos y - \sin x \sin y$$

This completes the proof of identity (1). ◆

➤ **EXAMPLE 1**

Application of Subtraction Formula for Cosine

Determine the value of $\cos \pi/12$.

Solution

Write $\pi/12$ in the form

$$\frac{\pi}{12} = \frac{\pi}{3} - \frac{\pi}{4}$$

then we have

$$\cos\frac{\pi}{12} = \cos\left(\frac{\pi}{3} - \frac{\pi}{4}\right)$$

Now apply the subtraction formula for the cosine to obtain

$$\cos\frac{\pi}{12} = \cos\frac{\pi}{3}\cos\frac{\pi}{4} + \sin\frac{\pi}{3}\sin\frac{\pi}{4}$$

We previously determined the values of the sine and cosine functions for angle $\pi/3$ and $\pi/4$. Inserting the previously determined values into the above expression gives us

$$\cos\frac{\pi}{12} = \frac{1}{2}\cdot\frac{\sqrt{2}}{2} + \frac{\sqrt{3}}{2}\cdot\frac{\sqrt{2}}{2}$$
$$= \frac{\sqrt{2}}{4} + \frac{\sqrt{6}}{4} = \frac{\sqrt{2} + \sqrt{6}}{4}$$

➤ **EXAMPLE 2**

Application of Addition Formula for Cosine

Use the addition formula for cosine to determine an alternate expression for $\cos(t + \pi/4)$.

Solution

$$\cos\left(t + \frac{\pi}{4}\right) = \cos t \cos\frac{\pi}{4} - \sin t \sin\frac{\pi}{4}$$
$$= \cos t \cdot \frac{\sqrt{2}}{2} - \sin t \cdot \frac{\sqrt{2}}{2}$$
$$= \frac{\sqrt{2}}{2}(\cos t - \sin t)$$

> **EXAMPLE 3**
Cosine of a Sum

Suppose that angles u and v are both in the third quadrant and that $\sin u = -\sqrt{3}/2$, and $\sin v = -1/2$. Determine the value of $\cos(u + v)$.

Solution

By the Pythagorean identity for the sine and cosine, we have

$$\cos u = \pm\sqrt{1 - \sin^2 u}$$

$$= \pm\sqrt{1 - \left(-\frac{\sqrt{3}}{2}\right)^2}$$

$$= \pm\sqrt{1 - \frac{3}{4}}$$

$$= \pm\frac{1}{2}$$

$$\cos v = \pm\sqrt{1 - \sin^2 v}$$

$$= \pm\sqrt{1 - \left(-\frac{1}{2}\right)^2}$$

$$= \pm\sqrt{\frac{3}{4}} = \pm\frac{\sqrt{3}}{2}$$

Because both u and v are given as lying in the third quadrant, the cosine of each angle is negative, so the $-$ sign prevails in the last two equations. Therefore, we have

$$\cos u = -\frac{1}{2}$$

$$\cos v = -\frac{\sqrt{3}}{2}$$

By the addition formula for the cosine, we have

$$\cos(u + v) = \cos v \cos u - \sin u \sin v$$

$$= \left(-\frac{\sqrt{3}}{2}\right) \cdot \left(-\frac{1}{2}\right) - \left(-\frac{\sqrt{3}}{2}\right) \cdot \left(-\frac{1}{2}\right)$$

$$= 0$$

Our next set of identities relates each of the functions sine, cosine, and tangent to its corresponding cofunction, cosine, sine, and cotangent, respectively.

COFUNCTION IDENTITIES

$$\cos\left(\frac{\pi}{2} - x\right) = \sin x \qquad \sin\left(\frac{\pi}{2} - x\right) = \cos x \qquad \tan\left(\frac{\pi}{2} - x\right) = \cot x$$

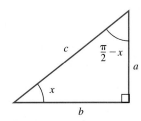

Figure 2

Proof   As motivation for these identities, consider the case where x is acute, as in the triangle of Figure 2. The second acute angle is $\pi/2 - x$ because the third angle is $\pi/2$ and the sum of the three angles is π. Using the right triangle definitions of the trigonometric functions, we see that

$$\sin x = \frac{\text{Opposite}}{\text{Hypotenuse}} = \frac{a}{c}$$

$$\cos\left(\frac{\pi}{2} - x\right) = \frac{\text{Adjacent}}{\text{Hypotenuse}} = \frac{a}{c}$$

This verifies the first cofunction identity. The proofs of the other identities for acute x may be given similarly. In case of a nonacute angle x, we can give the following argument. To prove the first identity, apply the subtraction formula for the cosine with x replaced with $\pi/2$ and y replaced with x. This gives

$$\cos\left(\frac{\pi}{2} - x\right) = \cos\frac{\pi}{2}\cos x + \sin\frac{\pi}{2}\sin x$$

Now use the facts $\cos(\pi/2) = 0$ and $\sin(\pi/2) = 1$ to obtain

$$\cos\left(\frac{\pi}{2} - x\right) = 1 \cdot \sin x$$

$$= \sin x$$

This proves the first cofunction identity.

To prove the second identity, start from the first and replace x everywhere with $\pi/2 - x$ to obtain

$$\cos\left[\frac{\pi}{2} - \left(\frac{\pi}{2} - x\right)\right] = \sin\left(\frac{\pi}{2} - x\right)$$

$$\cos x = \sin\left(\frac{\pi}{2} - x\right)$$

This proves the second identity.

The third identity follows from the first two. Its proof is left as an exercise. ◆

Using the cofunction identities and the addition and subtraction formulas for the cosine, we can deduce addition and subtraction formulas for the sine function:

ADDITION AND SUBTRACTION IDENTITIES FOR SINE

Let x and y be real numbers. Then we have the following formulas:

$$\sin(x + y) = \sin x \cos y + \cos x \sin y$$

$$\sin(x - y) = \sin x \cos y - \cos x \sin y$$

Proof To prove the subtraction formula for the sine, start with the addition formula for the cosine:

$$\cos(x + y) = \cos x \cos y - \sin x \sin y$$

Replace x with $\pi/2 - x$. This gives the identity

$$\cos\left[\left(\frac{\pi}{2} - x\right) + y\right] = \cos\left(\frac{\pi}{2} - x\right)\cos y - \sin\left(\frac{\pi}{2} - x\right)\sin y$$

$$\cos\left[\frac{\pi}{2} - (x - y)\right] = \sin x \cos y - \cos x \sin y \qquad \textit{(Cofunction formulas)}$$

Applying the first cofunction formula to the left side now gives us

$$\sin(x - y) = \sin x \cos y - \cos x \sin y$$

This is the subtraction formula for the sine.

To obtain the addition formula for the sine, start with the subtraction formula and replace y throughout with $-y$ to obtain

$$\sin[x - (-y)] = \sin x \cos(-y) - \cos x \sin(-y)$$

Now use the fact that $\sin(-x) = -\sin x$ and $\cos(-x) = \cos x$. This gives the identity

$$\sin(x + y) = \sin x \cos y - \cos x(-\sin y)$$
$$\sin(x + y) = \sin x \cos y + \cos x \sin y \qquad \blacklozenge$$

Here are two additional identities that follow simply from the addition formulas for the sine and cosine.

SYMMETRY IDENTITIES

$$\sin x = \sin(\pi - x)$$
$$\cos x = -\cos(\pi - x)$$

Proof   By the addition formula for the sine, we have

$$\sin(\pi - x) = \sin \pi \cos(-x) + \cos \pi \sin(-x)$$
$$= -\sin(-x) \qquad \text{(Because } \sin \pi = 0, \cos \pi = -1)$$
$$= \sin x \qquad \text{(Because } \sin(-x) = -\sin x)$$

This proves the first symmetry identity. To prove the second, apply the addition formula for the cosine:

$$\cos(\pi - x) = \cos \pi \cos(-x) - \sin \pi \sin(-x)$$
$$= -\cos(-x) \qquad \text{(Because } \cos \pi = -1, \sin \pi = 0)$$
$$= -\cos x \qquad \text{(Because } \cos(-x) = \cos x) \quad \blacklozenge$$

➤ **EXAMPLE 4**
Trigonometric Functions
of Special Angles

Determine the value of the following:

1. $\sin 75°$
2. $\sin 15°$

Solution

1. We have $75° = 30° + 45°$. Therefore, by the addition formula for the sine, we have

$$\sin 75° = \sin(30° + 45°)$$
$$= \sin 30° \cos 45° + \cos 30° \sin 45°$$

Now use the value for the trigonometric functions at $30°$ and $45°$.

$$\sin 75° = \frac{1}{2} \cdot \frac{\sqrt{2}}{2} + \frac{\sqrt{3}}{2} \cdot \frac{\sqrt{2}}{2}$$
$$= \frac{\sqrt{2} + \sqrt{6}}{4}$$

2. We first note that $15° = 45° - 30°$. In this case, we apply the subtraction formula for the sine:

$$
\begin{aligned}
\sin 15° &= \sin(45° - 30°) \\
&= \sin 45° \cos 30° - \sin 30° \cos 45° \\
&= \frac{\sqrt{2}}{2} \cdot \frac{\sqrt{3}}{2} - \frac{1}{2} \cdot \frac{\sqrt{2}}{2} \\
&= \frac{\sqrt{6}}{4} - \frac{\sqrt{2}}{4} = \frac{\sqrt{6} - \sqrt{2}}{4}
\end{aligned}
$$

In many applications, it is most convenient to deal with functions of the form $C \sin(x + D)$. The following example shows that every linear combination of the form $A \sin x + B \cos x$ may be written in this form by using the addition formula for the sine.

> **EXAMPLE 5**
Linear Combinations
of Sine and Cosine

Let A and B be real numbers. Express

$$A \sin x + B \cos x$$

in the form

$$C \sin(x + D)$$

where C and D are suitable real numbers.

Solution

The trick is to choose angle D correctly. Plot the point (A, B) on a Cartesian coordinate system. The distance of this point from the origin is $\sqrt{A^2 + B^2}$ (see Figure 3). Draw a line from the origin to the point (A, B) and let D be the angle with this line as terminal side and the positive x-axis as the initial side.

Then from Figure 3, we see that

$$\sin D = \frac{B}{\sqrt{A^2 + B^2}} \quad \text{and} \quad \cos D = \frac{A}{\sqrt{A^2 + B^2}}$$

We then have

$$
\begin{aligned}
A \sin x + B \cos x &= \sqrt{A^2 + B^2}\left(\frac{A}{\sqrt{A^2 + B^2}} \sin x + \frac{B}{\sqrt{A^2 + B^2}} \cos x\right) \\
&= \sqrt{A^2 + B^2}(\cos D \sin x + \sin D \cos x)
\end{aligned}
$$

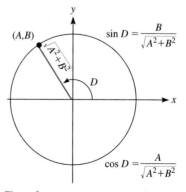

Figure 3

Now apply the addition formula for the sine to the expression in parentheses:

$$A \sin x + B \cos x = \sqrt{A^2 + B^2} \sin(x + D)$$

Therefore, we have

$$A \sin x + B \cos x = C \sin(x + D)$$

where D is as previously defined and $C = \sqrt{A^2 + B^2}$.

➤ **EXAMPLE 6**
A Particular Linear
Combination

Express $2 \sin x - 2 \cos x$ in the form $C \sin(x + D)$.

Solution

We apply the reasoning of the preceding example with $A = 2$, $B = -2$. In this case, we have

$$2 \sin x - 2 \cos x = C \sin(x + D) = 2\sqrt{2} \sin\left(x + \frac{7\pi}{4}\right)$$

$$C = \sqrt{A^2 + B^2} = \sqrt{2^2 + (-2)^2} = \sqrt{8} = 2\sqrt{2}$$

$$\sin D = \frac{B}{C} = -\frac{2}{2\sqrt{2}} = -\frac{\sqrt{2}}{2}$$

To determine D from the last equation, draw the angle D as in Figure 4. We see that $D = 2\pi - \pi/4 = 7\pi/4$, and thus

$$2 \sin x - 2 \cos x = C \sin(x + D) = 2\sqrt{2} \sin\left(x + \frac{7\pi}{4}\right)$$

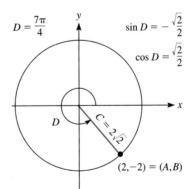

Figure 4

From the addition and subtraction formulas for the sine and cosine, we may deduce the following addition and subtraction formulas for the tangent function.

ADDITION AND SUBTRACTION IDENTITIES FOR THE TANGENT
$$\tan(x + y) = \frac{\tan x + \tan y}{1 - \tan x \tan y}$$
$$\tan(x - y) = \frac{\tan x - \tan y}{1 + \tan x \tan y}$$

Proof Let's prove the first identity. By the definition of the tangent function, we have

$$\tan(x + y) = \frac{\sin(x + y)}{\cos(x + y)}$$

Apply the addition formulas for sine and cosine to the right side to obtain

$$\tan(x + y) = \frac{\sin x \cos y + \cos x \sin y}{\cos x \cos y - \sin x \sin y}$$

Multiply the numerator and denominator on the right by $1/(\cos x \cos y)$ to obtain

$$\tan(x + y) = \frac{\dfrac{\sin x \cos y + \cos x \sin y}{1}}{\dfrac{\cos x \cos y - \sin x \sin y}{1}} \cdot \frac{\dfrac{1}{\cos x \cos y}}{\dfrac{1}{\cos x \cos y}}$$

Multiplying out the rational expressions in the numerators and denominators gives

$$\tan(x + y) = \frac{\dfrac{\sin x \cos y}{\cos x \cos y} + \dfrac{\cos x \sin y}{\cos x \cos y}}{\dfrac{\cos x \cos y}{\cos x \cos y} - \dfrac{\sin x \sin y}{\cos x \cos y}}$$

Rewrite this last expression using the definition of the tangent:

$$\tan(x + y) = \frac{\dfrac{\sin x}{\cos x} + \dfrac{\sin y}{\cos y}}{1 - \dfrac{\sin x}{\cos x} \cdot \dfrac{\sin y}{\cos y}}$$

$$= \frac{\tan x + \tan y}{1 - \tan x \tan y}$$

This completes the proof of the addition formula for the tangent function. The proof of the subtraction formula is left as an exercise. ◆

➤ **EXAMPLE 7**
Tangent of a Special Angle

Determine the exact value of $\tan \pi/12$. Do not use a calculator.

Solution

First note that $\pi/12 = \pi/3 - \pi/4$. Therefore, we may apply the subtraction formula for the tangent to obtain

$$\tan \frac{\pi}{12} = \frac{\tan \dfrac{\pi}{3} - \tan \dfrac{\pi}{4}}{1 + \tan \dfrac{\pi}{3} \tan \dfrac{\pi}{4}}$$

From calculations in the previous chapter, we saw that $\tan \pi/3 = \sqrt{3}$, $\tan \pi/4 = 1$. Inserting these values into the preceding formula, we have

$$\tan \frac{\pi}{12} = \frac{\sqrt{3} - 1}{1 + \sqrt{3} \cdot 1} = \frac{\sqrt{3} - 1}{\sqrt{3} + 1}$$

We may simplify this answer by rationalizing the denominator. We multiply both numerator and denominator by $-\sqrt{3}+1$. This yields

$$\tan\frac{\pi}{12} = \frac{\sqrt{3}-1}{\sqrt{3}+1} \cdot \frac{-\sqrt{3}+1}{-\sqrt{3}+1}$$

$$= \frac{-3+2\sqrt{3}-1}{-3+1}$$

$$= \frac{-4+2\sqrt{3}}{-2}$$

$$= 2-\sqrt{3}$$

➤ **EXAMPLE 8**

The Period of the Tangent

Prove that the function $\tan x$ has period π.

Solution

Apply the addition formula for the tangent to evaluate $\tan(x+\pi)$:

$$\tan(x+\pi) = \frac{\tan x + \tan \pi}{1 - \tan x \tan \pi}$$

$$= \frac{\tan x + 0}{1 - 0 \cdot \tan x}$$

$$= \tan x$$

➤ **EXAMPLE 9**

Formula for the Tangent of the Angle between Two Lines

Let L_1 and L_2 be straight lines having slopes m_1 and m_2, respectively, with L_2 having the larger angle of inclination (that is, the angle the line makes with the positive x-axis). Show that if θ denotes the angle between the two lines (see Figure 5), then

$$\tan\theta = \frac{m_2 - m_1}{1 + m_1 m_2}$$

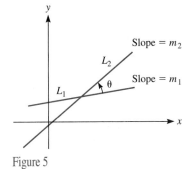

Figure 5

Solution

Let θ_1 and θ_2 denote the angle of inclination of L_1 and L_2, respectively. By the fundamental property of slope, if we start at any point on L_1 and proceed 1 unit in the positive x-direction, we must move m_1 units vertically in order to return to the line L_1. By referring to Figure 6 and using the definition of the tangent function, we see that

$$\tan\theta_1 = \frac{m_1}{1} = m_1$$

In a similar fashion, we have

$$\tan\theta_2 = m_2$$

Referring to Figure 6, we see that the angle θ between L_1 and L_2 is $\theta_2 - \theta_1$. Therefore, we have

$$\tan\theta = \tan(\theta_2 - \theta_1)$$

By the subtraction formula for the tangent function, we then have

$$\tan\theta = \frac{\tan\theta_2 - \tan\theta_1}{1 + \tan\theta_1 \tan\theta_2}$$

$$= \frac{m_2 - m_1}{1 + m_1 m_2}$$

This proves the desired formula.

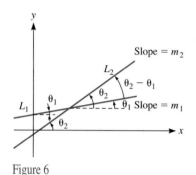

Figure 6

➤ **EXAMPLE 10**
Calculating Angle between Two Lines

Determine the angle between the lines with equations $y = 5x + 2, y = -x + 3$.

Solution

We apply the formula developed in the preceding exercise. The slopes of the lines are $m_1 = 5$ and $m_2 = -1$. If θ denotes the angle between the lines, then

$$\tan\theta = \frac{m_2 - m_1}{1 + m_1 m_2}$$

$$= \frac{-1 - 5}{1 + (-5)}$$

$$= \frac{-6}{-4}$$

$$= \frac{3}{2}$$

$$\theta = \tan^{-1}\left(\tfrac{3}{2}\right) \approx 56.310°$$

Exercises 6.2

Determine the exact value of the following without the use of a calculator or trigonometric tables.

1. $\cos\dfrac{7\pi}{12}$

2. $\sin\dfrac{5\pi}{12}$

3. $\tan\dfrac{17\pi}{12}$

4. $\cos\left(-\dfrac{13\pi}{12}\right)$

5. $\sin 255°$

6. $\tan 285°$

7. $\cos 15°$

8. $\sin\left(-\dfrac{11\pi}{12}\right)$

9. $\tan(-165°)$

10. $\cos 105°$

11. $\sin(-15°)$

12. $\tan\left(-\dfrac{5\pi}{12}\right)$

Use the addition and subtraction formulas to determine alternate expressions for each of the following.

13. $\cos\left(\dfrac{\pi}{2} + \theta\right)$

14. $\cos\left(\theta - \dfrac{\pi}{2}\right)$

15. $\sin\left(\theta + \dfrac{\pi}{3}\right)$

16. $\sin\left(\theta - \dfrac{\pi}{3}\right)$

17. $\tan\left(\theta - \dfrac{\pi}{4}\right)$

18. $\tan\left(\theta + \dfrac{\pi}{4}\right)$

19. $\cos(30° - \theta)$

20. $\cos(30° + \theta)$

21. $\sin(45° - \theta)$

22. $\sin(45° + \theta)$

23. $\tan(\theta - 135°)$

24. $\tan(\theta + 135°)$

Use the addition and subtraction formulas to determine the value of each of the following expressions.

25. $\cos\left(\dfrac{\pi}{2} + x\right)$ if

$\cos x = \dfrac{1}{2}, \qquad 0 < x < \dfrac{\pi}{2}$

26. $\cos\left(\theta - \dfrac{\pi}{2}\right)$ if

$\sin\theta = \dfrac{\sqrt{3}}{2}, \qquad \dfrac{\pi}{2} < \theta < \pi$

27. $\sin\left(\theta + \dfrac{\pi}{3}\right)$ if

$\sin\theta = \dfrac{\sqrt{2}}{2}, \qquad 0 < \theta < \dfrac{\pi}{2}$

28. $\sin\left(x + \dfrac{\pi}{6}\right)$ if

$\cos x = -\dfrac{1}{2}, \qquad \pi < x < \dfrac{3\pi}{2}$

29. $\sin\left(\dfrac{\pi}{3} - \theta\right)$ if

$\cos\theta = -\dfrac{1}{2}, \qquad \pi < \theta < \dfrac{3\pi}{2}$

30. $\tan\left(\theta + \dfrac{\pi}{4}\right)$ if

$\tan\theta = -1, \qquad \dfrac{3\pi}{2} < \theta < 2\pi$

31. $\tan\left(\theta - \dfrac{\pi}{4}\right)$ if

$\sin\theta = -\dfrac{1}{2}, \qquad \dfrac{3\pi}{2} < \theta < 2\pi$

32. $\tan\left(\dfrac{\pi}{6} + x\right)$ if

$\cos x = \dfrac{\sqrt{3}}{2}, \qquad 0 < x < \dfrac{\pi}{2}$

Determine the exact value of the indicated trigonometric expression, given the following. Do not use a calculator.

33. $\cos(x + y)$ if

$\sin x = \dfrac{\sqrt{2}}{2}, \qquad \dfrac{\pi}{2} < x < \pi$

$\sin y = \dfrac{1}{2}, \qquad 0 < y < \dfrac{\pi}{2}$

34. $\cos(x - y)$ if

$\sin x = -\dfrac{\sqrt{3}}{2}, \qquad \dfrac{3\pi}{2} < x < 2\pi$

$\sin y = -\dfrac{\sqrt{2}}{2}, \qquad \pi < y < \dfrac{3\pi}{2}$

35. $\sin(x + y)$ if

$\cos x = \dfrac{1}{2}, \qquad 0 < x < \dfrac{\pi}{2}$

$\cos y = -\dfrac{1}{2}, \qquad \dfrac{\pi}{2} < y < \pi$

36. $\sin(x - y)$ if

$\sin x = \dfrac{1}{2}, \qquad \dfrac{\pi}{2} < x < \pi$

$\sin y = -\dfrac{\sqrt{3}}{2}, \qquad \dfrac{3\pi}{2} < y < 2\pi$

37. $\cos 2x$ if

$\sin x = \dfrac{3}{5}, \qquad 0 < x < \dfrac{\pi}{2}$

(*Hint:* $2x = x + x$)

38. $\sin 2x$ if

$\cos x = -\dfrac{12}{13}, \qquad \pi < x < \dfrac{3\pi}{2}$

(*Hint:* $2x = x + x$)

39. $\tan(x + y)$ if

$\tan x = -1, \qquad \dfrac{3\pi}{2} < x < 2\pi$

$\tan y = -\sqrt{3}, \qquad \dfrac{3\pi}{2} < y < 2\pi$

40. $\tan(x - y)$ if

$\sin x = \dfrac{1}{2}, \qquad 0 < x < \dfrac{\pi}{2}$

$\cos y = -\dfrac{1}{2}, \qquad \dfrac{\pi}{2} < y < \pi$

41. $\cos(y + x)$ if

$\cos y = \dfrac{3}{5}, \qquad \cos x = \dfrac{4}{5},$

and x, y are both between 0 and $\dfrac{\pi}{2}$

42. $\cos(y - x)$ if

$\cos y = -\dfrac{3}{5}, \qquad \dfrac{\pi}{2} < y < \pi$

$\cos x = \dfrac{4}{5}, \qquad \dfrac{3\pi}{2} < x < 2\pi$

43. $\sin(y + x)$ if

$\sin y = \dfrac{4}{5}, \qquad 0 < y < \dfrac{\pi}{2}$

$\cos x = -\dfrac{3}{5}, \qquad \pi < x < \dfrac{3\pi}{2}$

44. $\sin(y - x)$ if

$\sin y = \dfrac{12}{13}, \qquad \dfrac{\pi}{2} < y < \pi$

$\cos x = -\dfrac{5}{13}, \qquad \dfrac{\pi}{2} < x < \pi$

45. $\tan(y + x)$ if

$\cos y = -\dfrac{12}{13}, \qquad \dfrac{\pi}{2} < y < \pi$

$\sin x = -\dfrac{5}{13}, \qquad \dfrac{3\pi}{2} < x < 2\pi$

46. $\tan(y - x)$ if

$\cos y = \dfrac{3}{5}, \qquad \dfrac{3\pi}{2} < y < 2\pi$

$\sin x = -\dfrac{4}{5}, \qquad \pi < x < \dfrac{3\pi}{2}$

Find the tangent of the smallest positive angle between the two given lines.

47. $2x - 2y = 6$ and $4x + 3y = 9$

48. $y = \frac{2}{3}x - 1$ and $y = -\frac{1}{2}x + 2$

49. $x = 3$ and $y = -1$ 50. $y = x$ and $y = -x$

51. $y = \frac{4}{5}x + 1$ and $2x + y = 0$

52. $y = \frac{1}{5}x - 2$ and $x - 2y = 3$

♦ **Applications**

53. A coordinate axis is placed with the center at the corner made by two roads that intersect. If a side of one road goes through the point $(4, 5)$ and a side of the other road goes through the point $(-3, -2)$, find the angle formed by the two roads at their point of intersection.

54. Two ships are on intersecting courses such that if a coordinate axis were placed with the center at their projected intersection, one ship would be at the point $(-3, 2)$ and the other would be at the point $(-5, -3)$. Find the angle between the two ships at their point of intersection.

Prove each of the following identities.

55. $\tan\left(\dfrac{\pi}{2} - x\right) = \cot x$ 56. $\cot\left(\dfrac{\pi}{2} - x\right) = \tan x$

57. $\sec\left(\dfrac{\pi}{2} - x\right) = \csc x$ 58. $\csc\left(\dfrac{\pi}{2} - x\right) = \sec x$

Express each of the following in the form $C \sin(Ex + D)$.

59. $\sin x + \cos x$ 60. $\sin x - \cos x$

61. $\sin x + \sqrt{3} \cos x$ 62. $4 \sin x - 3 \cos x$

63. $5 \sin x - 13 \cos x$ 64. $3 \sin x + 4 \cos x$

65. $\sin\dfrac{\pi}{4}x - \sqrt{3}\cos\dfrac{\pi}{4}x$ 66. $4 \sin 2x + 3 \cos 2x$

Without the use of tables or calculators, find the value of each of the following.

67. $\sin 27.35° \cos 2.65° + \cos 27.35° \sin 2.65°$

68. $\cos 34.721° \cos 10.279° - \sin 34.721° \sin 10.279°$

69. $\sin\dfrac{3\pi}{7} \cos\dfrac{\pi}{14} + \sin\dfrac{\pi}{14} \cos\dfrac{3\pi}{7}$

70. $\cos\dfrac{4\pi}{11} \cos\dfrac{7\pi}{11} - \sin\dfrac{4\pi}{11} \sin\dfrac{7\pi}{11}$

71. If $f(x) = \sin x$, prove that
$$\frac{f(x + h) - f(x)}{h} = \sin x\left(\frac{\cos h - 1}{h}\right) + \cos x\left(\frac{\sin h}{h}\right)$$

72. If $f(x) = \cos x$, prove that
$$\frac{f(x + h) - f(x)}{h} = -\sin x\left(\frac{\sin h}{h}\right) + \cos x\left(\frac{\cos h - 1}{h}\right)$$

73. If $f(x) = \tan x$, prove that
$$\frac{f(x + h) - f(x)}{h} = \frac{\sec^2 x}{\cos h - \sin h \tan x}\left(\frac{\sin h}{h}\right)$$

74. If $f(x) = \csc x$, prove that
$$\frac{f(x + h) - f(x)}{h}$$
$$= \frac{1}{h(\sin x \cos h + \cos x \sin h)} - \frac{1}{h \sin x}$$

Prove each of the following identities.

75. $2 \cos \alpha \cos \theta - \cos(\alpha - \theta) = \cos(\alpha + \theta)$

76. $\sin 2x = 2 \sin x \cos x$ (*Hint:* $2x = x + x$)

77. $\sin(x + y) - \sin(x - y) = 2 \sin y \cos y$

78. $\sin(x + y) \sec(x + y) = \dfrac{\tan x + \tan y}{1 - \tan x \tan y}$

79. $\cos(x + y) - \cos(x - y) = -2 \sin x \sin y$

80. $\sin(x - y) \sin(x + y) = \sin^2 x - \sin^2 y$

81. $\sin\left(\dfrac{\pi}{4} + x\right) - \sin\left(\dfrac{\pi}{4} - x\right) = \sqrt{2} \sin x$

82. $\cos\left(\dfrac{\pi}{6} + x\right)\cos\left(\dfrac{\pi}{6} - x\right)$
$$- \sin\left(\dfrac{\pi}{6} + x\right)\sin\left(\dfrac{\pi}{6} - x\right) = \dfrac{1}{2}$$

➡ **Technology**

83. Use a graphing calculator to check the plausibility of the addition formula for the sine by testing it for five particular values of y.

84. Use a graphing calculator to check the plausibility of the addition formula for the cosine by testing it for five particular values of y.

6.3 MULTIPLE- AND HALF-ANGLE IDENTITIES

For many applications of the trigonometric functions, especially for techniques of integration in calculus, it is necessary to use a variety of different identities to write trigonometric expressions in alternative forms. In this section, we will prove a number of such identities that give expressions of trigonometric functions for double angles and half angles. Here is a summary of the identities we will prove in this section.

SUMMARY OF IDENTITIES FOR SECTION 6.3

Double-Angle Identities

$$\sin 2u = 2 \sin u \cos u$$
$$\cos 2u = \cos^2 u - \sin^2 u$$
$$= 2 \cos^2 u - 1$$
$$= 1 - 2 \sin^2 u$$
$$\tan 2u = \frac{2 \tan u}{1 - \tan^2 u}$$

Square Identities

$$\sin^2 x = \frac{1 - \cos 2x}{2}, \qquad \cos^2 x = \frac{1 + \cos 2x}{2}, \qquad \tan^2 x = \frac{1 - \cos 2x}{1 + \cos 2x}$$

Half-Angle Identities

$$\sin \frac{x}{2} = \pm \sqrt{\frac{1 - \cos x}{2}}$$

$$\cos \frac{x}{2} = \pm \sqrt{\frac{1 + \cos x}{2}}$$

$$\tan \frac{x}{2} = \pm \sqrt{\frac{1 - \cos x}{1 + \cos x}} = \frac{1 - \cos x}{\sin x} = \frac{\sin x}{1 + \cos x}$$

Let's begin with the following identities for the trigonometric functions of a double angle.

DOUBLE-ANGLE IDENTITIES

$$\sin 2u = 2 \sin u \cos u, \qquad \cos 2u = \cos^2 u - \sin^2 u, \qquad \tan 2u = \frac{2 \tan u}{1 - \tan^2 u}$$

Proof   To prove the first formula, start with the addition formula for the sine and replace y with x throughout to obtain

$$\sin(x + x) = \sin x \cos x + \cos x \sin x = 2 \sin x \cos x$$
$$\sin 2x = 2 \sin x \cos x$$

To prove the second formula, start with the addition formula for the cosine and replace y with x throughout to obtain

$$\cos(x + x) = \cos x \cos x - \sin x \sin x = \cos^2 x - \sin^2 x$$
$$\cos 2x = \cos^2 x - \sin^2 x$$

In a similar fashion, we prove the third formula by replacing y with x in the addition formula for the tangent function. This completes the proof of the theorem. ♦

Starting with the double-angle formula for the cosine and replacing $\sin^2 x$ with $1 - \cos^2 x$, we obtain the following alternative form of the double-angle formula:

$$\cos 2x = 2 \cos^2 x - 1$$

Again starting with the double-angle formula for the cosine and replacing $\cos^2 x$ with $1 - \sin^2 x$, we arrive at the second alternative form of the identity:

$$\cos 2x = 1 - 2\sin^2 x$$

> EXAMPLE 1
Trigonometric Functions
of a Double Angle

Suppose that for a certain angle x in the first quadrant, we have $\sin x = 0.6$. Determine the value of $\sin 2x$, $\cos 2x$, and $\tan 2x$.

Solution

We have

$$\cos x = \pm \sqrt{1 - \sin^2 x} = \pm\sqrt{1 - (0.6)^2} = \pm 0.8$$

Because x is in the first quadrant, the plus sign applies, and

$$\cos x = 0.8$$
$$\tan x = \frac{\sin x}{\cos x} = \frac{0.6}{0.8} = 0.75$$

Now apply the double-angle formulas and these values of the trigonometric functions at x to obtain

$$\sin 2x = 2\sin x \cos x$$
$$= 2(0.6)(0.8)$$
$$= 0.96$$
$$\cos 2x = 2\cos^2 x - 1$$
$$= 2(0.8)^2 - 1$$
$$= 0.28$$
$$\tan 2x = \frac{2\tan x}{1 - \tan^2 x}$$
$$= \frac{2 \cdot (0.75)}{1 - (0.75)^2}$$
$$= 3.4286$$

> EXAMPLE 2
Triple-Angle Formula

Express $\sin 3x$ in terms of $\sin x$.

Solution

Write

$$\sin 3x = \sin(x + 2x)$$

Now apply the addition formula for the sine function along with the double-angle formula for the sine:

$$\sin(x + 2x) = \sin x \cos 2x + \cos x \sin 2x$$
$$= \sin x(1 - 2\sin^2 x) + \cos x \cdot 2\sin x \cos x$$
$$= -2\sin^3 x + \sin x + 2\sin x \cos^2 x$$
$$= -2\sin^3 x + \sin x + 2\sin x(1 - \sin^2 x)$$
$$= -2\sin^3 x + \sin x + 2\sin x - 2\sin^3 x$$
$$= -4\sin^3 x + 3\sin x$$

➤ **EXAMPLE 3**
Proving an Identity

Prove the following identity:

$$\sec 2x = \frac{\sec^2 x + \sec^4 x}{2 + \sec^2 x - \sec^4 x}$$

Solution

Factor the numerator and denominator of the right side to obtain

$$\frac{\sec^2 x + \sec^4 x}{2 + \sec^2 x - \sec^4 x} = \frac{\sec^2 x(1 + \sec^2 x)}{(2 - \sec^2 x)(1 + \sec^2 x)}$$

$$= \frac{\sec^2 x}{2 - \sec^2 x}$$

$$= \frac{\dfrac{1}{\cos^2 x}}{2 - \dfrac{1}{\cos^2 x}} \qquad \textit{(Definition of } \sec x\,)$$

$$= \frac{1}{2\cos^2 x - 1}$$

$$= \frac{1}{\cos 2x} \qquad \textit{(Because } \cos 2x = 2\cos^2 x - 1)$$

$$= \sec 2x \qquad \textit{(Definition of } \sec x\,)$$

This proves the identity.

➤ **EXAMPLE 4**
Nautical Mile

Latitude is used to measure north-south location on the earth between the equator and the poles. For example, the city of Hollywood, Florida, has latitude 26° N. (See Figure 1.) In Great Britain, the **nautical mile** is defined as the length of a minute of arc of the earth's radius. Because the earth is flattened at the poles, a British nautical mile varies with latitude. In fact, it is given, in feet, by the function $N(\phi) = 6066 - 31 \cos 2\phi$, where ϕ is the latitude in degrees.

1. What is the length of a British nautical mile at the city of Hollywood, Florida?
2. What is the length of a British nautical mile at the North Pole?
3. Express $N(\phi)$ in terms of $\cos \phi$ only (that is, eliminate the double angle).

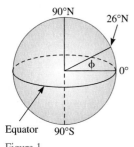

Figure 1

Solution

1. For Hollywood, Florida, we have $\phi = 26°$, so that

$$N(\phi) = 6066 - 31 \cos 2 \cdot 26°$$
$$= 6066 - 31 \cos 52°$$
$$\approx 6046.91 \text{ ft}$$

2. At the North Pole, we have $\phi = 90°$, so that

$$N(\phi) = 6066 - 31 \cos 2\phi$$
$$= 6066 - 31 \cos 2 \cdot 90°$$
$$= 6066 - 31 \cos 180°$$
$$= 6066 - 31(-1)$$
$$= 6097 \text{ ft}$$

3. Applying the double-angle formula for the cosine, we have

$$
\begin{aligned}
N(\phi) &= 6066 - 31\cos 2\phi \\
&= 6066 - 31(2\cos^2(\phi) - 1) \\
&= 6097 - 62\cos^2\phi
\end{aligned}
$$

> **EXAMPLE 5**
Acceleration Due to Gravity

The acceleration due to gravity is usually denoted by g. It has to do with the physics of motion near the earth's surface and is usually considered constant. In fact, however, g is not constant but varies slightly with latitude. If ϕ stands for latitude, in degrees, g is approximated by the formula

$$
g = 9.78049(1 + 0.005288\sin^2\phi - 0.000006\sin^2 2\phi)
$$

where g is measured in m/sec^2 at sea level.

1. Hollywood, Florida, has latitude 26°N. Find g.
2. Anchorage, Alaska, has latitude 61.8°N. Find g.
3. Express g in terms of $\sin\phi$ only (that is, eliminate the double angle).

Solution

1. In Hollywood, Florida, we have

$$
\begin{aligned}
g &= 9.78049(1 + 0.005288\sin^2\phi - 0.000006\sin^2 2\phi) \\
&= 9.78049(1 + 0.005288\sin^2 26° - 0.000006\sin^2 52°) \\
&\approx 9.79039 \text{ m/sec}^2
\end{aligned}
$$

2. In Anchorage, Alaska, we have

$$
\begin{aligned}
g &= 9.78049(1 + 0.005288\sin^2\phi - 0.000006\sin^2 2\phi) \\
&= 9.78049(1 + 0.005288\sin^2 61.8° - 0.000006\sin^2 123.6°) \\
&\approx 9.82062 \text{ m/sec}^2
\end{aligned}
$$

3. Applying the double-angle formula for the sine, we have

$$
\begin{aligned}
g &= 9.78049(1 + 0.005288\sin^2\phi - 0.000006\sin^2 2\phi) \\
&= 9.78049(1 + 0.005288\sin^2\phi - 0.000024\sin^2\phi\cos^2\phi) \\
&= 9.78049[1 + 0.005288\sin^2\phi - 0.000024\sin^2\phi(1 - \sin^2\phi)] \\
&\approx 9.78049(1 + 0.005264\sin^2\phi + 0.000024\sin^4\phi)
\end{aligned}
$$

Next, let's prove a series of identities that provide alternative expressions for the squares of the sine, cosine, and tangent functions. These expressions are useful in calculus as well as in proving the half-angle formulas later in this section.

SQUARE IDENTITIES
$\sin^2 x = \dfrac{1 - \cos 2x}{2}$
$\cos^2 x = \dfrac{1 + \cos 2x}{2}$
$\tan^2 x = \dfrac{1 - \cos 2x}{1 + \cos 2x}$

Proof   To prove the first identity, start from the identity for $\cos 2x$:

$$\cos 2x = 1 - 2\sin^2 x$$

This identity was established earlier. Solve this equation for $\sin^2 x$:

$$\cos 2x - 1 = -2\sin^2 x$$

$$-\tfrac{1}{2}(\cos 2x - 1) = \sin^2 x$$

$$\sin^2 x = \frac{-\cos 2x + 1}{2} = \frac{1 - \cos 2x}{2}$$

The second identity is proved in a similar fashion starting from the double-angle formula:

$$\cos 2x = 2\cos^2 x - 1$$

To prove the third identity, start from the definition of the tangent function $\tan^2 x = (\sin x / \cos x)^2$ and apply the first two identities to obtain

$$\tan^2 x = \left(\frac{\sin x}{\cos x}\right)^2$$

$$= \frac{\sin^2 x}{\cos^2 x}$$

$$= \frac{\dfrac{1 - \cos 2x}{2}}{\dfrac{1 + \cos 2x}{2}}$$

$$= \frac{1 - \cos 2x}{1 + \cos 2x}$$

This completes the proof of the third identity.   ♦

To close this section, let's derive formulas for the sine, cosine, and tangent for half an angle, namely $x/2$.

HALF-ANGLE IDENTITIES

Let x be a real number. Then

$$\sin \frac{x}{2} = \pm\sqrt{\frac{1 - \cos x}{2}}$$

$$\cos \frac{x}{2} = \pm\sqrt{\frac{1 + \cos x}{2}}$$

$$\tan \frac{x}{2} = \pm\sqrt{\frac{1 - \cos x}{1 + \cos x}} = \frac{1 - \cos x}{\sin x} = \frac{\sin x}{1 + \cos x}$$

where the sign that applies depends on the quadrant in which $x/2$ lies.

Proof   To prove the first identity, start from the square identity for $\sin^2 x$ and replace x with $x/2$ throughout. This gives us the identity

$$\sin^2 \frac{x}{2} = \frac{1 - \cos\left(2 \cdot \frac{x}{2}\right)}{2} = \frac{1 - \cos x}{2}$$

Now take the square roots of both sides to obtain

$$\sin \frac{x}{2} = \pm\sqrt{\frac{1 - \cos x}{2}}$$

This identity is valid for an appropriate choice of sign on the right. Moreover, it is clear that the sign of sin $x/2$ depends only on the quadrant of $x/2$. This establishes the first half-angle identity.

The proof of the second half-angle identity is similar to the proof of the first.

To prove the third half-angle identity, start from the first two identities to obtain

$$\tan \frac{x}{2} = \frac{\sin \dfrac{x}{2}}{\cos \dfrac{x}{2}} = \pm \sqrt{\frac{1 - \cos x}{1 + \cos x}}$$

which is the first equality of the third half-angle identity. To prove the second equality, note that

$$\tan \frac{x}{2} = \pm \sqrt{\frac{1 - \cos x}{1 + \cos x} \cdot \frac{1 - \cos x}{1 - \cos x}}$$

$$= \pm \sqrt{\frac{(1 - \cos x)^2}{1 - \cos^2 x}}$$

$$= \pm \sqrt{\frac{(1 - \cos x)^2}{\sin^2 x}}$$

$$= \pm \frac{1 - \cos x}{\sin x}$$

Note that $1 - \cos x \geq 0$ and that $\tan x/2$ and $\sin x$ have the same sign (both are positive for $0 < x < \pi$ and negative for $\pi < x < 2\pi$). Therefore, the + sign holds in the last equation and

$$\tan \frac{x}{2} = \frac{1 - \cos x}{\sin x}$$

which proves the second equality of the third half-angle identity. The proof of the final equality is similar. ◆

> **EXAMPLE 6**

Trigonometric Functions
for a Half Angle

Suppose that $\sin x = 3/5$ and that x lies in the second quadrant. Determine the values of sin $x/2$, cos $x/2$, and tan $x/2$.

Solution

Draw the angle x as shown in Figure 2. From the value of $\sin x$, we can calculate that $\cos x = -4/5$, and $\tan x = -3/4$. Because $\pi/2 < x < \pi$, we have $\pi/4 < x/2 < \pi/2$, so that $x/2$ lies in the first quadrant. Therefore, by the half-angle identities, we have

$$\sin \frac{x}{2} = +\sqrt{\frac{1 - \cos x}{2}}$$

$$= \sqrt{\frac{1 - \left(-\frac{4}{5}\right)}{2}}$$

$$= \sqrt{\frac{9}{10}}$$

$$= \frac{3\sqrt{10}}{10}$$

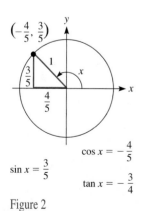

$\sin x = \dfrac{3}{5}$

$\cos x = -\dfrac{4}{5}$

$\tan x = -\dfrac{3}{4}$

Figure 2

$$\cos \frac{x}{2} = +\sqrt{\frac{1 + \cos x}{2}}$$

$$= \sqrt{\frac{1 + \left(-\frac{4}{5}\right)}{2}}$$

$$= \sqrt{\frac{1}{10}}$$

$$= \frac{\sqrt{10}}{10}$$

$$\tan \frac{x}{2} = +\sqrt{\frac{1 - \cos x}{1 + \cos x}}$$

$$= \sqrt{\frac{1 - \left(-\frac{4}{5}\right)}{1 + \left(-\frac{4}{5}\right)}}$$

$$= 3$$

➤ EXAMPLE 7
Using the Half-Angle Identities

Use the half-angle identities to determine the exact values of the following. Do not use a calculator.

1. $\sin \pi/8$
2. $\tan 105°$

Solution

1. The angle $\pi/8$ lies in the first quadrant. Therefore, the value of $\sin \pi/8$ is positive. By the half-angle identities, we have

$$\sin \frac{\pi}{8} = +\sqrt{\frac{1 - \cos \frac{\pi}{4}}{2}}$$

$$= \sqrt{\frac{1 - \frac{\sqrt{2}}{2}}{2}}$$

$$= \sqrt{\frac{2 - \sqrt{2}}{4}}$$

$$= \frac{\sqrt{2 - \sqrt{2}}}{2}$$

2. The angle $105°$ lies in the second quadrant, so the value of its tangent is negative. Twice this angle is $210°$. The value of $\cos 210°$ equals $-\sqrt{3}/2$. (See Figure 3.) By the half-angle identities, we have

$$\tan 105° = -\sqrt{\frac{1 - \cos 210°}{1 + \cos 210°}}$$

$$= -\sqrt{\frac{1 - \left(-\frac{\sqrt{3}}{2}\right)}{1 + \left(-\frac{\sqrt{3}}{2}\right)}}$$

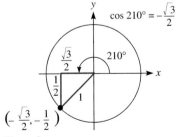

$\cos 210° = -\frac{\sqrt{3}}{2}$

$\left(-\frac{\sqrt{3}}{2}, -\frac{1}{2}\right)$

Figure 3

$$= -\sqrt{\frac{2 + \sqrt{3}}{2 - \sqrt{3}}}$$

$$= -\sqrt{\frac{2 + \sqrt{3}}{2 - \sqrt{3}} \cdot \frac{2 + \sqrt{3}}{2 + \sqrt{3}}}$$

$$= -\sqrt{7 + 4\sqrt{3}}$$

➤ **EXAMPLE 8**
A Half-Angle Identity
for Secant

Prove the identity

$$\sec^2 \frac{x}{2} = \frac{2}{1 + \cos x}$$

Solution

By the half-angle formula for $\cos x$, we have

$$\sec^2 \frac{x}{2} = \frac{1}{\cos^2 \frac{x}{2}}$$

$$= \frac{1}{\left(\pm\sqrt{\frac{1 + \cos x}{2}}\right)^2}$$

$$= \frac{1}{\frac{1 + \cos x}{2}}$$

$$= \frac{2}{1 + \cos x}$$

This proves the identity.

The fact that an expression involves half angles does not necessarily mean that the easiest way to simplify the expression is to use the half-angle formula. The following example shows that sometimes the easiest method is to apply the double-angle formula.

➤ **EXAMPLE 9**
Writing a Trigonometric
Expression in Terms of Sine

Write the following expression using only the sine function:

$$\sec \frac{x}{2} \csc \frac{x}{2}$$

Solution

We have

$$\sec \frac{x}{2} \csc \frac{x}{2} = \frac{1}{\cos \frac{x}{2}} \frac{1}{\sin \frac{x}{2}}$$

$$= \frac{1}{\sin \dfrac{x}{2} \cos \dfrac{x}{2}}$$

$$= \frac{2}{2\sin \dfrac{x}{2} \cos \dfrac{x}{2}}$$

$$= \frac{2}{\sin\left(2 \cdot \dfrac{x}{2}\right)} \qquad \text{(Double-angle formula)}$$

$$= \frac{2}{\sin x}$$

Note that if we had used the half-angle formula, it would have been necessary to determine the sign of the resulting expression, a nasty job.

➤ **EXAMPLE 10**
Proving an Identity
Involving Half Angles

Prove the identity

$$\cos x = \frac{1 - \tan^2 \dfrac{x}{2}}{1 + \tan^2 \dfrac{x}{2}}$$

Solution

Rewrite the right-hand side using the half-angle formula for the tangent function:

$$\frac{1 - \tan^2 \dfrac{x}{2}}{1 + \tan^2 \dfrac{x}{2}} = \frac{1 - \left(\pm\sqrt{\dfrac{1 - \cos x}{1 + \cos x}}\right)^2}{1 + \left(\pm\sqrt{\dfrac{1 - \cos x}{1 + \cos x}}\right)^2}$$

$$= \frac{1 - \dfrac{1 - \cos x}{1 + \cos x}}{1 + \dfrac{1 - \cos x}{1 + \cos x}}$$

$$= \frac{\dfrac{(1 + \cos x) - (1 - \cos x)}{1 + \cos x}}{\dfrac{(1 + \cos x) + (1 - \cos x)}{1 + \cos x}}$$

$$= \frac{\dfrac{2\cos x}{1 + \cos x}}{\dfrac{2}{1 + \cos x}}$$

$$= \frac{2\cos x}{2}$$

$$= \cos x$$

This proves the identity.

Exercises 6.3

Given the quadrant of angle x and one trigonometric function value of x, find (a) $\sin 2x$, (b) $\cos 2x$, and (c) $\tan 2x$.

1. $0 < x < \dfrac{\pi}{2}$, $\quad \sin x = \dfrac{1}{2}$

2. $0 < x < \dfrac{\pi}{2}$, $\quad \cos x = \dfrac{\sqrt{2}}{2}$

3. $\dfrac{\pi}{2} < x < \pi$, $\quad \cos x = -0.8$

4. $\dfrac{\pi}{2} < x < \pi$, $\quad \sin x = 0.6$

5. $\pi < x < \dfrac{3\pi}{2}$, $\quad \sin x = -\dfrac{\sqrt{3}}{2}$

6. $\pi < x < \dfrac{3\pi}{2}$, $\quad \cos x = -\dfrac{1}{2}$

7. $\dfrac{3\pi}{2} < x < 2\pi$, $\quad \tan x = -1$

8. $\dfrac{3\pi}{2} < x < 2\pi$, $\quad \cot x = -\dfrac{\sqrt{3}}{3}$

Express the following in terms of the indicated trigonometric function or functions.

9. $\cos 3x$ in terms of $\cos x$

10. $\tan 3x$ in terms of $\tan x$

11. $\sin 4x$ in terms of $\sin x$ and $\cos x$

12. $\cos 4x$ in terms of $\cos x$

Given the quadrant of angle x and one trigonometric function value of x, find (a) $\sin x/2$, (b) $\cos x/2$, and (c) $\tan x/2$.

13. $0 < x < \dfrac{\pi}{2}$, $\quad \cos x = \dfrac{4}{5}$

14. $\dfrac{3\pi}{2} < x < 2\pi$, $\quad \sin x = -\dfrac{3}{5}$

15. $\pi < x < \dfrac{3\pi}{2}$, $\quad \tan x = 1$

16. $\dfrac{\pi}{2} < x < \pi$, $\quad \cot x = -1$

17. $\pi < x < \dfrac{3\pi}{2}$, $\quad \sec x = -4$

18. $0 < x < \dfrac{\pi}{2}$, $\quad \csc x = \dfrac{5}{2}$

19. $\dfrac{3\pi}{2} < x < 2\pi$, $\quad \sin x = -\dfrac{2}{3}$

20. $\pi < x < \dfrac{3\pi}{2}$, $\quad \cos x = -\dfrac{12}{13}$

For the given information, find the exact value of the indicated trigonometric function.

21. $\cos 2x = \dfrac{1}{2}$, $\quad \pi < x < \dfrac{3\pi}{2}$;
    find $\sin x$.

22. $\cos 2x = -\dfrac{2}{9}$, $\quad \dfrac{\pi}{2} < x < \pi$;
    find $\sin x$.

23. $\cos 2x = -\dfrac{1}{8}$, $\quad \dfrac{\pi}{2} < x < \pi$;
    find $\cos x$.

24. $\cos 2x = -\dfrac{4}{25}$, $\quad 0 < x < \dfrac{\pi}{2}$;
    find $\cos x$.

Without using of tables or calculators, find each of the following.

25. $\sin 22.5°$ 26. $\cos 67.5°$ 27. $\tan \dfrac{\pi}{8}$

28. $\cot \dfrac{3\pi}{8}$ 29. $\sec\left(-\dfrac{3\pi}{8}\right)$ 30. $\csc \dfrac{\pi}{8}$

31. $\cos(-112.5°)$ 32. $\sin(-22.5°)$

Simplify the following expressions. Determine the exact value where possible.

33. $2\cos^2 20° - 1$ 34. $2\sin 50° \cos 50°$

35. $\dfrac{2\tan \dfrac{\pi}{5}}{1 - \tan^2 \dfrac{\pi}{5}}$ 36. $\cos^2 \dfrac{\pi}{8} - \sin^2 \dfrac{\pi}{8}$

37. $\sqrt{\dfrac{1 - \cos 140°}{1 + \cos 140°}}$ 38. $\dfrac{\sin \dfrac{\pi}{5}}{1 + \cos \dfrac{\pi}{5}}$

39. $-\sqrt{\dfrac{1 - \cos \dfrac{\pi}{4}}{2}}$ 40. $\sqrt{\dfrac{1 + \cos 130°}{2}}$

41. $\pm\sqrt{\dfrac{1 + \cos 14x}{2}}$ 42. $\dfrac{1 - \cos 40°}{\sin 40°}$

43. $\dfrac{\tan 80° + \tan 55°}{1 - \tan 80° \tan 55°}$

44. $\sin 12° \cos 42° - \sin 42° \cos 12°$

Simplify each expression.

45. $\sin 2a + (\sin a - \cos a)^2$

46. $\sin 2a - (\sin a + \cos a)^2$

47. $\cos 2x + 2\sin^2 x$

48. $\sin x \cos^2 x - \cos 2x \sin x$

49. $\dfrac{2\sin x}{\sin 2x}$ 50. $\dfrac{2\cos x}{\sin 2x}$

Prove each of the following identities.

51. $\tan x \sin 2x = 1 - \cos 2x$

52. $\left(\cos \dfrac{x}{2} + \sin \dfrac{x}{2}\right)^2 = 1 + \sin x$

53. $\csc x - \tan \dfrac{x}{2} = \cot x$

54. $\tan 2x = \dfrac{2 \cos x}{\csc x - 2 \sin x}$

55. $\tan x + \dfrac{\sin 3x}{\cos x} = 2 \sin 2x$

56. $\tan^2 \dfrac{x}{2} - 2 \csc x \tan \dfrac{x}{2} + 1 = 0$

57. $\dfrac{\sec^2 x}{2 - \sec^2 x} = \sec 2x$ 58. $3 - 4 \sin^2 x = \dfrac{\sin 3x}{\sin x}$

59. $2 \csc 2x = \tan x + \cot x$

60. $\dfrac{2 + \sin 2x}{2} = \dfrac{\sin^3 x - \cos^3 x}{\sin x - \cos x}$

61. $\cos^2 \dfrac{x}{2} = \dfrac{\sin x + \tan x}{2 \tan x}$ 62. $\csc 2x = \dfrac{1 + \tan^2 x}{2 \tan x}$

63. $\tan 2x = \dfrac{2}{\cot x - \tan x}$

64. $\cot 2x = \dfrac{\csc x - 2 \sin x}{2 \cos x}$

65. $4 \sin x \cos x \cos 2x = \sin 4x$

66. $1 + \cos 2x = \cot x \sin 2x$

67. $\tan \dfrac{x}{2} = \dfrac{\sin 2x - \sin x}{\cos 2x + \cos x}$

68. $\tan^2 \dfrac{x}{2} = \dfrac{2}{1 + \cos x} - 1$

69. $\csc x = \cot x + \tan \dfrac{x}{2}$ 70. $\sin x \sec \dfrac{x}{2} = 2 \sin \dfrac{x}{2}$

71. $\dfrac{1 - \sin^2 x}{1 - \cos^2 x} = \cot^2 x$

72. $\dfrac{\sin x \cos x}{1 - 2 \sin^2 x} = \dfrac{1}{\cot x - \tan x}$

73. $\csc x \sin 4x - \cos 3x = \sin 3x \cot x$

74. $\sin^2 2x - \cos^2 2x = \dfrac{\sin x}{\csc 3x} - \dfrac{\cos 3x}{\sec x}$

➡ **Technology**

75. Use a graphing calculator to determine the range of x for which the + sign pertains to the half-angle formula for the sine.

76. Use a graphing calculator to determine the range of x for which the + sign pertains to the half-angle formula for the cosine.

6.4 FURTHER TRIGONOMETRIC IDENTITIES

In this section, we will prove the following useful trigonometric identities:

PRODUCT AND FACTORING IDENTITIES

Product Identities (Product to Sum)

$$\sin x \cos y = \tfrac{1}{2}[\sin (x + y) + \sin (x - y)]$$

$$\cos x \sin y = \tfrac{1}{2}[\sin (x + y) - \sin (x - y)]$$

$$\cos x \cos y = \tfrac{1}{2}[\cos (x + y) + \cos (x - y)]$$

$$\sin x \sin y = \tfrac{1}{2}[\cos(x - y) - \cos(x + y)]$$

Factoring Identities (Sum to Product)

$$\sin x + \sin y = 2 \sin \left(\dfrac{x + y}{2}\right) \cos \left(\dfrac{x - y}{2}\right)$$

$$\sin x - \sin y = 2 \sin \left(\dfrac{x - y}{2}\right) \cos \left(\dfrac{x + y}{2}\right)$$

$$\cos x + \cos y = 2 \cos \left(\dfrac{x + y}{2}\right) \cos \left(\dfrac{x - y}{2}\right)$$

$$\cos x - \cos y = -2 \sin \left(\dfrac{x + y}{2}\right) \sin \left(\dfrac{x - y}{2}\right)$$

Let's begin by proving the first set of identities.

Proof   By the addition and subtraction formulas for the sine, we have

$$\sin(x + y) = \sin x \cos y + \cos x \sin y$$
$$\sin(x - y) = \sin x \cos y - \cos x \sin y$$

To obtain the first product identity, add these two equations together to obtain

$$\sin(x + y) + \sin(x - y) = 2 \sin x \cos y$$

from which the identity follows. The other three product formulas may be obtained similarly by using the addition and subtraction formulas for the cosine. The details of the proofs are left to the exercises.   ◆

➤ **EXAMPLE 1**
Reducing to a Sum of Sines

Write $\sin 3x \cos x$ as a sum or difference of trigonometric functions.

Solution

By the first product formula, we have

$$\sin 3x \cos x = \tfrac{1}{2}[\sin(3x + x) + \sin(3x - x)]$$
$$= \tfrac{1}{2}[\sin 4x + \sin 2x]$$

➤ **EXAMPLE 2**
Proving an Identity

Verify the identity

$$\frac{\cos 2x + \cos 2y}{\cos 2x - \cos 2y} = -\cot(x + y)\cot(x - y)$$

Solution

Rewrite the right side in terms of sines and cosines and apply the product identity:

$$-\cot(x + y)\cot(x - y) = -\frac{\cos(x + y)}{\sin(x + y)} \cdot \frac{\cos(x - y)}{\sin(x - y)}$$

$$= -\frac{\tfrac{1}{2}\{\cos[(x + y) + (x - y)] + \cos[(x + y) - (x - y)]\}}{\tfrac{1}{2}\{\cos[(x + y) - (x - y)] - \cos[(x + y) + (x - y)]\}}$$

$$= -\frac{\cos 2x + \cos 2y}{\cos 2y - \cos 2x}$$

$$= \frac{\cos 2x + \cos 2y}{\cos 2x - \cos 2y}$$

This proves the identity.

The product identities express products of sines and cosines as linear expressions in the trigonometric functions. The following factoring identities work in reverse. That is, they express a sum or difference of sines or cosines as products.

FACTORING IDENTITIES

$$\sin x + \sin y = 2\sin\left(\frac{x+y}{2}\right)\cos\left(\frac{x-y}{2}\right)$$

$$\sin x - \sin y = 2\sin\left(\frac{x-y}{2}\right)\cos\left(\frac{x+y}{2}\right)$$

$$\cos x + \cos y = 2\cos\left(\frac{x+y}{2}\right)\cos\left(\frac{x-y}{2}\right)$$

$$\cos x - \cos y = -2\sin\left(\frac{x+y}{2}\right)\sin\left(\frac{x-y}{2}\right)$$

Proof   Let's prove the first of these identities. The proofs of the others will be left to the exercises. Start from the first product formula:

$$\sin x \cos y = \tfrac{1}{2}[\sin(x+y) + \sin(x-y)]$$

Define new variables $t = x + y$, $u = x - y$. By adding these equations together, we have

$$t + u = 2x$$
$$x = \frac{t+u}{2}$$

Subtracting the second from the first yields

$$t - u = 2y$$
$$y = \frac{t-u}{2}$$

Writing the first product identity in terms of the variables t and u, we have

$$\sin\frac{t+u}{2}\cos\frac{t-u}{2} = \tfrac{1}{2}(\sin t + \sin u)$$

This is just the first factoring formula in which x is replaced with t and y is replaced with u.   ♦

➢ **EXAMPLE 3**
Reducing to a Product
of Trigonometric Functions

Express $\cos 3x - \cos 4x$ as a product of two trigonometric functions.

Solution

By the fourth factorization formula, in which x is replaced with $3x$ and y with $4x$, we have

$$\cos 3x - \cos 4x = -2\sin\left(\frac{3x+4x}{2}\right)\sin\left(\frac{3x-4x}{2}\right)$$

$$= -2\sin\frac{7x}{2}\sin\left(-\frac{x}{2}\right)$$

$$= -2\sin\frac{7x}{2}\cdot\left(-\sin\frac{x}{2}\right)$$

$$= 2\sin\frac{x}{2}\sin\frac{7x}{2}$$

Exercises 6.4

Write the following as a sum or difference of trigonometric functions.

1. $\sin 4x \cos 3x$ 2. $\sin 2a \cos a$ 3. $\sin y \sin 3y$

4. $\sin 5x \sin 2x$ 5. $\cos a \cos 3a$ 6. $\cos 4y \cos 5y$

7. $\sin 3x \cos 5x$ 8. $\sin 3y \sin(-2y)$

9. $\cos 5a \cos(-a)$ 10. $\cos(-2x)\cos(-3x)$

11. $6\sin \dfrac{\pi}{4}\cos \dfrac{\pi}{4}$ 12. $4\sin \dfrac{\pi}{3}\cos \dfrac{5\pi}{6}$

13. $\sin(x-y)\cos(x+y)$ 14. $\cos(x+y)\cos(x-y)$

Express the given sum or difference as a product.

15. $\sin 60° - \sin 20°$ 16. $\cos 40° - \cos 80°$

17. $\sin 7y + \sin 9y$ 18. $\sin x + \sin 5x$

19. $\cos \dfrac{3\pi}{4} - \cos \dfrac{\pi}{4}$ 20. $\sin \dfrac{\pi}{2} - \sin \dfrac{\pi}{6}$

21. $\sin(x+y) - \sin(x-y)$

22. $\cos(x+y) + \cos(x-y)$

23. $\cos 2x + \cos x$ 24. $\sin 8x + \sin 2x$

25. $\sin 50° + \sin 10°$ 26. $\sin 20° + \sin 80°$

Prove each of the following identities.

27. $\sin x \sin y = \frac{1}{2}[\cos(x-y) - \cos(x+y)]$

28. $\cos x \cos y = \frac{1}{2}[\cos(x+y) + \cos(x-y)]$

29. $\sin x - \sin y = 2\sin\left(\dfrac{x-y}{2}\right)\cos\left(\dfrac{x+y}{2}\right)$

30. $\cos x + \cos y = 2\cos\left(\dfrac{x+y}{2}\right)\cos\left(\dfrac{x-y}{2}\right)$

31. $\cos x - \cos y = -2\sin\left(\dfrac{x+y}{2}\right)\sin\left(\dfrac{x-y}{2}\right)$

32. $\sin 2x + \sin 2y = 2\sin(x+y)\cos(x-y)$

33. $\dfrac{\tan 4x}{\tan x} = \dfrac{\sin 5x + \sin 3x}{\sin 5x - \sin 3x}$

34. $\cot 5x = \dfrac{\cos x + \cos 9x}{\sin x + \sin 9x}$

35. $\tan \dfrac{x}{2} = \dfrac{\sin 2x - \sin x}{\cos 2x + \cos x}$

36. $\cot \dfrac{7x}{2} = \dfrac{\cos 5x + \cos 2x}{\sin 5x + \sin 2x}$

37. $\dfrac{\sin 4x - \sin 2x}{\cos 4x + \cos 2x} = \tan x$

38. $-\cot x = \dfrac{\sin 5x + \sin 3x}{\cos 5x - \cos 3x}$

39. $\sin 2x \sin 6x = \sin^2 4x - \sin^2 2x$

40. $\cos^2 y - \sin^2 x = \cos(x+y)\cos(x-y)$

41. $\dfrac{\sin x}{\sec 2x} + \dfrac{\cos x}{\csc 2x} = \sin 3x$

42. $\dfrac{\cos 2x}{\sec x} - \dfrac{\sin x}{\csc 2x} = \cos 3x$

43. $\tan \frac{1}{2}(x+y)\tan \frac{1}{2}(x-y) = -\dfrac{\cos x - \cos y}{\cos x + \cos y}$

44. $\tan \frac{1}{2}(x+y) = \dfrac{\sin x + \sin y}{\cos x + \cos y}$

45. $\dfrac{\cos 2x - \cos 6x}{\sin 6x - \sin 2x} = \tan 4x$

46. $\cot x = \dfrac{\sin 4x + \sin 6x}{\cos 4x - \cos 6x}$

Without the use of tables or calculators, find the value of each of the following expressions.

47. $\dfrac{\sin 75° + \sin 15°}{\cos 75° + \cos 15°}$ 48. $\dfrac{\cos 75° - \cos 15°}{\sin 75° - \sin 15°}$

49. $\dfrac{\cos 10° + \cos 80°}{\sin 10° + \sin 80°}$ 50. $\dfrac{\cos 10° - \cos 80°}{\sin 10° - \sin 80°}$

51. $\dfrac{\sin \dfrac{5\pi}{18} - \sin \dfrac{\pi}{12}}{\cos \dfrac{5\pi}{18} + \cos \dfrac{\pi}{12}}$ 52. $\dfrac{\cos \dfrac{5\pi}{18} + \cos \dfrac{\pi}{8}}{\sin \dfrac{5\pi}{18} + \sin \dfrac{\pi}{8}}$

53. $\dfrac{\sin \dfrac{4\pi}{9} - \sin \dfrac{\pi}{18}}{\cos \dfrac{4\pi}{9} - \cos \dfrac{\pi}{18}}$ 54. $\dfrac{\sin \dfrac{4\pi}{9} + \sin \dfrac{\pi}{18}}{\cos \dfrac{4\pi}{9} - \cos \dfrac{\pi}{18}}$

55. $2\sin 37.5° \cos 82.5°$ 56. $\sin \dfrac{\pi}{4}\cos \dfrac{\pi}{12}$

6.5 THE INVERSE TRIGONOMETRIC FUNCTIONS

In Chapter 2, we defined the notion of the inverse of a function. In this section, we will apply that discussion to define and investigate the properties of the inverse functions of the trigonometric functions.

To start with, let's summarize the discussion of Chapter 2. Suppose that $f: A \to B$ is a one-to-one function, where A and B are sets of real numbers and B is the range of f. Then we may define the inverse function $f^{-1}: B \to A$ by the formula

$$f^{-1}(y) = x \quad \text{if and only if} \quad f(x) = y \quad (x \text{ in } A, y \text{ in } B)$$

The following fundamental composition formulas then apply:

$$(f^{-1} \circ f)(x) = x \quad (x \in A)$$
$$(f \circ f^{-1})(y) = y \quad (y \in B)$$

The function $\sin x$ is a function with domain the real line **R** and range the interval $[-1, 1]$. However, it is *not* a one-to-one function, so that we may not directly apply the discussion of Chapter 2 to construct an inverse function for $\sin x$. To get around this difficulty, let's restrict the domain of $\sin x$ to be the interval $[-\pi/2, \pi/2]$. On this restricted domain, the function $\sin x$ is one-to-one. (See Figure 1.)

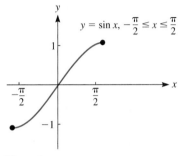

Figure 1

Definition 1	The inverse of the function $\sin x$ for x in the interval $[-\pi/2, \pi/2]$ is called the **arcsin**, denoted $\sin^{-1}$, and is defined by the property
Inverse Sine Function	$$\sin^{-1} y = x \quad \text{if and only if sin} \quad x = y \quad \left(-1 \le y \le 1, -\frac{\pi}{2} \le x \le \frac{\pi}{2}\right)$$

The domain of $\sin^{-1}$ is the interval $[-1, 1]$, and the range is the interval $[-\pi/2, \pi/2]$. The fundamental composition formulas in this case read

$$\sin^{-1}(\sin x) = x \quad (-\pi/2 \le x \le \pi/2)$$
$$\sin(\sin^{-1} y) = y \quad (-1 \le y \le 1)$$

The graph of the arcsine function is obtained by reflecting the graph of the portion of the sine graph $y = \sin x$, $-\pi/2 \le x \le \pi/2$, about the line $y = x$. The graph of the arcsine function is shown in Figure 2.

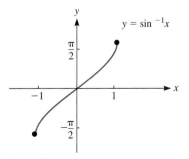

Figure 2

➤ **EXAMPLE 1**
Value of an Inverse Function

Calculate the exact value of $\sin^{-1}(\sqrt{3}/2)$. Do not use a calculator.

Solution

From the definition of the arcsine function, the equation

$$\sin^{-1}\frac{\sqrt{3}}{2} = y$$

is equivalent to

$$\sin y = \frac{\sqrt{3}}{2}, \qquad -\frac{\pi}{2} \le y \le \frac{\pi}{2}$$

There is a single value of y that satisfies this equation, namely $y = \pi/3$. That is,

$$\sin^{-1}\frac{\sqrt{3}}{2} = \frac{\pi}{3}$$

Note that there are an infinite number of values of y for which $\sin y = \sqrt{3}/2$. However, the condition $-\pi/2 \le y \le \pi/2$ serves to pick out exactly one among them.

For each remaining trigonometric function, it is possible to restrict the domain so that the function becomes one-to-one, just as for the sine. The **inverse trigonometric functions** are then defined as the inverses of the one-to-one functions. The restricted domain of the trigonometric function becomes the range of the inverse trigonometric function. The range of the trigonometric function becomes the domain of the inverse trigonometric function. The following table summarizes the definitions of the domains and ranges of the various inverse trigonometric functions.

Function	Range	Domain
$y = \sin^{-1} x$	$-\dfrac{\pi}{2} \le y \le \dfrac{\pi}{2}$	$-1 \le x \le 1$
$y = \cos^{-1} x$	$0 \le y \le \pi$	$-1 \le x \le 1$
$y = \tan^{-1} x$	$-\dfrac{\pi}{2} < y < \dfrac{\pi}{2}$	$-\infty < x < \infty$
$y = \sec^{-1} x$	$0 \le y < \dfrac{\pi}{2}$ or $\pi \le y < \dfrac{3\pi}{2}$	$x \ge 1$ or $x \le -1$
$y = \csc^{-1} x$	$0 < y \le \dfrac{\pi}{2}$ or $\pi < y \le \dfrac{3\pi}{2}$	$x \ge 1$ or $x \le -1$
$y = \cot^{-1} x$	$0 < y < \pi$	$-\infty < x < \infty$

Figures 3 and 4 show the graphs of the functions $y = \cos^{-1} x$ and $y = \tan^{-1} x$. We may summarize the fundamental composition properties of these functions as follows:

$$\cos^{-1}(\cos y) = y \qquad 0 \le y \le \pi$$
$$\cos(\cos^{-1} x) = x \qquad -1 \le y \le 1$$
$$\tan^{-1}(\tan y) = y \qquad -\frac{\pi}{2} < y < \frac{\pi}{2}$$
$$\tan(\tan^{-1} x) = x \qquad -\infty < x < \infty$$

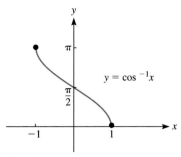

Figure 3

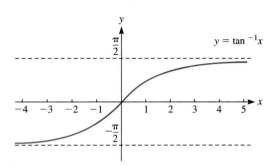

Figure 4

➢ **EXAMPLE 2**
Evaluating Inverse
Trigonometric Functions

Calculate the exact values of the following. Do not use a calculator.

1. $\cos^{-1}\frac{1}{2}$
2. $\tan^{-1} 1$

Solution

1. Set $y = \cos^{-1}(1/2)$. Then, by the definition of the inverse function, we have

$$\cos y = \tfrac{1}{2}, \qquad 0 \leq y \leq \pi$$

$$y = \frac{\pi}{3}$$

$$\cos^{-1}\tfrac{1}{2} = \frac{\pi}{3}$$

2. Set $y = \tan^{-1} 1$. Then, by the definition of the inverse function, we have

$$\tan y = 1, \qquad -\frac{\pi}{2} < y < \frac{\pi}{2}$$

$$y = \frac{\pi}{4}$$

$$\tan^{-1} 1 = \frac{\pi}{4}$$

➢ **EXAMPLE 3**
Composition Involving
Inverse Functions

Calculate the exact value of $\cos(\tan^{-1}(-5/4))$. Do not use a calculator.

Solution

Let $y = \tan^{-1}(-5/4)$. By the definition of the inverse function, we have

$$\tan y = -\tfrac{5}{4}$$

Because y is in the range of the inverse tangent function, it lies in the interval $(-\pi/2, \pi/2)$. Moreover, because the value of $\tan y$ is negative, y must actually lie in the interval $(-\pi/2, 0]$. According to the preceding equation, we have

$$\sec^2 y = \tan^2 y + 1$$

$$= \left(-\tfrac{5}{4}\right)^2 + 1$$

$$\sec^2 y = \tfrac{41}{16}$$

$$\frac{1}{\cos^2 y} = \frac{41}{16}$$

$$\cos^2 y = \frac{16}{41}$$

$$\cos y = \pm\sqrt{\frac{16}{41}} = \pm\frac{4}{\sqrt{41}}$$

Because y lies in the interval $[-\pi/2, 0]$, the value of $\cos y$ is positive and the $+$ sign in the last equation prevails. Thus, we have

$$\cos\left[\tan^{-1}\left(-\tfrac{5}{4}\right)\right] = \cos y = \frac{4\sqrt{41}}{41}$$

> **EXAMPLE 4**

Another Composition Problem

Prove the identity

$$\sin^{-1}(\cos x) = \frac{\pi}{2} - x, \quad 0 \le x \le \pi$$

Solution

Start from the cofunction identity

$$\cos x = \sin\left(\frac{\pi}{2} - x\right)$$

Because $0 \le x \le \pi$, we see that $0 \ge -x \ge -\pi$ and thus $\pi/2 \ge \pi/2 - x \ge -\pi/2$. Now apply the function $\sin^{-1}$ to both sides of the last equation:

$$\sin^{-1}(\cos x) = \sin^{-1}\left[\sin\left(\frac{\pi}{2} - x\right)\right]$$

$$= \frac{\pi}{2} - x \qquad \textit{(Definition of inverse function)}$$

This proves the identity.

> **EXAMPLE 5**

Solving a Trigonometric Equation

Solve the following equation for x in terms of y:

$$y = 4\tan^{-1}(3x - 1)$$

Solution

First divide both sides of the equation by 4 to obtain

$$\frac{y}{4} = \tan^{-1}(3x - 1)$$

Now apply the function tan to both sides of the equation:

$$\tan\left(\frac{y}{4}\right) = \tan(\tan^{-1}(3x - 1))$$

By the fundamental composition formula for inverse functions, the right side equals $(3x - 1)$. Therefore, we have

$$3x - 1 = \tan\left(\frac{y}{4}\right)$$

$$3x = \tan\left(\frac{y}{4}\right) + 1$$

$$x = \tfrac{1}{3}\tan\left(\frac{y}{4}\right) + \tfrac{1}{3}$$

➢ **EXAMPLE 6**
Simplification

Suppose that $0 \leq x \leq 1$. Write the expression $\cos(\sin^{-1} x)$ in a form that does not involve inverse trigonometric functions.

Solution

Draw an angle t whose sine is x. Let the opposite side be x and the hypotenuse be 1. (See Figure 5.) The adjacent side is then $\sqrt{1 - x^2}$, so that

$$\cos(\sin^{-1} x) = \cos t = \frac{\text{Adjacent}}{\text{Hypotenuse}} = \frac{\sqrt{1 - x^2}}{1} = \sqrt{1 - x^2}$$

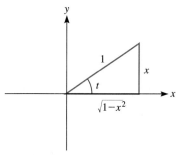

Figure 5

➢ **EXAMPLE 7**
Evaluation of an Inverse
Trigonometric Expression

Suppose that $y = \sin^{-1} x$, $0 \leq x \leq 1$. Express $\sec y$ in terms of x.

Solution

From the definition of the inverse function, we have

$$\sin y = \sin(\sin^{-1} x)$$
$$= x$$
$$\cos^2 y = 1 - \sin^2 y$$
$$= 1 - x^2$$
$$\cos y = \pm \sqrt{1 - x^2}$$

Because we are given that $0 \leq x \leq 1$, we know that $y = \sin^{-1} x$ lies in the interval $[0, \pi/2]$, so that $\cos y \geq 0$. Therefore, in the last equation, the $+$ sign prevails and we have

$$\cos y = \sqrt{1 - x^2}$$
$$\sec y = \frac{1}{\cos y}$$
$$= \frac{1}{\sqrt{1 - x^2}}$$

Exercises 6.5

Calculate the value of each of the following without a calculator or trigonometry table.

1. $\sin^{-1}\left(\frac{1}{2}\right)$

2. $\sin^{-1}\left(-\frac{\sqrt{3}}{2}\right)$

3. $\cos^{-1}\left(\frac{\sqrt{2}}{2}\right)$

4. $\cos^{-1}\left(\frac{1}{2}\right)$

5. $\tan^{-1}(-\sqrt{3})$

6. $\tan^{-1}(0)$

7. $\sec^{-1}(2)$

8. $\sec^{-1}(-\sqrt{2})$

9. $\csc^{-1}(\sqrt{2})$

10. $\csc^{-1}\left(\dfrac{2\sqrt{3}}{3}\right)$

11. $\cos^{-1}\left(-\dfrac{\sqrt{3}}{2}\right)$

12. $\cot^{-1}(\sqrt{3})$

13. $\cot^{-1}(-1)$

14. $\sin^{-1}\left(-\dfrac{\sqrt{2}}{2}\right)$

15. $\tan^{-1}\left(\dfrac{1}{\sqrt{3}}\right)$

16. $\sec^{-1}(\sqrt{2})$

17. $\csc\left(\csc^{-1}\dfrac{2}{\sqrt{3}}\right)$

18. $\sin\left(\csc^{-1}\left(-\dfrac{2}{\sqrt{3}}\right)\right)$

19. $\tan\left(\cot^{-1}\dfrac{1}{\sqrt{3}}\right)$

20. $\cot(\cot^{-1}(-1))$

21. $\cos\left(\cos^{-1}\dfrac{\sqrt{3}}{2}\right)$

22. $\csc^{-1}\left(\csc\dfrac{\pi}{2}\right)$

23. $\sin\left(\cos^{-1}-\tfrac{1}{2}\right)$

24. $\cos(\sin^{-1}0)$

25. $\sec\left(\cot^{-1}\tfrac{3}{5}\right)$

26. $\cos\left(\sin^{-1}\tfrac{12}{13}\right)$

27. $\cot^{-1}\left(\cot\dfrac{2\pi}{3}\right)$

28. $\sin^{-1}\left(\sin\dfrac{5\pi}{12}\right)$

29. $\cos[\tan^{-1}(-\sqrt{3})]$

30. $\sin(\cot^{-1}0)$

Prove each identity. (*Hint:* Use identities for sums when necessary.)

31. $\tan^{-1}(-x) = -\tan^{-1}x$

32. $\cos^{-1}(-x) = \pi - \cos^{-1}x$

33. $\sin(\cos^{-1}x) = \sqrt{1-x^2}$ for $|x| \le 1$

34. $\sin(\sin^{-1}x + \cos^{-1}x) = 1$ for $|x| \le 1$

35. $\sin^{-1}x + \cos^{-1}x = \dfrac{\pi}{2}$ for $|x| \le 1$

36. $\sin\left(\sin^{-1}\tfrac{4}{5} + \cos^{-1}\tfrac{4}{5}\right) = \dfrac{4 + 3\sqrt{3}}{10}$

37. $\cos[\sin^{-1}(-x)] = \cos(\sin^{-1}x)$ for $0 \le x \le 1$

38. $\sin[\cos^{-1}(-x)] = \sin(\cos^{-1}x)$ for $0 \le x \le 1$

39. $\tan^{-1}x + \cot^{-1}x = \dfrac{\pi}{2}$,   for $x \ge 0$

40. $\cos^{-1}\left[\sin\left(x + \dfrac{\pi}{2}\right)\right] = x$ for $0 \le x \le \pi$

41. $\sin^{-1}x = \tan^{-1}\dfrac{x}{\sqrt{1-x^2}}$ for $|x| < 1$

42. $\sin(2\tan^{-1}x) = \dfrac{2x}{1+x^2}$

43. $\cot^{-1}x = \dfrac{\pi}{2} - \tan^{-1}x$

44. $\cot^{-1}x = \tan^{-1}\left(\dfrac{1}{x}\right)$ for $x > 0$

Solve the following equations for x in terms of y.

45. $y = 2\sin^{-1}(x+1)$

46. $y = -\cos^{-1}(2x-1)$

47. $y = \tfrac{1}{2}\tan^{-1}(1-x)$

48. $y = 3\cot^{-1}(x+3)$

49. $y = -4\csc^{-1}(2x+1)$

50. $y = \tfrac{1}{3}\sec^{-1}(2-x)$

51. $y = -\tfrac{2}{3}\cos^{-1}(x-1)$

52. $y = -4\sin^{-1}(2x-3)$

53. $y = -\sec^{-1}(2-3x)$

54. $y = \tfrac{1}{5}\tan^{-1}(4x+1)$

Write the following expressions in a form that does not involve inverse trigonometric functions.

55. $\sin\left(\tan^{-1}\dfrac{x}{2}\right)$

56. $\sin(\cos^{-1}x)$

57. $\tan(\cos^{-1}x)$

58. $\sin\left(\sec^{-1}\dfrac{x}{2}\right)$

59. $\csc\left(\tan^{-1}\dfrac{x}{\sqrt{2}}\right)$

60. $\cos(2\tan^{-1}x)$

61. $\cos\left(\tan^{-1}\dfrac{x}{3}\right)$

62. $\cot\left(\cos^{-1}\dfrac{3}{x}\right)$

63. $\sin(\sec^{-1}x)$

64. $\sec(\tan^{-1}x)$

65. $\sin(\sin^{-1}x - \cos^{-1}x)$

66. $\sin(\cos^{-1}x - \sin^{-1}x)$

67. $\cos(\sin^{-1}x + \cos^{-1}x)$

68. $\cos(\sin^{-1}x - \cos^{-1}x)$

Find the six trigonometric functions of y in terms of x for each of the following.

69. $y = \sin^{-1}(-x)$, $0 \le x \le 1$

70. $y = \sin^{-1}(2x)$, $0 \le x \le \tfrac{1}{2}$

71. $y = \cos^{-1}(2x)$, $0 \le x \le \tfrac{1}{2}$

72. $y = \cos^{-1}\left(\tfrac{1}{2}x\right)$, $0 \le x \le 2$

73. $y = \csc^{-1}(x)$, $x \ge 1$ or $x \le -1$

74. $y = \sec^{-1}(x)$, $x \ge 1$ or $x \le -1$

75. $y = \tan^{-1}\left(\tfrac{1}{2}x\right)$, $x \ne 0$

76. $y = \cot^{-1}(x)$, $x \ne 0$

Graph each of the following.

77. $y = \cot^{-1}x$

78. $y = \sin^{-1}x$

79. $y = \sec^{-1}x$

80. $y = \csc^{-1}x$

Find the exact values of $\sin(x+y)$, $\cos(x+y)$, $\sin 2x$, and $\cos 2x$ for the following. Do not use a calculator.

81. $x = \sin^{-1}\left(\tfrac{1}{2}\right); y = \sin^{-1}\left(\dfrac{\sqrt{3}}{2}\right)$

82. $x = \sin^{-1}(-1); y = \sin^{-1}\left(-\dfrac{1}{2}\right)$

83. $x = \sin^{-1}\left(\tfrac{3}{5}\right); y = \tan^{-1}(-\sqrt{3})$

84. $x = \tan^{-1}(2); y = \sin^{-1}\left(\dfrac{\sqrt{2}}{2}\right)$

85. $x = \cos^{-1}(-1); y = \cos^{-1}\left(-\frac{1}{2}\right)$

86. $x = \cos^{-1}\left(\frac{1}{2}\right); y = \cos^{-1}\left(\frac{\sqrt{3}}{2}\right)$

Show that the following are *not* identities.

87. $\sin^{-1}(x) = \dfrac{1}{\sin x}$

88. $\cos^{-1}(2x) = 2\cos^{-1}(x)$

89. $(\sin^{-1}x)^2 + (\cos^{-1}x)^2 = 1$

90. $1 + (\tan^{-1}x)^2 = (\sec^{-1}x)^2$

91. $\sin^{-1}(2x) = 2\sin^{-1}x$

92. $\tan^{-1}x = \dfrac{1}{\tan x}$

➡ Technology

Use a graphing calculator to display graphs of the following functions. From each graph, determine the domain and range of the function. (Your calculator may not have keys for all of the inverse functions required. Inverse functions not represented by a key will need to be computed in terms of other functions.)

93. $y = \sin^{-1}x$

94. $y = \cos^{-1}x$

95. $y = \tan^{-1}x$

96. $y = x\tan^{-1}x$

97. $y = \sin^{-1}x + \cos^{-1}x$

98. $y = \tan(x + \tan^{-1}x)$

6.6 SOLVING TRIGONOMETRIC EQUATIONS

Many problems reduce to solving equations involving trigonometric functions. In this section, we will discuss some techniques for solving many of the most common of these equations. The simplest equations involving trigonometric functions are those that assert that a trigonometric function equals a constant. Some examples of such equations are

$$\sin x = 0.5$$
$$\tan x = 0.573$$
$$\sec x = -2$$

The following example illustrates how to solve such equations.

➤ EXAMPLE 1
A Trigonometric Equation

Determine all solutions to the equation $\sin x = 0.5$.

Solution

By examining the unit circle, we see that there are two points that have y-coordinates equal to 0.5. Corresponding to these points, the equation has two solutions

$$x = \frac{\pi}{6}, \frac{5\pi}{6}$$

By periodicity, we may add any integer multiple of 2π to these solutions and obtain other solutions. There are two infinite families of solutions to the equation:

$$x = \frac{\pi}{6} + 2k\pi \qquad (k = 0, \pm1, \pm2, \ldots)$$
$$x = \frac{5\pi}{6} + 2k\pi \qquad (k = 0, \pm1, \pm2, \ldots)$$

➤ EXAMPLE 2
Another Trigonometric Equation

Determine all solutions of the equation

$$\sin x = 0.452 \qquad (0 \le x \le 2\pi)$$

Solution

There are two points on the unit circle that have y-coordinate 0.452 (or any other y-coordinate in $(-1, 1)$, for that matter), so the equation has two solutions. One solution is obtained by taking the arcsine of both sides of the equation:

$$x = \sin^{-1}(\sin x) = \sin^{-1}0.452 \approx 0.46901$$

The second solution is $\pi - x \approx 2.67258$, because of the identity

$$\sin x = \sin(\pi - x)$$

Problems such as this are easily solved using the trace function on a graphing calculator. First, graph the two equations $y = \sin x$ and $y = 0.452$. Then, use the trace function to locate the x-coordinate of the point of intersection.

➢ **EXAMPLE 3**
Using Periodicity
in Solving an Equation

Determine all solutions of the following equation:

$$2\tan\left(3x - \frac{\pi}{4}\right) = 2, \qquad -\frac{\pi}{2} < x < \frac{\pi}{2}$$

Solution

Divide both sides of the equation by 2 to obtain

$$\tan\left(3x - \frac{\pi}{4}\right) = 1$$

An angle with tangent equal to 1 is $\pi/4$. Moreover, because the graph of the tangent function is increasing in the interval $(-\pi/2, \pi/2)$, we see that $\pi/4$ is the only angle in this interval for which the tangent has the value 1. Because the tangent function has period π, the following angles all have tangent 1:

$$\frac{\pi}{4} + n\pi, \qquad n = 0, \pm 1, \pm 2, \ldots$$

Therefore, the solutions of the equation are given by

$$3x - \frac{\pi}{4} = \frac{\pi}{4} + n\pi$$

$$3x = \frac{\pi}{2} + n\pi$$

$$x = \frac{\pi}{6} + \frac{n\pi}{3}, \qquad n = 0, \pm 1 \pm 2, \ldots$$

The condition

$$-\frac{\pi}{2} < x < \frac{\pi}{2}$$

imposes the inequality

$$-\frac{\pi}{2} < \frac{\pi}{6} + \frac{n\pi}{3} < \frac{\pi}{2}$$

$$-\frac{\pi}{6} - \frac{\pi}{2} < \frac{n\pi}{3} < \frac{\pi}{2} - \frac{\pi}{6}$$

$$-\frac{2\pi}{3} < \frac{n\pi}{3} < \frac{\pi}{3}$$

$$-2 < n < 1$$

The only possibilities for n are thus $n = -1, 0$, and the only solutions of the equation are

$$x = \frac{\pi}{6}, -\frac{\pi}{6}$$

Many trigonometric equations may be solved by regarding them as algebraic equations in which the variable is a particular trigonometric function, such as $\sin x$ or $\cos x$. The next two examples provide illustrations of equations that may be regarded as quadratic equations with a trigonometric function as variable.

➢ **EXAMPLE 4**
A Quadratic Equation
in sin x

Determine all solutions to the equation
$$\sin^2 x - \sin x = 0$$

Solution

The given equation can be regarded as a quadratic equation with variable $\sin x$. Factoring the quadratic expression on the left yields the equation
$$\sin x (\sin x - 1) = 0$$
Equating each of the factors to 0 yields

$\sin x = 0$ $\qquad\qquad$ $\sin x = 1$

$x = n\pi, \quad n = 0, \pm 1, \pm 2, \ldots$ $\qquad$ $x = \dfrac{\pi}{2} \pm 2n\pi, \quad n = 0, 1, 2, \ldots$

$x = \ldots, -\pi, 0, \pi, \ldots$ $\qquad\qquad$ $x = \ldots, -\dfrac{3\pi}{2}, \dfrac{\pi}{2}, \dfrac{5\pi}{2}, \ldots$

These two infinite families of numbers are the solutions of the given equation.

➢ **EXAMPLE 5**
Another Quadratic Equation
in sin x

Determine the solutions of the equation
$$2 \sin^2 x + \sin x - 1 = 0, \qquad 0 \le x < 2\pi$$

Solution

The given equation is a quadratic in the variable $\sin x$. Factor the equation:
$$2 \sin^2 x + \sin x - 1 = 0$$
$$(2 \sin x - 1)(\sin x + 1) = 0$$
Equate the factors to 0:
$$2 \sin x - 1 = 0 \qquad \sin x + 1 = 0$$
$$\sin x = \tfrac{1}{2} \qquad\quad \sin x = -1$$

There are an infinite number of values of x for which each of these equations holds. However, the only ones lying within the interval $[0, 2\pi)$ are $x = \pi/6, 5\pi/6$ (solutions of the first equation) and $x = 3\pi/2$ (solution of the second equation).

In solving trigonometric equations, it is often useful to simplify the equation using one of the various identities we have proved. The next two examples provide illustrations of some typical equations that may be solved in this way.

➢ **EXAMPLE 6**
Equation Involving Two
Trigonometric Functions

Find the solutions of the equation
$$\sin x \cos x = \frac{\sqrt{3}}{4}, \qquad 0 \le x < 2\pi$$

Solution

Let's apply the double-angle formula for the sine to rewrite the left side of the equation. To do so, multiply both sides by 2 to obtain
$$2 \sin x \cos x = \frac{\sqrt{3}}{2}$$
$$\sin 2x = \frac{\sqrt{3}}{2} \qquad (0 \le 2x < 4\pi)$$

$$2x = \frac{\pi}{3}, \frac{2\pi}{3}, \frac{7\pi}{3}, \frac{8\pi}{3}$$

$$x = \frac{\pi}{6}, \frac{\pi}{3}, \frac{7\pi}{6}, \frac{4\pi}{3}$$

➤ **EXAMPLE 7**

Equation Involving Trigonometric Functions of Several Angles

Solve the equation

$$\cos 2x - 2\cos^2 x = \cos x, \qquad 0 \le x < 2\pi$$

Solution

This equation involves one trigonometric function, cos, but it is evaluated at two different values. As a first step in solving the equation, let's use the double-angle formula for the cosine to get an equivalent form of the equation in which all the trigonometric functions are evaluated at the same value. We use the alternative form of the double-angle formula: $\cos 2x = 2\cos^2 x - 1$. This yields

$$(2\cos^2 x - 1) - 2\cos^2 x = \cos x$$
$$\cos x = -1$$
$$x = \pi$$

Exercises 6.6

Determine all solutions of the following equations.

1. $\sin x = 0$
2. $\cos x = 0$
3. $\tan x = 0$
4. $\cot x = 0$
5. $\sec x = 0$
6. $\csc x = 0$
7. $\sin x = -1$
8. $\cos x = -1$
9. $\tan x = 1$
10. $\cot x = -1$
11. $\sec x = 2$
12. $\csc x = -\dfrac{2\sqrt{3}}{3}$
13. $\tan x = -\dfrac{\sqrt{3}}{3}$
14. $\sin x = \dfrac{\sqrt{3}}{2}$
15. $\cos x = -\frac{1}{2}$
16. $\sec x = -2$
17. $\tan x = 1.621$
18. $\cos x = 0.9224$
19. $\cot x = -7.770$
20. $\sin x = 0.4318$
21. $\sin x = -3.2$
22. $\tan x = 4.449$
23. $2\cot\left(3x - \dfrac{\pi}{4}\right) = 2, \qquad -\dfrac{\pi}{2} < x < \dfrac{\pi}{2}$
24. $2\cot\left(3x - \dfrac{\pi}{4}\right) = -2, \qquad -\dfrac{\pi}{2} < x < \dfrac{\pi}{2}$
25. $2\sin\left(2x + \dfrac{\pi}{3}\right) = 1, \qquad -\dfrac{\pi}{2} < x < \dfrac{\pi}{2}$
26. $-2\cos(2x + \pi) = 1, \qquad -\dfrac{\pi}{2} < x < \dfrac{\pi}{2}$
27. $-\frac{1}{2}\sin\left(x - \frac{\pi}{4}\right) = \frac{\sqrt{3}}{4}, \qquad -\dfrac{\pi}{2} < x < \dfrac{\pi}{2}$
28. $-\sqrt{3}\tan\left(2x + \dfrac{\pi}{4}\right) = 1, \qquad -\dfrac{\pi}{2} < x < \dfrac{\pi}{2}$
29. $2\cos\left(\dfrac{\pi}{4} - x\right) = -\sqrt{2}, \qquad -\dfrac{\pi}{2} < x < \dfrac{\pi}{2}$
30. $3\sin\left(\dfrac{\pi}{3} - x\right) = 3, \qquad -\dfrac{\pi}{2} < x < \dfrac{\pi}{2}$

Determine all solutions to the following over the interval $0 \le x < 2\pi$, without a calculator or table.

31. $2\cos(x - \pi) = -\sqrt{3}$
32. $-\sin\left(x + \dfrac{\pi}{4}\right) = \frac{1}{2}$
33. $\tan^2 x - \tan x = 0$
34. $2\sin^2 x + 3\sin x + 1 = 0$
35. $\cos^3 x = \cos x$
36. $\tan^2 x - \sqrt{3}\tan x = 0$
37. $2\sec^2 x = 2 - 3\sec x$
38. $2\cos^2 x - \sqrt{3}\cot x = 0$
39. $2\cos^2 x - 3\cos x = -1$
40. $1 - 3\tan^2 x = 0$
41. $\csc^2 x + \csc x - 2 = 0$
42. $\sec^2 x + 2 = 3\sec x$
43. $\cos^2 x + \sin x + 1 = 0$
44. $\sqrt{3}\cot x + 1 = \csc^2 x$
45. $\sin^2 3x = 0$
46. $\cos^2 \dfrac{x}{2} = \frac{1}{4}$
47. $\sec x \csc x = 2\csc x$
48. $2\sin x + \csc x = 0$
49. $4\sin x \cos x = 1$
50. $\sin 2x \sin x - \cos x = 0$

Determine all solutions to the following over the interval $0 \le x < 2\pi$, with the use of a calculator or trigonometric table where necessary.

51. $\sec x + \tan x = 1$
52. $\cot x + 1 = \csc x$
53. $2\tan^2 x + 7\tan x = 15$
54. $4\cot^2 x - 4\cot x = 3$
55. $\sec^2 x = 6\sec x + 2$
56. $\sin^2 x = 20 - \sin x$
57. $\cos 3x = \cos x$
58. $\sin^2 3x - \sin^2 x = 0$

59. $3 \sin x = \sqrt{3} \cos x$

60. $3 \sin^2 x - \cos^2 x = 1$

61. $\cot x + \tan x = \sec x \csc x$

62. $\cos x = \sin 2x$

63. $\cos 2x + 1 = \cos x$

64. $\cos 2x = 2 \sin^2 x - 2$

65. $\sin 3x + \sin x = 0$

66. $\cos 3x \cos x - \sin x \sin 3x = \frac{1}{2}$

67. $2 \cot x \sin x - \cot x = 0$

68. $\cos x + 2 \cos^2 \frac{x}{2} = 2$

69. $\cos 3x = 1 - \sin 3x$

70. $\sec x + \tan x = 1$

71. $\tan\left(x + \frac{\pi}{4}\right) - \tan\left(x - \frac{\pi}{4}\right) = 4$

72. $\tan\left(x + \frac{\pi}{4}\right) + \tan\left(x - \frac{\pi}{4}\right) = 2$

73. $2 \sin x \tan x + \tan x - 2 \sin x - 1 = 0$

74. $2 \tan x \csc x + 2 \csc x + \tan x + 1 = 0$

75. $\sec^2 x = 5 - 3 \cot^2 x$

76. $\sin x = 3 \cos x$

Prove that each of the following equations is an identity.

77. $\sin\left(\frac{\pi}{4} + x\right) - \sin\left(\frac{\pi}{4} - x\right) = \sqrt{2} \sin x$

78. $(1 - 2 \sin^2 x) \sec 2x = 1$

79. $\sec x - \cos x - \sin x \tan x = 0$

80. $\dfrac{\sin x}{\sin x \cot x + \cos x} = \dfrac{\tan x}{2}$

81. $2 \tan x \csc 2x - \tan^2 x = 1$

82. $\sin^2 x = 1 - \cos^2 x$

Express each expression in terms of the specified function values.

83. $\sin(x + y - z)$; in terms of sines and cosines of x, y, and z.

84. $\tan(x - y - z)$; in terms of tangents of x, y, and z.

85. $\cos 4x$; in terms of $\cos x$.

86. $\sin 3x$; in terms of $\sin x$.

87. Suppose $\sin x = 7 \cos x$. Find $\sin x \cos x$.

➡ **Technology**

Use a computer software graphing package or a graphing calculator to solve each of the following equations over the interval $0 \le x \le 2\pi$.

88. $\sin x + \ln x = \dfrac{x}{7}$

89. $x + \cos x = 0.48$

90. $e^{-x/2} - \sin x = 0$

91. $\sin x + \cos x = \tan x$

6.7 CHAPTER REVIEW

Important Concepts, Properties, and Formulas—Chapter 6

Pythagorean identities	$\sin^2 x + \cos^2 x = 1$ $\tan^2 x + 1 = \sec^2 x$ $\cot^2 x + 1 = \csc^2 x$	p. 333
Addition and subtraction identities	$\sin(x + y) = \sin x \cos y + \cos x \sin y$ $\sin(x - y) = \sin x \cos y - \sin y \cos x$ $\cos(x - y) = \cos x \cos y + \sin x \sin y$ $\cos(x + y) = \cos x \cos y - \sin x \sin y$ $\tan(x + y) = \dfrac{\tan x + \tan y}{1 - \tan x \tan y}$ $\tan(x - y) = \dfrac{\tan x - \tan y}{1 + \tan x \tan y}$	pp. 341, 344, 347
Cofunction identities	$\cos\left(\dfrac{\pi}{2} - x\right) = \sin x$ $\sin\left(\dfrac{\pi}{2} - x\right) = \cos x$ $\tan\left(\dfrac{\pi}{2} - x\right) = \cot x$	p. 343
Angle between two lines	Tangent of the smallest angle formed by intersecting lines: $\tan \theta = \dfrac{m_2 - m_1}{1 + m_1 m_2}$	p. 349

Double-angle identities	$\sin 2x = 2 \sin x \cos x$ $\cos 2x = \cos^2 x - \sin^2 x$ $\tan 2x = \dfrac{2 \tan x}{1 - \tan^2 x}$ $\cos 2x = 1 - 2 \sin^2 x$ $\cos 2x = 2 \cos^2 x - 1$	p. 353
Tangent square identities	$\cos^2 x = \dfrac{1 + \cos 2x}{2}$ $\sin^2 x = \dfrac{1 - \cos 2x}{2}$ $\tan^2 x = \dfrac{1 - \cos 2x}{1 + \cos 2x}$	p. 356
Half-angle identities	$\sin \dfrac{x}{2} = \pm \sqrt{\dfrac{1 - \cos x}{2}}$ $\cos \dfrac{x}{2} = \pm \sqrt{\dfrac{1 + \cos x}{2}}$ $\tan \dfrac{x}{2} = \dfrac{1 - \cos x}{\sin x} = \dfrac{\sin x}{1 + \cos x}$	p. 357
Product identities	$\sin x \cos y = \frac{1}{2} [\sin(x + y) + \sin(x - y)]$ $\sin x \sin y = \frac{1}{2} [\cos(x - y) - \cos(x + y)]$ $\cos x \cos y = \frac{1}{2} [\cos(x + y) + \cos(x - y)]$ $\cos x \sin y = \frac{1}{2} [\sin(x + y) - \sin(x - y)]$	p. 363
Factoring identities	$\sin x + \sin y = 2 \sin \left[\dfrac{x + y}{2} \right] \cos \left[\dfrac{x - y}{2} \right]$ $\sin x - \sin y = 2 \sin \left(\dfrac{x - y}{2} \right) \cos \left(\dfrac{x + y}{2} \right)$ $\cos x + \cos y = 2 \cos \left(\dfrac{x + y}{2} \right) \cos \left(\dfrac{x - y}{2} \right)$ $\cos x - \cos y = -2 \sin \left(\dfrac{x + y}{2} \right) \sin \left(\dfrac{x - y}{2} \right)$	p. 365

Inverse trigonometric functions properties, pp. 367, 368

Function	Range	Domain
$y = \sin^{-1} x$	$-\dfrac{\pi}{2} \le y \le \dfrac{\pi}{2}$	$-1 \le x \le 1$
$y = \cos^{-1} x$	$0 \le y \le \pi$	$-1 \le x \le 1$
$y = \tan^{-1} x$	$-\dfrac{\pi}{2} < y < \dfrac{\pi}{2}$	$-\infty < x < \infty$
$y = \sec^{-1} x$	$0 \le y < \dfrac{\pi}{2}$ or $\pi \le y < \dfrac{3\pi}{2}$	$x \ge 1$ or $x \le -1$
$y = \csc^{-1} x$	$0 < y \le \dfrac{\pi}{2}$ or $\pi < y \le \dfrac{3\pi}{2}$	$x \ge 1$ or $x \le -1$
$y = \cot^{-1} x$	$0 < y < \pi$	$-\infty < x < \infty$

Cumulative Review Exercises—Chapter 6

Write the following in terms of a single trigonometric function.

1. $\cos x \csc x$

2. $\sec x \csc x - \tan x$

3. $\dfrac{\cot x + \tan x}{\csc x}$

4. $\sin x \sec x$

5. $\dfrac{\sec x}{\cot x + \tan x}$

6. $\dfrac{1 + \csc x}{\sec x} - \cot x$

Given the quadrant for angle x and one trigonometric function value, find the other five trigonometric function values.

7. x in fourth quadrant; $\sin x = -\frac{1}{2}$

8. x in second quadrant; $\tan x = -\sqrt{3}$

9. x in first quadrant; $\sec x = \frac{5}{3}$

10. x in third quadrant; $\cos x = -\frac{12}{13}$

Make the given substitution for t, with $0 < x < \pi/2$, and write the given expression in terms of x.

11. $t = 3 \sin x$; for $\sqrt{9 - t^2}$

12. $t = \sec x$; for $\sqrt{t^2 - 1}$

Prove each of the following identities.

13. $\dfrac{\tan^2 x + 1}{\cot^2 x + 1} = \tan^2 x$

14. $1 + \sin x = \dfrac{\cos x}{\sec x - \tan x}$

15. $\dfrac{1}{\sec x + \tan x} = \dfrac{1 - \sin x}{\cos x}$

16. $\sec^4 x - \tan^4 x = \dfrac{1 + \sin^2 x}{\cos^2 x}$

Determine the value of each of the following without using a calculator or trigonometric tables.

17. $\tan \dfrac{5\pi}{12}$

18. $\sin \left(-\dfrac{7\pi}{12}\right)$

19. $\cos(-15°)$

20. $\tan 255°$

Find an expression for each of the following involving only trigonometric functions evaluated at x.

21. $\sin \left(\dfrac{\pi}{3} + x\right)$

22. $\tan \left(x + \dfrac{\pi}{4}\right)$

23. $\tan(135° - x)$

24. $\cos(45° - x)$

Determine the value of the following expressions without using a calculator or trigonometric tables.

25. $\cos \left(x + \dfrac{\pi}{6}\right)$ if
$$\cos x = -\tfrac{1}{2}, \qquad \pi < x < \dfrac{3\pi}{2}$$

26. $\sin \left(\dfrac{\pi}{3} - x\right)$ if
$$\sin x = -\tfrac{3}{5}, \qquad \dfrac{3\pi}{2} < x < 2\pi$$

27. $\tan(45° - x)$ if
$$\sin x = -\tfrac{12}{13}, \qquad \pi < x < \dfrac{3\pi}{2}$$

28. $\cos(60° + x)$ if
$$\sec x = 2, \qquad 0 < x < \dfrac{\pi}{2}$$

29. $\cos(x + y)$ if
$$\sin x = \tfrac{3}{5}, \qquad 0 < x < \dfrac{\pi}{2}$$
and
$$\sin y = -\tfrac{4}{5}, \qquad \pi < y < \dfrac{3\pi}{2}$$

30. $\sin(x - y)$ if
$$\sin x = \tfrac{12}{13}, \qquad \dfrac{\pi}{2} < x < \pi$$
and
$$\cos y = -\tfrac{5}{13}, \qquad \pi < y < \dfrac{3\pi}{2}$$

31. $\tan(x + y)$ if
$$\tan x = 1, \qquad \pi < x < \dfrac{3\pi}{2}$$
and
$$\tan y = -\sqrt{3}, \qquad \dfrac{\pi}{2} < y < \pi$$

32. $\tan(y - x)$ if
$$\cos y = \tfrac{3}{5}, \qquad \dfrac{3\pi}{2} < y < 2\pi$$
and
$$\sin x = \tfrac{4}{5}, \qquad 0 < x < \dfrac{\pi}{2}$$

Find the value of the smallest positive angle between the two intersecting lines.

33. $x + 2y = 6$ and $2x - 3y = 4$

34. $y = \frac{1}{5}x + 2$ and $y = \frac{2}{3}x - 5$

35. If $f(x) = \cos x$, prove that
$$\dfrac{f(x + y) - f(x)}{y} = \cos x \left(\dfrac{\cos y - 1}{y}\right) - \sin x \left(\dfrac{\sin y}{y}\right)$$

Without a calculator or trigonometric tables, find the value of the following.

36. $\sin 25° \cos 65° + \cos 25° \sin 65°$

37. $\cos 125° \cos 55° - \sin 125° \sin 55°$

38. $\dfrac{\tan 15° + \tan 30°}{1 - \tan 15° \tan 30°}$

39. $\dfrac{\tan 240° - \tan 15°}{1 + \tan 240° \tan 15°}$

Prove the following identities.

40. $\cos(\pi + x)\cos(\pi - x) + \sin(\pi + x)\sin(\pi - x) = \cos 2x$

41. $\cot\left(\dfrac{\pi}{4} + x\right) = \dfrac{1 - \tan x}{\tan x + 1}$

42. $\cot(x + y) = \dfrac{\cot x \cot y - 1}{\cot x + \cot y}$

Given the quadrant where angle x is located and the trigonometric function value of x find: (a) $\sin 2x$, (b) $\cos 2x$, (c) $\tan 2x$, (d) $\sin x/2$, (e) $\cos x/2$ and (f) $\tan x/2$.

43. $0 < x < \dfrac{\pi}{2}$, $\cos x = \dfrac{3}{5}$

44. $\dfrac{\pi}{2} < x < \pi$, $\sin x = \dfrac{\sqrt{3}}{2}$

45. $\pi < x < \dfrac{3\pi}{2}$, $\tan x = \sqrt{3}$

46. $\dfrac{3\pi}{2} < x < 2\pi$, $\cot x = -\frac{3}{4}$

Express the following in terms of the indicated trigonometric function or functions.

47. $\sin 3x$ in terms of $\sin x$.

48. $\tan 14x$ in terms of $\tan 7x$.

49. $\cos 10x$ in terms of $\sin 5x$.

50. $\cos 3x$ in terms of $\cos x$.

Prove each of the following identities.

51. $\dfrac{1 - \cos 2x}{\sin 2x} = \tan x$

52. $\sec 2x = \dfrac{1}{1 - 2\sin^2 x}$

53. $\tan 3x = \dfrac{3\tan x - \tan^3 x}{1 - 3\tan^2 x}$

54. $1 + \cos x = 2\cos^2 \dfrac{x}{2}$

55. $\tan \dfrac{x}{2} = \dfrac{1 - \cos x}{\sin x}$

56. $\sec \dfrac{x}{2} = \pm\dfrac{\sqrt{2 + 2\cos x}}{1 + \cos x}$

Find the indicated trigonometric function value without using a calculator or tables.

57. Find $\cos x$ if
$$\cos 2x = \tfrac{1}{2}, \qquad \pi < x < \dfrac{3\pi}{2}$$

58. Find $\tan x$ if
$$\cos 2x = -\tfrac{2}{9}, \qquad \dfrac{\pi}{2} < x < \pi$$

59. Find $\cot x$ if
$$\cos 2x = -\dfrac{1}{8}, \qquad \dfrac{\pi}{2} < x < \pi$$

60. Find $\sin x$ if
$$\cos 2x = -\tfrac{4}{25}, \qquad 0 < x < \dfrac{\pi}{2}$$

Without the use of a calculator or a trigonometric table, find the value of the following.

61. $\sin 67.5°$

62. $\tan 22.5°$

63. $\cos -\dfrac{3\pi}{8}$

64. $\sec -\dfrac{\pi}{8}$

Write as a single trigonometric function value, or give the value.

65. $2\sin 15° \cos 15°$

66. $2\cos^2 22.5 - 1$

67. $\cos^2 \dfrac{\pi}{5} - \sin^2 \dfrac{\pi}{5}$

68. $\sqrt{\dfrac{1 + \cos \dfrac{\pi}{4}}{2}}$

69. $\dfrac{1 - \cos 35°}{\sin 35°}$

Simplify each expression.

70. $\dfrac{\sin 2x}{2\sin x}$

71. $\cos^2 x - \cos 2x$

72. $\sin 2x \cos x + \cos 2x \sin x$

73. $(\sin x + \cos x)^2 - \sin 2x$

Write as a sum or difference of trigonometric functions.

74. $2\sin x \sin 9x$

75. $2\cos 3x \cos 5x$

76. $2\sin 3x \cos 5x$

77. $2\sin 3x \cos x$

Express as a product.

78. $\sin 60° - \sin 20°$

79. $\sin 7y + \sin 9y$

80. $\cos 3x - \cos x$

81. $\cos 40° - \cos 80°$

Prove each of the following identities.

82. $\dfrac{\cos 4x + \cos 2x}{\sin 4x - \sin 2x} = \cot x$

83. $\dfrac{\cos x + \cos 3x}{\sin 3x - \sin x} = \cot x$

84. $\dfrac{\cos 2x - \cos 4x}{2\sin 3x} = \sin x$

85. $\dfrac{\sin x + \sin y}{\cos x + \cos y} = \tan\left(\dfrac{x + y}{2}\right)$

Calculate the value of the following.

86. $\cos^{-1}\left(\dfrac{\sqrt{3}}{2}\right)$

87. $\tan^{-1}(-\sqrt{3})$

88. $\sin(\csc^{-1} 2)$

89. $\cos\left(\tan^{-1} \tfrac{3}{5}\right)$

90. $\tan^{-1}(\sin \pi)$

91. $\sin^{-1}\left(\cot -\dfrac{\pi}{4}\right)$

Prove each of the following identities.

92. $\sin^{-1} x + \cos^{-1} x = \dfrac{\pi}{2}, \qquad 0 \le x \le 1$

93. $\tan^{-1} x + \cot^{-1} x = \dfrac{\pi}{2}, \qquad x \ge 0$

Solve for x in terms of y.

94. $y = 3\sin^{-1}(x + 1)$

95. $y = \tfrac{1}{2}\tan^{-1}(4x - 3)$

96. $y = -\tfrac{3}{2}\cos^{-1}(1 - x)$

97. $y = -2\csc^{-1}(2x + 1)$

Write in a form that does not involve inverse trigonometric functions.

98. $\sin\left(\tan^{-1}\dfrac{x}{3}\right)$

99. $\cos\left(\tan^{-1}\dfrac{2}{x}\right)$

100. $\cos\left(\sin^{-1}\dfrac{x}{5}\right)$

101. $\tan\left(\cos^{-1}x\right)$

Find the six trigonometric functions of y in terms of x.

102. $y = \sin^{-1}(x)$, $0 \le x \le 1$

103. $y = \cos^{-1}\left(\dfrac{x}{2}\right)$, $0 \le x \le 2$

104. $y = \csc^{-1}(-x)$, $x \ge 1$ or $x \le -1$

105. $y = \cot^{-1}(-x)$, $x \ne 0$

Find $\sin(x + y)$, $\cos(x + y)$, $\sin 2x$, $\cos 2x$ and $\tan 2x$. Do not use a calculator or tables.

106. $x = \sin^{-1}\left(\dfrac{\sqrt{3}}{2}\right)$, $y = \cos^{-1}\left(\dfrac{1}{2}\right)$

107. $x = \tan^{-1}\left(-\dfrac{3}{5}\right)$, $y = \tan^{-1}\dfrac{4}{3}$

Determine all solutions to the following.

108. $2\sin x = 1$

109. $-2\cos x = \sqrt{3}$

110. $\tan x = -1.621$

111. $\cot x = 7.770$

112. $3\cot\left(3x + \dfrac{\pi}{4}\right) = -3$, $-\dfrac{\pi}{2} < x < \dfrac{\pi}{2}$

113. $\sqrt{3}\tan\left(2x + \dfrac{\pi}{4}\right) = -1$, $-\dfrac{\pi}{2} < x < \dfrac{\pi}{2}$

Determine all solutions over the interval $0 \le x < 2\pi$.

114. $4\sin^2 x - 3 = 0$

115. $\sin^2 x = 3\cos^2 x$

116. $2\sin x + \csc x = 0$

117. $(\cos 2x + \sin 2x)^2 = 1$

118. $2\sin^2 x + 7\sin x - 15 = 0$

119. $4\cos^2 x = 4\cos x - 1$

120. $20\cos^2 x - \cos x = 1$

121. $\sec^2 x + \sec x + 1 = 0$

122. If $f(x) = \csc x$, prove that

$$\dfrac{f(x + h) - f(x)}{h}$$

$$= \dfrac{1}{h(\sin x \cos h + \cos x \sin h)} - \dfrac{1}{h \sin x}$$

CHAPTER 7

APPLICATIONS
OF TRIGONOMETRY

Trigonometry is of central importance in describing the motion of the heavenly bodies as seen from Earth.

T rigonometry has many applications in other fields as well as in mathematics itself. In this chapter, we will discuss a number of these applications.

7.1 THE LAW OF SINES

We have discussed the solution of right triangles using trigonometry (essentially the definitions of the trigonometric functions). It is often necessary to solve triangles that don't contain a right angle. Such triangles are called **oblique.** The following result is useful in solving oblique triangles.

> ### LAW OF SINES
>
> Suppose that a triangle has sides of length a, b, and c. Suppose that the angle opposite a is α, the angle opposite b is β, and the angle opposite c is γ. (See Figure 1.) Then the following equations hold:
>
> $$\frac{\sin \alpha}{a} = \frac{\sin \beta}{b} = \frac{\sin \gamma}{c}$$

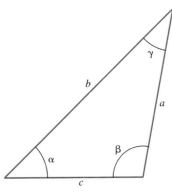

Figure 1

Proof   Let's prove the first equality. The proof of the second is similar and will be left for the exercises. Drop a perpendicular from the vertex opposite side c, as shown in Figure 2. Denote by d the length of the perpendicular. Then by the geometric definition of the sine, we have

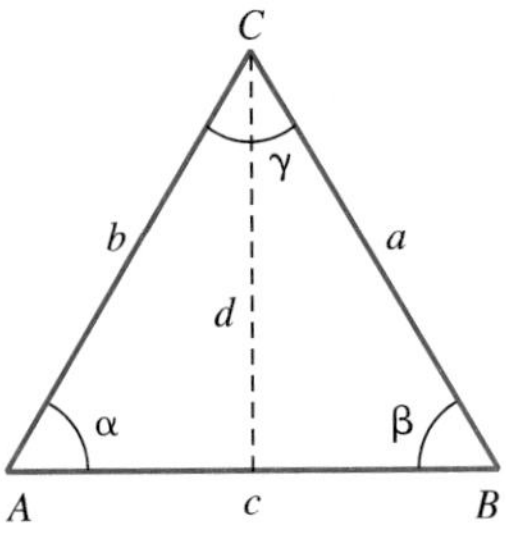

$$\sin \alpha = \frac{d}{b} \quad \text{so that} \quad d = b \sin \alpha$$

$$\sin \beta = \frac{d}{a} \quad \text{so that} \quad d = a \sin \beta$$

Equating the two expressions for d, we have

$$b \sin \alpha = a \sin \beta$$

This proves the first equality of the law of sines.   ♦

Figure 2

One application of the law of sines is to determine the sides and angles of a triangle given two angles and a side opposite one of them, as illustrated in the following example.

➤ **EXAMPLE 1**
Solving a Triangle

Suppose a triangle has its sides and angles labeled as in Figure 1, and that $a = 112$, $\alpha = 31°45'$, and $\beta = 72°37'$. Determine the remaining sides and angles.

Solution

The third angle is easily deduced from the geometric fact that the angles of a triangle add up to 180°:

$$\gamma = 180° - 31°45' - 72°37' = 75°38'$$

From the law of sines, we have

$$\frac{\sin \alpha}{a} = \frac{\sin \beta}{b}$$

$$\frac{\sin 31°45'}{112} = \frac{\sin 72°37'}{b}$$

$$b = \frac{112 \sin 72°37'}{\sin 31°45'} \approx 203.12$$

We may calculate the length of the third side c by applying the law of sines again:

$$\frac{\sin \alpha}{a} = \frac{\sin \gamma}{c}$$

$$c = \frac{a \sin \gamma}{\sin \alpha}$$

$$= \frac{112 \sin 75°38'}{\sin 31°45'}$$

$$\approx 206.19$$

The triangle with all its parts labeled is shown in Figure 3.

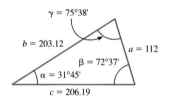

Figure 3

In the preceding example, we solved a triangle given two angles and a side opposite one of them. Another possibility is to be given sides a and b, and an angle opposite one of them, α. In such examples, we use the law of sines to solve the equation

$$\frac{\sin \alpha}{a} = \frac{\sin \beta}{b}$$

$$\sin \beta = \frac{b \sin \alpha}{a}$$

for β. There may be no, one, or two solutions. (See Figure 4.)

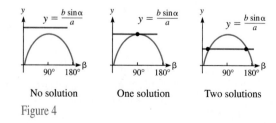

No solution   One solution   Two solutions

Figure 4

The various cases for α acute may be distinguished geometrically as follows: Let d be the length of the perpendicular to side c. As we have seen, $d = b \sin \alpha$.

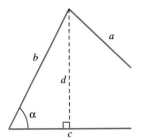

If $d > a$, there is no solution with the given data.

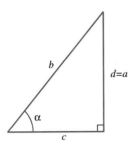

If $d = a$, there is one solution with the given data.

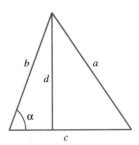

If $a > b$, there is one solution with the given data.

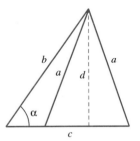

If $d < a < b$, there are two solutions with the given data.

For α obtuse, there are two cases:

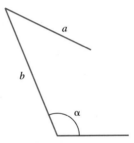

If $a \le b$, there is no solution with the given data.

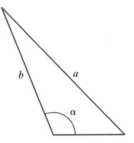

If $a > b$, there is one solution with the given data.

In case there are two distinct triangles with the given data, we say that the case is *ambiguous*. The following examples show how to recognize the three cases as they occur in practice.

➢ **EXAMPLE 2**
Data with No Solution

Determine all triangles for which $a = 50$, $b = 150$, and $\alpha = 60°$.

Solution

Applying the law of sines yields

$$\frac{\sin \alpha}{a} = \frac{\sin \beta}{b}$$

$$\frac{\sin 60°}{50} = \frac{\sin \beta}{150}$$

$$\sin \beta = \frac{150 \sin 60°}{50}$$

$$= \frac{3\sqrt{3}}{2} > 1$$

Because the sine of an angle must lie in the interval $[-1, 1]$, the above equation has no solution β. Thus, there are no triangles satisfying the given data.

➤ EXAMPLE 3
Data with Two Solutions

Determine all triangles for which $b = 520$, $c = 952$, and $\beta = 13°$.

Solution

By the law of sines, we have

$$\frac{\sin\beta}{b} = \frac{\sin\gamma}{c}$$

$$\frac{\sin 13°}{520} = \frac{\sin\gamma}{952}$$

$$\sin\gamma = \frac{952\sin 13°}{520} \approx 0.41183$$

Solving the last equation using the techniques of Section 6.5, we determine that there are two solutions, namely, $\gamma \approx 24.32°$, $155.68°$.

Let's consider the first solution. In this case, the value of α is given by

$$\alpha = 180° - \beta - \gamma$$

$$\approx 180° - 13° - 24.32°$$

$$= 142.68°$$

The value of a may then be determined from the law of sines:

$$\frac{\sin\alpha}{a} = \frac{\sin\beta}{b}$$

$$a = \frac{b\sin\alpha}{\sin\beta}$$

$$= \frac{520 \cdot \sin 142.68°}{\sin 13°}$$

$$\approx 1401.45$$

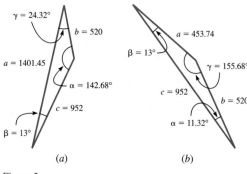

Figure 5

The triangle corresponding to this set of data is shown in Figure 5(a). In a similar fashion, we can determine the values of α and a corresponding to the second value of γ:

$$\alpha = 180° - \beta - \gamma$$

$$\approx 180° - 13° - 155.68°$$

$$= 11.32°$$

$$\frac{\sin \alpha}{a} = \frac{\sin \beta}{b}$$

$$a = \frac{b \sin \alpha}{\sin \beta}$$

$$\approx \frac{520 \cdot \sin 11.32°}{\sin 13°}$$

$$= 453.74$$

The triangle corresponding to the second set of data is shown in Figure 5(*b*).

➢ **EXAMPLE 4**
Data with One Solution

Determine all triangles for which $a = 10$, $c = 20$, and $\alpha = 30°$.

Solution

By the law of sines, we have

$$\frac{\sin \alpha}{a} = \frac{\sin \gamma}{c}$$

$$\sin \gamma = \frac{20 \sin 30°}{10} = 1$$

$$\gamma = 90°$$

In this case, there is a single value for γ, and the resulting triangle is a right triangle. The third angle is given by

$$\beta = 180° - \alpha - \gamma$$

$$= 180° - 30° - 90°$$

$$= 60°$$

To determine the third side, we may apply the Pythagorean theorem:

$$a^2 + b^2 = c^2$$

$$10^2 + b^2 = 20^2$$

$$b^2 = 300$$

$$b = 10 \sqrt{3}$$

The triangle corresponding to the given data is shown in Figure 6.

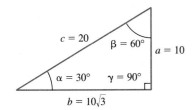

$c = 20$ $\beta = 60°$ $a = 10$ $\alpha = 30°$ $\gamma = 90°$ $b = 10\sqrt{3}$

Figure 6

➢ **EXAMPLE 5**
Navigation

A ship receives notification of a storm to its east. To avoid the storm, it first proceeds at a heading 48° north of east for 160 nautical miles. It then changes heading to 54° south of east (see Figure 7).

1. How far does the ship go until it intersects its original course?
2. How far out of its way did the ship go to avoid the storm?

Solution

1. Let angle α and side a be as shown in Figure 7. From geometry, $\alpha = 54°$ (alternate interior angles). Therefore, by the law of sines, we have

$$\frac{\sin 48°}{c} = \frac{\sin 54°}{160}$$

$$c = \frac{160 \sin 48°}{\sin 54°}$$

$$\approx 146.97 \text{ nautical miles}$$

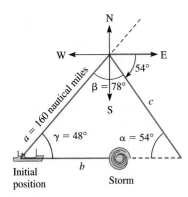

Figure 7

Therefore, the total distance the boat travels to avoid the storm is $146.97 + 160 = 306.97$ nautical miles.

2. If the ship did not avoid the storm, it would have traversed the third side of the triangle. By the law of sines, this side has length given by

$$\frac{\sin \beta}{b} = \frac{\sin \alpha}{a}$$

$$\frac{\sin 78°}{b} = \frac{\sin 54°}{160}$$

$$b \approx 193.45 \text{ nautical miles}$$

The distance the ship went out of its way equals $306.97 - 193.45 = 113.52$ nautical miles.

Exercises 7.1

Determine the remaining sides and angles.

1.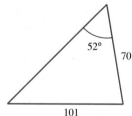

2.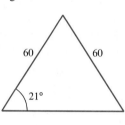

3.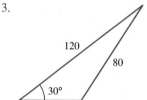

4.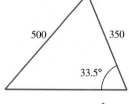

5.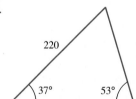

6.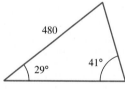

7. $\angle A = 115.5°, \angle C = 30°, a = 36$

8. $\angle A = 30°, \angle C = 52°, b = 20$

9. $b = 5, \alpha = 75°, \beta = 30°$

10. $c = 3, \beta = 37°, \gamma = 30°$

11. $a = 200, \alpha = 32°21', \gamma = 21°39'$

12. $a = 50, \alpha = 37°30', \beta = 71°10'$

13. $\angle A = 100°15', a = 48, b = 16$

14. $\beta = 2°45', b = 6.2, c = 5.8$

15. $\angle B = 15°30', a = 4.5, b = 6.8$

16. $\angle A = 145°, b = 4, a = 14$

Determine the number of solutions for the given triangle and find them.

17. $b = 125, c = 100, \beta = 110°$

18. $\alpha = 64°9', \beta = 13°, a = 12.3$

19. $b = 9, a = 25.6, \beta = 58°$

20. $c = 84.8, a = 50, \angle C = 58°$

21. $a = 10, b = 30, \alpha = 60°$

22. $b = 260, c = 476, \beta = 13°$

23. $a = 20, c = 40, \alpha = 30°$

24. $a = 15, b = 24, \beta = 36°$

♦ Applications

25. Two lighthouses at points X and Y are 40 kilometers apart. Each has a visual contact with a fishing boat at point Z. If $m\angle ZXY = 20°30'$ and $m\angle ZYX = 115°$, how far is the fishing boat from lighthouse Y?

26. Lookouts in two fire towers located 10 km apart sight a forest fire. Electronic equipment allows them to determine that the fire is at an angle of 71° from one tower and 100° from the other tower. Which tower is closer to the fire, and how far away is it?

27. A tree grows vertically on the side of a hill that slopes upward from the horizontal by 8°. When the angle of elevation of the sun from the side of the hill measures 20°, the shadow of the tree falls 42 m down the hill. How tall is the tree?

28. Two Coast Guard cutters on an east-west line and 4.6 km apart receive a distress call from a freighter that has a bearing of 47°40' from the cutter farther west and a

bearing of 302°30′ from the easternmost cutter. Which cutter has the shortest distance to travel to reach the freighter, and how far is it?

29. To find the distance across a swamp, *LM*, a distance *MN* meters is measured off on one side of the swamp. In the triangle *LMN*, the measure of angle *LMN* is 112°10′ and the measure of angle *LNM* is 15°20′. Find the distance across the swamp.

30. A tree stands at point *Z* across a river from point *X*. A base line *XY* is measured on one side of the river. The measure of *XY* is 80 meters. The measure of ∠*YXZ* is 54°20′, and that of ∠*ZYX* is 74°10′. The angle of elevation of the top of the tree from *X* measures 10°20′. How tall is the tree?

31. Two angles of a triangle are 55° and 30°, and the longest side measures 34 feet. Find the length of the shortest side.

32. From two points 400 meters apart on the bank of a river flowing due south, the bearings of a point on the opposite shore are 28°20′ and 148°20′, respectively. How wide is the river?

33. A railroad surveyor measures the angle from point *A* to point *B* as 63° west of south. He also measures the angle from point *A* to point *C* as 38° west of south. If the distance from *A* to *B* is 239 meters and the distance from *C* to *B* is 374 meters, find the distance from *C* to *A*.

34. An airplane was flying at 6500 ft altitude. When the plane was at point *A*, the angle of depression to a lighthouse at point *C* was 75°. When the plane reached point *B*, 10 seconds later, the angle of depression to the lighthouse was 50°30′. Find the ground speed of the plane in miles per hour.

35. A tree stands on a hill that has a 15° incline from the horizontal. The tree leans at a 20° angle down the hill from the vertical. The angle of elevation from the side of the hill to the sun over the top of the tree is 49° from the end of the tree's 50-foot shadow down the hill. How tall is the tree?

36. At 1 P.M., a train leaves Detroit at 30 miles per hour traveling due east, and a second train leaves Detroit at 40 miles per hour traveling south. At 3 P.M., the second train is 49.18° north of east from the first train. How far apart are the trains at 3 P.M.?

37. A battery commander *B* is ordered to shoot at a target *T* from a position *G*, from which *T* is visible. To check on the range *GT* as found by a range finder, *B* locates an observation point *H* from which *T* is visible. The bearing of *T* from *G* is 12°48′ east of north, the bearing of *T* from *H* is 6°23′ west of north, and the bearing of *H* from *G* is 82°53′ east of south. From a map, it is found that *GH* = 3250 meters. Find the range of the target from *G*.

38. A lighthouse stands at the top of a cliff. From a buoy at sea, out from the base of the cliff, the angle of elevation to the top of the lighthouse is 40° and the angle of elevation to the bottom of the lighthouse is 25°. From a ship 150 meters from the buoy, on line with the lighthouse, the angle of elevation to the top of the lighthouse is 32°. Find the height of the lighthouse.

39. Two fire towers are located 7 miles apart on a mountain ridge. If tower *A* is directly north of tower *B*, and a fire is spotted bearing 265° from tower *B* and 250° from tower *A*, how far is the fire from tower *B*?

40. A ship is sighted at point *C* from two observation posts, *A* and *B*, on shore. If points *A* and *B* are 24 kilometers apart, and if the measure of angle *CAB* is 41°40′ and the measure of angle *CBA* 36°10′, find the distance from observation post *A* to the ship.

41. By using the law of sines and the formula Area = $(\frac{1}{2})bh$ for finding the area of any triangle, develop the new formula Area = $(\frac{1}{2})bc \sin A$ for finding the area of any triangle. (*Hint*: Solve for *h* and use substitution.) (See figure.)

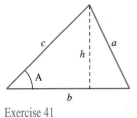

Exercise 41

Find the area for each of the following triangles.

42.

43.

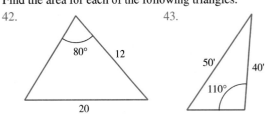

44. $m\angle A = 25°$, $a = 10$, $b = 25$

45. $m\angle C = 42°20'$, $m\angle A = 60°$, $c = 30$

46.

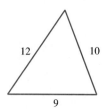

Prove that the following are true for any triangle ABC.

47. $\dfrac{a + b}{b} = \dfrac{\sin A + \sin B}{\sin B}$ 48. $\dfrac{a - b}{b} = \dfrac{\sin A - \sin B}{\sin B}$

49. $\dfrac{a - b}{a + b} = \dfrac{\sin A - \sin B}{\sin A + \sin B}$ 50. $\dfrac{a + b}{a - b} = \dfrac{\sin A + \sin B}{\sin A - \sin B}$

7.2 THE LAW OF COSINES

In this section, we discuss another set of identities useful in solving triangles, the **law of cosines**. In addition, we will derive several formulas for the area of a triangle that follow as consequences of the law of cosines.

LAW OF COSINES

In any triangle, the square of the length of any side equals the sum of the squares of the other two sides, minus twice the product of the lengths of the other two sides times the cosine of the angle between them. That is, suppose that a triangle has sides of length a, b, and c. Suppose that the angle opposite a is α, the angle opposite b is β, and the angle opposite c is γ. Then the following equations hold:

$$a^2 = b^2 + c^2 - 2bc\cos\alpha$$
$$b^2 = a^2 + c^2 - 2ac\cos\beta$$
$$c^2 = a^2 + b^2 - 2ab\cos\gamma$$

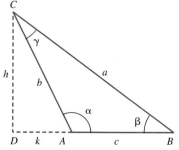

Figure 1

Proof   Let's prove the first formula. The proofs of the second and third are similar. Orient the triangle so that side c is horizontal. Draw a perpendicular to side c (or its extension), as in Figure 1. Let's assume that α is obtuse, as in the picture. At the end of the proof, we will indicate the modifications necessary if α is acute. Suppose that the length of $\overline{DA}$ is k and that the length of $\overline{DC}$ is h. From the definition of the sine, we have

$$\sin\alpha = \frac{h}{b}$$
$$h = b\sin\alpha$$

Because α is obtuse, $\cos\alpha$ is negative. Moreover, from the definition of the cosine, we have

$$\cos\alpha = -\frac{k}{b}$$
$$k = -b\cos\alpha$$

By the Pythagorean theorem applied to triangle BCD, we have

$$
\begin{aligned}
a^2 &= h^2 + (c + k)^2 \\
&= h^2 + c^2 + k^2 + 2ck \\
&= (b\sin\alpha)^2 + c^2 + (-b\cos\alpha)^2 + 2c(-b\cos\alpha) \\
&= b^2(\sin^2\alpha + \cos^2\alpha) + c^2 - 2bc\cos\alpha \\
&= b^2 + c^2 - 2bc\cos\alpha
\end{aligned}
$$

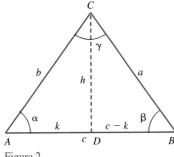

Figure 2

This proves the first equation of the law of cosines in case α is obtuse. In case α is acute, the reasoning is the same, except the first two equations in the argument read

$$\sin \alpha = \frac{h}{b}, \qquad \cos \alpha = \frac{k}{b}$$

$$h = b \sin \alpha, \qquad k = b \cos \alpha$$

(See Figure 2.) Moreover, in this case, the Pythagorean theorem yields

$$a^2 = h^2 + (c - k)^2$$

Working through the algebra gives the same result as in the first case.   ◆

The law of cosines may be used to solve a triangle given two sides and the angle between them. The following example illustrates how this may be done.

➤ **EXAMPLE 1**
Solving a Triangle Given
Two Sides and an Angle

Suppose that a triangle has sides $a = 300$, $b = 225$, and an angle $\gamma = 51°$. Determine the remaining parts of the triangle.

Solution

From the third equation in the law of cosines, we have

$$c^2 = a^2 + b^2 - 2ab \cos \gamma$$
$$= 300^2 + 225^2 - 2 \cdot 300 \cdot 225 \cos 51°$$
$$\approx 90,000 + 50,625 - 84,958$$
$$= 55,667$$
$$c \approx 235.94$$

We may determine the two unknown angles using the other two equations in the law of cosines. Applying the first equation, we have

$$a^2 = b^2 + c^2 - 2bc \cos \alpha$$
$$300^2 \approx 225^2 + (235.94)^2 - 2(225)(235.94) \cos \alpha$$
$$\cos \alpha \approx \frac{-16,292}{-106,172} = 0.15345$$
$$\alpha \approx 81.17°$$

We could determine the third angle using the second equation in the law of cosines. However, at this point, it is simplest to use the fact that the sum of the angles of a triangle is $180°$, so that

$$\beta = 180° - \alpha - \gamma$$
$$\approx 180° - 81.17° - 51°$$
$$\approx 47.83°$$

(The two missing angles could also have been found using the law of sines.)

The law of cosines may also be used to determine the angles of a triangle when the lengths of the sides are given.

➤ EXAMPLE 2
Solving a Triangle Given Three Sides

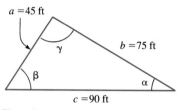

Figure 3

Suppose that a triangle has sides of length 45, 75, and 90 feet. Determine the angles of the triangle. (See Figure 3.)

Solution

From each of the three equations in the law of cosines, we may determine one of the angles. As in the preceding example, it is simplest to determine two of the angles in this way and then determine the remaining angle using the fact that the sum of the three angles is 180°. Here are the calculations:

$$a^2 = b^2 + c^2 - 2bc \cos \alpha$$
$$45^2 = 75^2 + 90^2 - 2 \cdot 75 \cdot 90 \cos \alpha$$
$$\cos \alpha \approx 0.86667$$
$$\alpha \approx 29.926°$$
$$b^2 = a^2 + c^2 - 2ac \cos \beta$$
$$75^2 = 45^2 + 90^2 - 2 \cdot 45 \cdot 90 \cos \beta$$
$$\cos \beta \approx 0.55556$$
$$\beta \approx 56.251°$$
$$\gamma \approx 180° - \alpha - \beta$$
$$\approx 180° - 29.926° - 56.251°$$
$$\approx 93.823°$$

Many applications require you to calculate the area of a triangle. In elementary geometry, you learned the following formula for the area:

$$\text{Area of a triangle} = \tfrac{1}{2} \cdot \text{Length of base} \cdot \text{Height}$$

This formula requires that you know (or can compute) the height and the length of the base. As an alternative, here is a formula that allows you to calculate the area from the lengths of two sides and the angle between them.

AREA OF A TRIANGLE

Suppose that a triangle has sides a and b and that the included angle is γ. (See Figure 4.) Then the area of the triangle is given by the formula

$$\text{Area} = \frac{ab}{2} \sin \gamma$$

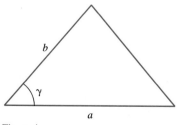

Figure 4

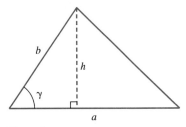

Figure 5

Proof   The proof is a simple application of trigonometry. Draw a perpendicular from the vertex opposite side a. (See Figure 5.) We have

$$\sin \gamma = \frac{h}{b}$$

$$h = b \sin \gamma$$

$$\text{Area of triangle} = \tfrac{1}{2} a h = \tfrac{1}{2} a b \sin \gamma$$

This completes the proof.   ◆

➤ **EXAMPLE 3**
Area of a Triangle from Two Sides and an Angle

Suppose that a triangle has sides of length 22 feet and 31 feet and that the angle between these sides is $37°$. What is the area of the triangle?

Solution

The area is given by the formula

$$\text{Area} = \frac{ab}{2} \sin \gamma$$

$$= \tfrac{1}{2}(22 \cdot 31 \sin 37°)$$

$$\approx 205.22 \text{ ft}^2$$

The area of a triangle may also be calculated from the lengths of the three sides using a particularly elegant result known as **Heron's formula.**

HERON'S FORMULA

Suppose that a triangle has sides of length a, b, and c. Then the area of the triangle is given by the formula

$$\text{Area} = \sqrt{s(s-a)(s-b)(s-c)}$$

where

$$s = \tfrac{1}{2}(a+b+c)$$

Proof   Let's rewrite the expression under the radical solely in terms of a, b, and c:

$$s(s-a)(s-b)(s-c) = \frac{a+b+c}{2} \cdot \frac{-a+b+c}{2} \cdot \frac{a-b+c}{2} \cdot \frac{a+b-c}{2}$$

$$= \left[\frac{(b+c)+a}{2} \cdot \frac{(b+c)-a}{2}\right]\left[\frac{a+(b-c)}{2} \cdot \frac{a-(b-c)}{2}\right]$$

$$= \left[\frac{(b+c)^2 - a^2}{4}\right]\left[\frac{a^2 - (b-c)^2}{4}\right]$$

$$= \left[\frac{b^2 + c^2 + 2bc - a^2}{4}\right]\left[\frac{a^2 - b^2 - c^2 + 2bc}{4}\right]$$

Let's now apply the law of cosines to replace a^2 in each of the expressions in the numerator:

$$\frac{b^2 + c^2 + 2bc - a^2}{4} = \frac{-(b^2 + c^2 - 2bc\cos\alpha) + 2bc + c^2 + b^2}{4}$$

$$= \frac{bc + bc\cos\alpha}{2}$$

$$= \frac{bc(1 + \cos\alpha)}{2}$$

$$\frac{a^2 - b^2 - c^2 + 2bc}{4} = \frac{(b^2 + c^2 - 2bc\cos\alpha) - b^2 - c^2 + 2bc}{4}$$

$$= \frac{-bc\cos\alpha + bc}{2}$$

$$= \frac{bc(1 - \cos\alpha)}{2}$$

This gives us

$$s(s - a)(s - b)(s - c) = \left[\frac{bc(1 + \cos\alpha)}{2}\right]\left[\frac{bc(1 - \cos\alpha)}{2}\right]$$

$$= b^2 c^2 \frac{(1 - \cos^2\alpha)}{4}$$

$$= \frac{b^2 c^2 \sin^2\alpha}{4}$$

$$\sqrt{s(s - a)(s - b)(s - c)} = \frac{|bc\sin\alpha|}{2}$$

However, the right side of the last equation equals the area of the triangle. This completes the proof of Heron's formula.   ◆

> **EXAMPLE 4**
Area of a Triangle
via Heron's Formula

Suppose that a triangle has sides of length 12 cm, 15 cm, and 11 cm. Use Heron's formula to determine the area of the triangle.

Solution

In this case, we have

$$s = \tfrac{1}{2}(12 + 15 + 11)$$

$$= 19$$

Therefore, by Heron's formula, we have

$$\text{Area} = \sqrt{s(s - a)(s - b)(s - c)}$$

$$= \sqrt{19 \cdot (19 - 12) \cdot (19 - 15) \cdot (19 - 11)}$$

$$= \sqrt{4256}$$

$$\approx 65.24 \text{ cm}^2$$

Exercises 7.2

Determine the remaining sides and angles. (Determine sides to the nearest tenth and angles to the nearest tenth of a degree.)

1. $a = 4$, $b = 5$, $\angle C = 30°$

2. $a = 1$, $b = 4$, $\angle C = 30°$

3. $c = 3$, $b = 2$, $\angle A = 60°$

4. $c = 4$, $b = \sqrt{3}$, $\angle A = 30°$

5. $b = 7$, $c = \sqrt{2}$, $\angle A = 135°$

6. $b = 9$, $c = \sqrt{5}$, $\angle A = 145°$

7. $a = 6$, $b = 9$, $\angle C = 70°$

8. $a = 4.5$, $c = 30.2$, $\angle B = 60°$

9. $b = 8$, $c = 12$, $\angle A = 53°$

10. $x = 27$, $z = 21$, $\angle Y = 112°$

11. $m = 10$, $n = 40$, $\angle N = 60°$

12. $a = 3.6$, $b = 7.2$, $\angle A = 32°20'$

13. $a = 3.2$, $b = 7.5$, $\angle C = 15°25'$

14. $b = 10.2$, $c = 15.5$, $\angle A = 72°50'$

15. $a = 14$, $b = 18$, $c = 12$

16. $a = 17$, $b = 25$, $c = 17$

17. $a = 2$, $b = 3$, $c = 4$

18. $a = 70$, $b = 240$, $c = 250$

19. $a = 16$, $b = 20$, $c = 28$

20. $a = 8.3$, $b = 16.4$, $c = 11.8$

21. $a = 7.2$, $b = 5.1$, $c = 11.4$

22. $a = 5$, $b = 12$, $c = 13$

23. $b = 13$, $c = 23$, $\angle A = 69°40'$

24. $a = .94$, $c = .35$, $\angle B = 72°$

◆ Applications

25. Pontiac is located approximately 14 miles due north of Southfield. Highland is located approximately 17 miles from Pontiac on a line that is 13° north of west from Pontiac. Find the distance from Southfield to Highland.

26. To measure the length of a lake, a surveyor measures the distances from point A, on one side of the lake, to points B and C, at the opposite ends of the lake, and finds them to be 52 meters and 47 meters, respectively. He then determines the measure of the angle BAC to be 55°. How long is the lake?

27. The distances from a sailboat, at point C, to two points A and B on the shore are known to be 100 meters and 80 meters, respectively. If the angle ACB measures 55°, find the distance between points A and B on the shore.

28. Two sides and a diagonal of a parallelogram are 7, 9, and 15 yards, respectively. Find the measures of the angles of the parallelogram.

29. A pilot is flying from Jackson, Michigan, to Chicago, Illinois, a distance of approximately 200 miles. As she leaves Jackson, she flies 20° off course for 50 miles. How far is she then from Chicago? By how much must she change her course to correct the error?

30. A gold mine is dug into the face of a hill that slopes upward from the horizontal by 12°. If a 50-foot shaft is dug into the hill so that the shaft makes an angle 12° down from the horizontal, how far is the end of the shaft vertically from the surface?

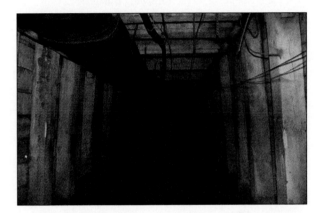

31. Two planes, one flying at 300 mph and the other flying at 400 mph, leave Cleveland at the same time. If their courses diverge by 40°, how far apart are the planes after 2 hours of flying?

32. In a storm cloud, an airplane meets an air current flowing vertically upward at a rate of 100 mph. The pilot aims her plane 58° from the horizontal downward, and her airspeed then reads 250 mph. If these conditions hold steady for 6 minutes, how far (the horizontal distance) will the plane have traveled? What would then be the ground speed of the plane?

33. Points X and Y are on opposite sides of Lake Fenton. From a third point Z, the angle between the sight lines to X and Y is 46°20'. If XZ is 350 meters and YZ is 286 meters, how far apart are X and Y?

34. Two planes leave St. Louis at 6 A.M. One plane travels 300 mph due east, while the other plane travels 400 mph 20° west of north. How far apart are the two planes at 8 A.M.?

35. Two lighthouses are 12 miles apart at points D and E. A freighter is observed between the two at point F, and an observer at lighthouse D notes that $\angle FDE$ measures $70°30'$. An observer at lighthouse E notes that $\angle FED$ measures $23°50'$. How far is the freighter from each lighthouse?

36. An isosceles triangle has a base of 24 cm. If the vertex angle measures $54°$, what is the perimeter of the triangle?

37. Two ships start from Pearl Harbor at the same time. If one cruises at 30 km per hour, the other cruises at 40 km per hour, and their courses diverge by $40°$, how far apart are they after 2 hours?

38. A navigator locates her position, C, on a map. She finds that she is 52 miles due east of lighthouse A, 43 miles from port B, and the distance between the lighthouse and the port is 22 miles. Find the heading she must take from the north to reach port B.

Find the area of each of the following triangles.

39. $a = 15, b = 8, \angle C = 30°$

40. $b = 6, c = 14, \angle A = 150°$

41. $a = 18, b = 25, \angle B = 148°40'$

42. $b = 8, \angle B = 16°40', \angle C = 145°$

43. $a = 26, b = 30, c = 13$

44. $a = 22, b = 10, c = 12\sqrt{2}$

45. Prove the following formula for the area of triangle ABC.

$$\text{Area} = \frac{1}{2}a^2 \frac{\sin B \sin C}{\sin A}$$

Prove the formulas in Exercises 46–48 for a general triangle ABC with opposite sides of respective lengths a, b, and c.

46. $A = \dfrac{1}{2}ab\sqrt{1 - \left(\dfrac{a^2 + b^2 - c^2}{2ab}\right)^2}$

47. $1 + \cos A = \dfrac{(b + c + a)(b + c - a)}{2bc}$

48. $1 - \cos A = \dfrac{(a - b + c)(a + b - c)}{2bc}$

49. Find the area of a quadrilateral $ABCD$ in terms of the lengths of the sides.

7.3 POLAR COORDINATES

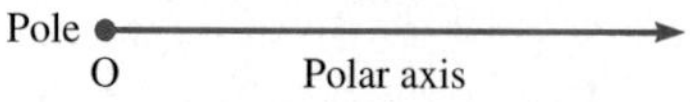

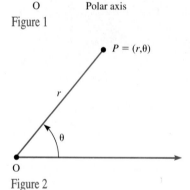

Pole •———————————————→
O   Polar axis

Figure 1

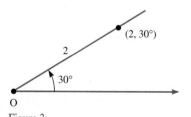

Figure 2

Throughout the book, we have graphed points and curves using rectangular coordinate systems. In many applications, especially in calculus and physics, polar coordinate systems provide a useful alternative. To define a polar coordinate system, we proceed as follows. Select a point O in the plane and a ray extending outward from the point along the positive x-axis. (See Figure 1.) The point is called the **pole**, and the ray is called the **polar axis.** The polar coordinates of a point P are a pair of numbers (r, θ), where r is the distance of P from the pole and θ is the angle the line segment $\overline{OP}$ makes with the polar axis. (See Figure 2.) The pole is assigned the coordinates $(0, \theta)$, where θ is any angle.

As defined above, the polar coordinates of a point P is a pair of real numbers (r, θ), with $r \geq 0$. We also assign a point to a pair (r, θ) with $r < 0$, namely, the point obtained by going $|r|$ units in the negative direction on the ray with angle θ to the polar axis.

➤ **EXAMPLE 1**
Plotting Points in Polar Coordinates

Plot the following points given in polar coordinates:

1. $(2, 30°)$ 2. $(3, 135°)$ 3. $(-1, -60°)$

Solution

1. To plot the point with polar coordinates $(2, 30°)$, we first draw an angle of $30°$ with the polar axis as its initial side and its vertex at the pole. We then start at the pole and proceed 2 units along the terminal side of the angle. The point we come to is the one with the given polar coordinates. (See Figure 3.)

Figure 3

2. To plot the point with polar coordinates (3, 135°), we first draw an angle of 135° with the polar axis as its initial side and its vertex at the pole. We then start at the pole and proceed 3 units along the terminal side of the angle. The point we come to is the one with the given polar coordinates. (See Figure 4.)

3. To plot the point with polar coordinates (−1, −60°), we first draw an angle of −60° with the polar axis as its initial side and its vertex at the pole. We then start at the pole and proceed 1 unit in the negative direction along the terminal side of the angle. The point we come to is the one with the given polar coordinates. (See Figure 5.)

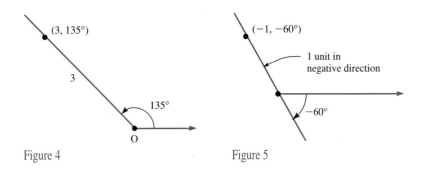

Figure 4   Figure 5

Polar coordinates are usually plotted on special coordinate paper (see Figure 6), which shows various rays from the origin corresponding to the most common angles θ, and various concentric circles centered at the origin corresponding to various values of r.

The polar coordinates (2, 30°) and (2, 390°) have angles that differ by 360° but have the same value of r. These polar coordinates correspond to the same point. More generally, two polar coordinates (r, θ), (r, θ') correspond to the same point provided that θ and θ' differ from one another by a multiple of 360°. In particular, any point P has an infinite number of different polar coordinate representations. Among this infinite collection of representations, it is most common to use the one for which $0 \leq \theta < 360°$.

An equation that expresses a relationship between r and θ is called a **polar equation**. By plotting the points (r, θ) that satisfy a polar equation, we obtain the graph of the polar equation. In the next four examples, we will graph a number of polar equations. As we shall see, many interesting curves arise as graphs of relatively simple polar equations.

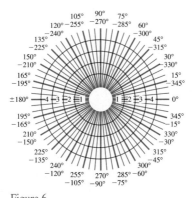

Figure 6

➤ EXAMPLE 2
A Polar Graph

Graph the polar equation $r = 2\cos\theta$.

Solution

Because the right side is periodic in θ with period 360°, it suffices to consider angles θ such that $0 \leq \theta < 360°$. We take a representative sample of angles in this range and compute the corresponding values of r. These are contained in the following table. Because $\cos\theta = \cos(360° - \theta)$ (the cosine function is even), the points corresponding to values of θ between 180° and 360° lead to repetition of the points obtained for values between $\theta = 0°$ and $\theta = 180°$. We plot these points and draw a smooth curve through them. (See Figure 7.) Note that the graph is a circle.

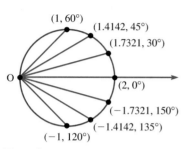

$r = 2\cos\theta$

θ	$r = 2\cos\theta$
0°	2
30°	1.7321
45°	1.4142
60°	1
90°	0
120°	−1
135°	−1.4142
150°	−1.7321
180°	−2

Figure 7

> **EXAMPLE 3**
Three-Leafed Rose

Graph the polar equation

$$r = 4\sin 3\theta$$

Solution

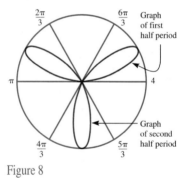

Figure 8

To sketch this graph, first recall that the sine function has period 2π. Therefore, the function $\sin 3\theta$ has period $2\pi/3$. As θ ranges over the interval $[0, 2\pi/6]$ (the first half period), the sine function first increases from 0 to 1 and then decreases from 1 to 0. So in this interval for θ, the value of r first increases from 0 to 4 and then decreases from 4 to 0. In the next half period, the value of r decreases from 0 to −4 and then increases from −4 to 0. Each half of the first period yields a portion of the graph that has the shape of a rose petal. Similarly, the portions of the graph corresponding to θ in the intervals $[2\pi/3, 4\pi/3]$ and $[4\pi/3, 2\pi]$ each correspond to two rose petals. You might think that because there are six half periods, there will be six rose petals. However, upon careful examination, you will see that each petal is traced out twice. For instance, the petal corresponding to the first half period is repeated as the petal for the fourth half period. The completed sketch is shown in Figure 8.

> **EXAMPLE 4**
Spiral of Archimedes

Graph the polar equation

$$r = \theta, \qquad \theta > 0, \qquad \theta \text{ in radians}$$

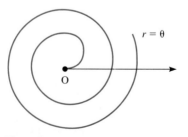

Figure 9

Solution

As θ increases, so does the value of r. Note, however, that the right side is not periodic, so we must consider all positive values of θ. With rotation through 360°, the value of r increases by 2π. The graph in this case is a spiral, as shown in Figure 9.

CONVERTING BETWEEN RECTANGULAR AND POLAR COORDINATES

Suppose that a polar coordinate system is superimposed on a rectangular coordinate system with the pole at the origin and the polar axis along the positive x-axis. Let P be a point with rectangular coordinates (x, y) and polar coordinates (r, θ). Then the coordinates are related by the equations

$$x = r \cos \theta$$
$$y = r \sin \theta$$
$$r = \sqrt{x^2 + y^2}$$
$$\tan \theta = \frac{y}{x}, \quad x \neq 0$$

Proof   Draw the coordinate systems and the point P as in Figure 10. From the definitions of the sine and cosine, we have

$$\cos \theta = \frac{d(O, A)}{d(O, P)} = \frac{x}{r}$$
$$x = r \cos \theta$$
$$\sin \theta = \frac{d(A, P)}{d(O, P)} = \frac{y}{r}$$
$$y = r \sin \theta$$

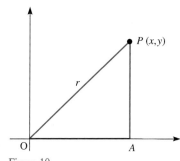

Figure 10

This proves the first two equations. To obtain the third proof, add the squares of the first two equations:

$$
\begin{aligned}
x^2 + y^2 &= (r \cos \theta)^2 + (r \sin \theta)^2 \\
&= r^2 \cos^2 \theta + r^2 \sin^2 \theta \\
&= r^2 (\sin^2 \theta + \cos^2 \theta) \\
&= r^2 \cdot 1 \\
&= r^2 \\
r &= \sqrt{x^2 + y^2}
\end{aligned}
$$

This proves the third equation. Finally, to prove the last equation, assume that $x \neq 0$ and divide the second equation by the first:

$$
\begin{aligned}
\frac{y}{x} &= \frac{r \sin \theta}{r \cos \theta} \\
&= \frac{\sin \theta}{\cos \theta} \\
&= \tan \theta
\end{aligned}
$$

This proves the fourth equation.   ◆

The next example provides some practice in converting rectangular coordinates to polar coordinates and vice versa.

➤ **EXAMPLE 5**
Rectangular to Polar
Coordinates

1. A point P has rectangular coordinates $(3, 6)$. Determine its polar coordinates.

2. Determine the rectangular coordinates of the point with polar coordinates $(2, 135°)$.

Solution

1. Let (r, θ) be the polar coordinates of P. We have

$$\tan \theta = \frac{y}{x} = \frac{6}{3} = 2$$

$$\theta = \tan^{-1} 2 = 63.435°$$

$$r = \sqrt{x^2 + y^2}$$

$$= \sqrt{3^2 + 6^2}$$

$$= \sqrt{45}$$

$$= 3\sqrt{5}$$

Therefore, the polar coordinates of the point are $(3\sqrt{5}, 63.435°)$.

2. From the first two equations, we have

$$x = r \cos \theta$$

$$= 2 \cos 135°$$

$$= 2 \cdot \left(-\frac{\sqrt{2}}{2} \right)$$

$$= -\sqrt{2}$$

$$y = r \sin \theta$$

$$= 2 \sin 135°$$

$$= 2 \cdot \frac{\sqrt{2}}{2}$$

$$= \sqrt{2}$$

The rectangular coordinates of the point are $(-\sqrt{2}, \sqrt{2})$.

The next two examples provide some practice in converting equations in rectangular coordinates into polar equations and vice versa.

> **EXAMPLE 6**
Converting an Equation
to Polar Coordinates

Write the equation $x - 2y = 3$ in terms of polar coordinates.

Solution

Apply the first two equations to obtain alternative expressions for x and y. Substitute these expressions into the given equation to obtain

$$x - 2y = 3$$

$$r \cos \theta - 2r \sin \theta = 3$$

$$r(\cos \theta - 2 \sin \theta) = 3$$

$$r = \frac{3}{\cos \theta - 2 \sin \theta}$$

> **EXAMPLE 7**
Converting a Polar Equation
to Rectangular Coordinates

Write the polar equation

$$r = \frac{1}{1 + 2 \sin \theta}$$

in terms of rectangular coordinates.

Solution

Multiply both sides of the equation by the denominator of the right side to obtain

$$r(1 + 2\sin\theta) = 1$$
$$r + 2r\sin\theta = 1$$
$$r + 2y = 1 \qquad (y = r\sin\theta)$$
$$\sqrt{x^2 + y^2} + 2y = 1$$
$$\sqrt{x^2 + y^2} = 1 - 2y$$
$$x^2 + y^2 = (1 - 2y)^2 \qquad (Squaring)$$
$$x^2 + y^2 = 1 - 4y + 4y^2$$
$$x^2 - 3y^2 + 4y = 1$$

Polar Coordinates and Graphing Technology

Most scientific calculators have built-in conversion from rectangular to polar coordinates and vice versa. The following example illustrates the procedure for the TI-81 calculator.

New Technology

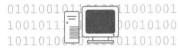

EXAMPLE 8
Converting to Polar Coordinates

1. Use the TI-81 calculator to determine the polar coordinates of the point with rectangular coordinates (3, 8).
2. Use the TI-81 calculator to determine the rectangular coordinates of the point with polar coordinates $(5, 0.1\pi)$.

Solution

1. Press the [MATH] key. You will see the menu in Figure 11. The first option is for conversion from rectangular to polar coordinates. Option 2 is for conversion from polar to rectangular coordinates. Selection option 1 and press [ENTER]. You will see the screen shown in Figure 12. Fill in the rectangular coordinates as shown in Figure 13. Be sure to include the comma and right parenthesis. (The comma is entered by pressing [ALPHA][,].) Press [ENTER]. The calculator will display 8.544003745. This is the r coordinate. Now press [ALPHA][θ] followed by [ENTER]. The calculator will now display the value of the θ coordinate, or 1.212025657 radians. The value of θ will be in degrees or radians, as specified by the current setting in the [MODE] menu for angle units. We have assumed that radians are specified.

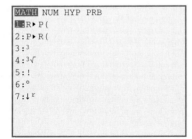

Figure 11

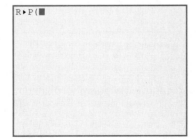

Figure 12

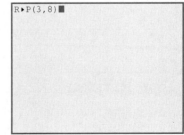

Figure 13

2. Press the [MATH] key. Select option 2 for conversion from polar to rectangular coordinates. Press [ENTER]. The display will be as shown in Figure 14. Fill in the polar coordinates as shown in Figure 15. (To enter π,

press $\boxed{\text{2nd}\,\boxed{\cdot}}$) Press $\boxed{\text{ENTER}}$. The calculator will display the x-coordinate, 4.755282581. Now press $\boxed{\text{ALPHA}}\boxed{\text{Y}}$. The calculator will display the y-coordinate, namely 1.545084972. Note that the angle $0.1 * \pi$ is interpreted as degrees or radians depending on the angle measurement setting in the $\boxed{\text{MODE}}$ menu. We have assumed radians are specified.

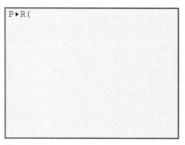

Figure 14

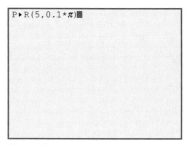

Figure 15

To graph in polar coordinates an equation of the form $r = f(\theta)$, we must graph the x and y coordinates of the point (r, θ) as

$$(x, y) = (r \cos\theta, r \sin\theta) = (f(\theta)\cos\theta, f(\theta)\sin\theta)$$

To graph separate equations for x and y in terms of θ, we switch to **parametric** plotting by selecting the option **param** from the $\boxed{\text{MODE}}$ menu (see Figure 16). To do this, display the $\boxed{\text{MODE}}$ menu, use the arrow keys to select **param**, and then press $\boxed{\text{CLEAR}}$. The $\boxed{\text{Y} =}$ menu now allows you to specify separate equations for X and Y. See Figure 17. Instead of θ, the variable is called T. Three separate sets of such equations are allowed, denoted, respectively, $X_{1T}, Y_{1T}, X_{2T}, Y_{2T}, X_{3T}, Y_{3T}$. The variable T is entered using the key $\boxed{\text{X}|\text{T}}$. The $\boxed{\text{RANGE}}$ menu allows you to set the range for T (see Figure 18).

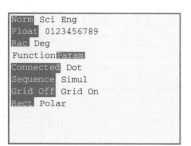

Figure 16

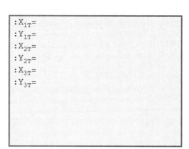

Figure 17

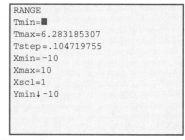

Figure 18   ◄

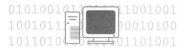

New Technology

EXAMPLE 9
Graphing in Polar
Coordinates

Use a graphing calculator to display the graph of the equation

$$r = 0.1\theta^2, \qquad 0 \le \theta \le 4\pi$$

Solution

The equations for x and y are shown in Figure 19. The range for T is from 0 to $4\pi \approx 12.58$. The graph is shown in Figure 20.

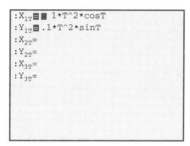

Figure 19

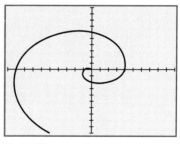

Figure 20

New Technology

010100101110011001001
1001011.[   ]00010100
1011010[   ]01101001

EXAMPLE 10
Graphing in Polar
Coordinates

Use a graphing calculator to display the graph of the equation

$$r = \sin 5\theta$$

Solution

The equations for x and y are shown in Figure 21. Because the equations have period 2π in T, we set the range to be $0 \le T \le 2\pi$. The graph is shown in Figure 22.

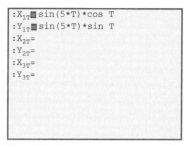

Figure 21

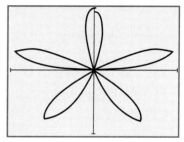

Figure 22

Exercises 7.3

If a graphing calculator is available, it may be used in Exercises 1–40. Plot the following points given in polar coordinates.

1. $A = (4, 30°)$

2. $A = (6, 45°)$

3. $A = \left(5, \dfrac{3\pi}{2}\right)$

4. $A = \left(4, -\dfrac{\pi}{2}\right)$

5. $A = \left(6, \dfrac{4\pi}{3}\right)$

6. $A = \left(5, \dfrac{2\pi}{3}\right)$

7. $A = (10, -120°)$

8. $A = (2, -135°)$

Graph the given polar equation.

9. $r = \sin\theta$

10. $r = 1 - \sin\theta$

11. $r = \cos 2\theta$ (four-leafed rose)

12. $r = \cos 4\theta$ (eight-leafed rose)

13. $r = 1 + 2\sin\theta$

14. $r = \sec\theta$

15. $r = \sin\dfrac{\theta}{2}$

16. $r = \sin 5\theta$ (lemniscate)

17. $r = 2\theta, \theta > 0, \theta$ in radians

18. $r = \sin 3\theta$ (three-leafed rose)

19. $r = 2 + 4\cos\theta$

20. $r = 2 - \cos\theta$

21. $r = 4 - 4\sin\theta$

22. $r = 5$

23. $r = 3 + \sin\theta$ (limaçon)

24. $r = 5 + \cos\theta$ (limaçon)

25. $r = e^\theta$ (logarithmic spiral)

26. $\theta = \dfrac{\log r}{2}$ (logarithmic spiral)

27. $r = \cos 2\theta \sec\theta$ (strophoid)

28. $r = \sin\theta \tan\theta$ (cissoid)

Determine polar coordinates for each of the given points. Give θ to the nearest tenth of a degree.

29. $(4, 4)$

30. $(-9, 9)$

31. $\left(-\frac{1}{2}, \frac{\sqrt{3}}{2}\right)$

32. $\left(\dfrac{\sqrt{2}}{2}, \dfrac{-\sqrt{2}}{2}\right)$

33. $(\sqrt{3}, 1)$

34. $(2\sqrt{3}, -2)$

35. $(3, 4)$ 36. $(-4, 3)$ 37. $(-2, -2)$

38. $(-\sqrt{3}, -1)$ 39. $(3, 2)$ 40. $(-2, 3)$

Write the given equation in terms of polar coordinates.

41. $3x + 2y = 4$ 42. $y = 2$ 43. $x^2 + y^2 = 16$

44. $y^2 = 25x$ 45. $y = 6$ 46. $x + y = 4$

47. $x^2 - 4y^2 = 4$ 48. $x = 5$

49. $x^2 + 9y^2 = 36$ 50. $3x + 4y = 5$

Write the given polar equation in rectangular form.

51. $r = 5$ 52. $r = 3\sin\theta + 2\cos\theta$

53. $r = -\dfrac{2}{\sin\theta}$ 54. $r = 2\sec\theta$

55. $r - 6\sin\theta = 0$ 56. $r(1 - \sin\theta) = 2$

57. $\theta = \dfrac{\pi}{4}$ 58. $r(\cos\theta + \sin\theta) = 2$

59. $r = 4\cos\theta$ 60. $r + 2\sin\theta = -2\cos\theta$

Graph.

61. $r = 2\cos 2\theta - 1$ (bow tie)

62. $r = \cos 2\theta - 2$ (peanut)

63. $r = \dfrac{\tan^2\theta \sec\theta}{4}$ (semicubical parabola)

64. $r = 2\theta + \cos\theta$ (twisted sister)

65. Match each statement in the first column with the correct statement in the second column about the graph of $r = f(\theta)$.

a. $f(\pi + \theta) = f(\theta)$ (1) The graph has a horizontal axis of symmetry.

b. $f(-\theta) = f(\theta)$ (2) The graph has a vertical axis of symmetry.

c. $f\left(\dfrac{\pi}{2} + \theta\right) = f\left(\dfrac{\pi}{2} - \theta\right)$ (3) The graph has a point of symmetry at the pole.

Determine coordinates of the points of intersection of the polar graphs for each given pair of equations.

66. $r = 2 + 4\sin\theta,\quad r = 2$

67. $r = 1 + \sin\theta,\quad r = 2\sin\theta$

7.4 TRIGONOMETRIC FORM OF A COMPLEX NUMBER

In Chapter 3, we introduced the complex number system and explored its elementary algebraic properties. Let's now discuss the connections between the complex numbers and the trigonometric functions. As we shall see, each complex number may be represented in a trigonometric form. Moreover, by using this form, we may give simple interpretations of various operations among complex numbers, including multiplication, division, and extraction of nth roots.

To introduce the trigonometric form of a complex number, let's first provide a geometric representation of a complex number. Recall that a complex number z has a unique representation in the form $z = a + bi$, where i is the imaginary unit and a and b are real. Let the point (a, b) correspond to $z = a + bi$ in the plane. (See Figure 1.) Then there is a one-to-one correspondence between points of the plane and complex numbers. When the plane is so identified, it is called the **complex plane**.

The points on the horizontal axis correspond to complex numbers for which $b = 0$, that is, real numbers. For this reason, the horizontal axis is also called the **real axis**. Similarly, the points on the vertical axis correspond to complex numbers for which $a = 0$, that is, the purely imaginary numbers. For this reason, the vertical axis is also called the **imaginary axis**.

Recall that we previously introduced the absolute value $|x|$ of a real number x. This number is equal to the distance of the point x from the origin of the number line. In a similar fashion, let's define the absolute value of a complex number.

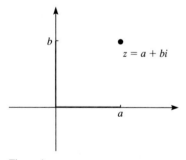

Figure 1

| **Definition 1**
Absolute Value of a Complex Number | Let $z = a + bi$. The **absolute value of the complex number** z, denoted $|z|$, is defined as the distance of the point z from the origin of the complex plane. (See Figure 2.) This value, $|z|$, is also referred to as the **modulus** of z. |

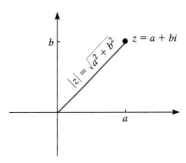

Figure 2

Referring to Figure 2, we see that if $z = a + bi$, then $|z|$ is equal to the length of the hypotenuse of a right triangle with legs of length a and b. So by the Pythagorean theorem, we have the formula

$$|z| = \sqrt{a^2 + b^2}$$

➤ **EXAMPLE 1**
Calculating Absolute Values

Evaluate the following:

1. $|3 + 4i|$
2. $|-2i|$
3. $\left| -\sqrt{2} \right|$

Solution

1. $|3 + 4i| = \sqrt{3^2 + 4^2} = \sqrt{25} = 5$
2. $|-2i| = |0 - 2i| = \sqrt{0^2 + (-2)^2} = \sqrt{4} = 2$
3. $\left| -\sqrt{2} \right| = \left| -\sqrt{2} + 0i \right| = \sqrt{(-\sqrt{2})^2 + 0^2} = \sqrt{2}$

➤ **EXAMPLE 2**
Absolute Value Identity

Let z be a complex number. Prove the identity:

$$z\bar{z} = |z|^2$$

Solution

Suppose that $z = a + bi$. Then

$$\bar{z} = \overline{a + bi} = a - bi$$
$$z\bar{z} = (a + bi)(a - bi)$$
$$= a^2 + b^2$$
$$= \left(\sqrt{a^2 + b^2} \right)^2$$
$$= |z|^2$$

This proves the given identity.

➤ **EXAMPLE 3**
Absolute Value of a Product

Let z and w be complex numbers. Prove the following identity:

$$|zw| = |z||w|$$

Solution

Let $z = a + bi$ and $w = c + di$. Then

$$|z| = \sqrt{a^2 + b^2}$$

$$|w| = \sqrt{c^2 + d^2}$$

$$|z||w| = \sqrt{a^2 + b^2}\,\sqrt{c^2 + d^2}$$

$$= \sqrt{(a^2 + b^2)(c^2 + d^2)} \qquad (1)$$

On the other hand, we have

$$zw = (a + bi)(c + di)$$

$$= (ac - bd) + i(ad + bc)$$

Therefore,

$$|zw| = \sqrt{(ad + bc)^2 + (ac - bd)^2}$$

$$= \sqrt{(a^2d^2 + b^2c^2 + 2adbc) + a^2c^2 + b^2d^2 - 2abcd}$$

$$= \sqrt{a^2d^2 + b^2c^2 + a^2c^2 + b^2d^2}$$

$$= \sqrt{(a^2 + b^2)(c^2 + d^2)}$$

$$= |z||w| \qquad \qquad \textit{(By equation (1))}$$

This completes the proof of the identity.

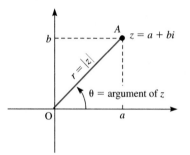

Figure 3

Suppose that $z = a + bi$ is a nonzero complex number. Represent z as a point in the complex plane. Draw a line $\overline{OA}$ from the origin to z as shown in Figure 3. Let $r = |z|$ denote the modulus of z. Let θ be the angle that $\overline{OA}$ makes with the positive "real" axis. The angle θ is called the **argument** of z.

The reason for assuming that z is nonzero is that for $z = 0$, the line $\overline{OA}$ reduces to a point, and the angle θ is undefined. That is, the argument of the complex number 0 is undefined.

Referring to Figure 3, the definitions of sine and cosine yield the relationships

$$\frac{a}{r} = \cos\theta, \qquad \frac{b}{r} = \sin\theta$$

$$a = r\cos\theta, \qquad b = r\sin\theta$$

$$z = a + bi$$

$$= r\cos\theta + ir\sin\theta$$

$$= r(\cos\theta + i\sin\theta)$$

That is, we have the following result.

TRIGONOMETRIC FORM OF A COMPLEX NUMBER

Let z be a nonzero complex number with modulus r and argument θ. Then z may be written in the following trigonometric form (also called **polar form**):

$$z = r(\cos\theta + i\sin\theta)$$

Again referring to Figure 3, we see that

$$a = r\cos\theta, \qquad b = r\sin\theta$$

$$\frac{b}{a} = \frac{r\sin\theta}{r\cos\theta} = \frac{\sin\theta}{\cos\theta} = \tan\theta$$

➤ EXAMPLE 4
Determining Polar Form

Determine the polar form of the following complex numbers. For each determine the modulus and argument. (Assume the argument is in $[0, 2\pi]$.)

1. $z = 1 - i$
2. $z = -3i$

Solution

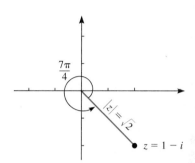

Figure 4

1. Draw the point corresponding to z, as in Figure 4. The radius from the origin to z makes a $-45°$ angle with the positive x-axis. The argument is $315°$ or

$$\theta = \frac{7\pi}{4}$$

The modulus r is given by

$$r = |z|$$
$$= \sqrt{1^2 + (-1)^2}$$
$$= \sqrt{2}$$

Therefore, the polar form of z is

$$z = \sqrt{2}\left(\cos \frac{7\pi}{4} + i \sin \frac{7\pi}{4}\right)$$

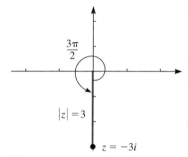

Figure 5

2. Draw the point corresponding to z as in Figure 5. From the figure, we see that the argument is $\theta = 3\pi/2$. The modulus is

$$r = |z|$$
$$= \sqrt{0^2 + (-3)^2}$$
$$= 3$$

Therefore, the polar form is given by

$$z = r(\cos \theta + i \sin \theta)$$
$$= 3\left(\cos \frac{3\pi}{2} + i \sin \frac{3\pi}{2}\right)$$

➤ EXAMPLE 5
Plotting from Polar Form

Describe the points corresponding to the following complex numbers in terms of their argument and modulus.

1. $2\left(\cos \frac{5\pi}{6} + i \sin \frac{5\pi}{6}\right)$

2. $3[\cos(-\pi) + i \sin(-\pi)]$

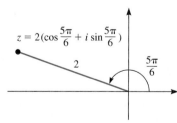

Figure 6

Solution

1. The argument is $5\pi/6$ radians, and the modulus is 2. Draw an angle of $5\pi/6$ radians in standard position. Start at the origin and move 2 units along the terminal side of the angle. (See Figure 6.) The point in rectangular form is $z = -\sqrt{3} + i$.

2. The argument is $-\pi$ radians, and the modulus is 3. Draw an angle of $-\pi$ radians in standard position. Start at the origin and move 3 units

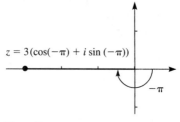

$z = 3(\cos(-\pi) + i \sin(-\pi))$

Figure 7

along the terminal side of the angle. The resulting point corresponds to the given polar form. (See Figure 7.) The point in rectangular form is $z = -3 + 0i$.

Suppose that the complex number z is written in the polar form

$$z = r(\cos\theta + i\sin\theta)$$

The argument θ is not uniquely determined by z. Indeed, because the sine and cosine functions have period 2π, you may add any integer multiple of 2π to θ and obtain another polar form for z:

$$z = r[\cos(\theta + 2n\pi) + i\sin(\theta + 2n\pi)], \qquad n = 0, \pm 1, \pm 2, \ldots$$

Another way of saying this is that z has an infinite number of arguments, differing from one another by integer multiples of 2π. A complex number has an infinite number of polar forms, one corresponding to each possible argument.

Among the arguments for a complex number z, there is precisely one lying in the interval $[0, 2\pi)$. This argument is called the **principal argument** of z. In the polar forms determined in the last example, the argument of part 1 is principal whereas the argument of part 2 is not.

Multiplication and division of complex numbers have simple interpretations in terms of their respective polar forms. Namely, we have the following result.

MULTIPLICATION AND DIVISION IN TERMS OF POLAR FORMS

Let the complex numbers z_1 and z_2 have polar forms

$$z_1 = r_1(\cos\theta_1 + i\sin\theta_1)$$
$$z_2 = r_2(\cos\theta_2 + i\sin\theta_2)$$

Then

$$z_1 z_2 = r_1 r_2[\cos(\theta_1 + \theta_2) + i\sin(\theta_1 + \theta_2)]$$

$$\frac{z_1}{z_2} = \frac{r_1}{r_2}[\cos(\theta_1 - \theta_2) + i\sin(\theta_1 - \theta_2)]$$

Proof   Let's prove the first formula. The proof of the second is similar. The key to the proof is to apply the addition formula for the sine and cosine functions:

$$\begin{aligned} z_1 z_2 &= [r_1(\cos\theta_1 + i\sin\theta_1)] \cdot [r_2(\cos\theta_2 + i\sin\theta_2)] \\ &= r_1 r_2[\cos\theta_1\cos\theta_2 - \sin\theta_1\sin\theta_2 + i(\cos\theta_1\sin\theta_2 + \cos\theta_2\sin\theta_1)] \\ &= r_1 r_2[\cos(\theta_1 + \theta_2) + i\sin(\theta_1 + \theta_2)] \qquad \text{\textit{Addition formulas}} \end{aligned}$$

This completes the proof.   ♦

➤ **EXAMPLE 6**
Multiplication and Division
from Polar Form

Suppose that z and w are given by the respective polar forms

$$z = 5\left(\cos\frac{3\pi}{5} + i\sin\frac{3\pi}{5}\right)$$

$$w = 7\left(\cos\frac{\pi}{3} + i\sin\frac{\pi}{3}\right)$$

Calculate a polar form of zw and z/w.

Solution

In this example, we have $r_1 = 5$, $r_2 = 7$, $\theta_1 = 3\pi/5$, $\theta_2 = \pi/3$. Therefore,

$$\theta_1 + \theta_2 = \frac{3\pi}{5} + \frac{\pi}{3}$$

$$= \frac{14\pi}{15}$$

$$\theta_1 - \theta_2 = \frac{3\pi}{5} - \frac{\pi}{3}$$

$$= \frac{4\pi}{15}$$

Thus, by the preceding formula, we have

$$zw = r_1 r_2[\cos(\theta_1 + \theta_2) + i\sin(\theta_1 + \theta_2)]$$

$$= 5 \cdot 7\left(\cos\frac{14\pi}{15} + i\sin\frac{14\pi}{15}\right)$$

$$= 35\left(\cos\frac{14\pi}{15} + i\sin\frac{14\pi}{15}\right)$$

$$\frac{z}{w} = \frac{r_1}{r_2}[\cos(\theta_1 - \theta_2) + i\sin(\theta_1 - \theta_2)]$$

$$= \frac{5}{7}\left(\cos\frac{4\pi}{15} + i\sin\frac{4\pi}{15}\right)$$

Exercises 7.4

Evaluate the following:

1. $|2 - 3i|$ 2. $|5 + 2i|$ 3. $\left|-\sqrt{3}\right|$

4. $|5|$ 5. $|-3i|$ 6. $\left|i\sqrt{5}\right|$

7. $\left|5 + i\sqrt{3}\right|$ 8. $|-3 - 5i|$

Show that the following are true.

9. $|z| + |w| \neq |z + w|$ 10. $|z| - |w| \neq |z + w|$

11. $\left|\frac{z}{w}\right| = \frac{|z|}{|w|}$ 12. $|z||\bar{z}| = |z \cdot \bar{z}|$

Determine the polar form of the following complex numbers; also determine the modulus and argument of each. (Assume the argument is in $[0, 2\pi)$.)

13. $z = 2 + 2i$ 14. $z = 3 + 3i$ 15. $z = 4i$

16. $z = 5i$ 17. $z = 3$ 18. $z = -10$

19. $z = 3 - 3i$ 20. $z = 2 - 2i$

21. $z = 4 + 4i$ 22. $z = -6 + 6i$

23. $z = \sqrt{3} - i$ 24. $z = -1 - i\sqrt{3}$

25. $z = 1 + i\sqrt{3}$ 26. $z = -\sqrt{3} + i$

27. $z = -2\sqrt{3} - 2i$ 28. $z = 4 + 4i\sqrt{3}$

29. $z = -2$ 30. $z = 4$

31. $z = 5i$ 32. $z = -2i$

Graph the points corresponding to the following complex numbers given in polar form.

33. $-2(\cos\pi + i\sin\pi)$ 34. $3\left(\cos\frac{\pi}{2} + i\sin\frac{\pi}{2}\right)$

35. $\frac{1}{2}\left(\cos\frac{\pi}{2} + i\sin\frac{\pi}{2}\right)$ 36. $4(\cos\pi + i\sin\pi)$

37. $\cos(-45°) + i\sin(-45°)$

38. $\frac{1}{3}[\cos(-60°) + i\sin(-60°)]$

39. $4\left[\cos\left(-\frac{5\pi}{6}\right) + i\sin\left(-\frac{5\pi}{6}\right)\right]$

40. $2\left[\cos\left(-\frac{3\pi}{2}\right) + i\sin\left(-\frac{3\pi}{2}\right)\right]$

41. $3(\cos 60° + i\sin 60°)$ 42. $(\cos 240° + i\sin 240°)$

43. $5(\cos 30° + i\sin 30°)$ 44. $3(\cos 150° + i\sin 150°)$

45. $2\left[\cos\left(-\frac{2\pi}{3}\right) + i\sin\left(-\frac{2\pi}{3}\right)\right]$

46. $2[\cos(-\pi) + i\sin(-\pi)]$

47. $3(\cos 270° + i \sin 270°)$

48. $-5(\cos 225° + i \sin 225°)$

For the following complex numbers in polar form, z and w, calculate a polar form of zw and $\dfrac{z}{w}$.

49. $z = 5(\cos 180° + i \sin 180°)$

$w = 2(\cos 45° + i \sin 45°)$

50. $z = 2(\cos 360° + i \sin 360°)$

$w = 3(\cos 135° + i \sin 135°)$

51. $z = \left[\cos\left(-\dfrac{\pi}{2}\right) + i \sin\left(-\dfrac{\pi}{2}\right)\right]$

$w = 4\left(\cos\dfrac{3\pi}{2} + i \sin\dfrac{3\pi}{2}\right)$

52. $z = 4\left[\cos\left(-\dfrac{\pi}{4}\right) + i \sin\left(-\dfrac{\pi}{4}\right)\right]$

$w = 3\left[\cos\left(-\dfrac{3\pi}{2}\right) + i \sin\left(-\dfrac{3\pi}{2}\right)\right]$

53. $z = 2\left[\cos\left(-\dfrac{5\pi}{6}\right) + i \sin\left(-\dfrac{5\pi}{6}\right)\right]$

$w = 3\left[\cos\left(-\dfrac{\pi}{2}\right) + i \sin\left(-\dfrac{\pi}{2}\right)\right]$

54. $z = 3\left[\cos\left(-\dfrac{5\pi}{3}\right) + i \sin\left(-\dfrac{5\pi}{3}\right)\right]$

$w = 2\left[\cos\left(-\dfrac{4\pi}{3}\right) + i \sin\left(-\dfrac{4\pi}{3}\right)\right]$

55. $z = 5\left(\cos\dfrac{5\pi}{6} + i \sin\dfrac{5\pi}{6}\right)$

$w = 2\left[\cos\left(-\dfrac{3\pi}{4}\right) + i \sin\left(-\dfrac{3\pi}{4}\right)\right]$

56. $z = 8(\cos 2\pi + i \sin 2\pi)$

$w = 4\left[\cos\left(-\dfrac{\pi}{4}\right) + i \sin\left(-\dfrac{\pi}{4}\right)\right]$

57. $z = 6(\cos 60° + i \sin 60°)$

$w = 2(\cos 90° + i \sin 90°)$

58. $z = 2(\cos 60° + i \sin 60°)$

$w = 6(\cos 135° + i \sin 135°)$

59. $z = 9[\cos(-225°) + i \sin(-225°)]$

$w = 3[\cos(-60°) + i \sin(-60°)]$

60. $z = 10[\cos(-90°) + i \sin(-90°)]$

$w = 2[\cos(-120°) + i \sin(-120°)]$

61. $z = 4\left(\cos\dfrac{5\pi}{6} + i \sin\dfrac{5\pi}{6}\right)$

$w = 8\left[\cos\left(-\dfrac{\pi}{2}\right) + i \sin\left(-\dfrac{\pi}{2}\right)\right]$

62. $z = \left[\cos\left(-\dfrac{\pi}{4}\right) + i \sin\left(-\dfrac{\pi}{4}\right)\right]$

$w = 6\left(\cos\dfrac{4\pi}{3} + i \sin\dfrac{4\pi}{3}\right)$

Given the following complex numbers in polar form, change them to $a + bi$ form (standard form).

63. $2\left(\cos\dfrac{3\pi}{4} + i \sin\dfrac{3\pi}{4}\right)$

64. $5[\cos(-\pi) + i \sin(-\pi)]$

65. $6\left[\cos\left(-\dfrac{\pi}{2}\right) + i \sin\left(-\dfrac{\pi}{2}\right)\right]$

66. $3\left(\cos\dfrac{2\pi}{3} + i \sin\dfrac{2\pi}{3}\right)$

67. $\cos(-150°) + i \sin(-150°)$

68. $2[\cos(-60°) + i \sin(-60°)]$

69. $\sqrt{3}[\cos(-30°) + i \sin(-30°)]$

70. $\sqrt{2}(\cos 135° + i \sin 135°)$

Write the following complex numbers in $a + bi$ form.

71. $[2(\cos 90° + i \sin 90°)][5(\cos 45° + i \sin 45°)]$

72. $[3(\cos 120° + i \sin 120°)][2(\cos 30° + i \sin 30°)]$

73. $[5(\cos \pi + i \sin \pi)]\left[3\left(\cos\left(-\dfrac{\pi}{2}\right) + i \sin\left(-\dfrac{\pi}{2}\right)\right)\right]$

74. $\left[2\left(\cos\dfrac{3\pi}{2} + i \sin\dfrac{3\pi}{2}\right)\right][\cos(2\pi) + i \sin(2\pi)]$

75. $[4(\cos 45° + i \sin 45°)] + [3(\cos(-30°) + i \sin(-30°))]$

76. $[3(\cos 60° + i \sin 60°)] + [6(\cos(-45°) + i \sin(-45°))]$

77. $\left[2\left(\cos\left(-\dfrac{3\pi}{2}\right) + i \sin\left(-\dfrac{3\pi}{2}\right)\right)\right]$
$+ [4(\cos(-2\pi) + i \sin(-2\pi))]$

78. $\left[12\left(\cos\dfrac{5\pi}{4} + i \sin\dfrac{5\pi}{4}\right)\right] + [3(\cos \pi + i \sin \pi)]$

Use trigonometric tables or calculators to find zw and z/w.

79. $z = 3(\cos 40° + i \sin 40°)$

$w = 2(\cos 130° + i \sin 130°)$

80. $z = 2[\cos(-20°) + i \sin(-20°)]$

$w = 4(\cos 305° + i \sin 305°)$

81. $z = 5\left[\cos\left(-\dfrac{\pi}{5}\right) + i \sin\left(-\dfrac{\pi}{5}\right)\right]$

$w = 2\left(\cos\dfrac{5\pi}{9} + i \sin\dfrac{5\pi}{9}\right)$

82. $z = 10\left(\cos\dfrac{\pi}{10} + i \sin\dfrac{\pi}{10}\right)$

$w = 2\left(\cos\dfrac{17\pi}{15} + i \sin\dfrac{17\pi}{15}\right)$

7.5 THE *n*TH ROOT OF A COMPLEX NUMBER

De Moivre

Let z be a complex number and n a positive integer. An nth root of z is a number w whose nth power is z. That is, w is an nth root of z provided that

$$w^n = z$$

In this section, we will use the trigonometric form of a complex number to determine its nth roots. As a first step in this direction, let's prove the following result for determining a polar form for the nth power of a complex number.

DE MOIVRE'S THEOREM

Let n be an integer and $z = r(\cos\theta + i\sin\theta)$ be a nonzero complex number in polar form. Then

$$z^n = [r(\cos\theta + i\sin\theta)]^n = r^n(\cos n\theta + i\sin n\theta)$$

That is, the modulus of z^n equals the nth power of the modulus of z and the argument of z^n equals n times the argument of z.

Proof   From the formula for multiplying terms in polar form from the preceding section, we see that

$$
\begin{aligned}
z^2 &= z \cdot z \\
&= r \cdot r[\cos(\theta + \theta) + i\sin(\theta + \theta)] \\
&= r^2(\cos 2\theta + i\sin 2\theta)
\end{aligned}
$$

This is the desired result for $n = 2$. For $n = 3$, note that $z^3 = z \cdot z^2$, so applying the result for $n = 2$, we have

$$
\begin{aligned}
z^3 &= z \cdot z^2 \\
&= [r(\cos\theta + i\sin\theta)][r^2(\cos 2\theta + i\sin 2\theta)] \\
&= r^3[(\cos(\theta + 2\theta) + i\sin(\theta + 2\theta)] \\
&= r^3(\cos 3\theta + i\sin 3\theta)
\end{aligned}
$$

This is the desired result for $n = 3$. Following the same reasoning, the result for each value of n may be deduced from the result for the value preceding it. We will fill in the details of the proof once we introduce the method of mathematical induction.   ◆

➤ **EXAMPLE 1**
Powers of a Complex Number

Use De Moivre's theorem to calculate the following powers:

1. z^3, $z = 2\left(\cos\dfrac{\pi}{5} + i\sin\dfrac{\pi}{5}\right)$

2. $(1 - i)^{10}$

Solution

1. By De Moivre's theorem, we have

$$
\begin{aligned}
z^3 &= 2^3\left[\cos\left(3\cdot\dfrac{\pi}{5}\right) + i\sin\left(3\cdot\dfrac{\pi}{5}\right)\right] \\
&= 8\left(\cos\dfrac{3\pi}{5} + i\sin\dfrac{3\pi}{5}\right)
\end{aligned}
$$

2. We previously determined the polar form of $1 - i$, namely

$$1 - i = \sqrt{2}\left(\cos\frac{7\pi}{4} + i\sin\frac{7\pi}{4}\right)$$

Therefore, by De Moivre's theorem, we have

$$(1 - i)^{10} = (\sqrt{2})^{10}\left[\cos\left(10 \cdot \frac{7\pi}{4}\right) + i\sin\left(10 \cdot \frac{7\pi}{4}\right)\right]$$

$$= 32\left(\cos\frac{35\pi}{2} + i\sin\frac{35\pi}{2}\right)$$

Because the sine and cosine functions have period 2π, we may rewrite the last expression as

$$(1 - i)^{10} = 32\left[\cos\left(\frac{3\pi}{2} + 16\pi\right) + i\sin\left(\frac{3\pi}{2} + 16\pi\right)\right]$$

$$= 32\left(\cos\frac{3\pi}{2} + i\sin\frac{3\pi}{2}\right)$$

$$= 32[0 + i(-1)]$$

$$= -32i$$

Suppose we are given a nonzero complex number z. Let's determine a formula for its nth roots. Let a polar form of z be

$$z = r(\cos\theta + i\sin\theta)$$

For each of the integers $j = 0, 1, 2, \ldots, n - 1$, set

$$t_j = r^{1/n}\left(\cos\frac{\theta + 2\pi j}{n} + i\sin\frac{\theta + 2\pi j}{n}\right)$$

By De Moivre's theorem, the nth power of t_j is equal to

$$t_j^n = \left(r^{1/n}\right)^n\left[\cos\left(n \cdot \frac{\theta + 2\pi j}{n}\right) + i\sin\left(n \cdot \frac{\theta + 2\pi j}{n}\right)\right]$$

$$= r[\cos(\theta + 2\pi j) + i\sin(\theta + 2\pi j)]$$

$$= r(\cos\theta + i\sin\theta) \qquad\qquad\qquad \textit{(Periodicity)}$$

$$= z$$

That is, each of the numbers t_j is an nth root of z. Moreover, any two consecutive nth roots have arguments differing by $2\pi/n$. In particular, the values of t_j are all different. Thus, z has n nth roots, and we have proved the following result:

nTH ROOTS OF A COMPLEX NUMBER

Let $z = r(\cos\theta + i\sin\theta)$ be a nonzero complex number and n a positive integer. Then there are n distinct complex numbers that are nth roots of z. These numbers are given by the formula

$$t_j = r^{1/n}\left[\cos\left(\frac{\theta + 2\pi j}{n}\right) + i\sin\left(\frac{\theta + 2\pi j}{n}\right)\right] \qquad (j = 0, 1, \ldots, n - 1)$$

> ## EXAMPLE 2
Complex Cube Roots

Calculate the cube roots of 2.

Solution

The polar form of 2 is

$$2(\cos 0 + i \sin 0)$$

Therefore, the cube roots of 2 are given by

$$t_j = 2^{1/3}\left(\cos \frac{0 + 2\pi j}{3} + i \sin \frac{0 + 2\pi j}{3}\right)$$

$$= 2^{1/3}\left(\cos \frac{2\pi j}{3} + i \sin \frac{2\pi j}{3}\right) \qquad (j = 0, 1, 2)$$

More explicitly, we have

$$t_0 = 2^{1/3}(\cos 0 + i \sin 0) = 2^{1/3}(1 - 0i) = 2^{1/3}$$

$$t_1 = 2^{1/3}\left(\cos \frac{2\pi}{3} + i \sin \frac{2\pi}{3}\right)$$

$$= 2^{1/3}\left(-\frac{1}{2} + i \frac{\sqrt{3}}{2}\right)$$

$$t_2 = 2^{1/3}\left(\cos \frac{4\pi}{3} + i \sin \frac{4\pi}{3}\right)$$

$$= 2^{1/3}\left(-\frac{1}{2} - i \frac{\sqrt{3}}{2}\right)$$

(See Figure 1.)

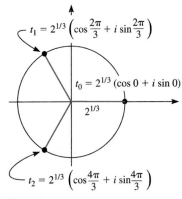

Figure 1

> ## EXAMPLE 3
Square Roots
of a Complex Number

Determine the square roots of *i*.

Solution

A polar form of *i* is given by

$$i = 1 \cdot \left(\cos \frac{\pi}{2} + i \sin \frac{\pi}{2}\right)$$

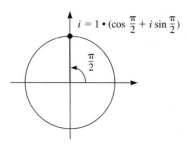

$i = 1 \cdot (\cos \frac{\pi}{2} + i \sin \frac{\pi}{2})$

$\frac{\pi}{2}$

Figure 2

(See Figure 2.) Therefore, the square roots of i are equal to t_0, t_1, where

$$t_j = 1 \cdot \left(\cos \frac{\frac{\pi}{2} + 2\pi j}{2} + i \sin \frac{\frac{\pi}{2} + 2\pi j}{2} \right), \quad (j = 0, 1)$$

$$t_0 = \cos \frac{\pi}{4} + i \sin \frac{\pi}{4} = \frac{\sqrt{2}}{2} + \frac{\sqrt{2}}{2} i$$

$$t_1 = \cos \frac{5\pi}{4} + i \sin \frac{5\pi}{4} = -\frac{\sqrt{2}}{2} - \frac{\sqrt{2}}{2} i$$

(See Figure 3.)

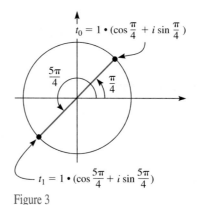

$t_0 = 1 \cdot (\cos \frac{\pi}{4} + i \sin \frac{\pi}{4})$

$\frac{5\pi}{4}$ $\frac{\pi}{4}$

$t_1 = 1 \cdot (\cos \frac{5\pi}{4} + i \sin \frac{5\pi}{4})$

Figure 3

➢ **EXAMPLE 4**
Complex Roots of an Equation

Determine the complex solutions of the equation $x^4 - 16 = 0$.

Solution

The solutions of the equation satisfy

$$x^4 = 16$$

That is, they are fourth roots of 16. The polar form of 16 is

$$16 = 16(\cos 0 + i \sin 0)$$

Therefore, its fourth roots are t_0, t_1, t_2, and t_3, where

$$t_j = 16^{1/4} \left(\cos \frac{2\pi j}{4} + i \sin \frac{2\pi j}{4} \right), \quad (j = 0, 1, 2, 3)$$

$$t_0 = 2 \left(\cos \frac{2\pi \cdot 0}{4} + i \sin \frac{2\pi \cdot 0}{4} \right)$$

$$= 2(\cos 0 + i \sin 0)$$

$$= 2$$

$$t_1 = 2 \left(\cos \frac{2\pi \cdot 1}{4} + i \sin \frac{2\pi \cdot 1}{4} \right)$$

$$= 2 \left(\cos \frac{\pi}{2} + i \sin \frac{\pi}{2} \right)$$

$$= 2i$$

$$t_2 = 2\left(\cos\frac{2\pi \cdot 2}{4} + i\sin\frac{2\pi \cdot 2}{4}\right)$$

$$= 2(\cos\pi + i\sin\pi)$$

$$= -2$$

$$t_3 = 2\left(\cos\frac{2\pi \cdot 3}{4} + i\sin\frac{2\pi \cdot 3}{4}\right)$$

$$= 2\left(\cos\frac{3\pi}{2} + i\sin\frac{3\pi}{2}\right)$$

$$= -2i$$

Exercises 7.5

Use De Moivre's theorem to calculate the following. Express results in $a + bi$ form.

1. z^{20}, $z = \cos 210° + i\sin 210°$

2. z^{12}, $z = \cos 180° + i\sin 180°$

3. z^{10}, $z = 2\left(\cos\frac{\pi}{4} + i\sin\frac{\pi}{4}\right)$

4. z^{30}, $z = 2\left(\cos\frac{5\pi}{4} + i\sin\frac{5\pi}{4}\right)$

5. z^3, $z = 5[\cos(-30°) + i\sin(-30°)]$

6. z^4, $z = 6[\cos(-60°) + i\sin(-60°)]$

7. $\left(-\frac{\sqrt{3}}{2} - \frac{1}{2}i\right)^6$ 8. $\left(\frac{\sqrt{3}}{2} + \frac{1}{2}i\right)^4$

9. $(5 + 5i)^{14}$ 10. $\left(-\frac{\sqrt{2}}{2} - \frac{\sqrt{2}}{2}i\right)^{13}$

11. $(-2\sqrt{3} + 2i)^8$ 12. $(-1 + \sqrt{3}i)^{10}$

13. $[3(\cos 15° + i\sin 15°)]^4$

14. $[2(\cos 10° + i\sin 10°)]^{18}$

15. $[2(\cos 20° + i\sin 20°)]^6$ 16. $[3(\cos 50° + i\sin 50°)]^6$

17. $\left\{2\left[\cos\left(-\frac{\pi}{18}\right) + i\sin\left(-\frac{\pi}{18}\right)\right]\right\}^3$

18. $\left\{2\left[\cos\left(-\frac{\pi}{12}\right) + i\sin\left(-\frac{\pi}{12}\right)\right]\right\}^4$

19. $\left\{2\left[\cos\frac{\pi}{8} + i\sin\frac{\pi}{8}\right]\right\}^{10}$

20. $\left[3\left(\cos\frac{\pi}{10} + i\sin\frac{\pi}{10}\right)\right]^{20}$

Find the indicated roots. Where possible, express the roots in $a + bi$ form.

21. Cube roots of -8.   22. Cube roots of 8.

23. Tenth roots of $-\sqrt{3} - i$.

24. Sixth roots of $-\sqrt{3} + i$.

25. Cube roots of i.   26. Fourth roots of -1.

27. Square roots of $9i$.   28. Fifth roots of $1 - i$.

29. Fifth roots of $1 + i$.

30. Square roots of $-16i$.

31. Fourth roots of $8 - 8i\sqrt{3}$.

32. Square roots of $-2 + 2\sqrt{3}i$.

33. Cube roots of $27(\cos 180° + i\sin 180°)$.

34. Fifth roots of $32(\cos 150° + i\sin 150°)$.

35. Fourth roots of $81(\cos 120° + i\sin 120°)$.

36. Cube roots of $125(\cos 135° + i\sin 135°)$.

37. Square roots of $16\left[\cos\left(\frac{\pi}{2}\right) + i\sin\left(\frac{\pi}{2}\right)\right]$.

38. Square roots of $100\left[\cos\left(-\frac{\pi}{3}\right) + i\sin\left(-\frac{\pi}{3}\right)\right]$.

Determine the complex solutions for the following.

39. $a^5 = 243$ 40. $m^6 - 64 = 0$

41. $m^4 + 81i = 0$ 42. $x^5 = -243$

43. $z^4 = -8\sqrt{3} + 8i$ 44. $x^3 + 4\sqrt{3} - 4i = 0$

45. $x^5 - 1 = 0$ 46. $z^5 - i = 0$ 47. $x^3 + 27 = 0$

48. $x^4 = -16$ 49. $x^2 + 25 = 0$ 50. $x^6 - 1 = 0$

51. $x^3 = 1 - \sqrt{3}i$ 52. $x^5 = \sqrt{2} - \sqrt{2}i$

53. $y^5 + \sqrt{2} - \sqrt{2}i = 0$ 54. $a^3 - 1 - \sqrt{3}i = 0$

Compute the following.

55. $\left(-\frac{1}{3} - \frac{1}{3}i\right)^{-3}$ 56. $(2\sqrt{3} - 2i)^{-4}$

57. $\dfrac{(1 + i)^3}{(\sqrt{3} + i)^4}$

58. $\dfrac{(2 - 2i)^4}{(1 + i)^3}$

59. $(\sqrt{3} + i)^4 (1 - i)^3$

60. $(2 - 2i)^3 (1 + i)^4$

61. $\left[2\left(\cos\dfrac{\pi}{2} + i\sin\dfrac{\pi}{2}\right)\right]^{-2}\left[3\left(\cos\dfrac{\pi}{2} + i\sin\dfrac{\pi}{2}\right)\right]^2$

62. $\left[3\left(\cos\dfrac{2\pi}{3} + i\sin\dfrac{2\pi}{3}\right)\right]^3\left[2\left(\cos\dfrac{\pi}{3} + i\sin\dfrac{\pi}{3}\right)\right]^2$

63. $\dfrac{\left[4\left(\cos\left(-\dfrac{\pi}{4}\right) + i\sin\left(-\dfrac{\pi}{4}\right)\right)\right]^3}{[2(\cos 2\pi + i\sin 2\pi)]^4}$

64. $\dfrac{\left[3\left(\cos\dfrac{3\pi}{4} + i\sin\dfrac{3\pi}{4}\right)\right]^4}{\left[9\left(\cos\left(-\dfrac{\pi}{6}\right) + i\sin\left(-\dfrac{\pi}{6}\right)\right)\right]^2}$

7.6 VECTORS

Another application of trigonometry is in elementary vector analysis, which is of critical importance in physics and engineering. In this section, we will provide a brief introduction to vectors in the plane and to the related applications of trigonometry.

In modeling the world around us, we describe some quantities by their magnitudes alone. Such quantities are called **scalar quantities**. Examples of such quantities are length, weight, area, and volume. Scalar quantities are represented by real numbers. Models also use quantities that are described by both a magnitude and a direction. Such quantities are called **vector quantities**. Examples of vector quantities abound in physics and include velocity, acceleration, and force.

It is customary to represent a vector geometrically as an arrow. The length of the arrow represents the magnitude of the vector, and the direction of the arrow represents its direction. For instance, the vector in Figure 1 shows a velocity of 50 miles per hour in a direction 30° north of east. The starting point of a vector is called its **initial point** and the ending point its **terminal point.**

Two vectors are equal if and only if they have the same magnitude and direction. For example, the vectors **A** and **B** in Figure 2 are equal to one another. However, the vectors **C** and **D** are not equal because they have the same direction but different magnitudes. Similarly, the vectors **E** and **F** are not equal because they have the same magnitude but different directions. The location of a vector in the plane is immaterial: Two vectors with the same magnitude and direction are equal, even though they are located in different parts of the plane.

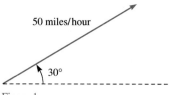

50 miles/hour

30°

Figure 1

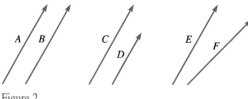

Figure 2

If **A** is a vector, then its magnitude is denoted $|\mathbf{A}|$. The **zero vector**, denoted **0**, is the vector with magnitude 0. By convention, the direction of the zero vector is undefined.

To distinguish between scalars and vectors, it is customary to write scalars in lower-case, nonbold letters, such as a, b, x, y. Vectors are denoted using bold letters, either upper- or lower-case, such as **A, B, i, j**.

Suppose that **A** and **B** are given vectors. We may form the sum **A** + **B** as follows: Place the initial point of **B** on the terminal point of **A**, as shown in Figure 3. The sum is the vector that connects the initial point of **A** to the terminal point of **B**.

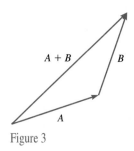

$A + B$ B

A

Figure 3

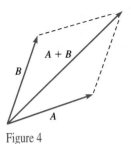

Figure 4

Another way of interpreting the sum is to draw a copy of **B** extending from the initial point of **A** and draw a parallelogram with the two vectors as sides. Then the sum **A** + **B** is just the diagonal of the parallelogram. (See Figure 4.) For this reason, vector addition is often called the **parallelogram rule**.

Let **A** be a vector. Then the vector −**A** is defined as the vector with the same length as **A** but with the opposite direction (see Figure 5). If k is a scalar, then the product k**A** is defined as the vector with magnitude $\left|k\right|$ times the magnitude of **A**, and pointing in the same direction as **A** if $k > 0$ and in the opposite direction if $k < 0$ (see Figure 6).

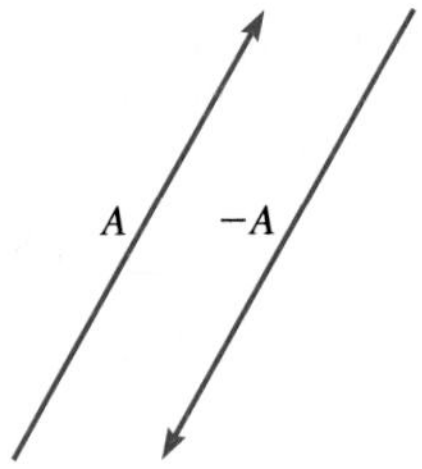

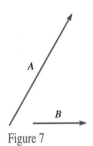

Figure 5

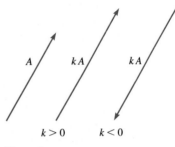

Figure 6

> **EXAMPLE 1**
Vector Arithmetic

Let **A** and **B** be as drawn in Figure 7. Use the parallelogram rule to describe the vectors **A** + (−**B**), $-\frac{1}{2}$**A**.

Solution

See Figure 8.

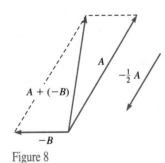

Figure 8

Figure 7

Let **i** be a vector of length 1 in the direction of the positive horizontal axis, and let **j** be a vector of length 1 in the direction of the positive vertical axis. (See Figure 9.) Then **i** and **j** are called **unit vectors** for the coordinate system.

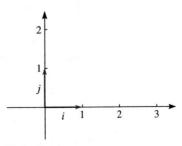

Figure 9

Any vector with horizontal direction is a scalar multiple of the unit vector **i**; any vector with vertical direction is a scalar multiple of the unit vector **j**. In particular, the horizontal component of a vector is a scalar multiple of **i** and the vertical component a scalar multiple of **j**.

Suppose that we are given a vector **A**. Then, referring to Figure 10 and the parallelogram rule, we see that **A** may be written in the form

$$\mathbf{A} = a\mathbf{i} + b\mathbf{j} \tag{1}$$

where a and b are scalars. Equation (1) expresses the vector **A** as the sum of a vector in the horizontal direction ($a\mathbf{i}$) and a vector in the vertical direction ($b\mathbf{j}$). The first vector is called the **horizontal component** of **A**, and the second is called the **vertical component** of **A** (with respect to the given coordinate system).

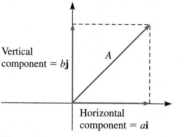

Figure 10   Figure 11

Let's make the last equation more explicit by determining formulas for a and b. Suppose that vector **A** makes an angle θ with the positive horizontal axis, as shown in Figure 11. Then the right triangle definitions of the sine and cosine imply that

$$a = |\mathbf{A}| \cos \theta \tag{2}$$
$$b = |\mathbf{A}| \sin \theta \tag{3}$$

> **EXAMPLE 2**
Length and Components of a Vector

Consider the vector **A** with initial point the origin and terminal point $(5, -3)$. Determine

1. $|\mathbf{A}|$
2. The horizontal and vertical components of **A**.

Solution

1. $|\mathbf{A}| = \sqrt{5^2 + (-3)^2} = \sqrt{34}$
2. When written in terms of the unit vectors **i** and **j**, the vector **A** is given by $\mathbf{A} = 5\mathbf{i} - 3\mathbf{j}$. The horizontal component is $5\mathbf{i}$ and the vertical component is $-3\mathbf{j}$.

> **EXAMPLE 3**
Inclined Plane

A block on a 41° inclined plane is acted on by two forces: Gravity exerts a force of 100 pounds downward, and a person exerts a force of 50 pounds at an angle

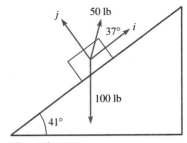

Figure 12

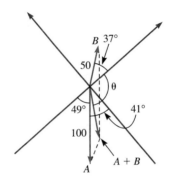

Figure 13

of 37° from the plane. (See Figure 12.) Determine the magnitude and direction of the sum of the two forces.

Solution

Let's use a coordinate system whose horizontal axis is parallel to the direction of the plane with positive direction pointing upward and whose vertical axis is perpendicular to the plane with positive direction pointing out of the plane. (See Figure 13.) Let **i** and **j** be, respectively, unit vectors parallel and perpendicular to the plane. Then referring to Figure 13, we see that

$$\mathbf{A} = -100\cos 49°\mathbf{i} - 100\sin 49°\mathbf{j}$$
$$= -65.606\mathbf{i} - 75.471\mathbf{j}$$
$$\mathbf{B} = 50\cos 37°\mathbf{i} + 50\sin 37°\mathbf{j}$$
$$= 39.932\mathbf{i} + 30.091\mathbf{j}$$
$$\mathbf{A} + \mathbf{B} = (-65.606 + 39.932)\mathbf{i} + (-75.471 + 30.091)\mathbf{j}$$
$$= -25.674\mathbf{i} - 45.38\mathbf{j}$$
$$|\mathbf{A} + \mathbf{B}| = \sqrt{(-25.674)^2 + (-45.38)^2}$$
$$= 52.139 \text{ lb}$$

Moreover, if θ denotes the angle that the sum vector makes with the positive x-axis, then we have

$$\cos\theta = \frac{-25.674}{52.139} \approx -0.492415$$
$$\theta = 240.5°$$

➤ EXAMPLE 4
Circular Motion

A race car is traveling at 180 miles per hour on a circular track. It hits a patch of oil that changes the direction of the car by 30°. (See Figure 14.) We can write the new velocity as a sum of two vectors, one directed along a radius of the circle, the radial component, and one along the tangent line to the circle, the tangential component. What are the magnitudes of the radial and tangential components of the car's velocity?

Solution

Referring to Figure 14, we see that the magnitude of the radial component of the car's velocity is equal to

$$180\sin 30° = 180 \cdot \tfrac{1}{2}$$
$$= 90 \text{ mph}$$

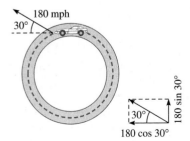

Figure 14

The magnitude of the tangential component of the velocity vector is equal to

$$180 \cos 30° = 180 \cdot \frac{\sqrt{3}}{2}$$
$$= 155.88 \text{ mph}$$

We have seen that a vector in the plane may be written in terms of a linear combination of unit vectors:

$$\mathbf{A} = a\mathbf{i} + b\mathbf{j}$$

It is often convenient to identify the vector $\mathbf{A}$ with the pair $\langle a, b \rangle$ of real numbers. (Of course, this identification depends on the particular coordinate system we are using.) In terms of this identification, vector addition and scalar multiplication can be carried out in simple terms: If $\mathbf{A} = \langle a_1, b_1 \rangle$ and $\mathbf{B} = \langle a_2, b_2 \rangle$, then

$$\mathbf{A} + \mathbf{B} = \langle a_1 + a_2, b_1 + b_2 \rangle$$
$$k\mathbf{A} = \langle ka_1, kb_1 \rangle$$

That is, addition and scalar multiplication may be carried out componentwise. Here is an illustration of how such calculations may be carried out:

➢ **EXAMPLE 5**
Vector Arithmetic

Suppose that $\mathbf{A} = \langle 3, 4 \rangle$, $\mathbf{B} = \langle -1, 10 \rangle$. Determine the vector $3\mathbf{A} - 2\mathbf{B}$.

Solution

We have

$$3\mathbf{A} = \langle 9, 12 \rangle$$
$$-2\mathbf{B} = \langle 2, -20 \rangle$$
$$3\mathbf{A} - 2\mathbf{B} = \langle 9, 12 \rangle + \langle 2, -20 \rangle$$
$$= \langle 11, -8 \rangle$$

SCALAR PRODUCT OF TWO VECTORS

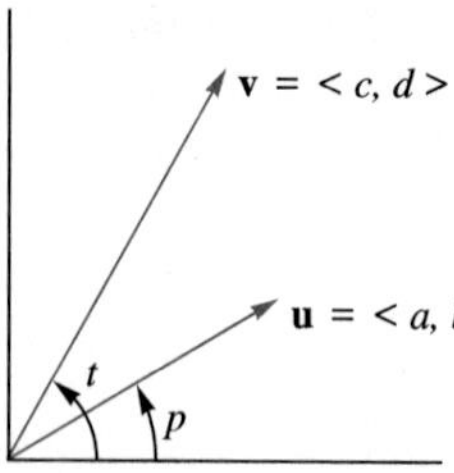

Figure 15

Let $\mathbf{u} = \langle a, b \rangle$, $\mathbf{v} = \langle c, d \rangle$ be two nonzero vectors based at the origin and let s and t be, respectively, the angles these vectors make with the positive x-axis (see Figure 15). From the definitions of the sine and cosine, we have the identities

$$\sin s = \frac{b}{\sqrt{a^2 + b^2}} = \frac{b}{|\mathbf{u}|}, \qquad \cos s = \frac{a}{\sqrt{a^2 + b^2}} = \frac{a}{|\mathbf{u}|}$$

$$\sin t = \frac{d}{\sqrt{c^2 + d^2}} = \frac{d}{|\mathbf{v}|}, \qquad \cos t = \frac{c}{\sqrt{c^2 + d^2}} = \frac{c}{|\mathbf{v}|}$$

Multiplying the first and third identities and the second and fourth, then adding the result gives

$$\frac{ac + bd}{|\mathbf{u}||\mathbf{v}|} = \cos s \cos t + \sin s \sin t = \cos(s - t) \qquad (4)$$

However, $s - t$ is just the angle between the two vectors, which we will denote by θ. The last identity may be written in the form

$$ac + bd = |\mathbf{u}||\mathbf{v}| \cos \theta \qquad (5)$$

This identity was proven under the assumption that the vectors were nonzero. However, it clearly holds as well if one or both vectors are zero. Let's assign a name to the expression on the left side of the equation.

Definition 1 Dot product	Let $\mathbf{u} = \langle a, b \rangle$, $\mathbf{v} = \langle c, d \rangle$ be two vectors. Then their **dot product** (also called the **scalar product**), denoted $\mathbf{u} \cdot \mathbf{v}$, is the real number defined by $$\mathbf{u} \cdot \mathbf{v} = ac + bd \qquad (6)$$

WARNING: Don't confuse the dot product of two vectors with the scalar product of a real number and a vector. The result of a dot product is a real number, whereas the result of a scalar product is a vector.

We can restate equation (4) as follows:

GEOMETRIC INTERPRETATION OF THE DOT PRODUCT

Let $\mathbf{u} = \langle a, b \rangle$, $\mathbf{v} = \langle c, d \rangle$ be two vectors and let θ denote the positive angle from $\mathbf{u}$ to $\mathbf{v}$. Then

$$\mathbf{u} \cdot \mathbf{v} = |\mathbf{u}||\mathbf{v}| \cos \theta \qquad (7)$$

If $\mathbf{u} \neq \mathbf{0}$, $\mathbf{v} \neq \mathbf{0}$, then $|\mathbf{u}| \neq 0$, $|\mathbf{v}| \neq 0$, so that the preceding relation may be written in the form

$$\cos \theta = \frac{\mathbf{u} \cdot \mathbf{v}}{|\mathbf{u}| \cdot |\mathbf{v}|}, \qquad \mathbf{u} \neq \mathbf{0}, \mathbf{v} \neq \mathbf{0} \qquad (8)$$

➤ **EXAMPLE 6**
Determining the Angle Between Two Vectors

Determine the angle between the vectors $\mathbf{u} = \langle 1, -1 \rangle$, $\mathbf{v} = \langle 3, 0 \rangle$.

Solution

Let θ be the positive angle with $\mathbf{u}$ as its initial side and $\mathbf{v}$ as its terminal side. We have:

$$\cos \theta = \frac{\mathbf{u} \cdot \mathbf{v}}{|\mathbf{u}||\mathbf{v}|}$$

$$= \frac{3}{\sqrt{1^2 + (-1)^2} \cdot \sqrt{3^2 + 0^2}}$$

$$= \frac{\sqrt{2}}{2}$$

$$\theta = \frac{\pi}{4}$$

Because two vectors are perpendicular if and only if $\cos \theta = 0$, we have:

DOT PRODUCT CRITERION FOR PERPENDICULARITY

The vectors $\mathbf{u}$ and $\mathbf{v}$ are perpendicular if and only if $\mathbf{u} \cdot \mathbf{v} = 0$.

Exercises 7.6

Given vectors **A**, **B**, and **C** as shown in the figure, draw the following.

1. −**A**
2. −**B**
3. **A** + **B**
4. **A** − **B**
5. **A** − **C**
6. **B** + **C**
7. 2**A**
8. −2**C**
9. **A** + 2**C**
10. −**A** − **C**

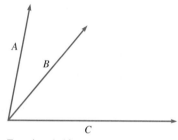

Exercises 1–10

For the following vectors **A** with initial point at the origin and the given terminal point, determine: (a) |**A**| and (b) the horizontal and vertical components of **A**.

11. Vector **A** with terminal point $(1, -1)$.
12. Vector **A** with terminal point $(-2, -2)$.
13. Vector **A** with terminal point $(3, 3)$.
14. Vector **A** with terminal point $(-5, 5)$.
15. Vector **A** with terminal point $(-1, \sqrt{3})$.
16. Vector **A** with terminal point $(4\sqrt{3}, 4)$.
17. Vector **A** with terminal point $(-3\sqrt{3}, -3)$.
18. Vector **A** with terminal point $(2, -2\sqrt{3})$.
19. Vector **A** with terminal point $(3, 4)$.
20. Vector **A** with terminal point $(-5, 12)$.
21. Vector **A** with terminal point $(10, -24)$.
22. Vector **A** with terminal point $(-6, -8)$.
23. Vector **A** with terminal point $\left(-\frac{1}{2}, -1\right)$.
24. Vector **A** with terminal point $(5, 3)$.

♦ **Applications**

25. A sled on an inclined plane weighs 500 lb, and the plane makes an angle of 50° with the horizontal. What force, perpendicular to the plane, is exerted on the plane by the sled?

26. A 150 lb box is dragged up a runway inclined 42° to the horizontal. Find the pressure of the box against the runway and the force required to drag the box.

27. If a force of 25 lb is required to move an 80 lb sled up a hill, what angle does the hill make with the horizontal?

28. What would be the force required to push a 100 lb object up a ramp that is inclined 10° with the horizontal?

29. A weight of 15 lb is placed on a smooth plane inclined at an angle of 24° with the horizontal. What force pushing along the plane will just prevent the weight from slipping?

30. A painting weighing 25 lb is supported at the top corners by a taut wire that passes around a nail embedded in the wall. If the wire forms an angle of 120° at the nail, what is the total pull on the wire?

31. When a girl pulls her sled with a rope, the rope makes an angle of 35° with the horizontal. If a pull of 16 lb on the rope is needed to move the sled, what is the horizontal component force?

32. A truck weighing 6875 lb moves up a bridge inclined 7°32′ from the horizontal. Find the pressure of the truck against the bridge.

33. What would be the largest weight a person could drag up a slope inclined 35° from the horizontal, if that person is able to pull with a force of 125 lb?

34. If a force of 52.1 lb is needed to keep a 75 lb block from sliding down an incline, what is the angle that the incline makes with the horizontal?

35. A ship heads through the water on a compass heading of 30° at a speed of 20 knots. It encounters a current that causes the ship to move on a path with a heading of 45° at 15 knots. Find the speed of the current if it is flowing directly from the north.

36. A plane that was flying due south at 250 mph was blown off course by the wind, which was blowing at 40 mph from the east. What is the angle that the plane is blown off course, and what is the actual ground speed of the plane?

37. A steamer sails 100 miles east and then 40 miles on a heading of 120°. How far is the steamer and what is its bearing from its starting point?

38. A sailboat is headed east at 18 mph relative to the water. An ocean current is carrying the water south at 3 mph. Find the course of the ship and its speed with respect to the ocean floor.

39. An airplane must fly at a ground speed of 450 mph on a course 170° to be on schedule. The wind velocity is 25 mph in the direction 40°. Find the necessary heading to the nearest degree and the necessary air speed to the nearest unit that the plane must fly to be on schedule.

40. A plane flies 650 mph on a bearing of 175°. A 25-mph wind, from a direction of 266°, blows against the plane. Find the resulting bearing of the plane.

41. At what bearing and speed should a pilot head if he wants to fly due south at 520 mph when a 40 mph east wind is blowing?

42. An airplane is headed at 230° at 320 knots when a wind of 15 knots is blowing from 50°. Find the ground speed of the plane and the wind correction angle.

43. A river is flowing at 2.4 mph when a girl rows across it. If the girl rows at a still-water speed of 3.1 mph and heads the boat perpendicular to the direction of the current, find the ground speed of the boat.

44. The velocity of a ship on a heading of 157° has a magnitude of 35.2 km/hr. The water current has a velocity of 8 km/hr in the direction of 213°. Find the magnitude of the actual velocity (ground speed) of the ship.

45. Two ships leave a harbor, one traveling at 20 knots on a course of 80° and the other at 24 knots on a course of 140°. How far apart are the ships after 2 hours? What is the bearing from the first ship to the second at that time?

46. An airplane flies on a compass heading of 90° at 200 mph. The wind affecting the plane is blowing from 300° at 30 mph. What is the true course and ground speed of the airplane?

Given the vectors $\mathbf{A} = \langle 3, 2 \rangle$, $\mathbf{B} = \langle -2, 4 \rangle$, $\mathbf{C} = \langle -1, -2 \rangle$, and $\mathbf{D} = \langle 4, -3 \rangle$, determine the following vectors.

47. 2**A**   48. −3**B**

49. **A** + **B**   50. **B** − 2**D**

51. 3**C** + 2**A**   52. 3**A** − 5**B**

53. 2**A** − 3**D**   54. 4**C** + 2**B**

55. **A** − 2**C**   56. 2**B** + 3**A**

Find the inner products of each pair of vectors.

57. $\mathbf{u} = \langle 1, \sqrt{2} \rangle$, $\mathbf{v} = \langle -3, -5 \rangle$

58. $\mathbf{u} = \langle -23, 35 \rangle$, $\mathbf{v} = \langle 12, -50 \rangle$

Prove the following for inner products.

59. $c_1(\mathbf{v}_1 \cdot \mathbf{v}_2) = (c_1\mathbf{v}_1) \cdot \mathbf{v}_2$

60. $(c_1\mathbf{v}_1) \cdot (c_2\mathbf{v}_2) = (c_1c_2)(\mathbf{v}_1 \cdot \mathbf{v}_2)$

61. $\mathbf{v}_1 \cdot (\mathbf{v}_2 + \mathbf{v}_3) = \mathbf{v}_1 \cdot \mathbf{v}_2 + \mathbf{v}_1 \cdot \mathbf{v}_3$

62. $\mathbf{v}_2 \cdot \mathbf{v}_1 = \mathbf{v}_1 \cdot \mathbf{v}_2$

Determine which of the following pairs of vectors are perpendicular to one another.

63. $\mathbf{i} + \mathbf{j}, \mathbf{i} - \mathbf{j}$

64. $\langle 3, 4 \rangle, \langle 4, -3 \rangle$

65. $\langle 1, -2 \rangle, \langle 2, 5 \rangle$

66. $2\mathbf{i} + 3\mathbf{j}, -4\mathbf{i} + \mathbf{j}$

Determine the angle between the following vectors.

67. $\mathbf{u} = \langle 1, 2 \rangle$, $\mathbf{v} = \langle 1, 1 \rangle$

68. $\mathbf{u} = \langle 1, 1 \rangle$, $\mathbf{v} = \langle 4, -4 \rangle$

69. $\mathbf{u} = \langle 0, 3 \rangle$, $\mathbf{v} = \langle -2, 1 \rangle$

70. $\mathbf{u} = \langle 5, -\frac{5}{2} \rangle$, $\mathbf{v} = \langle -1, 0 \rangle$

7.7 CHAPTER REVIEW

Important Concepts, Properties, and Formulas—Chapter 7

Law of sines	$\dfrac{\sin \alpha}{a} = \dfrac{\sin \beta}{b} = \dfrac{\sin \gamma}{c}$	p. 383
Law of cosines	$a^2 = b^2 + c^2 - 2bc \cos \alpha$ $b^2 = a^2 + c^2 - 2ac \cos \beta$ $c^2 = a^2 + b^2 - 2ab \cos \gamma$	p. 390

Area of triangle	$A = \frac{1}{2}ab\sin\gamma$ $A = \frac{1}{2}ac\sin\beta$ $A = \sqrt{s(s-a)(s-b)(s-c)}$, where $s = \dfrac{a+b+c}{2}$	pp. 392, 393
Conversion between rectangular and polar coordinates	$x = r\cos\theta$ $y = r\sin\theta$ $r = \sqrt{x^2 + y^2}$ $\tan\theta = \dfrac{y}{x}, \quad x \neq 0$	p. 399
Polar form	$z = r(\cos\theta + i\sin\theta)$ where $r = \sqrt{a^2 + b^2}$ and $\tan\theta = \dfrac{b}{a}$	p. 406
Polar products	$z_1 z_2 = r_1 r_2[\cos(\theta_1 + \theta_2) + i\sin(\theta_1 + \theta_2)]$	p. 408
Polar division	$\dfrac{z_1}{z_2} = \dfrac{r_1}{r_2}[\cos(\theta_1 - \theta_2) + i\sin(\theta_1 - \theta_2)]$	p. 408
De Moivre's theorem	$z^n = r^n(\cos n\theta + i\sin n\theta)$	p. 411
nth roots	$t_j = r^{1/n}\left[\cos\dfrac{\theta + 2\pi j}{n} + i\sin\dfrac{\theta + 2\pi j}{n}\right]$ $(j = 0, 1, \ldots, n-1)$	p. 412
Vectors	$|\mathbf{A}| = \sqrt{a^2 + b^2}$	pp. 416–421
	Horizontal component: $a = |\mathbf{A}|\cos\theta$	
	Vertical component: $b = |\mathbf{A}|\sin\theta$	
	Sum of $A = \langle a_1, b_1\rangle$ and $\mathbf{B} = \langle a_2, b_2\rangle$ $\mathbf{A} + \mathbf{B} = \langle a_1 + a_2, b_1 + b_2\rangle$	
	Scalar product of k and $\mathbf{A}$ $k\mathbf{A} = \langle ka_1, kb_1\rangle$	
	Dot product $\langle a, b\rangle \cdot \langle c, d\rangle = ac + bd$ $\mathbf{u} \cdot \mathbf{v} = |\mathbf{u}||\mathbf{v}|\cos\theta,$ $\theta = $ the angle between $\mathbf{u}$ and $\mathbf{v}$	

Cumulative Review Exercises—Chapter 7

Determine the remaining sides and angles.

1. $a = 15$, $b = 3$, $\angle A = 30°$
2. $c = 10$, $\angle C = 45°$, $\angle B = 30°$

Determine the number of solutions and find them for the given triangle.

3. $a = 260$, $b = 476$, $\angle A = 13°$

4. $b = 20$, $a = 40$, $\angle C = 30°$

◆ Applications

5. How long is a pole if, when it leans away from the sun at an angle of $7°$ to the vertical, the angle of elevation of the sun is $51°$ and the pole casts a shadow 47 ft long on level ground?

6. A tree grows vertically on the side of a hill that slopes upward from the horizontal by $8°$. When the angle of elevation of the sun measures $20°$, the shadow of the tree falls 42 ft down the hill. How tall is the tree?

7. One end of a 20-ft ladder is placed on the ground at a point 8 ft from the start of a $41°$ incline. If the other end rests on the incline, how far up the incline is this other end?

8. Two angles of a triangle are $30°$ and $55°$, and the longest side is 34 m. Find the length of the shortest side.

9. Points A and B are on opposite sides of Lake Tyrone. From a third point C, the angle between the line of sight to A and B is $46.3°$. If AC is 350 ft long and BC is 286 ft long, find AB.

10. A parallelogram has adjacent sides of length B and 10 inches, and the measure of one included angle is $35°$. Find the length of each diagonal of the parallelogram.

11. Two submarines, one cruising at 25 knots and the other at 20 knots, left a naval base at the same time. Three

hours later they were 100 nautical miles apart. What was the measure of the angle between their courses?

12. Two planes leave an airport together, traveling on courses that have an angle of $135°40'$ between them. If they each travel 402 miles, how far apart are they?

Determine the remaining sides and angles of the following triangles.

13. $a = 10$, $b = 12$, $\angle C = 45°$

14. $c = 20$, $a = 30$, $\angle C = \dfrac{\pi}{6}$

Find the area of the following triangles.

15. $a = 6$, $b = 14$, $\angle C = 5\pi/6$

16. $a = 26$, $b = 30$, $c = 8$

17. $a = 18$, $b = 25$, $\angle A = 120°$

18. $a = 10$, $b = 15$, $\angle C = \dfrac{\pi}{3}$

Evaluate the following.

19. $|1 - 2i|$

20. $|3 + 3i|$

21. $\left|-1 - \sqrt{3}i\right|$

22. $\left|-\sqrt{2} + \sqrt{2}i\right|$

For each of the following complex numbers, determine: (a) the polar form, (b) the modulus, and (c) the argument.

23. $z = 2i$

24. $z = 3 - 3i$

25. $z = -4 - \sqrt{3}i$

26. $z = 5i$

Graph the point corresponding to the following:

27. $3(\cos 30° + i \sin 30°)$

28. $4[\cos(-45°) + i \sin(-45°)]$

29. $2\left[\cos\left(-\dfrac{\pi}{3}\right) + i \sin\left(-\dfrac{\pi}{3}\right)\right]$

30. $5\left[\cos \dfrac{2\pi}{3} + i \sin \dfrac{2\pi}{3}\right]$

Compute zw and z/w for the following complex numbers in polar form.

31. $z = 5(\cos 90° + i \sin 90°)$
    $w = 2(\cos 45° + i \sin 45°)$

32. $z = 2[\cos(-30°) + i \sin(-30°)]$
    $w = 3(\cos 60° + i \sin 60°)$

33. $z = 3\left[\cos\left(-\dfrac{2\pi}{3}\right) + i \sin\left(-\dfrac{2\pi}{3}\right)\right]$
    $w = 2(\cos 2\pi + i \sin 2\pi)$

34. $z = 4\left[\cos\left(-\dfrac{\pi}{3}\right) + i \sin\left(-\dfrac{\pi}{3}\right)\right]$
    $w = 2\left[\cos\left(-\dfrac{5\pi}{6}\right) + i \sin\left(-\dfrac{5\pi}{6}\right)\right]$

Change from polar form to complex (standard) form.

35. $\sqrt{2}(\cos 45° + i \sin 45°)$

36. $\sqrt{3}[\cos(-60°) + i \sin(-60°)]$

37. $5[\cos(-2\pi) + i \sin(-2\pi)]$

38. $2\left(\cos \dfrac{\pi}{6} + i \sin \dfrac{\pi}{6}\right)$

Calculate the following and express results in $a + bi$ form.

39. $z^{10}; z = \left(\cos \dfrac{7\pi}{6} + i \sin \dfrac{7\pi}{6}\right)$

40. $z^{12}; z = \cos \pi + i \sin \pi$

41. $z^{30}; z = 2(\cos 225° + i \sin 225°)$

42. $z^3; z = 3\left[\cos\left(-\dfrac{\pi}{6}\right) + i \sin\left(-\dfrac{\pi}{6}\right)\right]$

43. $(2\sqrt{3} + 2i)^8$

44. $(3\sqrt{3} - 3i)^4$

Find the indicated roots and express them in $a + bi$ form if they are exact roots; otherwise, leave them in polar form.

45. Cube roots of 27.

46. Square roots of $16i$.

47. Fifth roots of $2 + 2i$.

48. Fourth roots of $8 + 8i\sqrt{3}$.

49. Square roots of $100[\cos(-60°) + i \sin(-60°)]$.

50. Cube roots of $125\left[\cos\left(\dfrac{3\pi}{4}\right) + i \sin\left(\dfrac{3\pi}{4}\right)\right]$.

Determine the complex solutions for the following.

51. $x^5 + 1 = 0$
52. $a^3 = 8$
53. $a^3 = 1 - \sqrt{3}i$
54. $x^4 + 8\sqrt{3} - 8i = 0$

Given vectors **A** and **B** in the figure, draw the following vectors.

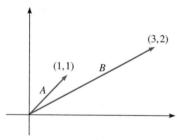

Exercises 55–58

55. **A** + **B**   56. **2A** − **B**   57. **3B** − **A**   58. **A** + **2B**

For the given vector **A**, with initial point at the origin and the given terminal point, determine (a) $|\mathbf{A}|$ and (b) the horizontal and vertical components of **A**.

59. vector **A** terminal point (2, 2)
60. vector **A** terminal point $(-1, -\sqrt{3})$
61. vector **A** terminal point $(-\sqrt{2}, \sqrt{2})$
62. vector **A** terminal point (3, −3)

♦ **Applications**

63. For a boat traveling at 30 mph to travel directly north across a river, it must aim at a point that has a bearing of 15°. If the current is flowing directly west, find the rate of the current.

64. At what compass heading and air speed should an aircraft fly if a wind of 40 mph is blowing from the north, and the pilot wants to maintain a ground speed of 200 mph on a true course of 90°?

65. A cable is attached to the top of a pole, and a 100-lb ball at the other end of the cable swings with a constant speed in a circle parallel to the ground. As it swings around, the cable makes an angle of 30° with its support. What is the centrifugal force on the ball and the force exerted on the ball by the cable?

66. A loading ramp makes an angle of 22° with the horizontal. A box weighing 130 lb slides down the ramp at a constant velocity. What is the force of friction acting on the box?

Determine the following vectors, given $\mathbf{A} = \langle 2, 5 \rangle$, $\mathbf{B} = \langle -2, -3 \rangle$, and $\mathbf{C} = \langle 1, -2 \rangle$.

67. **3A**   68. **A** + **B** + **C**
69. **2B** − **C**   70. **2C** − **3B**

Plot the following points given in polar coordinates:

71. $(-1, 180°)$ 72. $(3, -30°)$

73. $\left(-3, -\dfrac{3\pi}{4}\right)$ 74. $\left(3, -\dfrac{7\pi}{4}\right)$

Graph the given polar equation.

75. $r = 5$ 76. $r = 4\cos \theta$
77. $r = -6 - 6\cos \theta$ 78. $r = 2 + 4\sin \theta$

Determine the polar coordinates for each of the following points.

79. $(-2, 2)$ 80. $(1, -\sqrt{3})$
81. $(3, -3\sqrt{3})$ 82. $(-4, -4)$

Write the given equations in terms of polar coordinates.

83. $x = -3$

84. $y = 6$

85. $x^2 + y^2 = 4$

86. $x^2 + y^2 + 2y = 0$

Write the given polar equations in rectangular form.

87. $r = 2$

88. $r = 2 \sec \theta$

89. $r \cos \theta = 5$

90. $r = \tan \theta$

91. $r^2 = \dfrac{1}{\cos 2\theta}$

92. $r^2(4 \sin^2 \theta - 9 \cos^2 \theta) = 36$

93. The center of a regular hexagon is at the origin, and one vertex is at the point (5, 0°). Find the polar coordinates of the other vertices.

CHAPTER 8

LINEAR AND NONLINEAR SYSTEMS

*Applied mathematicians often must determine how to maximize
production of industrial goods in situations where a set of restrictions
on production are described by a system of inequalities.*

I n preceding chapters, it has often been necessary to solve equations and inequalities. Our attention has been concentrated, for the most part, on expressions in a single variable. In this chapter, we will turn to equations and inequalities involving several variables.

8.1 SYSTEMS OF EQUATIONS

In our applications thus far, the problems we have considered boiled down to solving a single equation in one unknown. However, in many applied problems, it is necessary to simultaneously satisfy several equations in several unknowns. For instance, consider the following problem. Suppose that a pension fund seeks to invest $1 million. In order to satisfy current and future obligations of the fund, it must earn $90,000 per year. Suppose that the fund has two sorts of investments open to it: stocks paying 8 percent per year and corporate bonds paying 12 percent per year. How much should the pension fund invest in each in order to achieve exactly the desired income?

To solve the problem, let $x = $ the amount invested in stocks and $y = $ the amount invested in bonds. Denote money amounts in multiples of $1 million. For instance, the required annual income from the portfolio is $0.09 million. The requirement that $1 million be invested may be expressed through the equation

$$x + y = 1$$

The requirement that the income of the portfolio be 0.09 may be expressed as

$$0.08x + 0.12y = 0.09$$

To solve the problem, we must determine x and y so that the two equations are *simultaneously* satisfied. The pair of equations is an example of a **system of equations** in the variables x and y.

In general, a system of equations is a list of equations in two or more variables. In this section, we will restrict ourselves to systems of two equations in two variables. For such systems, let us make the following definition.

Definition 1 Solution of a System of Equations	A **solution of a system in two variables** x and y is an ordered pair of numbers (a, b) such that all equations of the system are satisfied for $x = a$ and $y = b$.

For example, consider the system

$$\begin{cases} x - y = -1 \\ 5x - 4y = 3 \end{cases}$$

Then $(x, y) = (7, 8)$ is a solution to the system because when we substitute $x = 7$ and $y = 8$, both equations are satisfied:

$$\begin{cases} 7 - 8 = -1 \\ 5(7) - 4(8) = 3 \end{cases}$$

As we shall soon see, this is the only solution of the system. However, as we will illustrate in this chapter, some systems have a single solution, some a finite number of solutions, others infinitely many solutions, and still others no solutions at all.

Our goal in this chapter is to develop techniques to determine the solutions of systems of equations (if there are any). Actually, this is quite an ambitious goal because as the number of equations and number of variables increase, it becomes much more difficult to determine the solutions. However, we will go part way toward achieving our goal in developing techniques that can be used to solve a wide variety of systems.

The Method of Substitution

Very often, it is possible to solve one equation of a system for one variable in terms of another. When this is possible, we can often determine the solutions of the system by using the **method of substitution**, described as follows:

THE METHOD OF SUBSTITUTION (FOR TWO EQUATIONS IN TWO VARIABLES)

1. Solve an equation for one variable in terms of the other.
2. Substitute the expression found in step 1 into the other equation.
3. Solve the resulting equation for the possible values of one variable.
4. For each value found in step 3, substitute into the expression of step 1 to determine the corresponding value of the other variable.

The method of substitution is illustrated in the following example.

➢ EXAMPLE 1
Applying Substitution to a System of Two Linear Equations in Two Variables

Determine all solutions of the following system:

$$\begin{cases} x - y = 1 \\ 3x + 2y = 7 \end{cases}$$

Solution

Let's begin by solving the first equation for y in terms of x:

$$x - y = 1$$
$$y = x - 1 \qquad (1)$$

Next, we substitute the expression for y just obtained into the second equation:

$$3x + 2y = 7$$
$$3x + 2(x - 1) = 7$$
$$3x + 2x - 2 = 7$$
$$5x = 9$$
$$x = \tfrac{9}{5}$$

To obtain the value of y, we substitute the value for x into equation (1), which gives y in terms of x.

$$y = x - 1$$
$$= \tfrac{9}{5} - 1$$
$$= \tfrac{4}{5}$$

Thus, $x = \frac{9}{5}$ and $y = \frac{4}{5}$, so the system has a single solution, namely, $\left(\frac{9}{5}, \frac{4}{5}\right)$. We may check this solution by substituting the values for x and y into each equation:

$$
\begin{array}{c|c}
x - y = 1 & 3x + 2y = 7 \\
\frac{9}{5} - \frac{4}{5} = 1 & 3\left(\frac{9}{5}\right) + 2\left(\frac{4}{5}\right) = 7 \\
\frac{5}{5} = 1 & \frac{27}{5} + \frac{8}{5} = 7 \\
1 = 1 & \frac{35}{5} = 7 \\
 & 7 = 7
\end{array}
$$

Let's apply the method of substitution to solve the pension fund problem stated at the beginning of this section.

> **EXAMPLE 2**
Solving the Pension Fund Problem

Determine all solutions of the system

$$
\begin{cases}
x + y = 1 \\
0.08x + 0.12y = 0.09
\end{cases}
$$

Solution

Recall the pension fund problem posed at the beginning of the section. We set $x =$ the amount of money invested in stocks (in millions) and $y =$ the amount invested in bonds (in millions). We wished to determine x and y, which are a solution to the given system. Solve the first equation for y in terms of x:

$$y = 1 - x$$

Now substitute this expression into the second equation:

$$0.08x + 0.12(1 - x) = 0.09$$
$$-0.04x + 0.12 = 0.09$$
$$0.04x = 0.03$$
$$x = 0.75$$

Now substitute the value of x into the expression for y:

$$y = 1 - x = 1 - 0.75 = 0.25$$

That is, the pension fund should invest $750,000 in stocks and $250,000 in bonds.

Examples 1 and 2 applied the method of substitution to systems whose equations were linear. Here is an example of the method applied to solve a nonlinear system.

> **EXAMPLE 3**
Applying Substitution to a Nonlinear System

Solve the following system of equations in two variables:

$$
\begin{cases}
x^2 + 3y^2 = 28 \\
x + y = 4
\end{cases}
$$

Solution

We apply the method of substitution. The second equation may be simply solved for y in terms of x:

$$x + y = 4$$
$$y = 4 - x$$

Substitute the last expression for y into the first equation:

$$x^2 + 3y^2 = 28$$
$$x^2 + 3(4 - x)^2 = 28$$
$$x^2 + 3(16 - 8x + x^2) = 28$$
$$x^2 + 3x^2 - 24x + 48 = 28$$
$$4x^2 - 24x + 20 = 0$$
$$4(x^2 - 6x + 5) = 0$$
$$4(x - 5)(x - 1) = 0$$
$$x = 5, 1$$

In this example, there are two values of x. For each of these values, we go back to the equation expressing y in terms of x and obtain a corresponding value of y:

$$y = 4 - x \qquad\qquad y = 4 - x$$
$$y = 4 - 5 = -1 \qquad y = 4 - 1 = 3$$

There are two pairs that are solutions of the system:

$$(5, -1) \quad \text{and} \quad (1, 3)$$

We may check the solution by substituting each pair into the original equations of the system.

➢ **EXAMPLE 4**
A System with No Solutions

Solve the following system of linear equations:

$$\begin{cases} 3x - y = 1 \\ -6x + 2y = 4 \end{cases}$$

Solution

Solve the first equation for y in terms of x:

$$y = 3x - 1$$

Substitute this expression into the second equation:

$$-6x + 2y = 4$$
$$-6x + 2(3x - 1) = 4$$
$$-6x + 6x - 2 = 4$$
$$-2 = 4$$

The last equation is inconsistent and so has no solutions. Therefore, the system has no solutions.

Graphical Analysis of Systems in Two Variables

As we have seen in preceding chapters, an equation in two variables has a graph that is a curve in the x–y plane. The points on the curve correspond to the ordered pairs (x, y) that satisfy the equation.

We may graph a system of equations by graphing separately each equation of the system. The solutions of the system correspond to the points where the graphs intersect. For example, Figure 1 shows the graph of the system in Example 1. Each equation has a straight line as a graph. The intersection point of the two lines has coordinates that correspond to the solution of the system. Figure 2 shows the graph of the system in the pension fund problem. Again the graph of each equation is a straight line, and the intersection of the lines corresponds to the solution of the system. Figure 3 shows the graph of the nonlinear system in Example 3. In this case, there are two solutions, one corresponding to each intersection point of the two graphs. Finally, Figure 4 shows the graph of the system in Example 4. The fact that the two lines don't intersect reflects the result of Example 4 that the system has no solutions.

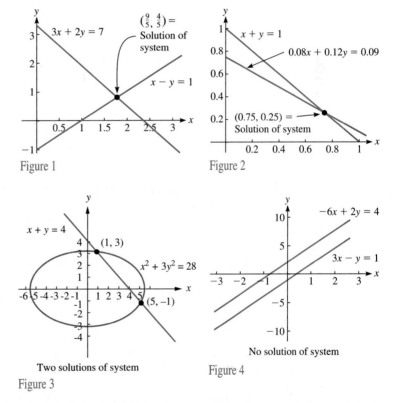

Figure 1

Figure 2

Two solutions of system

Figure 3

Figure 4

Sometimes it is difficult to solve a system using algebraic methods. In such cases, the solutions may usually be approximated numerically using a graphing calculator, as shown in the following example.

➤ EXAMPLE 5
Graphical Solution of a System

Solve the following system graphically:

$$\begin{cases} y = x^2 \\ y = \sin x \end{cases}$$

Solution

To solve this system by substitution would require that we solve an equation of the form

$$\sin x = x^2$$

which is impossible to do exactly. However, we can approximate the solutions of the system numerically by drawing the graph, as shown in Figure 5. It is clear from the graph that the system has two solutions. The first is $(0, 0)$, and the second corresponds to a value of x between 0 and 1. By enlarging the portion of the graph corresponding to $0.7 \le x \le 1.0$ (see Figure 6), we see that with single-digit accuracy, we have $(x, y) = (0.9, 0.8)$. By further enlarging portions of the graph, we could read off the second solution with greater accuracy.

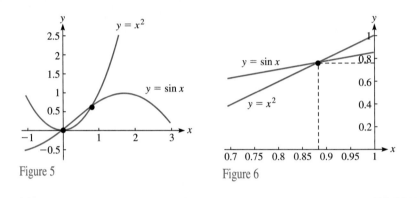

Figure 5   Figure 6

Applications of Systems in Two Variables

Here are some applied problems that give rise to systems of two equations in two variables.

➤ EXAMPLE 6
Farming

A rectangular cow pen is fenced on three sides, with the fourth side a barn. The fence opposite the barn cost $4 per foot. The other sides cost $3 per foot. The amount of fencing needed is 400 feet, and the cost of the fencing is $1400. (See Figure 7.) What are the dimensions of the pen?

Solution

In the figure, we have labeled the side opposite the barn as x and the other sides as length y. The amount of fencing required is $x + 2y$, so we have the equation

$$x + 2y = 400$$

The cost of the side opposite the barn is $4x$, and the cost of the other sides is $3y + 3y = 6y$. Therefore, the total cost of the fencing is $4x + 6y$, so that we have the equation

$$4x + 6y = 1400$$

Thus, we must solve the system of linear equations

$$\begin{cases} x + 2y = 400 \\ 4x + 6y = 1400 \end{cases}$$

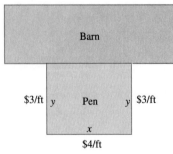

Figure 7

To do so, we use the method of substitution:

$$x = 400 - 2y$$
$$4x + 6y = 1400$$
$$4(400 - 2y) + 6y = 1400 \qquad \text{(Substitution in second equation)}$$
$$-2y = -200$$
$$y = 100$$
$$x = 400 - 2 \cdot 100$$
$$= 200$$

That is, the side opposite the barn is 200 ft, and the other two sides are each 100 ft. A simple check shows that these lengths satisfy the conditions set forth.

➤ **EXAMPLE 7**
Architectural Design

A cylindrical water cistern is to contain 10,000 cubic feet of water, and its sides are to be constructed of 5000 square feet of aluminum. What are the dimensions of the cistern?

Solution

Let r denote the radius of the cistern and h its height. (See Figure 8.) By the formula for the surface area of a cylinder, the area of the side of the cistern is equal to $2\pi rh$. Therefore, we have the equation

$$2\pi rh = 5000$$

By the formula for the volume of a cylinder, the volume of the cistern is equal to $\pi r^2 h$. Therefore, we have the second equation

$$\pi r^2 h = 10,000$$

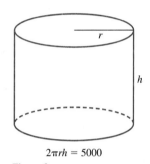

$2\pi rh = 5000$

Figure 8

To solve the problem, we must solve the system of equations

$$\begin{cases} 2\pi rh = 5000 \\ \pi r^2 h = 10,000 \end{cases}$$

We may solve the second equation for h in terms of r:

$$h = \frac{10,000}{\pi r^2} \qquad (2)$$

Substituting this expression for h into the first equation gives us

$$2\pi rh = 5000$$
$$2\pi r \cdot \frac{10,000}{\pi r^2} = 5000$$
$$\frac{20,000}{r} = 5000$$
$$r = 4 \text{ feet}$$

Substituting this value into equation (2), we have

$$h = \frac{10,000}{\pi r^2}$$
$$= \frac{10,000}{16\pi}$$
$$= 198.94 \text{ ft}$$

Exercises 8.1

Determine all solutions of the following systems.

1. $\begin{cases} 3x - 6y = -5 \\ 2x + 9y = 1 \end{cases}$

2. $\begin{cases} 3x + 8y = 4 \\ 12y + 6x = -1 \end{cases}$

3. $\begin{cases} x - 3y = 17 \\ 2x + y = 6 \end{cases}$

4. $\begin{cases} x - 3y = 10 \\ 3x + 2y = 2 \end{cases}$

5. $\begin{cases} 3x + 2y = 4 \\ 5x + 3y = 2 \end{cases}$

6. $\begin{cases} \frac{3}{2}x = 2 + \frac{5}{4}y \\ \frac{1}{2}x = \frac{3}{2} - \frac{5}{3}y \end{cases}$

7. $\begin{cases} x^2 + y^2 = 16 \\ y - 2x = 3 \end{cases}$

8. $\begin{cases} x + y = 1 \\ x^2 - y^2 = 4 \end{cases}$

9. $\begin{cases} x + 3 = y \\ 4y^2 - 16 = x^2 \end{cases}$

10. $\begin{cases} x + 2y = 3 \\ y^2 = 9 - x^2 \end{cases}$

11. $\begin{cases} 4y = 1 + 3x \\ 2x^2 - 8y^2 + 3x + 2y = 2 \end{cases}$

12. $\begin{cases} 5x - 2y = 6 \\ 4x^2 + 4x - y^2 - 4y = 12 \end{cases}$

13. $\begin{cases} xy + x^2 + 2y = 3 \\ 3x - 2y + 2 = 0 \end{cases}$

14. $\begin{cases} 5xy = 4x + y \\ x = 5 - 4y \end{cases}$

15. $\begin{cases} 3x^2 - 2y^2 = 9 \\ x^2 - y^2 = -1 \end{cases}$

16. $\begin{cases} x^2 + y^2 = 4 \\ x^2 - 2y^2 = 1 \end{cases}$

17. $\begin{cases} xy = 8 \\ y = -\frac{1}{2}x^2 + 2x + 2 \end{cases}$

18. $\begin{cases} xy = 8 \\ \frac{1}{x} + \frac{1}{y} = \frac{1}{4} \end{cases}$

◆ Applications

19. Tickets for a pre-Broadway showing of a play sold at $40.00 for the main floor and $27.50 for the balcony. If the receipts from the sale of 1600 tickets were $55,250.00, how many tickets were sold at each price?

20. Tickets to a class play were $2.50 and $5.00. In all, 275 tickets were sold, and the total receipts were $1187.50. Find the number of tickets sold at each price.

21. The atomic number of antimony is 1 less than four times the atomic number of aluminum. If twice the atomic number of aluminum is added to that of antimony, the result is 77. What are the atomic numbers of aluminum and antimony?

22. Two temperature scales are established. In the x scale, water freezes at 10° and boils at 300°, and in the y scale, water freezes at 30° and boils at 1430°. If x and y are linearly related, find an expression for any temperature y in terms of a temperature x.

23. You have $5000 to invest for one year. Part of your investment went into a risky stock that paid 12 percent annual interest, and the rest went into a savings account that paid 6 percent annual interest. If your annual return was $540 from both investments, how much was invested at each rate?

24. Suppose you invested a certain sum of money at 8.5 percent and another sum at 9.5 percent. If your total investment was $17,000 and your total interest for one year of both investments was $1535, how much did you invest at each rate?

25. Find two numbers whose sum is 10 and the sum of whose reciprocals is $\frac{5}{12}$.

26. Find two numbers whose difference is 6 and the difference of whose reciprocals is $\frac{2}{9}$.

27. Find the value of C for which the graph of $y = 2x + C$ is tangent to the graph of $y = x^2 + 1$. (*Hint:* There is only one point of intersection.)

28. Find the value of C for which the graph of $y = 2x + C$ is tangent to $x^2 + y^2 = 25$. (*Hint:* There is only one point of intersection.)

29. The sum of the tens digit and the ones digit of a three-digit number is 5. The product of the three digits is 24, and the hundreds digit is 2 less than twice the ones digit. Find the number.

30. The sum of the digits of a three-digit number is 8, the product of the digits is 10, and if the hundreds digit is subtracted from the ones digit, the result is 4. Find the number.

31. Juan wants to make a window box from a rectangular sheet of metal whose area is 680 cm². He cuts 6-cm squares from each corner and folds the sides to make a box with base area of 176 cm². Find the dimensions of the original metal sheet.

32. An art dealer bought some paintings for $360. When a thief stole two of them, he sold the rest for $380, thus making a profit of $8 on the original cost of each picture. How many pictures did he buy?

33. Find two numbers such that the square of their sum is 20 more than the square of their difference, and the difference of their squares is 24.

34. Nguyen made a round trip between two cities 270 miles apart. On his return trip, heavy snow reduced his average rate by 5 miles per hour and increased his traveling time by $\frac{3}{4}$ hours. What was his original rate?

35. A number of people shared the rental price of a skiing chalet equally. If two more people had shared the chalet, each would have paid $12 less. If there had been three fewer persons, each would have paid $24 more. How much did each pay?

36. **East–west air travel.** Airline schedules allow 3 hrs, 31 min (terminal to terminal) to fly from Chicago to Phoenix, and 3 hrs, 9 min to fly from Phoenix to Chicago. Of this, about $\frac{1}{2}$ hr might be used in getting off the ground and getting back down, so the actual flying times at cruising speed might be 3 hrs in one direction and 2.7 hrs in the other. The difference is in the jet stream, a wind that usually blows west to east.

   a. If there were no wind, or jet stream, how much time would you expect a flight to take?

   b. The airline distance between Chicago and Phoenix is 1453 mi. Estimate the average speed of the jet stream and the average speed of a plane.

37. **Catering costs.** A caterer finds that the expenses for a breakfast consist of fixed costs and costs per guest. A meal for 40 guests costs $150 and for 100 guests costs $225. Find the fixed costs and the cost per guest.

8.2 LINEAR SYSTEMS IN TWO VARIABLES

In this section, we will concentrate on a second method for solving systems of two linear equations in two unknowns.

The Idea of Elimination

Linear equations in one variable are simple to solve. One method, called **elimination**, of solving a system of two linear equations in two variables is to eliminate one of the variables from the system to obtain a single equation in one unknown. This procedure is illustrated in the following example.

➤ EXAMPLE 1
Solving Using Elimination

Use elimination to solve the following system:

$$\begin{cases} 2x + 3y = 10 \\ x - y = 5 \end{cases}$$

Solution

Let's eliminate x by creating equations with equal x-coefficients and subtracting:

$$\begin{cases} -2x - 3y = -10 & \textit{Multiply first equation by } -1 \\ 2x - 2y = 10 & \textit{Multiply second equation by 2} \end{cases}$$

$$-5y = 0 \qquad \textit{Add second equation to first equation}$$

$$y = 0$$

$$2x + 3(0) = 10 \qquad \textit{Substitute into first equation}$$

$$2x = 10$$

$$x = 5 \qquad \textit{Solve for x}$$

We may summarize the process of elimination as follows:

THE METHOD OF ELIMINATION
(TWO LINEAR EQUATIONS IN TWO VARIABLES)

1. Multiply one or both equations of the system by suitable numbers to obtain a new system in which the coefficients of one of the variables are negatives of one another.
2. Add the equations to obtain a linear equation in one variable.
3. Solve the equation obtained in step 2.
4. Substitute the value obtained into either of the original system equations and solve for the second variable.

Graphical Analysis of the Solutions

The graph of each equation in a linear system in two variables is a straight line. The solutions of the system correspond to points where the lines intersect. There are three possible configurations:

1. The graphs of the equations correspond to two distinct lines. In this case, the system has a single solution, corresponding to the point of intersection of the lines. (See Figure 1.)
2. The two straight lines coincide. In this case, the system has an infinite number of solutions, corresponding to all points on the line. (See Figure 2.)
3. The two straight lines are different but parallel. In this case, the system has no solutions. (See Figure 3.)

Figure 4 provides an illustration of a system with a single solution. It shows the graphs of the two equations and the solution (intersection point). The next example provides an example corresponding to the configuration in Figure 2.

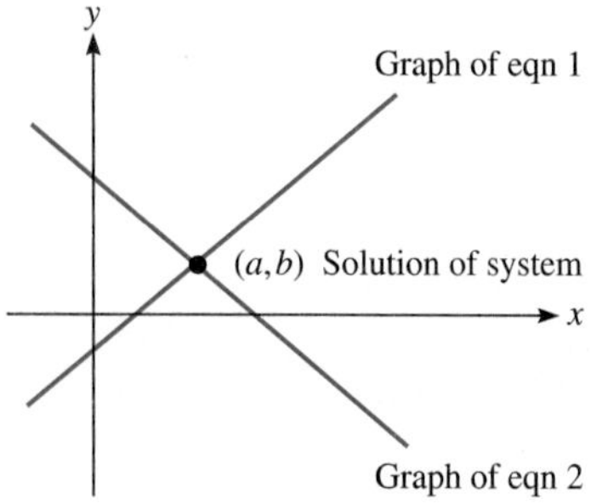

Figure 1

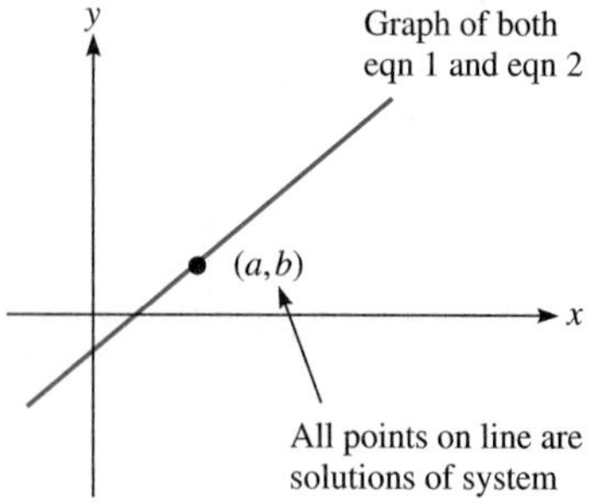

Figure 2

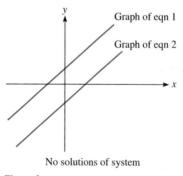

No solutions of system

Figure 3

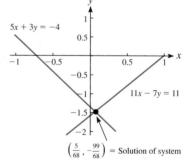

$\left(\frac{5}{68}, -\frac{99}{68}\right)$ = Solution of system

Figure 4

➤ EXAMPLE 2
A System in Two Variables with Infinitely Many Solutions

Solve the following system of linear equations:

$$\begin{cases} 3x + 8y = 12 \\ \frac{3}{4}x + 2y = 3 \end{cases}$$

Solution

Eliminate x:

$$\begin{cases} 3x + 8y = 12 \\ 3x + 8y = 12 \qquad \textit{Multiply equation 2 by 4} \\ 0 = 0 \qquad \textit{Subtract equation 2 from equation 1} \end{cases}$$

This equation is satisfied by any value of y. For a particular value of y, the corresponding value of x is obtained by solving one of the equations for x in terms of y:

$$3x = 12 - 8y$$
$$x = 4 - \tfrac{8}{3}y$$

Figure 5 shows the graphs of the two equations of the system. Both equations have the same graph, because one is four times the other. Any point on the line shown is a solution to the system.

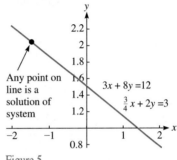

Figure 5

➤ **EXAMPLE 3**
A System in Two Variables
with No Solutions

Determine all solutions of the system

$$\begin{cases} 2x - 10y = 1 \\ 3x - 15y = 7 \end{cases}$$

Solution

Eliminate x:

$$\begin{cases} 6x - 30y = 3 \qquad \textit{Multiply equation 1 by 3} \\ 6x - 30y = 14 \qquad \textit{Multiply equation 2 by 2} \\ 0 = -11 \qquad \textit{Subtract equation 2 from equation 1} \end{cases}$$

Clearly, the last equation has no solutions, so the same can be said of the system. In this case, the graphs of the two equations are parallel, as shown in Figure 6.

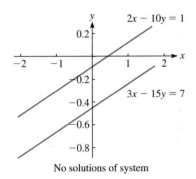

No solutions of system

Figure 6

Application

Let's apply the method of elimination to solve a linear system that arises in an application.

➤ **EXAMPLE 4**
Aviation

A plane can fly against the wind and cover a 500-mile distance in 2 hours. The same plane can fly with the wind and cover the same route in the opposite direction in 1.5 hours. What is the speed of the plane in still air, and what is the speed of the wind?

Solution

Let v denote the speed of the plane in still air, and let w denote the speed of the wind. Against the wind, the speed of the plane is $v - w$. Because the plane covers 500 miles in 2 hours, we have

$$\text{Distance} = \text{Rate} \cdot \text{Time}$$
$$500 = (v - w) \cdot 2$$
$$2v - 2w = 500$$

With the wind, the speed of the plane is $v + w$. Because it takes 1.5 hours to cover the same 500 miles in this case, we have

$$500 = 1.5(v + w)$$
$$1.5v + 1.5w = 500$$

To solve the problem, we must solve the following system:

$$\begin{cases} 2v - 2w = 500 \\ 1.5v + 1.5w = 500 \end{cases}$$

$$\begin{cases} 6v - 6w = 1,500 & \textit{Multiply equation 1 by 3} \\ 6v + 6w = 2,000 & \textit{Multiply equation 2 by 4} \end{cases}$$

$$12v = 3,500 \qquad \textit{Add equations}$$
$$v = 291.67 \text{ mph} \qquad \textit{Solve}$$
$$2 \cdot 291.67 - 2w = 500 \qquad \textit{Substitute in equation 1}$$
$$w = 41.67 \text{ mph}$$

Thus, the airplane goes 291.67 mph in still air, and the wind speed is 41.67 mph.

Exercises 8.2

Use elimination to determine all solutions of the following systems of equations.

1. $\begin{cases} 2x - y = 0 \\ 3x + y = 5 \end{cases}$

2. $\begin{cases} 3x + y = 7 \\ x - y = 1 \end{cases}$

3. $\begin{cases} 3x + 4y = -3 \\ 2x + 2y = 2 \end{cases}$

4. $\begin{cases} 3x - y = 3 \\ 2x + 2y = -6 \end{cases}$

5. $\begin{cases} 2x + 4y = 0 \\ 3x + 6y = 6 \end{cases}$

6. $\begin{cases} 3x - 6y = -5 \\ 3x + 9y = 1 \end{cases}$

7. $\begin{cases} 3x + 8y = 4 \\ 12y + 6x = -1 \end{cases}$

8. $\begin{cases} x - 3y = 17 \\ 2x + y = 6 \end{cases}$

9. $\begin{cases} x - 3y = 10 \\ 3x + 2y = 2 \end{cases}$

10. $\begin{cases} 3x + 2y = 4 \\ 5x + 3y = 2 \end{cases}$

11. $\begin{cases} \frac{3}{2}x = 2 + \frac{5}{4}y \\ \frac{1}{2}x = \frac{3}{2} - \frac{5}{3}y \end{cases}$

12. $\begin{cases} \frac{9}{2}x - 4y = -3 \\ \frac{4}{3}x - \frac{1}{2}y = \frac{7}{6} \end{cases}$

♦ Applications

13. The perimeter of a rectangular pool is 84 meters. The width of the pool is $\frac{3}{4}$ of the length. What are the length and the width of the pool?

14. The perimeter of a rectangular pool is 70 meters. The length is $\frac{4}{3}$ the width of the pool. What are the width and the length of the pool?

15. A bank contains $37 in quarters and dimes. The number of dimes is 10 more than twice the number of quarters. How many dimes and how many quarters are there in the bank?

16. Tickets to a high school play cost $4.50 and $6.00. If 450 tickets were sold and the receipts were $2550, how many tickets were sold at each price?

17. A lawn service company has two mowers, a new one and an old one. With both mowers working together, a certain job is done in 2 hours and 24 minutes. On another job of the same kind, the old mower worked alone for 3 hours and then was joined by the new mower, and the two mowers finished the job in an additional 1 hour and 12 minutes. How long would it take each mower operating alone to do the second job?

18. One bricklayer can build the wall to a standard basement in 6 hours less than a second bricklayer. If the faster bricklayer works for 2 hours alone on a standard

basement and is then joined by the second bricklayer, they complete the basement in an additional $1\frac{3}{7}$ hours working together. How long would it take each bricklayer working alone to build a standard basement?

19. Traveling downstream, the riverboat *New Baltimore* can travel 12 miles in 2 hours. Going up the same river, it can travel $\frac{2}{3}$ of this distance in twice the time. What is the rate of the river?

20. Flying with the wind, a SAC bomber can cover a distance of 3360 miles to its target in 5 hours and 15 minutes. On its return against the same wind, the trip would require 6 hours. Find the rate of the wind.

Solve the following systems for x and y in terms of a and b.

21. $\begin{cases} x + y = 2a \\ 2x + y = 3a + b \end{cases}$

22. $\begin{cases} 2x + y = \dfrac{a}{b} \\ x + y = 0 \end{cases}$

8.3 LINEAR SYSTEMS IN ANY NUMBER OF VARIABLES

In the preceding section, we learned to solve systems of two linear equations in two variables using elimination. Let's consider systems of linear equations in any number of variables. We will develop a method called **Gauss elimination** for determining all solutions of such a system.

Systems in Triangular Form

Let's begin by introducing a type of linear system that is very easy to solve. Consider a linear system in the variables $x, y, z, w, \ldots$. We say that the system is in **triangular form** (also called **echelon form**) provided that:

1. The second equation does not involve x, the third equation does not involve x and y, the fourth equation does not involve x, y, and z, and so forth.

2. The first nonzero coefficient of each equation is 1.

Here are two systems in triangular form:

$$\begin{cases} x + y + z = -1 \\ y - 2z = 4 \\ z = 4 \end{cases}$$

$$\begin{cases} x - 5y + z = -1 \\ z = 6 \end{cases}$$

In each system, the second equation does not involve x and the third equation does not involve x and y. Note that in the second system, the second equation has all 0 coefficients and is omitted. However, in terms of triangular form, we consider it as part of the system.

A triangular system is simple to solve. Just start from the last equation and determine the value of the last variable, then use the next-to-last equation to determine the value of the next-to-last variable, and so forth. The next example illustrates how this is done.

➤ **EXAMPLE 1**
Solving a Triangular System

Solve the following system of equations in triangular form:

$$\begin{cases} x + 2y - 5z = -1 \\ y - 4z = 7 \\ z = -2 \end{cases}$$

Solution

Start from the bottom equation and work upward. The bottom equation gives the value of z, namely, $z = -2$. Proceed upward to the next equation and substitute the value of z just obtained:

$$y - 4z = 7$$
$$y - 4(-2) = 7$$
$$y = -1$$

Now we have determined the values of y and z. Move up to the first equation and substitute these values:

$$x + 2y - 5z = -1$$
$$x + 2(-1) - 5(-2) = -1$$
$$x = -9$$

Thus, the system has the single solution

$$(x, y, z) = (-9, -1, -2)$$

Back substitution, the method just illustrated, may be used to solve any system in triangular form.

➤ **EXAMPLE 2**
Triangular System with an Infinite Number of Solutions

Solve the triangular system:

$$\begin{cases} x - y + z = 5 \\ y - z = 4 \end{cases}$$

Solution

The equation that would allow you to determine the value of z is missing. Let z be any real number: In terms of this choice for z, we may find y from the second equation:

$$y = 4 + z$$

We may find x from the first equation:

$$x = y - z + 5$$
$$= (4 + z) - z + 5$$
$$= 9$$

The solution of the system is

$$z = \text{any value}, \quad y = 4 + z, \quad x = 9$$

This system has an infinite number of solutions.

➤ **EXAMPLE 3**
Another System with an Infinite Number of Solutions

Determine all solutions of the triangular system:

$$\begin{cases} x + 3y - 2z = 1 \\ z = 4 \end{cases}$$

Solution

The equation used to determine the value of y is missing. Let y have any value. In terms of this value, we obtain the solution

$$z = 4$$
$$x = -3y + 2z + 1$$
$$= -3y + 2 \cdot 4 + 1$$
$$= -3y + 9$$

The solutions of the system are

$$y = \text{any value}, \quad x = -3y + 9, \quad z = 4$$

Elementary Operations on a Linear System

We can transform a linear system into triangular form using a sequence of the following operations, which don't change the solutions of the system.

ELEMENTARY OPERATIONS ON LINEAR SYSTEMS

1. Interchange two equations.
2. Multiply an equation by a nonzero real number.
3. Add to an equation a multiple of another equation.

These elementary operations are similar to the operations we used to generate equivalent equations in Section 1.3.

The next example illustrates how these operations can be used to transform a system into triangular form.

➤ **EXAMPLE 4**
Reducing a System in Two Variables to Triangular Form

Consider the following system of linear equations:
$$\begin{cases} 3x - y = -5 \\ x + y = 2 \end{cases}$$

1. Use a sequence of elementary row operations to transform the system into echelon form.
2. Determine all solutions of the system.

Solution

1. Start by interchanging the first equation with the second (to avoid fractions where possible):
$$\begin{cases} x + y = 2 \\ 3x - y = -5 \end{cases}$$

Multiply the first equation by -3 and add it to the second equation:
$$\begin{cases} x + y = 2 \\ -4y = -11 \end{cases}$$

Multiply the second equation by $-\frac{1}{4}$:
$$\begin{cases} x + y = 2 \\ y = \frac{11}{4} \end{cases}$$

2. Back-substituting $y = \frac{11}{4}$ into the first equation, we have
$$x + \frac{11}{4} = 2$$
$$x = -\frac{3}{4}$$

Therefore, the system has the single solution
$$\left(-\frac{3}{4}, \frac{11}{4}\right)$$

The main advantage of using the echelon form in solving a linear system is that the method works for any number of equations in any number of variables, and the method, although tedious, is simple to describe in a computer program. This is the preferred method for solving large systems. In the next section, we will discuss in detail how it works for solving linear systems in three or more variables.

➤ EXAMPLE 5
Reducing a System with Three Variables to Triangular Form

Use a sequence of elementary operations to transform the following system into triangular form. Determine all solutions of the system.

$$\begin{cases} \quad\quad 3y + 4z = 15 \\ x - y - z = -4 \\ 2x + 5y + 9z = 39 \end{cases}$$

Solution

Because the first equation has no x term, we interchange the first and second equations to obtain the system

$$\begin{cases} x - y - z = -4 \\ 3y + 4z = 15 \\ 2x + 5y + 9z = 39 \end{cases}$$

Next, add (-2) times the first equation to the third:

$$\begin{cases} x - y - z = -4 \\ 3y + 4z = 15 \\ 7y + 11z = 47 \end{cases}$$

Next, multiply the second equation by $\frac{1}{3}$ to obtain

$$\begin{cases} x - y - z = -4 \\ y + \frac{4}{3}z = 5 \\ 7y + 11z = 47 \end{cases}$$

Next, add (-7) times the second equation to the third equation:

$$\begin{cases} x - y - z = -4 \\ y + \frac{4}{3}z = 5 \\ \frac{5}{3}z = 12 \end{cases}$$

Finally, multiply the third equation by $\frac{3}{5}$ to obtain the triangular form

$$\begin{cases} x - y - z = -4 \\ y + \frac{4}{3}z = 5 \\ z = \frac{36}{5} \end{cases}$$

We now use back-substitution to solve the triangular system:

$$z = \frac{36}{5}$$
$$y = 5 - \frac{4}{3}z = 5 - \frac{48}{5} = -\frac{23}{5}$$
$$x = y + z - 4 = -\frac{23}{5} + \frac{36}{5} - 4 = -\frac{7}{5}$$

Matrices and Gauss Elimination

The elementary row operations on a system of linear equations involve only the numbers that appear on both sides of the equations; they don't involve the variable names in any way. Therefore, as a shorthand, it is convenient to write the system

$$\begin{cases} 3x - 2y = 4 \\ x + 5y = 0 \end{cases}$$

as the rectangular array of numbers:

$$\begin{bmatrix} 3 & -2 & 4 \\ 1 & 5 & 0 \end{bmatrix}$$

Actually, rectangular arrays such as this one occur often in mathematics and are called **matrices**. An extensive theory of matrices has been developed over the last century, and matrices have been used to describe diverse phenomena, including elementary particles in physics and the interaction of various segments of an economy.

Definition 1 **Matrix, Entry of a Matrix**	A **matrix** is a rectangular array $$\begin{bmatrix} a_{11} & a_{12} & \cdots & a_{1n} \\ a_{21} & a_{22} & \cdots & a_{2n} \\ \vdots & \vdots & \vdots & \vdots \\ a_{m1} & a_{m2} & \cdots & a_{mn} \end{bmatrix}$$ where $a_{11}, a_{21}, \ldots$ are algebraic expressions, called the **elements** or **entries** of the matrix.

The size of a matrix is measured by the numbers of rows and columns it contains. If a matrix contains m rows and n columns, it is called an $m \times n$ (read: m by n) matrix. For example, the matrix $A = \begin{bmatrix} 1 & 2 \\ 3 & 4 \end{bmatrix}$ is 2×2, whereas the matrix $B = \begin{bmatrix} -1 & 0 & \pi \\ \sqrt{2} & e & -3 \end{bmatrix}$ is 2×3. Note that in specifying the size of a matrix, the number of rows always comes first. The rows and columns are numbered beginning from 1, with the top row and the left-hand column each numbered 1.

The elements of a matrix are often denoted, as in the example, using subscript notation a_{ij}. In this notation, the first subscript, i, refers to the row number, and the second subscript, j, refers to the column number. For example, a_{32} refers to the element in row 3, column 2. Consider the matrix

$$\begin{array}{cccc} & \text{column 1} & \text{column 2} & \text{column 3} \\ & \downarrow & \downarrow & \downarrow \\ \text{row 1} \rightarrow & 3 & 5 & 4 \\ \text{row 2} \rightarrow & 1 & -2 & 0 \\ \text{row 3} \rightarrow & 2 & -7 & -3 \end{array}$$

element a_{32} is -7.

The elements $a_{11}, a_{22}, \ldots$, whose row and column numbers are equal, are said to form the **main diagonal** of the matrix. A matrix with an equal number of rows and columns is called a **square matrix**.

Suppose that we are given a system of linear equations:

$$\begin{cases} a_{11}x_1 + a_{12}x_2 + \cdots + a_{1n}x_n = b_1 \\ a_{21}x_1 + a_{22}x_2 + \cdots + a_{2n}x_n = b_2 \\ \vdots \\ a_{m1}x_1 + a_{m2}x_2 + \cdots + a_{mn}x_n = b_m \end{cases}$$

The matrix

$$\begin{bmatrix} a_{11} & a_{12} & \cdots & a_{1n} \\ a_{21} & a_{22} & \cdots & a_{2n} \\ \vdots & \vdots & \vdots & \vdots \\ a_{m1} & a_{m2} & \cdots & a_{mn} \end{bmatrix}$$

is called the **coefficient matrix** of the system. The matrix

$$\begin{bmatrix} a_{11} & a_{12} & \dots & a_{1n} & b_1 \\ a_{21} & a_{22} & \dots & a_{2n} & b_2 \\ \vdots & \vdots & \vdots & \vdots & \vdots \\ a_{m1} & a_{m2} & \dots & a_{mn} & b_m \end{bmatrix}$$

is called the **augmented matrix** of the system. For example, consider the system

$$\begin{cases} 5x - 2y = 3 \\ 3x + 9y = 1 \end{cases}$$

The coefficient matrix is

$$\begin{bmatrix} 5 & -2 \\ 3 & 9 \end{bmatrix}$$

and the augmented matrix is

$$\begin{bmatrix} 5 & -2 & 3 \\ 3 & 9 & 1 \end{bmatrix}$$

We may perform elementary row operations on the augmented matrix rather than on the system in order to solve the corresponding system of linear equations. Here is a summary of the elementary row operations in matrix form:

ELEMENTARY ROW OPERATIONS ON MATRICES

$R_i \leftrightarrow R_j$: Interchange rows i and j.

kR_i: Multiply row i by a nonzero real number k.

$kR_i + R_j$: Add k times row i to row j.

The notations introduced in this list are a compact way of describing elementary row operations without using long phrases. The next example provides some practice in using this notation.

➢ **EXAMPLE 6**
Elementary Row Operations

Consider the matrix

$$\begin{bmatrix} 3 & 0 & 1 \\ -1 & 5 & 8 \\ 0 & -4 & 7 \end{bmatrix}$$

Perform the following elementary row operations:

1. $R_1 \leftrightarrow R_3$
2. $\frac{1}{2}R_2$
3. $3R_1 + R_2$

Solution

1. The elementary row operation specified requires us to interchange rows 1 and 3. The result is

$$\begin{bmatrix} 0 & -4 & 7 \\ -1 & 5 & 8 \\ 3 & 0 & 1 \end{bmatrix}$$

2. The elementary row operation specified requires us to multiply row 2 by $\frac{1}{2}$. The result is

$$\begin{bmatrix} 3 & 0 & 1 \\ -\frac{1}{2} & \frac{5}{2} & 4 \\ 0 & -4 & 7 \end{bmatrix}$$

3. The elementary row operation specified requires us to add three times row 1 to row 2. The result is

$$\begin{bmatrix} 3 & 0 & 1 \\ 8 & 5 & 11 \\ 0 & -4 & 7 \end{bmatrix}$$

Corresponding to the echelon form of a linear system, we make the following definition:

Definition 2
Echelon Form

Let A be a matrix. We say that A is in **echelon form** provided that:

1. The first nonzero element of each nonzero row is a 1.
2. The position of the first 1 in a row is to the right of the first 1 in the row above it.
3. Any all-zero rows are at the bottom of the matrix.

Here are some matrices in echelon form:

$$\begin{bmatrix} 1 & 1 & 2 & -5 \\ 0 & 1 & 3 & 2 \\ 0 & 0 & 1 & 4 \end{bmatrix}$$

$$\begin{bmatrix} 1 & -1 & 5 & 4 \\ 0 & 0 & 1 & 9 \\ 0 & 0 & 0 & 0 \end{bmatrix}$$

By using a sequence of elementary operations, we may transform a matrix into echelon form. The next example illustrates a typical calculation.

$\succ$ **EXAMPLE 7**
Transforming a Matrix Into Echelon Form

Use a sequence of elementary operations to transform the following matrix into echelon form:

$$\begin{bmatrix} 4 & 1 & 0 & -1 \\ 1 & 0 & 1 & 0 \\ 2 & -1 & 3 & 0 \end{bmatrix}$$

Solution

We transform a column at a time, proceeding from the top of the column down, taking subsequent columns in left-to-right order:

$$\begin{bmatrix} 4 & 1 & 0 & -1 \\ 1 & 0 & 1 & 0 \\ 2 & -1 & 3 & 0 \end{bmatrix} \xrightarrow{\frac{1}{4}R_1} \begin{bmatrix} 1 & \frac{1}{4} & 0 & -\frac{1}{4} \\ 1 & 0 & 1 & 0 \\ 2 & -1 & 3 & 0 \end{bmatrix}$$

$$(-1)R_1 + R_2 \atop \longrightarrow \begin{bmatrix} 1 & \frac{1}{4} & 0 & -\frac{1}{4} \\ 0 & -\frac{1}{4} & 1 & \frac{1}{4} \\ 2 & -1 & 3 & 0 \end{bmatrix}$$

$$(-2)R_1 + R_3 \atop \longrightarrow \begin{bmatrix} 1 & \frac{1}{4} & 0 & -\frac{1}{4} \\ 0 & -\frac{1}{4} & 1 & \frac{1}{4} \\ 0 & -\frac{3}{2} & 3 & \frac{1}{2} \end{bmatrix}$$

$$(-4)R_2 \atop \longrightarrow \begin{bmatrix} 1 & \frac{1}{4} & 0 & -\frac{1}{4} \\ 0 & 1 & -4 & -1 \\ 0 & -\frac{3}{2} & 3 & \frac{1}{2} \end{bmatrix}$$

$$\left(\tfrac{3}{2}\right)R_2 + R_3 \atop \longrightarrow \begin{bmatrix} 1 & \frac{1}{4} & 0 & -\frac{1}{4} \\ 0 & 1 & -4 & -1 \\ 0 & 0 & -3 & -1 \end{bmatrix}$$

$$\left(-\tfrac{1}{3}\right)R_3 \atop \longrightarrow \begin{bmatrix} 1 & \frac{1}{4} & 0 & -\frac{1}{4} \\ 0 & 1 & -4 & -1 \\ 0 & 0 & 1 & \frac{1}{3} \end{bmatrix}$$

The last matrix is the echelon form for the given matrix.

Once we have the echelon form of the matrix of a linear system, we may determine the solutions of the system using back-substitution. This method of solving a linear system is called **Gauss elimination**, named for the 19th-century mathematician Karl Friedrich Gauss.

THE METHOD OF GAUSS ELIMINATION

To solve a system of linear equations in any number of variables:

1. Form the augmented matrix of the system.
2. Use elementary row operations to write the augmented matrix in echelon form.
3. Write the linear system corresponding to the echelon form. This system is triangular.
4. Use back-substitution to solve the triangular system.

The following examples illustrate the computational procedures of Gauss elimination.

➤ **EXAMPLE 8**

Solving a System
in Three Variables
Using Gauss Elimination

Solve the following system of linear equations:

$$\begin{cases} 4x + y & = -1 \\ x + & z = 0 \\ 2x - y + 3z = 0 \end{cases} \quad .$$

Solution

The augmented matrix of the system is precisely the matrix considered in the preceding example. We found that the echelon form of the matrix is

$$\begin{bmatrix} 1 & \frac{1}{4} & 0 & -\frac{1}{4} \\ 0 & 1 & -4 & -1 \\ 0 & 0 & 1 & \frac{1}{3} \end{bmatrix}$$

The corresponding linear system is

$$\begin{cases} x + \frac{1}{4}y & = -\frac{1}{4} \\ \quad y - 4z = -1 \\ \quad\quad z = \frac{1}{3} \end{cases}$$

Using back-substitution, we find that the solution is $(-\frac{1}{3}, \frac{1}{3}, \frac{1}{3})$.

Nonsquare Systems

In the preceding examples, we considered only square linear systems, that is, systems in which the number of equations equals the number of variables. However, Gauss elimination works just as well on nonsquare systems, as the next example shows.

➤ EXAMPLE 9

A System with an Infinite Number of Solutions

Solve the following system of linear equations:

$$\begin{cases} 2x - 3y + z - w = 4 \\ 2x - \quad\quad 9z - 3w = -2 \end{cases}$$

Solution

The matrix of the system is

$$\begin{bmatrix} 2 & -3 & 1 & -1 & 4 \\ 2 & 0 & -9 & -3 & -2 \end{bmatrix}$$

We transform the first column of the matrix, proceeding as in the preceding example:

$$\begin{bmatrix} 2 & -3 & 1 & -1 & 4 \\ 2 & 0 & -9 & -3 & -2 \end{bmatrix} \xrightarrow{\frac{1}{2}R_1} \begin{bmatrix} 1 & -\frac{3}{2} & \frac{1}{2} & -\frac{1}{2} & 2 \\ 2 & 0 & -9 & -3 & -2 \end{bmatrix}$$

$$\xrightarrow{(-2)R_1 + R_2} \begin{bmatrix} 1 & -\frac{3}{2} & \frac{1}{2} & -\frac{1}{2} & 2 \\ 0 & 3 & -10 & -2 & -6 \end{bmatrix}$$

$$\xrightarrow{\frac{1}{3}R_2} \begin{bmatrix} 1 & -\frac{3}{2} & \frac{1}{2} & -\frac{1}{2} & 2 \\ 0 & 1 & -\frac{10}{3} & -\frac{2}{3} & -2 \end{bmatrix}$$

The two rows are both in the proper form, so the last matrix is the echelon form. The corresponding system is

$$\begin{cases} x - \frac{3}{2}y + \frac{1}{2}z - \frac{1}{2}w = 2 \\ \quad y - \frac{10}{3}z - \frac{2}{3}w = -2 \end{cases}$$

The equations from which the values of z and w would be determined are not present in this echelon form. Accordingly, the values of z and w may be any real numbers. The echelon form may be used to determine the values of x and y in terms of the values chosen for z and w. From the second equation, we have

$$y = \frac{10z}{3} + \frac{2}{3}w - 2$$

Therefore, from the first equation, we have

$$x = \tfrac{3}{2}y - \tfrac{1}{2}z + \tfrac{1}{2}w + 2$$
$$= \tfrac{3}{2}\left(\tfrac{10}{3}z + \tfrac{2}{3}w - 2\right) - \tfrac{1}{2}z + \tfrac{1}{2}w + 2$$
$$= \tfrac{9}{2}z + \tfrac{3}{2}w - 1$$

Therefore, the solution of the system is

$$z = \text{any value}, \qquad w = \text{any value}$$
$$x = \tfrac{9}{2}z + \tfrac{3}{2}w - 1, \qquad y = \tfrac{10}{3}z + \tfrac{2}{3}w - 2$$

In particular, we see that the system has an infinite number of solutions. Here are three particular solutions of the system

$$\left(\tfrac{7}{2}, \tfrac{4}{3}, 1, 0\right), \left(\tfrac{1}{2}, -\tfrac{4}{3}, 0, 1\right), (-1, -2, 0, 0)$$

➤ **EXAMPLE 10**
A System with No Solution

Determine all solutions of the following system of linear equations:

$$\begin{cases} x + y - 2z = 1 \\ \quad\quad y - 3z = 4 \\ 2x + 3y - 7z = 5 \end{cases}$$

Solution

The first two rows are already in the proper echelon form. We use row operations on the system matrix to transform the third row, proceeding from left to right:

$$\begin{bmatrix} 1 & 1 & -2 & 1 \\ 0 & 1 & -3 & 4 \\ 2 & 3 & -7 & 5 \end{bmatrix} \overset{(-2)R_1 + R_3}{\longrightarrow} \begin{bmatrix} 1 & 1 & -2 & 1 \\ 0 & 1 & -3 & 4 \\ 0 & 1 & -3 & 3 \end{bmatrix}$$

$$\overset{(-1)R_2 + R_3}{\longrightarrow} \begin{bmatrix} 1 & 1 & -2 & 1 \\ 0 & 1 & -3 & 4 \\ 0 & 0 & 0 & -1 \end{bmatrix}$$

Note that the last row of the final matrix corresponds to the equation

$$0 = -1$$

This equation is inconsistent. Therefore, the system has no solutions. Whenever the echelon form of a system contains a row consisting of zeros in all columns except the rightmost one, the system has no solutions. Such systems are called **inconsistent**.

Exercises 8.3

Solve the following systems of equations in echelon form:

1. $\begin{cases} x + 2y - z = 4 \\ 5y - 2z = 8 \\ z = 1 \end{cases}$

2. $\begin{cases} 2x + y + 2z = 12 \\ 4y - 3z = 17 \\ z = 15 \end{cases}$

3. $\begin{cases} 3x + 2y - z = 15 \\ y - 3z = 5 \\ z = -2 \end{cases}$

4. $\begin{cases} x + 3y - z = 11 \\ 4y - 7z = 14 \\ z = 2 \end{cases}$

5. $\begin{cases} x + y + 5z = 13 \\ y - 4z = -10 \\ z = 3 \end{cases}$

6. $\begin{cases} 3x - 5y - z = 7 \\ 6y - 2z = 8 \\ z = -4 \end{cases}$

Consider the matrix

$$\begin{bmatrix} 4 & 1 & 0 \\ -1 & 2 & 3 \\ 0 & -1 & 5 \end{bmatrix}$$

Perform the following elementary row operations:

7. $R_1 \leftrightarrow R_2$ 8. $R_1 \leftrightarrow R_3$

9. $\frac{1}{2}R_2$ 10. $-2R_1$

11. $2R_1 - R_2$ 12. $4R_3 + 2R_2$

13. $-1R_1 + 2R_2 + R_3$ 14. $R_2 - R_3 + 2R_1$

Transform the following matrices into echelon form.

15. $\begin{bmatrix} 1 & 2 & 0 & 3 \\ 4 & 5 & 1 & 0 \\ -2 & 0 & 1 & -1 \end{bmatrix}$ 16. $\begin{bmatrix} 1 & -1 & 5 & -6 \\ 3 & 3 & -1 & 10 \\ 1 & 3 & 2 & 5 \end{bmatrix}$

17. $\begin{bmatrix} 1 & 1 & 2 \\ 2 & 2 & 5 \end{bmatrix}$ 18. $\begin{bmatrix} 2 & 6 & 18 \\ 4 & -3 & -19 \end{bmatrix}$

19. $\begin{bmatrix} 1 & 3 & 2 & 1 \\ 2 & 1 & -1 & 2 \\ 1 & 1 & 1 & 2 \end{bmatrix}$ 20. $\begin{bmatrix} 2 & 1 & 1 & 3 \\ 3 & -4 & 2 & -7 \\ 1 & 1 & 1 & 2 \end{bmatrix}$

21. $\begin{bmatrix} 1 & 5 & 6 \\ 0 & 1 & 1 \end{bmatrix}$ 22. $\begin{bmatrix} 2 & 7 & 1 \\ 5 & 0 & -15 \end{bmatrix}$

23. $\begin{bmatrix} 3 & 2 & 1 & 1 \\ 0 & 2 & 4 & 22 \\ -1 & -2 & 3 & 15 \end{bmatrix}$ 24. $\begin{bmatrix} 4 & 2 & 6 & 24 \\ 4 & -3 & 0 & 10 \\ 5 & 0 & -4 & -11 \end{bmatrix}$

25. $\begin{bmatrix} 1 & 2 & 3 & 0 \\ 1 & 0 & -2 & 3 \\ 0 & 1 & 1 & 1 \end{bmatrix}$ 26. $\begin{bmatrix} 2 & 1 & 2 & 1 \\ 1 & 2 & -3 & 4 \\ 3 & -1 & 1 & 0 \end{bmatrix}$

Solve the following systems of linear equations.

27. $\begin{cases} x + 2y + 3z = 0 \\ x \quad\quad - 2z = 3 \\ \quad\quad y + z = 1 \end{cases}$ 28. $\begin{cases} x + y + 4z = 1 \\ -2x - y + z = 2 \\ 3x - 2y + 3z = 1 \end{cases}$

29. $\begin{cases} 3x + 2y - z = 4 \\ 2x - y + 3z = 5 \\ x + 3y + 2z = -1 \end{cases}$ 30. $\begin{cases} 4x - 2y + 3z = 4 \\ 5x - y + 4z = 7 \\ 3x + 5y + z = 7 \end{cases}$

31. $\begin{cases} 2x + 3y + z = 1 \\ x - y + 2z = 3 \end{cases}$ 32. $\begin{cases} -5x + 3y - z = 0 \\ 2x + y - z = 4 \end{cases}$

33. $\begin{cases} 2x \quad\quad - z = 0 \\ \quad y + 2z = -2 \\ x + y \quad\quad = 1 \end{cases}$ 34. $\begin{cases} x \quad\quad + z = 1 \\ x + y \quad\quad = -1 \\ \quad y + z = 4 \end{cases}$

35. $\begin{cases} x + y + z + w = -2 \\ -2x + 3y + 2z - 3w = 10 \\ 3x + 2y - z + 2w = -12 \\ 4x - y + z + 2w = 1 \end{cases}$

36. $\begin{cases} x - y - 3z - 2w = 2 \\ 4x + y + z + 2w = 2 \\ x + 3y - 2z - w = 9 \\ 3x + y - z + w = 5 \end{cases}$

37. $\begin{cases} x + 2y + 3z = 0 \\ x \quad\quad - 2z = 3 \\ \quad\quad y + z = 1 \end{cases}$ 38. $\begin{cases} 3x - 5y = 6 \\ -6x + 10y = 7 \end{cases}$

39. $\begin{cases} 3x - 6y = 12 \\ -2x + 4y = -8 \end{cases}$ 40. $\begin{cases} x - y + z = 0 \\ 5x + y + 3z = 1 \end{cases}$

41. $\begin{cases} -5x + 3y - z = 1 \\ 2x + y - z = 4 \end{cases}$ 42. $\begin{cases} x - y \quad\quad = 5 \\ \quad y - z \quad\quad = 6 \\ \quad\quad z - w = 7 \\ x \quad\quad + w = 8 \end{cases}$

43. $\begin{cases} x + y - z + w = 4 \\ 2x + 3y + z - w = -1 \\ -x - 3y + 2z + 3w = 3 \\ 3x + 2y - 3z + w = 8 \end{cases}$

44. $\begin{cases} x + 2y + 3z - w = 7 \\ x - 4y + z = 3 \\ 2y - 3z + w = 4 \end{cases}$

45. $\begin{cases} x + 2y = 5 \\ 2x - 5y = -8 \end{cases}$

♦ Applications

46. Flying with the wind, an airplane can travel 1080 miles in six hours, but flying against the same wind it goes only $\frac{1}{3}$ this distance in half the time. Find the speed of the plane in still air and the wind speed.

47. To start a record company, Barry G. borrowed $50,000 from three different banks. He borrowed some money at 8 percent, and he borrowed $3000 more than that amount at 6.9 percent. The rest was borrowed at 12 percent. If his annual interest totals $4755 from all three loans, how much was borrowed at each rate?

48. Barry G.'s record company makes two types of records, 45s and $33\frac{1}{3}$s. Both require time on two machines: 45s, 1 hour on the recorder and 2 hours on the cutter; $33\frac{1}{3}$s, 3 hours on the recorder and 1 hour on the cutter. Both the recorder and the cutter operate 15 hours a day. How many of each record can be produced in a 5-day work week under these conditions?

49. A famous national appliance store sells both Sony and Sanyo stereos. Sony sells for $280, and Sanyo sells for $315. During a one-day sale, a total of 85 Sonys and Sanyos were sold for a total of $23,975. How many of each brand were sold during this one-day sale?

8.4 SYSTEMS OF INEQUALITIES

In Chapter 1, we introduced inequalities involving a single variable and showed how to graph such inequalities on the number line. Let's turn our attention to inequalities in two variables x and y. Here are some typical inequalities of this sort:

$$3x + 5y < 4$$
$$x^2 + 3xy \le 10$$
$$x + \frac{1}{y} > 1$$

Suppose that we are given an inequality in two variables. An ordered pair (a, b) is said to **satisfy the inequality** (or **is a solution of the inequality**) provided that the inequality is true when x is replaced by a and y is replaced by b. For instance, the ordered pair $(-1, 2)$ is not a solution of the first inequality shown because when x is replaced by -1 and y is replaced by 2, we have

$$3x + 5y = 3(-1) + 5(2)$$
$$= -3 + 10$$
$$= 7$$

Because 7 is not less than 4, we see that $(-1, 2)$ is not a solution. However, $(-4, 1)$ is a solution of the inequality, because when x is replaced with -4 and y is replaced with 1, we have

$$3x + 5y = 3(-4) + 5(1)$$
$$= -7$$
$$< 4 \qquad \textit{(True)}$$

The set of ordered pairs that are solutions of an inequality is called the **solution set** of the inequality. The solution set may be represented graphically as a region in the x–y plane. The next few examples determine the solution sets of some typical inequalities in two variables and the corresponding plane regions.

> **EXAMPLE 1**
Determining Solution Set

Graph the inequality $x^2 + y^2 < 4$.

Solution

Note that $x^2 + y^2$ equals the square of the distance of point (x, y) from the origin. The inequality specifies that the square of the distance of (x, y) be less than 4; that is, the distance is less than 2. These are the points in the interior of the circle shown in Figure 1. The dotted line indicates that the points on the circle are excluded from the shaded region.

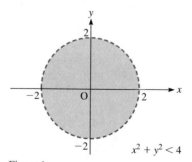

Figure 1

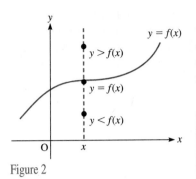

Figure 2

In Figure 2, we have drawn the graph of an equation of the form $y = f(x)$, where $f(x)$ is a function of the variable x. A point (x, y) for which $y = f(x)$ lies on the graph. A point (x, y) for which $y > f(x)$ lies above the graph. A point (x, y) for which $y < f(x)$ lies below the graph. That is, we have the following result.

GRAPHING SOLUTION SETS

The graph of the inequality $y > f(x)$ consists of the region above the graph of the equation $y = f(x)$. (See Figure 3(*a*).) The graph of the inequality $y < f(x)$ consists of the region below the graph of the equation $y = f(x)$. (See Figure 3(*b*).)

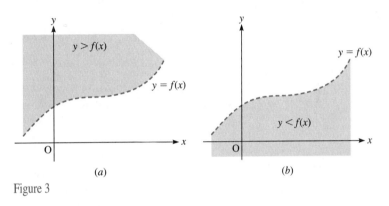

(*a*)   (*b*)

Figure 3

The next example gives an application of this result.

➤ **EXAMPLE 2**
Graphing a Solution Set

Graph the inequality $y \geq x^2 - 3$.

Solution

In Figure 4, we have drawn the graph of the equation $y = x^2 - 3$. The graph of the given inequality is the region above and on the graph. This is the shaded region in Figure 4.

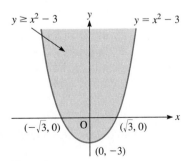

Figure 4

In Chapter 1, we discussed the rules that allow us to add, subtract, multiply, and divide inequalities. It is often possible to apply these rules to transform an inequality into one of the forms $y < f(x)$ or $y > f(x)$, to which we may apply the rules for graphing solution sets to determine the graph. The next two examples illustrate how this may be done.

➤ **EXAMPLE 3**
Solution Set
of a Linear Inequality

Graph the inequality $5x - 3y < 12$.

Solution

Let's transform the inequality so that y is isolated on the left side. First, subtract $5x$ from both sides to obtain

$$-3y < 12 - 5x$$

Next, multiply by $-\frac{1}{3}$. Because this quantity is negative, we must reverse the sign of the inequality:

$$y > -\tfrac{1}{3}(12 - 5y)$$

$$y > -4 + \tfrac{5}{3}x$$

We now graph the equation $y = -4 + \frac{5}{3}x$, as shown in Figure 5. The graph of the given inequality is the region above the line in the figure.

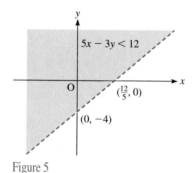

Figure 5

➤ **EXAMPLE 4**
Solution Set
of a Nonlinear Inequality

Graph the solution set of the inequality

$$xy \geq 1$$

Solution

As in the preceding example, we apply the laws of inequalities to isolate y on the left side. Basically, we wish to multiply the inequality by $1/x$. However, this quantity is positive if $x > 0$ and negative if $x < 0$. In the first case, the inequality is equivalent to

$$y \geq \frac{1}{x}, \qquad x > 0$$

In the second case, the inequality is equivalent to

$$y \leq \frac{1}{x}, \qquad x < 0$$

For the first case, draw the graph $y = 1/x$, $x > 0$, as shown in Figure 6(a). The points corresponding to solutions of the given inequality are those lying on or above the graph. These are the shaded points in the figure. For the second case, draw the graph $y = 1/x$, $x < 0$, as shown in Figure 6(b). The point corresponding to solutions of the given inequality are those lying on or below the graph. These are the shaded points in the figure. The solution set of the given inequality consists of the points belonging to the solution sets for either of the cases. This set is the shaded region shown in Figure 6(c).

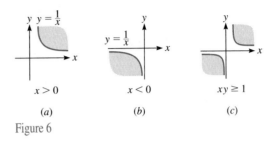

Figure 6

Another method of graphing the inequality is to replace the inequality sign by equal signs and graph the resulting equation. The graph of the inequality consists of all points lying to one side of the graph. To determine which side, choose a point (a, b) not on the graph of the equation and substitute this point into the inequality. If the result is a true inequality, then the point lies on the same side as the graph of the inequality. Otherwise, the point lies on the opposite side from the graph of the inequality. The point (a, b) is called a **test point**.

For example, consider the inequality $x^2 + y^2 < 4$ of Example 1 and the test point $(2, 2)$. The equation $x^2 + y^2 = 4$ has a circle as its graph, and the test point lies outside the circle. At the test point, we have

$$x^2 + y^2 = 2^2 + 2^2$$
$$= 8$$

That is, the inequality is false at the test point because $8 < 4$ is false. This means that the region containing the test point, the exterior of the circle, does not belong to the graph of the inequality. The graph consists of the points within the circle. Because the sign of the inequality is $<$ rather than $\leq$, the circle itself is not included in the graph.

In the previous sections of this chapter, we studied systems of equations and their solutions. By analogy, let's turn to systems of *inequalities* and their solutions. A **system of inequalities** is a collection of one or more inequalities. We will concentrate on systems of inequalities in two variables. A **solution** of a system of inequalities is an ordered pair that is a solution for each inequality of the system. The set of solutions of a system of inequalities is called the **feasible set** (or **graph**) of the system. One method for determining the feasible set is as follows:

1. Solve separately each inequality of the system.
2. Graph on the same coordinate system the solution set of each inequality. Shade each solution set with a different sort of shading (hatching).
3. The solution set of the system consists of the region that is shaded with all possible shading types.

The following examples illustrate how to solve a simple system of inequalities.

➢ **EXAMPLE 5**

Graph a Feasible Set

Determine the feasible set of the following system of inequalities:

$$\begin{cases} y \geq x^2 - 3 \\ 5x + 3y < 12 \end{cases}$$

Solution

In Figure 7(a), we have sketched the solution of the first inequality of the system. In Figure 7(b), we have sketched the solution set of the second inequality. Figure 7(c) shows the two solution sets superimposed on a single coordinate system. The solution set of the system consists of the region that is shaded twice. This is the shaded region in Figure 7(d).

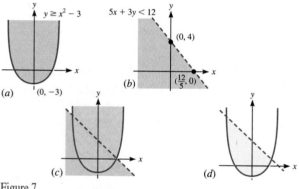

Figure 7

➢ **EXAMPLE 6**

Feasible Set
of a Linear System

Sketch the graph of the following system of linear inequalities:

$$\begin{cases} x \leq 8 \\ y \leq 3x - 7 \\ y > x + 4 \end{cases}$$

Solution

In this case, there are three solution sets to determine, one for each inequality. Figure 8(a) shows the three solution sets, each shaded with a different hatching. The region that is shaded three times is the solution set of the system. (See Figure 8(b).)

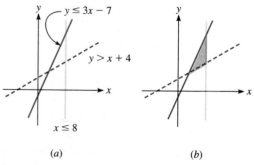

Figure 8

➢ EXAMPLE 7
Manufacturing

A pharmaceutical company manufactures two drugs using the same ingredients. Each case of drug 1 requires 300 grams of ingredient A and 600 grams of ingredient B. Each case of drug 2 requires 500 grams of ingredient A and 100 grams of ingredient B. Market conditions dictate that at most 100,000 grams of ingredient A and 50,000 grams of ingredient B are available each week. Determine a system of inequalities describing the production limitations for the two drugs and determine the feasible set of the system of inequalities.

Solution

Let

$$x = \text{number of cases of drug 1}$$
$$y = \text{number of cases of drug 2}$$

First, all the values of x and y must be nonnegative. That is,

$$x \geq 0, \qquad y \geq 0$$

The condition of ingredient A is

$$\text{total ingredient A} \leq 100,000$$
$$\text{ingredient A in drug 1} + \text{ingredient A in drug 2} \leq 100,000$$
$$300x + 500y \leq 100,000$$

Similarly, the condition on ingredient B is

$$\text{total of ingredient B} \leq 50,000$$
$$\text{ingredient B in drug 1} + \text{ingredient B in drug 2} \leq 50,000$$
$$600x + 100y \leq 50,000$$

Thus, the production limitations are described by the following system of four inequalities:

$$\begin{cases} x \geq 0, & y \geq 0 \\ 3x + 5y \leq 1000 \\ 6x + 1y \leq 500 \end{cases}$$

In Figure 9(a), we have graphed the system of inequalities. The region that is shaded four times is the feasible set for the system. This is the shaded region in Figure 9(b).

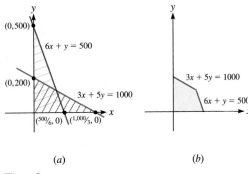

(a)   (b)

Figure 9

Exercises 8.4

Graph the inequality.

1. $x + 2y \le 4$ 2. $3x - y > 10$

3. $x^2 + y^2 < 4$ 4. $x^2 + 9y^2 \le 36$

5. $x < -y^2$ 6. $x + y^2 \ge -4$

7. $xy > 4$ 8. $x \le \frac{16}{y}$

9. $x^2 + 4y^2 - 10x + 24y + 57 \le 0$

10. $x^2 - y^2 + 2x - 2y \ge 0$

11. $3x - 2y > 6$ 12. $x < y$

13. $y \le x^2 - 4x + 2$ 14. $y \ge -1 - x^2$

Determine the feasible set of each of the following systems of inequalities.

15. $\begin{cases} x + 2y > 6 \\ x^2 > 2y \end{cases}$ 16. $\begin{cases} x^2 + y^2 \le 9 \\ \dfrac{x^2}{16} + \dfrac{y^2}{4} \ge 1 \end{cases}$

17. $\begin{cases} 4x + 3y < 12 \\ 4x + y > -4 \end{cases}$ 18. $\begin{cases} x^2 + y^2 \le 36 \\ -4 \le x \le 4 \end{cases}$

19. $\begin{cases} 9x^2 - y^2 \ge 0 \\ x^2 + y \le 25 \end{cases}$ 20. $\begin{cases} x^2 + 4y^2 < 16 \\ x^2 - 2xy \ge 0 \end{cases}$

21. $\begin{cases} y \ge (x - 2)^2 + 3 \\ 3x - 4y \le -16 \end{cases}$ 22. $\begin{cases} y \le -(x - 3)^2 + 2 \\ 3x + 5y \ge -20 \end{cases}$

23. $\begin{cases} |x| \ge 6 \\ |y| \le 1 \end{cases}$ 24. $\begin{cases} y > |x + 2| \\ 3 \ge |x| \end{cases}$

Sketch the graphs of the following systems of linear inequalities.

25. $\begin{cases} x + y > 6 \\ 2x - y \le 0 \\ y \ge -2 \end{cases}$ 26. $\begin{cases} 2x - y \le 6 \\ x + y > 6 \\ x \ge -1 \end{cases}$

27. $\begin{cases} x - y \le 2 \\ x + 2y \ge 8 \\ y \le 4 \end{cases}$ 28. $\begin{cases} 2x - y \ge -1 \\ 2x + y \ge 1 \\ x \le 2 \end{cases}$

29. $\begin{cases} x \ge 0 \\ y \ge 0 \\ 2x + 3y \le 12 \\ 3x + y \le 6 \end{cases}$ 30. $\begin{cases} x \ge 0 \\ y \ge 0 \\ 2x + y \le 4 \\ 2x - 3y \le 6 \end{cases}$

31. $\begin{cases} x \ge 2 \\ y \ge 5 \\ 3x - y \ge 12 \\ x + y \le 15 \end{cases}$ 32. $\begin{cases} x \ge 10 \\ y \ge 20 \\ 2x + 3y \le 100 \\ 5x + 4y \le 200 \end{cases}$

♦ Applications

For Exercises 33–39, (a) determine a system of inequalities describing the given problems, and (b) determine the feasible set of the systems of inequalities.

33. A manufacturer of skis can produce as many as 60 pairs of recreational skis and as many as 45 pairs of racing skis

per day. It takes 3 hours of labor to produce a pair of recreational skis and 4 hours of labor to produce a pair of racing skis. The company has up to 240 hours of labor available for ski production each day.

34. General Mills wants to make a cereal from a mixture of oats and wheat so that it will contain at least 88 g of protein and at least 36 mg of iron per kg. Given that wheat contains 80 g of protein and 40 mg of iron per kg, and that oats contain 100 g of protein and 30 mg of iron per kg, write the system of inequalities.

35. You are taking a test in which items of type x are worth 10 points and items of type y are worth 15 points. It takes 3 minutes for each item of type x and 6 minutes for each item of type y. Total time allowed is 60 minutes and you may not answer more than 16 questions.

36. Every day Martha B. needs a dietary supplement of 4 mg of vitamin A, 11 mg of vitamin B, and 100 mg of vitamin C. Either of two brands of vitamin pills can be used: brand x, at 6 cents a pill, or brand y, at 8 cents a pill. Brand x supplies 2 mg of vitamin A, 3 mg of vitamin B, and 25 mg of vitamin C. Likewise, a brand y pill supplies 1, 4, and 50 mg, respectively, of vitamins A, B, and C.

37. **Hockey.** A hockey team figures that it needs at least 60 points for the season to make the playoffs. A win (w) is worth 2 points in the standings, and a tie (t) is worth 1 point.

38. **Elevators.** Many elevators have a capacity of 1 metric ton (1000 kg). Suppose c children, each weighing 35 kg and a adults, each weighing 75 kg, are on an elevator.

39. **Width of a basketball floor.** Sizes of basketball floors vary considerably due to building dimensions and other constraints, such as cost. The length L is to be at most 74 ft, and the width w is to be at most 50 ft.

8.5 LINEAR PROGRAMMING

Linear programming is an important application involving systems of linear inequalities. To introduce this important subject, let's return to the pharmaceutical company example at the end of the preceding section. Suppose that the company earns \$50 for each case of drug 1 and \$75 for each case of drug 2. How many cases of each drug should it produce in order to maximize its profit?

Let's state the problem in algebraic form: Suppose that the drug company produces x cases of drug 1 and y cases of drug 2. The profit P it earns is given by

$$P = 50x + 75y$$

At first, you might ask: Why not just manufacture a huge number of each type of drug and thereby make an equally huge profit? The answer is that production has a number of constraints imposed on it, as stated in the example of the preceding section. Namely, production is subject to the inequalities

$$300x + 500y \le 100,000$$
$$600x + 100y \le 50,000$$
$$x \ge 0, \qquad y \ge 0$$

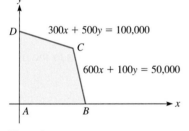

Figure 1

The four inequalities are called **constraints**, because they constrain the choice of x and y. As we saw, the set of ordered pairs (x, y) that satisfy the system of inequalities is represented by the points in the shaded region in Figure 1. Our problem is to choose the point (x, y) of the feasible set at which the value of P is a maximum.

A problem of this sort is called a **linear programming problem**. More precisely, suppose that we are given a function

$$P = ax + by$$

of two variables x and y and a system of linear inequalities in the same variables.

LINEAR PROGRAMMING PROBLEM

Determine the ordered pair (x, y) that is a solution of the system of inequalities and for which the value of P is maximized (respectively minimized).

The function P is called the **objective function**, and the inequalities are called **constraints**. The corners of the feasible set are called **vertices**. Then we have the following result.

FUNDAMENTAL THEOREM OF LINEAR PROGRAMMING

Suppose that the feasible set is a polygon.[1] Then the objective function assumes its maximum (respectively minimum) value at a vertex of the feasible set.

We will omit the proof of this theorem.

---

[1] In more advanced treatments of linear programming, the case of feasible sets that are "unbounded" (that is, those that extend to infinity in either the horizontal or vertical direction, or both) is also treated. For simplicity, we will not discuss problems with such feasible sets.

➢ EXAMPLE 1
Manufacturing

Determine the number of cases of each type of drug that the pharmaceutical company must produce in order to maximize its profit.

Solution

According to the Fundamental Theorem of Linear Programming, the maximum profit occurs at a vertex of the feasible set. Let's determine the vertices. In Figure 1, we have drawn the feasible set and labeled each of the vertices. Also, each of the bounding lines of the feasible set is labeled with its equation. Vertex A is the origin $(0,0)$. Vertex B is the intersection of the lines with equations

$$y = 0$$
$$600x + 100y = 50,000$$

The solution of this system is obtained by substituting $y = 0$ into the second equation:

$$600x + 100(0) = 50,000$$
$$x = \tfrac{250}{3}$$
$$B = \left(\tfrac{250}{3}, 0\right)$$

Vertex C is the intersection of the lines with equations

$$3x + 5y = 1000$$
$$6x + 1y = 500$$

To solve this system, multiply the second equation by -5 and add it to the first to obtain:

$$3x + 5y = 1000$$
$$+ \quad \underline{-30x - 5y = -2500}$$
$$-27x = -1500$$
$$x = \tfrac{1500}{27} = \tfrac{500}{9}$$
$$y = 500 - 6x$$
$$= 500 - 6\left(\tfrac{500}{9}\right)$$
$$= \tfrac{1500}{9}$$
$$= \tfrac{500}{3}$$
$$C = \left(\tfrac{500}{9}, \tfrac{500}{3}\right)$$

The final vertex, D, is the intersection of the lines with equations

$$x = 0$$
$$300x + 500y = 100,000$$
$$300 \cdot 0 + 500y = 100,000$$
$$y = \tfrac{1000}{5} = 200$$
$$D = (0, 200)$$

Let's now make a table of the vertices and the value of the function P at each of them:

Vertex	$P = 50x + 75y$
$A : (0, 0)$	0
$B : \left(\dfrac{250}{3}, 0\right)$	$\dfrac{12,500}{3} = 4166.67$
$C : \left(\dfrac{500}{9}, \dfrac{500}{3}\right)$	$15,277.78 \leftarrow$ Maximum
$D : (0, 200)$	$15,000$

From the table, we see that the maximum profit occurs for vertex C. By the Fundamental Theorem of Linear Programming, the maximum value of the function P for (x, y) in the feasible set is 15,277.78 and this value occurs for $x = \frac{500}{9}$, $y = \frac{500}{3}$. That is, the company should produce $\frac{500}{9}$ cases of drug 1 and $\frac{500}{3}$ cases of drug 2 per week.

➢ **EXAMPLE 2**
Minimization
of a Linear Function

Determine the minimum value of the function $P = 3x - 5y$ if x and y are constrained to lie in the triangle with vertices $(1, 1)$, $(3, 2)$, and $(1, 5)$. (See Figure 2.)

Solution

By the Fundamental Theorem of Linear Programming, the minimum point must occur at one of the vertices of the triangle. Let's tabulate the value of the function at each of the vertices:

Vertex	$P = 3x - 5y$
$(1, 1)$	-2
$(3, 2)$	-1
$(1, 5)$	-22

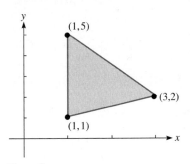

Figure 2

The minimum value of $3x - 5y$ is -22, which occurs for $x = 1$, $y = 5$.

Exercises 8.5

Determine the maximum and minimum values of the function and the values of x and y where they occur, given a feasible set with vertices as follows:

1. $f(x, y) = 3x + 5y; (4, 8), (2, 4), (1, 1), (5, 2)$
2. $f(x, y) = x + 4y; (0, 7), (0, 0), (6, 2), (5, 4)$
3. $Q = x + 3y$ subject to
$$\begin{cases} x \geq 0 \\ y \geq 0 \\ 5x + 2y \leq 20 \\ 2y \geq x \end{cases}$$
4. $F = 2x - 3y$ subject to
$$\begin{cases} x \geq 0 \\ y \geq 0 \\ 3x + 5y \leq 15 \\ 5x + 3y \geq 15 \end{cases}$$

5. $f(x, y) = 10x + 12y$ subject to
$$\begin{cases} 2x + 5y \geq 22 \\ 4x + 3y \geq 28 \\ 2x + 2y \leq 17 \\ x \geq 0 \\ y \geq 0 \end{cases}$$

6. $f(x, y) = -3x + y$ subject to
$$\begin{cases} x + 2y \leq 10 \\ 2x + y \geq 12 \\ x + y \leq 8 \\ x \geq 0 \\ y \geq 0 \end{cases}$$

Find the maximum and minimum values of the given function in the indicated region.

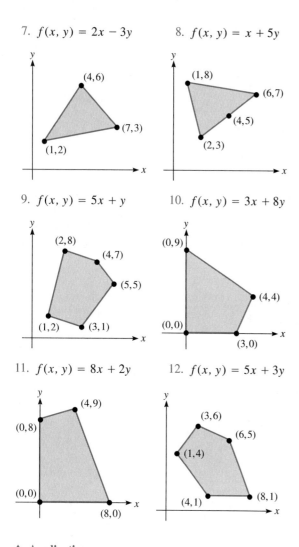

7. $f(x, y) = 2x - 3y$

8. $f(x, y) = x + 5y$

9. $f(x, y) = 5x + y$

10. $f(x, y) = 3x + 8y$

11. $f(x, y) = 8x + 2y$

12. $f(x, y) = 5x + 3y$

♦ **Applications**

13. RCA Manufacturing Company in Fort Wayne, Indiana, makes a $60 profit on each 19″ TV it produces and a $40 profit on each 13″ TV. A 19″ TV requires 1 hour on machine X, 1 hour on machine Y, and 4 hours on machine Z. The 13″ TV requires 2 hours on X, 1 hour on Y, and 1 hour on Z. In a given day, machines X, Y, and Z can work a maximum of 16, 9, and 24 hours, respectively. How many 19″ TVs and how many 13″ TVs should be produced per day to maximize the profit?

14. The Buick manufacturing plant in Flint, Michigan, must fill orders for Park Avenues from two dealers. The first dealer, Farmington Buick, has ordered 20 Park Avenues, and the second dealer, Brighton Buick, has ordered 30. The manufacturer has the cars stored in two different areas: southeast Flint and southwest Flint. There are 40 cars in southeast Flint and only 15 cars in southwest Flint. The shipping costs, per car, are: $15 from south-

east Flint to Brighton; $13 from southeast Flint to Farmington; $14 from southwest Flint to Brighton; $16 from southwest Flint to Farmington. With these conditions, find the number of cars to be shipped from each area to each dealer if the total shipping cost is to be a minimum. What is the cost?

15. Kellogg's of Battle Creek, Michigan, is going to produce a new cereal from a mixture of bran and rice so that it contains at least 88 g of protein and at least 36 mg of iron. Knowing that bran contains 80 g of protein and 40 mg of iron per kilogram, and that rice contains 100 g of protein and 30 mg of iron per kilogram, find the minimum cost of producing this new cereal, "Rice Bran," if bran costs 50 cents per kilogram and rice costs 40 cents per kilogram.

16. Sanyo makes stereo receivers. It produces a 30-watt receiver that it sells for a $100 profit and a 50-watt receiver that it sells for a $150 profit. In manufacturing, the 30-watt receiver requires 3 hours, and the 50-watt receiver takes 5 hours. The cabinet shop spends 1 hour on a 30-watt receiver and 3 hours on a 50-watt receiver. Packing takes 2 hours for both types of receivers. Per week, Sanyo has available 3900 work hours for manufacturing, 2100 hours for cabinet making, and 2200 hours for packing. How many receivers of each type should Sanyo produce per week to maximize its profit, and what is the maximum profit per week?

17. Almosttexas makes two types of calculators: Deluxe sells for $12 and Top of the Line sells for $10. It costs Almosttexas $9 to produce a Deluxe and $8 to produce a Top of the Line calculator. In one week, Almosttexas can produce between 200 and 300 (inclusive) Deluxe calculators, and between 100 and 250 (inclusive) Top of the Line calculators, but no more than 500 total calculators. How many of each type should be produced per week to maximize the profits for Almosttexas, Inc.?

18. You have $40,000 to invest in stocks and bonds. The least you are allowed to invest in stocks is $6000, and you cannot invest more than $22,000 in stocks. You may also invest no more than $30,000 in bonds. The interest

on stocks is 8 percent tax-free, and the interest on bonds is $7\frac{1}{2}$ percent tax-free. How much should you invest in each type to maximize your income? What is your income from the $40,000 invested?

19. The Glodfelty's 312-acre farm grows corn and beans. The work of growing and picking the corn takes 35 labor-hours per acre, and the work of growing and picking the beans takes 27 labor-hours per acre. There are only 9500 labor-hours available for these crops. The profit per acre on corn is $173, and the profit per acre on beans is $152. With this in mind, how many acres should be planted in corn to maximize the Glodfelty's profits? What are the maximum profits?

20. Fenton Cement Company has been contracted to haul 360 tons of cement per day for a highway construction job on U.S. 23. Fenton has seven 6-ton trucks, four 10-ton trucks, and nine drivers to haul cement per day. The 6-ton trucks can make eight trips per day and the 10-ton trucks can make only six trips per day, with the costs being $15 and $24, respectively, per day. Using all 9 drivers, how many trucks of each type should be used to minimize the cost per day? What is that cost?

21. The North Manchester building code requires that the area of the windows be at least $\frac{1}{8}$ the area of the walls and roof in any new building. The annual heating cost of a new building is $3 per square foot of windows and $1 per square foot of walls and roof. To the nearest square foot, what is the largest surface area a new building can have if its annual heating cost cannot exceed $1000?

22. Gary Smelting Company receives a monthly order for at least 40 tons of iron, 60 tons of copper, and 40 tons of lead. It can fill this order by smelting either alloy A or alloy B. Each railroad car of A will produce 1 ton of iron, 3 tons of copper, and 4 tons of lead after smelting. Each railroad car of B will produce 2 tons of iron, 2 tons of copper, and 1 ton of lead after smelting. If the cost of smelting the alloy in railroad car A is $350 and the cost of smelting the alloy in railroad car B is $200, then how many carloads of each should be used to fill the order at a minimum cost to Gary Smelting? What is that minimum cost?

23. A lumber company can convert logs into either lumber or plywood. In a given week, the mill can turn out 400 units of production, of which 100 units of lumber and 150 units of plywood are required by regular customers. The profit on a unit of lumber is $20, and on a unit of plywood it is $30. How many units of each type should the mill produce per week in order to maximize profit?

24. A farm consists of 240 acres of cropland. The farmer wishes to plant part or all of the acreage in corn and/or oats. Profit per acre in corn production is $40, and in oats it is $30. An additional restriction is that the total hours of labor during the production be no more than 320. Each acre of land in corn production uses 2 hours of labor during the production period; production of oats requires only 1 hour per acre. How many acres of land should be planted in corn and how many in oats in order to maximize profit?

8.6 CHAPTER REVIEW

Important Concepts, Properties, and Formulas—Chapter 8

Method of substitution for systems of two equations in two variables	1. Solve an equation for one of the variables in terms of the other.	p. 430
	2. Substitute the expression found in step 1 into the other equation.	
	3. Solve the resulting equation for the possible values of one variable.	
	4. For each value found in step 3, substitute into the expression of step 1 to determine the corresponding value of the other variable.	

Method of elimination for systems of two linear equations in two variables	1. Multiply one or both equations of the system by suitable numbers to obtain a new system in which the coefficients of one of the variables are equal. 2. Subtract the equations to obtain a linear equation in one variable. 3. Solve the equation obtained in step 2. 4. Substitute the value obtained into either of the original system equations and solve for the second variable.	p. 438
Graphical analysis of solutions of two linear equations in two variables.	1. The graphs of the equations correspond to two distinct lines. In this case, the system has a single solution, corresponding to the point of intersection of the lines. 2. The two straight lines coincide. In this case, the system has an infinite number of solutions, corresponding to all points on the line. 3. The two straight lines are different but parallel. In this case, the system has no solutions.	p. 438
Elementary operations on linear systems	1. Interchange two equations. 2. Multiply an equation by a nonzero real number. 3. Add to an equation a multiple of another equation.	p. 443
Determining the feasible set of a system of inequalities	1. Solve separately each inequality of the system. 2. Graph on the same coordinate system the solutions sets of each of the inequalities. Shade each solution set with a different sort of shading (hatching). 3. The solution set of the system consists of the region that is shaded with all possible shading types.	p. 455
Fundamental theorem of linear programming	The objective function assumes its maximum (respectively minimum) value at a vertex of the feasible set.	p. 459

Cumulative Review Exercises—Chapter 8

Determine all solutions of the following systems of equations.

1. $\begin{cases} 3x + y = 11 \\ 2x - 5y = 13 \end{cases}$

2. $\begin{cases} 4x + 3y = 7 \\ 2x + 4y = 16 \end{cases}$

3. $\begin{cases} x + y = 1 \\ x^2 + y^2 = 25 \end{cases}$

4. $\begin{cases} x = 3 - 2y \\ x^2 = 9 - y^2 \end{cases}$

5. $\begin{cases} x + y + z = 6 \\ xyz = 6 \\ 2x - y = 0 \end{cases}$

6. $\begin{cases} x - y - z = 0 \\ xyz = 6 \\ 3x + 4y = 5 \end{cases}$

♦ Applications

7. I have a total of 30 coins in my pocket. Some are dimes, and the rest are quarters. If I have a total of $4.50, then how many of the coins are dimes?

8. Tickets to the Fisher Theater cost $20 for the main floor and $16 for the balcony. If the receipts from the sale of 1420 tickets were $26,060, how many tickets were sold at each price?

9. Find two numbers whose sum is 14 and the sum of whose reciprocals is $\frac{14}{33}$.

10. The sum of the digits of a three-digit number is 11, the product of the digits is 40, and if the hundreds digit is subtracted from the ones digit the result is 1. Find the number.

Solve the following in echelon form.

11. $\begin{cases} x + y + z = 9 \\ 3y + 2z = 7 \\ z = -1 \end{cases}$ 12. $\begin{cases} x - y + 5z = 13 \\ y - 4z = -10 \\ z = 3 \end{cases}$

13. $\begin{cases} 3x - 2y = 1 \\ 4x + y = 5 \end{cases}$ 14. $\begin{cases} x + 2y = 3 \\ 2x + 3y = 4 \end{cases}$

15. $\begin{cases} 2x - 3y = -12 \\ -10x + 15y = 60 \\ 6x - 9y = -36 \end{cases}$ 16. $\begin{cases} 3x - 4y = -1 \\ x + 6y = -4 \\ -2x + 8y = -2 \end{cases}$

♦ **Applications**

17. Suppose you invested a certain sum of money at 12.5 percent and another sum at 14 percent. If your total investment was $60,000 and your total yearly interest for one year on both investments was $8190, how much did you invest at each rate?

18. The perimeter of a rectangular pool is 68 meters. If the width of the pool is 6 m less than the length, what are the dimensions of the pool?

Graph the inequality.

19. $x + 3y \geq 12$ 20. $x - y < 4$

21. $x^2 + y^2 < 25$ 22. $x^2 + 4y^2 \leq 36$

23. $xy > 16$ 24. $x + y^2 < -4$

25. $y > x^2 + 16$ 26. $xy \leq 25$

Determine the feasible set for each of the following systems of inequalities.

27. $\begin{cases} x \geq 0 \\ y \geq 0 \\ x + y \leq 8 \end{cases}$ 28. $\begin{cases} x \geq 1 \\ y < 4 \\ x + y > 8 \end{cases}$

29. $\begin{cases} x^2 + y^2 \leq 36 \\ -2 \leq x \leq 2 \end{cases}$ 30. $\begin{cases} 9x^2 - y^2 \geq 0 \\ x^2 + y^2 \leq 25 \end{cases}$

Sketch the graph of the following systems of linear inequalities.

31. $\begin{cases} 4x + 7y \leq 28 \\ 2x - 3y \geq -6 \\ y \geq -2 \end{cases}$ 32. $\begin{cases} 3x + 9y \leq 18 \\ 6x - 4y \geq 8 \\ 3y \geq 0 \end{cases}$

33. $\begin{cases} x + y \leq 10 \\ x - 3y \geq -18 \\ x \geq 0 \\ y \geq 0 \end{cases}$ 34. $\begin{cases} 2x + 3y \geq -120 \\ 6x - 3y \geq -150 \\ x \leq 0 \\ y \leq 0 \end{cases}$

Determine the maximum and minimum values of the functions and the values of x and y where they occur, given a feasible set with vertices as follows:

35. $f(x, y) = 2x + 3y; (3, 1), (5, 2), (1, 4), (8, 1)$

36. $f(x, y) = 4x + y; (2, 1), (1, 9), (3, 5), (2, 7)$

37. $Q = 2x + 3y$ subject to
$$\begin{cases} x \geq 0 \\ y \geq 0 \\ x + 2y \leq 7 \\ 5x - 8y \leq -3 \end{cases}$$

38. $F = \frac{3}{2}x + y$ subject to
$$\begin{cases} x \leq 6 \\ y \geq 0 \\ x + 2y \geq 4 \\ x - 2y \geq 2 \end{cases}$$

Find the maximum and the minimum values of the given function in the indicated region.

39. $f(x, y) = x + 4y$ 40. $f(x, y) = 3x + y$

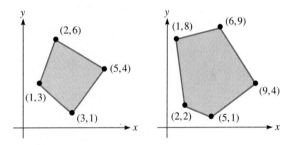

♦ **Applications**

For each of the following linear programming problems, (a) determine a system of inequalities describing the given problem, (b) determine the feasible set of the systems of inequalities, and (c) solve the problem.

41. A cabinet company makes two types of cabinet drawers, one plain and one fancy. Each plain drawer takes 2 hours of work to assemble and 1 hour of sanding. Each fancy drawer takes 1 hour of work to assemble and 4 hours of sanding. The four assembly workers and six sanding workers each work 12 hours per day. Each plain drawer nets a $3 profit, and each fancy drawer nets a $5 profit. If the company can sell all the drawers it makes, how many of each kind should be produced each day in order to maximize profit?

42. Seiberling Tire Company of Akron, Ohio, has 1000 units of raw rubber to use in producing radial tires for cars and for tractor tires. Each car tire requires 5 units of rubber, and each tractor tire requires 20 units of rubber. Labor costs are $8 for a car tire and $12 for a tractor tire. If Seiberling does not want to pay more than $1500 in labor costs and wishes to make a profit of $10 per car tire and $25 per tractor tire, how many of each type of tire should be made in order to maximize profits?

CHAPTER 9

APPLICATIONS OF LINEAR SYSTEMS

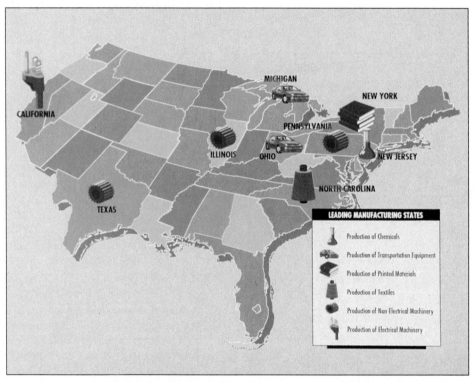

*The various segments of an economy are interrelated in the sense that they use the outputs of other segments. These interrelationships may be described using **input–output analysis**, a mathematical technique using matrices. Using input–output analysis, the effect of a change in one segment of the economy on the other segments may be calculated.*

I   n the preceding chapter, we developed the mathematics of systems of linear equations and inequalities. In this chapter, we apply this mathematics to partial fractions, matrices, and determinants.

9.1 PARTIAL FRACTIONS

We have learned to add rational expressions in a single variable. We found that the sum of rational expressions is another rational expression. Moreover, the denominator of the sum is the product of the denominators of the summands (at least before any reduction is carried out). For example, we have

$$\frac{1}{x - 2} + \frac{2}{x - 1} = \frac{3x - 5}{(x - 2)(x - 1)}$$

In many contexts (most notably in performing integration in calculus), an expression such as the one on the left is far easier to work with than one like the expression on the right. The method of **partial fractions** is a procedure for writing rational expressions (such as the one on the right) in terms of simple rational expressions (such as the ones on the left). As we will see, this method often leads to a system of linear equations.

Suppose that you are given a rational expression $f(x)/g(x)$, where $f(x)$ and $g(x)$ are polynomials with real coefficients. By the division algorithm for polynomials, we can divide $f(x)$ by $g(x)$ to obtain

$$\frac{f(x)}{g(x)} = q(x) + \frac{h(x)}{g(x)}$$

where $q(x)$ is the quotient and $h(x)$ is the remainder and has degree less than the degree of $g(x)$. We then write the rational expression $h(x)/g(x)$ as a sum of simple expressions as follows.

As we discussed in Chapter 3, the polynomial $g(x)$ may be factored into a product of linear and irreducible quadratic factors, all with real coefficients. A typical linear factor is of the form $(ax + b)^n$. Corresponding to each such factor, the expression for $h(x)/g(x)$ has a sum of simple expressions of the form

$$\frac{A_1}{ax + b} + \frac{A_2}{(ax + b)^2} + \cdots + \frac{A_n}{(ax + b)^n}$$

for suitable real numbers $A_1, A_2, \ldots, A_n$. A typical quadratic factor is of the form $\left(cx^2 + dx + e\right)^n$. Corresponding to each such factor, the expression for $h(x)/g(x)$ has a sum of simple expressions of the form

$$\frac{A_1 + B_1 x}{cx^2 + dx + e} + \frac{A_2 + B_2 x}{(cx^2 + dx + e)^2} + \cdots + \frac{A_n + B_n x}{(cx^2 + dx + e)^n}$$

where $A_1, B_1, A_2, B_2, \ldots, A_n, B_n$ are suitable real numbers.

The expression of $f(x)/g(x)$ as the sum of polynomials and rational expressions of the preceding form is called a **partial fraction expansion**. We may summarize the discussion as follows:

DETERMINING THE PARTIAL FRACTION EXPANSION OF $f(x)/g(x)$

1. Use long division to write $f(x)/g(x)$ in the form

$$\frac{f(x)}{g(x)} = q(x) + \frac{h(x)}{g(x)}, \quad \deg(h) < \deg(g)$$

2. Factor $g(x)$ into a product of distinct real factors, some linear and some quadratic.
3. For a linear factor $(ax + b)^n$, the partial fraction expansion has a component

$$\frac{A_1}{ax + b} + \frac{A_2}{(ax + b)^2} + \cdots + \frac{A_n}{(ax + b)^n}$$

4. For an irreducible quadratic factor $(cx^2 + dx + e)^n$, the partial fraction expansion has a component

$$\frac{A_1 + B_1 x}{cx^2 + dx + e} + \frac{A_2 + B_2 x}{(cx^2 + dx + e)^2} + \cdots + \frac{A_n + B_n x}{(cx^2 + dx + e)^n}$$

5. The partial fraction expansion of $f(x)/g(x)$ is the sum of $q(x)$, from step 1, and each of the components from steps 3 and 4.

The principal difficulty in determining the partial fraction expansion of a given rational expression is in determining the values of the various coefficients. The following examples show how this may be done when $g(x)$ is a product of nonrepeating linear factors.

➤ **EXAMPLE 1**
Simple Partial Fraction Expansion

Determine the partial fraction expansion of the rational expression

$$\frac{3x - 5}{(x - 2)(x - 1)}$$

Solution

Because the degree of the numerator is less than the degree of the denominator, it is unnecessary to carry out the division indicated in the previous discussion. (The result would indicate that $q(x)$ is the zero polynomial and $h(x)$ equals $f(x)$.) The denominator is already factored. According to the previous discussion, there is a component corresponding to each factor in the denominator. The component corresponding to the factor $x - 1$ is of the form

$$\frac{A}{x - 1}$$

The component corresponding to the factor $x - 2$ is of the form

$$\frac{B}{x - 2}$$

That is, we must have

$$\frac{3x - 5}{(x - 2)(x - 1)} = \frac{A}{x - 1} + \frac{B}{x - 2}$$

Multiply both sides by $(x - 2)(x - 1)$ to obtain

$$3x - 5 = A(x - 2) + B(x - 1)$$

This last equation must hold for all values of x. Let's choose two particular values for x to use in determining the values of A and B. The most convenient values are, respectively, $x = 1$ and $x = 2$. Substituting these values into the last equation, we see that when $x = 1$,

$$3(1) - 5 = A(1 - 2) + B(0)$$
$$-A = -2$$
$$A = 2$$

When $x = 2$,

$$3(2) - 5 = A(0) + B(2 - 1)$$
$$B = 1$$

Thus, the partial fraction expansion of the given rational expression is

$$\frac{3x - 5}{(x - 2)(x - 1)} = \frac{2}{x - 1} + \frac{1}{x - 2}$$

The calculation just performed reversed the addition of rational functions we performed at the beginning of the section.

➤ **EXAMPLE 2**
Partial Fractions
with Three Factors

Determine the partial fraction expansion of the following rational expression:

$$\frac{2x^4 - 1}{x^3 - x}$$

Solution

Begin by dividing the numerator by the denominator. The quotient is $q(x) = 2x$, and the remainder is $r(x) = 2x^2 - 1$. That is, we have

$$\frac{2x^4 - 1}{x^3 - x} = 2x + \frac{2x^2 - 1}{x^3 - x}$$

Next, factor the denominator:

$$x^3 - x = x(x^2 - 1)$$
$$= x(x - 1)(x + 1)$$

The partial fraction expansion has the form

$$\frac{2x^2 - 1}{x^3 - x} = \frac{A}{x} + \frac{B}{x - 1} + \frac{C}{x + 1}$$

Multiply both sides by $x(x - 1)(x + 1)$ to obtain the identity

$$2x^2 - 1 = A(x - 1)(x + 1) + Bx(x + 1) + Cx(x - 1)$$

To obtain the values of A, B, and C, we substitute three different values for x into the above identity. The three most convenient values are $x = 0$, $x = 1$, and $x = -1$, because these values result in various terms being equal to 0. When $x = 0$, we have

$$2(0)^2 - 1 = A(0 - 1)(0 + 1) + B \cdot 0(0 + 1) + C \cdot 0(0 - 1)$$
$$-1 = -A$$
$$A = 1$$

When $x = 1$, we have

$$2(1)^2 - 1 = A(1 - 1)(1 + 1) + B \cdot 1(1 + 1) + C \cdot 1(1 - 1)$$
$$1 = 2B$$
$$B = \tfrac{1}{2}$$

When $x = -1$, we have

$$2(-1)^2 - 1 = A(-1 - 1)(-1 + 1) + B \cdot (-1)(-1 + 1) + C \cdot (-1)(-1 - 1)$$
$$1 = 2C$$
$$C = \tfrac{1}{2}$$

Thus, the partial fraction expansion is given by

$$\frac{2x^4 - 1}{x^3 - x} = 2x + \frac{1}{x} + \frac{\tfrac{1}{2}}{x - 1} + \frac{\tfrac{1}{2}}{x + 1}$$

$$= 2x + \frac{1}{x} + \frac{1}{2(x - 1)} + \frac{1}{2(x + 1)}$$

In the preceding two examples, we were able to determine the unknown coefficients in the partial fraction expansion one at a time. However, in some instances, it is necessary to determine the coefficients by solving a system of linear equations. This phenomenon is illustrated in the next example.

> **EXAMPLE 3**
Partial Fractions
with a Multiple Factor

Determine the partial fraction expansion of the following rational expression:

$$\frac{3x^3 - 5x + 1}{x^4 + x^2}$$

Solution

The degree of the numerator is less than the degree of the denominator, so the initial division step is unnecessary. Factoring the denominator gives us

$$x^4 + x^2 = x^2(x^2 + 1)$$

Note that the quadratic factor

$$x^2 + 1$$

cannot be factored further because the factorization is assumed to be factored into polynomials with real coefficients. The general form of the partial fraction expansion is then

$$\frac{3x^3 - 5x + 1}{x^4 + x^2} = \frac{A}{x} + \frac{B}{x^2} + \frac{Cx + D}{x^2 + 1}$$

Multiplying both sides by $x^4 + x^2$ yields the identity

$$3x^3 - 5x + 1 = Ax(x^2 + 1) + B(x^2 + 1) + (Cx + D)x^2$$

Substituting $x = 0$ yields the value of B

$$1 = 0 + B(0^2 + 1) + 0$$
$$B = 1$$

Inserting this value back into the identity gives us

$$3x^3 - 5x + 1 = Ax(x^2 + 1) + (x^2 + 1) + (Cx + D)x^2$$
$$= (Ax^3 + Ax) + (x^2 + 1) + (Cx^3 + Dx^2)$$
$$= (A + C)x^3 + (D + 1)x^2 + Ax + 1$$

We now have an identity between two polynomials in x. For them to be equal, the coefficients of their respective powers of x must be equal. That is, we have the system of equations

$$A + C = 3 \qquad \textit{Coefficient of } x^3$$
$$D + 1 = 0 \qquad \textit{Coefficient of } x^2$$
$$A = -5 \qquad \textit{Coefficient of } x$$

Solving this system gives us

$$D = -1$$
$$A = -5$$
$$-5 + C = 3$$
$$C = 8$$

Therefore, the desired partial fraction decomposition is

$$\frac{3x^3 - 5x + 1}{x^4 + x^2} = \frac{-5}{x} + \frac{1}{x^2} + \frac{8x - 1}{x^2 + 1}$$

Note that in our solution we compared like coefficients on both sides of the identity rather than substituting assorted values of x to arrive at a system of equations. Both approaches are valid. For instance, we could just as well have arrived at a (different) system by substituting for x the values $x = 0, 1, 2, 3$. Why not carry out the calculations and verify that the resulting system leads to the same partial fraction expansion?

Exercises 9.1

Determine the partial fractional expansion of the given rational expressions.

1. $\dfrac{6x - 2}{(x + 1)(x - 1)}$

2. $\dfrac{4x + 1}{(x - 2)(x + 1)}$

3. $\dfrac{4x + 2}{x(x + 1)(x + 2)}$

4. $\dfrac{2x^2 - 6x - 2}{x(x + 2)(x - 1)}$

5. $\dfrac{7x - 10}{(2x - 1)(x - 2)}$

6. $\dfrac{2x + 8}{(x - 3)(3x - 2)}$

7. $\dfrac{x^2 - 17x + 35}{(x^2 + 1)(x - 4)}$

8. $\dfrac{7x^2 - 29x + 24}{(2x - 1)(x - 2)^2}$

9. $\dfrac{13x + 5}{3x^2 - 7x - 6}$

10. $\dfrac{4x - 13}{2x^2 + x - 6}$

11. $\dfrac{3x^3 + 4x^2 - x}{x^2 + x - 2}$

12. $\dfrac{2x^3 - x^2 - 13x + 6}{x^2 - x - 6}$

13. $\dfrac{3x^3 - 8x^2 + 9x - 6}{x^2 - 3x + 2}$

14. $\dfrac{4x^3 - 16x^2 + 9x - 15}{2x^2 - 9x + 4}$

15. $\dfrac{5x + 5}{x^2 - x - 6}$

16. $\dfrac{11x + 2}{2x^2 + x - 1}$

17. $\dfrac{-2x^2 + 7x + 2}{x^3 - 2x^2 + x}$

18. $\dfrac{-3x^2 - 10x - 4}{x(x + 1)^2}$

19. $\dfrac{-x^2 + 26x + 6}{(2x - 1)(x + 2)^2}$

20. $\dfrac{5x^2 + 9x - 56}{(x - 4)(x - 2)(x + 1)}$

21. $\dfrac{6x^3 + 29x^2 - 6x - 5}{2x^2 + 9x - 5}$

22. $\dfrac{-9x^2 + 7x - 4}{x^3 - 3x^2 - 4x}$

23. $\dfrac{x^3 + x^2 + 2}{(x^2 + 2)^2}$

24. $\dfrac{2x^3 + 8x^2 + 2x + 4}{(x + 1)^2(x^2 + 3)}$

25. $\dfrac{6x^3 + 5x^2 - 7}{3x^2 - 2x - 1}$

26. $\dfrac{3x^2 - 11x - 26}{(x^2 - 4)(x + 1)}$

27. $\dfrac{3x^2 + 10x + 9}{(x + 2)^3}$

28. $\dfrac{x^2 - x - 4}{(x - 2)^3}$

29. $\dfrac{5}{(x + 2)^2 (x + 1)}$

30. $\dfrac{-8x + 23}{2x^2 + 5x - 12}$

35. $y = \dfrac{2x^2 + 4x - 1}{x^2 + x}$

36. $y = \dfrac{9}{x^2 + x - 2}$

Decompose into partial fractions.

31. $\dfrac{x}{x^4 - a^4}$

32. $\dfrac{2x^2 + 2ax + 3x + 2a^2 - 3a}{x^3 - a^3}$

Decompose into partial fractions and graph by addition of ordinates.

33. $y = \dfrac{x - 11}{x^2 - 1}$

34. $y = \dfrac{5x - 1}{x^2 - x - 2}$

Find constants x, y, and z so that the following statements are true.

37. $\dfrac{2a + 1}{(a + 1)(a^2 - 3a + 5)} = \dfrac{x}{(a + 1)} + \dfrac{ya + z}{a^2 - 3a + 5}$

38. $\dfrac{a + 2}{a^3 + 2a^2 + a} = \dfrac{x}{a} + \dfrac{ya + z}{a^2 + 2a + 1}$

■ 9.2 MATRICES

In Chapter 8, we introduced the notion of a matrix and showed how elementary row operations on a matrix can be used to solve systems of linear equations. Actually, matrices are an important mathematical notion in their own right, and their applications extend throughout pure and applied mathematics. For instance, matrices are used by physicists extensively in describing mechanics problems and in representing elementary particles. Economists use matrices to describe the interactions of various sectors of an economy. Matrices are even used by computer scientists in describing computer graphics calculations. Mathematicians have developed an extensive algebraic theory of matrices that parallels (and draws its motivation from) the elementary algebraic notions for real numbers. In this section, we will provide an introduction to the algebra of matrices.

Recall that a matrix is a rectangular array of real numbers. Matrices are usually denoted with capital letters, such as A, B, C, X, Y, and Z. The entries of a matrix are typically denoted using the lowercase version of the matrix name, with subscripts to indicate position. For instance, the entries of the matrix A are denoted a_{ij}, where i denotes the row number and j denotes the column number. We write

$$A = \begin{bmatrix} a_{ij} \end{bmatrix}$$

to indicate that a typical entry of A is a_{ij}.

Suppose that $A = \begin{bmatrix} a_{ij} \end{bmatrix}$ and $B = \begin{bmatrix} b_{ij} \end{bmatrix}$ are matrices. We say that the two matrices are **equal**, denoted $A = B$, provided that the two matrices are of the same size and their corresponding entries are equal, that is, $a_{ij} = b_{ij}$ for all possible i and j.

A matrix whose entries are all 0 is called a **zero matrix** and is denoted **0**. Actually, there is a zero matrix corresponding to each possible matrix size. It is customary to determine the size of a zero matrix from context.

■ Definition 1
Sum of Two Matrices,
Product of a Matrix
and a Real Number

Suppose that A and B are matrices of the same size. The **sum of A and B**, denoted $A + B$, is the matrix obtained by adding corresponding entries of A and B. Suppose that k is a real number. The **product kA** is the matrix whose entries are the entries of A each multiplied by k.

> **EXAMPLE 1**
> Matrix Arithmetic

Let

$$A = \begin{bmatrix} 5 & 0 \\ 3 & -1 \end{bmatrix}, \qquad B = \begin{bmatrix} -2 & 4 \\ 1 & 3 \end{bmatrix}$$

Calculate the following:

    1. $A + B$ 2. $(-2)A$ 3. $3A + 4B$

Solution

1. $A + B = \begin{bmatrix} 5 & 0 \\ 3 & -1 \end{bmatrix} + \begin{bmatrix} -2 & 4 \\ 1 & 3 \end{bmatrix}$

$$= \begin{bmatrix} 5 - 2 & 0 + 4 \\ 3 + 1 & -1 + 3 \end{bmatrix}$$

$$= \begin{bmatrix} 3 & 4 \\ 4 & 2 \end{bmatrix}$$

2. $(-2)A = (-2)\begin{bmatrix} 5 & 0 \\ 3 & -1 \end{bmatrix}$

$$= \begin{bmatrix} (-2) \cdot 5 & (-2) \cdot 0 \\ (-2) \cdot 3 & (-2) \cdot (-1) \end{bmatrix}$$

$$= \begin{bmatrix} -10 & 0 \\ -6 & 2 \end{bmatrix}$$

3. $3A + 4B = 3\begin{bmatrix} 5 & 0 \\ 3 & -1 \end{bmatrix} + 4\begin{bmatrix} -2 & 4 \\ 1 & 3 \end{bmatrix}$

$$= \begin{bmatrix} 3 \cdot 5 + 4 \cdot (-2) & 3 \cdot 0 & + & 4 \cdot 4 \\ 3 \cdot 3 + 4 \cdot 1 & 3 \cdot (-1) & + & 4 \cdot 3 \end{bmatrix}$$

$$= \begin{bmatrix} 7 & 16 \\ 13 & 9 \end{bmatrix}$$

Addition of matrices obeys both the commutative and associative laws. That is, we know the following formulas hold for matrices of the same size:

$$A + B = B + A \qquad \textit{Commutative law}$$
$$A + (B + C) = (A + B) + C \qquad \textit{Associative law}$$

Moreover, the zero matrix (of appropriate size) acts as the identity for the operation of addition. That is, we have

$$A + \mathbf{0} = \mathbf{0} + A = A \qquad \textit{Identity}$$

The **additive inverse of the matrix** A, denoted $-A$, is the matrix whose entries are the negatives of the corresponding entries of A. The additive inverse has the property

$$A + (-A) = (-A) + A = \mathbf{0}$$

The difference $A - B$ of two matrices is defined via the formula

$$A - B = A + (-B)$$

All of these formulas should remind you of the corresponding properties of real numbers with respect to the operations of addition and subtraction. In fact, from the previously mentioned definitions, we can deduce analogues of many of the elementary algebraic properties of real numbers. The exercises contain a number of these for you to deduce.

A matrix consisting of a single row is called a **row matrix**, and a matrix consisting of a single column is called a **column matrix.** Suppose that we are given a row matrix and a column matrix of the same size. The product of the row times the column is the 1×1 matrix formed by multiplying corresponding entries of the two matrices and adding. That is, we have

$$[a_1 \quad a_2 \quad \cdots \quad a_n] \begin{bmatrix} b_1 \\ b_2 \\ \vdots \\ b_n \end{bmatrix} = \begin{bmatrix} a_1 b_1 + a_2 b_2 + \cdots + a_n b_n \end{bmatrix}$$

Thus, for example, we have

$$[2 \quad 1 \quad -3] \begin{bmatrix} 5 \\ 2 \\ -4 \end{bmatrix} = [2 \cdot 5 + 1 \cdot 2 + (-3) \cdot (-4)]$$

$$= [24]$$

Using this way of multiplying a row matrix by a column matrix, we may define multiplication of matrices as follows:

Definition 2
Matrix Multiplication

Suppose that A is an $m \times n$ matrix and B is an $n \times k$ matrix. Then the product AB is the $m \times k$ matrix with entries as follows: The entry in the ith row and jth column of AB is obtained by multiplying the ith row of A by the jth column of B.

Suppose that the ith row of A is equal to

$$[a_{i1} \quad a_{i2} \quad \ldots \quad a_{in}]$$

and that the jth column of B is equal to

$$\begin{bmatrix} b_{1j} \\ b_{2j} \\ \vdots \\ b_{nj} \end{bmatrix}$$

as shown by the following product:

$$\begin{bmatrix} a_{11} & a_{12} & \cdots & a_{1n} \\ \vdots & \vdots & \vdots & \vdots \\ a_{i1} & a_{i2} & \cdots & a_{in} \\ \vdots & \vdots & \vdots & \vdots \\ a_{m1} & a_{m2} & \cdots & a_{mn} \end{bmatrix} \begin{bmatrix} b_{11} & \cdots & b_{1j} & \cdots & b_{1k} \\ b_{21} & \cdots & b_{2j} & \cdots & b_{2k} \\ \vdots & \vdots & \vdots & \vdots & \vdots \\ b_{n1} & \cdots & b_{nj} & \cdots & b_{nk} \end{bmatrix} = i \rightarrow \begin{bmatrix} \cdots & \overset{\overset{\textstyle j}{\downarrow}}{} & \cdots \\ \cdots & c_{ij} & \cdots \\ \cdots & \vdots & \cdots \end{bmatrix}$$

Then the entry c_{ij} in the ith row and jth column of the product is given by

$$c_{ij} = a_{i1} b_{1j} + a_{i2} b_{2j} + \cdots + a_{in} b_{nj}$$

The next example shows how this calculation works in practice.

Note that the definition of matrix multiplication is based on the fact that we can form the product of rows of the first matrix times columns of the second matrix. In order for this to happen, the number of columns of the first matrix must equal the number of rows of the second. If this is not the case, the product is not defined.

➢ **EXAMPLE 2**
Product of 2×2 Matrices

Compute the following matrix product:

$$\begin{bmatrix} 5 & -1 \\ 2 & 0 \end{bmatrix} \begin{bmatrix} 3 & 1 \\ 0 & 9 \end{bmatrix}$$

Solution

The number of columns of the left matrix equals the number of rows of the right matrix, so the product is defined. Moreover, the number of rows in the product equals the number of rows (2) in the left matrix, and the number of columns in the product equals the number of columns (2) in the right matrix. That is, the product is a 2×2 matrix. Let's work out the entries in the product. The entry in row 1, column 1 is obtained by multiplying the first row on the left by the first column on the right. The product is

$$5 \cdot 3 + (-1) \cdot 0 = 15$$

This gives us

$$\begin{bmatrix} 5 & -1 \\ 2 & 0 \end{bmatrix} \begin{bmatrix} 3 & 1 \\ 0 & 9 \end{bmatrix} = \begin{bmatrix} 15 & * \\ * & * \end{bmatrix}$$

The entry in row 1, column 2 of the product is obtained as the product of the first row on the left times the second column on the right. This product is

$$5 \cdot 1 + (-1) \cdot 9 = -4$$

That is,

$$\begin{bmatrix} 5 & -1 \\ 2 & 0 \end{bmatrix} \begin{bmatrix} 3 & 1 \\ 0 & 9 \end{bmatrix} = \begin{bmatrix} 15 & -4 \\ * & * \end{bmatrix}$$

Next, we compute the entry in row 2, column 1 as the product of the second row times the first column:

$$\begin{bmatrix} 5 & -1 \\ 2 & 0 \end{bmatrix} \begin{bmatrix} 3 & 1 \\ 0 & 9 \end{bmatrix} = \begin{bmatrix} 15 & -4 \\ 6 & * \end{bmatrix}$$

Finally, we compute the entry in row 2, column 2 as the product of the second row times the second column:

$$\begin{bmatrix} 5 & -1 \\ 2 & 0 \end{bmatrix} \begin{bmatrix} 3 & 1 \\ 0 & 9 \end{bmatrix} = \begin{bmatrix} 15 & -4 \\ 6 & 2 \end{bmatrix}$$

➢ **EXAMPLE 3**
Which Products
Are Defined?

Determine which of the following products are defined and the size of those products.

1. A is 2×5; B is 5×7.
2. A is 3×4; B is 2×4.
3. A is 4×4; B is 4×3.

Solution

1. The product AB is defined because the inner dimensions of A and B match. That is, the number of columns of A (5) equals the number of rows of B (5). The size of the product is obtained from the outer dimensions and is 2×7.
2. The product is undefined because the number of columns of A (4) is unequal to the number of rows of B (2).
3. The product AB is defined and is of size 4×3.

Matrix multiplication obeys the distributive and associative laws. That is, the following formulas hold provided the products are defined:

$$A(BC) = (AB)C$$
$$A(B + C) = AB + AC$$

We will omit the proofs of these facts.

Unlike multiplication of real numbers, matrix multiplication does not necessarily obey the commutative law, as the following example illustrates.

➢ **EXAMPLE 4**

Matrix Multiplication Is Noncommutative

Suppose that

$$A = \begin{bmatrix} 5 & 1 \\ 2 & 3 \end{bmatrix}, \qquad B = \begin{bmatrix} 1 & 4 \\ 3 & 0 \end{bmatrix}$$

Show that $AB \neq BA$.

Solution

We may determine the values of AB and BA by direct calculation:

$$AB = \begin{bmatrix} 5 & 1 \\ 2 & 3 \end{bmatrix} \cdot \begin{bmatrix} 1 & 4 \\ 3 & 0 \end{bmatrix}$$
$$= \begin{bmatrix} 8 & 20 \\ 11 & 8 \end{bmatrix}$$
$$BA = \begin{bmatrix} 1 & 4 \\ 3 & 0 \end{bmatrix} \cdot \begin{bmatrix} 5 & 1 \\ 2 & 3 \end{bmatrix}$$
$$= \begin{bmatrix} 13 & 13 \\ 15 & 3 \end{bmatrix}$$

Two matrices are equal only when their corresponding entries are the same. By examining the entries here, we see that they are certainly not the same, so $AB \neq BA$.

For each positive integer n, let I_n denote the $n \times n$ matrix with 1s down the main diagonal and 0s everywhere else. For instance,

$$I_1 = [1] = 1$$
$$I_2 = \begin{bmatrix} 1 & 0 \\ 0 & 1 \end{bmatrix}$$
$$I_3 = \begin{bmatrix} 1 & 0 & 0 \\ 0 & 1 & 0 \\ 0 & 0 & 1 \end{bmatrix}$$

I_n is called the **identity matrix of size n**. Indeed, I_n acts as the identity for multiplication with square matrices of size $n \times n$. That is, if A is an $n \times n$ matrix, then one can prove that

$$I_n A = A I_n = A \tag{1}$$

In case $n = 2$, this statement follows from the equations

$$I_2 \cdot \begin{bmatrix} a & b \\ c & d \end{bmatrix} = \begin{bmatrix} 1 & 0 \\ 0 & 1 \end{bmatrix} \begin{bmatrix} a & b \\ c & d \end{bmatrix}$$

$$= \begin{bmatrix} 1 \cdot a + 0 \cdot c & 1 \cdot b + 0 \cdot d \\ 0 \cdot a + 1 \cdot c & 0 \cdot b + 1 \cdot d \end{bmatrix}$$

$$= \begin{bmatrix} a & b \\ c & d \end{bmatrix}$$

Similarly,

$$\begin{bmatrix} a & b \\ c & d \end{bmatrix} \cdot I_2 = \begin{bmatrix} a & b \\ c & d \end{bmatrix}$$

The proof of (1) for general n will be omitted.

Matrices and Technology

Matrix arithmetic can be quite tedious to do manually. Much of the tedium can be reduced by appropriate application of technology. In this section, we will indicate how to use a calculator to perform various matrix calculations.

A TI-81 calculator may store three matrices, denoted [A], [B], and [C]. You will find these indicated on keys at the bottom of the calculator.

Let's begin our discussion by learning to enter matrices into the calculator. To enter [A], press (MATRX). You will see the menu shown in Figure 1. Select *EDIT* from the first row and press (ENTER). You will see the display shown in Figure 2. Select option 1 to edit the matrix [A] and press (ENTER). You will see the display shown in Figure 3. First, enter the size of the matrix in the first row by changing 6×6 to 3×3. (Use the arrow keys to select each 6 and key in a 3.) Next, key in the entries of A. The entries are specified in the following rows using their subscript notation. That is, "1, 2" refers to row 1, column 2. When you have finished keying in the entries, press (ENTER). Follow a similar procedure to enter matrix B. To leave the matrix editing screen, press (2nd) (QUIT).

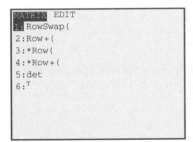

Figure 1

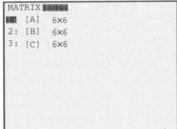

Figure 2

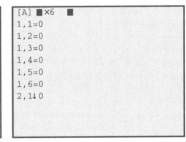

Figure 3

New Technology

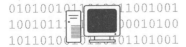

EXAMPLE 5
Multiplying Matrices
Using a Calculator

Let

$$A = \begin{bmatrix} 5 & -1 & 4 \\ 3 & 0 & 9 \\ 2 & -5 & 8 \end{bmatrix},$$

$$B = \begin{bmatrix} 0.1 & 0.15 & 3.0 \\ 1.1 & 2.7 & 4.0 \\ 0.3 & 1.0 & 0.9 \end{bmatrix}$$

Use a calculator to calculate

1. $A + B$
2. AB
3. BA

Solution

```
[A]+[B]
[ 5.1 -.85 7  ]
[ 4.1 2.7  13 ]
[ 2.3 -4   8.9]
```

Figure 4

First, enter the matrices into the calculator using the procedure described above.

1. Enter the expression for the sum $A + B$ as [2nd] [[A]] [+] [2nd] [[B]] [ENTER]. You will see the sum as shown in Figure 4.

2. Enter AB using the keystrokes [2nd] [[A]] [×] [2nd] [[B]] [ENTER]. You will see the product shown as in Figure 5. Note that some of the matrix is off the screen to the right. This is indicated by the dots. To see the portion off-screen, press the right arrow key several times. To scroll the screen back to the position shown, use the left arrow key.

3. See Figure 6.

```
[A]*[B]
[ .6   2.05 14....
[ 3    9.45 17....
[ -2.9 -5.2 -6....
```

Figure 5

```
[B]*[A]
[ 6.95 -15.1 25...
[ 21.6 -21.1 60...
[ 6.3  -4.8  17...
```

Figure 6 ◄

Exercises 9.2

Given

$$A = \begin{bmatrix} 2 & 0 \\ 1 & 3 \end{bmatrix}, \qquad B = \begin{bmatrix} -1 & 3 \\ 0 & -2 \end{bmatrix}, \qquad C = \begin{bmatrix} 4 & 2 \\ -1 & 0 \end{bmatrix},$$

$$D = \begin{bmatrix} 1 & 0 \\ 0 & 1 \end{bmatrix}, \qquad E = \begin{bmatrix} 2 & -1 \\ 0 & 0 \end{bmatrix}, \qquad F = \begin{bmatrix} 1 & 3 \\ 2 & 6 \end{bmatrix},$$

calculate the following.

1. $A + B$ 2. $A + E$ 3. $2F$ 4. $3C$

5. $B - C$ 6. $C - F$ 7. $2A + 3D$ 8. $4F - 2B$

9. $B + D$ 10. $E + D$ 11. $F - 2A$ 12. $E + 2B$

13. $-5B$ 14. $-3E$ 15. AB 16. EF

17. CD 18. BC 19. $(AB)C$ 20. $A(BF)$

Given

$$A = \begin{bmatrix} 1 & 2 \\ 3 & 4 \end{bmatrix}, \qquad B = \begin{bmatrix} -1 \\ 7 \end{bmatrix}, \qquad C = \begin{bmatrix} 3 & -4 & 1 \\ 5 & 0 & 2 \end{bmatrix},$$

$$D = \begin{bmatrix} -2 & -3 & -4 \\ 2 & -1 & 0 \\ 4 & -2 & 3 \end{bmatrix}, \qquad E = \begin{bmatrix} -1 & 2 \\ 3 & -2 \end{bmatrix}, \qquad F = \begin{bmatrix} 5 \\ -2 \end{bmatrix},$$

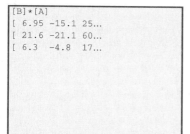

$$G = \begin{bmatrix} -1 & 2 & 3 \\ 0 & 1 & 0 \end{bmatrix}, \qquad H = \begin{bmatrix} 0 & 1 & 4 \\ 1 & 2 & -1 \\ 3 & 2 & -2 \end{bmatrix},$$

$$I = \begin{bmatrix} 4 & 0 & -1 \end{bmatrix}$$

find each of the following matrices, whenever possible.

21. AB 22. DH 23. CD 24. EF

25. FG 26. GH 27. CH 28. BI

29. DG 30. EA 31. IH 32. IF

33. FE 34. AF

Determine which of the following products are defined and the size of those products.

35. A is 3×2; B is 2×5.   36. A is 4×5; B is 5×2.

37. A is 2×2; B is 1×2.   38. A is 5×7; B is 5×7.

39. A is 1×4; B is 4×3.   40. A is 5×1; B is 1×3.

Find the values of $w, x, y,$ and z in the following matrix equations.

41. $\begin{bmatrix} 3 & 1 \\ 4 & 5 \end{bmatrix} = \begin{bmatrix} x & y \\ z & 5 \end{bmatrix}$

42. $\begin{bmatrix} 3 & 5 & x \\ 2 & y & 3 \end{bmatrix} = \begin{bmatrix} z & 5 & 2 \\ 2 & 7 & w \end{bmatrix}$

43. $\begin{bmatrix} x - 7 & 4y & 8z \\ 6w & 2 & 5 \end{bmatrix} + \begin{bmatrix} -9 & 8y & 3 \\ 2 & 5 & 4 \end{bmatrix} = \begin{bmatrix} 2 & 36 & 27 \\ 20 & 7 & 9 \end{bmatrix}$

44. $\begin{bmatrix} x + 2 & 3y + 1 & 5w \\ 4z & 0 & 18 \end{bmatrix}$

$\quad + \begin{bmatrix} 3x & 2y & 5w \\ 2z & 7 & -6 \end{bmatrix} = \begin{bmatrix} 10 & -14 & 80 \\ 10 & 7 & 12 \end{bmatrix}$

Prove the following for

$$A = \begin{bmatrix} a_{11} & a_{12} \\ a_{21} & a_{22} \end{bmatrix}, \quad B = \begin{bmatrix} b_{11} & b_{12} \\ b_{21} & b_{22} \end{bmatrix},$$

$$C = \begin{bmatrix} c_{11} & c_{12} \\ c_{21} & c_{22} \end{bmatrix}, \quad I = \begin{bmatrix} 1 & 0 \\ 0 & 1 \end{bmatrix}$$

45. $IA = AI = A$

46. $A + B = B + A$

47. $k(A + B) = kA + kB$

48. $(A + B) + C = A + (B + C)$

■ 9.3 DETERMINANTS

Let A be a square matrix. Associated with A is a real number, its determinant. The determinant plays a significant role in matrix theory, as we will see in the next section, where we discuss Cramer's rule for calculating the inverse of a matrix. Note that the concept of a determinant is defined only for square matrices. Throughout this section, all matrices will be assumed to be square.

The determinant of a square matrix A is denoted $|A|$ and is a real number. The concept of a determinant is defined for a matrix of a given size in terms of determinants of matrices of smaller sizes.

For a 1×1 matrix, the determinant is defined as follows: Suppose that $A = [a]$ is a 1×1 matrix. Then $|A|$ is defined as a.

■ Definition 1
Determinant of
a 2 × 2 Matrix

Suppose that A is the 2×2 matrix

$$A = \begin{bmatrix} a & b \\ c & d \end{bmatrix}$$

Then $|A|$ is defined by the formula

$$|A| = ad - bc$$

That is, $|A|$ is computed by forming the product of the elements along the diagonal from top left to bottom right, minus the product of the elements along the diagonal from bottom left to top right. (See Figure 1.)

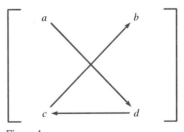

Figure 1

➤ **EXAMPLE 1**
Determinant of a
2 × 2 Matrix

Calculate $|A|$, where A is the matrix

$$A = \begin{bmatrix} 2 & -3 \\ 4 & 1 \end{bmatrix}$$

Solution

From the definition just given, the value of the determinant equals

$$|A| = 2 \cdot 1 - 4 \cdot (-3)$$
$$= 14$$

Definition 2
Minor of a 3×3 Matrix

By eliminating a row and a column of the matrix, we can arrive at a 2×2 matrix. The determinant of such a matrix is called a **minor** of A corresponding to the eliminated row and column.

For instance, suppose that

$$A = \begin{bmatrix} -1 & 0 & 3 \\ 2 & 1 & 0 \\ 3 & 4 & -1 \end{bmatrix}$$

The minor obtained by eliminating the second row and third column is denoted M_{23}, given by

$$M_{23} = \begin{vmatrix} -1 & 0 \\ 3 & 4 \end{vmatrix} = (-1) \cdot 4 - 3 \cdot 0 = -4$$

The subscript in the notation M_{23} indicates that the minor was formed from A by eliminating the second row and the third column. Note that the row comes first, then the column. More generally, we denote by M_{ij} the minor obtained by eliminating the ith row and jth column.

Definition 3
Cofactor of a Minor M_{ij}

Suppose that A is a 3×3 matrix. Then the **cofactor** A_{ij} is defined through the formula

$$A_{ij} = (-1)^{i+j} M_{ij}$$

For instance, let A be the matrix defined above. The cofactor A_{23} is equal to

$$A_{23} = (-1)^{2+3} M_{23}$$
$$= -(-4)$$
$$= 4$$

The cofactor of any entry of A is the minor of that entry multiplied by $+1$ or -1 according to the scheme:

$$\begin{bmatrix} + & - & + \\ - & + & - \\ + & - & + \end{bmatrix}$$

The determinant of a 3×3 matrix may be defined in terms of cofactors as follows:

Definition 4
Determinant of a 3×3 Matrix

Let A be a 3×3 matrix. Its determinant $|A|$ is defined by the formula

$$|A| = \begin{vmatrix} a_{11} & a_{12} & a_{13} \\ a_{21} & a_{22} & a_{23} \\ a_{31} & a_{32} & a_{33} \end{vmatrix} = a_{11}A_{11} + a_{12}A_{12} + a_{13}A_{13}$$

That is, to calculate the determinant of a 3×3 matrix, form the sum of products of each element of the first row times its corresponding cofactor, that is, the cofactor formed by eliminating the row and column containing the element.

We may derive an alternative formula for the determinant of a 3×3 matrix by using the definitions of the cofactors. We have

$$A_{11} = (-1)^{1+1} M_{11}$$

$$= \begin{vmatrix} a_{22} & a_{23} \\ a_{32} & a_{33} \end{vmatrix}$$

$$= a_{22} a_{33} - a_{23} a_{32}$$

$$A_{12} = (-1)^{1+2} M_{12}$$

$$= (-1) \begin{vmatrix} a_{21} & a_{23} \\ a_{31} & a_{33} \end{vmatrix}$$

$$= -(a_{21} a_{33} - a_{31} a_{23})$$

$$A_{13} = (-1)^{1+3} M_{13}$$

$$= \begin{vmatrix} a_{21} & a_{22} \\ a_{31} & a_{32} \end{vmatrix}$$

$$= a_{21} a_{32} - a_{31} a_{22}$$

Therefore, by the definition of a determinant for a 3×3 matrix, this gives us

$$|A| = a_{11} A_{11} + a_{12} A_{12} + a_{13} A_{13}$$

$$= a_{11} (a_{22} a_{33} - a_{32} a_{23}) - a_{12} (a_{21} a_{33} - a_{31} a_{23})$$

$$+ a_{13} (a_{21} a_{32} - a_{31} a_{22})$$

$$= a_{11} a_{22} a_{33} - a_{11} a_{32} a_{23} - a_{12} a_{21} a_{33} + a_{12} a_{31} a_{23}$$

$$+ a_{13} a_{21} a_{32} - a_{13} a_{31} a_{22}$$

When calculating determinants, it is usually most convenient to go back to the definition rather than use this last formula.

> **EXAMPLE 2**
Determinant of a
3×3 Matrix

Calculate the determinant of the matrix

$$A = \begin{bmatrix} 4 & -1 & 5 \\ 0 & -1 & 1 \\ 1 & -1 & 1 \end{bmatrix}$$

Solution

We form the sum of products of elements of the first row times the corresponding cofactor:

$$|A| = 4 \cdot (-1)^{1+1} \begin{vmatrix} -1 & 1 \\ -1 & 1 \end{vmatrix} + (-1) \cdot (-1)^{1+2} \begin{vmatrix} 0 & 1 \\ 1 & 1 \end{vmatrix} + 5 \cdot (-1)^{1+3} \begin{vmatrix} 0 & -1 \\ 1 & -1 \end{vmatrix}$$

$$= 4 \cdot [(-1) \cdot 1 - (-1) \cdot 1] + 1 \cdot (0 \cdot 1 - 1 \cdot 1) + 5 \cdot [0 \cdot (-1) - 1 \cdot (-1)]$$

$$= 4 \cdot 0 + 1 \cdot (-1) + 5 \cdot 1$$

$$= 4$$

This definition of a determinant involves products of elements of the first row times the corresponding cofactor. Actually, the determinant can also be computed in analogous fashion using any row or column. We describe this computation as **expanding the determinant about a given row or column.** For example, here is the calculation of the determinant of Example 2 expanding about the second column:

$$|A| = (-1) \cdot (-1)^{1+2} \begin{vmatrix} 0 & 1 \\ 1 & 1 \end{vmatrix} + (-1) \cdot (-1)^{2+2} \begin{vmatrix} 4 & 5 \\ 1 & 1 \end{vmatrix} + (-1) \cdot (-1)^{3+2} \begin{vmatrix} 4 & 5 \\ 0 & 1 \end{vmatrix}$$

$$= 1 \cdot (0 \cdot 1 - 1 \cdot 1) - 1 \cdot (4 \cdot 1 - 1 \cdot 5) + 1 \cdot (4 \cdot 1 - 0 \cdot 5)$$

$$= -1 + 1 + 4$$

$$= 4$$

Note that this agrees with the result derived in Example 2. You will find that we arrive at the same result in this case no matter what row or column we use. This observation is a special case of the following result.

CALCULATING A DETERMINANT BY COFACTORS

Let A be a 3×3 matrix. Then the determinant of A can be computed as the sum of the products of the elements in a particular row or column times the corresponding cofactors.

We will omit the proof of this result, which properly belongs in a course in linear algebra.

In our discussion so far, we have defined the determinants for square matrices of size 1, 2, and 3. Let's now define determinants for square matrices of any size. The general idea is to define the determinant for matrices of a given size in terms of determinants of matrices of smaller size.

Definition 5
Determinant of
an $n \times n$ Matrix

Let A be an $n \times n$ matrix. The determinant $|A|$ is defined by the formula

$$|A| = a_{11}A_{11} + a_{12}A_{12} + \cdots + a_{1n}A_{1n}$$

where A_{ij} denotes the cofactor determined by eliminating the ith row and jth column.

Note that the size of the determinants involved in the cofactors is $n - 1$. This means, for example, that this formula defines the determinant of a 4×4 matrix in terms of determinants of 3×3 matrices, the determinant of a 5×5 matrix in terms of determinants of 4×4 matrices, and so forth.

As with 3×3 matrices, the definition given involves expansion about the first row. However, an analogue of this definition holds for determinants of any size. That is, the determinant may be calculated by expanding about any row or column. The next example illustrates how this definition may be applied to calculate the determinant of a 4×4 matrix.

> **EXAMPLE 3**
Determinant of a
4 × 4 Matrix

Calculate the determinant of the matrix

$$A = \begin{bmatrix} 2 & 0 & 1 & 0 \\ 0 & 4 & 2 & 0 \\ 0 & 1 & 0 & 1 \\ 0 & -1 & 1 & 0 \end{bmatrix}$$

Solution

Note that the first column has a single nonzero element. The computation will be simplified if we expand the determinant about the first column because all but the first product are 0. We have

$$|A| = 2 \cdot \begin{vmatrix} 4 & 2 & 0 \\ 1 & 0 & 1 \\ -1 & 1 & 0 \end{vmatrix}$$

The simplest calculation for expanding the determinant on the right is about the third column because two of the resulting products are 0. The result is

$$|A| = 2 \cdot 1 \cdot (-1)^{2+3} \begin{vmatrix} 4 & 2 \\ -1 & 1 \end{vmatrix}$$

$$= -2 \cdot [4 \cdot 1 - (-1) \cdot 2] = -2(4 + 2) = -2(6)$$

$$= -12$$

Cramer's Rule

Cramer's rule gives a general formula for the solution of a system of linear equations in terms of determinants formed from the coefficients of the system. To motivate the general statement of Cramer's rule, let's consider the specialized case of a linear system in two variables:

$$\begin{cases} ax + by = S \\ cx + dy = T \end{cases}$$

Use the addition and subtraction method. Multiply the first equation by c and the second equation by a and subtract:

$$\begin{array}{rl} cax + cby &= cS \\ -(cax + ady) &= aT \\ \hline cby - ady &= cS - aT \end{array}$$

$$(cb - ad)y = cS - aT$$

$$y = \frac{aT - cS}{ad - bc}, \qquad ad - bc \neq 0$$

Substituting this value for y into the first equation of the system, we have

$$ax + by = S$$

$$ax + b\left(\frac{aT - cS}{ad - bc}\right) = S$$

$$ax = S - \frac{baT - bcS}{ad - bc}$$

$$= \frac{adS - baT}{ad - bc}$$

$$x = \frac{Sd - Tb}{ad - bc}, \qquad ad - bc \neq 0$$

The numerator and denominator of x and y may each be expressed as a determinant, namely,

$$x = \frac{\begin{vmatrix} S & b \\ T & d \end{vmatrix}}{\begin{vmatrix} a & b \\ c & d \end{vmatrix}}, \qquad y = \frac{\begin{vmatrix} a & S \\ c & T \end{vmatrix}}{\begin{vmatrix} a & b \\ c & d \end{vmatrix}}$$

We may describe these formulas as follows:

1. The denominator in each case is the determinant of the matrix of coefficients A of the system. We have assumed that this determinant is nonzero and shown that with this assumption the system has a unique solution.

2. In the formula for x, the column of coefficients $\begin{smallmatrix} a \\ b \end{smallmatrix}$ of x is replaced by the column of numbers $\begin{smallmatrix} S \\ T \end{smallmatrix}$ on the right side of the system. The matrix appearing in the numerator is denoted A_1, so that

$$x = \frac{|A_1|}{|A|}$$

3. In the formula for y, the column of coefficients $\begin{smallmatrix} c \\ d \end{smallmatrix}$ of y is replaced by the column of numbers $\begin{smallmatrix} S \\ T \end{smallmatrix}$ on the right side of the system. The matrix appearing in the numerator is denoted A_2, so that

$$y = \frac{|A_2|}{|A|}$$

The formulas

$$x = \frac{|A_1|}{|A|}, \qquad y = \frac{|A_2|}{|A|}$$

are a special case of **Cramer's rule**.

➤ **EXAMPLE 4**
Application of
Cramer's Rule

Use Cramer's rule to solve the system

$$\begin{cases} 5x - 2y = 8 \\ 3x + 4y = -1 \end{cases}$$

Solution

The determinant of the coefficients is

$$|A| = \begin{vmatrix} 5 & -2 \\ 3 & 4 \end{vmatrix} = 5 \cdot 4 - 3 \cdot (-2) = 26$$

This value is nonzero, so the system has a unique solution. Applying Cramer's rule, we have

$$x = \frac{|A_1|}{|A|} = \frac{\begin{vmatrix} 8 & -2 \\ -1 & 4 \end{vmatrix}}{26} = \frac{30}{26} = \frac{15}{13}$$

$$y = \frac{|A_2|}{|A|} = \frac{\begin{vmatrix} 5 & 8 \\ 3 & -1 \end{vmatrix}}{26} = \frac{-29}{26} = -\frac{29}{26}$$

The system has the solution $(x, y) = \left(\frac{15}{13}, -\frac{29}{26}\right)$.

Cramer's rule may be generalized to systems of linear equations in n variables. Suppose that we are given a system with coefficient matrix A. For the nth variable, let A_n denote the matrix obtained by replacing the nth column of A with the column of numbers on the right side of the system. Then we have the following result.

CRAMER'S RULE FOR $n \times n$ LINEAR SYSTEMS

If a linear system of n equations in n variables $x, y, z, \ldots$ has a nonzero determinant $|A|$, then the system has a unique solution given by the formulas

$$x = \frac{|A_1|}{|A|}, \qquad y = \frac{|A_2|}{|A|}, \qquad \ldots$$

Note that Cramer's rule applies only to systems in which the number of equations equals the number of variables. Contrast this with Gauss elimination, which can be used to solve any linear system. The proof of Cramer's rule is beyond the scope of this book and will be omitted. With the advent of computers, Cramer's rule is rarely used to solve systems of linear equations, because it is far less efficient than Gauss elimination. However, Cramer's rule is still useful as a theoretical tool for proving facts about solutions of linear systems.

Properties of Determinants

Here are some properties of determinants that are useful to know, but whose proofs will be omitted. They are useful in shortening the task of evaluating determinants using cofactors, which, as we have seen, can be a tedious task.

ROW OR COLUMN INTERCHANGE

If two rows or two columns of a matrix are interchanged, then the determinant changes sign.

For example, the determinant of the matrix

$$\begin{bmatrix} 2 & 3 \\ -1 & 8 \end{bmatrix}$$

is 19. Interchanging the first and second columns gives the matrix

$$\begin{bmatrix} 3 & 2 \\ 8 & -1 \end{bmatrix}$$

which has determinant -19.

MULTIPLYING A ROW BY A CONSTANT

If all elements of a row (column) are multiplied by k, then the determinant is multiplied by k.

For instance, if we multiply each element of the second row of the matrix

$$\begin{bmatrix} 2 & 3 \\ -1 & 8 \end{bmatrix}$$

by 3, we get the matrix

$$\begin{bmatrix} 2 & 3 \\ -3 & 24 \end{bmatrix}$$

which has determinant 57, which is $3 \cdot 19$.

ADDING ONE ROW TO ANOTHER

The determinant is unchanged if a row (column) is replaced by adding to it or subtracting from it any multiple of another row (column).

For instance, if we add twice the first row to the second of the matrix

$$\begin{bmatrix} 2 & 3 \\ -1 & 8 \end{bmatrix}$$

we get the matrix

$$\begin{bmatrix} 2 & 3 \\ -1 + 2 \cdot 2 & 8 + 2 \cdot 3 \end{bmatrix} = \begin{bmatrix} 2 & 3 \\ 3 & 14 \end{bmatrix}$$

which has determinant 19, the same as the original matrix.

Equations of Lines and Areas of Triangles

Determinants can be used to write many of the formulas in analytic geometry and physics in a simple and elegant form. For example, consider the equation of a line passing through the points (x_1, y_1) and (x_2, y_2). We have seen that this equation may be written in the form

$$y - y_1 = \frac{y_2 - y_1}{x_2 - x_1} (x - x_1)$$

Clear the fraction and write the equation in the form

$$(y - y_1)(x_2 - x_1) - (y_2 - y_1)(x - x_1) = 0$$

$$x_2 y - x_1 y - x_2 y_1 + x_1 y_1 - x y_2 + x_1 y_2 + x y_1 - x_1 y_1 = 0$$

$$(x y_1 - x_1 y) - (x y_2 - x_2 y) + (x_1 y_2 - x_2 y_1) = 0$$

Each of the differences is a determinant:

$$\begin{vmatrix} x & y \\ x_1 & y_1 \end{vmatrix} - \begin{vmatrix} x & y \\ x_2 & y_2 \end{vmatrix} + \begin{vmatrix} x_1 & y_1 \\ x_2 & y_2 \end{vmatrix} = 0$$

It is now easy to recognize this expression as the expansion of the 3×3 determinant

$$\begin{vmatrix} x & y & 1 \\ x_1 & y_1 & 1 \\ x_2 & y_2 & 1 \end{vmatrix} = 0$$

in minors about the third column. Thus, we have proven the following result:

DETERMINANT FORM OF THE EQUATION OF A LINE

The equation of a line passing through the points (x_1, y_1) and (x_2, y_2) is given in determinant form by

$$\begin{vmatrix} x & y & 1 \\ x_1 & y_1 & 1 \\ x_2 & y_2 & 1 \end{vmatrix} = 0$$

Suppose we are given three points (x_1, y_1), (x_2, y_2), and (x_3, y_3). They lie on a common line, provided that the third point lies on the line determined by the first two. Therefore, by the preceding result, we have:

CONDITION FOR COLLINEARITY

Three points (x_1, y_1), (x_2, y_2), and (x_3, y_3) are collinear provided that

$$\begin{vmatrix} x_1 & y_1 & 1 \\ x_2 & y_2 & 1 \\ x_3 & y_3 & 1 \end{vmatrix} = 0$$

It is possible to calculate the area of a triangle in terms of a determinant according to the following result:

DETERMINANT FORMULA FOR AREA OF A TRIANGLE

The area of the triangle with vertices (x_1, y_1), (x_2, y_2), and (x_3, y_3) is given by the formula

$$\text{Area} = \pm \tfrac{1}{2} \begin{vmatrix} x_1 & y_1 & 1 \\ x_2 & y_2 & 1 \\ x_3 & y_3 & 1 \end{vmatrix}$$

➤ **EXAMPLE 5**
Area of a Triangle

Verify the last result in case the triangle is a right triangle with the right angle at the origin.

Solution

In this case, we may set $(x_1, y_1) = (0, 0)$, $(x_2, y_2) = (x_2, 0)$, and $(x_3, y_3) = (0, y_3)$. The determinant formula then gives

$$\text{Area} = \pm \tfrac{1}{2} \begin{vmatrix} 0 & 0 & 1 \\ x_2 & 0 & 1 \\ 0 & y_3 & 1 \end{vmatrix}$$

$$= \pm \tfrac{1}{2} \left[0 \cdot \begin{vmatrix} 0 & 1 \\ y_3 & 1 \end{vmatrix} - 0 \cdot \begin{vmatrix} x_2 & 1 \\ 0 & 1 \end{vmatrix} + 1 \cdot \begin{vmatrix} x_2 & 0 \\ 0 & y_3 \end{vmatrix} \right]$$

$$= \pm \tfrac{1}{2} x_2 y_3$$

$$= \pm \tfrac{1}{2} \cdot \text{Base} \cdot \text{Altitude}$$

Graphing Calculators and Determinants

Many graphing calculators have built-in determinant functions. Using such a function avoids the considerable calculation involved in evaluating determinants, especially those of larger sizes.

New Technology

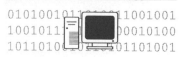

Let

$$B = \begin{bmatrix} 0.1 & 0.15 & 3.0 \\ 1.1 & 2.7 & 4.0 \\ 0.3 & 1.0 & 0.9 \end{bmatrix}$$

EXAMPLE 6
Using a Calculator to Evaluate Determinants

Use a calculator to calculate $|B|$.

Solution

First enter the matrix B. (If you haven't changed matrix $[B]$ from Example 5 in Section 9.2, it is still stored in your calculator.) Press (MATR X). Choose the option *det* from the menu shown (see Figure 2). Now indicate the matrix B by pressing (2nd) (B). The display is as shown in Figure 3. Now press (ENT ER). The determinant, 0.7445, is displayed.

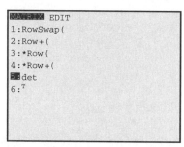

Figure 2

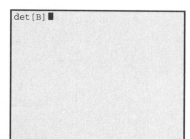

Figure 3   ◄

Exercises 9.3

Calculate $|A|$, where A is the matrix given.

1. $\begin{bmatrix} 1 & 2 \\ 3 & -1 \end{bmatrix}$ 2. $\begin{bmatrix} 4 & -2 \\ 1 & 3 \end{bmatrix}$ 3. $\begin{bmatrix} -5 & 0 \\ -2 & 1 \end{bmatrix}$

4. $\begin{bmatrix} 0 & -2 \\ 1 & 3 \end{bmatrix}$ 5. $\begin{bmatrix} 3 & -1 \\ -2 & -4 \end{bmatrix}$ 6. $\begin{bmatrix} -1 & -3 \\ -2 & -5 \end{bmatrix}$

Use

$$A = \begin{bmatrix} 4 & -1 & -6 \\ 3 & 0 & 7 \\ 1 & 2 & -1 \end{bmatrix} \qquad B = \begin{bmatrix} 2 & -1 & 0 & 3 \\ -1 & 4 & 1 & 3 \\ 0 & 1 & 0 & 2 \\ -1 & 3 & -2 & 1 \end{bmatrix}$$

for the following exercises.

7. Find A_{21}.   8. Find A_{12}.

9. Find B_{43}.   10. Find B_{31}.

11. Find A_{11}.   12. Find A_{23}.

13. Find B_{22}.   14. Find B_{44}.

Calculate $|A|$, where A is the given matrix.

15. $\begin{bmatrix} 1 & 2 & -1 \\ 2 & -1 & 1 \\ 4 & 0 & 2 \end{bmatrix}$ 16. $\begin{bmatrix} 1 & 2 & -3 \\ 4 & 5 & -9 \\ 0 & 0 & 1 \end{bmatrix}$

17. $\begin{bmatrix} 3 & 3 & -1 \\ 2 & 6 & 0 \\ -6 & -6 & 2 \end{bmatrix}$ 18. $\begin{bmatrix} 1 & 2 & -1 \\ 3 & 2 & 1 \\ 1 & 0 & -2 \end{bmatrix}$

19. $\begin{bmatrix} -2 & 0 & 4 & 2 \\ 3 & 6 & 0 & 4 \\ 0 & 0 & 0 & 3 \\ 9 & 0 & 2 & -1 \end{bmatrix}$ 20. $\begin{bmatrix} 7 & -8 & 1 & 2 \\ 21 & 4 & 3 & -1 \\ -35 & 8 & 3 & -2 \\ 14 & 16 & 0 & 1 \end{bmatrix}$

21. $\begin{bmatrix} 4 & 2 & 1 & 0 \\ -2 & 4 & -1 & 7 \\ -5 & 2 & 3 & 1 \\ 6 & 4 & -3 & 2 \end{bmatrix}$ 22. $\begin{bmatrix} 1 & -1 & 0 & 2 \\ 0 & 1 & -1 & 0 \\ 2 & 1 & 0 & -1 \\ -2 & 2 & 1 & 1 \end{bmatrix}$

23. $\begin{bmatrix} 11 & -15 & 20 \\ 16 & 24 & -8 \\ 6 & 9 & 15 \end{bmatrix}$ 24. $\begin{bmatrix} 2 & 1 & 1 \\ 2 & -3 & -1 \\ -4 & 5 & 2 \end{bmatrix}$

$$25. \begin{bmatrix} -3 & 0 & 2 & 6 \\ 2 & 4 & 0 & -1 \\ -1 & 0 & -5 & 2 \\ 0 & -1 & -2 & -3 \end{bmatrix} \qquad 26. \begin{bmatrix} 4 & 0 & 0 & 2 \\ -1 & 0 & 3 & 0 \\ 2 & 4 & 0 & 1 \\ 0 & 0 & 1 & 2 \end{bmatrix}$$

Use Cramer's rule to solve the following linear systems.

27. $\begin{cases} 4x - y = 10 \\ 3x + 5y = 19 \end{cases}$

28. $\begin{cases} 2x + 5y = 18 \\ 3x + 4y = 7 \end{cases}$

29. $\begin{cases} 2x + 3y = -9 \\ 3x + 5y = -13 \end{cases}$

30. $\begin{cases} 4x - y = 3 \\ 2x - 3y = -1 \end{cases}$

31. $\begin{cases} x + 2y = 5 \\ 2x - 5y = -8 \end{cases}$

32. $\begin{cases} 2x + 3y + z = 1 \\ x - y + 2z = 3 \end{cases}$

33. $\begin{cases} 3x - 5y = 6 \\ -6x + 10y = 7 \end{cases}$

34. $\begin{cases} 3x - 6y = 12 \\ -2x + 4y = -8 \end{cases}$

35. $\begin{cases} x + 2y + 3z = 0 \\ x - 2z = 3 \\ y + z = 1 \end{cases}$

36. $\begin{cases} x + y + 4z = 1 \\ -2x - y + z = 2 \\ 3x - 2y + 3z = 5 \end{cases}$

37. $\begin{cases} 3x + 2y - z = 4 \\ 2x - y + 3z = 5 \\ x + 3y + 2z = -1 \end{cases}$

38. $\begin{cases} 4x - 2y + 3z = 4 \\ 5x - y + 4z = 7 \\ 3x + 5y + z = 7 \end{cases}$

39. $\begin{cases} x + 2y + 3z - w = 7 \\ x - 4y + z = 3 \\ 2y - 3z + w = 4 \end{cases}$

40. $\begin{cases} x + y + z + w = -2 \\ -2x + 3y + 2z - 3w = 10 \\ 3x + 2y - z + 2w = -12 \\ 4x - y + z + 2w = 1 \end{cases}$

41. $\begin{cases} x - y - 3z - 2w = 2 \\ 4x + y + z + 2w = 2 \\ x + 3y - 2z - w = 9 \\ 3x + y - z + w = 5 \end{cases}$

42. $\begin{cases} x - y = 5 \\ y - z = 6 \\ z - w = 7 \\ x + w = 8 \end{cases}$

Determine the determinant form of the equation of the line passing through the following points.

43. (3, 4), (4, 8)

44. (1, 1), (0, −1)

45. (0, 0), (5, 2)

46. (10, 0), (0, 8)

Determine if the following points are collinear.

47. (1, 1), (4, 4), (8, 11)

48. (0, 0), (10, −3), (30, −9)

49. (3, 2), (2, 1), (1, −1)

50. (0, 0), (1, 0), (0, 1)

Calculate the area of the triangle with the following vertices.

51. (5, 3), (3, 1), (1, −2)

52. (10, 0), (1, 1), (4, 8)

53. (−5, 8), (4, 1), (0, 0)

54. Prove the determinant formula for the area of a triangle.

Use the appropriate results from this section to solve the following.

$$55. \begin{vmatrix} -20 & 2 & -3 \\ 6 & -4 & 1 \\ -1 & -1 & 2 \end{vmatrix} = 12x$$

$$56. \begin{vmatrix} 2 & -20 & -3 \\ 1 & 6 & 1 \\ 4 & -1 & 2 \end{vmatrix} = 2x + 1$$

$$57. \begin{vmatrix} a - 4 & 0 & 0 \\ 0 & a + 4 & 0 \\ 0 & 0 & a + 1 \end{vmatrix} = 0$$

$$58. \begin{vmatrix} 1 & m & m^2 \\ 1 & 1 & 1 \\ 4 & 5 & 0 \end{vmatrix} = 0$$

Calculate $|A|$, where A is the given matrix and x, y, and z are real numbers.

$$59. \begin{bmatrix} x & y & z \\ 0 & -4 & 2 \\ -1 & 3 & 1 \end{bmatrix} \qquad 60. \begin{bmatrix} x & -1 & 4x \\ y & 2 & 4y \\ z & -3 & 4z \end{bmatrix}$$

Show that

$$61. \begin{vmatrix} x_{11} & x_{12} & x_{13} & x_{14} \\ x_{21} & x_{22} & x_{23} & x_{24} \\ 0 & 0 & x_{33} & x_{34} \\ 0 & 0 & x_{43} & x_{44} \end{vmatrix} = \begin{vmatrix} x_{11} & x_{12} \\ x_{21} & x_{22} \end{vmatrix} \cdot \begin{vmatrix} x_{33} & x_{34} \\ x_{43} & x_{44} \end{vmatrix}$$

$$62. \begin{vmatrix} x_1 & y_1 & z_1 \\ x_2 & y_2 & z_2 \\ x_3 & y_3 & z_3 \end{vmatrix} = - \begin{vmatrix} x_2 & y_2 & z_2 \\ x_1 & y_1 & z_1 \\ x_3 & y_3 & z_3 \end{vmatrix}$$

9.4 INVERSE OF A SQUARE MATRIX

In this section, we will introduce the inverse of a matrix and provide a technique for calculating the inverse. As motivation for the inverse of a matrix, consider a real number a. We say that a^{-1} is the inverse of a under multiplication provided that a^{-1} satisfies

$$a a^{-1} = a^{-1} a = 1$$

Also, recall that the number 1 is the identity for multiplication.

Definition 1

Inverse of a Square Matrix

Suppose that A is an $n \times n$ matrix. By analogy with the situation for real numbers, we say that A^{-1} is an **inverse** of A, provided that A^{-1} satisfies the equations

$$AA^{-1} = A^{-1}A = I_n$$

Here I_n is the identity matrix of size n, which is the identity for multiplication for square matrices of size $n \times n$.

For instance, suppose that

$$A = \begin{bmatrix} 5 & 7 \\ 2 & 3 \end{bmatrix}$$

An inverse of A is given by

$$A^{-1} = \begin{bmatrix} 3 & -7 \\ -2 & 5 \end{bmatrix}$$

Indeed, a simple computation shows that

$$\begin{bmatrix} 5 & 7 \\ 2 & 3 \end{bmatrix} \begin{bmatrix} 3 & -7 \\ -2 & 5 \end{bmatrix} = \begin{bmatrix} 1 & 0 \\ 0 & 1 \end{bmatrix}$$

$$AA^{-1} = I_2$$

and similarly that

$$A^{-1}A = I_2$$

This shows that A^{-1} is an inverse of A.

The inverse of a 2×2 matrix may be determined from the following result.

INVERSE OF A 2 × 2 MATRIX

Let

$$A = \begin{bmatrix} a & b \\ c & d \end{bmatrix}$$

be a 2×2 matrix for which $ad - bc \neq 0$. Then A has an inverse given by the formula

$$A^{-1} = \begin{bmatrix} \dfrac{d}{\Delta} & -\dfrac{b}{\Delta} \\ -\dfrac{c}{\Delta} & \dfrac{a}{\Delta} \end{bmatrix}$$

where $\Delta = |A| = ad - bc$.

Proof   Let's calculate the product AA^{-1} and show that it equals I_2:

$$AA^{-1} = \begin{bmatrix} a & b \\ c & d \end{bmatrix} \begin{bmatrix} \dfrac{d}{\Delta} & -\dfrac{b}{\Delta} \\ -\dfrac{c}{\Delta} & \dfrac{a}{\Delta} \end{bmatrix}$$

$$= \begin{bmatrix} a \cdot \dfrac{d}{\Delta} + b \cdot \dfrac{-c}{\Delta} & a \cdot -\dfrac{b}{\Delta} + b \cdot \dfrac{a}{\Delta} \\ c \cdot \dfrac{d}{\Delta} + d \cdot -\dfrac{c}{\Delta} & c \cdot -\dfrac{b}{\Delta} + d \cdot \dfrac{a}{\Delta} \end{bmatrix}$$

$$= \begin{bmatrix} \dfrac{ad - bc}{\Delta} & 0 \\ 0 & \dfrac{ad - bc}{\Delta} \end{bmatrix}$$

$$= \begin{bmatrix} 1 & 0 \\ 0 & 1 \end{bmatrix}$$

$$= I_2 \quad \text{because } ad - bc = \Delta$$

Similarly, $A^{-1}A = I_2$.   ◆

It is possible to prove that a matrix has at most one inverse. Instead of referring to *an* inverse, we may refer to *the* inverse—provided, of course, that the matrix has an inverse.

➢ **EXAMPLE 1**
Matrix Inverse

Calculate the inverse of the matrix

$$A = \begin{bmatrix} 5 & -1 \\ 4 & 3 \end{bmatrix}$$

Solution

In this case, we have

$$\Delta = |A| = 5 \cdot 3 - 4 \cdot (-1) = 19$$

therefore, by the previous result, we have

$$A^{-1} = \begin{bmatrix} \dfrac{3}{19} & -\left(-\dfrac{1}{19}\right) \\ -\dfrac{4}{19} & \dfrac{5}{19} \end{bmatrix} = \begin{bmatrix} \dfrac{3}{19} & \dfrac{1}{19} \\ -\dfrac{4}{19} & \dfrac{5}{19} \end{bmatrix}$$

The previous result provides a formula for the inverse of a 2×2 matrix in case Δ is nonzero. It is possible to show that if $\Delta = 0$, then the matrix has no inverse. We will prove this in particular cases in the exercises.

There is a straightforward computational procedure for determining the inverse of any square matrix A (provided that the inverse exists). The procedure relies on performing elementary row operations and is described as follows:

1. Form the matrix

$$\begin{bmatrix} a_{11} & a_{12} & \cdots & a_{1n} & \vline & 1 & 0 & \cdots & 0 \\ a_{21} & a_{22} & \cdots & a_{2n} & \vline & 0 & 1 & \cdots & 0 \\ \vdots & \vdots & & \vdots & \vline & \vdots & \vdots & & \vdots \\ a_{n1} & a_{n2} & \cdots & a_{nn} & \vline & 0 & 0 & \cdots & 1 \end{bmatrix}$$

where the right half is the identity matrix I_n.

2. If possible, perform elementary row operations on this matrix to transform the left half into the identity matrix I_n.

3. After the transformation, the matrix in the right half is A^{-1}.

The proof that this method works belongs in a more advanced book and won't be included here. The next example illustrates this technique for calculating the inverse.

➤ **EXAMPLE 2**

Inverse of a 3 × 3 Matrix

Determine the inverse of the matrix

$$\begin{bmatrix} 4 & 8 & 0 \\ 3 & 1 & 4 \\ 0 & 0 & 2 \end{bmatrix}$$

Solution

We first form the matrix

$$\left[\begin{array}{ccc|ccc} 4 & 8 & 0 & 1 & 0 & 0 \\ 3 & 1 & 4 & 0 & 1 & 0 \\ 0 & 0 & 2 & 0 & 0 & 1 \end{array}\right]$$

Now we perform a sequence of elementary row operations to this matrix to transform the left half into the identity matrix I_3:

$$\left[\begin{array}{ccc|ccc} 4 & 8 & 0 & 1 & 0 & 0 \\ 3 & 1 & 4 & 0 & 1 & 0 \\ 0 & 0 & 2 & 0 & 0 & 1 \end{array}\right] \xrightarrow{-2R_3 + R_2} \left[\begin{array}{ccc|ccc} 4 & 8 & 0 & 1 & 0 & 0 \\ 3 & 1 & 0 & 0 & 1 & -2 \\ 0 & 0 & 2 & 0 & 0 & 1 \end{array}\right]$$

$$\xrightarrow{-8R_2 + R_1} \left[\begin{array}{ccc|ccc} -20 & 0 & 0 & 1 & -8 & 16 \\ 3 & 1 & 0 & 0 & 1 & -2 \\ 0 & 0 & 2 & 0 & 0 & 1 \end{array}\right]$$

$$\xrightarrow{-\frac{1}{20}R_1} \left[\begin{array}{ccc|ccc} 1 & 0 & 0 & -\frac{1}{20} & \frac{8}{20} & -\frac{16}{20} \\ 3 & 1 & 0 & 0 & 1 & -2 \\ 0 & 0 & 2 & 0 & 0 & 1 \end{array}\right]$$

$$\xrightarrow{-3R_1 + R_2} \left[\begin{array}{ccc|ccc} 1 & 0 & 0 & -\frac{1}{20} & \frac{8}{20} & -\frac{16}{20} \\ 0 & 1 & 0 & \frac{3}{20} & -\frac{4}{20} & \frac{8}{20} \\ 0 & 0 & 2 & 0 & 0 & 1 \end{array}\right]$$

$$\xrightarrow{\frac{1}{2}R_3} \left[\begin{array}{ccc|ccc} 1 & 0 & 0 & -\frac{1}{20} & \frac{8}{20} & -\frac{16}{20} \\ 0 & 1 & 0 & \frac{3}{20} & -\frac{4}{20} & \frac{8}{20} \\ 0 & 0 & 1 & 0 & 0 & \frac{1}{2} \end{array}\right]$$

The left side of the matrix is now the identity matrix I_3. The inverse may then be read off the right side:

$$A^{-1} = \begin{bmatrix} -\frac{1}{20} & \frac{2}{5} & -\frac{4}{5} \\ \frac{3}{20} & -\frac{1}{5} & \frac{2}{5} \\ 0 & 0 & \frac{1}{2} \end{bmatrix}$$

This result may be checked by verifying that A^{-1} satisfies the equations

$$AA^{-1} = A^{-1}A = I_3$$

We leave the calculations as an exercise.

Our original motivation for introducing matrices was to have a convenient way of performing row operations on a system of linear equations. Actually, the connection between matrices and systems of linear equations is much more fundamental than our previous discussion indicates. Let's now explore this connection in greater depth. Suppose that we are given the system of linear equations

$$\begin{cases} a_{11}x_1 + a_{12}x_2 + \cdots + a_{1n}x_n = b_1 \\ a_{21}x_1 + a_{22}x_2 + \cdots + a_{2n}x_n = b_2 \\ \qquad\qquad \vdots \\ a_{m1}x_1 + a_{m2}x_2 + \cdots + a_{mn}x_n = b_m \end{cases}$$

Define the matrices

$$A = \begin{bmatrix} a_{11} & a_{12} & \cdots & a_{1n} \\ a_{21} & a_{22} & \cdots & a_{2n} \\ \vdots & \vdots & & \vdots \\ a_{m1} & a_{m2} & \cdots & a_{mn} \end{bmatrix}, \qquad B = \begin{bmatrix} b_1 \\ b_2 \\ \vdots \\ b_m \end{bmatrix}, \qquad X = \begin{bmatrix} x_1 \\ x_2 \\ \vdots \\ x_n \end{bmatrix}$$

From the definition of matrix multiplication, we may write the system of linear equations in the compact form

$$AX = B$$

Indeed, by multiplying out the matrices on the left and equating the resulting entries to those of the matrix on the right, we arrive precisely at the given system of linear equations.

At first, the previous equation might seem like just a notational convenience, a shorthand way of writing the equations of the system. However, it can be used to arrive at a solution to the system in case A has an inverse. Indeed, multiplying both sides of the equation by A^{-1}, we have

$$\begin{aligned} AX &= B \\ A^{-1}(AX) &= A^{-1}B \\ (A^{-1}A)X &= A^{-1}B \qquad \text{(Associative law)} \\ I_n X &= A^{-1}B \qquad \text{(Definition of inverse)} \\ X &= A^{-1}B \qquad \text{(} I_n \text{ is the identity for multiplication)} \end{aligned}$$

Thus, we have the following equation:

$$X = A^{-1}B$$

The matrix X is just the column matrix of variables. The right side of the equation is a matrix whose value may be calculated in terms of the constants of the system. By equating corresponding entries on both sides of the equation, we arrive at a solution of the system. We have now derived a new method for solving a system of linear equations:

1. Calculate A^{-1}.
2. Form the product $A^{-1}B$.
3. Use the equation $X = A^{-1}B$ to determine the values of the variables.

The next example illustrates how to solve a system using this method.

➤ EXAMPLE 3
Matrix Solution
of a Linear System

Solve the following system using inverse matrices:

$$\begin{cases} 3x - 8y = 7 \\ -x + 4y = 6 \end{cases}$$

Solution

Let

$$A = \begin{bmatrix} 3 & -8 \\ -1 & 4 \end{bmatrix}, \qquad B = \begin{bmatrix} 7 \\ 6 \end{bmatrix}, \qquad X = \begin{bmatrix} x \\ y \end{bmatrix}$$

Then the system may be written in the matrix form

$$AX = B$$

$$X = A^{-1}B$$

Using the rule for determining the inverse of a 2×2 matrix, we can calculate A^{-1}:

$$\Delta = 3 \cdot 4 - (-1) \cdot (-8) = 4$$

$$A^{-1} = \begin{bmatrix} \frac{4}{4} & -(-\frac{8}{4}) \\ -(-\frac{1}{4}) & \frac{3}{4} \end{bmatrix}$$

$$= \begin{bmatrix} 1 & 2 \\ \frac{1}{4} & \frac{3}{4} \end{bmatrix}$$

Thus,

$$X = A^{-1}B$$

$$= \begin{bmatrix} 1 & 2 \\ \frac{1}{4} & \frac{3}{4} \end{bmatrix}\begin{bmatrix} 7 \\ 6 \end{bmatrix}$$

$$= \begin{bmatrix} 1 \cdot 7 + 2 \cdot 6 \\ \frac{1}{4} \cdot 7 + \frac{3}{4} \cdot 6 \end{bmatrix}$$

$$= \begin{bmatrix} 19 \\ \frac{25}{4} \end{bmatrix}$$

Thus, the solution of the system is $x = 19$, $y = \frac{25}{4}$.

Note that the method employed in the preceding example makes use of the inverse matrix. We should emphasize that not every matrix has an inverse; therefore, the previous method is not applicable to solving all systems of linear equations, just those for which the matrix of coefficients has an inverse. It can be proved that these are precisely the systems with a single solution. As we have previously seen, there are also systems that have no solutions and systems with an infinite number of solutions. For such systems, we can't use the method of the preceding example. Instead, we must rely on the method of Gauss elimination discussed earlier.

Graphing Calculators and Matrix Inverses

In this section, we will show how to use a TI-81 calculator to determine the inverse of a matrix.

New Technology

EXAMPLE 4
Using a Calculator to Determine the Inverse of a Matrix

Let

$$A = \begin{bmatrix} 5 & -1 & 4 \\ 3 & 0 & 9 \\ 2 & -5 & 8 \end{bmatrix}$$

Use a calculator to evaluate A^{-1}.

Solution

Enter the matrix A. (If you haven't changed matrix [A] from Example 5 in Section 9.2, it is still stored in your calculator.) Indicate the inverse of A with the keystrokes (2nd) (1) X^{-1}. On the display, you will see [A^{-1}]. Now press (ENTER). The inverse will be displayed as shown in Figure 1.

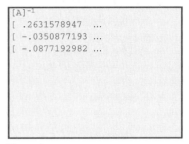

Figure 1

Exercises 9.4

If a graphing calculator is available, it may be used to solve the exercise in this set.

Calculate the inverse of the given matrix, if it exists.

1. $\begin{bmatrix} 3 & 5 \\ -2 & -14 \end{bmatrix}$ 2. $\begin{bmatrix} 3 & 8 \\ 2 & 5 \end{bmatrix}$ 3. $\begin{bmatrix} -3 & 4 \\ 1 & -2 \end{bmatrix}$

4. $\begin{bmatrix} 2 & 4 \\ 1 & -1 \end{bmatrix}$ 5. $\begin{bmatrix} 2 & -4 \\ 1 & -2 \end{bmatrix}$ 6. $\begin{bmatrix} 2 & -3 \\ -3 & 5 \end{bmatrix}$

7. $\begin{bmatrix} 2 & -1 & 1 \\ 1 & -2 & 3 \\ 4 & 1 & 2 \end{bmatrix}$ 8. $\begin{bmatrix} 1 & -4 & 8 \\ 1 & -3 & 2 \\ 2 & -7 & 10 \end{bmatrix}$

9. $\begin{bmatrix} -1 & -1 & -1 \\ 4 & 5 & 0 \\ 0 & 1 & -3 \end{bmatrix}$ 10. $\begin{bmatrix} 2 & 4 & 6 \\ -1 & -4 & -3 \\ 0 & 1 & -1 \end{bmatrix}$

11. $\begin{bmatrix} 3 & 1 & 0 \\ 1 & 1 & 1 \\ 1 & -1 & 2 \end{bmatrix}$ 12. $\begin{bmatrix} 2 & -1 & 0 \\ 3 & 0 & 1 \\ -2 & 4 & 0 \end{bmatrix}$

13. $\begin{bmatrix} 1 & -2 & 3 & 0 \\ 0 & 1 & -1 & 1 \\ -2 & 2 & -2 & 4 \\ 0 & 2 & -3 & 1 \end{bmatrix}$ 14. $\begin{bmatrix} 1 & 2 & 3 & 4 \\ 0 & 1 & 3 & -5 \\ 0 & 0 & 1 & -2 \\ 0 & 0 & 0 & 1 \end{bmatrix}$

Determine if the given matrices are inverses of each other.

15. $\begin{bmatrix} 2 & -1 & 1 \\ 1 & -2 & 3 \\ 4 & 1 & 2 \end{bmatrix}$ and $\begin{bmatrix} \frac{7}{15} & -\frac{1}{5} & \frac{1}{15} \\ -\frac{2}{3} & 0 & \frac{1}{3} \\ -\frac{3}{5} & \frac{2}{5} & \frac{1}{5} \end{bmatrix}$

16. $\begin{bmatrix} 1 & -1 & 2 \\ 0 & 1 & 3 \\ 2 & 1 & -2 \end{bmatrix}$ and $\begin{bmatrix} \frac{1}{3} & 0 & \frac{1}{3} \\ -\frac{2}{5} & \frac{2}{5} & \frac{1}{5} \\ \frac{2}{15} & \frac{1}{5} & -\frac{1}{15} \end{bmatrix}$

17. $\begin{bmatrix} 11 & 3 \\ 7 & 2 \end{bmatrix}$ and $\begin{bmatrix} 2 & -3 \\ -7 & 11 \end{bmatrix}$

18. $\begin{bmatrix} 3 & 2 \\ 5 & 3 \end{bmatrix}$ and $\begin{bmatrix} -3 & -2 \\ 5 & -3 \end{bmatrix}$

19. $\begin{bmatrix} 1 & 3 & 3 \\ 1 & 4 & 3 \\ 1 & 3 & 4 \end{bmatrix}$ and $\begin{bmatrix} 7 & -3 & -3 \\ -1 & 1 & 0 \\ -1 & 0 & 1 \end{bmatrix}$

20. $\begin{bmatrix} -1 & 0 & 2 \\ 3 & 1 & 0 \\ 0 & 2 & -3 \end{bmatrix}$ and $\begin{bmatrix} -\frac{1}{5} & \frac{4}{15} & -\frac{2}{15} \\ \frac{3}{5} & \frac{1}{5} & \frac{2}{5} \\ \frac{2}{5} & \frac{2}{15} & -\frac{1}{15} \end{bmatrix}$

Solve the following systems of equations using inverse matrices.

21. $\begin{cases} x + 2y = 6 \\ 2x + y = 9 \end{cases}$

22. $\begin{cases} x + 3y = 8 \\ 3x + 4y = 9 \end{cases}$

23. $\begin{cases} 3x - 2y = 9 \\ 2x + 5y = 8 \end{cases}$

24. $\begin{cases} 2x + 4y = 6 \\ x - 3y = 3 \end{cases}$

25. $\begin{cases} 4x - y = 3 \\ 6x + 4y = -1 \end{cases}$

26. $\begin{cases} 3x - 6y = 1 \\ -5x + 9y = -1 \end{cases}$

27. $\begin{cases} 4x - 3y = -23 \\ -3x + 2y = 16 \end{cases}$

28. $\begin{cases} 2x + 3y = 13 \\ x + 2y = 8 \end{cases}$

29. $\begin{cases} 4x + z = 1 \\ 2x + 2y = 3 \\ x - y + z = 4 \end{cases}$

30. $\begin{cases} x + 2y + 3z = -1 \\ 2x - 3y + 4z = 2 \\ -3x + 5y - 6z = 4 \end{cases}$

31. $\begin{cases} x + y - 3z = 4 \\ 2x + 4y - 4z = 8 \\ -x + y + 4z = -3 \end{cases}$

32. $\begin{cases} 2x - 2y = 5 \\ 4y + 8z = 7 \\ kx + 2z = 1 \end{cases}$

In each of the following, state the conditions under which A^{-1} exists, and then find a formula for A^{-1}.

33. $A = \begin{bmatrix} a & 0 \\ 0 & b \end{bmatrix}$

34. $A = \begin{bmatrix} x & 0 & 0 \\ 0 & y & 0 \\ 0 & 0 & z \end{bmatrix}$

35. If $A = \begin{bmatrix} 1 & 2 \\ 3 & 4 \end{bmatrix}$ and $B = \begin{bmatrix} -1 & -1 \\ 2 & 3 \end{bmatrix}$, show that $(AB)^{-1} = B^{-1}A^{-1}$.

36. Suppose A is a 2×2 matrix. Prove that if $\Delta = 0$, then the inverse of A does not exist.

37. Suppose A is a 3×3 matrix. Prove that if $|A| = 0$, then the inverse of A does not exist.

◼ 9.5 CHAPTER REVIEW

Important Concepts, Properties, and Formulas—Chapter 9

Determining a partial fraction expansion	1. Use long division to write $\dfrac{f(x)}{g(x)}$ in the form $$\frac{f(x)}{g(x)} = q(x) + \frac{h(x)}{g(x)}, \qquad \deg(h) < \deg(g)$$ 2. Factor $g(x)$ into a product of distinct real factors, some linear and some quadratic. 3. For a linear factor $(ax + b)^n$, the partial fraction expansion has a component: $$\frac{A_1}{ax + b} + \frac{A_2}{(ax + b)^2} + \cdots + \frac{A_n}{(ax + b)^n}$$ 4. For a quadratic factor $(cx^2 + dx + e)^n$, the partial fraction expansion has a component $$\frac{A_1 + B_1x}{cx^2 + dx + e} + \frac{A_2 + B_2x}{(cx^2 + dx + e)^2}$$ $$+ \cdots + \frac{A_n + B_nx}{(cx^2 + dx + e)^n}$$ 5. The partial fraction expansion is the sum of $q(x)$ in step 1 and each of the components in steps 3 and 4.	p. 468

Matrices	Subscript notation for matrix entries.	pp. 472, 474, 476
	Zero matrix: matrix of all zeros.	
	Sum of two matrices: Add corresponding entries.	
	Row matrix: only one row.	
	Column matrix: only one column.	
	Matrix multiplication: Suppose that A is an $m \times n$ matrix and B is an $n \times k$ matrix. The product AB is the $m \times k$ matrix with entries as follows: The entry in the ith row and jth column of AB is obtained by multiplying the ith row of A by the jth column of B.	
	Identity matrix I_n of size n.	
Determinant of a square matrix	Let A be an $n \times n$ matrix. Then the determinant $|A|$ is defined by the formula $$|A| = a_{11}A_{11} + a_{12}A_{12} + \cdots + a_{1n}A_{1n},$$ where A_{ij} denotes the cofactor determined by eliminating the ith row and jth column.	p. 482
Cramer's rule	If a linear system of n equations in n variables $x, y, z, \ldots$ has a nonzero determinant $|A|$, then the system has a unique solution given by the formulas $$x = \frac{|A_1|}{|A|}, \qquad y = \frac{|A_2|}{|A|}, \ldots$$	p. 485
Determinant form of the equation of a line	The equation of a line passing through points (x_1, y_1) and (x_2, y_2) is given in determinant form: $$\begin{vmatrix} x & y & 1 \\ x_1 & y_1 & 1 \\ x_2 & y_2 & 1 \end{vmatrix} = 0$$	p. 487
Determinant condition for collinearity	Three points (x_1, y_1), (x_2, y_2), and (x_3, y_3) are collinear provided that $$\begin{vmatrix} x_1 & y_1 & 1 \\ x_2 & y_2 & 1 \\ x_3 & y_3 & 1 \end{vmatrix} = 0$$	p. 487
Determinant formula for the area of a triangle	The area of the triangle with the vertices (x_1, y_1), (x_2, y_2), and (x_3, y_3) is given by the formula $$\text{Area} = \pm\frac{1}{2}\begin{vmatrix} x_1 & y_1 & 1 \\ x_2 & y_2 & 1 \\ x_3 & y_3 & 1 \end{vmatrix}$$	p. 487
Inverse of a square matrix	A^{-1} is an inverse of A provided that A^{-1} satisfies the equations $$AA^{-1} = A^{-1}A = I_n$$	p. 490

Computing the inverse of a square matrix	1. Form the matrix $$\left[\begin{array}{cccc|cccc} a_{11} & a_{12} & \cdots & a_{1n} & 1 & 0 & \cdots & 0 \\ a_{21} & a_{22} & \cdots & a_{2n} & 0 & 1 & \cdots & 0 \\ \vdots & & & & \vdots & & & \\ a_{n1} & a_{n2} & \cdots & a_{nn} & 0 & 0 & \cdots & 1 \end{array}\right]$$ where the matrix on the right is the identity matrix I_n. 2. If possible, perform elementary row operations on this matrix to transform the left half into the identity matrix I_n. 3. After the transformation, the matrix in the right half is A^{-1}.	p. 495
Matrix form of a linear system	The solution of the matrix equation $AX = B$ is $X = A^{-1}B$.	p. 493

Cumulative Review Exercises—Chapter 9

Determine the partial fraction expansion of the given rational expressions.

1. $\dfrac{-3a^2 - 10a - 4}{a(a+1)^2}$

2. $\dfrac{13y + 5}{2y^2 - 7y + 6}$

3. $\dfrac{7x - 1}{6x^2 - 5x + 1}$

4. $\dfrac{5x^2 + 9x - 56}{(x-4)(x-2)(x+1)}$

Given

$$A = \begin{bmatrix} 4 & 2 \\ 1 & -3 \end{bmatrix}, \quad B = \begin{bmatrix} 0 \\ 3 \end{bmatrix}, \quad C = \begin{bmatrix} 2 & -1 & 4 \\ 3 & 0 & 5 \end{bmatrix},$$

$$D = \begin{bmatrix} -2 & 4 & 5 \\ 1 & -1 & 2 \\ 3 & 2 & -3 \end{bmatrix}, \quad E = \begin{bmatrix} 3 & 4 \\ -1 & 5 \end{bmatrix}, \quad F = \begin{bmatrix} 5 \\ -2 \end{bmatrix},$$

$$G = \begin{bmatrix} -1 & 3 & 4 \\ 0 & 2 & -1 \end{bmatrix}, \quad H = \begin{bmatrix} 0 & 4 & -2 \\ 2 & -1 & 3 \\ 3 & 5 & 2 \end{bmatrix},$$

$$I = \begin{bmatrix} 4 & 1 & 0 \end{bmatrix}$$

find each of the following matrices, whenever possible.

5. DH
6. EF
7. $4A + 3E$
8. $2E - 3A$
9. GH
10. CH
11. AB
12. CD
13. BI
14. FG

Determine which of the following products are defined and the size of those products.

15. A is 3×5; B is 4×5.
16. A is 3×4; B is 4×3.
17. A is 1×3; B is 3×1.
18. A is 2×2; B is 4×2.

Find the values of w, x, y, and z in the following matrix equations.

19. $\begin{bmatrix} x & 5 & 2 \\ 2 & 7 & y \end{bmatrix} = \begin{bmatrix} 3 & 5 & z \\ 2 & w & 3 \end{bmatrix}$

20. $\begin{bmatrix} 4 & 1 \\ 2 & 4 \end{bmatrix} = \begin{bmatrix} x & w \\ y & 4 \end{bmatrix}$

Calculate $|A|$, where A is the given matrix.

21. $\begin{bmatrix} 1 & -1 \\ 2 & 3 \end{bmatrix}$

22. $\begin{bmatrix} 4 & 3 \\ -1 & -2 \end{bmatrix}$

23. $\begin{bmatrix} -1 & -2 \\ -4 & 5 \end{bmatrix}$

24. $\begin{bmatrix} 0 & 3 \\ 2 & -5 \end{bmatrix}$

Use

$$A = \begin{bmatrix} 1 & 3 & -1 \\ 4 & 5 & 1 \\ 3 & -1 & 2 \end{bmatrix}, \quad B = \begin{bmatrix} 7 & 2 & 0 & 4 \\ 2 & -1 & 2 & 2 \\ 1 & 4 & 3 & -1 \\ -3 & 5 & 1 & 5 \end{bmatrix}$$

for the following exercises.

25. Find A_{12}.
26. Find A_{11}.
27. Find B_{44}.
28. Find B_{22}.

Calculate $|A|$, where A is the given matrix.

29. $\begin{bmatrix} 1 & 3 & -1 \\ 2 & 4 & 5 \\ -1 & 2 & 0 \end{bmatrix}$

30. $\begin{bmatrix} 0 & 3 & 1 \\ 4 & -2 & -1 \\ -1 & 4 & 2 \end{bmatrix}$

31. $\begin{bmatrix} 2 & 0 & 1 & 4 \\ 1 & 2 & 3 & 0 \\ 4 & -1 & 0 & 2 \\ 0 & -2 & 3 & 1 \end{bmatrix}$ 32. $\begin{bmatrix} -3 & 0 & 2 & 1 \\ 2 & 4 & 0 & -1 \\ 3 & 0 & 1 & 2 \\ 0 & 2 & -1 & 4 \end{bmatrix}$

Solve the following for x.

33. $\begin{bmatrix} 3 & -1 \\ 2 & 5 \end{bmatrix} = 2x - 1$ 34. $\begin{bmatrix} 2 & -1 & 1 \\ 3 & 4 & 0 \\ 1 & 0 & 2 \end{bmatrix} = x - 5$

Calculate the inverse of the given matrix, if it exists.

35. $\begin{bmatrix} 3 & 5 \\ 1 & 2 \end{bmatrix}$ 36. $\begin{bmatrix} 4 & -3 \\ 1 & 2 \end{bmatrix}$

37. $\begin{bmatrix} 1 & -4 & 8 \\ 1 & -3 & 2 \\ 2 & -7 & 10 \end{bmatrix}$ 38. $\begin{bmatrix} 1 & -1 & 2 \\ 0 & 1 & 3 \\ 2 & 1 & -2 \end{bmatrix}$

39. $\begin{bmatrix} 1 & 2 & 3 & 4 \\ 0 & 1 & 3 & -5 \\ 0 & 0 & 1 & -2 \\ 0 & 0 & 0 & -1 \end{bmatrix}$ 40. $\begin{bmatrix} -2 & -3 & 4 & 1 \\ 0 & 1 & 1 & 0 \\ 0 & 4 & -6 & 1 \\ -2 & -2 & 5 & 1 \end{bmatrix}$

Determine if the given matrices are inverses of each other.

41. $\begin{bmatrix} 1 & 2 & -1 \\ 3 & 4 & 2 \\ 1 & 1 & 0 \end{bmatrix}$ and $\begin{bmatrix} 1 & 1 & 0 \\ 3 & 4 & 2 \\ 1 & 2 & -1 \end{bmatrix}$

42. $\begin{bmatrix} 2 & 3 \\ 1 & 5 \end{bmatrix}$ and $\begin{bmatrix} \frac{1}{2} & \frac{1}{3} \\ 1 & \frac{1}{5} \end{bmatrix}$

Solve the following systems of equations using inverse matrices.

43. $\begin{cases} 2x + y = 7 \\ 3x + 2y = 9 \end{cases}$ 44. $\begin{cases} 6x + y = 0 \\ 3x + 5y = 9 \end{cases}$

45. $\begin{cases} x + y = 4 \\ y + z = 2 \\ 2x - z = 1 \end{cases}$ 46. $\begin{cases} x + z = 1 \\ 2x + y = 3 \\ x - y + z = 4 \end{cases}$

SEQUENCES, SERIES,
AND PROBABILITY

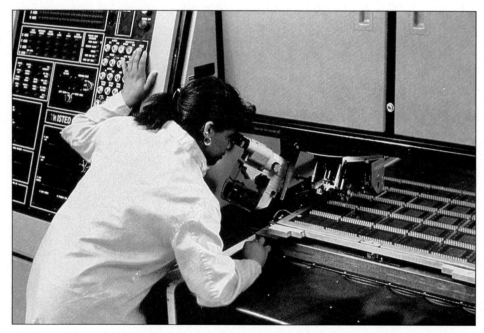

Counting and probability concepts form the groundwork for statistical quality control. Controlling quality and preventing defects is a key concern in manufacturing environments, particularly those with sensitive electronic parts, such as these computer chips at IBM.

I n this chapter, we discuss three topics that are very important for more advanced courses, especially calculus. Our first main topic is the **binomial theorem**, which states how to raise a binomial expression to any positive integer power. Second, we discuss **mathematical induction**, a valuable technique for proving many theorems, including the binomial theorem. Finally, we introduce **sequences** and **series**, which play an important role in calculus, and apply them to permutations, combinations, and probability.

10.1 THE BINOMIAL THEOREM

A **binomial** is an algebraic expression that is a sum of two terms. Here are some examples:

$$a + b, \qquad 2x^2 - y, \qquad \frac{5}{t} - 2t, \qquad \sqrt{x} + 3y$$

Note that we have been working with such expressions throughout the book. In many applications, it is necessary to raise a binomial to a nonnegative integer power. Here are some examples of such powers:

$$(a + b)^2$$
$$(a + b)^3$$
$$(2x^2 - y)^5$$
$$(4\sqrt{x} + 3y)^{18}$$

It is possible to compute such powers by repeated multiplication. For example, we did just that in the case of the first two powers above and came up with the following formulas in Chapter 1:

$$(a + b)^2 = a^2 + 2ab + b^2$$
$$(a + b)^3 = a^3 + 3a^2b + 3ab^2 + b^3$$

For higher powers, such results are tedious to obtain by direct multiplication. However, the **binomial theorem** provides a shortcut.

The binomial theorem allows you to write out the terms of the power $(a+b)^n$, where n is a nonnegative integer. To illustrate the binomial theorem, let's examine some particular expansions:

$$(a + b)^0 = 1$$
$$(a + b)^1 = a + b$$
$$(a + b)^2 = a^2 + 2ab + b^2$$
$$(a + b)^3 = a^3 + 3a^2b + 3ab^2 + b^3$$
$$(a + b)^4 = a^4 + 4a^3b + 6a^2b^2 + 4ab^3 + b^4$$
$$(a + b)^5 = a^5 + 5a^4b + 10a^3b^2 + 10a^2b^3 + 5ab^4 + b^5$$

Based on these expansions, we can guess at the general pattern for $(a + b)^n$.

1. The first term is

$$a^n b^0 = a^n \cdot 1 = a^n$$

2. Each successive term is formed by decreasing the exponent of a by 1 and increasing the exponent of b by 1.
3. The sum of the exponents of a and b in any term is n.
4. The last term is

$$a^0 b^n = 1 \cdot b^n = b^n$$

5. Each term has a numerical coefficient that must be determined.

If we denote the coefficients by $c_0, c_1, \ldots, c_n$, then we have the following result:

BINOMIAL THEOREM (FIRST FORM)

The expansion of $(a + b)^n$ can be written in the form

$$(a + b)^n = c_0 a^n + c_1 a^{n-1} b + c_2 a^{n-2} b^2 + \cdots + c_{n-1} a b^{n-1} + c_n b^n$$

where the coefficients on the right are calculated as follows:

1. The first coefficient, c_0, equals 1.
2. Multiply the preceding coefficient by the exponent of a in the preceding term.
3. Divide the result by the exponent of b in the present term.

We will prove the binomial theorem in the next section, once we have developed the method of mathematical induction.

➢ **EXAMPLE 1**
Cube of a Binomial

Use the binomial theorem to verify the cube of a binomial formula cited earlier.

Solution

We calculate the coefficients beginning with that for the term 0 coefficient:

$$c_0 = 1$$

Term 1 coefficient:

$$c_1 = c_0 \cdot \tfrac{3}{1} = 1 \cdot \tfrac{3}{1} = 3$$

Term 2 coefficient: The preceding term has exponent 2 on a. This term has exponent 2 on b.

$$c_2 = c_1 \cdot \tfrac{2}{2} = 3 \cdot \tfrac{2}{2} = 3$$

Term 3 coefficient: The preceding term has exponent 1 on a. This term has exponent 3 on b.

$$c_3 = c_2 \cdot \tfrac{1}{3} = 3 \cdot \tfrac{1}{3} = 1$$

These coefficients agree with those cited in the expansion given at the beginning of the section, namely,

$$(a + b)^3 = a^3 + 3a^2 b + 3ab^2 + b^3$$

➢ **EXAMPLE 2**
Sixth Power of a Binomial

1. Calculate the coefficients in the expansion of $(a + b)^6$.
2. Write out the expansion of $(a + b)^6$.

Solution

1. We follow the calculational procedure described earlier:

$$c_0 = 1$$

$$c_1 = c_0 \cdot \tfrac{6}{1} = 6$$

$$c_2 = c_1 \cdot \tfrac{5}{2} = 6 \cdot \tfrac{5}{2} = 15$$

$$c_3 = c_2 \cdot \tfrac{4}{3} = 15 \cdot \tfrac{4}{3} = 20$$

$$c_4 = c_3 \cdot \tfrac{3}{4} = 20 \cdot \tfrac{3}{4} = 15$$

$$c_5 = c_4 \cdot \tfrac{2}{5} = 15 \cdot \tfrac{2}{5} = 6$$

$$c_6 = c_5 \cdot \tfrac{1}{6} = 6 \cdot \tfrac{1}{6} = 1$$

2. From the previous calculation of the coefficients, the expansion of $(a + b)^6$ is equal to

$$a^6 + 6a^5b + 15a^4b^2 + 20a^3b^3 + 15a^2b^4 + 6ab^5 + b^6$$

Factorials, Binomial Coefficients, and Applications

The method just described allows us to calculate the coefficients of any binomial expansion. However, the method suffers from a serious difficulty: it requires that we compute the coefficients in order. To determine the 10th coefficient, we must compute the first 9. In many applications, such as probability and statistics, it is necessary to calculate single terms in binomial expansions. For such applications, the method we have given is very cumbersome. Let's present another method that can be used to compute any specified term of a binomial expansion.

Definition 1 Factorial	Let n be a positive integer. Then n **factorial**, denoted $n!$, is the product of all positive integers from 1 to n. That is, $$n! = n(n - 1)(n - 2) \cdots 2 \cdot 1$$

Here are some calculations of factorials:

$$1! = 1$$
$$2! = 2 \cdot 1 = 2$$
$$3! = 3 \cdot 2 \cdot 1 = 6$$
$$4! = 4 \cdot 3 \cdot 2 \cdot 1 = 24$$
$$5! = 5 \cdot 4 \cdot 3 \cdot 2 \cdot 1 = 120$$

➤ **EXAMPLE 3**
Factorials

Calculate the value of $\frac{11!}{6!5!}$.

Solution

This type of quotient arises in calculating the coefficients in binomial expansions (as we shall see later). In computing such a quotient, it is best to write out the definitions of all the factorials and do whatever simplification is possible:

$$\frac{11!}{6!5!} = \frac{11 \cdot 10 \cdot 9 \cdot 8 \cdot 7 \cdot 6 \cdot 5 \cdot 4 \cdot 3 \cdot 2 \cdot 1}{(6 \cdot 5 \cdot 4 \cdot 3 \cdot 2 \cdot 1)(5 \cdot 4 \cdot 3 \cdot 2 \cdot 1)}$$

$$= \frac{11 \cdot 10 \cdot 9 \cdot 8 \cdot 7}{5 \cdot 4 \cdot 3 \cdot 2 \cdot 1}$$

$$= 11 \cdot 3 \cdot 2 \cdot 7$$

$$= 462$$

The preceding definition of factorials defines $n!$ for a positive integer n. For use in connection with the binomial theorem, it is convenient to define $0!$ according to the formula

$$0! = 1$$

Let's introduce a new notation. The coefficient of the $(k + 1)$th term in the binomial expansion of $(a + b)^n$ is customarily denoted with the symbol

$$\binom{n}{k}$$

This symbol is called a **binomial coefficient** and is read "binomial n over k." (Note, however, that "n over k" does not mean the quotient of n divided by k. Rather, the term is just a convenient way of reminding us of the notation for the binomial coefficient.)

From the binomial expansion for $(a + b)^3$, for example, we can read off the values of the following binomial coefficients:

$$\binom{3}{0} = 1, \qquad \binom{3}{1} = 3, \qquad \binom{3}{2} = 3, \qquad \binom{3}{3} = 1$$

We can calculate binomial coefficients using the following result.

FACTORIAL FORMULA FOR BINOMIAL COEFFICIENTS

The binomial coefficients may be computed using the formula

$$\binom{n}{k} = \frac{n!}{k!(n - k)!}$$

We will delay proving this result until the next section. However, in the meantime, we may use it to evaluate binomial coefficients, as illustrated in the next example.

➤ **EXAMPLE 4**
Calculating Binomial Coefficients

Use the factorial formula to compute the following binomial coefficients:

1. $\binom{3}{2}$

2. $\binom{4}{3}$

3. $\binom{7}{2}$

Solution

1. $\binom{3}{2} = \dfrac{3!}{2!(3 - 2)!} = \dfrac{3!}{2! \cdot 1!} = \dfrac{3 \cdot 2 \cdot 1}{2 \cdot 1} = 3$

Note that this result agrees with the value of the coefficient stated earlier.

2. $\binom{4}{3} = \dfrac{4!}{3!(4 - 3)!} = \dfrac{4!}{3! \cdot 1!} = \dfrac{4 \cdot 3 \cdot 2 \cdot 1}{3 \cdot 2 \cdot 1 \cdot 1} = 4$

3. $\dbinom{7}{2} = \dfrac{7!}{2!(7-2)!}$

$= \dfrac{7!}{2! \cdot 5!}$

$= \dfrac{7 \cdot 6 \cdot 5 \cdot 4 \cdot 3 \cdot 2 \cdot 1}{(2 \cdot 1)(5 \cdot 4 \cdot 3 \cdot 2 \cdot 1)}$

$= \dfrac{7 \cdot 6}{2 \cdot 1}$

$= 7 \cdot 3 = 21$

Using the binomial coefficient notation, we may compute the $(k + 1)$th term in the expansion of the binomial $(a + b)^n$ according to the formula

$$(k + 1)\text{th term} = \dbinom{n}{k} a^{n-k} b^k$$

Using this formula, we can restate the expansion of a binomial in a slightly different form.

BINOMIAL THEOREM (SECOND FORM)

Let n be a nonnegative integer. Then

$$(a + b)^n = \dbinom{n}{0} a^n + \dbinom{n}{1} a^{n-1} b + \dbinom{n}{2} a^{n-2} b^2 + \cdots + \dbinom{n}{n} b^n$$

Using the factorial formula for the binomial coefficients, we may easily calculate any particular term of a binomial expansion, as illustrated in the following example.

> **EXAMPLE 5**
> Calculating a Term
> in a Binomial Expansion

Determine the fifth term in the expansion of $(x + y)^9$.

Solution

We have

$$n = 9$$
$$k + 1 = 5$$
$$k = 4$$

The fifth term of the expansion is

$$\dbinom{9}{4} x^{9-4} y^4 = \dfrac{9!}{4!(9-4)!} x^5 y^4$$

$$= \dfrac{9!}{4!5!} x^5 y^4$$

$$= \dfrac{9 \cdot 8 \cdot 7 \cdot 6}{4 \cdot 3 \cdot 2 \cdot 1} x^5 y^4$$

Simplifying the fractional coefficient gives the equivalent expression

$$3 \cdot 7 \cdot 6x^5y^4 = 126x^5y^4$$

➤ EXAMPLE 6
Medicine

The recovery rate for a certain serious viral infection is 80 percent. A group of 100 persons are infected with the virus. The probability that exactly k will recover is given by the $(k + 1)$th term of binomial expansion $(0.8 + 0.2)^{100}$. Determine the probability that exactly three persons will recover from the infection.

Solution

The described probability is given by the fourth term in the expansion, which is equal to

$$\binom{100}{3}(0.8)^{97}(0.2)^3 = \frac{100!}{3!(100-3)!}(0.8)^{97}(0.2)^3$$

$$= \frac{100!}{3!97!}(0.8)^{97}(0.2)^3$$

$$= \frac{100 \cdot 99 \cdot 98}{3 \cdot 2 \cdot 1}(0.8)^{97}(0.2)^3$$

To compute a numerical value of this last expression, you should use a calculator with a power function. (Computing a 97th power by hand is not fun!) You will find that the value of the above expression is approximately $5.1467 \times 10^{-7} = 0.00000051467$.

Expansions Using the Binomial Theorem

The binomial theorem gives the expansion of a binomial $(a + b)^n$. By replacing a and b with algebraic expressions, we may use the binomial theorem to compute the expansion of any binomial to a nonnegative integer power. The following examples illustrate some typical expansions.

➤ EXAMPLE 7
Expansion of a More
Complex Binomial

Determine the expansion of the following binomial:

$$(2x^2 - y)^3$$

Solution

We apply the binomial theorem with a replaced by $2x^2$ and b replaced by $-y$. The result is

$$(2x^2 - y)^3 = \binom{3}{0}(2x^2)^3(-y)^0 + \binom{3}{1}(2x^2)^2(-y)^1$$

$$+ \binom{3}{2}(2x^2)^1(-y)^2 + \binom{3}{3}(2x^2)^0(-y)^3$$

Now we replace the binomial coefficients by their respective values and simplify the various terms to obtain the expansion

$$(2x^2)^3 + 3(2x^2)^2(-y)^1 + 3(2x^2)^1(-y)^2 + (-y)^3 = 8x^6 - 12x^4y + 6x^2y^2 - y^3$$

➤ EXAMPLE 8
A Binomial Involving
a Rational Expression

Determine the expansion of the following binomial:

$$\left(\frac{5}{t} - 2t\right)^4$$

Solution

By the binomial theorem, with a replaced with $5/t$ and b replaced with $-2t$, we have

$$\left(\frac{5}{t} - 2t\right)^4 = \binom{4}{0}\left(\frac{5}{t}\right)^4 + \binom{4}{1}\left(\frac{5}{t}\right)^3(-2t) + \binom{4}{2}\left(\frac{5}{t}\right)^2(-2t)^2$$

$$+ \binom{4}{3}\left(\frac{5}{t}\right)^1(-2t)^3 + \binom{4}{4}(-2t)^4$$

Inserting the values of the binomial coefficients, we have the following expansion:

$$\left(\frac{5}{t}\right)^4 + 4\left(\frac{5}{t}\right)^3(-2t) + 6\left(\frac{5}{t}\right)^2(-2t)^2 + 4\left(\frac{5}{t}\right)(-2t)^3 + (-2t)^4$$

Simplifying each of the terms, we obtain the following expansion:

$$\frac{625}{t^4} - \frac{1000}{t^2} + 600 - 160t^2 + 16t^4$$

➢ EXAMPLE 9
A Binomial Involving a Radical

Determine term 5 in the expansion of the following binomial:

$$(\sqrt{x} + 3y)^{18}$$

Solution

The desired term is given by the expression:

$$\binom{18}{4}(\sqrt{x})^{18-4}(3y)^4 = \frac{18!}{4!(18-4)!}(\sqrt{x})^{14}(3y)^4$$

$$= \frac{18 \cdot 17 \cdot 16 \cdot 15}{4 \cdot 3 \cdot 2 \cdot 1}x^7 \cdot 81y^4$$

$$= 247,860x^7y^4$$

New Technology

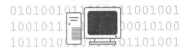

EXAMPLE 10
Using a Calculator to Determine Binomial Coefficients

Use a calculator to determine $\binom{20}{9}$.

Solution

For large values of n, calculation of binomial coefficients may be very tedious. The TI-81 allows you to calculate the binomial coefficient $\binom{n}{r}$ using the notation $_nC_r$ (which we will introduce later in the chapter). To calculate $\binom{20}{9}$, use these keystrokes: Enter 20 and then press (MATH) to get the menu shown in Figure 1. Select the option *PRB* in the first row. You will see the display in Figure 2. Use the

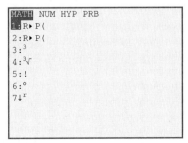

Figure 1

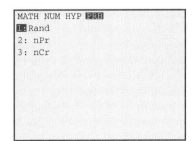

Figure 2

arrow keys to select $_nC_r$ from the menu. Now press (ENTER). You will be returned to the main screen. Then key in 9. See Figure 3. Press (ENTER). The value of the binomoal coefficient will be displayed. See Figure 4.

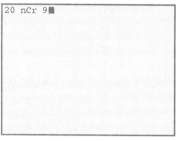

Figure 3

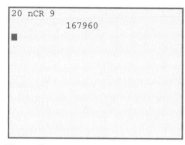

Figure 4   ◀

Pascal's Triangle

Pascal

The 17th-century mathematician and philosopher Blaise Pascal discovered a simple method for computing binomial coefficients using only addition (as opposed to the methods just presented, which involve multiplying numbers that tend to be rather large).

To understand this method, list the binomial coefficients corresponding to $n = 1, 2, 3, \ldots$, with the coefficients corresponding to a given value of n written across a row, as follows:

$$
\begin{array}{c}
1 \\
1 \; 1 \\
1 \; 2 \; 1 \\
1 \; 3 \; 3 \; 1 \\
1 \; 4 \; 6 \; 4 \; 1 \\
1 \; 5 \; 10 \; 10 \; 5 \; 1
\end{array}
$$

Each row of the table may be computed from the preceding row as follows: The first element of the row is always 1. Each subsequent element is obtained by adding the two elements immediately above. For example, consider row 3. The first element is 1. The next element, 2, is obtained by adding the two elements immediately above in the preceding row, namely, $1 + 1$. The next element, 1, is obtained by adding a blank and a 1.

Consider now row 4. The first element is 1. The next is obtained by adding $1 + 2$ from the row above. The next element is obtained by adding $2 + 1$ from the row above. The final element is obtained by adding a blank and 1 from the row above.

In this manner, we may calculate the rows of the table one after another. Each row gives the set of binomial coefficients corresponding to one particular value of n. This method of computing the binomial coefficients is called **Pascal's triangle**.

The advantage of Pascal's triangle is that it allows you to calculate the binomial coefficients using addition rather than multiplication. The disadvantage, however, is that to apply Pascal's triangle, you must calculate all rows preceding the one you are interested in. If you need a row corresponding to a large value of n, then a considerable amount of calculation will be involved.

Exercises 10.1

Evaluate.

1. $0!$
2. $1!$
3. $8!$
4. $5!$
5. $\dfrac{6!}{4!2!}$
6. $\dfrac{8!}{5!3!}$
7. $\dfrac{7!}{2!5!}$
8. $\dfrac{10!}{6!4!}$
9. $\dbinom{6}{1}$
10. $\dbinom{8}{0}$
11. $\dbinom{9}{5}$
12. $\dbinom{12}{3}$

Determine the indicated term of the expansion.

13. Fourth, $(x + y)^7$
14. Sixth, $(a + b)^8$
15. 11th, $(t - 3)^{15}$
16. 10th, $(y - 4)^{13}$
17. Fourth, $(\sqrt{x} + 4t^3)^9$
18. Fifth, $\left(\dfrac{a}{5} - \dfrac{1}{a^2}\right)^{10}$
19. Middle two, $(\sqrt{t} - \sqrt{2})^9$
20. Middle, $(5a^2 + 7b^3)^8$
21. Last, $(a - b^2)^{100}$

Expand.

22. $(a + b)^4$
23. $(p - q)^5$
24. $(4a - b)^6$
25. $(a^2 + 3b)^5$
26. $(\sqrt{2} + 2\sqrt{3})^4$
27. $\left(t - \dfrac{t}{2}\right)^8$
28. $\left(\dfrac{a^2}{b} - b^3\right)^6$
29. $\left(2 + \dfrac{t}{2}\right)^8$
30. $(3x^2 + 2xy^3)^5$
31. $\left(\dfrac{3a^2}{2b} - \dfrac{2b^2}{3a}\right)^4$
32. $\left(\dfrac{1}{\sqrt{x}} + \sqrt{x}\right)^6$
33. $(a^3 - a^{-3})^5$
34. $(2 - 2)^n$
35. $(1 + 1)^n$
36. $(1 + \sqrt{2})^5 - (1 + \sqrt{2})^5$
37. $(1 - \sqrt{3})^4 + (1 - \sqrt{3})^4$
38. $(a + \sqrt{5})^6$
39. $(a - \sqrt{3})^4$
40. $(a^{2/3} + b^{1/3})^3$
41. $(2a - b)^{10}$, to four terms.
42. $(2p + 3q)^9$, to four terms.
43. $\left(\dfrac{1}{\sqrt{x}} - 3x^2\right)^8$, to three terms.
44. $(\sqrt{a} + \sqrt{b})^{10}$, to five terms.
45. $\left(3x - \dfrac{1}{y}\right)^{12}$, last four terms.
46. $\left(\dfrac{1}{a} + \dfrac{1}{3}a\sqrt{b}\right)^{10}$, last three terms.

47. Find the term that does not contain a in the expansion of $\left(\dfrac{3a^{-2}}{2} + \dfrac{1}{3}a\right)^{12}$.

48. Find the term containing $\dfrac{1}{t^3}$ in $\left(4 - \dfrac{5}{2}t\right)^9$.

♦ Applications

49. Before the advent of modern calculators and computers, mathematicians were always looking for ways to approximate the results of difficult or lengthy computations. For example, consider 1.01^{17}. An approximation might have been found by considering this as $1.01^{17} = (1.00 + 0.01)^{17}$. Use the sum of the first four terms of this binomial expansion to approximate 1.01^{17} to six decimal places. Approximate 1.01^{17} to six decimal places on your calculator. Compare the two approximations.

50. About 9 percent of the female population will contract breast cancer. There are 100 women at a sales meeting. The probability that exactly 4 of them will contract breast cancer in their lifetimes is given by the fifth term of the binomial expansion of $(0.91 + 0.09)^{100}$. Find this term and use your calculator to approximate the probability.

51. An electrical company manufactures computer chips. About 2 percent of the chips are defective. A quality control inspector pulls 50 chips at random off an assembly line. The probability that exactly 10 of them are defective is given by the 11th term of the binomial expansion of $(0.98 + 0.02)^{50}$. Find this term and use your calculator to approximate the probability.

52. At one point in a recent season, Eric Davis of the Los Angeles Dodgers was batting .320. Suppose he came to bat five times in one game. The probability of his getting at most four hits in the game is found by adding the first five terms of the binomial expansion of

$(0.680 + 0.320)^5$. Find these terms and use your calculator to approximate the probability.

53. Use Pascal's triangle to find the coefficients of the binomial expansion of $(a + b)^n$ for:

a. $n = 7$

b. $n = 8$

c. $n = 9$

d. $n = 10$

Then find each expansion.

54. Divide the fourth term of $(x^2 + 3yx^{1/3})^5$ by the third term and simplify.

55. Expand and simplify $\dfrac{(x + h)^5 - x^5}{h}$

56. Expand and simplify $\dfrac{(x + h)^n - x^n}{h}$

57. Find a formula for $(a + b + c)^n$.

58. Expand $(a^2 + b^2 + c^2)^4$.

In calculus, it can be shown that the formula for the binomial expansion can be altered in a way that includes powers other than nonnegative integers. The formula is as follows:

$$(1 + x)^n = 1 + nx + \frac{n(n - 1)}{2!}x^2 + \frac{n(n - 1)(n - 2)}{3!}x^3 + \cdots$$

where n is any real number, x is any real number such that $|x| < 1$, and the dots $\cdots$ mean that the expansion continues indefinitely. Find each of the following expansions to eight terms, assuming $|x| < 1$:

59. $(1 + x)^{-1}$ 60. $(1 - x)^{-2}$

61. $(1 + x)^{1/2}$ 62. $(1 + x)^{1/3}$

Simplify.

63. $\dfrac{\dfrac{(n + 2)^2 2^{n+1}}{(n + 1)!}}{\dfrac{n^2 2^n}{n!}}$ 64. $\dfrac{\dfrac{(3x + 6)^{n+1}}{(n + 1)!}}{\dfrac{(3x + 6)^n}{n!}}$

■ 10.2 MATHEMATICAL INDUCTION

How would you calculate the sum

$$1 + 2 + 3 + \cdots + 100?$$

As a youngster, Gauss was given this problem to keep him busy for a while. Within seconds, he gave the correct answer: 5050. We can generalize Gauss' calculation in the following way. Let n be a positive integer. Consider the following formula, which we will prove later in this section:

$$1 + 2 + \cdots + n = \frac{n(n + 1)}{2}$$

It can be viewed as a series of statements, one corresponding to each value of n. Let's denote the preceding statement by P_n. Then P_1 is the statement

$$1 = \frac{1 \cdot (1 + 1)}{2}$$

P_2 is the statement

$$1 + 2 = \frac{2 \cdot (2 + 1)}{2}$$

and so forth. The following result can be used to prove such a series of statements.

PRINCIPLE OF MATHEMATICAL INDUCTION

Suppose that for each positive integer n there is given a statement P_n. Furthermore, suppose that the following two conditions are fulfilled:

1. P_1 is true.
2. By assuming that P_n is true, we can deduce the truth of P_{n+1}. (Here n is any positive integer.)

Then P_n is true for every positive integer n.

The operation of the principle of mathematical induction is analogous to a row of dominoes. The truth of statement P_n is represented by the falling of the nth domino. Condition 1 is analogous to the first domino falling. Condition 2 is analogous to domino n knocking down domino $n + 1$. Conditions 1 and 2 together are analogous to all the dominoes falling. That is, the statement P_n is true for every value of n.

In more advanced courses, the principle of mathematical induction (in a slightly different form) is used as the fundamental tool for proving many of the most important properties of the natural numbers. In this text, we will assume its truth. The next three examples show how the principle of mathematical induction may be used as a technique of proof, called **proof by mathematical induction.**

➢ **EXAMPLE 1**

Sum of the First
n Integers

Let n be a positive integer. Use mathematical induction to prove the formula

$$1 + 2 + \cdots + n = \frac{n(n + 1)}{2}$$

Solution

According to the principle of mathematical induction, we may prove the formula by verifying conditions 1 and 2. Condition 1 is simple, because P_1 is the statement

$$1 = \frac{1 \cdot (1 + 1)}{2}$$

and this is a true statement. As for condition 2, let's assume that statement P_n is true. That is, we assume that

$$1 + 2 + \cdots + n = \frac{n(n + 1)}{2}$$

Let's deduce from this statement the truth of statement P_{n+1}. To do so, start from the last equation and add $n + 1$ to each side to obtain

$$1 + 2 + \cdots + n + (n + 1) = \frac{n(n + 1)}{2} + (n + 1)$$

Let's simplify the right side by combining the two expressions found there:

$$1 + 2 + \cdots + n + (n + 1) = \frac{n(n + 1) + 2(n + 1)}{2}$$

Factoring $n + 1$ from each term in the numerator, we have the equation

$$1 + 2 + \cdots + n + (n + 1) = \frac{(n + 1)(n + 2)}{2}$$

This is precisely the statement P_{n+1}. Indeed, if we take the statement P_n and replace n with $n + 1$ everywhere, we arrive at the last equation. Thus, starting from the truth of P_n, we have deduced the truth of P_{n+1}, so condition 2 is verified. Because conditions 1 and 2 hold, the principle of mathematical induction says that P_n holds for all n. That is, the given formula holds for every positive integer n. This completes the proof.

Condition 2 requires that you prove that P_n implies P_{n+1}. This part of a proof by induction is often called the **induction step.**

The next example provides further practice in constructing proofs by induction.

➤ **EXAMPLE 2**

Sum of First
n Odd Integers

Use mathematical induction to prove that for each positive integer n, the following formula holds:

$$1 + 3 + 5 + \cdots + (2n - 1) = n^2$$

Solution

In this example, the statement P_1 is the formula

$$1 = 1^2$$

which is clearly true, so condition 1 holds. To prove the inductive step, assume that P_n holds. That is,

$$1 + 3 + 5 + \cdots + (2n - 1) = n^2$$

We must deduce P_{n+1} from this formula. That is, we must deduce from the last formula the result

$$1 + 3 + 5 + \cdots + [2(n + 1) - 1] = (n + 1)^2$$

To see how to proceed, let's rewrite what we wish to deduce. Explicitly write out the next-to-last term to obtain

$$1 + 3 + 5 + \cdots + (2n - 1) + [2(n + 1) - 1] = (n + 1)^2$$

Note that all the terms but the last form the expression on the left side of the statement P_n. This suggests that we start from P_n and add $2(n + 1) - 1 = 2n + 1$ to both sides of the equation to obtain

$$1 + 3 + \cdots + (2n - 1) + (2[n + 1] - 1) = n^2 + [2n + 1]$$

Factoring the right side yields

$$1 + 3 + \cdots + (2[n + 1] - 1) = (n + 1)^2$$

This is the statement P_{n+1}. That is, we have deduced P_{n+1} from P_n. This is the inductive step and completes the proof of the formula.

Recall that in the preceding section, we stated and applied the binomial theorem, but did not supply a proof. Let's remedy that omission.

➤ **EXAMPLE 3**

Binomial Theorem

Let n be a positive integer. Use mathematical induction to prove the binomial theorem

$$(a + b)^n = a^n + \binom{n}{1}a^{n-1}b + \binom{n}{2}a^{n-2}b^2 + \cdots + \binom{n}{n-1}ab^{n-1} + b^n$$

Solution

Here, P_1 is the statement

$$(a + b)^1 = a^1 + b^1$$

This is clearly true, so let's prove the inductive step. Assume that P_n is true. In other words, assume that

$$(a + b)^n = a^n + \binom{n}{1}a^{n-1}b + \binom{n}{2}a^{n-2}b^2 + \cdots + \binom{n}{n-1}ab^{n-1} + b^n$$

Let's multiply both sides of the equation by $a + b$. On the left, we obtain $(a + b)^{n+1}$, so that the equation reads

$$(a+b)^{n+1} = (a+b)\left[a^n + \binom{n}{1}a^{n-1}b + \binom{n}{2}a^{n-2}b^2 + \cdots + \binom{n}{n-1}ab^{n-1} + b^n\right]$$

Apply the distributive law to the right side to obtain

$$(a + b)^{n+1} =$$

$$\left[\binom{n}{0}a^{n+1} + \binom{n}{1}a^n b + \binom{n}{2}a^{n-1}b^2 + \cdots + \binom{n}{n-1}a^2 b^{n-1} + \binom{n}{n}ab^n\right]$$

$$+ \left[\binom{n}{0}a^n b + \binom{n}{1}a^{n-1}b^2 + \binom{n}{2}a^{n-2}b^3 + \cdots + \binom{n}{n-1}ab^n + \binom{n}{n}b^{n+1}\right]$$

Collecting like terms, we then have

$$(a + b)^{n+1} = \binom{n}{0}a^{n+1} + \left[\binom{n}{0} + \binom{n}{1}\right]a^n b + \left[\binom{n}{1} + \binom{n}{2}\right]a^{n-1}b^2$$

$$+ \left[\binom{n}{2} + \binom{n}{3}\right]a^{n-2}b^3 + \cdots + \left[\binom{n}{n-1} + \binom{n}{n}\right]ab^n + \binom{n}{n}b^{n+1}$$

$$= a^{n+1} + \left[\binom{n}{0} + \binom{n}{1}\right]a^n b + \left[\binom{n}{1} + \binom{n}{2}\right]a^{n-1}b^2$$

$$+ \left[\binom{n}{2} + \binom{n}{3}\right]a^{n-2}b^3 + \cdots + \left[\binom{n}{n-1} + \binom{n}{n}\right]ab^n + b^{n+1}$$

To complete simplification of the right side, let's obtain an alternate formula for the sums of binomial coefficients that appear. Let j be one of the integers $0, 1, \ldots, n - 1$. Then we have

$$\binom{n}{j} + \binom{n}{j+1} = \frac{n!}{j!(n-j)!} + \frac{n!}{(j+1)![n-(j+1)]!}$$

$$= n!\left[\frac{1}{j!(n-j)!} + \frac{1}{j! \cdot (j+1) \cdot (n-j-1)!}\right]$$

$$= \frac{n!}{j!}\left[\frac{1}{(n-j-1)! \cdot (n-j)} + \frac{1}{(j+1) \cdot (n-j-1)!}\right]$$

$$= \frac{n!}{j!(n-j-1)!}\left(\frac{1}{n-j} + \frac{1}{j+1}\right)$$

$$= \frac{n!}{j!(n-j-1)!} \cdot \frac{n+1}{(n-j)(j+1)}$$

$$= \frac{(n+1)!}{(j+1)!(n-j)!}$$

$$= \binom{n+1}{j+1}$$

Using this formula, we replace the sums of the binomial coefficients in the preceding equation to obtain

$$(a + b)^{n+1} = a^{n+1} + \binom{n+1}{1}a^n b + \binom{n+1}{2}a^{n-1}b^2 + \binom{n+1}{3}a^{n-2}b^3$$

$$+ \cdots + \binom{n+1}{n}ab^n + b^{n+1}$$

This equation is precisely statement P_{n+1}. This completes the inductive step and, with it, the proof of the binomial theorem.

Note that in the first two examples, we added a term to each side of the formula to obtain the inductive step. In the third example, the inductive step was more complicated and involved first multiplying both sides of the equation by $a + b$, then performing considerable simplification, and finally proving an identity about binomial coefficients. This is fairly typical of the ingenuity that must sometimes be employed to prove the inductive step.

All the preceding examples used induction to prove formulas. Induction may also be used to prove inequalities, as the next example shows.

> **EXAMPLE 4**

Inductive Proof
of an Inequality

Let n be a positive integer. Prove that $2n + 1 \le 3^n$.

Solution

In this case, the statement P_1 is the true inequality

$$2 \cdot 1 + 1 \le 3^1$$
$$3 \le 3$$

Let's now prove the inductive step. We assume that P_n holds. That is, we assume that

$$2n + 1 \le 3^n$$

We must prove (somehow) that

$$2(n + 1) + 1 \le 3^{n+1}$$

To prove this, let's start from the assumed inequality and add 2 to both sides to obtain

$$(2n + 1) + 2 \le 3^n + 2$$
$$2(n + 1) + 1 \le 3^n + 2$$

Because n is a positive integer, we have

$$3^n \ge 3 > 2$$

Therefore, applying this fact to the preceding inequality, we have

$$2(n + 1) + 1 \le 3^n + 2$$
$$< 3^n + 3^n \qquad \textit{because } 3^n > 2$$
$$= 2 \cdot 3^n$$
$$\le 3 \cdot 3^n = 3^{n+1}$$
$$2(n + 1) + 1 \le 3^{n+1}$$

This last inequality is the assertion for $n + 1$. This proves the inductive step and completes the proof of the inequality.

Exercises 10.2

Use mathematical induction to prove that the following formulas hold for all positive integers n.

1. $1 + 5 + 9 + \cdots + (4n - 3) = 2n^2 - n$

2. $1 + 4 + 7 + \cdots + (3n - 2) = \dfrac{n(3n - 1)}{2}$

3. $1 \cdot 2 + 3 \cdot 4 + 5 \cdot 6 + \cdots + (2n - 1)2n = \dfrac{n(n + 1)(4n - 1)}{3}$

4. $1^2 + 3^2 + 5^2 + \cdots + (2n - 1)^2 = \dfrac{n(4n^2 - 1)}{3}$

5. $\dfrac{1}{2} + \dfrac{1}{4} + \dfrac{1}{8} + \cdots + \dfrac{1}{2^n} = \dfrac{2^n - 1}{2^n}$

6. $3 + 3^2 + 3^3 + \cdots + 3^n = \dfrac{3^{n+1} - 3}{2}$

Prove by mathematical induction that the following are true for all positive integers n.

7. $1^3 + 2^3 + 3^3 + \cdots + n^3 = \left[\dfrac{n(n + 1)}{2} \right]^2$

8. $\dfrac{1}{1 \cdot 2} + \dfrac{1}{2 \cdot 3} + \dfrac{1}{3 \cdot 4} + \cdots + \dfrac{1}{n(n + 1)} = \dfrac{n}{n + 1}$

9. $1^2 + 2^2 + 3^2 + \cdots + n^2 = \dfrac{n(n + 1)(2n + 1)}{6}$

10. $2 + 2^2 + 2^3 + \cdots + 2^n = 2(2^n - 1)$

11. $2 + 4 + 6 + \cdots + 2n = n^2 + n$

12. $1 + 3 + 5 + \cdots + (2n - 1) = n^2$

13. $1 + 2 + 3 + \cdots + n = \dfrac{n(n + 1)}{2}$

14. $1 + 2 + 3 + \cdots + n < \dfrac{(2n + 1)^2}{8}$

15. $n(n^2 + 5)$ is a multiple of 6.

16. $3^{2n} - 1$ is a multiple of 8.

17. 3 is a factor of $n^3 - n + 3$.

18. 4 is a factor of $5^n - 1$.

19. $n^2 + n$ is divisible by 2.   20. $n^3 - n$ is divisible by 6.

21. 7 is a factor of $11^n - 4^n$.   22. 3 is a factor of $n^3 + 2n$.

23. $4^n > n^4$ for $n \geq 5$. (*Hint:* Start by proving P_5 true.)

24. $3^n < 3^{n+1}$

25. $n^2 < 2^n$ for $n > 4$.   26. $2n < 2^n$ for $n \geq 3$.

For the following $f(n)$: (*a*) complete the table; (*b*) on the basis of the results in the table, what would you believe to be the value of $f(7)$? Compute $f(7)$; (*c*) make a conjecture about the value of $f(n)$.

27. $f(n) = 1 + 2 + 4 + \cdots + 2^{n-1}$

n	1	2	3	4	5	6
$f(n)$						

28. $f(n) = 3 + 7 + 11 + \cdots + (4n - 1)$

n	1	2	3	4	5	6
$f(n)$						

29. $f(n) = 2 + 9 + 16 + \cdots + (7n - 5)$

n	1	2	3	4	5	6
$f(n)$						

30. $f(n) = 1 + 3 + 5 + \cdots + (2n - 1)$

n	1	2	3	4	5	6
$f(n)$						

31. Prove that it is possible to pay any debt of \$4, \$5, \$6, \$7, ..., \$$n$ by using only \$2 and \$5 bills.

32. Prove that it is possible to pay any debt of \$8, \$9, \$10, ..., \$$n$ by using only \$3 and \$5 bills.

33. Prove that the number of diagonals of an n-sided convex polygon is $n(n - 3)/2$ for $n \geq 3$.

34. Prove that the sum of the measures of the angles in an n-sided convex polygon is $180°(n - 2)$ for $n \geq 3$.

35. If, in a room with n people ($n \geq 2$), every person shakes hands once with every other person, prove that there will be $n(n - 1)/2$ handshakes.

36. Use mathematical induction to prove $[r(\cos \theta + i \sin \theta)]^n = r^n(\cos n\theta + i \sin n\theta)$ (De Moivre's theorem) for a positive integer n.

37. Prove: $\sin(\theta + n\pi) = (-1)^n \sin \theta$ for all positive integers n.

38. Prove: $\cos(\theta + n\pi) = (-1)^n \cos \theta$ for all positive integers n.

Find the smallest positive integer n for which the given statement is true. Then prove that the statement is true for all integers greater than or equal to that smallest value.

39. $n + 5 < 2^n$ 40. $\log n < n$

41. $3^n > 2^n + 20$

42. $(1 + x)^n \geq 1 + nx$, if $x \geq -1$

For the following problems, tell what you can conclude from the given information about the sequence of statements. (*Example:* Given that P_4 is true and that P_k implies P_{k+1} for any k, then you may conclude that P_n is true for every integer $n \geq 4$.)

43. P_{18} is true, and P_k implies P_{k+1}.

44. P_{18} is not true, and P_k implies P_{k+1}.

45. P_1 is true, but P_k does not imply P_{k+1}.

46. P_1 is true, and P_k implies P_{k+2}.

47. P_1 and P_2 are true, and P_k and P_{k+1} together imply P_{k+2}.

48. P_1 is true, and P_k implies P_{4k}.

10.3 SEQUENCES AND SERIES

Many applications involve quantities that are reported at discrete intervals. For instance, the balance on a mortgage is computed monthly, the unemployment rate is reported monthly, and the temperature is reported hourly by the weather bureau. To record and manipulate such data, mathematicians use sequences.

Definition 1
Sequence, Terms

A **sequence** is a function f from the positive integers to the real numbers. A sequence may be viewed as an ordered collection of numbers

$$f(1), f(2), f(3), \ldots$$

The number $f(n)$ is called the nth **term of the sequence**. Rather than use the function notation to denote sequences, it is tradition to denote the variable n using a subscript such as $a_n = f(n)$. In terms of this notation, the above sequence may be written as

$$a_1, a_2, a_3, \ldots$$

Here are some examples of sequences:

$$a_1 = 1, \quad a_2 = \tfrac{1}{2}, \quad a_3 = \tfrac{1}{4}, \quad a_4 = \tfrac{1}{8}, \ldots$$
$$a_1 = 1, \quad a_2 = -1, \quad a_3 = 1, \quad a_4 = -1, \ldots$$
$$a_1 = 5, \quad a_2 = 4, \quad a_3 = 3, \quad a_4 = 2, \quad a_5 = 1$$

The first two sequences above contain an infinite number of terms and are called **infinite sequences**. The third sequence contains only a finite number of terms and is therefore called a **finite sequence**. The terms of a sequence are customarily labeled using positive integer subscripts to indicate the term's position. For example, here is how a typical sequence is written using subscript notation:

$$a_1, a_2, a_3, \ldots, a_n, \ldots$$

That is, a_1 denotes the first term of the sequence, a_2 the second term, and a_n the nth term.

One method for defining a sequence (finite or infinite) is to specify a formula for computing a_n in terms of n. In the next example, we calculate terms of sequences specified in this fashion.

➤ **EXAMPLE 1**
Determining Terms of Sequences

Determine terms 1, 2, 3, and 7 of the sequences defined by

1. $a_n = \dfrac{1}{n}$

2. $a_n = 1 + \dfrac{(-1)^n}{n}$

3. $a_n = \dfrac{n^n}{n!}$

Solution

In each case, we calculate a_n by substituting the value of n in the given formula for a_n.

1. $a_1 = \frac{1}{1} = 1$

   $a_2 = \frac{1}{2}$

   $a_3 = \frac{1}{3}$

   $a_7 = \frac{1}{7}$

2. $a_1 = 1 + \frac{(-1)^1}{1} = 1 + \frac{-1}{1} = 0$

   $a_2 = 1 + \frac{(-1)^2}{2} = 1 + \frac{1}{2} = \frac{3}{2}$

   $a_3 = 1 + \frac{(-1)^3}{3} = 1 + \frac{-1}{3} = \frac{2}{3}$

   $a_7 = 1 + \frac{(-1)^7}{7} = 1 + \frac{-1}{7} = \frac{6}{7}$

3. $a_1 = \frac{1^1}{1!} = \frac{1}{1} = 1$

   $a_2 = \frac{2^2}{2!} = \frac{2 \cdot 2}{1 \cdot 2} = 2$

   $a_3 = \frac{3^3}{3!} = \frac{3 \cdot 3 \cdot 3}{1 \cdot 2 \cdot 3} = \frac{9}{2}$

   $a_7 = \frac{7^7}{7!} = \frac{7^6}{1 \cdot 2 \cdot 3 \cdot 4 \cdot 5 \cdot 6} = \frac{117,649}{720}$

All the sequences of the preceding example were defined by formulas for calculating a_n in terms of n. In many applications, it is most convenient to define a sequence by specifying how to calculate a_n in terms of preceding terms of the sequence $a_1, a_2, \ldots, a_{n-1}$. Such sequences are said to be defined **recursively**.

For instance, let a_n denote the amount of money in a bank account at the end of n months, where the account earns 0.5 percent interest per month. Then the amount a_n is related to the amount at the end of the preceding month a_{n-1} by the formula

$$a_n = 1.005 a_{n-1}$$

If the amount in the account at the end of month 1 is $1000, then $a_1 = 1000$. The two equations, $a_n = 1.005 a_{n-1}$ and $a_1 = 1000$, suffice to define the sequence of bank balances recursively.

➤ **EXAMPLE 2**
Calculating Terms
from One Another

Determine the first five terms of the previously defined bank balance sequence.

Solution

The value of a_1 gives the first term. Each successive term may be computed from its predecessor using the given formula. Here are the calculations for the first five terms:

$$a_1 = 1000$$
$$a_2 = 1.005 \cdot a_1 = 1.005 \cdot 1000 = 1005$$
$$a_3 = 1.005 \cdot a_2 = 1.005 \cdot 1005 = 1010.025$$
$$a_4 = 1.005 \cdot a_3 = 1.005 \cdot 1010.025 = 1015.0751$$
$$a_5 = 1.005 \cdot a_4 = 1.005 \cdot 1015.0751 = 1020.1505$$

The sum of the first n terms of a sequence is called the nth **partial sum** and is denoted S_n. That is,

$$S_n = a_1 + a_2 + \cdots + a_n$$

Thus, for example, if we consider the sequence $a_n = 2^{-n}$, then

$$S_1 = a_1 = 2^{-1} = \tfrac{1}{2}$$
$$S_2 = a_1 + a_2$$
$$= 2^{-1} + 2^{-2}$$
$$= \tfrac{1}{2} + \tfrac{1}{4} = \tfrac{3}{4}$$
$$S_3 = a_1 + a_2 + a_3$$
$$= 2^{-1} + 2^{-2} + 2^{-3}$$
$$= \tfrac{1}{2} + \tfrac{1}{4} + \tfrac{1}{8} = \tfrac{7}{8}$$

In dealing with sums of consecutive terms of a sequence, it is often convenient to use a special notation:

Definition 2
Summation Notation,
Summation Index

Suppose that A and B are integers with $A < B$. The symbol

$$\sum_{n=A}^{B} a_n$$

is another way of denoting the sum

$$a_A + a_{A+1} + \cdots + a_B$$

The notation

$$\sum_{n=A}^{B} a_n$$

says to add up the terms a_n, where the variable n starts at A (written underneath the $\sum$) and ends with B (written above the $\sum$). The variable n is called the **summation index**.

The next example provides some practice in using summation notation.

> **EXAMPLE 3**
Evaluating Sums Given by
Summation Notation

Evaluate the following expressions:

1. $\displaystyle\sum_{n=1}^{5} n$ 2. $\displaystyle\sum_{n=2}^{5} n^2$ 3. $\displaystyle\sum_{n=0}^{3} \binom{3}{n}$

Solution

1. We write out the terms of the summation by substituting, in succession, the values $n = 1, 2, 3, 4, 5$:

$$\sum_{n=1}^{5} n = 1 + 2 + 3 + 4 + 5 = 15$$

2. The terms of the summation are obtained by evaluating n^2 in succession, for $n = 2, 3, 4, 5$:

$$\sum_{n=2}^{5} n^2 = 2^2 + 3^2 + 4^2 + 5^2 = 54$$

3. The terms of the summation are obtained by substituting, in succession, $n = 0, 1, 2, 3$:

$$\sum_{n=0}^{3} \binom{3}{n} = \binom{3}{0} + \binom{3}{1} + \binom{3}{2} + \binom{3}{3}$$
$$= 1 + 3 + 3 + 1$$
$$= 8$$

Summation notation provides a convenient way of writing many mathematical facts in a shorthand form. For instance, the definition of the nth partial sum of a sequence $a_1, a_2, \ldots, a_n, \ldots$ may be written in summation notation as

$$S_n = \sum_{j=1}^{n} a_j$$

Note that in this formula, we used j as the summation index. The right side of the previous equation says to add up all the terms a_j for j beginning with 1 and ending with n. Note that we couldn't use n to denote the summation index, because this letter is already in use as the index of the last term to sum. You may use any variable name you wish for the summation index as long as it can't be confused with a variable in use for some other purpose (such as n in the previous equation). For instance, we could have just as easily denoted the summation index k and written the last formula in the form

$$\sum_{k=1}^{n} a_k$$

Because the summation index may be replaced by any unused variable, it is often referred to as a **dummy variable**. The most common are i, j, and k.

Following are some further applications of summation notation. Here is a formula we proved by induction in the preceding section:

$$1 + 2 + 3 + \cdots + N = \frac{N(N + 1)}{2}$$

In summation notation, this formula may be written

$$\sum_{n=1}^{N} n = \frac{N(N+1)}{2}$$

The binomial theorem states that for a positive integer N, we have

$$(a + b)^N = a^N + \binom{N}{1}a^{N-1}b + \binom{N}{2}a^{N-2}b^2 + \cdots + \binom{N}{N-1}ab^{N-1} + b^N$$

In terms of summation notation, this theorem may be written

$$(a + b)^N = \sum_{n=0}^{N} \binom{N}{n}a^{N-n}b^n$$

Indeed, the first term in the previous summation corresponds to $n = 0$:

$$\binom{N}{0}a^{N-0}b^0 = 1 \cdot a^N \cdot 1 = a^N$$

The second term corresponds to $n = 1$:

$$\binom{N}{1}a^{N-1}b^1 = \binom{N}{1}a^{N-1}b$$

The final term corresponds to $n = N$:

$$\binom{N}{N}a^{N-N}b^N = 1 \cdot a^0 b^N = b^N$$

As you can see, the terms in the summation notation correspond exactly to the terms in the right side of the binomial theorem. As another example of an application of the summation notation, consider the polynomial

$$a_0 + a_1 x + a_2 x^2 + \cdots + a_N x^N$$

In terms of the summation notation, this polynomial may be written

$$\sum_{n=0}^{N} a_n x^n$$

Exercises 10.3

Determine terms 1, 2, 3, and 8 of the sequences defined by the following:

1. $a_n = n + 2$ 2. $a_n = n - 4$ 3. $a_n = \frac{n+1}{n-2}$

4. $a_n = \frac{n-1}{n+1}$ 5. $a_n = 2^{n-3}$ 6. $a_n = 3^{n-4}$

7. $a_n = n^{-2}$ 8. $a_n = n^3$

9. $a_n = (-2)\sqrt[n]{n}$ 10. $a_n = (-1)^{n-1}\sqrt{2n}$

11. $a_n = 3^{n-1} - 3^n$ 12. $a_n = 2^{n+1} - 2^{n-1}$

13. $a_n = \frac{n^2-1}{n^2-2}$ 14. $a_n = \frac{n^3+1}{n^3-2}$

15. $a_n = -n\cos n\pi$ 16. $a_n = -n\sin n\pi$

17. $a_n = \sin\left(\frac{-3n\pi}{2}\right)$ 18. $a_n = \cos\left(\frac{5n\pi}{2}\right)$

Find the first five terms for the following sequences:

19. $a_1 = 2, a_n = a_{n-1} + 5, n > 1$

20. $a_1 = -3, a_n = 2a_{n-1}, n > 1$

21. $a_1 = 9, a_n = \frac{1}{3}a_{n-1}, n > 1$

22. $a_1 = 1, a_n = \left(-\frac{1}{3}\right)^n a_{n-1}, n > 1$

23. $a_n = n + \frac{1}{n}$ 24. $a_n = n^3 + 1$

25. $a_n = \left(-\frac{1}{2}\right)^n (n^{-1})$ 26. $a_n = \frac{1}{(n+1)!}$

27. $a_n = \frac{1}{2n}\log 1000^n$ 28. $a_n = n\log 10^{n-1}$

29. $a_1 = 3, a_{n+1} = \frac{1}{a_n}, n \geq 1$

30. $a_1 = -2$, $a_{n+1} = -a_n^{-1}$, $n \geq 1$

31. $a_1 = -2$, $a_{n+1} = \log 100^{a_n}$, $n \geq 1$

32. $a_n = -n \sin(-n\pi)$

Evaluate the following expressions:

33. $\displaystyle\sum_{n=1}^{4}(n - 2)$ 34. $\displaystyle\sum_{n=1}^{3}(2n + 1)$ 35. $\displaystyle\sum_{n=1}^{3}(n^2 + 1)$

36. $\displaystyle\sum_{n=1}^{5}(-2)^n$ 37. $\displaystyle\sum_{n=2}^{4}n^{-1}$ 38. $\displaystyle\sum_{n=0}^{3}(4n - 1)$

39. $\displaystyle\sum_{n=0}^{7}\log 100^n$ 40. $\displaystyle\sum_{n=1}^{6}\log 1000^{n-1}$

41. $\displaystyle\sum_{n=3}^{10}\left(-\cos\frac{n\pi}{2}\right)$ 42. $\displaystyle\sum_{n=0}^{6}\cos(-n\pi)$

43. $\displaystyle\sum_{n=1}^{50}3$ 44. $\displaystyle\sum_{n=2}^{200}-1$

45. $\displaystyle\sum_{n=1}^{10}[1 + (-1)^n]$ 46. $\displaystyle\sum_{n=1}^{20}(-2n)$

For each of the following sequences, find the general or nth term. Answers may vary.

47. $1, 3, 5, 7, 9, \ldots$

48. $3, 9, 27, 81, 243, \ldots$

49. $2, -4, 8, -16, 32, -64, \ldots$

50. $1, -\frac{1}{2}, \frac{1}{3}, -\frac{1}{4}, \frac{1}{5}, -\frac{1}{6}, \ldots$

51. $\frac{1}{2}, \frac{1}{8}, \frac{1}{32}, \frac{1}{128}, \ldots$ 52. $17, 21, 25, 29, \ldots$

Rewrite each of the following summations using sigma notation.

53. $1 + \frac{1}{2} + \frac{1}{3} + \frac{1}{4} + \frac{1}{5}$

54. $3 + 6 + 12 + 24 + 48 + 96$

55. $1 + 2 + 3 + 4 + 5 + 6 + 7 + 8 + 9 + 10$

56. $-2 + 4 - 8 + 16 - 32 + 64 - 128 + 256$

57. $-3 + 9 - 27 + 81 - 243 + 729$

58. $1 + 5 + 9 + 13 + 17$

Solve the following equations for x:

59. $\displaystyle\sum_{n=4}^{6}nx = 30$ 60. $\displaystyle\sum_{n=1}^{12}(nx - 5) = 64$

61. $\displaystyle\sum_{n=3}^{5}(nx + 3) = 45$ 62. $\displaystyle\sum_{n=1}^{7}(-1)^n x = -2$

63. $\displaystyle\sum_{n=0}^{5}n(n - x) = 25$ 64. $\displaystyle\sum_{n=1}^{6}(x - 3n) = -3$

10.4 ARITHMETIC SEQUENCES

In this section and the next, we study two particular types of sequences that play an important role in both pure and applied problems, namely the arithmetic and geometric sequences.

| Definition 1 | An **arithmetic sequence** is a sequence of real numbers in which the difference |
| Arithmetic Sequence | between consecutive terms is constant. |

➤ **EXAMPLE 1**
Proving a Sequence Is Arithmetic

Prove that the following sequence is arithmetic:

$$6, 11, 16, \ldots, 5n + 1, \ldots$$

Solution

Let's compute the difference between two typical consecutive terms, namely a_{n+1} and a_n. We have

$$a_{n+1} - a_n = [5(n + 1) + 1] - (5n + 1)$$
$$= (5n + 6) - (5n + 1)$$
$$= 5$$

This computation shows that the differences between consecutive terms are constant and, in fact, all equal to 5. Thus, the sequence is arithmetic.

Suppose that an arithmetic sequence has first term a_1 and that the difference between consecutive terms is d. Then the terms of the sequence are

$$a_1$$
$$a_2 = a_1 + d$$
$$a_3 = a_2 + d$$
$$= (a_1 + d) + d$$
$$= a_1 + 2d$$
$$a_4 = a_3 + d$$
$$= (a_1 + 2d) + d$$
$$= a_1 + 3d$$

These formulas generalize to the following result.

---

nTH TERM OF AN ARITHMETIC SEQUENCE

Suppose that in an arithmetic sequence, the first term is a_1 and the difference between consecutive terms is d. The nth term of the sequence is given by the formula

$$a_n = a_1 + (n - 1)d$$

Proof   We can proceed by induction. We have shown that the formula holds for $n = 1, 2, 3, 4$. Assume that the formula holds for n. That is, assume that

$$a_n = a_1 + (n - 1)d$$

We want to prove the induction step, namely that

$$a_{n+1} = a_1 + [(n + 1) - 1]d = a_1 + nd$$

Because the sequence is arithmetic with difference d, we see that

$$a_{n+1} - a_n = d$$
$$a_{n+1} = a_n + d$$
$$= [a_1 + (n - 1)d] + d \qquad \text{(By assumption)}$$
$$= a_1 + nd$$
$$= a_1 + [(n + 1) - 1]d$$

This last equation is the assertion of the formula for $n + 1$. This completes the induction step and the proof of the formula.   ♦

➢ **EXAMPLE 2**

Evaluation of the nth Term of an Arithmetic Sequence

Suppose that the first term of an arithmetic sequence is 5 and the difference between consecutive terms is 7. What is the value of the 30th term of the sequence?

Solution

By the formula for the nth term of an arithmetic sequence, with $a_1 = 5$ and $d = 7$, we see that

$$a_{30} = a_1 + (30 - 1)d$$
$$= 5 + 29 \cdot 7$$
$$= 208$$

> **EXAMPLE 3**

Calculating the First Term
and Difference
of an Arithmetic Sequence

Suppose that $a_1, a_2, \ldots$ is an arithmetic sequence. Suppose that $a_{15} = 40$ and $a_{20} = 50$. Determine the first term and the difference d between consecutive terms of the sequence.

Solution

According to the formula for the nth term of an arithmetic sequence and the given information, we have

$$a_{15} = 40$$
$$a_1 + (15 - 1)d = 40 \tag{1}$$
$$a_1 + 14d = 40$$
$$a_{20} = 50$$
$$a_1 + 19d = 50 \tag{2}$$

Subtracting equation (1) from equation (2), we have

$$(19 - 14)d = 50 - 40$$
$$5d = 10$$
$$d = 2$$

Substituting this value of d into equation (1) gives us

$$a_1 + 14d = 40$$
$$a_1 + 14 \cdot 2 = 40$$
$$a_1 = 12$$

Thus, the sequence has first term 12 and difference 2.

The following result provides a simple formula for calculating the sum of an arithmetic sequence.

SUM OF AN ARITHMETIC SEQUENCE

Let $a_1, a_2, a_3, \ldots$ be an arithmetic sequence with difference d, and let S_n be its nth partial sum. Then

$$S_n = \frac{n}{2}(a_1 + a_n)$$

$$= na_1 + \frac{n(n-1)}{2}d$$

Proof   From the definition of the nth partial sum, we have

$$S_n = a_1 + a_2 + \cdots + a_n$$

Let's substitute the arithmetic sequence formula for each term on the right side:

$$S_n = a_1 + (a_1 + d) + (a_1 + 2d) + \cdots + [a_1 + (n - 1)d]$$
$$= (a_1 + a_1 + \cdots + a_1) + [d + 2d + \cdots + (n - 1)d]$$
$$= na_1 + [1 + 2 + \cdots + (n - 1)]d \tag{3}$$

In section 2, we proved the formula

$$1 + 2 + \cdots + n = \frac{n(n + 1)}{2}$$

Replacing n with $n - 1$ in this formula, we see that

$$1 + 2 + \cdots + (n - 1) = \frac{(n - 1)(n - 1 + 1)}{2} = \frac{n(n - 1)}{2}$$

Substituting this formula into equation (3), we have

$$S_n = na_1 + \frac{n(n - 1)}{2}d$$

This yields the second formula stated. To get the first formula, rewrite equation (4) as follows:

$$S_n = \frac{n}{2}[2a_1 + (n - 1)d]$$

$$= \frac{n}{2}[a_1 + a_1 + (n - 1)d]$$

$$= \frac{n}{2}(a_1 + a_n)$$

This is the desired statement.   ◆

➤ EXAMPLE 4
Calculating Sum of an Arithmetic Sequence

Compute the sum of the first 50 terms of the arithmetic sequence 1, 4, 7, 10, ..., $3n - 2, \ldots$.

Solution

The sequence has $a_1 = 1$, $d = 3$. Therefore, by the preceding result, we have

$$S_n = na_1 + \frac{n(n - 1)}{2}d$$

$$S_{50} = 50 \cdot 1 + \frac{50 \cdot (50 - 1)}{2} \cdot 3 = 3725$$

➤ EXAMPLE 5
Seats in an Auditorium

A concert auditorium has 30 rows of seats. The first row contains 50 seats. As you move to the rear of the auditorium, each row has two more seats than the previous one. How many seats are in the auditorium?

Solution

Let a_n denote the number of seats in the nth row. Then the sequence $a_1, a_2, \ldots, a_{30}$ is an arithmetic sequence with first term 50 and difference 2. The number of seats in the auditorium is the sum of the first 30 terms of the sequence, or S_{30}. By the formula for the sum of an arithmetic sequence, this quantity equals

$$S_n = na_1 + \frac{n(n - 1)}{2}d$$

$$S_{30} = 30 \cdot 50 + \frac{30 \cdot 29}{2} \cdot 2 = 2370$$

That is, the auditorium has 2370 seats.

➤ **EXAMPLE 6**

Evaluating the Sum of Terms in an Arithmetic Sequence

Evaluate the sum

$$\sum_{n=1}^{10}(5n+3)$$

Solution

The numbers being summed are

$$5 \cdot 1 + 3, 5 \cdot 2 + 3, \ldots, 5 \cdot 10 + 3$$

an arithmetic sequence with first term 8 and last term 53. By the formula for the sum of an arithmetic sequence, the sum of these 10 terms is equal to

$$S_{10} = \tfrac{10}{2}(a_1 + a_{10})$$

$$= 5 \cdot (8 + 53)$$

$$= 305$$

Exercises 10.4

Prove that the following sequences are arithmetic.

1. $1, 3, 5, 7, \ldots, 2n - 1, \ldots$

2. $4, 5, 6, \ldots, n + 3, \ldots$ 3. $4, 7, 10, \ldots, 1 + 3n, \ldots$

4. $0, 2, 4, \ldots, 2n - 2, \ldots$ 5. $7, 8, 9, \ldots, n + 6, \ldots$

6. $9, 14, 19, \ldots, 4 + 5n, \ldots$

7. $1, -1, -3, \ldots, 3 - 2n, \ldots$

8. $3, -1, -5, \ldots, 7 - 4n, \ldots$

Find the indicated term for the given arithmetic sequences.

9. $a_1 = 2, d = -3$; find a_{10}.

10. $a_1 = -5, d = 7$; find a_{12}.

11. $a_1 = 0, d = 9$; find a_{30}.

12. $a_1 = 4, d = -5$; find a_{40}.

13. $a_1 = 9, d = -4$; find a_8.

14. $a_1 = -7, d = -4$; find a_{20}.

15. $a_1 = -3, d = -9$; find a_{19}.

16. $a_1 = 0, d = \frac{1}{2}$; find a_{35}.

17. $a_1 = 3, d = -\frac{1}{2}$; find a_{40}.

18. $a_1 = -\frac{1}{2}, d = 2$; find a_8.

For each of the following arithmetic sequences, find the first term, a_1, and the difference d between consecutive terms of the sequence.

19. $a_{10} = -26, a_{14} = -7$ 20. $a_{18} = 49, a_{20} = 28$

21. $a_{15} = 16, a_{19} = -12$ 22. $a_{10} = 3, a_{16} = 21$

23. $a_5 = 3, a_{50} = 30$ 24. $a_5 = \frac{1}{2}, a_{20} = \frac{7}{8}$

25. $a_{17} = -40, a_{28} = -73$ 26. $a_{10} = -11, a_{40} = -71$

27. $a_{20} = 66, a_{59} = 222$

28. $a_{16} = 60\frac{1}{2}, a_{41} = 160\frac{1}{2}$

The following exercises refer to arithmetic sequences.

29. If $d = 4$ and $a_8 = 33$, what is a_1?

30. If $a_1 = 8$ and $a_{11} = 26$, what is d?

31. If $a_7 = \frac{7}{3}$ and $d = -\frac{2}{3}$, what is a_{15}?

32. If $d = -7$ and $a_3 = 37$, what is a_{17}?

33. If $a_1 = -2, d = 7$, and $a_n = 138$, what is n?

34. If $a_3 = -7, d = -5$, and $a_n = -142$, what is n?

35. If $a_2 = 5$ and $a_4 = 1$, what is a_{10}?

36. If $a_4 = 5$ and $d = -2$, what is a_1?

37. If $a_3 = 5, d = -3$, and $a_n = -76$, what is n?

38. If $a_3 = 25, d = -14$, and $a_n = -507$, what is n?

Compute the indicated sums for the following arithmetic sequences.

39. $3, 5, 7, \ldots, 2n + 1, \ldots$; find S_{20}.

40. $2, 5, 8, \ldots, 3n - 1, \ldots$; find S_{30}.

41. $-1, -4, -7, \ldots, 2 - 3n, \ldots$; find S_{19}.

42. $7, 9, 11, \ldots, 5 + 2n, \ldots$; find S_{35}.

43. $2, \frac{21}{2}, 3, \ldots, \dfrac{3+n}{2}, \ldots$; find S_{44}.

44. $\frac{11}{2}, \frac{1}{2}, -\frac{1}{2}, \ldots, \dfrac{5-2n}{2}, \ldots$; find S_{50}.

45. $-4, -1, 2, \ldots, 30 - 7, \ldots$; find S_7.

46. $3, 9, 15, \ldots, 6n - 3, \ldots$; find S_{15}.

47. $a_1 = 10, d = -3$; find S_{30}.
48. $a_1 = 15, d = 9$; find S_{28}.
49. $a_5 = -15, d = 6$; find S_{45}.
50. $a_3 = -12, d = -2$; find S_{89}.

♦ Applications

51. A log pyramid has 16 logs on the bottom row, 15 on the next row, and so on, until there is just one log on the top row. How many logs are in the pyramid?

52. A student doing some typing finds that he can type 5 words per minute faster each half hour that he types. If he starts at 35 words per minute at 6:30 P.M., how fast is he typing at 10:00 P.M.?

53. A well-drilling firm charges $2.50 to drill the first foot, $2.65 for the second foot, and so on in arithmetic sequence. At this rate, what would be the cost to drill the last foot of a well 350 feet deep?

54. The developer of a housing development finds that the sale of the first house gives her a $200 profit. Her profit on the second sale is $325, and she finds that each additional house sold increases her profit per house by $125. How many houses did she sell if her profit on the last house sold was $6450 ?

Solve the following arithmetic sequences for the indicated terms.

55. If $a_1 = 1, a_n = 408$, and $S_n = 2454$, find n and d.

56. If $a_1 = 4$ and $S_{13} = 247$, find d.
57. If $d = \frac{1}{2}, n = 81$, and $S_n = 2025$, find a_n.
58. If $n = 14, a_n = 53$, and $S_n = 378$, find d.

Write the following expressions in summation notation.

59. $4 + 7 + 10 + 13$
60. $3 + \frac{5}{2} + 2 + \frac{3}{2} + 1$
61. $-9 - 3 + 3 + 9 + 15 + 21 + 27$
62. $-2 - 8 - 14 - 20 - 26 - 32$

63. Show that if a and b are real numbers and r and s are integers such that $r \le s$, then $\sum_{i=r}^{s}(ai + b)$ is an arithmetic sequence having $s - r + 1$ terms; a is the common difference when $r < s$.

64. Consider the sequence for which $a_n = n^2$ starting with $n = 0$. If $b_n = a_n - a_{n-1}$, show that $b_1, b_2, b_3, \ldots$ is an arithmetic sequence.

65. Prove that an expression for the nth term of an arithmetic sequence defines a linear function, given a_1 and d.

66. Prove that for any three consecutive terms of any arithmetic sequence, the middle term is the average of the other two terms.

67. Given that r, s, t, and u are the first four terms of an arithmetic sequence, show that $r + u - s = t$.

■ 10.5 GEOMETRIC SEQUENCES AND SERIES

Let's proceed to the second special type of sequence that arises most commonly in applications, namely the geometric sequence.

■ Definition 1 Geometric Sequence	A sequence $a_1, a_2, \ldots, a_n$ is called a **geometric sequence** provided that each term is a fixed multiple of its immediate predecessor. That is, there is a fixed real number r such that for each positive integer n, we have

$$a_{n+1} = ra_n$$

The number r is called the **ratio**.

An example of a geometric sequence is

$$\frac{1}{2}, \frac{1}{2^2}, \frac{1}{2^3}, \cdots$$

Here, $r = 1/2$. Indeed, we have $a_n = 1/2^n$, so that

$$a_{n+1} = \frac{1}{2^{n+1}}$$

$$= \frac{1}{2 \cdot 2^n}$$

$$= \frac{1}{2} \cdot \frac{1}{2^n}$$

$$= \frac{1}{2} \cdot a_n$$

The next example provides another illustration of a geometric sequence as well as practice in proving that a sequence is geometric.

> **EXAMPLE 1**
Proving a Sequence Is Geometric

Consider the sequence given by

$$a_n = (-1)^{n-1}\left(\tfrac{4}{3}\right)^n$$

Prove that this is a geometric sequence.

Solution

In looking for a possible value of r, we need to compute the ratio

$$\frac{a_{n+1}}{a_n} = \frac{(-1)^{(n+1)-1}\left(\tfrac{4}{3}\right)^{n+1}}{(-1)^{n-1}\left(\tfrac{4}{3}\right)^n}$$

$$= (-1)\left(\tfrac{4}{3}\right)^{(n+1)-n}$$

$$= -\left(\tfrac{4}{3}\right)^1$$

$$= -\tfrac{4}{3}$$

Because this ratio is a constant (it does not depend on the value of n), the sequence is geometric with ratio $-\tfrac{4}{3}$.

Our development of geometric sequences parallels our discussion of arithmetic sequences of the preceding section, in which we derived a formula for the nth term of an arithmetic sequence in terms of the first term, the index, and the difference. Here is the analogous result for geometric sequences.

---

nTH TERM OF A GEOMETRIC SEQUENCE

Suppose that $a_1, a_2, \ldots, a_n, \ldots$ is a geometric sequence with ratio r. Then

$$a_n = a_1 r^{n-1}$$

Proof Let's proceed by induction. The initial case corresponds to $n = 1$, in which case the result asserts that

$$a_1 = a_1 r^{1-1}$$

$$= a_1 r^0$$

$$= a_1$$

The result is true in case $n = 1$. Let's assume that the result holds for a particular value of n. That is, we assume that

$$a_n = a_1 r^{n-1}$$

We must deduce from this the truth of the result for case $n + 1$. That is, we must prove the induction step, namely that

$$a_{n+1} = a_1 r^{[(n+1)-1]} = a_1 r^n$$

To do so, let's first recall that the sequence is geometric, so that

$$a_{n+1} = r a_n$$

Into this last equation, substitute the assumed formula for a_n. This gives us

$$a_{n+1} = r a_n$$
$$= r \cdot a_1 r^{n-1}$$
$$= a_1 r^n$$
$$= a_1 r^{(n+1)-1}$$

This is the statement of the result for $n + 1$, so the induction step, and thus the general formula is proved. ◆

The next two examples provide applications of the preceding result.

➤ **EXAMPLE 2**
Calculating a Term
of a Geometric Sequence

Suppose that a geometric sequence has first term 3 and ratio $\frac{2}{3}$. Determine the fifth term of the sequence.

Solution

By the preceding result, the nth term a_n of the sequence is given by the formula

$$a_n = a_1 r^{n-1}$$
$$= 3 \cdot \left(\frac{2}{3}\right)^{n-1}$$

In particular, for $n = 5$, this equation yields

$$a_5 = 3 \cdot \left(\frac{2}{3}\right)^{5-1}$$
$$= 3 \cdot \left(\frac{2}{3}\right)^4$$
$$= 3 \cdot \frac{16}{81}$$
$$= \frac{16}{27}$$

➤ **EXAMPLE 3**
Formula for the nth Term

Suppose that $a_1, a_2, \ldots$ is a geometric sequence. Further, suppose that $a_2 = \frac{5}{4}$ and $a_5 = \frac{5}{32}$. Determine a formula for a_n.

Solution

Let r denote the ratio of the sequence. Then we have

$$a_2 = \frac{5}{4}$$
$$= a_1 r^{2-1}$$
$$\frac{5}{4} = a_1 r \tag{1}$$

and

$$a_5 = \frac{5}{32}$$
$$\frac{5}{32} = a_1 r^4 \tag{2}$$

Dividing equation (2) by the corresponding terms of equation (1), we have

$$\frac{\frac{5}{32}}{\frac{5}{4}} = \frac{a_1 r^4}{a_1 r}$$

$$\tfrac{1}{8} = r^3$$

$$r = \tfrac{1}{2}$$

Inserting this value of r into equation (1) gives

$$\tfrac{5}{4} = a_1 r$$

$$= a_1 \cdot \left(\tfrac{1}{2}\right)$$

$$a_1 = \tfrac{5}{2}$$

Inserting the values of a_1 and r into the formula for the nth term of a geometric sequence, we find the desired formula for a_n, namely

$$a_n = a_1 r^{n-1}$$

$$= \tfrac{5}{2} \cdot \left(\tfrac{1}{2}\right)^{n-1}$$

➤ **EXAMPLE 4**
Compound Interest

Suppose that a bank account earns compound interest at a rate of i per period (year, month, day). Further, suppose that at the start of period 1, the account has balance P. Then the balance a_n in the account at the end of period n is given by the formula

$$a_n = P(1 + i)^n$$

1. Show that the sequence $a_0, a_1, a_2, \ldots, a_n$ is a geometric sequence with ratio $1 + i$.

2. Suppose that \$400 is deposited in a bank account paying 8 percent interest compounded monthly. Calculate the amount in the account at the end of 5 years.

Solution

1. Note that the first index of this sequence is $n = 0$. In dealing with financial calculations, it is usually most convenient to begin sequences with the 0th term rather than the first. This allows the sequence to start its description with time 0 (the beginning of the first time period). Note that if n is any nonnegative integer, then

$$a_{n+1} = P(1 + i)^{n+1}$$
$$= P(1 + i)^n \cdot (1 + i)$$
$$= (1 + i)a_n$$

Thus, the sequence is geometric with ratio $1 + i$.

2. The 8 percent interest refers to an annual rate. Because the interest is compounded monthly, the length of a single financial period is a month, and the amount of interest for one month is $i = \frac{0.08}{12} \approx 0.00667$. Five years

corresponds to 60 months, that is, 60 financial periods, so the balance in the account at the end of five years is

$$a_{60} \approx 400 \, (1 + 0.00667)^{60}$$
$$\approx \$595.94$$

(We have used a scientific calculator to calculate the numerical value of the expression.)

Let's turn to the problem of calculating the partial sum S_n of a geometric sequence. We have the following formula:

PARTIAL SUM OF A GEOMETRIC SEQUENCE

Suppose that $a_1, a_2, \ldots, a_n, \ldots$ is a geometric sequence and that S_n is its nth partial sum. Then

$$S_n = a_1 \frac{1 - r^n}{1 - r} \quad (r \neq 1) \qquad (3)$$

$$= na_1 \quad (r = 1) \qquad (4)$$

Proof   We have

$$S_n = a_1 + a_2 + a_3 + \cdots + a_n$$
$$= a_1 + a_1 r + a_1 r^2 + \cdots + a_1 r^{n-1}$$
$$= a_1(1 + r + r^2 + \cdots + r^{n-1})$$

In case $r = 1$, each term within the parentheses on the right is equal to 1, and the formula for S_n reads

$$S_n = a_1(1 + 1 + 1 + \cdots + 1) = na_1$$

Let's assume that $r \neq 1$. We may now write the right side in the form

$$S_n = a_1(1 + r + r^2 + \cdots + r^{n-1})$$

$$= a_1(1 + r + r^2 + \cdots + r^{n-1})\frac{1 - r}{1 - r}$$

$$= a_1 \frac{1 - r^n}{1 - r}$$

This last formula is the result claimed in equation (3) and completes the proof.   ◆

➤ **EXAMPLE 5**
Sum of a Geometric
Sequence

Determine the value of the following sum:

$$3 + 3\left(\tfrac{3}{4}\right) + 3\left(\tfrac{3}{4}\right)^2 + \cdots + 3\left(\tfrac{3}{4}\right)^9$$

Solution

The sum is the partial sum S_{10} of a geometric series with $a_1 = 3$, $r = \tfrac{3}{4}$. Therefore, by equation (3), we have

$$S_{10} = a_1 \frac{1 - r^{10}}{1 - r}$$

$$= 3 \cdot \frac{1 - \left(\frac{3}{4}\right)^{10}}{1 - \frac{3}{4}}$$

$$= 12 \cdot \left[1 - \left(\frac{3}{4}\right)^{10}\right] = 11.324$$

➤ **EXAMPLE 6**
Value of a Pension Fund

Each year for 20 years, a woman deposits $5000 into a pension fund. The money earns 9 percent compounded annually. How much money is in the pension fund after 20 years?

Solution

There are 20 deposits to the pension fund, each earning interest for a different length of time. The last deposit earns interest for 0 years, the next-to-last deposit for 1 year, and the first deposit for 19 years. Let a_1 denote the amount of money generated by the last deposit, a_2 the amount of money generated by the next-to-last, and a_{20} the amount generated by the first deposit. The amount of money represented by a_n earns interest for $n - 1$ years at 9 percent compounded annually. By our previous discussion, we have

$$a_n = 5000(1 + 0.09)^{n-1}$$
$$= 5000 \cdot 1.09^{n-1}$$

The amount in the pension fund at the end of 20 years equals the sum of the amounts attributable to each of the deposits. This equals

$$a_1 + a_2 + a_3 + \cdots + a_{20}$$

This sum is the partial sum S_{20} of a geometric series with $a_1 = 5000$, $r = 1.09$. By equation (3), this sum equals

$$a_1 \frac{1 - r^n}{1 - r} = 5000 \cdot \frac{1 - 1.09^{20}}{1 - 1.09}$$

$$= -\frac{5000}{0.09}(1 - 1.09^{20})$$

$$= \$255,800.60$$

That is, the pension fund balance after 20 years equals $255,800.60.

Now let us consider the infinite geometric sequence

$$1, \frac{1}{2}, \frac{1}{2^2}, \frac{1}{2^3}, \cdots$$

Here are its first few partial sums:

$$S_1 = 1$$
$$S_2 = \frac{3}{2}$$
$$S_3 = \frac{7}{4}$$
$$S_4 = \frac{15}{8}$$
$$S_5 = \frac{31}{16}$$

As more and more terms of the sequence are included, the partial sums approach the value 2. The nth partial sum may be obtained as the sum of a geometric progression.

$$1 + \frac{1}{2} + \frac{1}{2^2} + \cdots + \frac{1}{2^{n-1}} = \frac{1 - \frac{1}{2^n}}{1 - \frac{1}{2}}$$

As n increases, $(\frac{1}{2})^n$ approaches 0, so that the expression on the right approaches

$$\frac{1}{1 - \frac{1}{2}} = 2$$

Similar reasoning works to obtain the sum of any infinite geometric series with ratio r for which $|r| < 1$. Namely we have:

SUM OF AN INFINITE GEOMETRIC SEQUENCE

Suppose that $|r| < 1$. The partial sums of the infinite geometric sequence a, ar, ar^2, ..., ar^n approach the limiting value $a/(1 - r)$. This fact is expressed using the notation

$$\sum_{n=0}^{\infty} ar^n = \frac{a}{1 - r}$$

➤ **EXAMPLE 7**

Calculating the Sum of an Infinite Geometric Series

Calculate the sum of the infinite geometric sequence

$$3, -3 \cdot \frac{1}{5}, 3 \cdot \left(\frac{1}{5}\right)^2, -3 \cdot \left(\frac{1}{5}\right)^3, \ldots$$

Solution

The geometric sequence has $a_1 = 3$, $r = -\frac{1}{5}$. Therefore, by the formula for the sum of an infinite geometric sequence, the partial sums of the sequence approach a limiting value, and this value equals

$$\frac{a}{1 - r} = \frac{3}{1 - \left(-\frac{1}{5}\right)}$$

$$= \frac{3}{\frac{6}{5}}$$

$$= \frac{5}{2}$$

Note that if $|r| \geq 1$, then the partial sums of the geometric series do not approach a limiting value.

Infinite sums are discussed in detail as part of a calculus course. The preceding discussion provides a taste of what you will learn later.

Exercises 10.5

Prove that the following are geometric sequences.

1. $a_n = 3\left(-\frac{1}{3}\right)^{n-1}$

2. $a_n = 24\left(\frac{1}{2}\right)^{n-1}$

3. $a_n = 5(4)^{n-1}$

4. $a_n = 2^{n-1}$

5. $a_n = 3^{-n}$

6. $a_n = \left(-\frac{1}{2}\right)^{n+1}$

Find the indicated term for each of the following geometric sequences.

7. $a_1 = \frac{1}{2}$, $r = \frac{1}{2}$; find a_{10}.

8. $a_1 = \frac{2}{3}$, $r = \frac{2}{3}$; find a_9.

9. $a_1 = 1$, $r = -\sqrt{2}$; find a_7.

10. $a_1 = 81$, $r = -\frac{1}{3}$; find a_{12}.

11. 54, 36, 24, ...; find a_8.

12. 32, 16, 8, ...; find a_{15}.

13. If $a_2 = 3$ and $r = 2$, find a_{13}.

14. If $a_4 = 64$ and $r = -4$, find a_{10}.

15. If $a_4 = 6$ and $a_5 = 12$, find a_{14}.

16. If $a_6 = 9$ and $a_7 = 3$, find a_2.

17. 300, -30, 3, ...; find a_4.

18. $-\sqrt{5}$, 5, $-5\sqrt{5}$, ...; find a_{20}.

19. 2, $\frac{2}{3}$, $\frac{2}{9}$, ...; find a_5.

20. -162, 54, -18, 6, ...; find a_8.

♦ **Applications**

21. On a merit salary plan, yearly raises are determined as a percent of the present salary. If a worker begins with a \$15,000 yearly salary and receives merit increases of 8.5 percent for nine years, what will be her yearly salary for her ninth year?

22. In place of a \$1 million salary, a baseball pitcher agreed to accept \$10 for his first win, \$20 for his second,

\$40 for his third, and so on. How many games must he win before his earnings exceed the \$1 million salary? What would his salary be if he were to win one more game?

23. If a new printing press cost \$50,000 and had a yearly depreciation of 25 percent of its value at the beginning of the year, what is its value after five years?

24. Highway Appliance Stores have CD players that can be purchased on a daily installment plan. You pay 1 cent the first day, 3 cents the second day, 9 cents the third day, and so on for 10 days. How much will the CD player cost?

25. A girl states that one of her 10th-generation ancestors was on the Mayflower. How many descendants does this ancestor have in the 10th generation, assuming each descendent has exactly two children?

26. If the interest on a 10-year certificate of deposit was set at 9 percent compounded semiannually, what would be the value after 10 years of an investment of \$8000?

27. The city of Akron, Ohio, lost $\frac{1}{100}$ of its population each year from 1972 through 1980. If the population in 1972 was 350,000, what was the population at the end of 1980?

28. If the federal government were to put an extra \$1 billion into the economy such that each business and individual saves 25 percent of its income and spends the rest, so that of the \$1 billion, 75 percent is respent by individuals and businesses, then 75 percent of this amount is respent, and so forth, what is the total increase in spending due to the government action?

29. Which term of the geometric sequence 2, -6, 18, ... is 162?

30. If $a_7 = 256$ and $a_1 = 4$, find a_5.

31. If $r = 3$ and $a_5 = 324$, find a_1.

32. Which term of the geometric sequence -243, 81, -27, ... is $\frac{1}{9}$?

33. Fill in the blanks for this geometric sequence:
_____, -5, _____, _____, _____, -405.

34. Fill in the blanks for this geometric sequence:
_____, _____, 1, _____, _____, 729.

In the following geometric sequences, three of the five real numbers a_1, a_n, r, n, and S_n are given. Find the other two.

35. $a_1 = 5$, $r = -2$, $S_n = -25$

36. $a_1 = 5$, $a_n = 320$, $r = 2$

37. $a_1 = 4$, $a_n = 324$, $r = 3$

38. $n = 9$, $r = 2$, $S_n = 1022$

39. $n = 5$, $r = -3$, $S_n = 244$

40. $a_1 = 10$, $r = 3$, $n = 5$

41. $a_n = 384, n = 7, r = 2$
42. $a_n = 729, n = 7, r = \frac{3}{2}$
43. $a_1 = 162, r = -\frac{1}{3}, n = 8$
44. $a_1 = 4, r = -\frac{3}{2}, n = 6$
45. $a_n = 9, r = -\sqrt{3}, S_n = 4 - 4\sqrt{3}$
46. $a_n = -625\sqrt{5}, r = -\sqrt{5}, S_n = 780 - 781\sqrt{5}$

Find a formula for the following geometric sequences.

47. $a_2 = -4, a_5 = \frac{4}{27}$ 48. $a_3 = 6, a_6 = 18\sqrt{3}$
49. $a_3 = 50, a_6 = -6550$ 50. $a_2 = 3, a_5 = 54$

♦ Applications

51. To begin a new business, a woman must borrow $80,000 at 12 percent interest compounded monthly. If the loan is to be paid off in one lump sum at the end of two years, what will be the amount of the payment?

52. To buy a used car, you borrow $1000 at 10 percent compounded monthly. You also elect to pay off this loan in one lump-sum payment at the end of four years. What will be the amount of your payment?

53. For retirement, you elect to deposit $1000 a year in a savings account that pays 11 percent compounded annually. How much will be in your account at the end of 40 years?

54. A couple with a 1-year-old decides to put $500 a year into a savings account paying 12 percent compounded annually for their son's college money. At the end of 18 years, what will be the amount in this account?

Solve the following infinite sequence problems, if possible.

55. Find the sum of $1 - \frac{1}{2} + \frac{1}{4} - \frac{1}{8} + \frac{1}{16} - \cdots$.
56. Find the sum of $1 - \frac{1}{5} + \frac{1}{25} - \frac{1}{125} + \cdots$.
57. Find the sum of $-\frac{5}{3} - \frac{10}{9} - \frac{20}{27} - \frac{40}{81} \cdots$.
58. Find the sum of $2 + \frac{2}{3} + \frac{2}{9} + \frac{2}{27} + \frac{2}{81} + \cdots$.
59. Find the sum of $9/10 + 9/100 + 9/1000 + 9/10,000 + \cdots$.
60. Find the sum of $\frac{1}{2} + \frac{1}{3} + \frac{2}{9} + \frac{4}{27} + \cdots$.
61. Find the sum of $1 + 2 + 4 + 8 + \cdots$.
62. Find the sum of $-1 + 1 - 1 + 1 - 1 + \cdots$.
63. If $s = 14$ and $a_1 = 21$, then $r = ?$
64. If $r = \frac{2}{3}$ and $s = 18$, then $a_1 = ?$
65. If $r = -\frac{1}{2}$ and $s = 20$, then $a_1 = ?$

66. If $s = 35$ and $a_1 = 7$, then $r = ?$
67. If $a_1 = 4$ and $r = \frac{1}{2}$, then $s = ?$
68. If $a_1 = \frac{1}{3}$ and $s = \frac{2}{3}$, then $r = ?$
69. A superball is dropped 40 feet and rebounds on each bounce $\frac{2}{5}$ of the height from which it fell. How far will it travel before coming to rest?
70. If a ball is dropped from a height of 12 meters and rebounds $\frac{1}{3}$ of the distance it fell on each bounce, how far will it travel before coming to rest?
71. Here is a diagram of steps that might be taken in getting some grapes from the grower to you:

    Grower → Trucker → Regional market → Trucker → Wholesaler → Trucker → Retailer → You

    If each person or organization in this chain makes a 25 percent profit, if all grapes are sold, and if the grower sells her grapes for $0.20 per pound, how much would you have to pay per pound?
72. If a dentist earns $40,000 during his first year of practice, and if each succeeding year his salary increases 10 percent, what would be the total of the dentist's salary over the first nine years?
73. In Exercise 71, if the grower receives $10,000 for the grapes she sold to the first trucker, then what is the sum of all the profits made once you buy these grapes?
74. In Exercise 72, how many years would the dentist have to work if the salary total is to exceed $1 million?
75. Show that there is no infinite series for which $a_1 = 2$ and $s = 1$.

✍ In Your Own Words

76. Can the sum of an infinite geometric series be less than the first term? Justify your answer.
77. Prove that every rational number a/b can be expressed as a repeating decimal.
78. Prove that every repeating decimal represents a rational number.
79. True or false: Suppose that to proceed to the door of the room, you walk half the distance to the door, then stop, then move half the remaining distance to the door, then stop, and so forth. You never cover the total distance to the door and hence can never leave the room. Explain your answer.
80. Explain the differences between arithmetic and geometric sequences.

▍ 10.6 PERMUTATIONS AND COMBINATIONS

Many applied problems require you to count the number of possible occurrences of a particular type. Here are some examples:

1. A communications network consists of 500 cables from point A to point B and 300 cables from point B to point C. How many ways are there to choose a transmission path from point A to point C?

2. A state lottery requires you to pick a sequence of five digits. How many different outcomes of the lottery are possible?

3. A certain genetic combination consists of eight traits, for each of which there are four possibilities. How many genetic combinations are possible?

The field of mathematics that deals with counting problems is called **combinatorics**. In this section, we provide a brief introduction to combinatorics and consider a few of the simplest types of counting problems.

Elementary Counting Problems

One of the basic techniques for solving counting problems is to break the problem down into a sequence of independent choices, each of whose possibilities can be counted. In counting such sequences, the following basic result is used:

FUNDAMENTAL PRINCIPLE OF COUNTING

Suppose that tasks $A_1, A_2, \ldots, A_n$ are independent of one another. Further, suppose that it is possible to perform task A_1 in m_1 ways, task A_2 in m_2 ways, ..., and task A_n in m_n ways. Then the number of possible ways to perform the sequence of tasks $A_1, A_2, \ldots, A_n$, is equal to $m_1 \cdot m_2 \cdots \cdots m_n$.

We will assume the truth of the Fundamental Principle of Counting. The next examples illustrate how it may be applied in solving counting problems.

➢ EXAMPLE 1
Counting License Plates

Car license plates in a certain state consist of three letters followed by 3 digits. How many different license plates are possible?

Solution

The process of selecting a license plate may be viewed as a sequence of six tasks, corresponding to the selection of the three letters and three digits. Each of the first three tasks may be accomplished in 26 different ways. Each of the last three tasks may be accomplished in 10 different ways. By the fundamental principle of counting, the sequence of six tasks may then be accomplished in

$$26 \cdot 26 \cdot 26 \cdot 10 \cdot 10 \cdot 10 = 17,576,000$$

different ways. That is, there are 17,576,000 different possible license plates.

➢ EXAMPLE 2
Hexadecimal Numbers

In the *hexadecimal number system*, used extensively in computer work, there are 16 digits, namely, 0–9 and A–F. A typical hexadecimal number consists of a sequence of hexadecimal digits. How many four-digit hexadecimal numbers are there whose first digit is not 0 or 1 ?

Solution

The specified four-digit hexadecimal number may be viewed as a sequence of four independent tasks, corresponding to choosing each of the four letters. The first task may be chosen in 14 ways (the hexadecimal digit can't be 0 or 1). Each of the remaining tasks may be accomplished in 16 different ways. The sequence of tasks may be accomplished in

$$14 \cdot 16 \cdot 16 \cdot 16 = 57,344$$

ways. That is, there are 57,344 such hexadecimal numbers.

Permutations

One of the most common types of counting problems involves permutations, which are defined as follows:

Definition 1 **Permutation**	Suppose that we are given n different objects. A selection of r of them arranged in a particular order is called a **permutation of n objects taken r at a time.**

For instance, suppose that $n = 3$ and that the objects are the three letters A, B, and C. Then there are six permutations of these objects taken two at a time, namely, AB, BA, AC, CA, BC, and CB.

The number of permutations of n objects taken r at time is denoted $_nP_r$. From the previous example, we see that $_3P_2 = 6$. We may compute $_nP_r$ from n and r using the following result:

FUNDAMENTAL THEOREM OF PERMUTATIONS
Suppose that n and r are positive integers with $r \leq n$. Then $$_nP_r = n(n-1)(n-2)\cdots(n-r+1) \quad \text{(a total of } r \text{ factors)}$$

Proof   Let us consider the task of choosing a permutation of n objects taken r at a time. This can be viewed as a sequence of r independent tasks. Select the first object in the permutation; this may be done in n ways. Next, select the second object in the permutation; because one object has already been used, this second task may be done in $n - 1$ ways. Next, select the third object in the permutation; because two objects have already been used, the second task may be done in $n - 2$ ways, and so on. The final task is selecting the rth object. The preceding tasks have used up $r - 1$ objects, so the final task may be done in $n - (r - 1) = n - r + 1$ ways. By the fundamental theorem of counting, the sequence of r tasks may be accomplished in

$$n(n-1)(n-2)\cdots(n-r+1)$$

ways. This proves the desired result.   ♦

Here is a useful formula that follows from the Fundamental Theorem of Permutations:

$$
\begin{aligned}
_nP_r &= n(n-1)(n-2)\cdots(n-r+1) \\
&= \frac{n(n-1)(n-2)\cdots(n-r+1)(n-r)(n-r-1)\cdots 2\cdot 1}{(n-r)(n-r-1)\cdots 2\cdot 1} \\
&= \frac{n!}{(n-r)!}
\end{aligned}
$$

➢ **EXAMPLE 3**
Calculating Numbers
of Permutations

Determine the values of the following:

1. $_5P_3$ 2. $_{10}P_4$ 3. $_nP_3$

Solution

1. We have

$$_5P_3 = 5\cdot 4\cdot 3 = 60$$

2. We have
$$_{10}P_4 = 10 \cdot 9 \cdot 8 \cdot 7 = 5040$$

3. The second subscript, 3, determines the number of factors. So we have
$$_{n}P_3 = n(n-1)(n-2)$$

➢ **EXAMPLE 4**
Assignment of Tasks

There are 30 students in a certain class. In how many ways can the instructor assign different students to perform four tasks, labeled *A–D*?

Solution

Each assignment of students to tasks may be viewed as a permutation of 30 objects (students) taken 4 at a time. The ordering of the permutation creates the assignment to the tasks: The first student is assigned to task *A*, the second student to task *B*, and so forth. The number of ways of assigning the students is $_{30}P_4$. The numerical value of this symbol is
$$_{30}P_4 = 30 \cdot 29 \cdot 28 \cdot 27 = 657{,}720$$

That is, there are 657,720 ways to assign 30 students to four tasks.

➢ **EXAMPLE 5**
Communications
Engineering

In a communications network, there are 13 cables connecting points 1 and 2, and there are five messages, labeled *A–E*, to be sent from point 1 to point 2. In how many ways can the cables for the transmission be selected if each cable can carry a single message?

Solution

Each assignment of messages to cables is a permutation of 13 objects (the cables) taken 5 at a time. The number of such permutations is equal to
$$_{13}P_5 = 13 \cdot 12 \cdot 11 \cdot 10 \cdot 9 = 154{,}440$$

The messages may be sent in 154,440 different ways.

Combinations

Let's consider a second common type of counting problem, namely one involving combinations, defined as follows:

Definition 2 **Combination**	Suppose that we are given *n* distinct objects. A selection of *r* of them (without regard to order) is called a **combination of *n* objects taken *r* at a time**.

Note that a combination differs from a permutation in that it does not take into account the order of arrangement. For instance, we previously saw that the permutations of the three letters *A*, *B*, and *C* taken two at a time are

$$AB, \ BA, \ AC, \ CA, \ BC, \ CB$$

However, with combinations of these same three letters taken two at a time, order does not count. For instance, the two permutations *AB* and *BA* are regarded as the same combination. There are only three combinations in this example, namely:

$$AB, \ AC, \ BC$$

The number of combinations of *n* objects taken *r* at a time is denoted $_{n}C_r$. This number may be calculated using the following result:

> ### FUNDAMENTAL THEOREM OF COMBINATIONS
>
> The number of combinations of n objects taken r at a time is equal to
>
> $$_nC_r = \frac{n!}{r!(n-r)!}$$

Proof Let's begin by forming all permutations of the n objects taken r at a time. The number of such permutations is $_nP_r$. Each permutation corresponds to a combination. However, permutations that are obtained by rearranging the same objects in differing order correspond to the same combination. How many such rearrangements are there? We can choose the first rearrangement in r ways, the second in $r-1$ ways, the third in $r-2$ ways, and so forth. By the fundamental principle of counting, we see that each permutation may be rearranged in

$$r(r-1)(r-2)\cdots\cdot 2\cdot 1 = r!$$

ways, so each combination is counted $r!$ times. That is, we have

$$_nP_r = r!\,_nC_r$$

$$_nC_r = \frac{_nP_r}{r!}$$

$$= \frac{n(n-1)\cdots(n-r+1)}{r!} \qquad \textit{(Fundamental theorem of permutations)}$$

$$= \frac{n(n-1)\cdots(n-r+1)(n-r)(n-r-1)\cdots\cdot 1}{r!(n-r)(n-r-1)\cdots\cdot 1}$$

$$= \frac{n!}{r!(n-r)!}$$

This proves the theorem.   ◆

Note that the right side of the formula in the theorem is just a binomial coefficient. That is, another way of stating the fundamental theorem of combinations is

$$_nC_r = \binom{n}{r}$$

Let's use the fundamental theorem of combinations to solve some applied counting problems.

➤ **EXAMPLE 6**
Calculating Number
of Combinations

Calculate the number of combinations of 10 objects taken 4 at a time.

Solution

The specified number equals $_{10}C_4$. By the fundamental theorem of combinations, this symbol has the value

$$_{10}C_4 = \frac{10!}{4!(10-4)!}$$

$$= \frac{10\cdot 9\cdot 8\cdot 7\cdot 6\cdot 5\cdot 4\cdot 3\cdot 2\cdot 1}{(4\cdot 3\cdot 2\cdot 1)\cdot(6\cdot 5\cdot 4\cdot 3\cdot 2\cdot 1)}$$

$$= \frac{10 \cdot 9 \cdot 8 \cdot 7}{4 \cdot 3 \cdot 2 \cdot 1}$$

$$= 210$$

Note that in the computation of the preceding example, we performed cancellation between the denominator and numerator in order to ease the burden of working with large numbers. Actually, we may perform such cancelation generally in working with $_nC_r$, as the following result shows:

FORMULA FOR COMBINATIONS

Let n and r be positive numbers with $r \leq n$. Then

$$_nC_r = \frac{n(n-1)\cdots(n-r+1)}{r!}$$

Proof   Start from the preceding formula for $_nC_r$:

$$_nC_r = \frac{n!}{r!(n-r)!}$$

$$= \frac{n(n-1)(n-2)\cdots(n-r+1)(n-r)(n-r-1)\cdots 2 \cdot 1}{r!(n-r)(n-r-1)\cdots 2 \cdot 1}$$

$$= \frac{n(n-1)(n-2)\cdots(n-r+1)}{r!}$$

This proves the desired formula.   ♦

➤ **EXAMPLE 7**
Communications
Engineering

In a communications network, there are 13 cables connecting points 1 and 2. In how many ways can 4 cables be selected to send messages from point 1 to point 2?

Solution

Each selection of cables is a combination of 13 objects (the cables) taken 4 at a time. The number of such combinations equals

$$_{13}C_4 = \frac{13 \cdot 12 \cdot 11 \cdot 10}{4!}$$

$$= 715$$

➤ **EXAMPLE 8**
Baseball

A baseball team has 14 players. How many different 9-player teams (disregarding order) are possible?

Solution

The number of such teams equals the number of combinations of 14 objects taken 9 at a time. This number equals

$$_{14}C_9 = \frac{14 \cdot 13 \cdot 12 \cdot 11 \cdot 10 \cdot 9 \cdot 8 \cdot 7 \cdot 6}{9 \cdot 8 \cdot 7 \cdot 6 \cdot 5 \cdot 4 \cdot 3 \cdot 2 \cdot 1}$$

$$= \frac{14 \cdot 13 \cdot 12 \cdot 11 \cdot 10}{5 \cdot 4 \cdot 3 \cdot 2 \cdot 1}$$

$$= 2002$$

That is, the team may be selected in 2002 ways.

➤ EXAMPLE 9
Poker

A standard deck of playing cards contains 52 cards. A poker hand contains 5 cards. How many different poker hands are possible?

Solution

A poker hand is a combination of 52 objects taken 5 at a time. The number of such objects equals

$$_{52}C_5 = \frac{52 \cdot 51 \cdot 50 \cdot 49 \cdot 48}{5!} = 2,598,960$$

That is, there are 2,598,960 possible poker hands.

Exercises 10.6

1. How many numbers consisting of one, two, or three digits, without repetitions, can you form using the digits 1, 2, 3, 4, 5, and 6?

2. How many truck license plates may be made if each plate will have three numbers followed by two letters followed by one number?

3. How many five-digit numbers begin with an even digit and end with an even digit?

4. In how many different orders can you arrange four books on a shelf?

5. How many possibilities are there for a license plate with three letters and two numbers?

6. If a student has six different slacks and 10 different shirts, how many different slacks-shirt outfits are possible?

7. If six people run a race, in how many different orders can they all finish if there are no ties?

8. For a nine-person baseball team, how many different batting orders are possible?

9. Marv's Pizza Palace offers pepperoni, onions, sausage, mushrooms, and anchovies as toppings for its pizza. How many different pizzas can be made?

10. A club consists of 30 men and 70 women. In how many ways can a president, vice-president, secretary, and sergeant-at-arms be chosen if the president must be a woman, the sergeant-at-arms must be a man, and a person cannot hold more than one office?

Determine the following.

11. $_4P_4$ 12. $_7P_4$ 13. $_6P_2$ 14. $_9P_P$

15. $_3P_1$ 16. $_9P_6$ 17. $_nP_5$ 18. $_nP_4$

19. $_7P_3$ 20. $_6P_1$

♦ Applications

Use permutations to solve the following.

21. How many "words" can be formed from the letters of the word *fragments*, taken four at a time? (Any four-letter group is considered a word.)

22. A Navy signal officer has six different flags. How many different signals can be sent by flying three flags at a time, one above the other, on a flagpole?

23. On leaving a train, a man finds that he has a nickel, a dime, a quarter, and a half-dollar in his pocket. In how many ways can he tip the porter?

24. In how many ways can three books be chosen from a group of seven different books and arranged in three spaces on a bookshelf?

25. How many permutations are there of five cards taken from a bridge deck of 52 cards? (Order is to be considered.)

26. How many arrangements can be formed from the letters of the word *hyperbola,* with all nine letters taken at a time?

27. A concert is to consist of three songs and two violin selections. In how many ways can the program be arranged so that the concert begins and ends with a song, and neither violin selection follows immediately after the other?

28. For Exercise 26, in how many of these words will the letters *h* and *y* occur next to each other?

Determine the following.

29. $_8C_3$ 30. $_9C_2$ 31. $_5C_4$ 32. $_6C_1$

33. $_{10}C_2$ 34. $_{12}C_8$ 35. $_nC_{n-1}$ 36. $_nC_4$

37. $_4C_0$ 38. $_nC_n$

♦ Applications

39. If five blue, three yellow, two black, and two white balls are to be arranged in a row, find the number of possible color arrangements for the balls.

40. In how many ways can a hand of 13 cards be selected from a standard bridge deck of 52 cards?

41. In how many ways can a committee of three be chosen from four married couples if all are equally eligible?

42. In how many ways can a selection of one or more books be made from five distinct algebra books and four distinct geometry books?

43. In Exercise 41, what would be the number if the committee must consist of two women and one man?

44. A contractor needs 4 carpenters, and 10 apply for the job. In how many ways can he pick the four that he needs?

45. Your math professor has five basic algebra books, four advanced algebra books, and eight geometry books. In how many ways can they be arranged on her shelf if books in the same category are kept next to each other?

46. How many lines are determined by 10 points, no 3 of which are collinear?

47. From a group of six persons, how many committees of three can be formed if two of the six people cannot be on the same committee?

48. In how many ways can you choose a committee of 5 girls and 7 boys from a group of 10 girls and 11 boys?

49. A math history club has a membership of four women and six men. A research committee of three is formed. In how many ways can this be done: (*a*) If there are to be two women and one man on the committee? (*b*) If there is to be at least one woman on the committee? (*c*) If all three are to be of the same sex?

50. Using three letters of the word *example* and two letters of *boot,* how many arrangements of five different letters are possible?

51. How many five-letter words, each consisting of three consonants and two vowels, can be formed from the letters of the word *equations*?

52. In how many ways can a coach choose a team of 5 from among 10 girls: (*a*) If 2 specified girls must be included? (*b*) If there are no restrictions?

53. In how many ways can 15 different objects be divided among *A*, *B*, and *C* if *A* must receive 2 objects, *B* must receive 3 objects, and *C* must receive 10 objects?

54. How many diagonals does a 20-sided polygon have? How many sides does a polygon with 35 diagonals have?

55. A student leadership of seven wishes to send a two-person committee to the dean to discuss new dorm policies. (*a*) How many different committees are possible? (*b*) If there are two women and five men student leaders, how many of the committees would include a woman? (*c*) If the leadership president must be on the committee, how many different committees are possible?

56. Prove $_{n+1}C_r = \,_nC_{r-1} + \,_nC_r$.

57. Prove $_nC_r = \,_nC_{n-r}$.

Solve the following for *n*.

58. $_nC_2 = \,_{100}C_{98}$ 59. $_nC_5 = \,_nC_3$

60. $_nP_4 = 4(_nP_3)$ 61. $_nP_3 = (_{n-1}P_2)$

62. $_nP_4 = 8(_{n-1}P_3)$

Prove the following.

63. $_nP_4 - \,_nP_3 = (n-4)(_nP_3)$

64. $_5P_r = 5(_4P_{r-1})$

✍ In Your Own Words

65. Describe the differences between permutations and combinations.

66. In what ways are permutations and combinations similar?

■ 10.7 PROBABILITY

The theory of probability is a branch of mathematics that deals with processes exhibiting uncertainty. Examples of such processes are throwing a pair of dice, playing a hand of blackjack, or observing the size of leaves on a tree. Probability theory developed in attempts to give a mathematical theory explaining games of chance. Initial attempts go back to the works of Cardano (16th century), Pascal and Fermat (17th century), and Bermoulli and Laplace (18th century). In the past two centuries, probability theory has developed into a major discipline that supports applications in biology, physics, psychology, and business, to mention only a few! In this section, we will provide a very brief taste of what probability theory is about. Our discussion will give us a chance to use the counting techniques we learned in the preceding section.

Laplace

Sample Spaces and Events

Any process whose result is subject to chance is called an **experiment**. The result of an experiment is called an **outcome**. One example of an experiment consists of flipping a coin and observing the outcome, heads or tails. A second experiment consists of rolling a die and observing the number that appears (the possibilities are the digits 1–6). A third experiment is to observe the height of a three-year-old. The outcomes in this case are positive real numbers.

The third example of an experiment differs from the first two because it has an infinite number of possible outcomes (positive real numbers), whereas the first has only two possible outcomes and the second has only six possible outcomes. In this section, we will deal only with experiments that have a finite number of outcomes, such as the first two. Experiments with an infinite number of outcomes require more advanced mathematics to deal with and are beyond the scope of this book.

We will make a further simplifying assumption by dealing only with experiments in which all outcomes are equally likely. For instance, the first and second experiments just mentioned satisfy this condition: In tossing a fair coin, heads and tails are equally likely to occur; and in rolling a fair die, the six possible outcomes are equally likely to occur. However, in selecting a ball from an urn containing 100 white balls and 1 black ball, the possible outcomes are not equally likely.

Let's use sets to describe experiments in mathematical terms.

Definition 1 **Sample Space**	The sample space of an experiment is the set that consists of all possible outcomes.

For example, the experiment of tossing a coin has the sample space
$$\{H,\ T\}$$
where H indicates heads and T indicates tails. The experiment of rolling a die consists of the sample space
$$\{1, 2, 3, 4, 5, 6\}$$
where the digits are the results of rolling the die.

➤ EXAMPLE 1
Determining a Sample Space

Suppose an experiment consists of flipping a coin three times. Describe the associated sample space.

Solution

Each outcome of the experiment consists of a sequence of three letters indicating, in order, the results of the three coin tosses. There are eight possible outcomes, as follows:

Sample Space = $\{HHH, \ HHT, \ HTH, \ HTT, \ THH, \ THT, \ TTH, \ TTT\}$

When performing an experiment, it is often necessary to determine whether or not a particular event occurs. For instance, in rolling a die, you might wish to know whether the number rolled is even. In this case, the event is "even number occurs." In terms of the sample space, this event is equivalent to the rolled number occurring in the subset $\{2, 4, 6\}$ of the sample space. In a similar fashion, we may define an event in terms of the sample space as follows:

Definition 2 **Event**	Suppose that we are given an experiment and its associated sample space. An **event** is a subset of the sample space.

➢ **EXAMPLE 2**
Exhibiting Events

Suppose that an experiment consists of rolling two dice.

1. Describe the sample space.
2. Describe the event "the dice add up to 7."

Solution

1. Each outcome of the experiment consists of an ordered pair of numbers (n, m), where n and m are integers from 1 to 6. The sample space consists of the following set of 36 ordered pairs:

$$\{(1, 1), \quad (1, 2), \quad (1, 3), \quad (1, 4), \quad (1, 5), \quad (1, 6)$$
$$(2, 1), \quad (2, 2), \quad (2, 3), \quad (2, 4), \quad (2, 5), \quad (2, 6)$$
$$(3, 1), \quad (3, 2), \quad (3, 3), \quad (3, 4), \quad (3, 5), \quad (3, 6)$$
$$(4, 1), \quad (4, 2), \quad (4, 3), \quad (4, 4), \quad (4, 5), \quad (4, 6)$$
$$(5, 1), \quad (5, 2), \quad (5, 3), \quad (5, 4), \quad (5, 5), \quad (5, 6)$$
$$(6, 1), \quad (6, 2), \quad (6, 3), \quad (6, 4), \quad (6, 5), \quad (6, 6)\}$$

2. The event "dice add up to 7" consists of those ordered pairs of the sample space in which the two entries sum to equal 7. Here is the subset E corresponding to this event:

$$E = \big\{(1, 6), (2, 5), (3, 4), (4, 3), (5, 2), (6, 1)\big\}$$

Assigning Probabilities to Events

Suppose that we are given a sample space S and an associated event E. We assign to the event a number, called its probability, as follows:

Definition 3 **Probability of an Event**	Suppose that $n(S)$ denotes the number of outcomes in the sample space S and that $n(E)$ denotes the number of outcomes in the event E. Then the **probability of E** is defined as the number $$p(E) = \frac{n(E)}{n(S)}$$

For instance, consider the sample space $S = \{H, T\}$ for the experiment of flipping a coin. The probability of the event $E = \{H\}$ (heads) is then

$$p(E) = \frac{n(\{H\})}{n(\{H, T\})} = \frac{1}{2}$$

➤ EXAMPLE 3
Dice

Suppose that you roll two dice. What is the probability that the dice add up to 7?

Solution

We have previously seen that the sample space for the experiment of throwing two dice is given by

$$
\begin{array}{cccccc}
\{(1, 1), & (1, 2), & (1, 3), & (1, 4), & (1, 5), & (1, 6) \\
(2, 1), & (2, 2), & (2, 3), & (2, 4), & (2, 5), & (2, 6) \\
(3, 1), & (3, 2), & (3, 3), & (3, 4), & (3, 5), & (3, 6) \\
(4, 1), & (4, 2), & (4, 3), & (4, 4), & (4, 5), & (4, 6) \\
(5, 1), & (5, 2), & (5, 3), & (5, 4), & (5, 5), & (5, 6) \\
(6, 1), & (6, 2), & (6, 3), & (6, 4), & (6, 5), & (6, 6)\}
\end{array}
$$

Moreover, the event $E = dice\ add\ up\ to\ 7$ is equal to

$$E = \{(1, 6), (2, 5), (3, 4), (4, 3), (5, 2), (6, 1)\}$$

Then from the definition of probability, we have

$$p(E) = \frac{n(E)}{n(S)}$$

$$= \frac{6}{36} = \frac{1}{6}$$

That is, the probability that the dice add up to 7 is $\frac{1}{6}$.

The probability of an event is a measure of likelihood that the event will occur. If the experiment is repeated over and over, then the proportion of repetitions in which the outcome will belong to an event E is approximately $p(E)$. Of course, the proportion will not be exactly $p(E)$, but it will usually approximate $p(E)$ very closely. To make these intuitive interpretations of the probability of an event more exact requires more advanced mathematics, usually found in courses on probability and statistics. For now, however, this inexact, intuitive approach will suffice.

In calculating probabilities, counting methods such as introduced in the preceding section are often useful, as is shown in the next example.

➤ EXAMPLE 4
Coin Flipping

Suppose that you flip a coin six times in succession. What is the probability that exactly two heads will occur?

Solution

A typical outcome in the sample space for this experiment consists of a sequence of six letters, where each letter is either H or T. We could list these outcomes, but it would be quite tedious to do so. Actually, we don't need a list of the outcomes. Rather, we need to know how many there are, and this can be deduced from the fundamental principle of counting. There are six letters to be chosen, and each can be chosen in two ways. So the number of choices is

$$n(S) = 2 \cdot 2 \cdot 2 \cdot 2 \cdot 2 \cdot 2 = 64$$

Let's count the number of outcomes in the event $E = $ "exactly two heads occur." An event with two heads is completely determined once we specify which two of the six coin tosses resulted in heads. If we represent the coin tosses by the numbers 1, 2, 3, 4, 5, 6, then an outcome in E is determined by a combination of these six objects taken two at a time. For instance, the combination $\{2, 5\}$ corresponds to the event $THTTHT$, which has heads in positions 2 and 5. The number of outcomes in E is thus the number of combinations of six objects taken two at a time. That is,

$$n(E) = {}_6C_2 = \frac{6 \cdot 5}{1 \cdot 2} = 15$$

Thus, the probability of the event E is given by

$$p(E) = \frac{n(E)}{n(S)} = \frac{15}{64}$$

That is, the probability that exactly two heads occur is $\frac{15}{64}$.

Exercises 10.7

Describe the associated sample space for each of the following.

1. A coin is tossed, and then a die is thrown.

2. Two letters are randomly chosen, one after another, from the word *tack*.

3. You are making a survey of two-children families. You want to record the sex of each child, in the order of their births.

4. You survey four-children families and record the sex of each child in the order of birth. How many of these points correspond to families having three boys and one girl?

5. Two dice, one green and one white, are tossed, and the numbers of dots on their upper faces are noted.

6. From five different cards, A, B, C, D, and E, three are selected.

7. Five cards are drawn from a deck of 52 playing cards.

8. In Exercise 6, how many of the sample points correspond to a selection including A?

9. In a game, three balls are rolled down an incline into slots numbered 1–9.

10. Four cards are drawn from a standard pinochle deck of 48 cards. .

11. Two dice are viewed from an angle so that three sides are visible.

12. Four coins are flipped.

13. Two letters from the name *William* are selected.

14. Three balls are drawn one at a time from a bag of eight green and seven red balls.

15. Of 20 cartons of milk in a grocery store cooler, 5 are spoiled. You choose 3 at random.

16. Two faulty batteries are placed on a shelf with a dozen reliable ones. Two are chosen by a customer.

17. You have a penny, a nickel, a dime, and a quarter in your pocket. You take out two coins, one after the other.

18. In a group of three men and two women, a two-person committee is chosen.

19. Three dice are tossed, and the number on the top surface of each die is written down.

20. Three letters are chosen one at a time from the word *jumping*.

♦ **Applications**

21. For a chronic disease, there are five standard treatments: 1, 2, 3, 4, and 5. A research doctor is going to study three of the treatments. If the three are chosen at random, what is the probability that treatment 1 will be chosen for study?

22. If two cards are drawn at random from a 52-card standard deck, what is the probability of drawing one king?

23. Two balls are drawn at random from a bag containing 6 red and 10 green balls. What is the probability of drawing 2 red balls?

24. In the game in which three balls are rolled down an incline into slots numbered 1–9, what is the probability that two of the balls will fall into the same slot?

25. From a standard deck of 52 cards, what is the probability of a 5-card hand having all cards from the same suit?

26. If four coins are flipped, find the probability of not obtaining two heads and two tails.

27. In tossing two dice, what is the probability of their sum totaling 5 ?

28. Four enlistees, Ed, José, Marvin, and Leroy, are to be assigned at random to either the infantry or the artillery. What is the probability that Ed and Marvin will be assigned to the same service?

29. A poker hand of 5 cards is dealt from a 52-card pack of standard cards. What is the probability of getting a full house (2 of one denomination and 3 of another)?

30. In a poker hand of 7 cards taken at random from a standard 52-card deck, what is the probability of getting 4 of a kind?

31. You and your best friend are placed in a group of 10 persons. From this group, a committee of 3 will be picked for an all-expense-paid trip to Disney World. What is the probability that both you and your best friend will be chosen?

32. You and your best friend join a softball team of 12 persons. In a game, 10 of the group are picked at random to play. What is the probability that you will be picked to play but your best friend will not be picked?

33. From Exercise 31, what is the probability that your best friend will be chosen but you will not be chosen?

34. From Exercise 32, what is the probability that neither you nor your friend will be picked?

35. A slot machine in Reno contains three reels with each reel containing 20 symbols. If the first reel contains five

apples, the middle reel has four apples, and the third reel has only two apples, what is the probability of getting three apples in a row in order to win $5000 from this slot machine?

36. To complete your college physical education requirements you must take two classes from the following list: bowling, volleyball, tennis, basketball, softball, swimming, and soccer. If you decide that you will choose your two at random by drawing classes from a hat one at a time, what is the probability that you will take tennis and swimming?

37. If a single die is tossed and then a single card is drawn from a standard 52-card deck, what is the probability of a 5 on the die and the 2 of spades on the card?

38. In a carnival game, 20 colored balls are placed in a bag. Of this 20, 6 are white, 4 are blue, 3 are red, 3 are green, 3 are yellow and 1 is black. In order to win, 3 balls are drawn at random, and of these, exactly 2 must be the same color. Also, if the black ball is drawn, you automatically lose. What is the probability of your winning?

39. Mrs. Robinson invites 10 relatives to a family party: her mother, two uncles, three brothers, and four cousins. If the chances of any one of her guests arriving first are equally likely, what is the probability that the first guest will be an uncle?

40. At a credit union dinner, a door prize of $300 is to be given to the person whose ticket stub is drawn from a bowl. If 35 tickets numbered 1–35 were sold, what is the probability that the winning ticket bears a number that is less than 20 ?

41. Of the 500,000 state income tax returns that one branch of the state tax bureau receives, 50,000 will be analyzed thoroughly. What is the probability that a return will not be thoroughly analyzed?

42. Ten cards, the 2 through the 6 of clubs and the 2 through the 6 of diamonds, are shuffled thoroughly and then taken one by one from the top of the deck and placed on a table. What is the probability that each card is next to a card bearing the same numeral?

▌ 10.8 CHAPTER REVIEW

Important Concepts, Properties, and Formulas—Chapter 10

n factorial	Let n be a positive integer. Then **n factorial,** denoted $n!$, is the product of all positive integers from 1 to n. That is, $$n! = n(n-1)(n-2)\cdots 2 \cdot 1$$	p. 503

Binomial coefficient	$$\binom{n}{k} = \frac{n!}{k!(n-k)!}$$	p. 504
Binomial theorem	Let n be a nonnegative integer. Then $$(a+b)^n = \binom{n}{0}a^n + \binom{n}{1}a^{n-1}b$$ $$+ \binom{n}{2}a^{n-2}b^2 + \cdots + \binom{n}{n}b^n$$	p. 505
Principle of mathematical induction	Suppose that for each positive integer k there is given a statement P_k. Furthermore, suppose that the following two conditions are fulfilled: a. P_1 is true. b. By assuming that P_k is true, we can deduce the truth of P_{k+1}. (Here k is any positive integer.) Then P_k is true for every positive integer k.	p. 510
Summation notation	$$\sum_{n=A}^{B} a_n$$ is another way of writing $$a_A + a_{A+1} + \cdots + a_B$$	p. 518
Arithmetic sequences	$$a_n = a_1 + d(n-1)$$ $$s_n = \frac{n}{2}(a_1 + a_n)$$ $$s_n = \frac{n}{2}[2a_1 + d(n-1)]$$	pp. 522, 523
Geometric series	$$a_n = a_1 r^{n-1}$$ $$s_n = \begin{cases} a_1\dfrac{1-r^n}{1-r}, & r \neq 1 \\ na_1, & \cdot r = 1 \end{cases}$$	pp. 527, 530
Infinite geometric series	$$a + ar + ar^2 + \cdots$$ $$S = \frac{a}{1-r} \text{ if } |r| < 1$$	p. 532
Number of permutations of n things taken k at a time	$$_nP_k = n(n-1)(n-2)\cdots(n-k+1)$$ $$= \frac{n!}{(n-k)!}$$	p. 536

Number of combinations of n things taken r at a time	$_nC_r = \dfrac{n!}{r!(n-r)!}$	p. 538

Cumulative Review Exercises—Chapter 10

Expand.

1. $(a + t)^4$

2. $(3ab^2 - 4a^2b)^3$

3. $(3ab^2 - 4a^2b)^7$

4. $\left(\sqrt{a} - \dfrac{1}{\sqrt{a}}\right)^6$

5. Find the middle term of
$$\left(t - \frac{1}{t}\right)^{10}$$

6. Find the fourth term of
$$\left(\frac{4a}{5} - \frac{5}{2a}\right)^{10}$$

Use mathematical induction to prove that the following hold for all positive integers n.

7. $2^3 + 4^3 + 6^3 + \cdots + (2n)^3 = 2n^2(n + 1)^2$

8. $1 + 2 + 3 + \cdots + n = \dfrac{n^2 + n}{2}$

9. $1 + 4 + 7 + \cdots + (3n - 2) = \dfrac{n}{2}(3n - 1)$

10. $\dfrac{1}{2} + \dfrac{1}{2^2} + \dfrac{1}{2^3} + \cdots + \dfrac{1}{2^n} = 1 - \dfrac{1}{2^n}$

11. $x^n - y^n$ is divisible by $x - y$.

12. $x^{2n+1} + y^{2n+1}$ is divisible by $x + y$.

For the given $f(n)$: (a) Complete the table. (b) On the basis of the results in the table, what would you believe to be the value of $f(9)$? (c) Compute $f(9)$. (d) Make a conjecture about the value of $f(n)$.

13. $f(n) = 2 + 5 + 8 + \cdots + 3n - 1$

n	1	2	3	4	5	6
$f(n)$						

14. $f(n) = 1 + 3 + 9 + \cdots + 3^{n-1}$

n	1	2	3	4	5	6
$f(n)$						

15. Prove that if a is any real number greater than 1, then $a^n > 1$ for every positive integer n.

16. Prove that 5 is a factor of $n^4 - 1$ for all natural numbers $n \geq 2$ not divisible by 5.

Determine the first three terms and the 10th term of the sequences defined by the following:

17. $a_n = n - 2$

18. $a_n = 2n + 3$

19. $a_n = \dfrac{4}{n + 1}$

20. $a_n = n^2$

Determine the first five terms for the following defined sequences.

21. $a_n = n \log 10^{n+1}$

22. $a_n = \log 100^{n-1}$

23. $a_1 = 2,\ a_n = \dfrac{3}{a_{n-1}},\ n > 1$

24. $a_1 = -2,\ a_n = \frac{1}{2}a_{n-1},\ n > 1$

Evaluate the following expressions.

25. $\displaystyle\sum_{n=4}^{10}(n - 5)$

26. $\displaystyle\sum_{n=1}^{6}(2n - 1)$

27. $\displaystyle\sum_{n=2}^{6}\left(\frac{1}{n - 1}\right)^2$

28. $\displaystyle\sum_{n=4}^{10}\left[-\cos\left(\frac{n\pi}{2}\right)\right]$

Prove the following sequences are arithmetic.

29. $2, 3, 4, \ldots, 1 + n, \ldots$

30. $2, -2, -6, -10, \ldots, 6 - 4n, \ldots$

31. $2, 5, 8, 11, \ldots, 3n - 1, \ldots$

32. $1, \dfrac{3}{2}, 2, \dfrac{5}{2}, \ldots, \dfrac{n + 1}{2}, \ldots$

Find the indicated term for the given arithmetic sequences.

33. $a_1 = 2,\ d = 3$; find a_6.

34. $a_1 = -1,\ d = 5$; find a_9.

35. $a_1 = 4,\ d = -2$; find a_{10}.

36. $a_1 = 5,\ d = -2$; find a_7.

For each of the following arithmetic sequences, find the first term, a_1, and the difference d between consecutive terms of the sequence.

37. $a_9 = 6,\ a_{12} = 21$

38. $a_3 = 1$, $a_8 = -14$

39. $a_{10} = -32$, $a_{40} = -122$

40. $a_{17} = 96$, $a_{28} = 141$

For the following arithmetic sequences, answer the given question.

41. If $a_1 = 2$, $d = 2$, and $a_n = 32$, what is n?

42. If $a_3 = 9$, $d = -2$, and $a_n = -27$, what is n?

43. If $a_6 = 12$ and $d = -2$, what is a_1?

44. If $a_{12} = -20$ and $d = 4$, what is a_3?

Compute the indicated sums for the following arithmetic sequences.

45. $-2, -1, 0, \ldots, n - 3, \ldots$; find s_{20}.

46. $-4, -2, 0, \ldots, 2n - 6, \ldots$; find s_{30}.

47. $2, 0, -2, \ldots, 4 - 2n, \ldots$; find s_{35}.

48. $5, 7, 9, \ldots, 3 + 2n, \ldots$; find s_{25}.

♦ Applications

49. In building a tower, a child uses 15 cubes for the first row, and 2 fewer cubes for each row thereafter. How many blocks would there be in the eighth row?

50. How many meters would a skydiver fall during the ninth second if he falls 10 meters during the first second, 20 meters during the second second, 30 meters during the third second, and so on?

51. The population of St. John, Michigan, is 42,000. If the zoning commission permits an increase of 250 people each year, what will be the maximum population in six years?

52. How much material will be needed for the rungs of a ladder of 31 rungs if the rungs taper from 18 cm to 28 cm and the lengths form an arithmetic sequence?

Prove that the following are geometric sequences.

53. $a_n = 4\left(\frac{1}{4}\right)^{n-1}$

54. $a_n = -24\left(\frac{1}{2}\right)^{n-1}$

55. $a_n = -\frac{1}{2}(2^{n+1})$

56. $a_n = (-4)^{n+1}$

Find the indicated term for each of the following geometric sequences.

57. $a_1 = 2$, $r = 4$; find a_{10}.

58. $a_1 = \frac{4}{5}$, $r = -\frac{1}{5}$; find a_8.

59. If $a_6 = -9$ and $a_7 = 3$; find a_3.

60. If $a_{10} = 12$ and $a_{11} = 18$; find a_7.

61. $4, 8, 16, \ldots$; find a_n.

62. $-3, 2, -\frac{4}{3}, \ldots$; find a_n.

Solve the following for geometric sequences.

63. Which term of the geometric sequence $2, -6, 18, \ldots$, is -54?

64. Which term of the geometric sequence $81, -27, 9, \ldots$, is $-\frac{1}{27}$?

♦ Applications

65. If the interest of a five-year certificate of deposit was set at 8% compounded semiannually, what would be the value after five years if you invested $2000?

66. To buy the new car of your dreams, you borrow $4000 at 10% compounded monthly. You also elect to pay off your loan in one lump sum payment at the end of three years. What will be the amount of your payment?

Given three of the values: a_1, a_n, r, n, and S_n of a geometric sequence, find the missing two.

67. $a_n = 324$, $n = 5$, $r = 3$

68. $a_n = -128$, $a_1 = -2$, $n = 6$

Find a formula for the following geometric sequence.

69. $a_2 = 10$, $a_5 = -1250$

70. $a_1 = 4$, $a_5 = 324$ $(r > 0)$

Solve the following problems for infinite sequences (if possible).

71. Find the sum of $10 - 5 + \frac{5}{2} - \frac{5}{4} + \cdots$.

72. Find the sum of $12 + 18 + 27 + \cdots$.

73. Find the sum of $24 + 16 + \frac{32}{3} + \frac{64}{9} + \cdots$.

74. Find the sum of $16 - 12 + 9 - \frac{27}{4} + \cdots$.

♦ Applications

75. A ball is dropped 20 feet and rebounds on each bounce $\frac{2}{5}$ of the height from which it fell. How far will it travel before coming to rest?

76. A ball is tossed from the ground to a height of 60 feet. If it will rebound on each bounce to a height that is $\frac{3}{10}$ of the height from which it fell, then how far will it travel before coming to rest?

Solve the following.

77. In how many different orders can you arrange five books on a shelf?

78. How many numbers consisting of two or three digits, without repetitions, can you form from the digits 1, 2, 3, 4, 5 ?

79. How many possibilities are there for a license plate with two letters followed by four numbers?

80. If four people run a race, how many different orders can they all finish in if there are no ties?

Determine the following.

81. $_3P_2$ 82. $_{10}P_3$ 83. $_9P_9$ 84. $_6P_1$

Use permutations to solve the following.

85. How many "words" of four letters can be formed from the letters of the word *empty*? (Any four-letter group is considered a word.)

86. How many permutations are there of 13 cards taken from a bridge deck of 52 cards?

87. How many 3-digit numbers can be formed from the 10 digits 0, 1, ..., 9?

88. In arranging five pictures on a wall such that a certain one must be second, how many arrangements are possible?

Determine the following.

89. $_{10}C_4$ 90. $_7C_6$ 91. $_9C_6$ 92. $_{10}C_2$

93. You are one of 12 people in your class. How many 5-member committees can be formed that include you?

94. In a group of nine Republicans and eight Democrats, how many eight-member committees can be formed with six Republicans and two Democrats?

95. In a group of 30 members, how many different committees of 4 people can be formed?

96. How many five-letter "words," each having three consonants and two vowels, can be formed from the letters of the word *equations?*

Describe the associated sample space for each of the following.

97. Five coins are flipped.

98. Six cards are drawn from a deck of 52 different cards.

99. Three balls are drawn, one at a time, from a bag of seven blue and six red balls.

100. Two dice are thrown.

Solve the following.

101. In the Indian game of Tong, two players simultaneously show their right hands to each other, exhibiting either one, two, or three extended fingers. If each player is equally likely to extend one, two, or three fingers, then what is the probability that the total number of fingers extended is odd?

102. The numbers from 1 through 15 are printed on 15 balls, one number per ball. If one of these balls is drawn at random, what is the probability that the number of it is divisible by 5?

103. A committee of two is to be selected from Bill, Ted, Carol, Alice, and Dave. What is the probability that Dave will not be on the committee?

104. If two cards are drawn one at a time from a standard bridge deck, what is the probability that the first card is an ace and the second card is a jack?

TOPICS IN ANALYTIC GEOMETRY

A cooling tower at an electrical power generation plant, such as the one shown here, uses a hyperbolic design to efficiently dissipate excess heat.

I n this final chapter, let's return to the subject of analytic geometry and discuss topics that are often useful in other courses, especially calculus and physics.

11.1 CONIC SECTIONS, PARABOLAS

A **conic section** is a curve obtained as the intersection of a cone and a plane. There are four different types of conic sections: circles, ellipses, parabolas, and hyperbolas. Exactly which type of curve is generated depends on the relative position of the cone and the plane. Each of the four types of conic sections is shown in Figure 1.

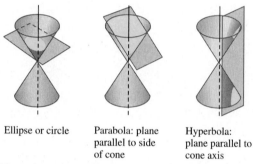

Ellipse or circle   Parabola: plane parallel to side of cone   Hyperbola: plane parallel to cone axis

Figure 1

The conic sections were first studied by the ancient Greek geometers, who were interested in them for their geometric properties. However, beginning in the 17th century, conic sections were studied in terms of their equations and were the testing ground for the new ideas of analytic geometry and calculus. Although the conic sections were originally studied from the point of view of pure mathematics, they were found to have many applications in physics, optics, and astronomy.

Rather than define the conic sections in terms of the intersection of a plane and a cone, we will give definitions of each of the conics as a **locus**, that is, as the set of points in the plane satisfying a specific geometric property. From the locus definitions, we will derive the equations and properties of each of the types of conics.

Definition and Standard Form for Parabolas

Let us start our discussion of the conics with a locus definition of a parabola.

Definition 1 **Parabola, Directrix, Focus**	A **parabola** is the set of all points whose distance from a fixed line D equals the distance[1] from a fixed point F not on the line. The line D is called the **directrix** and the point F is called the **focus**.

---

[1]The distance from a point P to a line L is the length of the perpendicular from P to L.

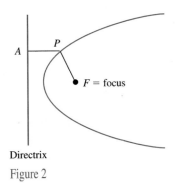

Directrix

Figure 2

In Figure 2, we have drawn the focus and directrix of a parabola. The parabola consists of all points P for which the distance $\overline{PA}$ equals the distance $\overline{PF}$.

The line through the focus perpendicular to the directrix is shown as a dotted line. This line is called the **axis** of the parabola. (See Figure 3.) If a point is on the parabola, then its reflection in the axis is also on the parabola. That is, the parabola is symmetric about its axis. The point on the axis halfway between the directrix and the focus is clearly on the parabola. This point is called the **vertex**. The directed distance from the vertex to the focus is called the **focal length** and is usually denoted p. If the focus is either to the right of or above the vertex, then the value of p is positive; if the focus is either to the left of or below the vertex, then the value of p is negative.

Parabolas for which the directrix is either horizontal or vertical have especially simple equations, as the following result shows:

STANDARD FORMS OF THE EQUATION OF A PARABOLA

Suppose that a parabola has vertex at (h, k) and that the focal length is p.

1. If the directrix is vertical, the parabola has equation
$$(y - k)^2 = 4p(x - h)$$
   (See Figure 4.)
2. If the directrix is horizontal, the parabola has equation
$$(x - h)^2 = 4p(y - k)$$
   (See Figure 5.)

In general, the graph of a parabola is bowl-shaped. The focus is within the bowl. The directrix is outside the bowl and perpendicular to the axis of the parabola.

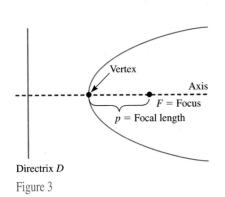

Directrix D

Figure 3

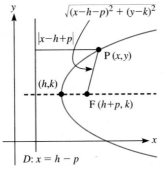

Figure 4

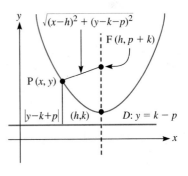

Figure 5

Proof   Let's prove statement 1. For simplicity, assume the vertex is at the origin, so $(h, k) = (0, 0)$. The general case then follows by translation. Let $P(x, y)$ be a point on the parabola. Because the directrix is vertical, the distance from P to the directrix is $d_1 = |x + p|$. The focus is the point (p), so that the distance from P to the focus is

$$d_2 = \sqrt{(x - p)^2 + y^2}$$

(See Figure 4.) Because P is on the parabola, we must have $d_1 = d_2$:

$$\sqrt{(x - p)^2 + y^2} = (x + p)$$
$$(x - p)^2 + y^2 = (x + p)^2$$
$$x^2 - 2xp + p^2 + y^2 = x^2 + 2xp + p^2$$
$$y^2 = 4xp$$

This gives us the standard form of the equation of a parabola in case the directrix is vertical. ◆

Parabolas and Their Equations

➤ **EXAMPLE 1**
Determining Equation
from Focus and Directrix

Determine the equation of a parabola with focus at $(3, -1)$ and directrix $y = 10$.

Solution

The directrix is horizontal, so the axis of the parabola is vertical. The focus is on the axis so that the axis has equation $x = 3$ (see Figure 6). The vertex is midway between the focus and directrix and is therefore $(3, \frac{9}{2})$. The directed distance from the focus to the vertex is given by

$$p = -1 - \frac{9}{2} = -\frac{11}{2}$$

Because the directrix is horizontal, standard form 2 applies. The equation of the parabola is

$$(x - 3)^2 = 4\left(-\frac{11}{2}\right)\left(y - \frac{9}{2}\right)$$
$$(x - 3)^2 = -22\left(y - \frac{9}{2}\right)$$

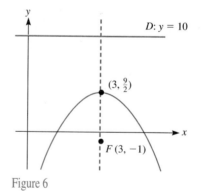

Figure 6

A parabola whose directrix is parallel to one of the coordinate axes has an equation of either of the following forms:

$$y = ax^2 + bx + c \qquad (1)$$
$$x = ay^2 + by + c \qquad (2)$$

(To get this form of the equations, just multiply out the standard form and rearrange terms.) Conversely, if we are given an equation in one of these forms, we may complete the square and write the equation in one of the standard forms. By reading the proof of statement 1 in reverse, we see that the graph of the equation is a parabola. The next two examples illustrate how to sketch the parabola from an equation of the form (1) or (2).

> **EXAMPLE 2**
Determine Focus and Directrix from Equation

Determine the focus and the directrix of the parabola with equation

$$y = 3x^2 - 6x$$

Sketch the parabola.

Solution

Complete the square on the right to obtain:

$$y = 3x^2 - 6x$$
$$y = 3(x^2 - 2x + 1) - 3$$
$$y + 3 = 3(x - 1)^2$$
$$(x - 1)^2 = \tfrac{1}{3}(y + 3)$$
$$(x - 1)^2 = 4 \cdot \tfrac{1}{12}(y + 3)$$

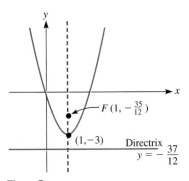

Figure 7

From the last form of the equation, we see that the graph is a parabola with vertex $(h, k) = (1, -3)$ and $p = \tfrac{1}{12}$. The last equation is in the second standard form, so the directrix is horizontal and the axis is vertical. The x-coordinate of the focus is the same as that of the vertex, namely 1. The y-coordinate of the focus is obtained by adding p to the y-coordinate of the vertex: $-3 + p = -3 + \tfrac{1}{12} = -\tfrac{35}{12}$, so the focus is at $(1, -\tfrac{35}{12})$. The directrix is

$$y = -3 - p = -3 - \tfrac{1}{12} = -\tfrac{37}{12}$$

The graph of the equation is sketched in Figure 7.

> **EXAMPLE 3**
Reducing to Standard Form

Determine the focus and the directrix of the parabola with equation

$$y^2 + 2y = -x - 5$$

Sketch the parabola.

Solution

In this example, y appears in a term of the second degree, so we complete the square on the left:

$$y^2 + 2y + 1 = (-x - 5) + 1$$
$$(y - (-1))^2 = 4 \cdot \left(-\tfrac{1}{4}\right) \cdot (x - (-4))$$

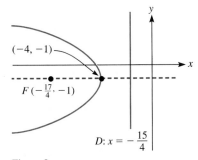

Figure 8

From the equation, we read off the vertex: $(-4, -1)$. Also from the equation, we see that $p = -\tfrac{1}{4}$. Because the y-term is the one of second degree, the directrix is vertical and the axis horizontal. Because the value of p is negative, the focus is to the left of the vertex, at

$$\left(-4 - \tfrac{1}{4}, -1\right) = \left(-\tfrac{17}{4}, -1\right)$$

The directrix lies a directed distance p to the right of the vertex and is therefore the line $x = -4 - p = -4 + \tfrac{1}{4} = -\tfrac{15}{4}$. Figure 8 shows the graph of the equation.

The Reflection Property of the Parabola

When a light or sound wave hits a surface, it is reflected in such a way that the angle made by the incoming wave equals the angle made by the reflected wave. (See Figure 9.) This is a physical law that was first observed in the 17th century. The parabola has a geometric property that, when combined with the law of reflection of light and sound waves, allows for many interesting and useful applications. Figure 10 shows a parabola with vertical axis. A series of light (or sound) waves moves parallel to the axis and is reflected off the surface of the parabola, so it can be proved that the reflected waves all pass through the focus. This result is known as the **reflection property** of the parabola.

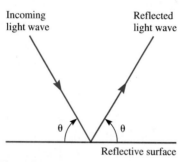

Figure 9

Figure 10

The reflection property may be used to construct a telescope. The observed light waves travel the length of the telescope tube and are reflected off a parabolic mirror. The reflected light rays all pass through the focus of the parabola, where the eyepiece of the telescope is located. In this case, the reflection property of the parabola states that the observed image is displayed at the eyepiece. (See Figure 10.)

Another application using a similar design principle is the parabolic microphone, often used in sporting events. This device consists of a parabolic dish to reflect sound to a microphone located at the focus. By concentrating diverse sound waves at the focus, the parabolic dish can pick up and amplify sounds quite a distance away.

To prove the reflection property of the parabola requires calculus.

Exercises 11.1

Find the vertex, focus, and directrix of the following parabolas.

1. $y = 9x^2$
2. $x = 4y^2$
3. $y^2 = -2x$
4. $x^2 = -12y$
5. $y + 1 = 2(x - 1)^2$
6. $3y - 4 = 6(x + 2)^2$
7. $y^2 - (y + 1)^2 = x^2$
8. $y = 4x^2 - 4x + 1$
9. $(3x + 1)^2 = 4y + 12$
10. $(2y - 1)^2 = 3(x + 5) - 1$

11. $y^2 + 3y - 6x = 8$
12. $y^2 + 3y - 6x = 0$
13. $25x^2 + 10x - y + 5 = 0$
14. $2x^2 + 6x + 5y - 20 = 0$

Determine the equation of the parabola with the following properties.

15. vertex: (0, 0)
    directrix: $x = 2$
16. vertex: (0, 0)
    directrix: $x = -3$

17. vertex: (2, 1)
    directrix: $y = -1$

18. vertex: (1, 1)
    directrix: $y = 2$

19. vertex: (0, 0)
    focus: (1, 0)

20. vertex: (0, 0)
    focus: $(-2, 0)$

21. vertex: $(3, -2)$
    focus: (3, 4)

22. vertex: (4, 2)
    focus: (4, 0)

23. focus: (1, 2)
    directrix: $y = -2$

24. focus: (1, 2)
    directrix: $y = 5$

25. focus: $(-1, -4)$
    directrix: $x = 3$

26. focus: (1, 0)
    directrix: $x = 0$

27. Axis vertical and passes through the three points (0, 0), $(-2, 4)$, (3, 6).

28. Axis vertical, vertex at the origin, and passes through the point (3, 10).

29. Axis horizontal and passes through the three points (0, 0), (2, 5), (3, 12).

30. Axis horizontal, vertex at (2, 1), and passes through origin.

♦ **Applications**

31. The main span of a suspension bridge is 4000 feet long. Suppose that the towers are 500 feet high and the roadway is 300 feet above the surface of the water. The cables form a parabola. Assuming a coordinate system with horizontal axis along the roadway and vertical axis along the left tower, determine the equation of the parabola.

32. The main span of a suspension bridge is 5200 feet long. Suppose that the towers hold the ends of the cable 300 feet above the roadway. The cables form a parabola. Assuming a coordinate system with horizontal axis along the roadway and vertical axis along the left tower, determine the equation of the parabola.

33. The mirror of a telescope has diameter 6 inches and depth .01 inches at the center. Determine the distance from the center of the mirror to the focus.

34. The wall and ceiling in back of the orchestra of a symphony hall has a cross-section that is a parabola. Suppose that the conductor stands at the focus, which is 50 feet from the wall. The height of the wall above the conductor is 60 feet. Determine the equation of the parabola. Explain what the reflection property of the parabola has to do with the design of the wall.

35. The headlight of a car has a cross-section that is a parabola. The light source is located at the focus. Suppose that the light source is 2 inches from the back of the headlight and that the radius of the headlight is 6 inches. Determine the equation of the parabola. Explain what the reflection property of the parabola has to do with the operation of the headlight.

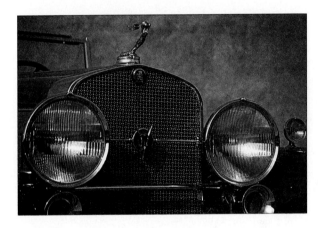

In calculus, it is proved that the slope m of the tangent line to the parabola

$$4p(y - k) = (x - h)^2$$

at the point where $x = a$ is given by the formula

$$m = \frac{1}{2p}(a - h)$$

Determine the equation of:

36. The tangent line to $y = x^2$ at (2, 4).

37. The tangent line to $y = 2(x - 3)^2$ at the point where $x = 1$.

38. The line perpendicular to the tangent line to $y = 2x^2 + 3$ at the point where $x = -2$ and passing through the point of tangency.

39. The line parallel to the tangent line to $y = x^2$ at $x = 3$ and passing through the origin.

40. Use the formula for the slope of the tangent line to prove the reflection property of the parabola.

Match the equations in Exercises 41–46 with their graphs.

41. $y = 9x^2$

42. $x = 4y^2$

43. $y^2 = -2x$

44. $x^2 = -12y$

45. $y + 1 = 2(x - 1)^2$

46. $(2y - 1)^2 = 3(x + 5) - 1$

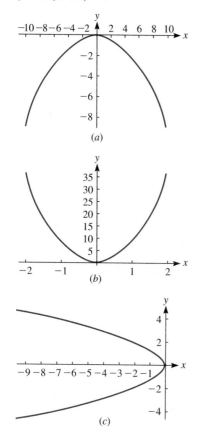

(a)

(b)

(c)

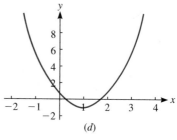

(d)

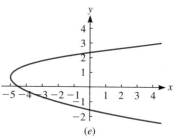

(e)

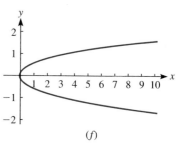

(f)

➡ **Technology**

47. **Graphing calculator.** On one set of axes, draw parabolas with vertex at the origin and focus at $(p, 0)$ for $p = 0.5, 0.75, 1, 2, 3$. Describe what happens to the graph as the focal length p increases.

Use a graphing utility to graph the parabolas in Exercises 48–51. (For some of them, you will need to graph two separate equations, obtained by solving for y in terms of x.)

48. $3y^2 = 4x + 1$

49. $-\dfrac{y^2}{2} = \dfrac{4x}{7}$

50. $(y - 3)^2 = 2(x + 5) + 1$

51. $(3y + 2)^2 = x - 3$

■ 11.2 ELLIPSES

Let's consider two additional conic sections: circles and ellipses. As we shall see, circles are special types of ellipses, so let's start the section by concentrating on ellipses.

Definition and Standard Equation of an Ellipse

We may define an ellipse in geometric terms as follows.

Definition 1 **Ellipse, Foci**	An **ellipse** is the set of all points P, the sum of whose distances from two fixed points is a constant. The two points are called the **foci** (the plural of focus).

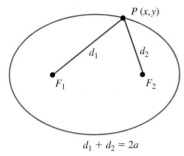

Figure 1

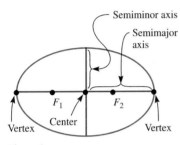

Figure 2

Refer to Figure 1. The two foci are F_1 and F_2. A typical point on the ellipse is $P(x, y)$. The distance of P to F_1 is denoted d_1, and the distance of P to F_2 is denoted d_2. The definition of an ellipse says that there is a constant $2a$, independent of the point P such that

$$d_1 + d_2 = 2a \qquad (1)$$

(The reason for writing $2a$ rather than a will become apparent shortly.)

Draw a line through the foci. (See Figure 2.) It intersects the ellipse in two points, called **vertices**. The line segment connecting the vertices is called the **major axis**. Half the length of the major axis is called the **semimajor axis** of the ellipse. The midpoint of the major axis is called the **center** of the ellipse. It is clear that the center is midway between the two foci. Draw a line segment perpendicular to the major axis through the center, with both endpoints on the ellipse. This line segment is called the **minor axis**. Half the length of the minor axis is called the **semiminor axis** of the ellipse.

Assume that the center of the ellipse is at (h, k) and that the distance from the center to each focus is c. The number c is called the **focal length** of the ellipse. Let us assume that the major axis is horizontal. Then the foci are at $(h \pm c, k)$. Suppose that the right vertex is at (l, m). It is clear that $m = k$. Moreover, the distance of the right vertex from F_1 is $(l - h) + c$, and the distance from F_2 is $(l - h) - c$. By equation (1), we have

$$(l - h + c) + (l - h - c) = 2a$$
$$2(l - h) = 2a$$
$$l = h + a$$

That is, the coordinates of the right vertex are $(h + a, k)$. Similarly, the coordinates of the left vertex are $(h - a, k)$. Moreover, from these coordinates, we see that $2a$ is the length of the major axis of the ellipse.

Let's write equation (1) in terms of the coordinates of the foci and the point $P(x, y)$:

$$d(F_1, P) + d(F_2, P) = 2a$$

$$\sqrt{(x - [h - c])^2 + (y - k)^2} = 2a - \sqrt{(x - [h + c])^2 + (y - k)^2}$$

$$(x - [h - c])^2 + (y - k)^2 = 4a^2 + (x - [h + c])^2 + (y - k)^2$$
$$\qquad\qquad - 4a\sqrt{(x - [h + c])^2 + (y - k)^2}$$

$$4a\sqrt{(x - [h + c])^2 + (y - k)^2} = 4a^2 - 4c(x - h)$$

$$a\sqrt{(x - [h + c])^2 + (y - k)^2} = a^2 - c(x - h)$$

$$a^2(x - [h + c])^2 + a^2(y - k)^2 = (a^2 - c(x - h))^2$$
$$a^2(x - h)^2 + a^2c^2 + a^2(y - k)^2 = a^4 + c^2(x - h)^2$$
$$(a^2 - c^2)(x - h)^2 + a^2(y - k)^2 = a^2(a^2 - c^2)$$
$$\frac{(x - h)^2}{a^2} + \frac{(y - k)^2}{a^2 - c^2} = 1 \qquad (2)$$

From Figure 3, we see that if b denotes half the length of the minor axis, then by equation (1), we have

$$2\sqrt{b^2 + c^2} = 2a$$
$$a^2 = b^2 + c^2$$

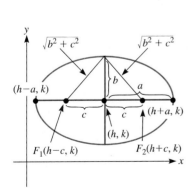

Figure 3

Therefore, we may rewrite equation (2) in the form

$$\frac{(x - h)^2}{a^2} + \frac{(y - k)^2}{b^2} = 1 \qquad \text{(major axis horizontal)} \qquad (3)$$

This is the standard form of the equation of an ellipse whose major axis is horizontal.

If we assume that the major axis is vertical, we arrive at the equation

$$\frac{(x - h)^2}{b^2} + \frac{(y - k)^2}{a^2} = 1 \qquad \text{(major axis vertical)} \qquad (4)$$

In either case, the foci are a distance c from the center along the major axis, where

$$a^2 = c^2 + b^2 \qquad (5)$$

A circle is an ellipse in which the semimajor and semiminor axes are equal (to the radius). In this case, the above standard equations of an ellipse reduce to the familiar equation of a circle. Indeed, if an ellipse has $a = b = r$ and center (h, k), then the equation of the ellipse is

$$\frac{(x - h)^2}{r^2} + \frac{(y - k)^2}{r^2} = 1$$
$$(x - h)^2 + (y - k)^2 = r^2$$

the standard form of a circle with center at (h, k) and radius r.

Illustrations

➤ EXAMPLE 1
Determine Equation from Foci and Minor Axis

Determine the equation of the ellipse with foci at (3, 5) and (9, 5) and minor axis of length 10.

Solution

Because the foci have the same y-coordinate, the major axis is horizontal. The center of the ellipse is midway between the two foci and is therefore (6, 5), and the distance from the center to the foci is given by $c = 3$. The semiminor axis b is given by

$$2b = 10$$
$$b = 5$$
$$a^2 = b^2 + c^2$$
$$a^2 = 5^2 + 3^2$$
$$= 25 + 9$$
$$= 34$$
$$a = \sqrt{34}$$

Therefore, the equation of the ellipse is

$$\frac{(x - 6)^2}{\left(\sqrt{34}\right)^2} + \frac{(y - 5)^2}{5^2} = 1$$

$$\frac{(x-6)^2}{34} + \frac{(y-5)^2}{25} = 1$$

The graph is sketched in Figure 4.

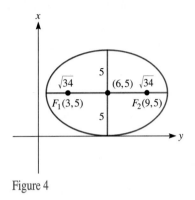

Figure 4

➤ EXAMPLE 2
Determine Parameters from Equation

Sketch the graph of the ellipse with equation

$$4x^2 + 25y^2 = 100$$

Determine the foci, vertices, semimajor axis, and semiminor axis.

Solution

Because both x and y appear to the second power and the coefficients of both have the same sign, the graph of the equation is an ellipse. In the standard form of the equation of an ellipse, the constant term is 1. To reduce the equation to this form, we divide both sides by 100:

$$\frac{4x^2}{100} + \frac{25y^2}{100} = 1$$

$$\frac{x^2}{25} + \frac{y^2}{4} = 1$$

$$\frac{x^2}{5^2} + \frac{y^2}{2^2} = 1$$

From the equation, we see that the ellipse is centered at the origin with semimajor axis along the x-axis with length 5 and semiminor axis along the y-axis with length 2. The vertices are $(\pm 5, 0)$. The foci are at a distance c from the center, where

$$c^2 + b^2 = a^2$$
$$c^2 = 5^2 - 2^2$$
$$= 21$$
$$c = \sqrt{21}$$

Therefore, the foci are $(\pm \sqrt{21}, 0)$. The graph is sketched in Figure 5.

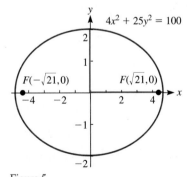

Figure 5

➤ EXAMPLE 3
Reducing to Standard Form

Sketch the graph of the ellipse with equation

$$5x^2 + 10x + 4y^2 + 8y - 5 = 0$$

Determine the foci, vertices, semimajor axis, and semiminor axis.

Solution

Note that the equation involves both x and y raised to the second power and that the coefficients of the square terms have the same sign. This means that the graph of the equation is an ellipse. To transform the equation into standard form, we complete the square in both x and y:

$$5x^2 + 10x + 4y^2 + 8y - 5 = 0$$

$$5(x^2 + 2x) + 4(y^2 + 2y) = 5$$

$$5(x^2 + 2x + 1) + 4(y^2 + 2y + 1) = 5 + 5 + 4$$

$$5(x + 1)^2 + 4(y + 1)^2 = 14$$

$$\frac{5(x + 1)^2}{14} + \frac{4(y + 1)^2}{14} = 1$$

$$\frac{(x + 1)^2}{\frac{14}{5}} + \frac{(y + 1)^2}{\frac{7}{2}} = 1$$

$$\frac{(x + 1)^2}{\left(\sqrt{\frac{14}{5}}\right)^2} + \frac{(y + 1)^2}{\left(\sqrt{\frac{7}{2}}\right)^2} = 1$$

This last equation has a graph that is an ellipse with center $(-1, -1)$. Because $\frac{7}{2} = 3.5$ is greater than $\frac{14}{5} = 2.8$, we see that the semimajor axis lies in the vertical direction and the semiminor axis in the horizontal direction. The lengths of the axes are

$$\text{semimajor axis} = \sqrt{\frac{7}{2}} = \frac{\sqrt{14}}{2}, \text{ vertical}$$

$$\text{semiminor axis} = \sqrt{\frac{14}{5}} = \frac{\sqrt{70}}{5}, \text{ horizontal}$$

The foci are at a distance c from the center, where

$$c^2 = a^2 - b^2$$

$$= \left(\sqrt{\frac{7}{2}}\right)^2 - \left(\sqrt{\frac{14}{5}}\right)^2$$

$$= \frac{7}{2} - \frac{14}{5}$$

$$= \frac{7}{10}$$

$$c = \sqrt{\frac{7}{10}} = \frac{\sqrt{70}}{10}$$

Therefore, the foci are $(-1, -1 \pm \sqrt{70}/10)$. The graph is sketched in Figure 6.

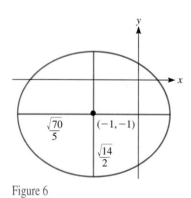

Figure 6

Eccentricity of an Ellipse

The eccentricity e measures the *ovalness* of an ellipse. The eccentricity always lies between 0 and 1. The closer the eccentricity is to 0, the closer the ellipse is to a circle. The closer the eccentricity is to 1, the more elongated the ellipse.

Definition 2 **Eccentricity**	Suppose that an ellipse has semimajor axis a and focal length c. Then the **eccentricity** e of the ellipse is defined as $$e = \frac{c}{a}$$

Note that because the focal length is less than the semimajor axis, the eccentricity as defined by the formula lies between 0 and 1.

➤ EXAMPLE 4
Calculating Eccentricity

Determine the eccentricity of the ellipse of Example 3.

Solution

In Example 3, we showed that $a = \sqrt{14}/2$ and $c = \sqrt{70}/10$. Therefore, we have

$$e = \frac{c}{a} = \frac{\frac{\sqrt{70}}{10}}{\frac{\sqrt{14}}{2}} = \frac{\sqrt{5}}{5}$$

➤ EXAMPLE 5
Eccentricity of a Circle

Show that an ellipse of eccentricity 0 is a circle.

Solution

If $e = 0$, then $c = 0$, so that

$$0 = a^2 - b^2$$
$$a^2 = b^2$$
$$a = b \qquad \text{(because a and b are positive)}$$

That is, the semimajor and semiminor axes are equal and the ellipse is a circle.

One of the major accomplishments of physics is Kepler's laws, which state that the planets travel in elliptical orbits around the sun, with the sun located at a focus. Using Newton's law of gravitation, it is possible to deduce the precise equations of the orbits in terms of the masses of the planets, the mass of the sun, and observed distances and velocities of the planets at particular times.

➤ EXAMPLE 6
Earth's Orbit

The eccentricity of the earth's orbit about the sun is approximately 0.0167. The closest distance between the earth and sun is approximately 93 million miles. What is the furthest distance between the earth and the sun? (See Figure 7.)

Solution

Let a be the semimajor axis of the orbit. Assume that the center of the orbit is $(0, 0)$ and the sun is at the focus $(c, 0)$. Then

$$a = c + 93$$

$$e = \frac{c}{a} = \frac{c}{c + 93} = 0.0167$$

$$c = 0.0167c + 1.5531$$

$$c = 1.5795$$

The furthest distance of the earth from the sun is

$$a + c = 2c + 93$$
$$= 96.159 \text{ million miles}$$

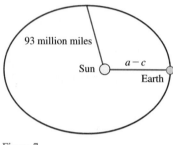

Figure 7

Exercises 11.2

Determine the equation of the ellipse having the following properties:

1. vertices: $(0, \pm 3)$
   foci: $(0, \pm 2)$

2. vertices: $(5, 1), (-5, 1)$
   foci: $(3, 1), (-3, 1)$

3. vertices: $(-1, 4), (-1, 10)$
   foci: $(-1, 5), (-1, 9)$

4. vertices: $(-3, 2), (5, 2)$
   foci: $(-1, 2), (3, 2)$

5. vertices: $(0, \pm 7)$
   semiminor axis: 3

6. vertices: $(\pm 5, 0)$
   semiminor axis: 2

7. vertices: $(-1, 2), (3, 2)$
   semiminor axis: 1

8. vertices: $(5, -3), (5, 7)$
   semiminor axis: 3

9. vertices: $(5, -3), (5, 7)$
   eccentricity: $\frac{2}{5}$

10. vertices: $(\pm 3, 0)$
    eccentricity: $\frac{3}{4}$

11. vertices: $(0, \pm 1)$
    foci: $\left(0, \pm \frac{1}{3}\right)$

12. vertices: $(\pm 5, 0)$
    foci: $(\pm 1, 0)$

13. vertices: $(5, 1), (15, 1)$
    foci: $(7, 1), (13, 1)$

14. vertices: $(3, 7), (3, -5)$
    foci: $(3, 5), (3, -3)$

15. vertices: $(0, \pm 4)$
    passing through: $(2, 0)$

16. vertices: $(\pm 3, 0)$
    passing through: $(0, 2)$

17. vertices: $(\pm 3, 0)$
    passing through: $(2, 1)$

18. vertices: $(0, \pm 5)$
    passing through: $(-2, 3)$

Graph each of the following ellipses. Determine the semimajor and semiminor axes, the foci, and the eccentricity.

19. $\dfrac{x^2}{9} + \dfrac{y^2}{4} = 1$

20. $\dfrac{x^2}{4} + \dfrac{y^2}{9} = 1$

21. $2x^2 + 4y^2 = 16$

22. $4x^2 + 2y^2 = 16$

23. $\dfrac{(x + 2)^2}{4} + \dfrac{(y - 2)^2}{16} = 1$

24. $\dfrac{(x + 2)^2}{16} + \dfrac{(y - 2)^2}{4} = 1$

25. $3(x - 1)^2 + 6(y - 2)^2 = 108$

26. $6(x - 1)^2 + 3(y - 2)^2 = 108$

27. $x^2 - 4x + 5y^2 + 30y + 24 = 0$

28. $4x^2 + 8x + 6y^2 + 24y + 4 = 0$

29. $36x^2 + 4y^2 - 144 = 0$ 30. $x^2 + 9y^2 - 9 = 0$

31. $9x^2 + y^2 - 54x - 4y + 49 = 0$

32. $4x^2 + 25y^2 + 8x + 150y + 129 = 0$

33. $\dfrac{(x + 1)^2}{16} + \dfrac{(y + 1)^2}{25} = 1$

34. $16(x + 1)^2 + 25(y + 1)^2 = 400$

35. $x^2 + y^2 + 6y = 0$ 36. $3x^2 + 3y^2 + 6x = 1$

37. The **latus recta** of an ellipse are the line segments perpendicular to the major axis, drawn through the foci, and having endpoints on the ellipse. Determine the endpoints of the latus recta of the ellipse

$$\frac{x^2}{4} + \frac{y^2}{9} = 1$$

38. Show that the length of the latus recta are $2b^2/a$.

39. Suppose that an ellipse has semimajor axis a, semiminor axis b, and eccentricity e. Prove that $b^2 = a^2(1 - e^2)$.

40. A communications satellite is launched into an earth orbit with low point 22,000 miles and high point 24,000

miles. Determine the eccentricity of the orbit. Determine the equation of the orbit.

41. Suppose that the semimajor and semiminor axis of an ellipse are both multiplied by the same factor k. Show that the eccentricity is unchanged.

Match the equation in Exercises 42–47 with their graphs.

42. $x^2 + \dfrac{y^2}{4} = 1$ 　　　　　43. $\dfrac{x^2}{4} + y^2 = 1$

44. $\dfrac{x^2}{12.25} + \dfrac{y^2}{6.25} = 1$ 　　　45. $\dfrac{x^2}{6.25} + \dfrac{y^2}{12.25} = 1$

46. $\dfrac{(x+2)^2}{4} + \dfrac{(y+3)^2}{9} = 1$

47. $\dfrac{(x-2)^2}{9} + \dfrac{(y+2)^2}{9} = 1$

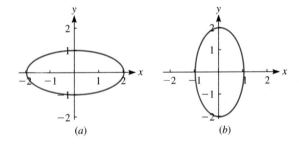

(a)　　　　　　　　(b)

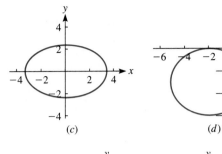

(c)　　　　　　　　(d)

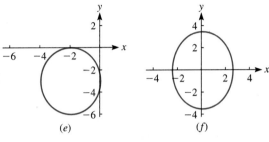

(e)　　　　　　　　(f)

Use a graphing utility to graph the ellipses in Exercises 48–51.

48. $\dfrac{x^2}{2.25} + \dfrac{y^2}{6.25} = 1$ 　　　49. $3x^2 + 5y^2 = 13$

50. $\dfrac{(x-2)^2}{9} + \dfrac{(y+2)^2}{20.25} = 1$

51. $7(x-2)^2 + 6(y-2)^2 = 17$

11.3 HYPERBOLAS

Let's turn our attention to the final conic section, the hyperbola. We can give a geometric definition of the hyperbola in terms of the distances to two foci, as follows:

Definition 1
Hyperbola, Foci

A **hyperbola** is the set of all points P, the difference of whose distances from two fixed points is a constant. The two points are called the **foci**.

Refer to Figure 1. The two foci are F_1 and F_2. A typical point on the hyperbola is $P(x, y)$. The distance of P to F_1 is denoted d_1, and the distance of P to

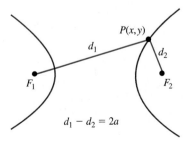

Figure 1

F_2 is denoted d_2. The definition of a hyperbola says that there is a constant $2a$, independent of the point P such that

$$d_1 - d_2 = 2a \qquad (1)$$

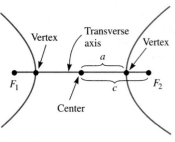

Draw a line connecting the foci. (See Figure 2.) It intersects the hyperbola in two points, called the **vertices**. The line connecting the vertices is called the **transverse axis** of the hyperbola. The midpoint of the transverse axis is called the **center** of the hyperbola. The distance from the center to either of the vertices is a. (Proof similar to the case of an ellipse.) Let c be the distance of the foci from the center. Then c is called the **focal length**. Define the number b by the equation

$$b^2 = c^2 - a^2 \qquad (2)$$

Then, by reasoning as in the case of the ellipse, we can deduce:

Figure 2

STANDARD FORMS FOR THE EQUATION OF A HYPERBOLA
$\dfrac{(x-h)^2}{a^2} - \dfrac{(y-k)^2}{b^2} = 1$ *(transverse axis horizontal)*   (3)
$\dfrac{(y-k)^2}{a^2} - \dfrac{(x-h)^2}{b^2} = 1$ *(transverse axis vertical)*   (4)

We will leave the derivation of these equations as an exercise.

➤ EXAMPLE 1
Determine Equation from Foci and Transverse Axis

Determine the equation of a hyperbola with foci at $(-3, 1)$ and $(5, 1)$, and transverse axis of length 4.

Solution

The transverse axis is horizontal, because the two foci lie on a horizontal line. The center is halfway between the foci, which is the point $(h, k) = (1, 1)$. Because the length of the transverse axis is 4, the vertices are $(3, 1)$ and $(-1, 1)$ and $a = 2$. The distance c of the foci from the center is 4. Thus, the value of b is given by

$$b^2 = c^2 - a^2$$
$$= 4^2 - 2^2$$
$$= 12$$
$$b = \sqrt{12} = 2\sqrt{3}$$

Thus, the equation of the hyperbola is

$$\frac{(x-1)^2}{2^2} - \frac{(y-1)^2}{(2\sqrt{3})^2} = 1$$
$$\frac{(x-1)^2}{4} - \frac{(y-1)^2}{12} = 1$$

Graphing Hyperbolas

Let's examine the graph of a hyperbola. Let's first consider the standard form corresponding to a horizontal transverse axis:

$$\frac{(x-h)^2}{a^2} - \frac{(y-k)^2}{b^2} = 1$$

Solve the equation for $y - k$:

$$\frac{(y-k)^2}{b^2} = \frac{(x-h)^2}{a^2} - 1$$

$$\frac{y-k}{b} = \pm\sqrt{\left(\frac{x-h}{a}\right)^2 - 1}$$

$$y - k = \pm\frac{b}{a}\sqrt{(x-h)^2 - a^2}$$

The square root on the right yields a real number only if $|x - h| \geq a$. Therefore, the x-coordinates of the points on the hyperbola must satisfy the inequality

$$|x - h| \geq a$$

which is equivalent to

$$x \leq h - a \quad \text{or} \quad x \geq h + a$$

That is to say, a point on the hyperbola is either to the right of the right vertex or to the left of the left vertex. As the value of x increases in absolute value, so does the value of

$$\frac{b}{a}\sqrt{(x-h)^2 - a^2}$$

Moreover, the $\pm$ indicates that for each value of x there are two points on the graph, symmetrically placed, above and below the line $y = k$. The graph is as shown in Figure 3. We may write the equation of the hyperbola in the form

$$y - k = \pm\frac{b}{a}\sqrt{(x-h)^2\left(1 - \frac{a^2}{(x-h)^2}\right)}$$

$$= \pm\frac{b}{a}(x - h)\sqrt{1 - \frac{a^2}{(x-h)^2}}$$

As x approaches either infinity or minus infinity, the expression under the square root approaches 1, so that the value of y approaches

$$k \pm \frac{b}{a}(x - h)$$

This means that as x approaches either infinity or minus infinity, the graph approaches the lines with equations

$$y = k + \frac{b}{a}(x - h)$$

$$y = k - \frac{b}{a}(x - h)$$

These lines are called **asymptotes** of the hyperbola. The lines are drawn in Figure 3. Note that if we draw a rectangle centered at (h, k) of width $2a$ and height $2b$, then the asymptotes are the diagonals of the rectangle. (See Figure 3.) The vertical line through the center extended b units above and below is called the **conjugate axis** of the hyperbola.

Repeating this reasoning for a hyperbola with vertical transverse axis, we see that the graph is as shown in Figure 4. In this case, the asymptotes are the equations

$$y = k + \frac{a}{b}(x - h)$$

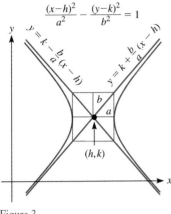

Figure 3

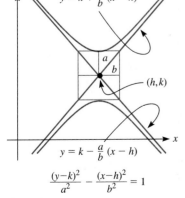

Figure 4

$$y = k - \frac{a}{b}(x - h)$$

The conjugate axis in this case is horizontal.

> ## EXAMPLE 2
> ### Graphing Hyperbolas

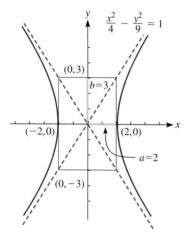

Figure 5

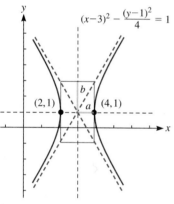

Figure 6

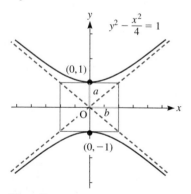

Figure 7

Sketch the graphs of the following equations:

1. $\dfrac{x^2}{4} - \dfrac{y^2}{9} = 1$ 2. $(x - 3)^2 - \dfrac{(y - 1)^2}{4} = 1$ 3. $y^2 - \dfrac{x^2}{4} = 1$

Solution

1. We first write the equation in one of the two standard forms:

$$\frac{x^2}{2^2} - \frac{y^2}{3^2} = 1$$

We see that $a = 2$ and $b = 3$. Moreover, because the positive sign is on the x-term, the axis is horizontal. The center is $(0, 0)$. The vertices are $(\pm a, 0) = (\pm 2, 0)$. Draw the transverse axis of length $2a = 4$ and the conjugate axis of length $2b = 6$, centered at $(0, 0)$. Around the axes, draw the associated rectangle and its diagonals. Next, draw the hyperbola, symmetric with respect to the transverse and conjugate axes and asymptotic to the diagonal lines. (See Figure 5.)

2. Write the equation in the standard form

$$\frac{(x - 3)^2}{1^2} - \frac{(y - 1)^2}{2^2} = 1$$

From this form, we determine that $a = 1$ and $b = 2$. Because the positive sign is on the x-term, the transverse axis of the hyperbola is horizontal. Moreover, the center of the hyperbola is the point $(3, 1)$. We plot the center and then draw the transverse axis of length $2a = 2$ and the conjugate axis of length $2b = 4$. We use the axes to draw a rectangle. The diagonals of the rectangle are the asymptotes. Next, we draw the hyperbola, symmetric with respect to both axes and asymptotic to the diagonal lines. (See Figure 6.)

3. We write the equation in the standard form

$$\frac{y^2}{1^2} - \frac{x^2}{2^2} = 1$$

In this case, we have $a = 1$ and $b = 2$. Because $h = 0$ and $k = 0$, the center of the hyperbola is the point $(0, 0)$. Moreover, because the positive sign is on the y-term, the transverse axis is vertical. The transverse axis has length $2a = 2$. The conjugate axis is of length $2b = 4$ and is horizontal. To sketch the graph, we plot the center and draw the axes. We use the axes to draw a rectangle. The diagonals of the rectangle are the asymptotes. Next, draw the hyperbola, symmetric with respect to both axes and asymptotic to the diagonal lines. (See Figure 7.)

> **EXAMPLE 3**
Reducing to Standard Form

Sketch the graphs of the following equations:

    1. $x^2 - 2y^2 = 1$ 2. $x^2 - 2x - y^2 + 6y - 7 = 0$

Solution

1. We write the equation in the standard form

$$x^2 - \frac{y^2}{\frac{1}{2}} = 1$$

$$\frac{x^2}{1^2} - \frac{y^2}{\left(\dfrac{1}{\sqrt{2}}\right)^2} = 1$$

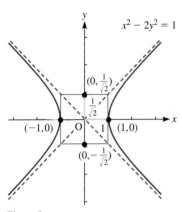

Figure 8

From the last equation, we see that the hyperbola has a horizontal transverse axis of length $2a = 2$, and a vertical conjugate axis of length $2b = 2/\sqrt{2} = \sqrt{2}$. To sketch the graph, we first draw in the rectangle determined by the vertices $(\pm 1, 0)$ and the points $(0, \pm \sqrt{2}/2)$. Next, we draw the diagonals of the rectangle. These are the asymptotes of the hyperbola. Finally, we draw the hyperbola. (See Figure 8.)

2. We begin by putting the equation into standard form. To do this, we complete the square with respect to both x and y:

$$(x^2 - 2x) - (y^2 - 6y) = 7$$
$$(x^2 - 2x + 1) - (y^2 - 6y + 9) = 7 + 1 - 9$$
$$(x - 1)^2 - (y - 3)^2 = -1$$
$$(y - 3)^2 - (x - 1)^2 = 1$$
$$\frac{(y - 3)^2}{1^2} - \frac{(x - 1)^2}{1^2} = 1$$

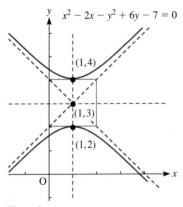

Figure 9

From this equation, we see that the hyperbola is centered at $(1, 3)$. Because the variable y appears with a positive sign, the transverse axis of the hyperbola is vertical. Moreover, the lengths of the two axes are both 2. This provides us with the data to sketch the graph. (See Figure 9.)

The Eccentricity of a Hyperbola

Paralleling the definition in the case of an ellipse, we define the eccentricity of a hyperbola as the focal length divided by the length of the semitransverse axis (half the length of the transverse axis), that is,

$$e = \frac{c}{a} \qquad (5)$$

Because the foci are located within the interior of the *wings* of the hyperbola, we see that $c > a$, so that $e > 1$. The eccentricity measures the *spread* of the wings of a hyperbola. The larger the eccentricity, the wider the spread of the wings.

    We have defined the eccentricity of a hyperbola and an ellipse. Using analogous reasoning, we may define the eccentricity of a parabola. The eccentricity is the ratio of the focal length to the semiaxis. In the case of a parabola, the semiaxis is the line segment from the vertex to the focus. The eccentricity of a parabola is therefore 1.

We can thus classify the conics by their respective eccentricities as follows:

Circle: $e = 0$
Ellipse: $0 < e < 1$
Parabola: $e = 1$
Hyperbola: $e > 1$

It can be proved, using calculus and elementary physics, that the orbit of a comet or an asteroid is always a conic section. The eccentricity may be interpreted as the total energy of the comet. Roughly speaking, the faster the comet travels, the greater the eccentricity. Thus, for example, a slow-moving comet (relatively speaking) will have eccentricity less than 1 and travel in an ellipse. With precisely the correct speed, the eccentricity becomes 1 and the orbit changes to a parabola. If the comet is moving faster still, the eccentricity is greater than 1 and the orbit is a hyperbola.

General Equations the Second Degree

In the first three sections of this chapter, we have studied the various conic sections. By completing the square, any equation of the form

$$Ax^2 + By^2 + Cx + Dy + E = 0, \qquad A, B \text{ not both } 0.$$

may be reduced to one of the standard forms for equations of conics. Here is how to tell which conic will result based on the coefficients of the equation:

GRAPHS OF SECOND DEGREE EQUATIONS

Parabola: Either A or B zero.
Ellipse: A and B nonzero and of the same sign, $A \neq B$.
Circle: $A = B$.
Hyperbola: A and B nonzero and of opposite signs.

➢ EXAMPLE 4
Equations of Second Degree

Determine which type of conic is the graph of each equation.

1. $5x^2 + 3y^2 - 7x + 3y = 1$
2. $-5x^2 + 3y^2 + 12x - 4 = 0$
3. $10x^2 + 10y^2 + 3x - 2y - 100 = 0$
4. $x + 7y^2 + 3y + 5 = 0$

Solution

1. In this case, we have $A = 5$ and $B = 3$. Because both coefficients are nonzero and of the same sign, the graph of the equation is an ellipse.
2. In this case, we have $A = -5$ and $B = 3$. Because the coefficients are nonzero and of opposite sign, the graph of the equation is a hyperbola.
3. In this case, $A = B = 10$. Because the coefficients are nonzero and equal, the graph of the equation is a circle.
4. In this case, $A = 0$ and $B = 7$. Because one of the coefficients is 0, the graph of the equation is a parabola.

Exercises 11.3

Determine the center, vertices, foci, and asymptotes of the following hyperbolas. Sketch the graph with the aid of this information.

1. $x^2 - y^2 = 1$

2. $y^2 - x^2 = 1$

3. $\dfrac{x^2}{4} - y^2 = 1$

4. $\dfrac{y^2}{9} - \dfrac{x^2}{4} = 1$

5. $\dfrac{x^2}{100} - \dfrac{y^2}{36} = 1$

6. $y^2 - \dfrac{x^2}{3} = 1$

7. $(x - 1)^2 - (y - 2)^2 = 1$

8. $4(x + 3)^2 - 9y^2 = 36$

9. $\dfrac{(x - 4)^2}{9} - \dfrac{(y - 1)^2}{25} = 1$

10. $\dfrac{x^2}{\frac{1}{3}} - \dfrac{y^2}{\frac{1}{2}} = 1$

11. $4x^2 - 9y^2 = 1$

12. $2x^2 - 3y^2 = 1$

13. $5(x + 6)^2 - 3(y - 4)^2 = 15$

14. $10y^2 - 3(x + 2)^2 = 30$

15. $x^2 - 6x - 4y^2 + 12y = 30$

16. $2y^2 - 3x^2 + 12x - y = 0$

17. $(x + 2)^2 - 3(y - 4)^2 = 1$

18. $16x^2 - 48x - 2y^2 - 10 = 0$

19. $9y^2 + 18y - 4x^2 + 24x - 28 = 0$

20. $100x^2 + 200x - y^2 = 200$

Determine the equation of the hyperbola having the following properties:

21. center: $(0, 0)$
    transverse axis: 4, horizontal
    conjugate axis: 2

22. center: $(0, 0)$
    transverse axis: 10, horizontal
    conjugate axis: 20

23. center: $(1, 1)$
    transverse axis: 2, vertical
    conjugate axis: 3

24. center: $(-1, -2)$
    transverse axis: 3, vertical
    conjugate axis: 1

25. vertices: $(\pm 3, 0)$
    foci: $(\pm 5, 0)$

26. vertices: $(\pm 8, 0)$
    foci: $(\pm 12, 0)$

27. vertices: $(-5, 1), (3, 1)$
    foci: $(-7, 1), (5, 1)$

28. vertices: $(-2, -2), (-2, 4)$
    foci: $(-2, -5), (-2, 7)$

29. vertices: $(\pm 2, 0)$
    asymptotes: $y = \pm 2x$

30. vertices: $(0, \pm 1)$
    asymptotes: $y = \pm 2x$

31. vertices: $(0, \pm 2)$
    passing through $(5, 3)$

32. vertices: $(1, -2), (1, 5)$
    passing through $(-1, 8)$

Without completing the square, determine which conic is the graph of the equation.

33. $x^2 + 3y^2 + 12x - 14y - 30 = 0$

34. $x + 12y^2 - 13y + 128 = 0$

35. $y^2 = 3x^2 + 10$

36. $100x^2 - 300x + 100y^2 - 200x + 50 = 0$

37. $-30x = y^2 - 3y + 100$

38. $y^2 + 10y = 100 - 12x^2$

39. Suppose that rescuers are searching a wooded area for a trapped camper. Two teams hear the camper's shouts. Suppose that the two rescue teams are 5 miles apart and hear the shouts of the camper 3 seconds apart. Assuming a coordinate system that has its origin at the rescue team that hears the call first and its positive x-axis extending from the first rescue team to the second, write an equation describing the possible location of the camper. (The equation is a hyperbola.) Assume that sound travels at 1100 feet per second.

40. (Continuation of the preceding exercise.) Suppose that a third rescue team, located at $(0, 3$ miles$)$, hears the camper 1 second after the first rescue team. Locate the camper.

41. Suppose that the transverse axis and conjugate axis are both multiplied by the same factor. Show that the eccentricity is unchanged.

42. Suppose that the transverse axis and the conjugate axis are both multiplied by the same factor. Show that the asymptotes remain unchanged.

43. Prove that a hyperbola never crosses its asymptotes.

✍ In Your Own Words

44. Describe how the shape of an ellipse changes as its eccentricity increases.

45. Describe how the shape of an ellipse changes as the foci get closer to the vertices.

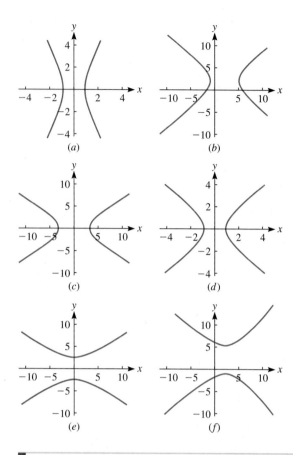

(a)   (b)

(c)   (d)

(e)   (f)

Match the equations in Exercises 46–51 with their graphs.

46. $x^2 - \dfrac{y^2}{4} = 1$

47. $\dfrac{x^2}{1} - \dfrac{y^2}{1} = 1$

48. $\dfrac{x^2}{12.25} - \dfrac{y^2}{6.25} = 1$

49. $\dfrac{y^2}{6.25} - \dfrac{x^2}{12.25} = 1$

50. $\dfrac{(x - 2)^2}{9} - \dfrac{(y - 2)^2}{16} = 1$

51. $\dfrac{(y - 2)^2}{9} - \dfrac{(x - 2)^2}{9} = 1$

➡ Technology

Use a graphing utility to graph the hyperbolas in Exercises 52–55. (You will need to graph them as two separate equations, obtained by solving for y in terms of x.)

52. $\dfrac{x^2}{2.25} - \dfrac{y^2}{6.25} = 1$

53. $3y^2 - 5x^2 = 13$

54. $\dfrac{(x - 2)^2}{9} - \dfrac{(y + 2)^2}{20.25} = 1$

55. $7(y - 2)^2 - 6(x - 2)^2 = 17$

11.4 TRANSLATION AND ROTATION OF AXES

In Chapter 1, we introduced the idea of a Cartesian coordinate system in the plane. As you will recall, the origin for the coordinate system was subject to an arbitrary choice, as were the perpendicular coordinate axes that met at the origin. By either translating or rotating a given set of coordinate axes, we arrive at a new set. In this section, we will study the relationship of coordinates with respect to the new and the original coordinate systems. As we shall see, the equation of a curve can often be written in a simpler form by choosing an appropriate coordinate system.

Translation of Coordinate Axes

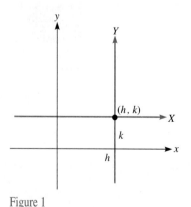

Figure 1

Suppose that we are given an x-y coordinate system. We can construct another coordinate system by translating the coordinate axes to another origin, say the point $P = (h, k)$ (where the coordinates (h, k) are given with respect to the original coordinate system). (See Figure 1.) Designate the coordinate with respect to the new coordinate system by (X, Y) and those with respect to the original coordinate system by (x, y). Then the coordinates with respect to the two systems are related to one another by the equations

$$T : \begin{cases} x = X + h \\ y = Y + k \end{cases}$$

> **EXAMPLE 1**
Coordinates of a Translated
Point

Suppose that a new coordinate system is defined by the translation

$$T : \begin{cases} x = X - 3 \\ y = Y + 4 \end{cases}$$

Calculate the coordinates of the point $P = (-1, 7)$ in the X-Y coordinate system.

Solution

To determine the new X-Y coordinates, we substitute $x = -1$ and $y = 7$ into the translation equations:

$$-1 = X - 3$$
$$7 = Y + 4$$

Solving for X and Y, we have

$$X = 2$$
$$Y = 3$$

That is, with respect to the new coordinate system, the point P has coordinates $(X, Y) = (2, 3)$.

> **EXAMPLE 2**
Equation of a Translated Curve

Suppose that a new coordinate system is defined by the translation

$$T : \begin{cases} x = X + 1 \\ y = Y - 5 \end{cases}$$

A curve C is the graph of the equation $x^2 - 2xy + 3y = 1$. Determine the equation of C in the new coordinate system.

Solution

To obtain the equation in the new coordinate system, substitute in the equation of C the expression $X + 1$ for x and $Y - 5$ for y:

$$(X + 1)^2 - 2(X + 1)(Y - 5) + 3(Y - 5) = 1$$
$$(X^2 + 2X + 1) - (2XY - 10X + 2Y - 10) + (3Y - 15) = 1$$
$$X^2 + 12X - 2XY + Y = 5$$

Rotation of Coordinate Axes

Another method for obtaining a new coordinate system from a given one is by rotating the axes. Suppose we are given an x-y coordinate system. We may obtain another one by rotating the axes about the origin through an angle θ. (See Figure 2.)

The relationship between the coordinates in the original coordinate system and those in the rotated coordinate system is given by the following result.

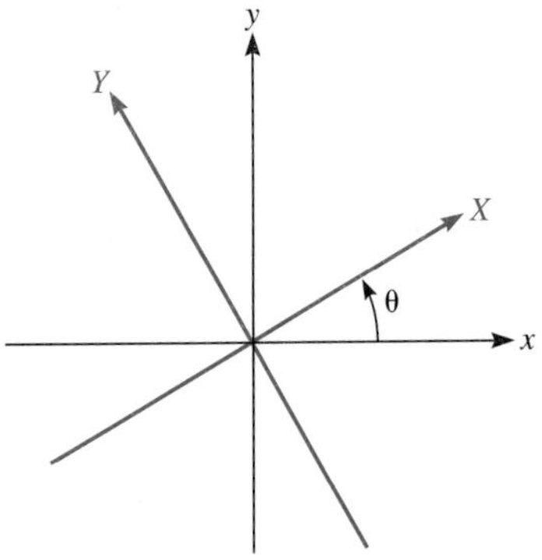

Figure 2

ROTATION OF COORDINATE AXES

Suppose that an x-y coordinate system is rotated through an angle θ. The coordinates (X, Y) in the rotated coordinate system are related to the coordinates (x, y) in the original coordinate system by the equations

$$T : \begin{cases} x = X \cos \theta - Y \sin \theta \\ y = X \sin \theta + Y \cos \theta \end{cases}$$

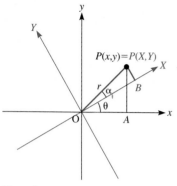

Figure 3

Proof   Draw right triangles POA and POB as shown in Figure 3. Denote by α the angle $\angle POB$. Then

$$\angle POA = \angle POB + \angle BOA$$
$$= \alpha + \theta$$

Using the coordinates in the first coordinate system, we see that $\overline{OA} = x$, and $\overline{AP} = y$. Therefore, by the definition of the trigonometric functions, we have

$$\cos(\theta + \alpha) = \frac{\overline{OA}}{\overline{OP}} = \frac{x}{r}$$

$$\sin(\theta + \alpha) = \frac{\overline{AP}}{\overline{OP}} = \frac{y}{r}$$

To these equations, apply the sum formulas for the sine and cosine functions to obtain

$$\begin{aligned}
x &= r\cos(\theta + \alpha) \\
&= r(\cos\theta\cos\alpha - \sin\theta\sin\alpha) \qquad (1) \\
&= r\cos\theta\cos\alpha - r\sin\theta\sin\alpha \\
y &= r\sin(\theta + \alpha) \\
&= r(\sin\theta\cos\alpha + \cos\theta\sin\alpha) \qquad (2) \\
&= r\sin\theta\cos\alpha + r\cos\theta\sin\alpha
\end{aligned}$$

Next, let's use the coordinates of P in the rotated coordinate system to obtain

$$\cos\alpha = \frac{\overline{OB}}{\overline{OP}} = \frac{X}{r}$$

$$\sin\alpha = \frac{\overline{BP}}{\overline{OP}} = \frac{Y}{r}$$

Therefore, we have

$$X = r\cos\alpha, \qquad Y = r\sin\alpha$$

Inserting these values for X and Y into equations (1) and (2), we have

$$\begin{aligned}
x &= r\cos\alpha\cos\theta - r\sin\alpha\sin\theta \\
&= X\cos\theta - Y\sin\theta \\
y &= r\cos\alpha\sin\theta + r\sin\alpha\cos\theta \\
&= X\sin\theta + Y\cos\theta
\end{aligned}$$

This completes the proof of the theorem.   ◆

➤ EXAMPLE 3
Equation of a Rotated Ellipse

Suppose that the x-y coordinate system is rotated $45°$. Determine the equation of the ellipse

$$x^2 + 4xy + 4y^2 = 4 \qquad (3)$$

in the rotated coordinate system.

Solution

In this case, $\theta = 45°$, so that the equations read

$$\begin{aligned}
x &= X\cos\theta - Y\sin\theta \\
&= \tfrac{\sqrt{2}}{2}X - \tfrac{\sqrt{2}}{2}Y \\
y &= X\sin\theta + Y\cos\theta \\
&= \tfrac{\sqrt{2}}{2}X + \tfrac{\sqrt{2}}{2}Y
\end{aligned}$$

Substituting these expressions for x and y in (3), we derive

$$x^2 + 4xy + 4y^2 = 4$$

$$\left(\tfrac{\sqrt{2}}{2}X - \tfrac{\sqrt{2}}{2}Y \right)^2 + 4\left(\tfrac{\sqrt{2}}{2}X - \tfrac{\sqrt{2}}{2}Y \right)\left(\tfrac{\sqrt{2}}{2}X + \tfrac{\sqrt{2}}{2}Y \right)$$

$$+ 4\left(\tfrac{\sqrt{2}}{2}X + \tfrac{\sqrt{2}}{2}Y \right)^2 = 4$$

$$\left(\tfrac{1}{2}X^2 - XY + \tfrac{1}{2}Y^2 \right) + 4\left(\tfrac{1}{2}X^2 - \tfrac{1}{2}Y^2 \right) + 4\left(\tfrac{1}{2}X^2 + XY + \tfrac{1}{2}Y^2 \right) = 4$$

$$\tfrac{9}{2}X^2 + 3XY + \tfrac{1}{2}Y^2 = 4$$

The last equation is the equation of the given ellipse in the rotated coordinate system.

➤ **EXAMPLE 4**

Equation of a Rotated Hyperbola

Suppose that the hyperbola with equation

$$4x^2 - 9y^2 = 36 \tag{4}$$

is rotated about the origin through an angle of $-30°$. Determine the equation of the rotated curve.

Solution

Let X-Y denote the rotated coordinate axes. Then the equations of rotation in this case read

$$x = X \cos(-30°) - Y \sin(-30°)$$

$$= \tfrac{\sqrt{3}}{2}X + \tfrac{1}{2}Y$$

$$y = X \sin(-30°) + Y \cos(-30°)$$

$$= -\tfrac{1}{2}X + \tfrac{\sqrt{3}}{2}Y$$

Substituting these expressions for x and y into (4) yields

$$4x^2 - 9y^2 = 36$$

$$4\left(\tfrac{\sqrt{3}}{2}X + \tfrac{1}{2}Y \right)^2 - 9\left(-\tfrac{1}{2}X + \tfrac{\sqrt{3}}{2}Y \right)^2 = 36$$

$$4\left(\tfrac{3}{4}X^2 + \tfrac{\sqrt{3}}{2}XY + \tfrac{1}{4}Y^2 \right) - 9\left(\tfrac{1}{4}X^2 - \tfrac{\sqrt{3}}{2}XY + \tfrac{3}{4}Y^2 \right) = 36$$

$$\tfrac{3}{4}X^2 + \tfrac{5\sqrt{3}}{2}XY - \tfrac{23}{4}Y^2 = 36$$

Because it is actually the hyperbola that is to be rotated, and not the axes, we replace X by x and Y by y in the last equation:

$$\tfrac{3}{4}x^2 + \tfrac{13\sqrt{3}}{2}xy - \tfrac{23}{4}y^2 = 36$$

This is the equation of the rotated hyperbola.

➤ **EXAMPLE 5**

Using Rotation to Eliminate the xy Term

Rotate the x-y coordinate axes so that in the new coordinate system, the equation

$$5x^2 + 3x + 4y^2 - xy + 10 = 0$$

has no XY term. Determine the angle θ of rotation.

Solution

Make the rotation specified by the equations

$$x = X \cos\theta - Y \sin\theta$$
$$y = X \sin\theta + Y \cos\theta$$

In the new coordinate system, the equation becomes

$$5x^2 + 3x + 4y^2 - xy + 10 = 0$$
$$5(X \cos\theta - Y \sin\theta)^2 + 3(X \cos\theta - Y \sin\theta) + 4(X \sin\theta + Y \cos\theta)^2$$
$$-(X \cos\theta - Y \sin\theta)(X \sin\theta + Y \cos\theta) + 10 = 0$$

Expanding this equation, we see that the coefficient of the term XY equals

$$2 \sin\theta \cos\theta + \cos^2\theta - \sin^2\theta = 0$$

We wish to choose θ so that this coefficient equals 0. That is, we wish to solve the equation

$$2 \sin\theta \cos\theta + \cos^2\theta - \sin^2\theta = 0$$

However, from the trigonometric identities we have proved, we have

$$\cos^2\theta - \sin^2\theta = \cos 2\theta$$
$$2 \sin\theta \cos\theta = \sin 2\theta$$

Substituting, we have the equation

$$\sin 2\theta + \cos 2\theta = 0$$
$$\frac{\sin 2\theta}{\cos 2\theta} = -1$$
$$\tan 2\theta = -1$$
$$2\theta = \tan^{-1}(-1)$$
$$2\theta = 135°$$
$$\theta = 67.5°$$

This choice of θ makes the coefficient of XY equal to 0.

Let's consider a generalization of the last example. Consider the curve with equation

$$Ax^2 + Bxy + Cy^2 + Dx + Ey + F = 0$$

Let's determine an angle θ, where $0° < \theta < 90°$, such that a rotation through this angle eliminates the xy term in the above equation. If we make the substitution

$$T : \begin{cases} x = X \cos\theta - Y \sin\theta \\ y = X \sin\theta + Y \cos\theta \end{cases}$$

in the equation of the curve, then the coefficient of the XY term is

$$-2A \sin\cos\theta + B(\cos^2\theta - \sin^2\theta) + 2C \sin\theta \cos\theta$$
$$= -A \sin 2\theta + B \cos 2\theta + C \sin 2\theta$$
$$= (-A + C)\sin 2\theta + B \cos 2\theta$$

Setting this coefficient equal to 0 gives us

$$B \cos 2\theta = (A - C)\sin 2\theta$$
$$\frac{\cos 2\theta}{\sin 2\theta} = \frac{A - C}{B}$$
$$\cot 2\theta = \frac{A - C}{B}$$

That is, we have proved the following result:

ELIMINATING THE XY TERM USING ROTATION OF COORDINATE SYSTEM

To eliminate the xy term in the equation
$$Ax^2 + Bxy + Cy^2 + Dx + Ey + F = 0$$
rotate the coordinate system through an angle θ satisfying
$$\cot 2\theta = \frac{A - C}{B} \qquad (0° < \theta < 90°)$$

Exercises 11.4

Suppose that a new coordinate system is defined as given in each of the following exercises. Calculate the coordinates of the given point P in the new coordinate system.

1. $x = X - 2,\ y = Y + 2,\ P = (1, -3)$
2. $x = X + 3,\ y = Y - 2,\ P = (-1, 2)$
3. $x = X + 3,\ y = Y + 5,\ P = (-1, -2)$
4. $x = X - 2,\ y = Y - 5,\ P = (4, 1)$
5. $x = X + 3,\ y = Y - 1,\ P = (-3, 3)$
6. $x = X - 1,\ y = Y - 3,\ P = (-1, -4)$
7. $x = X - 5,\ y = Y - 3,\ P = (-3, 0)$
8. $x = X + 2,\ y = Y + 3,\ P = (0, 0)$
9. $x = 2X + 1,\ y = Y - 5,\ P = (-2, -3)$
10. $x = X + 5,\ y = 2Y - 1,\ P = (-3, 4)$

Suppose that a new coordinate system is defined by the given translation. A curve C is the graph of the given equation. Determine the equation of C in the new coordinate system.

11. $x = X + 2,\ y = Y + 1,\ x^2 + 2xy - 3y = 1$
12. $x = X - 1,\ y = Y + 2,\ x^2 - 3x + 2y = 1$
13. $x = X - 2,\ y = Y - 3,\ x^2 + 4xy + 2y = 2$
14. $x = X - 5,\ y = Y - 1,\ 2x^2 + 3xy - 2y = 4$
15. $x = X + 3,\ y = Y + 1,\ x^2 + 2xy + y^2 = 1$
16. $x = X + 2,\ y = Y - 3,\ x^2 - 3xy + y^2 = 0$
17. $x = X - 5,\ y = Y + 2,\ 2x^2 - 3y^2 = 9$
18. $x = X + 3,\ y = Y + 2,\ x^2 + 3xy - y^2 = 4$
19. $x = 2X - 1,\ y = Y - 2,\ x + 2y = 6$
20. $x = X - 7,\ y = 3Y + 2,\ x - 3y = 6$

Suppose that a new coordinate system is determined by rotating the x-y coordinates by the given angle θ. Determine the equation of the rotated curve in this new coordinate system.

21. $\theta = 45°,\ x^2 + y^2 = 9$
22. $\theta = 60°,\ x^2 + y^2 = 16$
23. $\theta = 30°,\ x^2 - y^2 = 1$
24. $\theta = 45°,\ 4x^2 + 9y^2 = 36$
25. $\theta = 135°,\ x^2 + 4xy + 4y^2 = 4$
26. $\theta = -30°,\ y - 1 = \frac{1}{4}(x + 2)^2$
27. $\theta = -60°,\ y = x^2 + 2x - 3$
28. $\theta = 210°,\ 4x^2 + 4xy + y^2 = 4$
29. $\theta = -45°,\ 16x^2 - 25y^2 = 400$
30. $\theta = -45°,\ y = 3x - 5$

Determine the angle θ of rotation such that the rotated equation has no XY term in the new coordinate system.

31. $xy = 1$
32. $2xy + 9 = 0$
33. $x^2 = xy - 3$
34. $2x^2 + 4xy - 3y^2 - 4y - 2 = 0$
35. $x^2 - 5xy + 3y^2 + 2y + 10 = 0$
36. $2x^2 + 5x + 4y^2 - 3xy + 12 = 0$
37. $x^2 + 2xy + y^2 = 10$
38. $4x^2 - 6xy - 9y^2 = 36$
39. $y = 4x^2 - 6xy + 2$
40. $(x - y)^2 + 4 = y$

In Exercises 41–44, after a rotation of axes that eliminates the XY term, describe and sketch the graph of the equation.

41. $9x^2 - 16y^2 + 24xy - 16x + 60y + 25 = 0$
42. $4y^2 + 4xy = 18y \sqrt{5} - 6x \sqrt{5} - 45 - x^2$
43. $32x^2 + 53y^2 - 72xy - 80 = 0$
44. $5x^2 + 5y^2 - 8xy - 9 = 0$
45. Show that after a rotation of an angle θ, where $0° < \theta < 90°$ and
$$\cot 2\theta = \frac{A - C}{B}$$
the XY term has been eliminated from the equation
$$Ax^2 + Bxy + Cy^2 + Dx + Ey + F = 0$$

11.5 POLAR EQUATIONS OF CONICS

Definition of a Conic in Terms of Eccentricity

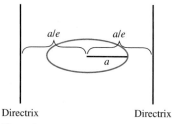

Directrix   Directrix

Figure 1

Recall that we defined the parabola in terms of a focus and a directrix. However, in defining an ellipse and a parabola, we did not make any mention of directrices. Actually, all conic sections have directrices. Let's now introduce them and use them to provide an alternative definition of the conics.

Consider first an ellipse with semimajor axis a and eccentricity e. The lines perpendicular to the major axis and at a distance a/e from the center are called the **directrices of the ellipse** (see Figure 1). Because $e < 1$, we see that $a/e > a$, so that the directrices don't intersect the ellipse. Each focus has an associated directrix, namely the directrix on the same side of the center. It is possible to verify the following property of the ellipse:

> ### DESCRIPTION OF AN ELLIPSE
>
> An ellipse is the set of all points P whose distance from the focus F is e times the distance from the associated directrix D, where $e < 1$.

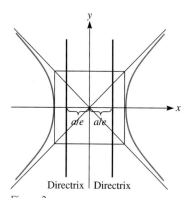

Figure 2

Let us now consider a hyperbola with semitransverse axis a and eccentricity e. The lines perpendicular to the transverse axis and at a distance a/e from the center are called **directrices of the hyperbola**. (See Figure 2.) In the case of a hyperbola, $e > 1$, so that $a/e < a$. Therefore, the directrices don't intersect the hyperbola. Each focus has an associated directrix, namely the directrix on the same side of the center. It is possible to verify the following property of the hyperbola:

> ### DESCRIPTION OF A HYPERBOLA
>
> A hyperbola is the set of all points P whose distance from the focus F is e times the distance from the associated directrix D, where $e > 1$.

Recall that the eccentricity of a parabola equals 1. From the definition of a parabola, we have the following property:

> ### DESCRIPTION OF A PARABOLA
>
> A parabola is the set of all points P whose distance from the focus F is e times the distance from the associated directrix D, where $e = 1$.

Combining the preceding properties, we arrive at the following result, which gives a unified description of all conic sections.

> ### ALTERNATIVE DESCRIPTION OF A CONIC
>
> Suppose that a conic has eccentricity e, focus F, and associated directrix D. Then the conic consists of all points P whose distance from F is e times the distance from D.

Equation of a Conic in Polar Coordinates

It is possible to give a very elegant form of the equation of a conic by using polar coordinates. This form of the equation makes use of the eccentricity and is independent of the type of conic involved.

DESCRIPTION OF A CONIC IN POLAR COORDINATES

Let e be the eccentricity of a conic and p the distance from the focus F to the directrix D. Set up a polar coordinate system with the origin at the focus F.

1. If the directrix is vertical, then the equation of the conic is

$$r = \frac{ep}{1 \pm e\cos\theta}$$

where the $+$ sign holds if the directrix is to the right of the focus, and the $-$ sign holds otherwise.

2. If the directrix is horizontal, then the equation of the conic is

$$r = \frac{ep}{1 \pm e\sin\theta}$$

where the $+$ sign holds if the directrix is above the focus, and the $-$ sign holds otherwise.

Conversely, each of the equations in statements 1 and 2 has a conic as its graph.

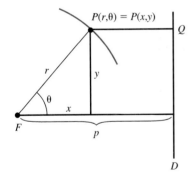

Figure 3

Proof   Let's prove statement 1 in the case $p > 0$. Draw the directrix and the focus as in Figure 3. Let $P(r, \theta) = P(x, y)$ be a typical point on the conic, given in both polar and rectangular coordinates. Because p is positive, the directrix lies to the right of the focus, as in the figure. According to the alternative description of a conic, we have

$$\text{Distance to focus} = e \cdot \text{Distance to directrix}$$
$$\overline{PF} = e\overline{PQ}$$
$$r = e(p - x)$$
$$= e(p - r\cos\theta)$$

Multiply out and solve for r:

$$r = ep - er\cos\theta$$
$$r + er\cos\theta = ep$$
$$r(1 + e\cos\theta) = ep$$
$$r = \frac{ep}{1 + e\cos\theta}$$

This gives the equation of the conic in polar coordinates. The proof in each of the other cases is similar.   ♦

Illustrations

➢ **EXAMPLE 1**
Graph of a Conic in Polar Coordinates

Sketch the graph of the conic with polar equation

$$r = \frac{20}{5 + 4\cos\theta}$$

Solution

First put the equation in the form of one of the standard equations of the preceding box. To do so, divide the numerator and denominator of the right side by 5 to obtain:

$$r = \frac{4}{1 + \frac{4}{5}\cos\theta}$$

From statement 1, we see that $e = \frac{4}{5}$. Therefore, the directrix is vertical and is located 5 units to the right of the focus. The conic is an ellipse with eccentricity $\frac{4}{5}$. Moreover,

$$pe = 4$$
$$p = \frac{4}{e} = \frac{4}{\frac{4}{5}} = 5$$

Because the major axis of the ellipse is horizontal, the vertices correspond to the points for which $\theta = 0, \pi$. That is, the vertices (in polar coordinates) are

$$(r, \theta) = \left(\frac{20}{5 + 4\cos 0}, 0\right), \qquad \left(\frac{20}{5 + 4\cos\pi}, \pi\right)$$
$$= \left(\frac{20}{9}, 0\right), \qquad (20, \pi)$$

In particular, the major axis is $\frac{20}{9} + 20 = \frac{200}{9}$, and the semimajor axis is $a = \frac{100}{9}$. Because $c = ea$, we have

$$b^2 = a^2 - c^2 = a^2 - (ea)^2 = a^2(1 - e^2)$$
$$= \left(\frac{100}{9}\right)^2 \left(1 - \left(\frac{4}{5}\right)^2\right)$$
$$= \frac{400}{9}$$
$$b = \frac{20}{3}$$

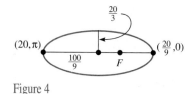

Figure 4

the graph is shown in Figure 4.

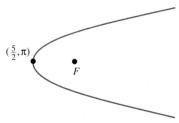

Figure 5

➤ EXAMPLE 2
Graph of a Conic in Polar Coordinates

Sketch the graph of the conic with polar equation

$$r = \frac{5}{2 - 2\cos\theta}$$

Solution

Divide both numerator and denominator by 2 to obtain the equation

$$r = \frac{\frac{5}{2}}{1 - \cos\theta}$$

According to statement 1, the directrix is vertical and lies to the left of the focus. Because $e = 1$, the graph is a parabola. Because the directrix is to the left of the focus, the parabola opens to the right. The point corresponding to $\theta = \pi$ is the vertex. This point is

$$(r, \theta) = \left(\frac{10}{2 - 2\cos\theta}, \pi\right) = \left(\frac{5}{2}, \pi\right)$$

The graph is shown in Figure 5.

➤ EXAMPLE 3
Determining the Equation of a
Conic in Polar Form

Give the polar form of the equation of an ellipse of eccentricity $\frac{1}{2}$ whose focus is at the origin and whose directrix is the line $x = -3$.

Solution
The directrix is vertical, and the distance from the focus to the directrix is 3. Therefore, the equation of the ellipse is

$$r = \frac{pe}{1 - e\cos\theta} = \frac{(3)\frac{1}{2}}{1 - (\frac{1}{2})\cos\theta} = \frac{3}{2 - \cos\theta}$$

We have used the $-$ sign in the denominator because the directrix lies to the left of the focus.

➤ EXAMPLE 4
The Orbit of the Earth

The orbit of the earth around the sun is an ellipse with the sun at the focus. The eccentricity of the orbit is 0.0167, and the major axis is 186 million miles. Find the equation of the orbit.

Solution
The semimajor axis of the ellipse is $\frac{186}{2} = 93$ million miles. Assume that the sun is located at the focus and that the directrices are vertical. Then the equation is of the form

$$r = \frac{ep}{1 + e\cos\theta}$$

We are given that $e = 0.0167$. Moreover, the vertices of the ellipse occur for $\theta = 0, \pi$. Therefore, if a denotes the semimajor axis, we have

$$2a = \frac{0.0167p}{1 + 0.0167} + \frac{0.0167p}{1 - 0.0167}$$

$$= 0.033409p$$

$$p = \frac{2a}{0.033509} = 5567.4$$

Therefore, the desired equation is

$$r = \frac{0.0167 \cdot 5567.4}{1 + 0.0167\cos\theta} = \frac{92.976}{1 + 0.0167\cos\theta}$$

Exercise 11.5

Determine an equation in polar form of the conic with focus at the origin and having the following properties:

1. Parabola, directrix $x = 1$
2. Parabola, directrix $y = 2$
3. Parabola, directrix $x = -2$
4. Parabola, directrix $y = 1$
5. Parabola, vertex $(5, 0)$
6. Parabola, vertex $(3, \pi)$
7. Parabola, vertex $\left(4, \frac{\pi}{2}\right)$
8. Parabola, vertex $\left(3, \frac{3\pi}{2}\right)$
9. Ellipse, $e = \frac{1}{3}$, directrix $x = 5$
10. Ellipse, $e = \frac{2}{3}$, directrix $y = 3$
11. Ellipse, vertices $(4, 0)$ and $(2, \pi)$
12. Ellipse, vertices $\left(2, \frac{\pi}{2}\right)$ and $\left(3, \frac{3\pi}{2}\right)$

13. Hyperbola, $e = 2$, directrix $x = 3$

14. Hyperbola, $e = 2$, directrix $y = -1$

15. Hyperbola, vertices $(2, 0)$ and $(4, 0)$

16. Hyperbola, vertices $\left(4, \dfrac{3\pi}{2}\right)$ and $\left(1, \dfrac{3\pi}{2}\right)$

Describe the conic represented by each of the following polar equations

17. $r = \dfrac{1}{1 + \cos\theta}$

18. $r = \dfrac{3}{1 - \cos\theta}$

19. $r = \dfrac{2}{1 + 2\sin\theta}$

20. $r = \dfrac{5}{1 - 3\cos\theta}$

21. $r = \dfrac{4}{1 + 10\cos\theta}$

22. $r = \dfrac{3}{2 + \cos\theta}$

23. $r = \dfrac{4}{1 + \sin\theta}$

24. $r = \dfrac{6}{2 - 3\sin\theta}$

25. $r = \dfrac{10}{25 + 6\cos\theta}$

26. $r = \dfrac{5}{10 + 25\sin\theta}$

27. $r = \dfrac{3}{1 - 3\cos\theta}$

28. $r = \dfrac{1}{3 - 3\sin\theta}$

Determine the polar equations of the conics with the following equations in rectangular coordinates. (Set up the polar coordinate system so that one focus is at the origin.)

29. $x^2 - y^2 = 1$

30. $y^2 - 2x^2 = 1$

31. $\dfrac{x^2}{4} - \dfrac{y^2}{16} = 1$

32. $\dfrac{y^2}{16} - \dfrac{x^2}{25} = 1$

♦ Applications

33. A satellite has an orbit with the earth as focus and with eccentricity 0.05. The closest distance of the orbit from the center of the earth is 12,000 miles. Determine the equation of the orbit.

34. Refer to the preceding exercise. Determine the furthest distance of the orbit from the center of the earth.

11.6 PARAMETRIC EQUATIONS

In the early chapters of the book, we introduced graphs of equations of the form $f(x, y) = 0$, where $f(x, y)$ is an expression in x and y. We have graphed many equations and arrived at an interesting collection of graphs, not the least of which are the conic sections considered in the preceding sections.

Rather than specifying a curve by an equation relating x and y, it is often expedient to relate both x and y to a third variable, which we will call t.

Definition 1 Parametric Equations for a Curve	A set of **parametric equations** in the **parameter** t is a pair of equations $$x = f(t), \qquad y = g(t) \tag{1}$$ where t ranges over an interval (finite or infinite). Each value of t corresponds to a point $(x, y) = (f(t), g(t))$. As t ranges over the interval, the set of points (x, y) traces out a curve in the plane.

In many examples, the parameter has a physical interpretation. For instance, t may represent time, so that the point $(f(t), g(t))$ may be interpreted as the position of a point at time t.

A set of parametric equations for a curve gives much more information than just the set of points describing the curve. As t runs through a specified set of values, the parametric equations specify the speed with which the curve is traced out, the direction in which it is traced, and even if certain portions of the curve are traced over themselves more than once.

➤ **EXAMPLE 1**
Motion of a Particle

The position of a particle at time t is given by the parametric equations

$$x = 2t, \qquad y = 3t, \qquad 0 \le t \le 5$$

Describe the motion of the particle.

Solution

At time $t = 0$, the particle starts at point $(x, y) = (2 \cdot 0, 3 \cdot 0) = (0, 0)$. The following table lists the location of the point at various times t:

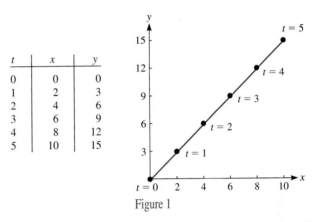

t	x	y
0	0	0
1	2	3
2	4	6
3	6	9
4	8	12
5	10	15

Figure 1

We plot the various points in Figure 1. We see that the graph is a line. As time runs from 0 to 5, the point moves along the line from $(0, 0)$ to $(10, 15)$.

➤ **EXAMPLE 2**
Graphing Parametric Equations

Graph the curve described by the parametric equations

$$x = t, \qquad y = \tfrac{3}{2}t, \qquad 0 \le t \le 10$$

Solution

The curve is the same curve shown in Figure 1. However, the present curve is traced out at half the speed. That is, for any given time t, the location of the present curve is the same as the location on the previous curve at time $t/2$. This example shows that there is no single set of parametric equations that describes a particular curve.

➤ **EXAMPLE 3**
Translating Parametric
Equations into
Rectangular Coordinates

Write the following set of parametric equations in rectangular coordinates:

$$x = \sqrt{t}, \qquad y = 3t + 1, \qquad t \ge 0$$

Solution

We may solve the first equation for t in terms of x:

$$x = \sqrt{t}$$
$$t = x^2$$

Now substitute the expression for t into the equation for y:

$$y = 3t + 1 = 3x^2 + 1$$

From the first equation, the values of x are all nonnegative. In fact, each nonnegative real number is a value of x for some value of t. Therefore, when we graph the nonparametric equation $y = 3x^2 + 1$, we must subject x to the restriction $x \geq 0$, which is inherited from the parametric equations.

➤ **EXAMPLE 4**
Parametric Form of a Circle

Graph the curve described by the parametric equations

$$x = 2 \cos t, \qquad y = 2 \sin t, \quad 0 \leq t \leq 2\pi$$

Solution

To solve this problem, let's pull a rabbit out of a hat. Draw a circle of radius 2 about the origin. From the definitions of the trigonometric functions sine and cosine, the coordinates of a point P on the circle determined by an angle t in standard position are $(\cos t, \sin t)$. So if we interpret the parameter t as the angle shown in Figure 2, we see that the parametric equations describe the point P. As t ranges over the interval $[0, 2\pi]$, the point described by the parametric equations traces out the circle shown, in the counterclockwise direction. This example illustrates how the parameter may often be interpreted in geometric terms, as an angle, length, or area.

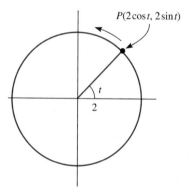

Figure 2

In some instances, it is possible to solve the parametric equation for t in terms of either x or y and then eliminate the parameter t, thereby obtaining an equation for the curve in nonparametric form.

➤ **EXAMPLE 5**
Translating Parametric
Equations into
Rectangular Coordinates

Write the following equations in rectangular coordinates:

$$x = e^t, \qquad y = t^2$$

Solution

Note that the value of x is positive. Moreover, solving the first equation for t in terms of x, we have

$$t = \ln x$$

Substituting this value into the equation for y, we have

$$y = t^2 = (\ln x)^2, \quad x > 0$$

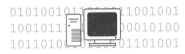

New Technology

Using a Graphing
Utility to Graph Parametric
Equations

Most graphing utilities allow you to graph parametric equations. For instance, the TI-81 can plot parametric equations by entering the [MODE] menu and selecting `Param` instead of `Function`. Or in *Mathematica*, you would use the `ParametricPlot` command. This is very useful because the graphs of some equations, such as those for many conic sections, are not easily produced with just one equation on most graphing utilities.

On the TI-81, after setting the [MODE], return to the [Y=] screen. The screen will now look like the one shown in Figure 3. At this screen you can enter up to three parametrically defined equations. For example, to graph

$$x = 3 \cos t, \qquad y = 3 \sin t, \qquad 0 \le t \le 2\pi$$

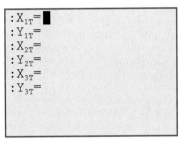

Figure 3

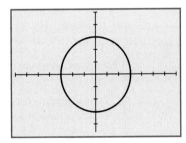

Figure 4

enter $3 \cos t$ for X_{1T} and $3 \sin t$ for Y_{1T}. Then graph the equation. You may need to adjust your viewing box using the `Square` command in the [ZOOM] screen to get a graph that looks like Figure 4. ◄

Exercises 11.6

Sketch the graphs of the following parametric equations.

1. $x = 3t, y = 5t - 1, \quad 0 \le t \le 1$

2. $x = 2t - 1, y = t + 1, \quad -1 \le t \le 1$

3. $x = 2t, y = e^t, \quad t \ge 0$

4. $x = 2t, y = \ln t, \quad 0 < t < 5$

5. $x = 2\cos\theta, y = 2\sin\theta, \quad 0 \le \theta \le 2\pi$

6. $x = 3 + 2\sin\theta, y = 1 + 2\cos\theta, \quad 0 \le \theta \le 2\pi$

7. $x = \sin\theta, y = \cos\theta, \quad 0 \le \theta \le 2\pi$

8. $x = \sin 3t, y = \cos 3t, \quad 0 \le t \le \pi$

9. $x = \cos\theta, y = \sin\theta, \quad -\pi \le \theta \le \pi$

10. $x = 3\sin\theta, y = -3\cos\theta, \quad 0 \le \theta \le 6\pi$

11. $x = 2t, y = 3t, \quad t = 1 \text{ to } t = 0$

12. $x = 2 + \cos t, y = -1 + 3\sin t, \quad 0 \le t \le 2\pi$

13. $x = \dfrac{1}{t}, y = t, \quad t > 0$

14. $x = \dfrac{1}{t^2 + 1}, y = \dfrac{t}{t^2 + 1}, \quad$ all real t

15. $x = \sqrt{t}, y = \dfrac{1}{t}, \quad t > 0$

16. $x = \cos t, y = \sin t, \quad 0 \le t \le \dfrac{\pi}{2}$

17–32. For each parametric equations in Exercises 1–16, write the equation in rectangular coordinates. Be careful to state the correct domain in each case.

Show that the conic sections may be written in the following parametric form.

33. Circle: $(x - h)^2 + (y - k)^2 = r^2$:

$$x = h + r\cos\theta, y = k + r\sin\theta, \quad 0 \le \theta \le 2\pi$$

34. Ellipse: $\dfrac{(x - h)^2}{a^2} + \dfrac{(y - k)^2}{b^2} = 1$:

$$x = h + a\cos\theta, y = k + b\sin\theta, \quad 0 \le \theta \le 2\pi$$

35. Hyperbola: $\dfrac{(x - h)^2}{a^2} - \dfrac{(y - k)^2}{b^2} = 1$:

$$x = h + a\sec\theta, y = k + b\tan\theta,$$

$$\begin{cases} -\dfrac{\pi}{2} < \theta < \dfrac{\pi}{2} & \text{(left branch)} \\[2mm] \dfrac{\pi}{2} < \theta < \dfrac{3\pi}{2} & \text{(right branch)} \end{cases}$$

36. Derive a set of parametric equations for a circle of radius 2 and center $(3, -1)$.

37. Derive a set of parametric equations for the unit circle.

38. Derive a set of parametric equations for an ellipse with equation $x^2/4 + y^2/25 = 1$, traced out twice in the clockwise direction, starting at the point for which $x = 0$.

39. Derive a set of parametric equations for an ellipse with center at the origin, horizontal semimajor axis 2, and semiminor axis 1.

40. Derive a set of parametric equations for an ellipse with center at $(5, -2)$, vertical semimajor axis 3, and semiminor axis 2.

41. Derive a set of parametric equations for an ellipse with foci at $(5, 1)$ and $(-3, 1)$ and semimajor axis 10.

42. Derive a set of parametric equations for the hyperbola $x^2 - y^2 = 1$.

43. Derive a set of parametric equations for the right branch of the hyperbola $(x - 1)^2/3 - (y - 1)^2/6 = 1$.

44. Prove that the line segment connecting (x_1, y_1) to (x_2, y_2) may be written parametrically in the form

$$x = x_1 + t(x_2 - x_1), \quad y = y_1 + t(y_2 - y_1), \; 0 \le t \le 1$$

45. Write parametric equations of a line segment going from $(2, 3)$ to $(-1, 4)$.

46. Describe the graph of the parametric equations

$$x = 3 + 4t, \, y = -1 + 2t, \quad 0 \le t \le 1$$

47. Describe the graph of the parametric equations obtained by replacing t by $1 - t$ in the preceding exercise.

Suppose that $x = f(t), y = g(t), 0 \le t \le 1$ is a set of parametric equations for a curve C. Describe the graph of the following equations.

48. $x = f(2t), y = g(2t), \quad 0 \le t \le \frac{1}{2}$

49. $x = f(1 - t), y = g(1 - t), \quad 0 \le t \le 1$

50. $x = g(t), y = f(t), \quad 0 \le t \le 1$

51. $x = f(t) + 1, y = g(t) + 2, \quad 0 \le t \le 1$

➡ **Technology**

Use the parametric plotting features of your graphing utility and the formulas from Exercises 33–35 to graph the conic sections in Exercises 52–59.

52. $\dfrac{x^2}{16} + \dfrac{y^2}{36} = 1$

53. $(x - 2)^2 + (y + 3)^2 = 11$

54. $\dfrac{(x - 2)^2}{49} + \dfrac{(y + 2)^2}{20} = 1$

55. $7(x - 2)^2 + 6(y - 2)^2 = 17$

56. $\dfrac{x^2}{2} - \dfrac{y^2}{9} = 1$

57. $7(y + 1)^2 - 5(x + 2)^2 = 13$

58. $\dfrac{(x - 2)^2}{9} - \dfrac{(y + 2)^2}{20} = 1$

59. $(y - 2)^2 + (x - 2)^2 = 17$

11.7 CHAPTER REVIEW

Important Concepts, Properties, and Formulas—Chapter 11

Parabolas	A parabola is the set of all points whose distance from the focus equals the distance from the directrix. Standard form of equations of parabolas: $\quad (x - h)^2 = 4p(y - k)$ axis vertical $\quad (y - k)^2 = 4p(x - h)$ axis horizontal where p = directed focal length.	p. 553
Ellipses	An ellipse is the set of all points the sum of whose distance from the two foci is a given constant. Standard form equation of an ellipse: $\dfrac{(x - h)^2}{a^2} + \dfrac{(y - k)^2}{b^2} = 1,$ If $a > b$, then major axis horizontal. If $a < b$, then major axis vertical If c is the focal length, then $\quad\quad a^2 = b^2 + c^2$	p. 560

Hyperbola	A hyperbola is the set of all points the difference of whose distances from the two foci is a given constant. Standard form equations of hyperbolas: $$\frac{(y-k)^2}{a^2} - \frac{(x-h)^2}{b^2} = 1, \quad \text{vertical transverse axis}$$ $$\frac{(x-h)^2}{a^2} - \frac{(y-k)^2}{b^2} = 1, \quad \text{horizontal transverse axis}$$ If c is the focal length, then $$c^2 = a^2 + b^2$$ Asymptotes are the lines: $$y - k = \pm\frac{b}{a}(x - h) \quad \text{(horizontal transverse axis)}$$ $$y - k = \pm\frac{a}{b}(x - h) \quad \text{(vertical transverse axis)}$$	pp. 566, 567
Translation of coordinate axes	Suppose that the origin of the X–Y coordinate system is at the point $P = (h, k)$ of the x–y coordinate system. Then the relationship between the coordinate systems is: $$T : \begin{cases} x = X + h \\ y = Y + k \end{cases}$$	p. 572
Rotation of coordinate axes	Suppose that an x–y coordinate system is rotated through an angle θ. The coordinates (X, Y) in the rotated coordinate system are related to the coordinates (x, y) in the original coordinate system by the equations: $$T : \begin{cases} x = X \cos\theta - Y \sin\theta \\ Y = X \sin\theta + Y \cos\theta \end{cases}$$	p. 573
Polar form of equations for conics	Let e be the eccentricity of a conic and p the distance from the focus F to the directrix D. Set up a polar coordinate system with the origin at the focus F. 1. If the directix is vertical, then the equation of the conic is $$r = \frac{ep}{1 \pm \cos\theta}$$ where the $+$ sign holds if the directrix is to the right of the focus and the $-$ sign holds otherwise. 2. If the directrix is horizontal, then the equation of the conic is $$r = \frac{ep}{1 \pm e\sin\theta}$$ where the $+$ sign holds if the directrix is above the focus and the $-$ sign holds otherwise.	p. 579
Parametric equation of a curve	$$x = f(t), \quad y = g(t), \quad a \le t \le b$$	p. 583

Cumulative Review Exercises—Chapter 11

Sketch the graphs of the following conics. For ellipses, determine the semimajor and semiminor axes. For hyperbolas, determine the transverse and conjugate axes and asymptotes.

1. $y^2 = 3x + 1$

2. $x^2 = 2y - 3$

3. $2x^2 - y^2 = 4$

4. $y^2 - 3x^2 = 10$

5. $x^2 + (y - 3)^2 = 10$

6. $2(x - 1)^2 + 5(y - 3)^2 = 25$

7. $\dfrac{x^2}{25} - y^2 = 6$

8. $\dfrac{(x - 1)^2}{3^3} + \dfrac{(y + 2)}{4^3} = 1$

9. $-x^2 + 3x + y = 1$

10. $x^2 = y^2 + 3$

11. $y^2 = -5x - 8$

12. $5x^2 - 3x + 2y^2 + 6y + 3 = 0$

13. $x^2 - 2y^2 + 3y = 0$

14. $(x + 1)^2 + 2(y - 3)^2 + 3x = 0$

Suppose that a new coordinate system is defined as given in each of the following exercises. Calculate the coordinates of the given point P in the defined coordinate system.

15. $x = X + 2$, $y = X - 3$, $P = (1, 4)$

16. $x = X - 3$, $y = Y + 4$, $P = (-2, 3)$

17. $x = X - 4$, $y = Y - 5$, $P = (4, -2)$

18. $x = X + 5$, $y = Y - 2$, $P = (-1, -3)$

19. $x = 3X + 1$, $y = Y + 2$, $P = (-2, -4)$

20. $x = 4X - 2$, $y = 2Y + 1$, $P = (0, 3)$

Suppose that a new coordinate system is defined by the given translation. A curve C is the graph of the given equation. Determine the equation of C in the new coordinate system.

21. $x = X - 1$, $y = Y + 2$, $x^2 + 3xy - 2y = 2$

22. $x = X + 2$, $y = Y - 3$, $x^2 + 2xy - 2y = 3$

23. $x = X + 3$, $y = Y - 1$, $x^2 + 4xy - y^2 = 1$

24. $x = X - 3$, $y = Y - 2$, $x^2 - 3xy - y^2 = 4$

25. $x = X - 3$, $y = Y - 2$, $2x + 3 = 6$

26. $x = X + 1$, $y = Y - 3$, $x^2 + 3xy = 6$

Suppose that a new coordinate system is determined by rotating the x-y coordinates by the given angle θ. Deter-

mine an equation of the rotated curve in this new coordinate system.

27. $\theta = 60°; 9x^2 - y^2 = 36$

28. $\theta = 45°; x^2 - y^2 = 25$

29. $\theta = -45°; x^2 - y^2 = 49$

30. $\theta = -30°; 4x^2 - 9y^2 = 36$

31. $\theta = 210°; y = x - 2$

32. $\theta = 135°; x^2 - 4xy + y^2 = 4$

Determine the angle θ of rotation such that the given equation has no XY term.

33. $5x^2 - 8xy + 5y^2 = 9$

34. $xy = 4$

35. $5x^2 + 6xy\sqrt{3} - y^2 + 8x - 8y\sqrt{3} - 12 = 0$

36. $25x^2 + 4y^2 = 100$

Determine the polar equations of the following conics.

37. Parabola, directrix $x = -2$

38. Ellipse, $e = \frac{1}{3}$, directrix $x = 3$

39. Hyperbola, vertices $(1, 0)$ and $(6, \pi)$

40. Hyperbola, $e = 2$, directrix $y = -3$

Graph the following conics.

41. $r = \dfrac{1}{1 + 2\cos\theta}$

42. $r = \dfrac{4}{2 - \cos\theta}$

43. $r = \dfrac{2}{1 - 2\sin\theta}$

44. $r = \dfrac{6}{1 + 3\cos\theta}$

Graph the following parametric equations.

45. $x = -t + 1$, $y = 2t$, $\quad 0 \le t \le 4$

46. $x = t^2$, $y = -t$, $\quad 0 \le t \le 1$

47. $x = -1 + 2\sin t$, $y = 4 + 2\cos t$, $\quad 0 \le t \le \pi$

48. $x = \sin t$, $y = t$, $\quad 0 \le t \le \pi$

Determine the rectangular form of the following parametric equations.

49. $x = 3z - 1$, $y = 2z + 3$, $\quad 0 \le z \le 5$

50. $x = t^2$, $y = \dfrac{1}{t - 1}$, $\quad t > 1$

51. $x = t$, $y = \sqrt{t - 1}$, $\quad t \ge 1$

52. $x = -1 + \cos t$, $y = 5 + \cos t$, $\quad 0 \le t \le 2\pi$

53. $x = 4\sin t$, $y = 3\cos t$, $\quad 0 \le t \le 2\pi$

54. $x = 5\ln(t - 1)$, $y = 3t + 1$, $\quad 1 < t \le 5$

ANSWERS TO ODD-NUMBERED EXERCISES

CHAPTER 1

Section 1.1, page 7

1. $24, 0, \sqrt[5]{32}, \sqrt{16}, -234$ 3. $2\pi, \sqrt{5}$

5. $0.\overline{285714}$ 7. $1.91\overline{6}$ 9. $\frac{1}{4}$

11. 11 13. 10 15. 0

17. $\sqrt{10}$ 19. $\sqrt{5} - 2$ 21. $>$

23. $<$ 25. 3 27. $\frac{7}{20}$

29. 27.5

31. $\{x : x < 0\}$

33. $\{x : x \geq -3\}$

Section 1.2, page 22

1. $\dfrac{1}{a^3}$ 3. q^4 5. $-12x^{11}$ 7. $\dfrac{36}{y^2}$

9. $9a^2b^8$ 11. $6^4x^4y^4$ 13. $\dfrac{1}{t^5}$ 15. $\dfrac{a^5}{b^4}$

17. $\dfrac{b^8}{9x^4}$ 19. $\dfrac{8b^3}{27a^3}$ 21. $\dfrac{yz^8}{2x^3}$ 23. $4\sqrt{2}(t+1)$

25. $3\sqrt{2x}$ 27. $2xy^2$ 29. $\dfrac{7\sqrt{2}a}{b^3}$ 31. $3\sqrt[3]{2}xy^{4/3}$

33. 3 35. $\dfrac{x^2\sqrt{6}}{3}$ 37. $5ab^{5/3}$ 39. $\dfrac{3ac^2}{b}$

41. 2

43. a^{20}

45. $a^3 - 8$

47. $6a^3b^4 + 4a^4b^3 - 15a^2b^3 - 10a^3b^2$

49. $25x^2 - 30xy + 9y^2$

51. $4a^2b^2 + 12ab^2c + 9b^2c^2$

53. $16x^4 - 56x^3y + 49x^2y^2$

55. $3x(1 - xy)$

57. $(x - 7)(x + 3)$

59. $2(2x - 3)(5x - 2)$

61. $2ab(7a - 6b)$

63. $(y - w)(x - z)$

65. $(x - 2)(x + 2)(2x + 1)$

67. $(23 - 18x)(23 + 18x)$

69. $4x(2a - 3x + a^2 - 5)$

71. $(x - 3)(x^2 + 3x + 9)$

73. $2a(2x + 3)^2$

75. $(a^n - b^n)(a^n + b^n)$

77. $(6t^n - 5)^2$

79. $\left(\dfrac{x^4}{10} + 1\right)\left(\dfrac{x^8}{100} - \dfrac{x^4}{10} + 1\right)$

81. 1

83. $\dfrac{t - 1}{t + 1}$

85. $\dfrac{(x + 2)(7x^2 + 3)}{6(x^2 + 2)(x^2 - 3)}$

87. $\dfrac{6x - 10}{2x + 1}$

89. $\dfrac{x - 7 - 3y}{x - 3}$

91. $\dfrac{3t^2 - 4t + 7}{t^2 + 3t - 5}$

93. $\dfrac{-2x^2 + 4x - 6}{x^2 - 3x - 8}$

95. $\dfrac{a^2 + b^2}{a + b}$

97. $\dfrac{y(x + z) + xz}{x + z}$

99. $-\dfrac{1}{x^2 + xh}$

101. $-\dfrac{3x^2 + 3xh + h^2}{x^3(x + h)^3}$

103. $\dfrac{\sqrt{3xy}}{2y}$

105. $4(\sqrt{6} - \sqrt{5})$

107. $\dfrac{\sqrt{a} + \sqrt{b}}{a - b}$

109. $\dfrac{(\sqrt{a} + \sqrt{2})(\sqrt{a} + \sqrt{b})}{a - b}$

111. $\dfrac{1}{\sqrt{a + h} + \sqrt{a}}$

113. $\dfrac{a - 2}{a - 2\sqrt{2a} + 2}$

115. 2×10^9

117. 1.6×10^{-18}

119. 3.56×10^{12}

121. $12,000,000,000

123. 0.000000000000000000000000383

125. 0.00000000000000000578

127. 1.29316×10^{19} miles

129. 2.387×10^5 miles

131. 66.779 ft

133. 11.355 sec

135. 4983 m/sec

137. 5.7557%

139. 329.67 m/sec

Section 1.3, page 31

1. $\pm\sqrt{3}$

3. No solutions

5. No solutions

7. 3 or 1

9. $\frac{2}{3}$ or $-\frac{4}{3}$

11. $-\frac{5}{2} \pm \frac{\sqrt{65}}{2}$

13. $-\frac{2}{3} \pm \frac{\sqrt{7}}{3}$

15. No solutions

17. $\dfrac{-1 \pm \sqrt{33}}{4}$

19. 0 or $\frac{5}{9}$

21. $\frac{3}{2}$ or $-\frac{7}{3}$

23. No solutions

25. $\dfrac{43 \pm 5\sqrt{85}}{6}$

27. $\dfrac{-\sqrt{2} \pm \sqrt{2 + 4\sqrt{3}}}{2}$

29. No solutions

31. $-\frac{2}{7}$ and $\frac{4}{3}$

33. $\dfrac{17}{2} \pm \dfrac{\sqrt{293}}{2}$

35. $\dfrac{5 \pm \sqrt{73}}{6}$

37. $\dfrac{-3 \pm \sqrt{329}}{40}$

39. $\dfrac{-15 \pm 5\sqrt{67}}{29}$

41. 100

43. No solutions

45. 9

47. 11

49. $0, -3, \frac{4}{5}$

51. 1

53. $-3, -1, 1$

55. 0,1

57. $-1,1$

59. 4

61. $\dfrac{-5 \pm \sqrt{221}}{14}$

63. $8\sqrt{10}$ ft

65. $400, $500

67. *a.* $t = 6$ sec ($t = -10$ rejected)

  b. $t = -2 + \frac{\sqrt{131}}{2}$ sec

69. 45

71. *a.* Young's : $C = 1$ cc; Cowling's : $C = 0.8\overline{3}$ cc;

  b. $A = \dfrac{11 + \sqrt{73}}{2} \approx 9.8$ yr or $A = \dfrac{11 - \sqrt{73}}{2} \approx$ 1.2 yr

73. 6.5%

75. 10.92%

77. 2 in. $\times$ 2 in.

79. *a.* 90   *b.* Yes, if there are 8 teams in the league.

Section 1.4, page 34

1. $[-2, 3)$

3. $[-4, 5]$

5. $[-2, 3)$

7. $[-2, \infty)$

9. $\{x : -1 \le x < 3\}$

11. $\{x : x > -2\}$

13. $\{x : x \le 2\}$

15. $[-1, \infty)$

17. $(-\infty, -\frac{3}{4})$

19. $(-\infty, 4]$

21. $[-\frac{4}{5}, \infty)$

23. $[4, \infty)$

25. $(-\infty, -2.6)$

27. $[-4, 5]$

29. $[1, \frac{13}{3}]$

31. $(-5, -2)$

33. $[25, \infty)$

35. $[-8, \frac{14}{3})$

37. $(\frac{14}{3}, \infty)$

39. $(-\infty, -\frac{3}{2})$

41. $(-\infty, -5) \cup (3, \infty)$

43. $(-\infty, 3] \cup [4, \infty)$

45. $(-5, 5)$

47. $(0, \frac{2}{3})$

49. $(-3, -1) \cup (2, \infty)$

51. $(-\infty, -5] \cup [0, 2]$

53. $(-\infty, -2) \cup (-1, 1)$

55. $(-\infty, -1) \cup (3, \infty)$

57. $[\frac{3}{2}, 4)$

59. $(-\infty, -4) \cup (-1, 2)$

61. $(-\infty, 0) \cup (0, \infty)$

63. $[-23, 23]$

65. $(-\infty, -14) \cup (14, \infty)$

67. $(-\infty, -2) \cup (6, \infty)$

69. $(-1, 5)$

71. $[-6, \frac{22}{3}]$

73. $x \le \frac{1}{2}; (-\infty, \frac{1}{2}]$

75. $(3, \infty)$

77. $-4 < m < 4$

79. No solution

81. Case 1 : $(t + 8)(t - 2) > 0$
    Case 2: $(t - 8)(t + 2) > 0$
    $(-\infty, -2) \cup (2, \infty)$

83. $[-\frac{11}{4}, -\frac{4}{3}) \cup (-\frac{4}{3}, -\frac{5}{8}]$

85. $[-3, 2] \cup [1, 2]$

87. $p \le \frac{1}{2}$

89. $(-\infty, 0) \cup (0, \infty)$

91. $(-\infty, -1) \cup (\frac{1}{5}, 2)$
    $\cup (2, \infty)$

93. $(0, 2\sqrt{2}) \cup (2\sqrt{2}, \infty)$

95. $0 \le h \le 33\frac{1}{3}$ in.

97. $52 < S < 92$

99. *a.* $x \ge 167$
    b. $x \ge 167$

101. $8 \le n \le 17$
     n an integer

103. [\$2.775 million, \$3.225 million]

105. $(6, 14)$

Section 1.5, page 49

1.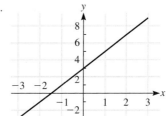

3. *a.* Quadrant I; *b.* Quadrant IV; *c.* y-axis; *d.* x-axis

5. $\sqrt{61}$ 7. $2\sqrt{2}$ 9. 6

11. $\sqrt{9a^2 + 4}$ 13. $\sqrt{a^2 + b^2}$ 15. $\sqrt{a + b}$

17. 1   19. $\sqrt{a^2 + b^2}$ 21. $\left(-1, -\frac{1}{2}\right)$

23. $(1, -4)$ 25. $(1, 5)$ 27. $\left(\frac{a}{2}, 4\right)$

29. $\left(\frac{a}{2}, \frac{b}{2}\right)$ 31. $\left(\frac{\sqrt{a}}{2}, \frac{\sqrt{b}}{2}\right)$ 33. $\left(0, -\frac{5}{3}\right)$

35. $(14.5, 0)$ 37. $-2 \pm \sqrt{33}$ 39. 21

41. $\left(0, -\frac{61}{20}\right)$ 43. No   45. Yes

47. $x - y = 2$

49. Radius $= \frac{5\sqrt{2}}{2}$; center $= (7, 10)$

55. The points $(a, 0)$ greater than $\sqrt{19}$ from $(0, 3)$ are $(a, 0)$,
    such that $|a| > \sqrt{10}$.

Section 1.6, page 59

1.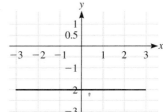

3.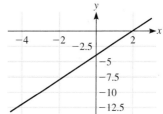

5.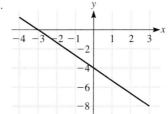

7.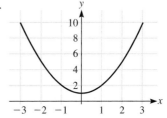

9.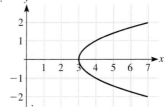

11.

13.

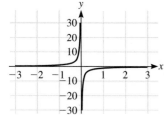

15.

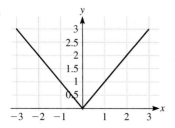

17.

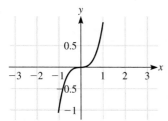

19.

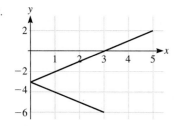

21.

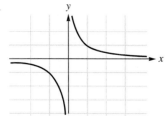

23.

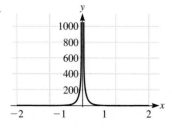

25.

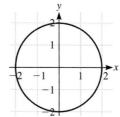

27.

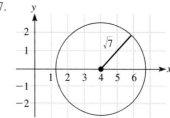

29.

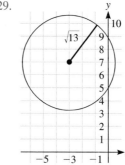

31.

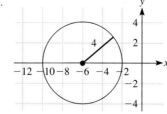

33.

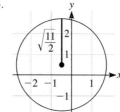

35. $(x + 6)^2 + (y - 1)^2 = 4$
37. $x^2 + y^2 = 39$
39. $(x + 2)^2 + (y - 3)^2 = 4$
41. $(x - 11)^2 + (y + 11)^2 = 121$
43. $(x + 1)^2 + (y - 6)^2 = 20$

45. $(x + 2)^2 + (y + 3)^2 = 9$
47. $(x - 7)^2 + (y - 10)^2 = \frac{25}{2}$
49. $(-5, 6), r = 9$
51. $\left(\frac{3}{2}, -2\right), r = \frac{15}{2}$
53. Yes   55. No   57. Yes   59. No
61. The solution set is the set of all points on and enclosed by the circle whose center is at $(-3, -5)$ and whose radius equals 6.
63. *a.*

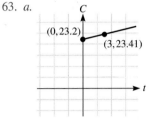

   b. The C intercept is the cost, in cents per mile, of driving a car in 1980. x-intercept: $t = 331.4$

Section 1.7, page 69

1. 0.3
3. −6
5. $\frac{1}{2}$
7. $\frac{2}{3}$
9. *a.* y decreases by 36
   b. y increases by 600
11. 2
13. −2
15. 0
17. 2
19. $y = -3x - 1$
21. $x = 2$
23. $y = -\frac{1}{5}x - 1$
25. $y = -\frac{2}{3}x + \frac{13}{3}$
27. $x = 9$
29. $m = -\frac{2}{3}; b = 1$
31. $m = \frac{5}{3}; b = -3$
33. $m = 0; b = \frac{3}{2}$
35. Parallel
37. Neither
39. Neither
41. Neither
43. $y = -x - 5$
45. $y = x - 4$
47. $y = \frac{2}{3}x$
49. $y = -\frac{3}{2}x - 2$
51. $y = -x$
53. $x = -2$

55. $m_{AB} = \frac{1}{3}$; parallel line: $y = \frac{1}{3}x - \frac{22}{3}$; perpendicular line: $y = -3x + 6$
59. $m_{AB} = -\frac{5}{2}$; midpoint $(4, -2)$ perpendicular bisector $y = \frac{2}{5}x - \frac{18}{5}$
61. No
63. $k = \frac{34}{3}$
65. *a.* Change in weekly orange juice sales with respect to price
   b. Increases sales by 3000 cans
67. *a.* $P = -\frac{12}{5}A + 82$
   b. 10%
69. $m = \frac{3}{5}$; *a.* $T = \frac{3}{5}V - 273$
   b. $T = -273°C$
71. $C = \$3.5x + \$43,000, x = \#$ of shirts
73. *a.* The slope is the amount the area increases each year.
   b. Area is increased by 18 sq. miles.
75. *a.* The slope is the price of one set.
   b. Revenue increases by \$818,000.
77. *a.* If the total length increases by 1 mm, the tail length increases by 0.143 mm.
   b. The longer-tailed snake is 559.44 mm longer than the other snake.

Section 1.8, Cumulative Review Exercises, page 74

1. $\dfrac{1}{9x^{7/4}}$
3. x^4
5. $\dfrac{b^{24}}{27a^{30}c^{18}}$
7. $|y|$
9. $t^{1/2}$
11. $\dfrac{1}{x^{25/6}}$
13. $\dfrac{y^2 + xy + x^2}{xy(y + x)}$
15. $\dfrac{x^2 - 4 + x^2}{x(x^4 - 4)}$
17. $2xz\sqrt{3yz}$
19. $5ab\sqrt[3]{b^2}$
21. $27a^3b^6$
23. $\dfrac{2xn^5}{(n + 1)^5}$
25. $(2x + 1)(2x - 3)2$
27. $(5a + 1)(25a^2 - 5a + 1)$
29. $(a - b)^3$
31. $(x^a + 3)(x^{2a} - 3x^a + 9)$
33. $2x(2x + 1)(x + 7)$
35. $-(x + 11)(x - 5)$

37. $(a + b)(x + y)^2$

39. $4(3t - 2m)(3t + 2m)$

41. $\dfrac{5\sqrt{6}}{6}$

43. $4(\sqrt{3} + \sqrt{2})$

45. $\dfrac{1}{5\sqrt[4]{t^3}}$

47. $-\dfrac{n}{\sqrt{n^2 - n} + n}$

49. $3xy + 4\sqrt{xy} + 6x\sqrt{y} + 2y\sqrt{x}$

51. $x^{2\sqrt{3}} + 2x^{\sqrt{2}+\sqrt{3}} + x^{2\sqrt{2}}$

53. $\dfrac{2}{\sqrt{2(x + h) - 3} + \sqrt{2x - 3}}$

55. $(-\infty, \infty)$

57. $-\dfrac{16}{17}$

59. $\dfrac{-5 \pm \sqrt{41}}{8}$

61. $\dfrac{1 \pm \sqrt{4001}}{20}$

63. $-\dfrac{1}{2}, -1$

65. $4, \dfrac{2}{3}$

67. No solutions

69. 16, 25, 0

71. -10

73. No solutions

75. $\pm\sqrt{5}, \pm\sqrt{2}$

77. $\left\{\dfrac{4}{3}, -12\right\}$

79. $\{\pm 3, -4\}$

81. $(-5, \infty)$

83. $[-9, 9)$

85. $\sqrt{34}$

87. $\dfrac{3}{5}$

89. $(x - \frac{1}{2})^2 + (y + \frac{11}{2})^2 = \frac{17}{2}$

91. $y + 6 = -\frac{3}{2}(x - 5)$

93. Neither

95. $\left(\dfrac{53}{12}, 0\right)$

97. $\left(\dfrac{5}{2}\sqrt{2}, -\dfrac{\sqrt{5}}{2}\right)$

99.

101.

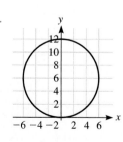

103. $y = \frac{4}{3}x - \frac{11}{3}$

105. $y = -\frac{1}{3}x + \frac{7}{3}$

107. 2.467×10^3

109. 0.0648

111. *a.* 1.674×10^{-2} g
    b. 4.482×10^{-23} g
    c. 3.442×10^{-22} g

113. $\sqrt{x^2 - \left(\dfrac{x}{2}\right)^2} = \sqrt{3} \cdot \dfrac{x}{2}$

115. $a\sqrt{2}$

117. 77.8%

119. $A = 2$ hr, $B = 6$ hr
    $C = 3$ hr

121. *a.* $\$74,837.01$
    b. $\$75,447.91$
    c. $\$75,986.81$
    d. $\$76,094.35$

123. For $\$5,000$; bank A
    For $\$10,000$; bank B

125. $3x + 6$

127. $.2$ acid = 152.727 l
    $.75$ acid = 87.273 l

129. 352.2 ft

131. *a.* 56.25 ft, 206.25 ft, 276.25 ft
    b. r = 40 mph
    c. $30 \le r \le 50$
    d. $r = -10 \times \sqrt{100 + 20D}$

CHAPTER 2
Section 2.1, page 88

1. -5 3. $-8a - 5$ 5. 309

7. $3(a - h)^2 + (a - h) - 1$ 9. 5 11. 3.56

13. 0 is not in the domain of g.

15. 11 17. -1

19. $-2h^3 + 9h^2 - 12h + 4$ 21. 27

23. $(1 + a + h)^3$ 25. 1

27. $4a^2 + 8ah + 4h^2 + 4a + 4h + 1$

29. 12 31. 12

33. Domain, range = all real numbers

35. Domain = all real numbers; range = $\{12\}$

37. Domain = all real numbers except 3; range = all real numbers except 0

39. Domain = $\left\{x : x \le \frac{3}{2}\right\}$; range = set of all nonnegative real numbers

41. Domain, range = all real numbers

43. Domain = all real numbers; range = set of all nonnegative real numbers

45. Domain = $\{x : -2 \le x \le 2\}$;
    range = $\{x : 0 \le x \le 2\}$
47. Domain = $\{x : x \le 2\} \cup \{x : x \ge 3\}$;
    range = $\{y : y \ge 0\}$
49. 2, 4   51. $2a - 5 + h$, $4a - 10$
53. $\dfrac{2}{\sqrt{2a + 2h - 3} + \sqrt{2a - 3}}$,
    $\dfrac{4}{\sqrt{2a + 2h - 3} + \sqrt{2a - 2h - 3}}$
55. 0, 0   57. $4a^3 + 6a^2h + 4ah^2 + h^3$, $8a^3 + 8ah^2$
59. Domain = all nonzero real numbers
61. Domain = $\{x : x \ne -1, -2\}$
63. Domain = all real numbers
65. Domain = all real numbers
67. Domain = $\{x : x \ne 1\}$
69. Domain = $\left\{ x : x \ne \frac{5}{4} \right\}$
71. Domain = all real numbers
73. $D(t) = 5t \sqrt{290}$ 75. $S(x) = x^2 + \dfrac{1200}{x}$
77. $A(x) = 14x - x^2 \left(\dfrac{1}{2} + \dfrac{\pi}{8} \right)$
79. $C(x) = 4000(3 - \sqrt{x^2 - 1}) + 6000x$
81. $S(t) = -16t^2 - 9.5t + 1800$

Section 2.2, page 102

1. Yes   3. No
5. 0, 2, −2, 5   7. −2, undefined, 1, 2, $\frac{1}{2}$
9. Domain = $[-2, 5]$; range = $[-2, 5]$
11. Domain = all reals $x \ne 0$; range = all reals $y \ne 0$
13. Domain = $[-1, 1]$;   15. Domain = $[-1, 1]$;
    range = $[-4, 12]$.   range = $[0, 1]$.

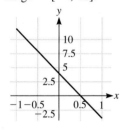   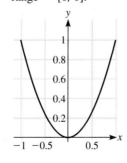

17. Domain = all reals;   19. Domain = $[-2, 2]$;
    range = $[-3, \infty)$.   range = $[-9, 7]$.

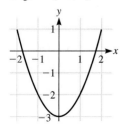   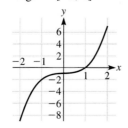

21. Domain = $[-2, 2]$; range = $[0, 2]$

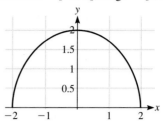

23. Domain = all positive reals; range = all positive reals.

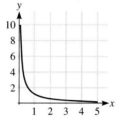

25.

27.

29.

31.

33.

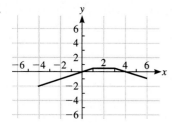

35. *a.*

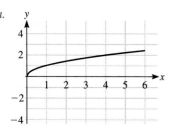

b.

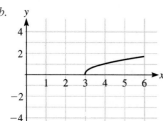

c.

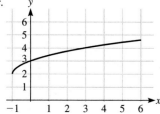

d.

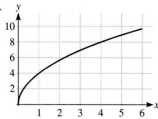

e.

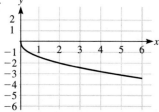

f. y

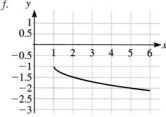

37. Both
39. Neither
41. Neither
43. x-axis
45. Yes
47. Yes
49. No
51. Yes
53. Neither
55. Neither
57. Even
59. Neither
61. Odd
63. Even

65.

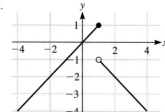

67.

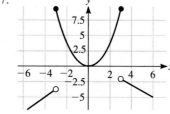

69.

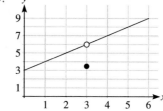

71.

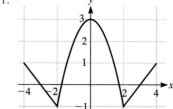

73.

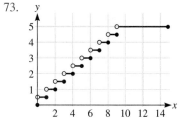

75.

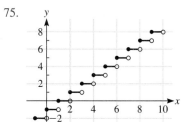

77.

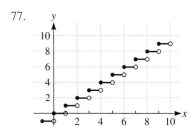

79. *a.* 0, 6, 2, 2
   b. Domain = all reals;
      range = {0, 1, 2, 3, 4, 5, 6, 7, 8, 9}
   c. The graph resembles an ascending staircase.
81. Symmetric with respect to the x-axis when $b = 0$; symmetric with respect to the y-axis when $a = 0$

Section 2.3, page 110

The answers depend on which graphing calculator you use. See your instructor or calculator manual.

Section 2.4, page 120

1. D_f = set of all real numbers, D_{ff} = set of all real numbers
   D_g = set of all real numbers, $D_{f/g}$ = set of all real numbers except -5
   D_{f+g} = set of all real numbers, $D_{g/f}$ = set of all real numbers except 4
   D_{f-g} = set of all real numbers, $D_{f \circ g}$ = set of all real numbers
   D_{fg} = set of all real numbers, $D_{g \circ f}$ = set of all real numbers
3. D_f = set of all real numbers, D_{ff} = set of all all real numbers
   D_g = set of all real numbers, $D_{f/g}$ = set of all real numbers except 0
   D_{f+g} = set of all real numbers, $D_{g/f}$ = set of all real numbers except 1 and $-\frac{3}{2}$

D_{f-g} = set of all real numbers, $D_{f \circ g}$ = set of all real numbers
D_{fg} = set of all real numbers, $D_{g \circ f}$ = set of all real numbers
5. -1
7. $x^2 - 2x - 4$
9. 72
11. undefined
13. $2x^3 + 3x^2 - 2x - 3$
15. $\dfrac{2x + 3}{x^2 - 1}$
17. $4x^2 + 12x + 8$
19. $2(2x + 3) + 3 = 4x + 9$
21. x, x
23. x, x
25. x, x
27. $|x|, x$
29. x, x
31. $3, -5$
33. $f(x) = x^5$
    $g(x) = 4x^3 - 1$
35. $f(x) = \dfrac{1}{x^4}$
    $g(x) = x + 5$
37. $f(x) = \dfrac{x + 1}{x - 1}$
    $g(x) = x^3$
39. $f(x) = \left(\dfrac{1 + x}{1 - x}\right)^4$
    $g(x) = x^3$
41. $f(x) = \sqrt{x}, \quad g(x) = \dfrac{x + 2}{x - 2}$
43. $f(x) = x^{63}, g(x) = x^5 + x^4 + x^3 - x^2 + x - 2$

45.

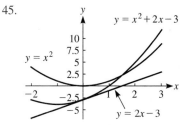

47.

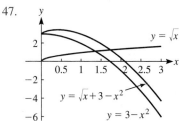

49.

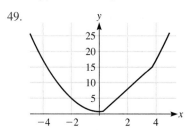

51.

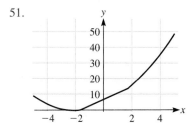

53.

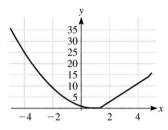

55.

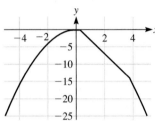

57.

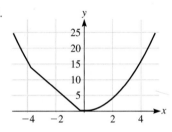

59.

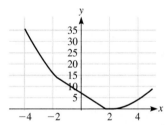

61. *a.* $a(t) = 125t; b.$ $p(a) = \sqrt{a^2 + 40,000}$

   c. $p(a(t)) = \sqrt{15,625t^2 + 40,000}$
   The distance from the tower after t hours.

63. *a.* $b(t) = 90 - 25t; b.$ $h(b) = \sqrt{b^2 + 8100}$

   c. $(h \circ b)(t) = \sqrt{9t^2 - 540t + 16,200}$

65. $P(x) = -0.5x^2 + 40x - 3$

67. $m = \frac{1}{3}, b = \frac{2}{3}$

Section 2.5, page 131

1. No inverse   3. No inverse

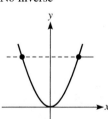

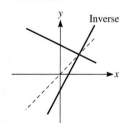

5. No inverse   7. $f^{-1}(x) = \dfrac{5}{8} - \dfrac{x}{8}$

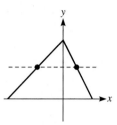

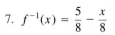

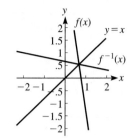

9. No inverse

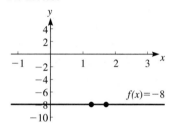

11. No inverse

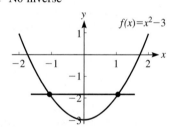

13. $f^{-1}(x) = \sqrt{4 - x^2}$,   15. No inverse
    $0 \leq x \leq 2$.

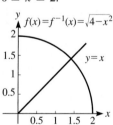

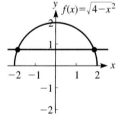

17. Yes   19. Yes   21. Yes
23. No   25. Yes   27. No

29. $f^{-1}(x) = \frac{1}{2}(x + 3)$ 31. $f^{-1}(x) = \frac{3}{2}\left(x - \frac{3}{5}\right)$

33. $f^{-1}(x) = \sqrt{x}, x \geq 0$ 35. $f^{-1}(x) = x^2 - 2, x \geq 0$

37. $f^{-1}(x) = \sqrt[5]{x}$ 39. $f^{-1}(x) = \dfrac{2}{x}$

41. $\dfrac{x + 3}{2 - 3x}$ 43. $f^{-1}(x) = -\sqrt{25 - x^2}, 0 \leq x \leq 5$

45. $x^3 - 8$

47. $(f \circ f^{-1})(739) = 739, (f^{-1} \circ f)(5.00023) = 5.00023$

51. $\dfrac{x - b}{m}$ 53. No

Section 2.6, page 137

1. *a.* $y = \frac{1}{70}x$; *b.* $y = 5$ 3. *a.* $y = 20x^2$
                                                b. $y = 273.8$

5. *a.* $y = \dfrac{.98}{x^2}$ 7. *a.* $y = \dfrac{xz}{w}(3.25)$

   b. $y = 98$ *b.* $y = 62.4$

9. *a.* $y = \dfrac{xz}{w^2}$ (22.62857143); *b.* $y = 15,713.28$

11. *a.* $y = \dfrac{z^3}{x^2}\left(\dfrac{2}{9}\right)$; *b.* $y = \dfrac{1}{6}$

13. $A = \pi r^2$
    Constant of variation is π.
    Varies directly as r^2.

15. $d = 65t$
    Constant of variation is 65.
    d varies directly as t.

17. 28.16 g   19. 22°C   21. 235.6942 ft

23. *a.* $s = \sqrt{30d}$, $d = 53.3333$ ft, $s = 65.03845$ mph

   b. $s = \sqrt{12d}$, $d = 252.08333$ ft, $s = 49.95998$ mph

25. 360°   27. 66.3158 mph

29. 17.320508 in.   31. -3.59375×10^{-11} dyn

33. *a.* y varies directly as $\sqrt{x}$; *b.* $\dfrac{1}{\sqrt{k}}$

Section 2.7, Cumulative Review Exercises, page 141

1. Yes   3. No   5. -6 7. -4
9. $(a - 1)^2 + 3(a - 1) - 4$
11. $2a + 3 + h$ 13. 1
15. Domain = all reals; range = all reals
17. Domain = $\{x : x \le 6\}$; range = $[0, \infty]$
19. Domain = $(-\infty, \infty)$; range = $\{56.7\}$
21. Domain = $\{x : |x| \le 8\}$; range = $[0, 8]$
23. Domain = $(-\infty, 0) \cup (0, \infty)$
25. Domain = $[-1, \infty)$ 27. Domain = $(-\infty, \infty)$

29.   31.

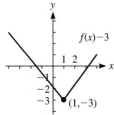

33. Symmetric with respect to y-axis
35. Symmetric with respect to y-axis
37. Symmetric with respect to y-axis

39. Neither   41. Even   43. Odd   45. Even
47. Even   49. All real numbers
51. *a.* $\{x : x \ne \pm 1\}$ 53. *a.* All reals

   b. $\dfrac{2x + 3}{x^2 - 1}$ *b.* $4x^2 + 12x + 8$

55. $(f \circ g)(x) = \dfrac{9}{x^4} - \dfrac{6}{x^2} + 1$

   $(g \circ f)(x) = \dfrac{3}{(x^2 - 2x + 1)^2}$

57. $(f \circ g)(x) = x^{33}$; $(g \circ f)(x) = x^{33}$

59. $(f \circ g)(x) = \dfrac{(10/x^2) + 7}{(15/x^2) - 4} = \dfrac{10 + 7x^2}{15 - 4x^2}$

   $(g \circ f)(x) = \dfrac{5(3x - 4)^2}{(2x + 7)^2}$

61. $f(x) = \sqrt[5]{x}$, $g(x) = \dfrac{x^3 + 1}{x^3 - 1}$

63. $f(x) = |x|$, $g(x) = \dfrac{x^2 - 5}{x^2 + 5}$

65. $3x + 4y = -23$ 67. $y - 3 = 3(x + 2)$

69. $f(x) = \begin{cases} 3 & (x = -3) \\ 1 & (-3 < x \le 0) \\ \sqrt{x} & (x > 0) \end{cases}$

71. $f \circ g(x) = \begin{cases} (2x - 7)^2 & (-3 \le x \le 0) \\ 1 - x^3 & (0 < x \le 2) \\ \dfrac{x}{3 - x} & (2 < x) \end{cases}$

   $g \circ f(x) = \begin{cases} 2x^2 - 7 & (-3 \le x \le 0) \\ (1 - x)^3 & (0 < x \le 2) \\ 3x - 5 & (2 < x) \end{cases}$

73. Yes   75. Yes

77. $\dfrac{4x + 4}{2x + 3}$ 79. $f^{-1}(x) = \sqrt[3]{x + 1}$

81. $L = \dfrac{R}{d^2}$

The loudness decreases by a factor of $\frac{1}{9}$ if d is tripled.

83. $G = N(700 - N)k$; or $G = (700N - N^2)k$

85. *a.* salary
   $= \begin{cases} 25,000 & (S \le 400,000) \\ 25,000 + .24(S - 400,000) & (S > 400,000) \end{cases}$
                     S = amount of
                     sales in
                     dollars

b.

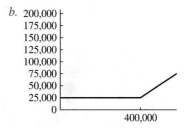

87. v would have to decrease by a factor of 0.959714.

89. $.55x^2 + \dfrac{246.4}{x}$

91. $\dfrac{\pi h^3}{9}$

CHAPTER 3

Section 3.1, page 154

1. (b)   3. (d)   5. (c)

7. $y = \frac{5}{2}x^2$ 9. $y = -7x^2$

11. $y = -\frac{1}{3}(x-2)^2 + 4$

13. a. $(0, 0)$; b. Downward;

c.
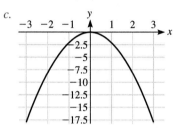

15. a. $(2, 0)$; b. Upward;

c.
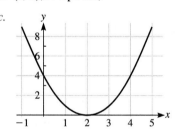

17. a. $(3, 0)$; b. Upward;

c.
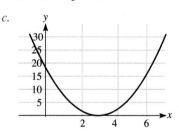

19. a. $(-3, -2)$; b. Downward;

c.
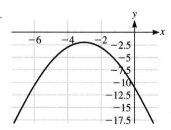

21. a. $f(x) = (x+3)^2 - 4$; b. $(-3, -4)$;

c. Upward; d. min $= -4$;

e.
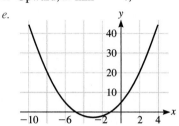

23. a. $f(x) = -\frac{1}{4}(x+6)^2 + 8$; b. $(-6, 8)$;

c. Downward; d. max $= 8$;

e.
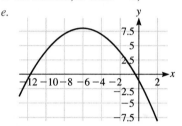

25. max $= 0$ 27. min $= 3$

29. max $= \frac{121}{8}$ 31. min $= -\frac{169}{12}$

33. min ≈ -435.99 35. max ≈ 5.84

37. 15 and 15   39. $b = h = 83$ cm

41. a. $A(x) = -2x^2 + 150x$; b. 37.5 yd; c. 2812.5 yd^2

43. Dimensions are 18.5 ft. $\times$ 18.5 ft.; area $= 342.25$ ft.^{2}

45. 1250

47. a. $R(x) = 1200 - 10x - x^2$; b. 0; c. 1200

49. a. $A(x) = \dfrac{4+\pi}{16\pi}x^2 - \dfrac{7}{2}x + 49$

b. $\dfrac{28\pi}{4+\pi} \approx 12.32$ in.; c. 27.44 in^2

51. a. He should wait 5 weeks.; b. \$6000

53. $-\frac{2}{7}$ 55. -4.5

57. maximum value $= \dfrac{(t^2 - 4)(t-2)}{4} - 16$

59. $\left(\dfrac{3}{2p}, \dfrac{4p^2 - 9}{4p} \right)$

61.
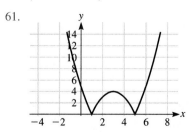

Section 3.2, page 163

1.
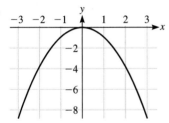

3.
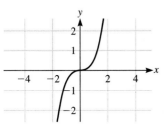

5.
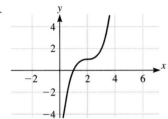

7.
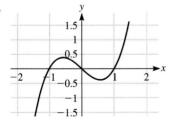

9.
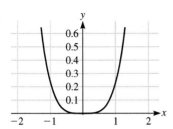

11.
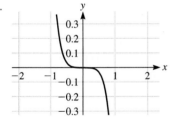

13.
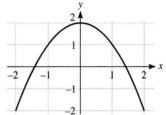

15.
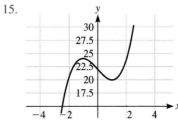

17.
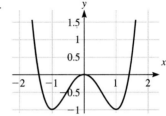

19.
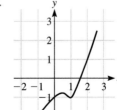

21.
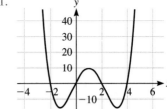

23.
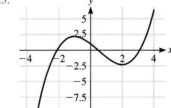

25.

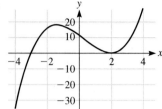

27. *a.* $V(x) = 4x^3 - 32x^2 + 64x$

b.

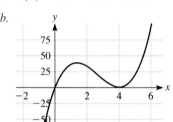

c. $x \approx 1.5$ gives the approximate maximum; max ≈ 38

29.

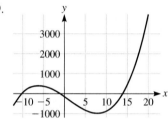

31.

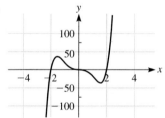

33. *a.* $P(x) = -\frac{1}{3}x^3 + 5x^2 + 11x - 100$

b.

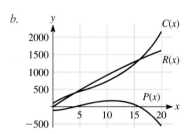

c. Max when $x = 11$; max $= 182.\overline{3}$.

35. *a.*

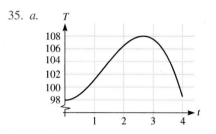

b. max ≈ 108; variable at max ≈ 2.65

37. (b) 39. (f) 41. (e)

Section 3.3, page 172

1. $\sqrt{3}i$ 3. $9i$ 5. $7i\sqrt{2}$ 7. $-7i$
9. $4 - 2i\sqrt{15}$ 11. $2i(1 + \sqrt{3})$ 13. $-i$
15. 1 17. -1 19. i
21. $8 + 0i = 8$ 23. $1 - 26i$ 25. $0 + 0i = 0$
27. 0 29. 1 31. $5 - 8i$
33. $2 - \frac{\sqrt{6}}{2}i$ 35. $\frac{4}{5}$ 37. $5 + 6i$
39. 6 41. $2 + 4i$ 43. $-7 + 10i$
45. $14 - 2i$ 47. 29 49. $6 + 8i$
51. $21 - 20i$ 53. $\frac{1}{10} + \frac{7}{10}i$ 55. $\frac{1}{2} - \frac{\sqrt{3}}{2}i$
57. $-3 - 5i$ 59. $-\frac{1}{2} + \frac{1}{2}i$ 61. $\frac{2}{25} + \frac{11}{25}i$
63. $\frac{1}{5} + \frac{2}{5}i$ 65. $\frac{1}{34} + \frac{2}{17}i$ 67. $\frac{3}{5} + \frac{1}{5}i$
69. $\frac{-1}{2}i$ 71. $\pm\sqrt{5}i$ 73. $\pm 8i$
75. $x^2 + 4$ 77. $x^2 - \frac{2}{3}x + \frac{1}{3}$
79. $(x - 2i)(x + 2i)$ 81. $(x - \sqrt{3}i)(x + \sqrt{3}i)$
83. $(a - bi)(a + bi)$ 85. $\left(-\frac{9}{5}, 8\right)$
87. $\left(3, \frac{3}{2}\right), \left(-3, \frac{3}{2}\right)$ 99. $4\overline{\alpha}^2 - 25$

Section 3.4, page 181

1. $q(x) = x^4 + x^2 + 2x + 4, \quad r(x) = 3$
3. $q(x) = 2x^2 - 3x - 7, \quad r(x) = 20$
5. $q(x) = x^3 + \frac{1}{3}x^2 + \frac{2}{9}x + \frac{22}{7}, \quad r(x) = -\frac{118}{27}$
7. $q(x) = x^3 - 2x^2 + 3x - 4, \quad r(x) = 6x - 13$
9. $q(x) = x^2 - 5x + 25, \quad r(x) = 0$
11. $q(x) = x^2 - 2, \quad r(x) = 6$
13. $q(x) = x, r(x) = 2x$
15. $q(x) = x^5 + yx^4 + y^2x^3 + y^3x^2 + y^4x + y^5, \quad r(x) = 0$
17. $-\frac{5}{2}(x + 1)(x - 2)$ 19. $\frac{6}{5}(x + i)(x - i)$
21. $4(x + 1)(x)(x - 1)$
23. $(-2)x(x + 2)(x + 1)(x - i)(x + i)$
25. $\frac{2}{25}(x - 1 + \sqrt{2})(x - 1 - \sqrt{2})(x - 4 + 3i)(x - 4 - 3i)$
27. $(x - 2 + i)$ and $(x - 2 - i)$ are its factors.
29. *a.* -31; *b.* -31; *c.* -1; *d.* -1
31. *a.* -2; *b.* -2
33. $q(x) = 4x^4 - 8x^3 + 18x^2 - 36x + 71, r(x) = -137$
35. $q(x) = 2x^2 + 5x - 3, \quad r(x) = 0$

37. $q(x) = x^2 - 5x + 25, \quad r(x) = 0$

39. $q(x) = 3x^3 - 2x^2 - \frac{61}{3}x - \frac{50}{9}, \quad r(x) = \frac{224}{27}$

41. $q(x) = x^5 + x^4y + x^3y^2 + x^2y^3 + xy^4 + y^5, \quad r(x) = 0$

43. $0, 30, 402$ 45. $4, 0, 34$

47. $41, -351, 2353, \frac{43}{27}$ 49. Yes; no

51. Yes; no   53. Yes; yes

55. $\{-2, -3, 1, 5\}$ 57. $\{-4, -2, 1, 3\}$

59. $\{-4 < x < -2\} \cup \{1 < x < 3\}$ 61. 1

63. $q(x) = x - 1, r(x) = 2$

65. $q(x) = 4(x+1)^2(x-5)^3 + 3(x-5)^4(x+1), \quad r(x) = 0$

67. $b.\ q(x) = \frac{5}{2}x^2 - \frac{7}{4}x + \frac{9}{8}, r(x) = -\frac{83}{8}$

Section 3.5, page 191

1. $x^2 \left[x - \dfrac{-5 + i\sqrt{95}}{12} \right]\left[x - \dfrac{-5 - i\sqrt{95}}{12} \right]$

3. $(x - i)(x + i)(x + 1 + \sqrt{2}i)(x + 1 - \sqrt{2}i)$

5. $(x - 1)\left(x - \dfrac{-1 + i\sqrt{3}}{2} \right)\left(x - \dfrac{-1 - i\sqrt{3}}{2} \right)$

7. $(x + 3)(x + \sqrt{5}i)(x - \sqrt{5}i)$

9. $\left(x - \dfrac{-1 + i\sqrt{11}}{2} \right)^2\left(x - \dfrac{-1 - i\sqrt{11}}{2} \right)^2$

11. 0, multiplicity $= 1$; 2, multiplicity $= 2$; $\frac{5}{2}$, multiplicity $= 3$

13. 2, multiplicity $= 2$; -2, multiplicity $= 5$

15. 2, multiplicity $= 3$; -1, multiplicity $= 1$

17. 0, multiplicity $= 1$; 3, multiplicity $= 2$ i, multiplicity $= 1$; $-i$, multiplicity $= 1$

19. 0, multiplicity $= 5$; $\dfrac{-1 + \sqrt{5}}{2}$, multiplicity $= 3$; $\dfrac{-1 - \sqrt{5}}{2}$, multiplicity $= 3$

21. 3 23. 2

25. $\pm 2, 1 \pm \sqrt{3}i, -1 \pm \sqrt{3}i$ 27. $1, \dfrac{-1 \pm \sqrt{3}i}{2}$

29. $-2, 1 \pm \sqrt{3}i$ 31. $\pm\sqrt{2}, \pm\sqrt{2}i$

33. $0, \pm i$ 35. $\pm 2i, \dfrac{1 \pm \sqrt{5}}{2}$

37. $\pm\sqrt{3}, \pm i$ 39. $2, -3$

41. At most 2 positive; at most 2 negative.

43. At most 2 positive; at most 2 negative; at least 6 nonreal.

45. At most 2 positive; at most 2 negative.

47. At most 3 positive; at most 2 negative.

49. At most 2 positive; at most 2 negative, at least 6 nonreal.

51. At most 2 positive; at most 2 negative; 0 a zero of multiplicity 2.

53. $-2 + i$ and -1 55. $0, \pm\sqrt{2}$

57. 3 59. $1 - 2i, 1,$ and 2

Section 3.6, page 200

1. $2, 3, \frac{1}{2}$ 3. None   5. $\frac{4}{5}$ 7. None

9. Lower bound $= -1$; upper bound $= 2$

11. Lower bound $= -4$; upper bound $= 1$

13. Lower bound $= -2$; upper bound $= 1$

15. Lower bound $= -3$; upper bound $= 3$

17. $1, \frac{1}{3}, \frac{2}{3}$ 19. $\pm 5, \frac{1}{4}$ 21. $\pm\frac{5}{2}$ 23. None.

25. $-2, -\frac{1}{3}, \frac{1}{2}$ 27. $f(-1) = -1; f(0) = 1$

29. $f(2) = -1; f(3) = 7$ 31. $f(0) = -2; f(1) = 9$

33. $f(-1) = -3; f(-2) = 9; f(1) = -3; f(2) = 9$

35. $-.6823$ 37. 2.2134

39. 0.2346 41. 1.3802 and -0.8191

43. $-0.6985, 1.3645, 1.8186$ 45. $a.\ 25; b.\ 13$

47. $x \approx 7.862$ 49. $t \approx .9507$ 51. $2, -1$

53. $2, \dfrac{-1 \pm \sqrt{5}}{2}$ 55. $1, 2, -1 \pm \sqrt{2}$

57. $\frac{1}{2}, -\frac{2}{3}, \pm i$ 59. $4, 2, -\frac{1}{2}, -1, \frac{3}{2}$

61. $\dfrac{39 \pm \sqrt{1305}}{6}$ 63. $3, 2, -1$

65. $\pm\sqrt{\dfrac{-3 \pm \sqrt{89}}{2}}$ 67. $0, \pm 2, \pm\frac{1}{2}$

Section 3.7, page 212

1. $a.\ x = 0, y = 0$

$b.$

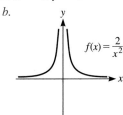

$f(x) = \dfrac{2}{x^2}$

3. $a.\ x = 0, y = 0$

$b.$

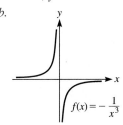

$f(x) = -\dfrac{1}{x^3}$

5. $a.\ x = -1, y = 0$

$b.$

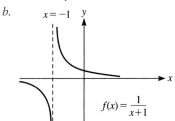

$x = -1$

$f(x) = \dfrac{1}{x+1}$

7. *a.* $x = -2, y = 0$

b.

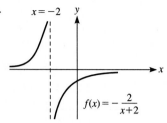

$$f(x) = -\frac{2}{x+2}$$

9. *a.* $x = -3, y = 0$

b.

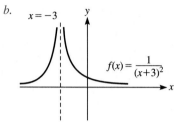

$$f(x) = \frac{1}{(x+3)^2}$$

11. *a.* $x = -2, y = -1$

b.

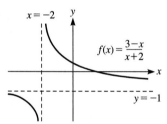

$$f(x) = \frac{3-x}{x+2}$$

13. *a.* $x = -\frac{10}{3}, y = -\frac{2}{3}$

b.

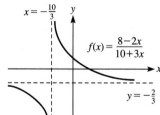

$$f(x) = \frac{8-2x}{10+3x}$$

15. *a.* $y = 0$

b.

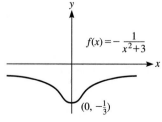

$$f(x) = -\frac{1}{x^2+3}$$

17. *a.* $x = \pm3, y = 0$

b.

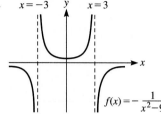

$$f(x) = -\frac{1}{x^2-9}$$

19. *a.* $x = \frac{1}{2}, x = -3, y = 0$

b.

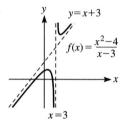

$$f(x) = \frac{x-3}{2x^2+5x-3}$$

21. *a.* $x = 3, y = x + 3$

b.

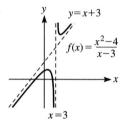

$$f(x) = \frac{x^2-4}{x-3}$$

23. *a.* $x = 2, x = -1, y = 1$

b.

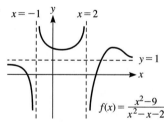

$$f(x) = \frac{x^2-9}{x^2-x-2}$$

25. *a.* $x = -4, y = x - 6$

b.

$$f(x) = \frac{x^2-2x-8}{x+4}$$

27. *a.* $x = 0, y = x$

b.
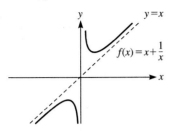

29. *a.* $x = \pm\dfrac{\sqrt{3}}{3}, y = \dfrac{1}{3}$

b.
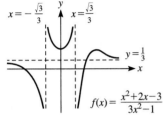

31. *a.* $x = 0$

b.
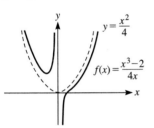

33. *a.* None

b.
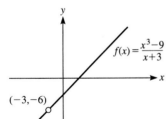

35. *a.* None

b.
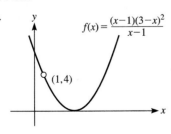

37. *a.* None

b.
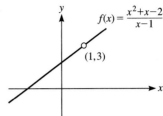

39. *a.* $x = \pm1, y = 4$

b.
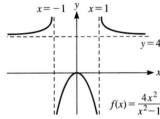

41. *a.* $y = x + 2, x = \pm5$

b.
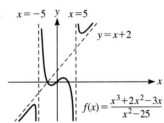

43. *a.*
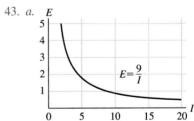

b.
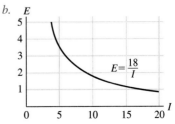

c.

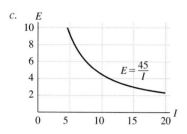

$E = \dfrac{45}{I}$

45.

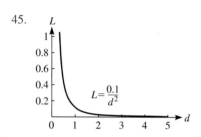

$L = \dfrac{0.1}{d^2}$

47. *a.* $W_h = 200\left(\dfrac{4000}{4000 + h}\right)^2$

b.

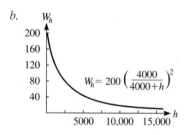

$W_h = 200\left(\dfrac{4000}{4000+h}\right)^2$

c. $4000(\sqrt{2} - 1)$ miles

49.

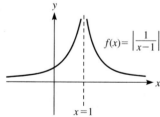

$f(x) = \left|\dfrac{1}{x-1}\right|$

$x = 1$

51.

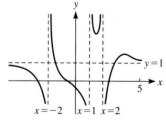

$y = 1$

$x = -2 \quad x = 1 \quad x = 2$

53.

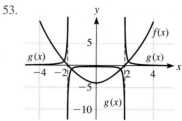

$g(x)$ $f(x)$ $g(x)$ $g(x)$

55.

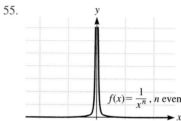

$f(x) = \dfrac{1}{x^n}$, n even

57.

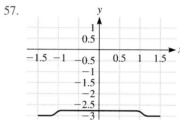

59.

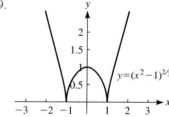

$y = (x^2 - 1)^{2/3}$

61. (*d*)   63. (*f*)   65. (*e*)

67.

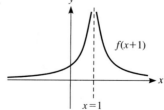

$f(x+1)$

$x = 1$

69.

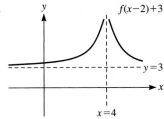

$f(x-2)+3$

$y=3$

$x=4$

71.

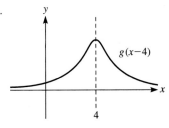

$g(x-4)$

4

Section 3.8, Cumulative Review Exercises, page 216

1. *a.* $f(x) = \left(x - \frac{3}{2}\right)^2 - \frac{9}{4}$; *b.* $\left(\frac{3}{2}, -\frac{9}{4}\right)$;
   c. Upward; *d.* min $= -\frac{9}{4}$

e.

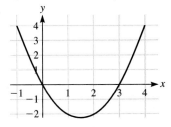

3. *a.* $f(x) = 2(x + 3)^2 - 16$; *b.*$(-3, -16)$
   c. Upward; *d.* min $= -16$

e.

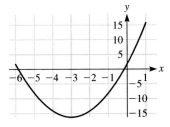

5. *a.* $f(x) = -\frac{1}{3}\left(x - \frac{3}{5}\right)^2 - \frac{17}{25}$; *b.* $\left(\frac{3}{5}, -\frac{17}{25}\right)$
   c. Downward; *d.* max $= -\frac{17}{25}$

e.

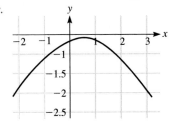

7. min $= -\frac{1198}{3}$ 9. max $= \frac{11}{4}\sqrt{2}$

11. \$0.35

13.

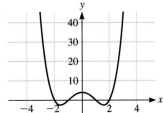

15.

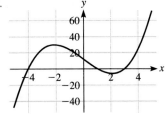

17.

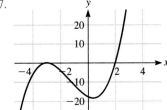

19.

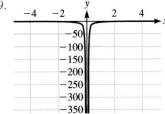

21.

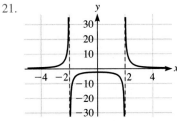

23.

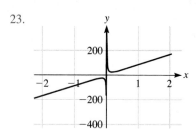

25.

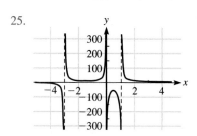

27.

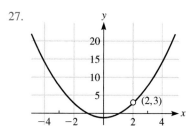

29.

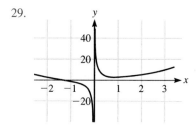

31. $q(x) = x^4 + 4x^3 + 4x^2 + 2x + 8$, $r(x) = 5$

33. $f(x) = 10x^2 + \frac{5}{3}x - \frac{10}{3}$

35. $f(x) = \frac{10}{9}x^2 - \frac{10}{9}x + \frac{25}{9}$

37. $f(x) = \frac{3}{20}x^5 - \frac{3}{20}x^4 - \frac{3}{20}x^3 - \frac{21}{20}x^2 - 3x - \frac{9}{5}$

39. $x^3 - 3x^2 - 2x + 1$ remainder 0

41. $x^3 + (-2 - i)x^2 + (2i - 1)x + i$ remainder 0

43. $-4, 41, 2 - 2i$

45. Yes; no; no   47. $-1, 2, 5$

49. 0 of multiplicity 2, 1 of multiplicity 4   51. 5

53. 1 positive real zero, 1 negative real zero, 2 nonreal zeros.

55. At most 3 positive, at most 2 negative real zeros.

57. $\pm\sqrt{\frac{3}{2}}$, $-1 + i\sqrt{2}$ 59. $2 \pm 2i\sqrt{3}$

61. ± 1, ± 2, ± 3, ± 6, $\pm\frac{1}{2}$, $\pm\frac{3}{2}$, $\pm\frac{1}{4}$, $\pm\frac{3}{4}$

63. $2, -2$ 65. None   67. None

69. $1, \dfrac{-1 \pm i\sqrt{3}}{2}$ 71. $1 \pm \sqrt{2}, 3$

73. $0, 1 \pm 2\sqrt{2}$ 75. $-1, \frac{1}{2}, 1 \pm i$

77. $f(-1) = 2 > 0, f(0) = -4 < 0$

79. -0.7320 81. $2, \dfrac{21 - \sqrt{177}}{4}$

83.

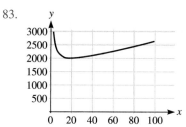

85. x intercepts are $2, 4, -\frac{2}{3}$.

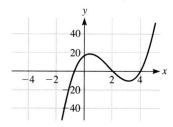

87. No x intercepts

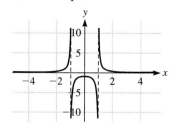

89. x intercept is 0.

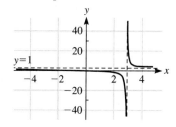

91. $c = 226.12305$

CHAPTER 4
Section 4.1, page 228

1. *a.* 64; *b.* 73.5; *c.* 77.71; *d.* 77.816;
   e. 77.8702; *f.* 77.87995; *g.* 77.880163; *h.* 77.88;

3. *a.* $f(1) = 3$; $f(2) = 9$; $f(3) = 27$; $f(0) = 1$
$f(-1) = \frac{1}{3}$; $f(-2) = \frac{1}{9}$; $f(-3) = \frac{1}{27}$;
$f\left(\frac{1}{2}\right) = \sqrt{3}$

b.

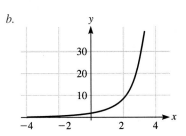

5.

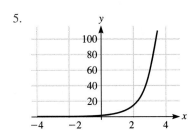

7.

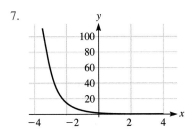

9.

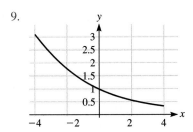

11.

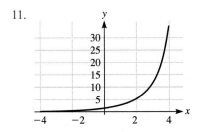

13.

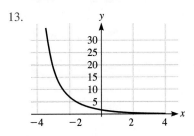

15.

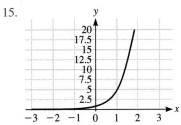

17.

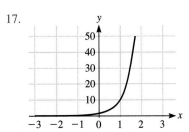

19.

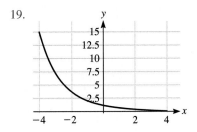

21.

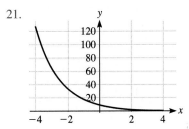

23.

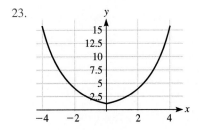

25.

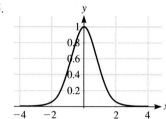

27.

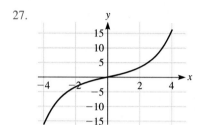

29.

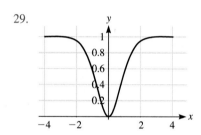

31. *a.* $10,000(1.08)^t$

b. 1 yr → \$10,800   3 yr → \$12,597.12
    2 yr → \$11,664   10 yr → \$21,589.25

c.

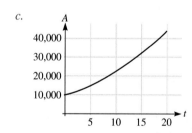

33. *a.* $A_0(.99997)^t$

b. 1 yr → 399.988 kg   200 yr → 397.607 kg
    2 yr → 399.976 kg

35. *a.* $16,370,000(1.0305955)^t$

b. 23 yr → 32,740,000   10 yr → 22,127,445
    46 yr → 65,480,000   15 yr → 25,726,019

37. *a.* 2 cm → 56.25%   10 cm → 5.63%
    3 cm → 42.19% x cm → $(.75)^x \cdot 100\%$

b. $(.75)^x \cdot (100\%)$

c.

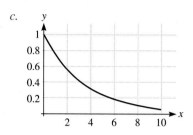

39. (*i*)   41. (*a*)   43. (*c*)
45. (*e*)   47. (*f*)

Section 4.2, page 237

1. $e^3 \approx 20.086$ 3. $e^3 + 2e^7 \approx 2213.4$
5. 0.9999092 7. $e^2 - 2e + 1 \approx 2.9525$
9. e^{4x} 11. $e^{2x} - 2 + e^{-2x}$
13. $e^{3x} - 3e^{2x} + 3e^x - 1$ 15. $4e^{2x-2} + 5e^{x-2}$
17. $e^{2x} - 9$ 19. e^x 21. 1.2761926
23. *a.* 1 sec → 29.003 mph   10 sec → 138.346 mph
        2 sec → 52.749 mph   20 sec → 157.069 mph
        5 sec → 101.139 mph

b.

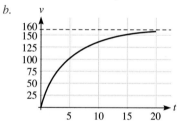

25. $P(t) = .5(1 - e^{-0.08t})$

a. 1 → .0384   10 → .2753   50 → .4908
    2 → .0739   20 → .399

b.

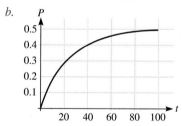

27. $P(t) = 1 - e^{-0.3t}$

a. 1 → 25.918%   5 → 77.687%   12 → 97.268%
    2 → 45.119%   10 → 95.021%

b.

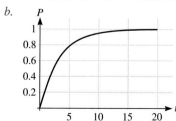

29. $Y(n) = 30(1 - e^{-0.04n})$

    a. $1 \to 1.1763 \quad 10 \to 9.8904$
         $2 \to 2.3065 \quad 20 \to 16.5201$
         $5 \to 5.4381 \quad 30 \to 20.9642$
         $6 \to 6.4012 \quad 40 \to 23.9431$

    b.

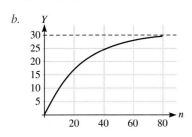

31. $N(t) = \dfrac{540}{1 + 4.4e^{-3t}}$

    a. $10 \to 540 \quad 30 \to 540$
        $20 \to 540 \quad 100 \to 540$

    b.

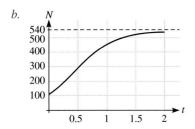

33. *a.* CE; *b.* $\approx 4.722 \times 10^{-5}$ coulombs

35. $\dfrac{4}{(e^x + e^{-x})^2}$

37. $\dfrac{(e^x + e^{-x-h})(e^h - 1)}{2h}$

39. $e^{-3x} \cdot \dfrac{e^{-3h} - 1}{h}$

Section 4.3, page 247

1. $2^3 = 8$ 3. $3^5 = 243$ 5. $2^{-3} = \frac{1}{8}$ 7. $x^{-12} = 3$

9. $\log_4 x = 2$ 11. $\log_{10} 0.001 = -3$

13. $\log_4 2 = \frac{1}{2}$ 15. $\log_x 0.01 = -3$

17. 1   19. 1   21. 0   23. 0

25. 4   27. -3 29. 0.8241   31. 3

33. $2x - 1$ 35. x^4

37. *a.*

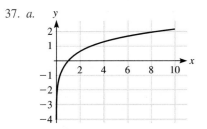

    b.

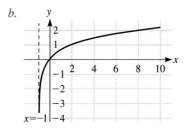

    c.

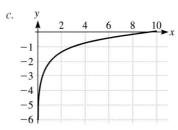

39.

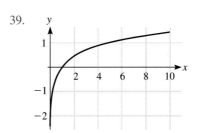

41.

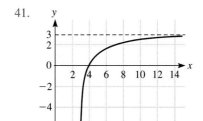

43.

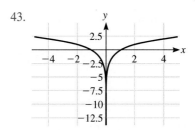

45.

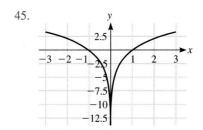

47.

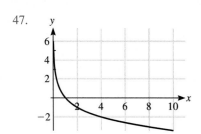

49. $-\log_{10} 3$ 51. $\log_a 3$ 53. 1
55. $\log_a(x + 3)$ 57. 1   59. $\log_a(x - y)$
61. $\log_b x + \frac{1}{2} \log_b y - 2 \log_b z$
63. $2 \log_2 x + 2 \log_2 y + 4 \log_2 z$
65. $4 \log_3 x + \log_3 y - 2 \log_3 z$
67. $2 \log_{10} x + \frac{2}{3} \log_{10} y - 4 \log_{10} z$
69. $\frac{3}{2} \log_{10} x + \frac{1}{4} \log_{10} y + \frac{5}{4} \log_{10} z$
71. $\log_a(m^3 - n^3) = \log_a(m - n) + \log_a(m^2 + mn + n^2)$
73. $\frac{1}{3} \left[\log_a(x - 1) + \log_a(x - 5) \right]$
75. $2 \log_b(x^3 + 8) = 2 \log_b(x + 2) + 2 \log_b(x^2 - 2x + 4)$

77.

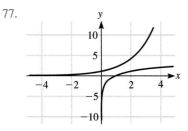

79. .7124   81. −.2084   83. .4354   85. −1
87. 2.0104   89. .41355   91. .32664
93. −.0360667   95. 17   97. 18
99. $-\dfrac{33}{\log_2 0.4} = 24.96354$ days
101. $\dfrac{9}{\log_2(21.5/14.5)} \approx 15.8$ yr
103. $x = 5$ 105. $x = 0$ 107. $x = 4$
109. $x = 2197$ 111. $x = 6$ 113. $x = \sqrt[3]{5}$
115. $x = \pm \sqrt{5 + \log_5 14}$ 117. $-\frac{1}{3}$
119. $x = \log_b 5$ 121. $x = \log_5 7$ 123. -2.5
125. $x = \frac{7}{4}$ 127. $\dfrac{-1 \pm \sqrt{7}}{2}$ 129. $x = \pm 3$
131. False   133. False   135. False   137. True
139. $x \geq 1$ 141. $-3 < x < -2$
143. $x < 0$ 145. $x < 2$
147. $3 \log_a(x + y)$ 151. $\log_3 x$

Section 4.4, page 255

1. 1   3. 3   5. −1   7. −3
9. 1.92153   11. .538573734   13. −.33913
15. 6.828144   17. 49.9615   19. −3.0654

21. −22.9164   23. 130 decibels
25. 10^{-15} watts per m^2
29. $I = I_0 \cdot 10^{D/10}$
31. a. 1; b. 2; c. 3; d. 4; e. 5; f. 8
33. The magnitude is 2 greater on the Richter scale.
37. $I = I_0 10^R$
39. a. 1.585×10^{-8} mol/l; b. 6.3096×10^{-5} mol/l
41. $H^+ < 10^{-7}$ 43. 0   45. −7
47. 1   9. 7k   51. $\dfrac{1}{625e^8}$
53. 3   55. 128
57. $\ln(x^2)^{x^{-1}} = \ln x^{2/x}$ 59. $\left[\ln(n^{3n}) \right]^{-1}$
61. $\dfrac{x}{y^2}$ 63. $\ln 4$ 65. $\ln(18)^{3^{2x} 2^x}$
67. $\ln \sqrt[3]{\dfrac{3 + \sqrt{10}}{2}}$ 69. 2.523 days
71. a. $1995 \to 260{,}249{,}015$ $2001 \to 274{,}688{,}828$
    b. 77.0163534 yr
73. a. 0.0150964837; b. $P(t) = 23{,}669{,}000 e^{0.015t}$
    c. 31,949,808
75. $1582.02   77. 28.2%
79. a. 9.477121255   81. a. $80,000
    b. 10.477121255   b. $1465.25
83. ≈ 48.008444 85. ≈ 2.010382
87. ≈ -1.5058883 89. $\dfrac{\ln 4578}{\ln 10}$
91. $\dfrac{\ln x^2}{\ln 10}$ 93. $\dfrac{\log 0.987}{\log e}$ 95. $\dfrac{1}{2} \dfrac{\log y}{\log e}$
97. e^{2e^x} 99. $\dfrac{1}{\sqrt{3}} \ln \left| \dfrac{z^2 - 3}{z^2 + 2\sqrt{3}z + 3} \right|$
101. 3.598   103. $3.3, -7.1$ 105. (c)   107. (a)

Section 4.5, page 262

1. $\pm \sqrt{\dfrac{\ln 4}{\ln 7}}$ 3. $\dfrac{2}{5 - \ln 4}$ 5. $\dfrac{1}{3e^6 - 1}$
7. $\ln 100$ 9. $\ln 100$ 11. $\dfrac{\ln 1.845}{10}$
13. $\frac{1}{10} \ln \frac{5}{3}$ 15. $\dfrac{-3 \pm \sqrt{13}}{2}$ 17. $\frac{2}{3}$
19. 1   21. 2   23. $\sqrt{41}$
25. $\frac{96}{31}$ 27. $\frac{1}{8}, 4$ 29. $\dfrac{1 + \sqrt{5}}{2}$
31. $10^{1/4}, 10^{-2/3}$ 33. $5, -5$ 35. 8
37. $\ln 2$ 39. No solutions   41. $\ln(8 + \sqrt{65})$
43. $\frac{1}{2} \ln 3$ 45. $9, \frac{1}{9}$ 47. 16
49. $1000, \frac{1}{1000}$ 51. $3^{1/(\log_2 3 - 1)}$ 53. $\dfrac{1}{k} \ln \dfrac{P}{P_0}$
55. $-\dfrac{1}{k} \ln \dfrac{P - N}{AN}$ 57. $\ln(x \pm \sqrt{x^2 - 1})$
59. $\ln \sqrt{\dfrac{1 + x}{1 - x}}$ 61. $-\dfrac{L}{R} \ln \left(1 - \dfrac{RI}{E} \right)$
69. $3e^{-2x/e}$ 71. $e^{-2/3} e^{-4x/3}$ 73. $-4e^{-2x/3}$

Section 4.6, Cumulative Review Exercises, page 264

1. $e^{0.6931} = 2$
3. $(0.5)^2 = 0.25$
5. $p = e^k$
7. $\log_2 2 = 1$
9. $\ln 7.3891 = 2$
11. *a.* $N(t) = 20 \cdot 2^{t/23}$
   b. 23 years $\rightarrow$ 40 million   1995 $\rightarrow$ 27,034,143
   46 years $\rightarrow$ 80 million   2000 $\rightarrow$ 31,430,689
   c. ≈ 7.4 yr

13.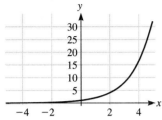

15.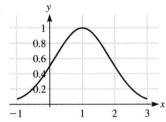

17.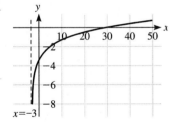

19.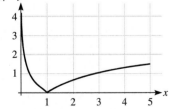

21. 2
23. $\dfrac{\sqrt{2}}{2}$
25. $\frac{1}{4}$
27. 2
29. 2
31. $-\frac{5}{6}$
33. $-\frac{1}{2}\ln 0.01$
35. 5
37. $4, \frac{1}{8}$
39. $10^4, 10^{-4}$
41. $16, \frac{1}{16}$
43. e^{2x}
45. $e^{2x} - e^{2y}$
47. *a.* $t = 0 \rightarrow 0, 1 \rightarrow 2.083317, 2 \rightarrow 3.9496,$
   $5 \rightarrow 8.461004, 10 \rightarrow 13.343$

b.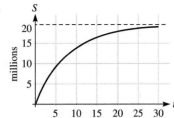

c. 6.3 yr
49. $5x - 2$
51. 1
53. 3
55. 4
57. $\log_a \dfrac{x^3 + x^2 y}{y^3}$
59. $\log_a \dfrac{\sqrt[3]{x}\, v^3}{\sqrt[6]{w}}$
61. $\log_3 x + \log_3(y + z)$
63. 2
65. b
67. $\frac{1}{3}$
69. $\frac{1}{2}$
71. 18
73. ≈ 40.3817 days
75. 11.086186
77. *a.* $1 \rightarrow 1.4816$ mg $4 \rightarrow 0.60239$ mg
   $2 \rightarrow 1.0976$ mg $6 \rightarrow 0.330598$ mg
   b. ≈ 2.3105 hr
79. 11.75 yr
81. *a.* 0.012; *b.* $P(t) = 2,286,000 e^{0.012t}$; *c.* 3,276,591
83. $\dfrac{1}{2}\ln\left(\dfrac{y+1}{y-1}\right)$
85. $\sqrt[n]{\dfrac{y}{a}}$
87. $\dfrac{\ln(A/P)}{n\ln(1 + i/n)}$
89. 5.005×10^{-4}
91. $\dfrac{1}{h}\ln\left(1 + \dfrac{h}{x}\right)$

CHAPTER 5
Section 5.1, page 276

1.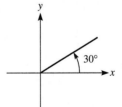

3.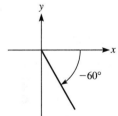

5.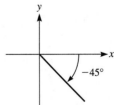

7.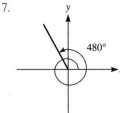

9. *a.* $53°36'$
   b. $143°36'$
11. *a.* $33°25'7''$
   b. $123°25'7''$
13. $36.4°$
15. $56.5814°$
17. $-35.8°$
19. $142.5667°$
21. $46°19'37''$
23. $-72°49'53''$
25. $364°2'42''$
27. $-1°39'20''$
29. $\dfrac{\pi}{5}$
31. $\dfrac{3\pi}{4}$
33. $-\dfrac{4\pi}{3}$
35. $\dfrac{1528}{1125}\pi$

37. 150°   39. −315°   41. 201.73844°

43. −390°   45. $\dfrac{7\pi}{6}$ 47. 135°, 45°

49. 45°, 45°, 90°; $\dfrac{\pi}{4}, \dfrac{\pi}{4}, \dfrac{\pi}{2}$ 51. 108.81 inches

53. 3.998 feet   55. 6.27 feet

57. 0.14545455 mph   59. $\dfrac{6600}{\pi}$ rev/sec

61. 24.43 meters   65. 1.041855 radians

67. −1.02654 radians   69. 109°2′37″

71. −114°35′30″

Section 5.2, page 286

	$\sin\theta$	$\cos\theta$	$\tan\theta$	$\cot\theta$	$\sec\theta$	$\csc\theta$
1.	$-\dfrac{\sqrt{2}}{2}$	$-\dfrac{\sqrt{2}}{2}$	1	1	$-\sqrt{2}$	$-\sqrt{2}$
3.	$-\dfrac{1}{2}$	$-\dfrac{\sqrt{3}}{2}$	$\dfrac{1}{\sqrt{3}}$	$\sqrt{3}$	$-\dfrac{2}{\sqrt{3}}$	-2
5.	$-\dfrac{\sqrt{3}}{2}$	$-\dfrac{1}{2}$	$\sqrt{3}$	$\dfrac{1}{\sqrt{3}}$	-2	$-\dfrac{2\sqrt{3}}{2}$
7.	$-\dfrac{\sqrt{2}}{2}$	$\dfrac{\sqrt{2}}{2}$	-1	-1	$\sqrt{2}$	$-\sqrt{2}$
9.	$-\dfrac{\sqrt{2}}{2}$	$-\dfrac{\sqrt{2}}{2}$	1	1	$-\sqrt{2}$	$-\sqrt{2}$
11.	$-\dfrac{\sqrt{3}}{2}$	$-\dfrac{1}{2}$	$\sqrt{3}$	$\dfrac{1}{\sqrt{3}}$	-2	$-\dfrac{2}{\sqrt{3}}$
13.	0	1	0	undef.	1	undef.
15.	$-\dfrac{\sqrt{3}}{2}$	$-\dfrac{1}{2}$	$\sqrt{3}$	$\dfrac{1}{\sqrt{3}}$	-2	$-\dfrac{2}{\sqrt{3}}$
17.	$-\dfrac{\sqrt{2}}{2}$	$\dfrac{\sqrt{2}}{2}$	-1	-1	$\sqrt{2}$	$-\sqrt{2}$
19.	1	0	undef.	0	undef.	1
21.	$-\dfrac{1}{2}$	$-\dfrac{\sqrt{3}}{2}$	$\dfrac{1}{\sqrt{3}}$	$\sqrt{3}$	$-\dfrac{2}{\sqrt{3}}$	-2
23.	$\dfrac{1}{2}$	$-\dfrac{\sqrt{3}}{2}$	$-\dfrac{1}{\sqrt{3}}$	$-\sqrt{3}$	$-\dfrac{2\sqrt{3}}{2}$	2
25.	$-\dfrac{\sqrt{2}}{2}$	$\dfrac{\sqrt{2}}{2}$	-1	-1	$\sqrt{2}$	$-\sqrt{2}$
27.	$-\dfrac{1}{2}$	$\dfrac{\sqrt{3}}{2}$	$-\dfrac{1}{\sqrt{3}}$	$-\sqrt{3}$	$\dfrac{2}{\sqrt{3}}$	-2
29.	0	-1	0	undef.	-1	undef.
31.	$\dfrac{4}{5}$	$\dfrac{3}{5}$	$\dfrac{4}{3}$	$\dfrac{3}{4}$	$\dfrac{5}{3}$	$\dfrac{5}{4}$
33.	$-\dfrac{12}{13}$	$\dfrac{5}{13}$	$-\dfrac{12}{5}$	$-\dfrac{5}{12}$	$\dfrac{13}{5}$	$-\dfrac{13}{12}$

35. 57.2900   37. 0.55920   39. 1.0888
41. 1.0002   43. 0.3065   45. 0.2507
47. 0.2399   49. 1.7251   51. 0.1700
53. 1.5508   55. −1.3764   57. 0.8253
59. 1.45846 radians = 83.56361°

61. 0.06112 radians = 3.50211°
63. 0.83770 radians = 47.99682°
65. 1.41315 radians = 80.96770°
67. 1.44547 radians = 82.81924°
69. 1.24507 radians = 71.33708°

Section 5.3, page 291

	$\sin\theta$	$\cos\theta$	$\tan\theta$	$\cot\theta$	$\sec\theta$	$\csc\theta$
1.	$\dfrac{12}{13}$	$\dfrac{5}{13}$	$\dfrac{12}{5}$	$\dfrac{5}{12}$	$\dfrac{13}{5}$	$\dfrac{13}{12}$
3.	$\dfrac{7}{25}$	$\dfrac{24}{25}$	$\dfrac{7}{24}$	$\dfrac{24}{7}$	$\dfrac{25}{24}$	$\dfrac{25}{7}$
5.	$\dfrac{3}{\sqrt{34}}$	$\dfrac{5}{\sqrt{34}}$	$\dfrac{3}{5}$	$\dfrac{5}{3}$	$\dfrac{\sqrt{34}}{5}$	$\dfrac{\sqrt{34}}{3}$
7.	$\dfrac{12}{13}$	$\dfrac{5}{13}$	2.4	$\dfrac{5}{12}$	2.6	$\dfrac{13}{12}$
9.	$\dfrac{3}{\sqrt{13}}$	$\dfrac{2}{\sqrt{13}}$	$\dfrac{3}{2}$	$\dfrac{2}{3}$	$\dfrac{\sqrt{13}}{2}$	$\dfrac{\sqrt{13}}{3}$
11.	$\dfrac{a}{\sqrt{1+a^2}}$	$\dfrac{1}{\sqrt{1+a^2}}$	a	$\dfrac{1}{a}$	$\sqrt{1+a^2}$	$\dfrac{\sqrt{1+a^2}}{a}$
13.	$\dfrac{2}{3}$	$\dfrac{\sqrt{5}}{3}$	$\dfrac{2}{\sqrt{5}}$	$\dfrac{\sqrt{5}}{2}$	$\dfrac{3}{\sqrt{5}}$	$\dfrac{3}{2}$
15.	$\sqrt{\dfrac{2}{3}}$	$\dfrac{\sqrt{3}}{3}$	$\sqrt{2}$	$\dfrac{1}{\sqrt{2}}$	$\sqrt{3}$	$\sqrt{\dfrac{3}{2}}$
17.	$\dfrac{24}{25}$	$\dfrac{7}{25}$	$\dfrac{24}{7}$	$\dfrac{7}{24}$	$\dfrac{25}{7}$	$\dfrac{25}{24}$
19.	$\sqrt{1-a^2}$	a	$\dfrac{\sqrt{1-a^2}}{a}$	$\dfrac{a}{\sqrt{1-a^2}}$	$\dfrac{1}{a}$	$\dfrac{1}{\sqrt{1-a^2}}$
21.	0.9844	0.1762	5.5866	0.179	5.6754	1.0159
23.	0.9893	0.1457	6.7899	0.1473	6.8631	1.0108
25.	$\dfrac{12}{13}$	$\dfrac{5}{13}$	$\dfrac{12}{5}$	$\dfrac{5}{12}$	$\dfrac{13}{5}$	$\dfrac{13}{12}$

27. By approximation, 0.980066578;
by calculation, .980066578

29. $x > 0.144$

Section 5.4, page 295

1. $\alpha = 37°$, $b = 19.9$, $c = 24.9$
3. $b = 7.5$, $\beta = 36.8699°$, $\alpha = 53.1301°$
5. $c = 7.647876$, $B = 62.765°$, $C = 90°$, $A = 27.235°$
7. $a = 30.25$, $A = 52.75°$, $B = 37.25°$, $C = 90°$
9. $B = 66°20'$, $a = 0.6224$, $c = 1.5504$, $C = 90°$
11. $A = 33°52'$, $b = 506.6098$, $c = 610.1258$, $C = 90°$
13. $c = 51.522$, $A = 38.54°$, $B = 51.46°$, $C = 90°$
15. $B = 52°17'$, $C = 90°$, $b = 1616.34$, $c = 2043.29$
17. $A = 30.84°$, $B = 59.16°$, $C = 90°$, $b = 54.44$
19. 95.395 ft   21. 272.56 ft
23. 1.11355 m   25. 12 ft, 53.1301°
27. 26,469.4 ft   29. 41.44°
31. Horizontal distance is 2674.3 ft. The jumpers travel 2931 ft while in the air.
33. $c = 58.097$, $b = 43.061$
35. 7498.6 m   37. 52.65 ft

39. 11.9 yd

41. 2677.6 ft

43. 110°

45. 34.034 miles

47. $h = \dfrac{s \tan \alpha \tan \beta}{\tan \beta - \tan \alpha}$

49. 19.438

Section 5.5, page 309

1. a. 2π; b. 1
   c.

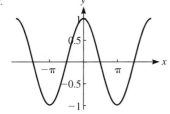

3. a. 4π; b. 1
   c.

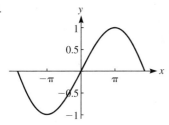

5. a. 1; b. 1
   c.

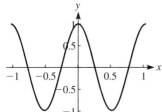

7. a. $\dfrac{2\pi}{3}$; b. 1
   c.

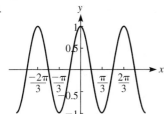

9. a. 2π; b. 2
   c.

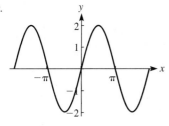

11. a. 2π; b. $\frac{1}{2}$
    c.

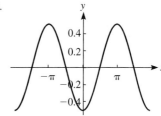

13. a. π; b. 2
    c.

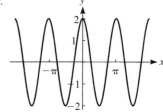

15. a. 1; b. 4
    c.

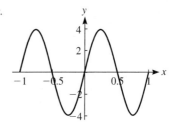

17. a. 2; b. 3
    c.

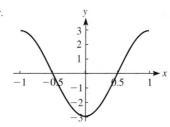

19. a. 3; b. 1
    c.

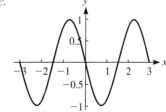

21. *a.* 2π; *b.* 1; *c.* $\dfrac{\pi}{2}$ right

d.

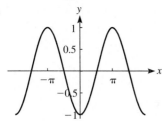

23. *a.* 2π; *b.* 3; *c.* π left

d.

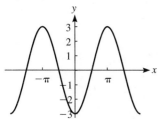

25. *a.* $\dfrac{\pi}{2}$; *b.* 2; *c.* $\dfrac{\pi}{4}$ left

d.

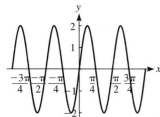

27. *a.* π; *b.* 3; *c.* $\dfrac{\pi}{6}$ right

d.

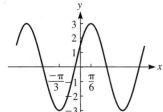

29. *a.* 1; *b.* $\frac{3}{2}$; *c.* $\dfrac{1}{2\pi}$ left

d.

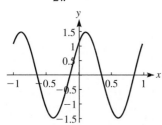

31. *a.* 2π; *b.* 1; *c.* None

d.

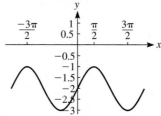

33. *a.* 2; *b.* 2; *c.* None

d.

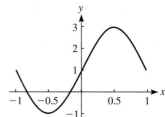

35. *a.* π; *b.* 6; *c.* $\dfrac{\pi}{2}$ right   37. *a.* $\dfrac{1}{2}$; *b.* 5; *c.* $\dfrac{1}{2\pi}$ left

39. *a.* $\dfrac{2\pi}{3}$; *b.* $\frac{2}{3}$; *c.* $\dfrac{\pi}{12}$ left

41. *a.* 2; *b.* 10 inches; *c.* 5 and -5

d.

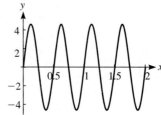

43. *a.* 104.6° and 98.6°

b.

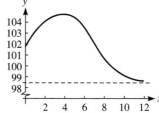

45. *a.*

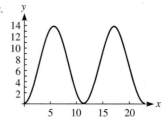

b. Minimum sales = \$0 at $t = 0, 12, 24$.

c. Maximum sales = $14,000 at $t = 6, 18$

d. Period = 12 months. Yes, given the cyclic nature of the sport.

47. a. 103 ft, 68.45 ft, 37.55 ft, 12.55 ft, 68.45 ft, 93.45 ft, 103 ft

b. 3 ft.; c. 103 ft.; d. 10

e.

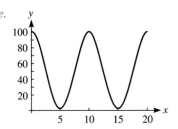

49. a.

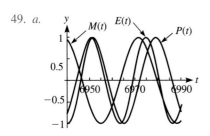

b. Best when $M(t) > 0$ and worst when $M(t) < 0$

51. a. $|A|$ = amplitude; $2\pi T$ = period; $\dfrac{xT}{\lambda}$ = phase shift

b. $|A|$ = amplitude; $2\pi\lambda$ = period; $\dfrac{t\lambda}{T}$ = phase shift

53. a. $|I|$ = amplitude; $\dfrac{2\pi}{w}$ = period; $-\dfrac{\alpha}{w}$ = phase shift

b.

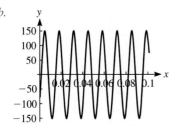

55. $y = -\cos x$ 57. $y = -4\cos(\pi x)$

59. $y = 2.8\cos(2\pi x - 12\pi)$

61. $x = -2.41549, 2.009285, -0.2031$

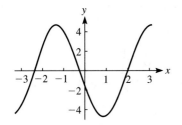

63. -0.36581

65. period = $\dfrac{2\pi}{3}$; amplitude = 1; phase shift = 0

67. period = 2π; amplitude = 2.2; phase shift = -1

71. (e)   73. (f)   75. (a)   77. (g)

Section 5.6, page 321

1.

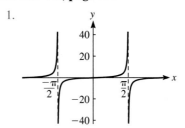

3.

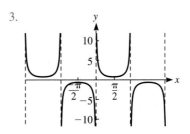

5.

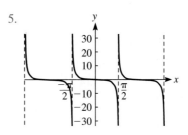

7.

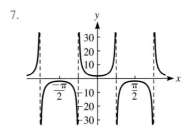

9.

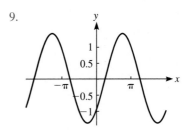

11.

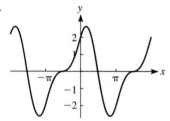

13.

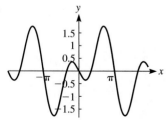

15.

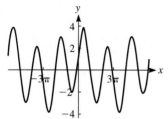

17.

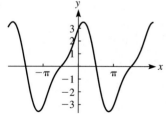

19.

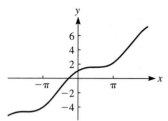

21.

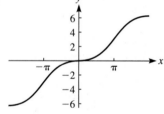

23.

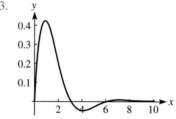

25.

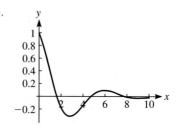

27.

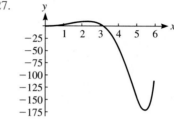

29.

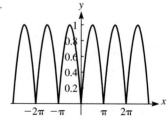

31.

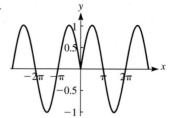

33.

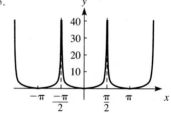

35.

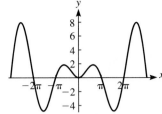

37.

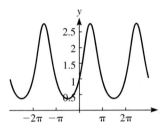

39. *a.* $D(0) = 7$ $D(1) = 6.3339$ $D(2) = 5.7311$
      $D(3) = 5.1857$ $D(4) = 4.6922$ $D(5) = 4.2457$
      $D(6) = 3.8417$ $D(7) = 3.4761$ $D(8) = 3.1453$
      $D(9) = 2.846$ $D(10) = 2.5752$

b.

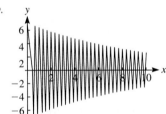

c. 7; *d.* -6.884397

41. *a.* $\theta = \dfrac{\pi t}{2}$; *b.* $d = 180 \tan \dfrac{\pi t}{2}$

c.

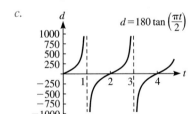

d. Spotlight beam is parallel to wall.
e. Spotlight is turned away from wall.

43.

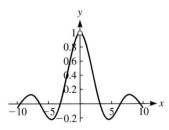

45.

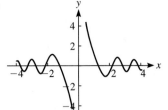

47. (*b*)   49. (*c*)   51. (*d*)

Section 5.7, page 327

	$\sin\theta$	$\cos\theta$	$\tan\theta$	$\cot\theta$	$\sec\theta$	$\csc\theta$
1.	$-\frac{2}{5}$	$-\dfrac{\sqrt{21}}{5}$	$\dfrac{2}{\sqrt{21}}$	$\dfrac{\sqrt{21}}{2}$	$-\dfrac{5}{\sqrt{21}}$	$-\frac{5}{2}$
3.	$\dfrac{\sqrt{5}}{3}$	$-\frac{2}{3}$	$-\dfrac{\sqrt{5}}{2}$	$-\dfrac{2}{\sqrt{5}}$	$-\frac{3}{2}$	$\dfrac{3}{\sqrt{5}}$
5.	$-\dfrac{\sqrt{5}}{4}$	$\dfrac{\sqrt{11}}{4}$	$-\dfrac{\sqrt{5}}{\sqrt{11}}$	$-\dfrac{\sqrt{11}}{\sqrt{5}}$	$\dfrac{4}{\sqrt{11}}$	$-\dfrac{4}{\sqrt{5}}$
7.	$\dfrac{2}{\sqrt{13}}$	$-\dfrac{3}{\sqrt{13}}$	$-\frac{2}{3}$	$-\frac{3}{2}$	$-\dfrac{\sqrt{13}}{3}$	$\dfrac{\sqrt{13}}{2}$

9. $2\sin^2 t - 1$ 11. $9 - 5\sin^2 t$

13. $2\cos^4 t - 2\cos^2 t + 1$ 15. $1 - 2\cos^2 t$

17. $\sin^5 t - 2\sin^3 t + \sin t$ 19. $\dfrac{2\cos^2 t - 1}{\cos^2 t - \cos^4 t}$

21. $\cot^4 t + \cot^2 t - 4\cot t - 3$

23. $\dfrac{1}{\tan^3 t} + \tan^3 t$ 25. $\cos t + 1 - \cos^2 t$

27. $\left(\dfrac{\sin^2 t - 1}{\sin t}\right)^2$ 29. $\tan t$ 31. $\dfrac{7\sin^3 \theta}{1 - \sin^2 \theta}$

33. $\sin x - 1$ 35. $\dfrac{1}{\cos\alpha + 5}$ 37. 1

39. $\dfrac{\cos y}{5}$ 41. 1

43. $\frac{1}{4}(7 - 3\cos^2 x)(\cos^2 x + 5)$

45. $\sec^8 x$ 47. $1 - \sin^2 t + \sin^4 t$

Section 5.8, Cumulative Review Exercises, page 328

1. $42°9'$, $132°9'$ 3. $47.85°$ 5. $-124.5667°$

7. $56°47'20''$ 9. $117°2'26''$ 11. $\dfrac{5\pi}{12}$ radians

13. $-\frac{5}{6}\pi$ radians   15. $\left(\dfrac{180}{\pi}\right)^{\circ}$ 17. $-2340°$

19. $60°$, $\dfrac{\pi}{3}$ radians   21. 0.3925 radians

23. 2.6 radians   25. $\dfrac{152\pi}{9}$ cm/sec

27.
$$\sin\theta = -\dfrac{2}{\sqrt{13}} \qquad \tan\theta = -\frac{2}{3}$$
$$\cos\theta = \dfrac{3}{\sqrt{13}} \qquad \cot\theta = -\frac{3}{2}$$
$$\csc\theta = -\dfrac{\sqrt{13}}{2} \qquad \sec\theta = \dfrac{\sqrt{13}}{3}$$

29. -1.527

31. -1.4422

33. 0.8445

35. -0.309

37. $\sin\theta = \dfrac{2}{\sqrt{13}}$ $\sec\theta = -\dfrac{\sqrt{13}}{3}$

   $\tan\theta = -\dfrac{2}{3}$ $\csc\theta = \dfrac{\sqrt{13}}{2}$

   $\cot\theta = -\dfrac{3}{2}$

39. $\sin\theta = -\dfrac{k}{\sqrt{1+k^2}}$ $\csc\theta = -\dfrac{\sqrt{1+k^2}}{k}$

   $\cos\theta = \dfrac{1}{\sqrt{1+k^2}}$ $\sec\theta = \sqrt{1+k^2}$

   $\tan\theta = -k$

41. $85.369°$, 1.49 radians   43. $7.1148°$, 0.1242 radians

	$\sin\theta$	$\cos\theta$	$\tan\theta$	$\cot\theta$	$\sec\theta$	$\csc\theta$
45.	$\dfrac{\sqrt{3}}{2}$	$-\dfrac{1}{2}$	$-\sqrt{3}$	$-\dfrac{1}{\sqrt{3}}$	-2	$\dfrac{2}{\sqrt{3}}$
47.	$-\dfrac{1}{2}$	$-\dfrac{\sqrt{3}}{2}$	$\dfrac{1}{\sqrt{3}}$	$\sqrt{3}$	$-\dfrac{2}{\sqrt{3}}$	-2
49.	0	-1	0	undef.	-1	undef.
51.	-1	0	undef.	0	undef.	-1

53. $t = \left\{\dfrac{4\pi}{3} + 2k\pi,\ \dfrac{5\pi}{3} + 2k\pi : k = 0, \pm1, \pm2, \ldots\right\}$

55. $t = \left\{\dfrac{\pi}{3} + k\pi : k = 0, \pm1, \pm2, \ldots\right\}$

57. $-\cot t$

59. $\csc^3 t - \csc^2 t - \csc t + 2$

61. 0

63. $\sec^2 x$

65. 1

67. $\dfrac{3\sin^4 t - 3\sin^2 t + 1}{(1 - \sin^2 t)(\sin^2 t)}$

69. 1 71. 144 73. $1 + \sqrt{3}$ 75. $\dfrac{8}{\sqrt{\sin\beta}}$

77. tan and cot $= \pi$; all others $= 2\pi$

79. Range of $\sin x$ and $\cos x$: $|y| \le 1$
   Range of $\sec x$, $\csc x$: $|y| \ge 1$
   Range of $\tan x$, $\cot x$: all real numbers

81. *a.* 2π; *b.* None; *c.* 1

   d.

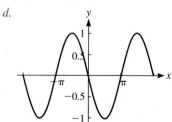

83. *a.* 4π; *b.* None; *c.* 2

   d.

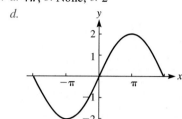

85.

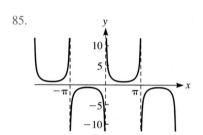

87.

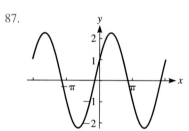

89.

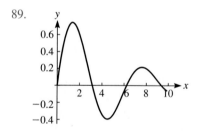

91. $y = -2\sin 2x$ 93. $y = 7.2\cos(2x - 2\pi)$

95. π 97. 0.6533 99. 85.94 in.

101. $18{,}229.54$ ft   103. 272.428 ft^2

105. 1489.6714 m

CHAPTER 6

Section 6.1, page 339

1. 1 3. $\sin^2 x$ 5. $\dfrac{1 - \sin^2 x}{\sin x}$

7. $\dfrac{1}{\sin^2 x}$ 9. $\sin^2 x$ 11. $\dfrac{1 - \cos^2 x}{\cos x}$

13. $\dfrac{1}{\cos^2 x}$ 15. $\dfrac{1}{1 - \cos^2 x}$ 17. $-\dfrac{1}{\cos^2 x}$

19. $\tan^2 x$ 21. $\cot x$ 23. $\cos x$

25. $a\sec x$ 27. $\dfrac{5(1 - \cos^2 x)}{\cos x}$

29. $a\tan x$ 31. $4\sec x$

Section 6.2, page 350

1. $\dfrac{\sqrt{2}-\sqrt{6}}{4}$

3. $2+\sqrt{3}$

5. $-\dfrac{\sqrt{2}+\sqrt{6}}{4}$

7. $\dfrac{\sqrt{2}+\sqrt{6}}{4}$

9. $2-\sqrt{3}$

11. $\dfrac{\sqrt{2}-\sqrt{6}}{4}$

13. $-\sin\theta$

15. $\dfrac{\sqrt{3}}{2}\cos\theta+\dfrac{1}{2}\sin\theta$

17. $\dfrac{\tan\theta-1}{\tan\theta+1}$

19. $\dfrac{\sqrt{3}}{2}\cos\theta+\dfrac{1}{2}\sin\theta$

21. $\dfrac{\sqrt{2}}{2}\cos\theta-\dfrac{\sqrt{2}}{2}\sin\theta$

23. $\dfrac{\tan\theta+1}{1-\tan\theta}$

25. $-\dfrac{\sqrt{3}}{2}$

27. $\dfrac{\sqrt{2}+\sqrt{6}}{4}$ or $\dfrac{\sqrt{2}-\sqrt{6}}{4}$

29. 0

31. $\dfrac{\sqrt{3}+1}{1-\sqrt{3}}$

33. $\dfrac{-\sqrt{6}-\sqrt{2}}{4}$

35. 0

37. $\frac{7}{25}$

39. $2+\sqrt{3}$

41. 0

43. $-\frac{24}{25}$

45. $-\frac{120}{119}$

47. 7

49. Undefined

51. $\frac{14}{3}$

59. $\sqrt{2}\sin\left(x+\dfrac{\pi}{4}\right)$

61. $2\sin\left(x+\dfrac{\pi}{3}\right)$

63. $\sqrt{194}\sin\left[x+\sin^{-1}\left(-\dfrac{13}{\sqrt{194}}\right)\right]$

65. $2\sin\left(\dfrac{\pi}{4}x+\dfrac{5\pi}{3}\right)$

67. $\frac{1}{2}$

69. 1

Section 6.3, page 362

1. $a.\ \dfrac{\sqrt{3}}{2}; b.\ \frac{1}{2}; c.\ \sqrt{3}$

3. $a.\ -0.96; b.\ 0.28; c.\ -\frac{24}{7}$

5. $a.\ \dfrac{\sqrt{3}}{2}; b.\ -\frac{1}{2}; c.\ -\sqrt{3}$

7. $a.\ -1; b.\ 0; c.$ Undefined

9. $4\cos^3 x-3\cos x$

11. $-4\sin x\cos x(-\cos^2 x+\sin^2 x)$

13. $a.\ \dfrac{1}{\sqrt{10}}; b.\ \dfrac{3}{\sqrt{10}}; c.\ \frac{1}{3}$

15. $a.\ \sqrt{\dfrac{2+\sqrt{2}}{4}}; b.\ -\sqrt{\dfrac{2-\sqrt{2}}{4}}; c.\ \dfrac{\sqrt{2}}{\sqrt{2}-2}$

17. $a.\ \sqrt{\frac{5}{8}}; b.\ -\sqrt{\frac{3}{8}}; c.\ -\sqrt{\frac{5}{3}}$

19. $a.\ \sqrt{\dfrac{3-\sqrt{5}}{6}}; b.\ -\sqrt{\dfrac{3+\sqrt{5}}{6}}; c.\ -\dfrac{2}{3+\sqrt{5}}$

21. $-\frac{1}{2}$

23. $-\dfrac{\sqrt{7}}{4}$

25. $\sqrt{\dfrac{2-\sqrt{2}}{4}}$

27. $\sqrt{\dfrac{\sqrt{2}-1}{\sqrt{2}+1}}$

29. $\dfrac{2}{\sqrt{2-\sqrt{2}}}$

31. $-\dfrac{\sqrt{\sqrt{2}-1}}{2^{3/4}}$

33. $\cos 40°$

35. $\tan\dfrac{2\pi}{5}$

37. $\tan 70°$

39. $-\sin\dfrac{\pi}{8}$

41. $\cos 7x$

43. $\tan 135°$

45. 1

47. 1

49. $\sec x$

Section 6.4, page 366

1. $\frac{1}{2}[\sin 7x+\sin x]$

3. $\frac{1}{2}[\cos 2y-\cos 4y]$

5. $\frac{1}{2}[\cos 4a+\cos 2a]$

7. $\frac{1}{2}[\sin 8x-\sin 2x]$

9. $\frac{1}{2}[\cos 4a+\cos 6a]$

11. $3\sin\dfrac{\pi}{2}$

13. $\frac{1}{2}[\sin 2x-\sin(2y)]$

15. $2\sin 20°\cos 40°$

17. $2\sin 8y\cos(y)$

19. $-2\sin\dfrac{\pi}{2}\sin\left(\dfrac{\pi}{4}\right)$

21. $2\sin y\cos x$

23. $2\cos\dfrac{3x}{2}\cos\dfrac{x}{2}$

25. $2\sin 30°\cos 20°$

47. 1

49. 1

51. $\tan\dfrac{7\pi}{72}$

53. -1

55. $\dfrac{\sqrt{3}-\sqrt{2}}{2}$

Section 6.5, page 371

1. $\dfrac{\pi}{6}$

3. $\dfrac{\pi}{4}$

5. $\dfrac{-\pi}{3}$

7. $\dfrac{\pi}{3}$

9. $\dfrac{\pi}{4}$

11. $\dfrac{5\pi}{6}$

13. $\dfrac{3\pi}{4}$

15. $\dfrac{\pi}{6}$

17. $\dfrac{2}{\sqrt{3}}$

19. $\sqrt{3}$

21. $\dfrac{\sqrt{3}}{2}$

23. $\dfrac{\sqrt{3}}{2}$

25. $\dfrac{\sqrt{34}}{3}$

27. $\dfrac{2\pi}{3}$

29. $\frac{1}{2}$

45. $\sin\dfrac{y}{2}-1$

47. $1-\tan 2y$

49. $\dfrac{\csc\left(-\dfrac{y}{4}\right)-1}{2}$

51. $\cos\left(-\frac{3}{2}y\right)+1$

53. $\dfrac{2-\sec(-y)}{3}$

55. $\dfrac{x}{\sqrt{4+x^2}}$

57. $\dfrac{\sqrt{1-x^2}}{x}$

59. $\dfrac{\sqrt{x^2+2}}{x}$

61. $\dfrac{3}{\sqrt{9+x^2}}$

63. $\dfrac{\sqrt{x^2-1}}{x}$

65. $2x^2-1$

67. 0

69. $\sin y=-x \quad \cos y=\sqrt{1-x^2} \quad \tan y=-\dfrac{x}{\sqrt{1-x^2}}$

$\sec y=\dfrac{1}{\sqrt{1-x^2}} \quad \csc y=-\dfrac{1}{x} \quad \cot y=-\dfrac{\sqrt{1-x^2}}{x}$

71. $\sin y=\sqrt{1-4x^2} \quad \cos y=2x \quad \tan y=\dfrac{\sqrt{1-4x^2}}{2x}$

$\sec y=\dfrac{1}{2x} \quad \csc y=\dfrac{1}{\sqrt{1-4x^2}} \quad \cot y=\dfrac{2x}{\sqrt{1-4x^2}}$

73. $\sin y=\dfrac{1}{x} \quad \cos y=\dfrac{\sqrt{x^2-1}}{x} \quad \tan y=\dfrac{1}{\sqrt{x^2-1}}$

$\sec y=\dfrac{x}{\sqrt{x^2-1}} \quad \csc y=x \quad \cot y=\sqrt{x^2-1}$

75. $\sin y = \dfrac{x}{\sqrt{4+x^2}}$ $\cos y = \dfrac{2}{\sqrt{4+x^2}}$ $\tan y = \dfrac{x}{2}$

$\sec y = \dfrac{\sqrt{4+x^2}}{2}$ $\csc y = \dfrac{\sqrt{4+x^2}}{x}$ $\cot y = \dfrac{2}{x}$

77.

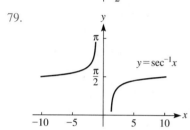

$y = \cot^{-1} x$

79.

$y = \sec^{-1} x$

81. $1, 0, \dfrac{\sqrt{3}}{2}, \dfrac{1}{2}$

83. $\dfrac{3-4\sqrt{3}}{10}, \dfrac{(4+3\sqrt{3})}{10}, \dfrac{24}{25}, \dfrac{7}{25}$

85. $\dfrac{-\sqrt{3}}{2}, \dfrac{1}{2}, 0, 1$ 87. Set $x = 1$

89. Set $x = 0$ 91. Set $x = \dfrac{1}{2}$

Section 6.6, page 376

1. $\{\pi k: k = 0, \pm 1, \pm 2, \pm 3, \ldots\}$

3. $\{\pi k: k = 0, \pm 1, \pm 2, \pm 3, \ldots\}$ 5. $\varnothing$

7. $\left\{\dfrac{3\pi}{2} + 2\pi k : k = 0, \pm 1, \pm 2, \pm 3, \ldots\right\}$

9. $\left\{\dfrac{\pi}{4} + \pi k : k = 0, \pm 1, \pm 2, \pm 3, \ldots\right\}$

11. $\left\{\dfrac{\pi}{3} + 2\pi k, \dfrac{5\pi}{3} + 2\pi k : k = 0, \pm 1, \pm 2, \ldots\right\}$

13. $\left\{\dfrac{5\pi}{6} + k\pi : k = 0, \pm 1, \pm 2, \ldots\right\}$

15. $\left\{\dfrac{2\pi}{3} + 2k\pi, \dfrac{4\pi}{3} + 2k\pi : k = 0, \pm 1, \pm 2, \ldots\right\}$

17. $\{\tan^{-1} 1.621 + k\pi : k = 0, \pm 1, \pm 2, \ldots\}$

19. $\{\cot^{-1}(-7.770) + k\pi : k = 0, \pm 1, \pm 2, \ldots\}$

21. No solutions   23. $\left\{-\dfrac{\pi}{6}, \dfrac{\pi}{6}\right\}$

25. $\left\{-\dfrac{\pi}{12}, \dfrac{\pi}{4}\right\}$ 27. $\left\{-\dfrac{5\pi}{12}, -\dfrac{\pi}{12}\right\}$

29. No solutions   31. $\left\{\dfrac{11\pi}{6}, \dfrac{\pi}{6}\right\}$

33. $\left\{0, \dfrac{\pi}{4}, \pi, \dfrac{5\pi}{4}\right\}$ 35. $\left\{0, \dfrac{\pi}{2}, \dfrac{3\pi}{2}, \pi\right\}$

37. $\left\{\dfrac{2\pi}{3}, \dfrac{4\pi}{3}\right\}$ 39. $\left\{\dfrac{\pi}{3}, \dfrac{5\pi}{3}, 0\right\}$

41. $\left\{\dfrac{\pi}{2}, \dfrac{7\pi}{6}, \dfrac{11\pi}{6}\right\}$ 43. $\left\{\dfrac{3\pi}{2}\right\}$

45. $\left\{0, \dfrac{\pi}{3}, \dfrac{2\pi}{3}, \pi, \dfrac{4\pi}{3}, \dfrac{5\pi}{3}\right\}$ 47. $\left\{\dfrac{\pi}{3}, \dfrac{5\pi}{3}\right\}$

49. $\left\{\dfrac{\pi}{12}, \dfrac{5\pi}{12}, \dfrac{13\pi}{12}, \dfrac{17\pi}{12}\right\}$ 51. $\{0\}$

53. 0.98279, 4.12439, 1.76819, 4.9098

55. 1.41182, 4.87137

57. $0, \dfrac{\pi}{2}, \pi, \dfrac{3\pi}{2}$ 59. $\dfrac{\pi}{6}, \dfrac{7\pi}{6}$

61. Equation is true for all x in $[0, 2\pi)$, except for $x = 0$, $\dfrac{\pi}{2}, \pi, \dfrac{3\pi}{2}$

63. $\dfrac{\pi}{3}, \dfrac{\pi}{2}, \dfrac{3\pi}{2}, \dfrac{5\pi}{3}$ 65. $0, \dfrac{\pi}{2}, \pi, \dfrac{3\pi}{2}$

67. $\dfrac{\pi}{2}, \dfrac{3\pi}{2}, \dfrac{\pi}{6}, \dfrac{5\pi}{6}$

69. $\left\{0, \dfrac{2\pi}{3}, \dfrac{4\pi}{3}, \dfrac{\pi}{6}, \dfrac{5\pi}{6}, \dfrac{3\pi}{2}\right\}$

71. $\dfrac{\pi}{6}, \dfrac{5\pi}{6}, \dfrac{7\pi}{6}, \dfrac{11\pi}{6}$ 73. $\dfrac{7\pi}{6}, \dfrac{11\pi}{6}, \dfrac{\pi}{4}, \dfrac{5\pi}{4}$

75. $\dfrac{\pi}{4}, \dfrac{3\pi}{4}, \dfrac{5\pi}{4}, \dfrac{7\pi}{4}, \dfrac{\pi}{3}, \dfrac{2\pi}{3}, \dfrac{4\pi}{3}, \dfrac{5\pi}{3}$

Section 6.7, Cumulative Review Exercises, page 379

1. $\cot x$ 3. $\sec x$ 5. $\sin x$

	$\sin x$	$\cos x$	$\tan x$	$\cot x$	$\sec x$	$\csc x$
7.	$-\dfrac{1}{2}$	$\dfrac{\sqrt{3}}{2}$	$-\dfrac{1}{\sqrt{3}}$	$-\sqrt{3}$	$\dfrac{2}{\sqrt{3}}$	-2
9.	$\dfrac{4}{5}$	$\dfrac{3}{5}$	$\dfrac{4}{3}$	$\dfrac{3}{4}$	$\dfrac{5}{3}$	$\dfrac{5}{4}$

11. $3\cos x$ 17. $\dfrac{\sqrt{3}+1}{\sqrt{3}-1}$

19. $\dfrac{1+\sqrt{3}}{2^{3/2}}$ 21. $\dfrac{\sqrt{3}}{2}\cos x + \dfrac{1}{2}\sin x$

23. $\dfrac{1+\tan x}{\tan x - 1}$ 25. 0

27. $-\dfrac{7}{17}$ 29. 0   31. $\dfrac{1-\sqrt{3}}{1+\sqrt{3}}$

33. $\tan^{-1}\dfrac{7}{4}$ 37. -1 39. 1

	$\sin 2x$	$\cos 2x$	$\tan 2x$	$\sin \dfrac{x}{2}$	$\cos \dfrac{x}{2}$	$\tan \dfrac{x}{2}$
43.	.96	$-.28$	$-\dfrac{24}{7}$	$\dfrac{1}{\sqrt{5}}$	$\dfrac{2}{\sqrt{5}}$	.5
45.	$\dfrac{\sqrt{3}}{2}$	$-\dfrac{1}{2}$	$-\sqrt{3}$	$\dfrac{\sqrt{3}}{2}$	$-\dfrac{1}{2}$	$-\sqrt{3}$

47. $-4\sin^3 x + 3\sin x$ 49. $1 - 2\sin^2 5x$

57. $-\dfrac{\sqrt{3}}{2}$ 59. $-\dfrac{\sqrt{7}}{3}$ 61. $\dfrac{\sqrt{2 + \sqrt{2}}}{2}$

63. $\dfrac{\sqrt{2 - \sqrt{2}}}{2}$ 65. $\frac{1}{2}$ 67. $\cos\dfrac{2\pi}{5}$

69. $\tan 17.5°$ 71. $\sin^2 x$

73. 1   75. $(\cos 8x) + \cos(-2x)$

77. $(\sin 4x) + \sin(2x)$ 79. $2\sin 8y \cos y$

81. $2\sin 60° \sin 20°$ 87. $-\dfrac{\pi}{3}$

89. $\dfrac{5}{\sqrt{34}}$ 91. $-\dfrac{\pi}{2}$

95. $x = \frac{1}{4}[\tan(2y) + 3]$ 97. $\frac{1}{2}\csc\left(-\frac{y}{2}\right) - \frac{1}{2}$

99. $\dfrac{x}{\sqrt{4 + x^2}}$ 101. $\dfrac{\sqrt{1 - x^2}}{x}$

103. $\cos y = \dfrac{x}{2}$ $\cot y = \dfrac{x}{\sqrt{4 - x^2}}$

$\sin y = \dfrac{\sqrt{4 - x^2}}{2}$ $\sec y = \dfrac{2}{x}$

$\tan y = \dfrac{\sqrt{4 - x^2}}{x}$ $\csc y = \dfrac{2}{\sqrt{4 - x^2}}$

105. $\cot y = -x$ $\cos y = -\dfrac{x}{\sqrt{x^2 + 1}}$

$\tan y = -\dfrac{1}{x}$ $\sec y = -\dfrac{\sqrt{x^2 + 1}}{x}$

$\sin y = \dfrac{1}{\sqrt{x^2 + 1}}$ $\csc y = \sqrt{x^2 + 1}$

107. $\dfrac{11}{5\sqrt{34}}$, $\dfrac{27}{5\sqrt{34}}$, $-\frac{15}{17}$, $\frac{8}{17}$, $-\frac{15}{8}$

109. $\left\{\dfrac{5\pi}{6} + 2k\pi, \dfrac{7\pi}{6} + 2k\pi : k = 0, \pm 1, \pm 2, \ldots\right\}$

111. $x = 0.1279965 + \pi k : k = 0, \pm 1, \pm 2, \ldots$

113. $\dfrac{7\pi}{24}$, $-\dfrac{5\pi}{24}$ 115. $\dfrac{\pi}{3}$, $\dfrac{2\pi}{3}$, $\dfrac{5\pi}{3}$, $\dfrac{4\pi}{3}$

117. $\dfrac{k\pi}{4} : k = 0, 1, 2, 3, 4, 5, 6, 7$

119. $\dfrac{\pi}{3}$, $\dfrac{5\pi}{3}$ 121. No solutions

CHAPTER 7

Section 7.1, page 388

1. $33.1°, 94.9°, 127.7$ 3. $30°, 120°, 138.56$
5. $90°, 275.5, 165.8$
7. $\angle B = 34.5°, c = 19.9, b = 22.6$
9. $\gamma = 75°, a = 9.66, c = 9.66$
11. $\beta = 126°, b = 302.39, c = 137.9$
13. $\angle B = 19.15°, \angle C = 60.6°, c = 42.5$

15. $\angle C = 154.31°, \angle A = 10.186°, c = 11.029$
17. $\gamma = 48.74°, \alpha = 21.26°, a = 48.23$ 19. None
21. None   23. $\gamma = 90°, \beta = 60°, b = 34.64$
25. 20 km   27. 9.3 m   29. $0.33\overline{MN}$ 31. 17.06 ft
33. 576.7 m   35. 32 ft   37. 9609.4 m
39. 25.4 miles   43. 363.3 ft^2 45. 565.33

Section 7.2, page 395

1. $c = 2.5, \angle B = 97.5°, \angle A = 52.5°$
3. $a = 2.6, \angle B = 40.9°, \angle C = 79.1°$
5. $a = 8.1, \angle B = 37.9°, \angle C = 7.1°$
7. $c = 8.9, \angle A = 39.1°, \angle B = 70.9°$
9. $a = 9.6, \beta = 41.6°, \gamma = 85.4°$
11. $o = 44.1, \angle M = 12.5°, \angle O = 107.5°$
13. $c = 4.5, \angle A = 10.9°, \angle B = 153.7°$
15. $\angle A = 51°, \angle B = 87.3°, \angle C = 41.8°$
17. $\angle A = 29°, \angle B = 46.6°, \angle C = 104.5°$
19. $\angle A = 34°, \angle B = 44.4°, \angle C = 101.5°$
21. $\angle A = 26.4°, \angle B = 18.4°, \angle C = 135.2°$
23. $a = 22.1, \angle B = 33.4°, \angle C = 76.9°$
25. 24.3 miles   27. 85 m
29. 154 miles; course should change to $26.4°$.
31. 514.4 miles   33. 257 m
35. 4.9 miles, 11.3 miles   37. 51.4 km
39. 30   41. 36.5   43. 168.4

Section 7.3, page 403

1.

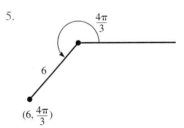

3. $\dfrac{3\pi}{2}$

$\left(5, \dfrac{3\pi}{2}\right)$

5.

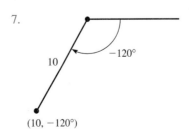

7.

9.
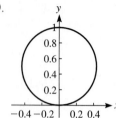

11.
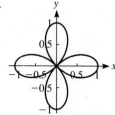

27.
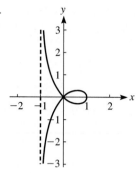

13.
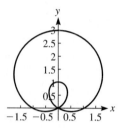

15.
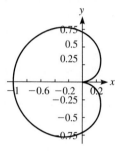

29. $(2^{5/2}, 45°)$ 31. $(1, 120°)$ 33. $(2, 30°)$
35. $(5, 53.1°)$ 37. $(2^{3/2}, 225°)$ 39. $(\sqrt{13}, 33.7°)$
41. $r = \dfrac{4}{3\cos\theta + 2\sin\theta}$ 43. $r = 4$
45. $r = 6\csc\theta$ 47. $r^2 = \dfrac{4}{\cos^2\theta - 4\sin^2\theta}$
49. $r^2 = \dfrac{36}{\cos^2\theta + 9\sin^2\theta}$ 51. $x^2 + y^2 = 25$
53. $y = -2$ 55. $x^2 + y^2 = 6y$
57. $y = x$ 59. $x^2 + y^2 = 4x$

17.
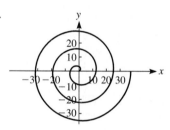

61.
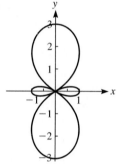

63.
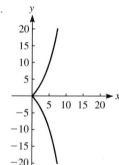

19.
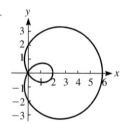

21.
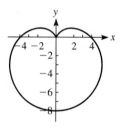

65. (a)-(3), (b)-(1), (c)-(2)
67. $\left(2, \dfrac{\pi}{2} + 2\pi k\right)$ $(k = 0, \pm 1, \pm 2, \dots)$

23.
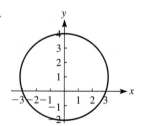

Section 7.4, page 409

1. $\sqrt{13}$ 3. $\sqrt{3}$ 5. 3 7. $2\sqrt{7}$
13. $2\sqrt{2}\left(\cos\dfrac{\pi}{4} + i\sin\dfrac{\pi}{4}\right)$ 15. $4\left(\cos\dfrac{\pi}{2} + i\sin\dfrac{\pi}{2}\right)$
17. $3(\cos 0 + i\sin 0)$

25.
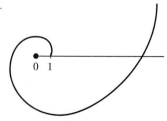

19. $3\sqrt{2}\left(\cos\dfrac{7\pi}{4} + i\sin\dfrac{7\pi}{4}\right)$

21. $4\sqrt{2}\left(\cos\dfrac{\pi}{4} + i\sin\dfrac{\pi}{4}\right)$ 23. $2\left(\cos\dfrac{11\pi}{6} + i\sin\dfrac{11\pi}{6}\right)$

25. $2\left(\cos\dfrac{\pi}{3} + i\sin\dfrac{\pi}{3}\right)$ 27. $4\left(\cos\dfrac{7\pi}{6} + i\sin\dfrac{7\pi}{6}\right)$

29. $2(\cos\pi + i\sin\pi)$ 31. $5\left(\cos\dfrac{\pi}{2} + i\sin\dfrac{\pi}{2}\right)$

33.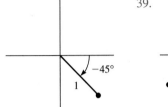

35.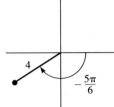

37.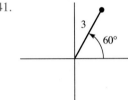

39.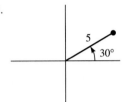

41.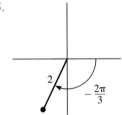

43.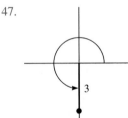

45.

47.

49. $10(\cos 225° + i \sin 225°)$, $\frac{5}{2}(\cos 135° + i \sin 135°)$

51. $4(\cos \pi + i \sin \pi)$, $\frac{1}{4}(\cos 2\pi + i \sin 2\pi)$

53. $6\left(\cos \frac{2\pi}{3} + i \sin \frac{2\pi}{3}\right)$, $\frac{2}{3}\left(\cos \frac{5\pi}{3} + i \sin \frac{5\pi}{3}\right)$

55. $10\left(\cos \frac{\pi}{12} + i \sin \frac{\pi}{12}\right)$, $\frac{5}{2}\left(\cos \frac{19\pi}{12} + i \sin \frac{19\pi}{12}\right)$

57. $12(\cos 150° + i \sin 150°)$, $3(\cos 330° + i \sin 330°)$

59. $27(\cos 75° + i \sin 75°)$, $3(\cos 195° + i \sin 195°)$

61. $32\left(\cos \frac{\pi}{3} + i \sin \frac{\pi}{3}\right)$, $\frac{1}{2}\left(\cos \frac{4\pi}{3} + i \sin \frac{4\pi}{3}\right)$

63. $-\sqrt{2} + \sqrt{2}i$ 65. $-6i$ 67. $-\frac{\sqrt{3}}{2} - \frac{1}{2}i$

69. $\frac{3}{2} - \frac{\sqrt{3}}{2}i$ 71. $-5\sqrt{2} + 5\sqrt{2}i$ 73. $15i$

75. $\left(2\sqrt{2} + \frac{3\sqrt{3}}{2}\right) + \left(-\frac{3}{2} + 2\sqrt{2}\right)i$

77. $4 + 2i$ 79. $-5.909 + 1.042i$; $-\frac{3}{2}i$

81. $4.384 + 8.988i$; $-1.798 - 1.737i$

Section 7.5, page 415

1. $-\frac{1}{2} - \frac{\sqrt{3}}{2}i$ 3. $1024i$

5. $-125i$ 7. -1

9. $-(5\sqrt{2})^{14}i$ 11. $4^8\left(-\frac{1}{2} + \frac{\sqrt{3}}{2}i\right)$

13. $\frac{81}{2} + \frac{81}{2}\sqrt{3}i$ 15. $-32 + 32\sqrt{3}i$

17. $4\sqrt{3} - 4i$ 19. $-512\sqrt{2} - 512\sqrt{2}i$

21. -2, $1 \pm \sqrt{3}i$

23. $2^{1/10}\left[\left(\cos \frac{7\pi/6 + 2\pi j}{10}\right) + \left(i \sin \frac{7\pi/6 + 2\pi j}{10}\right)\right]$
$(j = 0, 1, \ldots, 9)$

25. $\frac{\sqrt{3}}{2} + \frac{1}{2}i$, $-\frac{\sqrt{3}}{2} + \frac{1}{2}i$, $-i$

27. $\frac{3\sqrt{2}}{2} + \frac{3\sqrt{2}}{2}i$; $-\frac{3\sqrt{2}}{2} - \frac{3\sqrt{2}}{2}i$

29. $t_j = (\sqrt{2})^{1/5}\left[\cos \frac{\frac{\pi}{4} + 2\pi j}{5} + i \sin \frac{\frac{\pi}{4} + 2\pi j}{5}\right]$
$(j = 0, 1, 2, 3, 4)$

31. $t_j = 2\left[\cos\left(\frac{5\pi/3 + 2\pi j}{4}\right) + i \sin\left(\frac{5\pi/3 + 2\pi j}{4}\right)\right]$
$(j = 0, 1, 2, 3)$

33. $\frac{3}{2} + \frac{3\sqrt{3}}{2}i$; $\frac{3}{2} - \frac{3\sqrt{3}}{2}i$; -3

35. $\pm\left(\frac{3\sqrt{3}}{2} + \frac{3}{2}i\right)$, $\pm\left(\frac{3}{2} - \frac{3\sqrt{3}}{3}i\right)$

37. $2\sqrt{2} + 2\sqrt{2}i$, $-2\sqrt{2} - 2\sqrt{2}i$

39. $3\left[\cos \frac{2\pi j}{5} + i \sin \frac{2\pi j}{5}\right]$ $(j = 0, 1, 2, 3, 4)$

41. $t_j = 3\left[\cos\left(\frac{3\pi/2 + 2\pi j}{4}\right) + i \sin\left(\frac{3\pi/2 + 2\pi j}{4}\right)\right]$
$(j = 0, 1, 2, 3)$

43. $t_j = 2\left[\cos\left(\frac{5\pi/6 + 2\pi j}{4}\right) + i \sin\left(\frac{5\pi/6 + 2\pi j}{4}\right)\right]$
$(j = 0, 1, 2, 3)$

45. $t_j = \cos\left(\frac{2\pi j}{5}\right) + i \sin\left(\frac{2\pi j}{5}\right)$ $(j = 0, 1, 2, 3, 4)$

47. $\frac{3}{2} + \frac{3\sqrt{3}}{2}i$, $\frac{3}{2} - \frac{3\sqrt{3}}{2}i$, -3

49. $\pm 5i$

51. $t_j = 2^{1/3}\left[\cos\left(\frac{5\pi/3 + 2\pi j}{3}\right) + i \sin\left(\frac{5\pi/3 + 2\pi j}{3}\right)\right]$
$(j = 0, 1, 2)$

53. $t_j = 2^{1/5}\left[\cos\left(\dfrac{3\pi/4 + 2\pi j}{5}\right) + i\sin\left(\dfrac{3\pi/4 + 2\pi j}{5}\right)\right]$

   $(j = 0, 1, 2, 3, 4)$

55. $\frac{27}{4} + \frac{27}{4}i$ 57. $\dfrac{\sqrt{2}}{8}\left[\cos\dfrac{\pi}{12} + i\sin\dfrac{\pi}{12}\right]$

59. $16\sqrt{3} + 16 + (16 - 16\sqrt{3})i$

61. $\frac{9}{4}$ 63. $-2\sqrt{2} - 2\sqrt{2}i$

Section 7.6, page 422

1.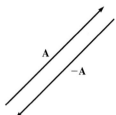

3.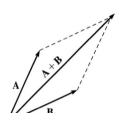

5.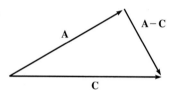

7.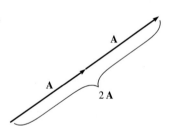

9.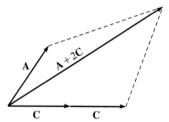

11. $a.\ \sqrt{2};\ b.\mathbf{i} - \mathbf{j}$
13. $a.3\sqrt{2};\ b.3\mathbf{i} + 3\mathbf{j}$
15. $a.2;\ b.-\mathbf{i} + \sqrt{3}\mathbf{j}$
17. $a.6;\ b.-3\sqrt{3}\mathbf{i} - 3\mathbf{j}$
19. $a.5;\ b.3\mathbf{i} + 4\mathbf{j}$
21. $a.26;\ b.10\mathbf{i} - 24\mathbf{j}$
23. $a.\dfrac{\sqrt{5}}{2}\ b.-\frac{1}{2}\mathbf{i} - \mathbf{j}$
25. 321.39 lbs
27. $\theta \approx 18.2°$
29. $15\sin 24° = 6.1$ lbs
31. 13.1 lbs
33. 217.93 lbs
35. $10\sqrt{3} - 10$ knots
37. 136 miles, 98.6°
39. 466 mph, bearing 97.45°

41. 521.54 mph, bearing 184.4°   43. 3.92 mph
45. 44.54 miles, bearing 11.05°
47. $\langle 6, 4\rangle$ 49. $\langle 1, 6\rangle$ 51. $\langle 3, -2\rangle$
53. $\langle -6, 13\rangle$ 55. $\langle 5, 6\rangle$ 57. $-3 - 5\sqrt{2}$
63. yes   65. no   67. $\cos^{-1}\left(\dfrac{3}{\sqrt{10}}\right)$
69. $\cos^{-1}\left(\dfrac{\sqrt{5}}{5}\right)$

Section 7.7, Cumulative Review Exercises, page 424

1. $c = 17.52, B = 5.74°, C = 144.26°$
3. $B = 24.32006°, C = 142.68°, c = 700.402$ or
   $B = 155.7°, C = 11.3°, c = 226.5$
5. 50.8 ft   7. 13.26 ft   9. 256.87 ft   11. 94.94°
13. $c = 8.62, A = 55.1°, B = 79.9°$
15. 21   17. No solution   19. $\sqrt{5}$
21. 2   23. $a.2\left(\cos\dfrac{\pi}{2} + i\sin\dfrac{\pi}{2}\right); b.2; c.\dfrac{\pi}{2}$
25. $a.\ \sqrt{19}(\cos 203.4° + i\sin 203.4°); b.\ \sqrt{19}; c.203.4°$

27. 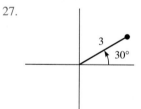   29.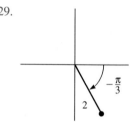

31. $zw = 10(\cos 135° + i\sin 135°),$
   $\dfrac{z}{w} = \frac{5}{2}(\cos 45° + i\sin 45°)$
33. $zw = 6\left(\cos\dfrac{4\pi}{3} + i\sin\dfrac{4\pi}{3}\right),$
   $\dfrac{z}{w} = \dfrac{3}{2}\left(\cos\dfrac{4\pi}{3} + i\sin\dfrac{4\pi}{3}\right)$
35. $1 + i$ 37. 5   39. $\dfrac{1 - \sqrt{3}i}{2}$
41. $-2^{30}i$ 43. $-2^{15} - \sqrt{3}2^{15}i$
45. $3, -\frac{3}{2} + \dfrac{3\sqrt{3}}{2}i, -\frac{3}{2} - \dfrac{3\sqrt{3}}{2}i$
47. $a.(2\sqrt{2})^{1/5}\left(\cos\dfrac{\pi}{20} + i\sin\dfrac{\pi}{20}\right)$
   $b.(2\sqrt{2})^{1/5}\left(\cos\dfrac{9\pi}{20} + i\sin\dfrac{9\pi}{20}\right)$
   $c.(2\sqrt{2})^{1/5}\left(\cos\dfrac{17\pi}{20} + i\sin\dfrac{17\pi}{20}\right)$
   $d.(2\sqrt{2})^{1/5}\left(\cos\dfrac{5\pi}{4} + i\sin\dfrac{5\pi}{4}\right)$
   $e.(2\sqrt{2})^{1/5}\left(\cos\dfrac{33\pi}{20} + i\sin\dfrac{33\pi}{20}\right)$

49. $5\sqrt{3} - 5i, -5\sqrt{3} + 5i$

51. $\cos\left(\dfrac{\pi + 2\pi j}{5}\right) + i\sin\left(\dfrac{\pi + 2\pi j}{5}\right)$ $(j = 0, 1, 2, 3, 4)$

53. $a.\ 2^{1/3}\left(\cos\dfrac{5\pi}{9} + i\sin\dfrac{5\pi}{9}\right)$

$b.\ 2^{1/3}\left(\cos\dfrac{11\pi}{9} + i\sin\dfrac{11\pi}{9}\right)$

$c.\ 2^{1/3}\left(\cos\dfrac{17\pi}{9} + i\sin\dfrac{17\pi}{9}\right)$

55.
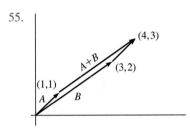

57.
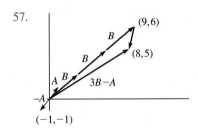

59. $a.\ 2\sqrt{2};\ b.\ 2\mathbf{i} + 2\mathbf{j}$
61. $a.\ 2;\ b.\ -\sqrt{2}\mathbf{i} + \sqrt{2}\mathbf{j}$
63. 8.03848 mph
65. Centrifugal force = 57.74 lbs; force by cable = 115.48 lbs
67. $\langle 6, 15\rangle$
69. $\langle -5, -4\rangle$

71.
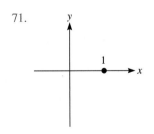

73.
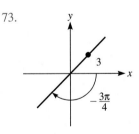

75.
77.
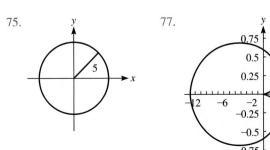

79. $\left(2\sqrt{2}, \dfrac{3\pi}{4}\right)$ 81. $(6, -60°)$

83. $r\cos\theta = -3$ 85. $r = 2$

87. $x^2 + y^2 = 4$ 89. $x = 5$ 91. $x^2 - y^2 = 1$

93. $\left(5, \dfrac{\pi}{3}\right), \left(5, \dfrac{2\pi}{3}\right), (5, \pi), \left(5, \dfrac{4\pi}{3}\right), \left(5, \dfrac{5\pi}{3}\right)$

CHAPTER 8
Section 8.1, page 436

1. $x = -1, y = \frac{1}{3}$ 3. $x = 5, y = -4$

5. $x = -8, y = 14$

7. $(x, y) = \left(\dfrac{-6 + \sqrt{71}}{5}, \dfrac{3 + 2\sqrt{71}}{5}\right),$

$\left(\dfrac{-6 - \sqrt{71}}{5}, \dfrac{3 - 2\sqrt{71}}{5}\right)$

9. $(x, y) = \left(\dfrac{-12 + 2\sqrt{21}}{3}, \dfrac{-3 + 2\sqrt{21}}{3}\right),$

$\left(\dfrac{-12 - 2\sqrt{21}}{3}, \dfrac{-3 - 2\sqrt{21}}{3}\right)$

11. No real solutions

13. $(x, y) = \left(\dfrac{-4 + \sqrt{26}}{5}, \dfrac{-2 + 3\sqrt{26}}{10}\right),$

$\left(\dfrac{-4 - \sqrt{26}}{5}, \dfrac{-2 - 3\sqrt{26}}{10}\right)$

15. $(x, y) = (\pm\sqrt{11}, \pm 2\sqrt{3})$ (all four sign combinations)
17. $(x, y) = (-2, -4), (2, 4), (4, 2)$
19. 700 balcony tickets, 900 main floor tickets
21. Aluminum = 13, antimony = 51
23. $4000 at 12%, 1000 at 6%
25. 4, 6   27. 0
29. 614 or 423   31. 20 cm × 34 cm
33. 5 and 1 or −5 and −1   35. $120
37. Fixed cost = $100; cost per guest = $1.25

Section 8.2, page 440

1. $x = 1, y = 2$ 3. $x = 7, y = -6$
5. No solutions   7. $x = -\frac{14}{3}, y = \frac{9}{4}$
9. $x = \frac{26}{11}, y = -\frac{28}{11}$ 11. $x = \frac{5}{3}, y = \frac{2}{5}$
13. Length = 24 m; width = 18 m
15. 170 dimes, 80 quarters
17. Old mower = 6 hours; new mower = 4 hours
19. 2 mph   21. $x = a + b, y = a - b$

Section 8.3, page 450

1. $x = 1, y = 2, z = 1$
3. $x = 5, y = -1, z = -2$
5. $x = -4, y = 2, z = 3$

7. $\begin{bmatrix} -1 & 2 & 3 \\ 4 & 1 & 0 \\ 0 & -1 & 5 \end{bmatrix}$

9. $\begin{bmatrix} 4 & 1 & 0 \\ -(\frac{1}{2}) & 1 & \frac{3}{2} \\ 0 & -1 & 5 \end{bmatrix}$

11. $\begin{bmatrix} 4 & 1 & 0 \\ 9 & 0 & -3 \\ 0 & -1 & 5 \end{bmatrix}$

13. $\begin{bmatrix} 4 & 1 & 0 \\ -1 & 2 & 3 \\ -6 & 2 & 11 \end{bmatrix}$

15. $\begin{bmatrix} 1 & 2 & 0 & 3 \\ 0 & 1 & -(\frac{1}{3}) & 4 \\ 0 & 0 & 1 & -(\frac{33}{7}) \end{bmatrix}$

17. $\begin{bmatrix} 1 & 1 & 2 \\ 0 & 0 & 1 \end{bmatrix}$

19. $\begin{bmatrix} 1 & 3 & 2 & 1 \\ 0 & 1 & 1 & 0 \\ 0 & 0 & 1 & 1 \end{bmatrix}$

21. $\begin{bmatrix} 1 & 5 & 6 \\ 0 & 1 & 1 \end{bmatrix}$

23. $\begin{bmatrix} 1 & 2 & -3 & -15 \\ 0 & 1 & 2 & 11 \\ 0 & 0 & 1 & 5 \end{bmatrix}$

25. $\begin{bmatrix} 1 & 2 & 3 & 0 \\ 0 & 1 & 1 & 1 \\ 0 & 0 & 1 & -(\frac{5}{3}) \end{bmatrix}$

27. $x = -\frac{1}{3}, y = \frac{8}{3}, z = -\frac{5}{3}$

29. $x = 2, y = -1, z = 0$

31. $x = 2 - \dfrac{7z}{5}, y = -1 + \dfrac{3z}{5}, z = $ any real number

33. $x = -1, y = 2, z = -2$

35. $x = 1, y = -2, z = 3, w = -4$

37. $x = -\frac{1}{3}, y = \frac{8}{3}, z = -\frac{5}{3}$

39. $x = 2y + 4, y = $ any number

41. $x = \frac{2}{11}z + 1, y = \frac{7}{11}z + 2, z = $ any number

43. $x = 1, y = 0, z = -1, w = 2$

45. $x = 1, y = 2$

47. $12,000 at 8%, $15,000 at 6.9%, $23,000 at 12%

49. 80 Sonys, 5 Sanyos

Section 8.4, page 458

1.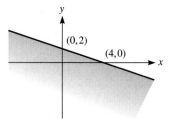

3.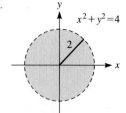

5.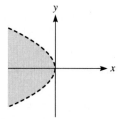

7.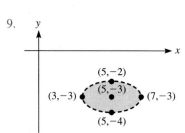

9.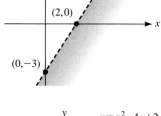

11.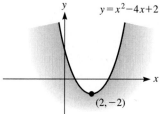

13.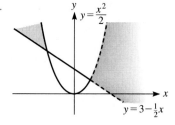

15.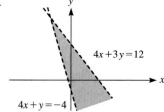

17.

19.

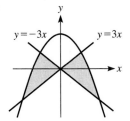

21.

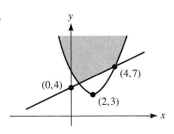

23.

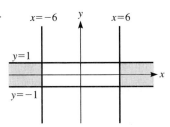

25.

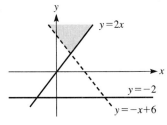

27.

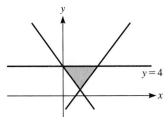

29.

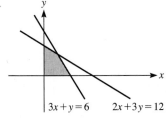

31.

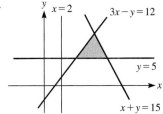

33. *a.* x = recreational skis, y = racing skis
$x \leq 60$, $y \leq 45$, $3x + 4y \leq 240$, $x \geq 0$, $y \geq 0$

b.

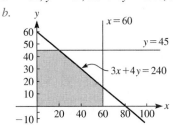

35. *a.* $x \geq 0$, $y \geq 0$, $x + y \leq 16$, $3x + 6y \leq 60$

b.

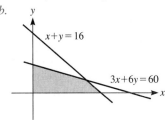

37. *a.* $2w + t \geq 60$, $t \geq 0$, $w \geq 0$

b.

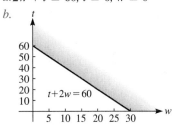

39. *a.* $0 \leq L \leq 74$, $0 \leq w \leq 50$

b.

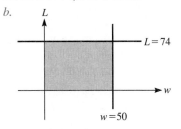

Section 8.5, page 461

1. $(x, y) = (4, 8)$ maximum value $= 52$
   $(x, y) = (1, 1)$ minimum value $= 8$

3. $(x, y) = (0, 10)$ minimum value $= 30$

5. $(x, y) = \left(\frac{5}{2}, 6\right)$ maximum value $= 97$

$(x, y) = \left(\frac{37}{7}, \frac{16}{7}\right)$ minimum value $= \frac{562}{7}$

7. $(x, y) = (7, 3)$ maximum value $= 5$
$(x, y) = (4, 6)$ minimum value $= -10$

9. $(x, y) = (5, 5)$ maximum value $= 30$
$(x, y) = (1, 2)$ minimum value $= 7$

11. $(x, y) = (8, 0)$ maximum value $= 64$
$(x, y) = (0, 0)$ minimum value $= 0$

13. Five 19-inch and four 13-inch televisions give a profit of $460.

15. 46¢ per bag with .6 kg bran and .4 kg rice

17. Deluxe $= 300$; top $= 200$;
maximum profit $= \$1300$/week

19. 134.5 acres in corn, 177.5 acres in beans, maximum profit $= \$50,248.50$

21. 818 ft^2

23. 100 units of lumber, 300 units of plywood, maximum profit $= \$11,000$

Section 8.6, Cumulative Review Exercises, page 464

1. $x = 4, y = -1$

3. $(x, y) = (4, -3), (-3, 4)$

5. $(x, y, z) = (1, 2, 3); \left(\dfrac{1 + \sqrt{5}}{2}, 1 + \sqrt{5}, \dfrac{9 - 3\sqrt{5}}{2}\right)$

$\left(\dfrac{1 - \sqrt{5}}{2}, 1 - \sqrt{5}, \dfrac{9 + 3\sqrt{5}}{2}\right)$

7. 20 dimes and 10 quarters

9. 3 and 11

11. $x = 7, y = 3, z = -1$

13. $x = 1, y = 1$

15. $x = -6 + \dfrac{3y}{2}, y = $ any real number

17. $14,000 at 12.5% and $46,000 at 14%

19.

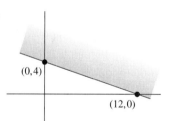

21.

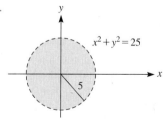

23.

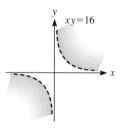

25.

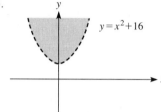

27.

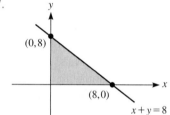

29.

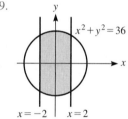

31.

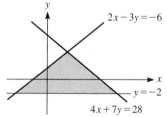

33.

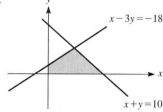

35. $(3, 1)$, minimum $= 9$; $(8, 1)$, maximum $= 19$
37. Minimum value $\frac{9}{8}$ at $x = 0$, $y = \frac{3}{8}$, maximum value $107/9$ at $x = \frac{25}{9}$, $y = \frac{19}{9}$.
39. $(2, 6)$, maximum $= 26$; $(3, 1)$, minimum $= 7$
41. 17 plain drawers, 13 fancy drawers

CHAPTER 9

Section 9.1, page 471

1. $\dfrac{2}{x - 1} + \dfrac{4}{x + 1}$

3. $\dfrac{1}{x} + \dfrac{2}{x + 1} - \dfrac{3}{x + 2}$

5. $\dfrac{\frac{4}{3}}{x - 2} + \dfrac{\frac{13}{3}}{2x - 1}$

7. $-\dfrac{1}{x - 4} + \dfrac{2x - 9}{x^2 + 1}$

9. $\dfrac{4}{x - 3} + \dfrac{1}{3x + 2}$

11. $1 + 3x + \dfrac{2}{x - 1} + \dfrac{2}{2 + x}$

13. $3x + 1 + \dfrac{4}{x - 2} + \dfrac{2}{x - 1}$

15. $\dfrac{4}{x - 3} + \dfrac{1}{x + 2}$

17. $\dfrac{2}{x} - \dfrac{4}{x - 1} + \dfrac{7}{(x - 1)^2}$

19. $\dfrac{3}{2x - 1} - \dfrac{2}{x + 2} + \dfrac{10}{(x + 2)^2}$

21. $1 + 3x$

23. $\dfrac{x + 1}{x^2 + 2} - \dfrac{2x}{(x^2 + 2)^2}$

25. $2x + 3 + \dfrac{5}{3x + 1} + \dfrac{1}{x - 1}$

27. $\dfrac{3}{x + 2} - \dfrac{2}{(x + 2)^2} + \dfrac{1}{(x + 2)^3}$

29. $\dfrac{5}{x + 1} - \dfrac{5}{x + 2} - \dfrac{5}{(x + 2)^2}$

31. $\dfrac{1}{4a^2(x + a)} + \dfrac{1}{4a^2(x - a)} - \dfrac{x}{2a^2(a^2 + x^2)}$

33. $\dfrac{6}{x + 1} - \dfrac{5}{x - 1}$

35. $2 - \dfrac{1}{x} + \dfrac{3}{x + 1}$

37. $x = -\frac{1}{9}$, $y = \frac{1}{9}$, $z = \frac{14}{9}$

Section 9.2, page 478

1. $\begin{bmatrix} 1 & 3 \\ 1 & 1 \end{bmatrix}$

3. $\begin{bmatrix} 2 & 6 \\ 4 & 12 \end{bmatrix}$

5. $\begin{bmatrix} -5 & 1 \\ 1 & -2 \end{bmatrix}$

7. $\begin{bmatrix} 7 & 0 \\ 2 & 9 \end{bmatrix}$

9. $\begin{bmatrix} 0 & 3 \\ 0 & -1 \end{bmatrix}$

11. $\begin{bmatrix} -3 & 3 \\ 0 & 0 \end{bmatrix}$

13. $\begin{bmatrix} 5 & -15 \\ 0 & 10 \end{bmatrix}$

15. $\begin{bmatrix} -2 & 6 \\ -1 & -3 \end{bmatrix}$

17. $\begin{bmatrix} 4 & 2 \\ -1 & 0 \end{bmatrix}$

19. $\begin{bmatrix} -14 & -4 \\ -1 & -2 \end{bmatrix}$

21. $\begin{bmatrix} 13 \\ 25 \end{bmatrix}$

23. $\begin{bmatrix} -10 & -7 & -9 \\ -2 & -19 & -14 \end{bmatrix}$

25. Undefined

27. $\begin{bmatrix} -1 & -3 & 14 \\ 6 & 9 & 16 \end{bmatrix}$

29. Undefined

31. $\begin{bmatrix} -3 & 2 & 18 \end{bmatrix}$

33. Undefined

35. 3×5

37. Undefined

39. 1×3

41. $x = 3$, $y = 1$, $z = 4$

43. $x = 18$, $y = 3$, $z = 3$, $w = 3$

Section 9.3, page 488

1. -7 3. -5 5. -14 7. -13

9. -5 11. -14 13. 8 15. -6

17. 0 19. 720 21. -410 23. 9072

25. -153 27. $x = 3$, $y = 2$

29. $x = -6$, $y = 1$ 31. $x = 1$, $y = 2$

33. Zero determinant; Cramer's rule not applicable

35. $x = -\frac{1}{3}$, $y = \frac{8}{3}$, $z = -\frac{5}{3}$

37. $x = 2$, $y = -1$, $z = 0$

39. Nonsquare matrix; Cramer's rule not applicable

41. $x = 0$, $y = 2$, $z = -2$, $w = 1$

43. $\begin{vmatrix} x & y & 1 \\ 3 & 4 & 1 \\ 4 & 8 & 1 \end{vmatrix} = 0$

45. $\begin{vmatrix} x & y & 1 \\ 0 & 0 & 1 \\ 5 & 2 & 1 \end{vmatrix} = 0$

47. No   49. No   51. 1 53. $\frac{37}{2}$

55. 12 57. $a = \pm 4, -1$ 59. $-10x - 2y - 4z$

Section 9.4, page 495

1. $\begin{bmatrix} \frac{7}{16} & \frac{5}{32} \\ \frac{1}{16} & \frac{3}{32} \end{bmatrix}$

3. $\begin{bmatrix} -1 & -2 \\ -(\frac{1}{2}) & -(\frac{3}{2}) \end{bmatrix}$

5. Inverse doesn't exist.

7. $\begin{bmatrix} \frac{7}{15} & -(\frac{1}{5}) & \frac{1}{15} \\ -(\frac{2}{3}) & 0 & \frac{1}{3} \\ -(\frac{3}{5}) & \frac{2}{5} & \frac{1}{5} \end{bmatrix}$

9. $\begin{bmatrix} 15 & 4 & -5 \\ -12 & -3 & 4 \\ -4 & -1 & 1 \end{bmatrix}$

11. $\begin{bmatrix} \frac{3}{8} & -(\frac{1}{4}) & \frac{1}{8} \\ -(\frac{1}{8}) & \frac{3}{4} & -(\frac{3}{8}) \\ -(\frac{1}{4}) & \frac{1}{2} & \frac{1}{4} \end{bmatrix}$

13. $\begin{bmatrix} \frac{1}{2} & \frac{1}{2} & -(\frac{1}{4}) & \frac{1}{2} \\ -1 & 4 & -(\frac{1}{2}) & -2 \\ -(\frac{1}{2}) & \frac{5}{2} & -(\frac{1}{4}) & -(\frac{3}{2}) \\ \frac{1}{2} & -(\frac{1}{2}) & \frac{1}{4} & \frac{1}{2} \end{bmatrix}$

15. Yes   17. Yes   19. Yes

21. $\begin{bmatrix} x \\ y \end{bmatrix} = \begin{bmatrix} 4 \\ 1 \end{bmatrix}$

23. $\begin{bmatrix} x \\ y \end{bmatrix} = \begin{bmatrix} \frac{61}{19} \\ \frac{6}{19} \end{bmatrix}$

25. $\begin{bmatrix} x \\ y \end{bmatrix} = \begin{bmatrix} \frac{1}{2} \\ -1 \end{bmatrix}$

27. $\begin{bmatrix} x \\ y \end{bmatrix} = \begin{bmatrix} -2 \\ 5 \end{bmatrix}$

29. $\begin{bmatrix} x \\ y \\ z \end{bmatrix} = \begin{bmatrix} -(\frac{9}{4}) \\ \frac{15}{4} \\ 10 \end{bmatrix}$

31. $\begin{bmatrix} x \\ y \\ z \end{bmatrix} = \begin{bmatrix} 0 \\ 1 \\ -1 \end{bmatrix}$

33. $A^{-1} = \begin{bmatrix} a^{-1} & 0 \\ 0 & b^{-1} \end{bmatrix}$ provided $ab \neq 0$

Section 9.5, Cumulative Review Exercises, page 498

1. $-\dfrac{4}{a} + \dfrac{1}{a+1} - \dfrac{3}{(a+1)^2}$

3. $-\dfrac{4}{(3x-1)} + \dfrac{5}{2x-1}$

5. $\begin{bmatrix} 23 & 13 & 26 \\ 4 & 15 & -1 \\ -5 & -5 & -6 \end{bmatrix}$

7. $\begin{bmatrix} 25 & 20 \\ 1 & 3 \end{bmatrix}$

9. $\begin{bmatrix} 18 & 13 & 19 \\ 1 & -7 & 4 \end{bmatrix}$

11. $\begin{bmatrix} 6 \\ -9 \end{bmatrix}$

13. $\begin{bmatrix} 0 & 0 & 0 \\ 12 & 3 & 0 \end{bmatrix}$

15. Not defined

17. 1×1

19. $x = 3, y = 3, w = 7, z = 2$

21. 5 23. -13 25. -5 27. -85

29. -33 31. 157

33. 9

35. $\begin{bmatrix} 2 & -5 \\ -1 & 3 \end{bmatrix}$

37. No inverse exists.

39. $\begin{bmatrix} 1 & -2 & 3 & 8 \\ 0 & 1 & -3 & 1 \\ 0 & 0 & 1 & -2 \\ 0 & 0 & 0 & -1 \end{bmatrix}$

41. Not inverses

43. $x = 5, y = -3$

45. $x = -1, y = 5, z = -3$

CHAPTER 10

Section 10.1, page 509

1. 1 3. $40{,}320$ 5. 15 7. 21
9. 6 11. 126 13. $35x^4 y^3$
15. $3^{10} \cdot 3003 t^5$ 17. $5376 x^3 t^9$
19. $504 t^{5/2}, -504\sqrt{2} t^2$ 21. b^{200}
23. $p^5 - 5p^4 q + 10 p^3 q^2 - 10 p^2 q^3 + 5 pq^4 - q^5$
25. $a^{10} + 15 a^8 b + 90 a^6 b^2 + 270 a^4 b^3 + 405 a^2 b^4 + 243 b^5$

27. $\dfrac{t^8}{256}$

29. $256 + 512t + 448t^2 + 224t^3 + 70t^4 + 14t^5$
$+ \frac{7}{4}t^6 + \frac{1}{8}t^7 + \frac{1}{256}t^8$

31. $\dfrac{81 a^8}{16 b^4} - \dfrac{9 a^5}{b} + 6 a^2 b^2 - \dfrac{16 b^5}{9a} + \dfrac{16 b^8}{81 a^4}$

33. $a^{15} - 5a^9 + 10a^3 - 10a^{-3} + 5a^{-9} - a^{-15}$

35. $\dbinom{n}{0} + \dbinom{n}{1} + \dbinom{n}{2} + \cdots + \dbinom{n}{n}$

37. $56 - 32\sqrt{3}$

39. $a^4 - 4a^3\sqrt{3} + 18a^2 - 12\sqrt{3}a + 9$

41. $1024 a^{10} - 5120 a^9 b + 11{,}520 a^8 b^2 - 15{,}360 a^7 b^3$

43. $x^{-4} - 24 x^{-3/2} + 252 x$

45. $\dfrac{-5940 x^3}{y^9} + \dfrac{594 x^2}{y^{10}} - \dfrac{36x}{y^{11}} + \dfrac{1}{y^{12}}$

47. $\dbinom{12}{8}\left(\dfrac{3a^{-2}}{2}\right)^4 \cdot \left(\dfrac{a}{3}\right)^8 = \dfrac{55}{144}$

49. 1.18428 by binomial theorem, 1.18430 by calculator

51. $\dbinom{50}{10}(.02)^{10}(.98)^{40}$

53. a. $a^7 + 7a^6 b + 21 a^5 b^2 + 35 a^4 b^3 + 35 a^3 b^4 + 21 a^2 b^5$
$+ 7ab^6 + b^7$

b. $a^8 + 8a^7 b + 28 a^6 b^2 + 56 a^5 b^3 + 70 a^4 b^4 + 56 a^3 b^5$
$+ 28 a^2 b^6 + 8ab^7 + b^8$

c. $a^9 + 9a^8 b + 36 a^7 b^2 + 84 a^6 b^3 + 126 a^5 b^4$
$+ 126 a^4 b^5 + 84 a^3 b^6 + 36 a^2 b^7 + 9ab^8 + b^9$

d. $a^{10} + 10 a^9 b + 45 a^8 b^2 + 120 a^7 b^3 + 210 a^6 b^4$
$+ 252 a^5 b^5 + 210 a^4 b^6 + 120 a^3 b^7 + 45 a^2 b^8 + 10 ab^9$
$+ b^{10}$

55. $5x^4 + 10 x^3 h + 10 x^2 h^2 + 5xh^3 + h^4$

57. $[a + (b+c)]^n = a^n + na^{n-1}(b+c)$
$+ \dbinom{n}{2} a^{n-2}(b+c)^2 + \cdots + (b+c)^n$

59. $1 - x + x^2 + \cdots - x^7$

61. $1 + \dfrac{1}{2}x - \dfrac{1}{8}x^2 + \dfrac{1}{16}x^3 - \dfrac{5}{128}x^4 + \dfrac{7}{256}x^5 - \dfrac{21}{1024}x^6$
$+ \dfrac{33}{2048}x^7$

63. $\dfrac{2(n+2)^2}{n^2(n+1)}$

Section 10.3, page 520

1. $3, 4, 5, 10$ 3. -2, undefined, $4, \frac{3}{2}$
5. $\frac{1}{4}, \frac{1}{2}, 1, 32$ 7. $1, \frac{1}{4}, \frac{1}{9}, \frac{1}{64}$
9. $-2, -2\sqrt{2}, -2\sqrt[3]{3}, -2\sqrt[8]{8}$
11. $-2, -6, -18, -4374$
13. $0, \frac{3}{2}, \frac{8}{7}, \frac{63}{62}$ 15. $1, -2, 3, -8$
17. $1, 0, -1, 0$ 19. $2, 7, 12, 17, 22$
21. $9, 3, 1, \frac{1}{3}, \frac{1}{9}$ 23. $2, \frac{5}{2}, \frac{10}{3}, \frac{17}{4}, \frac{26}{5}$
25. $-\frac{1}{2}, \frac{1}{8}, -\frac{1}{24}, \frac{1}{64}, -\frac{1}{160}$ 27. $\frac{3}{2}, \frac{3}{2}, \frac{3}{2}, \frac{3}{2}, \frac{3}{2}$
29. $3, \frac{1}{3}, 3, \frac{1}{3}, 3$ 31. $-2, -4, -8, -16, -32$
33. 2 35. 17 37. $\frac{13}{12}$ 39. 56
41. 0 43. 150 45. 10
47. $a_n = 2n - 1$ 49. $a_n = (-1)^{1+n} 2^n$
51. $a_n = 2^{-(2n-1)}$ 53. $\displaystyle\sum_{n=1}^{5} \dfrac{1}{n}$ 55. $\displaystyle\sum_{n=1}^{10} n$
57. $\displaystyle\sum_{n=1}^{6} (-3)^n$ 59. $x = 2$ 61. $x = 3$ 63. $x = 2$

Section 10.4, page 525

1. Difference is 2.   3. Difference is 3.
5. Difference is 1.   7. Difference is -2.
9. -25 11. 261 13. -19 15. -165
17. -16.5 19. $a_1 = -\frac{275}{4}, d = \frac{19}{4}$
21. $a_1 = 114, d = -7$ 23. $a_1 = \frac{3}{5}, d = \frac{3}{5}$
25. $a_1 = 8, d = -3$ 27. $a_1 = -10, d = 4$
29. 5 31. -3 33. 21 35. -11

37. 30   39. 440   41. −532   43. 561
45. 35   47. −1005   49. 4185   51. 136
53. $54.85   55. $n = 12, d = 37$

57. $a_{81} = 45$ 59. $\sum_{d=0}^{3}(4 + 3d)$ 61. $\sum_{d=0}^{6}(-9 + 6d)$

Section 10.5, page 532

1. $\dfrac{a_{n+1}}{a_n} = -\dfrac{1}{3}$ 3. $\dfrac{a_{n+1}}{a_n} = 4$ 5. $\dfrac{a_{n+1}}{a_n} = \dfrac{1}{3}$

7. $\dfrac{1}{1024}$ 9. 8   11. $\dfrac{256}{81}$

13. 6144   15. 6144   17. $-\dfrac{3}{10}$ 19. $\dfrac{2}{81}$

21. $28,809.07   23. $11,865.23   25. 1024

27. 322,960   29. 5   31. 4

33. $\dfrac{5}{3}, 15, -45, 135$ or $-\dfrac{5}{3}, -15, -45, -135$

35. $n = 4, a_n = -40$ 37. $n = 5, S_n = 484$
39. $a_n = 324, a_1 = 4$ 41. $a_1 = 6, S_n = 762$
43. $a_n = -\dfrac{2}{27}, S_n = \dfrac{3280}{27}$ 45. $a_1 = 3, n = 3$

47. $a_n = (12) \cdot \left(-\dfrac{1}{3}\right)^{n-1}$

49. $r = -\sqrt[3]{131}, a_n = -\dfrac{50(\sqrt[3]{-131})^n}{131}$

51. $101,578.77   53. $581,826.07   55. $\dfrac{2}{3}$

57. −5   59. 1   61. Sum does not exist.

63. $-\dfrac{1}{2}$ 65. 30   67. 8

69. $93\frac{1}{3}$ ft   71. $0.95   73. $28,146.97

Section 10.6, page 540

1. 156   3. 20,000   5. 17,576,000
7. 720   9. 32   11. 24
13. 30   15. 3
17. $n(n - 1)(n - 2)(n - 3)(n - 4)$
19. 210   21. 3024   23. 15
25. 311,875,200   27. 12   29. 56
31. 5   33. 45   35. n 37. 1

39. $\dfrac{12!}{5!3!2!2!} = 166,320$ 41. 56

43. 24   45. 696,729,600   47. 16
49. *a.* 36; *b.* 100; *c.* 24   51. 4800
53. 30,030   55. *a.* 21; *b.* 11; *c.* 6
59. 8   61. No solutions

Section 10.7, page 545

1. {(H, 1), (H, 2), (H, 3), (H, 4), (H, 5), (H, 6),
   (T, 1), (T, 2), (T, 3), (T, 4), (T, 5), (T, 6)}
3. {MM, MF, FF, FM}
5. {2, 3, 4, 5, 6, 7, 8, 9, 10, 11, 12}
7. $_{52}C_5 = 2,598,960$ hands   9. 729
11. Ordered pairs of the numbers:
   {653,654,623, 624,153,154,123,124}; 64 elements in the
   sample space.

13. {WI, WL, WA, WM, IW, IL, II, IA, IM, LW, LI, LL, LA,
    LM, AW, AI, AL, AM, MW, MI, ML, MA}
15. {GGG, BGG, BBG, BBB}
17. {PN, PD, PQ, NP, ND, NQ, DN, DP, DQ, QP,
    QN, QD}
19. {3, 4, ..., 18}
21. $\dfrac{3}{5}$ 23. $\dfrac{1}{8}$ 25. $\dfrac{33}{16,660}$ 27. $\dfrac{1}{9}$

29. $\dfrac{4}{465}$ 31. $\dfrac{1}{15}$ 33. $\dfrac{7}{30}$ 35. $\dfrac{1}{200}$

37. $\dfrac{1}{312}$ 39. $\dfrac{1}{5}$ 41. $\dfrac{9}{10}$

Section 10.8, Review Exercises, page 548

1. $a^4 + 4a^3t + 6a^2t^2 + 4at^3 + t^4$
3. $2187a^7b^{14} - 20,412a^8b^{13} + 81,648a^9b^{12}$
   $- 181,440a^{10}b^{11} + 241,920a^{11}b^{10} - 193,536a^{12}b^9$
   $+ 86,016a^{13}b^8 - 16,384a^{14}b^7$
5. −252

13. *a.*

n	1	2	3	4	5	6
$f(n)$	2	7	15	26	40	57

   b. 126; *c.* 126;
   d. $f(n) = 2n + \frac{3}{2}n(n - 1)$

17. −1, 0, 1, 8   19. $2, \frac{4}{3}, 1, \frac{4}{11}$
21. $a_n = n(n + 1), a_1 = 2, a_2 = 6, a_3 = 12, a_4 = 20,$
    $a_5 = 30$
23. $a_1 = 2, a_2 = \frac{3}{2}, a_3 = 2, a_4 = \frac{3}{2}, a_5 = 2$
25. 14   27. $\dfrac{5269}{3600}$
29. Common difference = 1
31. Common difference = 3
33. 17   35. −14
37. $d = 5, a_1 = -34$ 39. $d = -3, a_1 = -5$
41. 16   43. 22   45. 150
47. −1120   49. 1   51. 43,500
53. Common ratio = $\frac{1}{4}$.   55. Common ratio = 2.
57. 2^{19} 59. 243   61. 2^{n+1}
63. The fourth term   65. $2960.49
67. $a_1 = 4, s_5 = 484$ 69. $a_n = (-2) \cdot (-5)^{n-1}$
71. $\dfrac{20}{3}$ 73. 72   75. $46\frac{2}{3}$ ft   77. 120
79. 6,760,000   81. 6
83. $9! = 362,880$ 85. 120
87. 900 (0 cannot be first digit)
89. 210   91. 84   93. 330   95. 27,405
97. {HHHHH, HHHHT, HHHTH, ..., TTTTT},
    32 sample points
99. {BBB, BBR, BRB, RBB, BRR, RBR, RRB, RRR}

101. $\dfrac{4}{9}$ 103. $\dfrac{3}{5}$

CHAPTER 11
Section 11.1, page 556

1. vertex: (0, 0)
   focus: $(0, \frac{1}{36})$
   directrix: $y = -(\frac{1}{36})$

3. vertex: $(0, 0)$
   focus: $(-\frac{1}{2}, 0)$
   directrix: $x = \frac{1}{2}$

5. vertex: $(1, -1)$
   focus: $(1, -\frac{7}{8})$
   directrix: $y = -\frac{9}{8}$

7. vertex: $(0, -\frac{1}{2})$
   focus: $(0, -1)$
   directrix: $y = 0$

9. vertex: $(-\frac{1}{3}, -3)$
   focus: $(-(\frac{1}{3}), -(\frac{26}{9}))$
   directrix: $y = -(\frac{28}{9})$

11. vertex: $(-\frac{41}{24}, -\frac{3}{2})$
    focus: $(-(\frac{5}{24}), -\frac{3}{2})$
    directrix: $x = -\frac{77}{24}$

13. vertex: $(-(\frac{1}{5}), 4)$
    focus: $(-(\frac{1}{5}), (\frac{401}{100}))$
    directrix: $y = (\frac{399}{100})$

15. $y^2 = -8x$

17. $(x - 2)^2 = 8(y - 1)$

19. $y^2 = 4x$

21. $(x - 3)^2 = 24(y + 2)$

23. $(x - 1)^2 = 8y$

25. $(y + 4)^2 = -8(x - 1)$

27. $(x - \frac{1}{4})^2 = \frac{5}{4}(y + \frac{1}{20})$

29. $(y - \frac{71}{6})^2 = -\frac{140}{3}(x - \frac{5041}{1680})$

31. $(x - 2000)^2 = 8000y$ 33. 225 inches

35. $y^2 = 8x$ 37. $y - 8 = -8(x - 1)$

39. $y = 6x$ 41. (b) 43. (c) 45. (d)

Section 11.2, page 564

1. $\dfrac{x^2}{5} + \dfrac{y^2}{9} = 1$

3. $\dfrac{(x + 1)^2}{5} + \dfrac{(y - 7)^2}{9} = 1$

5. $\dfrac{x^2}{9} + \dfrac{y^2}{49} = 1$

7. $\dfrac{(x - 1)^2}{4} + (y - 2)^2 = 1$

9. $\dfrac{(x - 5)^2}{21} + \dfrac{(y - 2)^2}{25} = 1$

11. $\dfrac{9x^2}{8} + y^2 = 1$

13. $\dfrac{(x - 10)^2}{25} + \dfrac{(y - 1)^2}{16} = 1$

15. $\dfrac{x^2}{4} + \dfrac{y^2}{16} = 1$

17. $\dfrac{x^2}{9} + \dfrac{5y^2}{9} = 1$

19. $a = 3, b = 2, c = \sqrt{5}, e = \dfrac{\sqrt{5}}{3}$; foci: $(\pm \sqrt{5}, 0)$

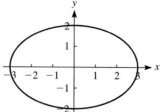

21. $a = 2\sqrt{2}, b = 2, c = 2, e = \dfrac{\sqrt{2}}{2}$; foci: $(\pm 2, 0)$

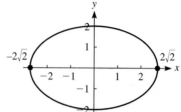

23. $a = 4, b = 2, c = 2\sqrt{3}, e = \dfrac{\sqrt{3}}{2}$; foci: $(-2, 2 \pm 2\sqrt{3})$

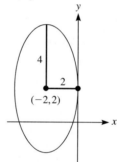

25. $a = 6, b = 3\sqrt{2}, c = 3\sqrt{2}, e = \dfrac{\sqrt{2}}{2}$; foci: $(1 \pm 3\sqrt{2}, 2)$

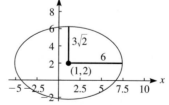

27. $a = 5, b = \sqrt{5}, c = 2\sqrt{5}, e = \dfrac{2\sqrt{5}}{5}$; foci: $(2 \pm 2\sqrt{5}, -3)$

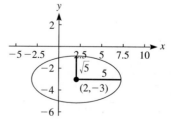

29. $a = 6$, $b = 2$, $c = 4\sqrt{2}$, $e = \dfrac{2\sqrt{2}}{3}$; foci: $(0, \pm 4\sqrt{2})$

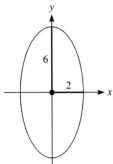

31. $a = 6$, $b = 2$, $c = 4\sqrt{2}$, $e = \dfrac{2\sqrt{2}}{3}$; foci: $(3, 2\pm 4\sqrt{2})$

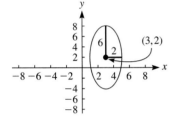

33. $a = 5$, $b = 4$, $c = 3$, $e = \frac{3}{5}$; foci: $(-1, -1 \pm 3)$

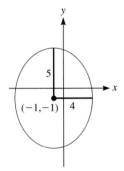

35. $a = b = 3$, $c = 0$, $e = 0$; foci coincide: $(0, -3)$

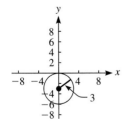

37. Endpoints of latus recta are $(\frac{4}{3}, \sqrt{5})$, $(\frac{4}{3}, -\sqrt{5})$, $(\frac{-4}{3}, \sqrt{5})$, and $(\frac{-4}{3}, -\sqrt{5})$.

43. (a)　　　　45. (f)　　　　47. (d)

Section 11.3, page 571

1. center: $(0,0)$, $e = \sqrt{2}$; vertices: $(\pm 1, 0)$; foci: $(\pm \sqrt{2}, 0)$; asymptotes: $y = \pm x$

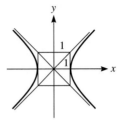

3. center: $(0,0)$, $e = \dfrac{\sqrt{5}}{2}$; vertices: $(\pm 2, 0)$; foci: $(\pm \sqrt{5}, 0)$; asymptotes: $y = \pm\left(\dfrac{x}{2}\right)$

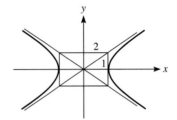

5. center: $(0,0)$, $e = \dfrac{\sqrt{34}}{5}$; vertices: $(\pm 10, 0)$; foci: $(\pm 2\sqrt{34}, 0)$; asymptotes: $y = \pm\dfrac{3x}{5}$

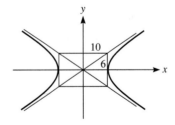

7. center: $(1,2)$, $e = \sqrt{2}$; vertices: $(1 \pm 1, 2)$; foci: $(1 \pm \sqrt{2}, 2)$; asymptotes: $y - 2 = \pm(x - 1)$

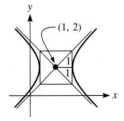

9. center: (4,1), $e = \frac{\sqrt{34}}{3}$; vertices: (1,1),(7,1); foci: (4 $\pm$ $\sqrt{34}$, 1); asymptotes: $y - 1 = \pm[\frac{5}{3}(x - 4)]$

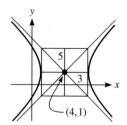

11. center: (0,0), $e = \frac{\sqrt{13}}{3}$; vertices: $(\frac{1}{2}, 0),(-\frac{1}{2}, 0)$; foci: $(\pm \frac{\sqrt{13}}{6}, 0)$; asymptotes: $y = \pm\frac{2x}{3}$

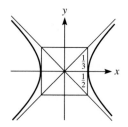

13. center: (−6, 4), $e = \frac{\sqrt{24}}{3}$; vertices: $(-6 \pm \sqrt{3}, 4)$; foci: $(-6 \pm \sqrt{8}, 4)$; asymptotes: $y - 4 = \pm\sqrt{\frac{5}{3}}(x + 6)$

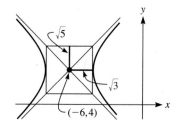

15. center: $(3, \frac{3}{2})$, $e = \frac{\sqrt{5}}{2}$; vertices: $(3 \pm \sqrt{30}, \frac{3}{2})$; foci: $(3 \pm 5\frac{\sqrt{6}}{2}, \frac{3}{2})$; asymptotes: $y - \frac{3}{2} = \pm(\frac{1}{2})(x - 3)$

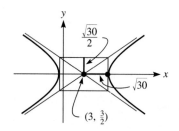

17. center: (−2, 4), vertices: (−2 $\pm$ 1, 4); foci: (−2 $\pm$ $2\frac{\sqrt{3}}{3}$, 4); asymptotes: $y - 4 = \pm\frac{1}{\sqrt{3}}(x + 2)$; $e = \frac{2\sqrt{3}}{3}$

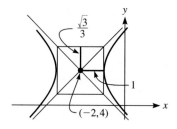

19. center: (3, −1), vertices: $(3, -1 \pm \frac{1}{3})$; foci: $(3, -1 \pm \frac{\sqrt{13}}{6})$; $e = \frac{\sqrt{13}}{2}$

asymptotes: $y + 1 = \pm(\frac{2}{3})(x - 3)$

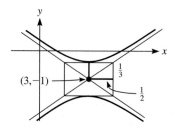

21. $\dfrac{x^2}{4^2} - \dfrac{y^2}{1^2} = 1$

23. $\dfrac{(y - 1)^2}{1^2} - \dfrac{(x - 1)^2}{(\frac{3}{2})^2} = 1$

25. $\dfrac{x^2}{9} - \dfrac{y^2}{16} = 1$

27. $\dfrac{(x + 1)^2}{16} - \dfrac{(y - 1)^2}{20} = 1$

29. $\dfrac{x^2}{4} - \dfrac{y^2}{16} = 1$

31. $\dfrac{y^2}{4} - \dfrac{x^2}{20} = 1$

33. Ellipse   35. Hyperbola   37. Parabola

39. $\dfrac{256(x - \frac{5}{2})^2}{25} - \dfrac{256y^2}{1575} = 1$

47. (d)   49. (e)   51. (f)

Section 11.4, page 577

1. (3, −5)   3. (−4, −7)   5. (−6, 4)   7. (2, 3)
9. $(-\frac{3}{2}, 2)$ 11. $X^2 + 2XY + 6X + Y = -4$
13. $X^2 + 4XY - 16X - 6Y = -20$
15. $X^2 + 2XY + Y^2 + 4X + Y = -3$
17. $2X^2 - 3Y^2 - 20X - 12Y = -29$
19. $2X + 2Y = 11$ 21. $X^2 + Y^2 = 9$
23. $X^2 - 2\sqrt{3}XY - Y^2 = 2$ 25. $X^2 - 6XY + 9Y^2 = 8$
27. $X^2 + 2\sqrt{3}XY + 3Y^2 + (4 + 2\sqrt{3})X + (4\sqrt{3} - 2)Y = 12$
29. $9X^2 - 82XY + 9Y^2 = -800$
31. $\theta = 45°$ 33. $\theta = 67.5°$ 35. $\theta \approx 34.1°$
37. $\theta = 45°$ 39. $\theta \approx 61.845°$
41. $16x^2 - 24xy + 100 = 60x = 8y - 9y^2$
43. $32x^2 + 53y^2 - 72xy - 80 = 0$

Section 11.5, page 582

1. $r = 1/(1 + \cos\theta)$ 3. $r = 2/(1 - \cos\theta)$
5. $r = 10/(1 + \cos\theta)$ 7. $r = 8/(1 + \sin\theta)$
9. $r = 5/(3 + \cos\theta)$ 11. $r = 8/(3 - \cos\theta)$
13. $r = 6/(1 + 2\cos\theta)$ 15. $r = 1/(1 + 3\cos\theta)$
17. parabola, directrix $x = 1$ 19. hyperbola, directrix $y = 1$
21. hyperbola, directrix $x = \frac{2}{5}$
23. parabola, directrix $y = 4$ 25. ellipse, directrix $x = \frac{5}{3}$
27. hyperbola, directrix $x = -1$
29. $r = 1/(1 + \sqrt{2}\cos\theta)$
31. $r = 8/(1 + \sqrt{5}\cos\theta)$
33. $r = 12{,}600/(1 + .05\cos\theta)$

Section 11.6, page 585

1.
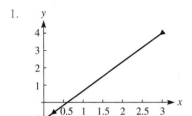

3.
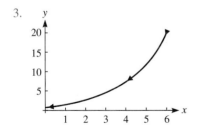

5.   7.
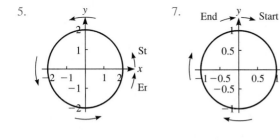

9.   11.
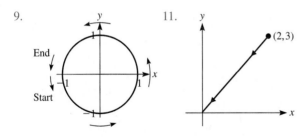

13.
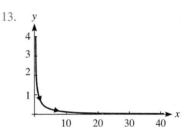

15.
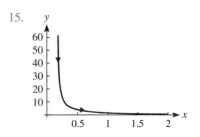

17. $y = \frac{5}{3}x - 1,\ 0 \le x \le 3$

19. $x = e^{x/2},\ x \ge 0$

21. $x^2 + y^2 = 4$, Domain: $-2 \le x \le 2$

23. $x^2 + y^2 = 1$, Domain: $-1 \le x \le 1$

25. $x^2 + y^2 = 1$, Domain: $-1 \le x \le 1$

27. $y = \dfrac{3x}{2},\ 0 \le x \le 2$

29. $y = \dfrac{1}{x},\ x > 0$ 31. $y = \dfrac{1}{x^2},\ x > 0$

37. $x = \cos t,\ y = \sin t,\ 0 \le t \le 2\pi$

39. $x = 2\cos t,\ y = \sin t,\ 0 \le t \le 2\pi$

41. $x = 1 + 10\cos t,\ y = 1 + \sqrt{84}\sin t,\ 0 \le t \le 2\pi$

43. $x = 1 + \sqrt{3}\sec t,\ y = 1 + \sqrt{6}\tan t,\ \dfrac{\pi}{2} \le t \le \dfrac{3\pi}{2}$

45. $x = 2 - 3t,\ y = 3 + t,\ 0 \le t \le 1$

47. Line segment from $(7, 1)$ to $(3, -1)$.

49. Same graph traced in the opposite direction.

51. Graph is translated horizontally 1 unit to the right and vertically 2 units upward.

Section 11.7, Cumulative Review Exercises, page 588

1.
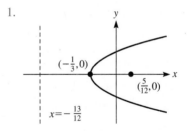

3.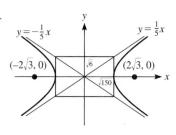
$y = -\sqrt{2}x$ $y = \sqrt{2}x$
$(-\sqrt{6}, 0)$ $(\sqrt{6}, 0)$

5.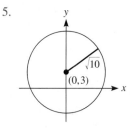
$\sqrt{10}$
$(0, 3)$

7.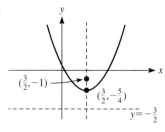
$y = -\frac{1}{5}x$ $y = \frac{1}{5}x$
$(-2\sqrt{3}, 0)$ $\sqrt{6}$ $(2\sqrt{3}, 0)$
$\sqrt{150}$

9.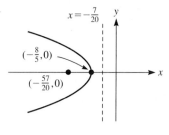
$(\frac{3}{2}, -1)$
$(\frac{3}{2}, -\frac{5}{4})$
$y = -\frac{3}{2}$

11.
$x = -\frac{7}{20}$
$(-\frac{8}{5}, 0)$
$(-\frac{57}{20}, 0)$

13.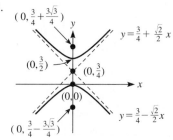
$(0, \frac{3}{4} + \frac{3\sqrt{3}}{4})$
$y = \frac{3}{4} + \frac{\sqrt{2}}{2}x$
$(0, \frac{3}{2})$ $(0, \frac{3}{4})$
$(0, 0)$
$y = \frac{3}{4} - \frac{\sqrt{2}}{2}x$
$(0, \frac{3}{4} - \frac{3\sqrt{3}}{4})$

15. $(-1, 7)$ 17. $(8, 3)$ 19. $(-1, -6)$

21. $X^2 + 3XY + 4X - 5Y = 11$

23. $X^2 + 4XY - Y^2 + 2X + 14Y = 5$

25. $2X = 9$

27. $3X^2 - 10\sqrt{3}XY + 13Y^2 = 72$

29. $2XY = 49$

31. $(\sqrt{3} - 1)X - (\sqrt{3} + 1)Y + 4 = 0$

33. $45°$ 35. $30°$

37. $r = 2/(1 - \cos\theta)$ 39. $r = 12/(5 + 7\cos\theta)$

41. hyperbola, $e = 2$, focus at origin, directrix $x = \frac{1}{2}$

43. hyperbola, $e = 2$, focus at origin, directrix $y = -1$

45.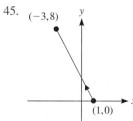
$(-3, 8)$
$(1, 0)$

47.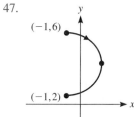
$(-1, 6)$
$(-1, 2)$

49. $y = (\frac{2}{3})x + \frac{11}{3}, \; -1 \leq x \leq 14$

51. $y = \sqrt{x - 1}, \; x \geq 1$ 53. $\dfrac{x^2}{16} + \dfrac{y^2}{9} = 1$

INDEX

ANSWERS TO EVEN-NUMBERED EXERCISES

CHAPTER 1
Section 1.1, page 7

2. $0, 24, \frac{2}{3}, -1.8, 93\%, \sqrt[5]{32}, \sqrt{16}, \sqrt{\frac{1}{4}}, -234, -\frac{16}{7},$
   0.0035

4. $0, 24, \frac{2}{3}, -1.8, 2\pi, 93\%, \sqrt[5]{32}, \sqrt{5}, \sqrt{16}, \sqrt{\frac{1}{4}}, -234,$
   $-\frac{16}{7}, 0.0035$

6. $4.\overline{09}$ 8. $3.\overline{7}$ 10. 2 12. 5

14. 6 16. $\frac{2}{3}$ 18. $\sqrt{10} - \pi$ 20. $\sqrt{10} - \pi$

22. $<$ 24. $=$ 26. 19 28. 89

30. $\pi - \sqrt{2}$ 32. $\{x : x \neq 2\}$ 34. $\{x : x > 0\}$

Section 1.2, page 22

2. 1 4. m^{12} 6. $48/y^{14}$ 8. $60p^2 t^4$

10. $-\dfrac{x^6}{125y^{12}}$ 12. $72a^5 b^5$ 14. a^2 16. $-\dfrac{5x^5}{y^7}$

18. $-\dfrac{p^9}{125q^{12}r^3}$ 20. $\dfrac{625q^{24}y^8}{256p^{24}}$ 22. $\dfrac{4a^{17}c^8}{625}$

24. $7x\sqrt{x}$ 26. $4y$ 28. $4z^2$

30. $-\dfrac{x^2}{y^3}$ 32. $2b\sqrt[4]{90a^2}$ 34. 4

36. $x\sqrt[3]{3y}$ 38. $x\sqrt[3]{20xy}$ 40. $2xy^2 \sqrt[5]{x^3}$

42. 1 44. $-5^5 x^{5a} y^{5b}$ 46. $m^3 + n^3$

48. $6p^3 q^3 - 10p^2 q^3 + 30p^3 q^2 - 50p^2 q^2$

50. $a^2 + 14at + 49t^2$ 52. $a^4 - b^4$

54. $81x^4 - t^4$ 56. $2a(4a - 1 + 6b)$

58. $(x - 6)(x - 5)$ 60. $(\sqrt{5} + \sqrt{7})(x - y)$

62. $3b(7b + 6x)$ 64. $(b + c)(a - m)$

66. $(x + 2y)(x^2 - 2xy + 4y^2)(1 - 2x)$

68. $(7 - 3y)(7 + 3y)$ 70. $x(3 - 10x)$

72. $2(z - 2)(z^2 + 2z + 4)$ 74. $2x(x^2 + 3)^2$

76. $(5x^y + 3)(2x^y - 1)$

78. $(a^{2n} - b^n)(a^{4n} + a^{2n}b^n + b^{2n})$

80. $(a^x + 5 - 6y^x)(a^x + 5 + 6y^x)$

82. $\dfrac{(x + 3)(x + 2)}{4(x - 2)(x - 3)}$ 84. $\dfrac{(a^3 + c^3)(a + c)}{2a^2(a + c)^2}$

86.
$$\dfrac{(x + y + 1)(15x^3 + 11x^2 y + 2xy^4 + 3xy^2 + 8x + 2y^5 + 4y)}{(7x^2 + 4)(x^2 + 3xy + 2y^4)}$$

88. $\dfrac{3a - 4}{2a + 7}$ 90. $\dfrac{9a - 3b - 4}{a - b}$

92. $\dfrac{-x^2 + 12x + 8}{x^2 - 3x - 8}$ 94. $\dfrac{-a^2 + a - 9}{a(a + 1)^2}$

96. $\dfrac{2(a + y)}{a(x + y)}$ 98. $\dfrac{x + y}{y + 1}$

100. $\dfrac{-2x - h}{x^2(x + h)^2}$ 102. $\dfrac{a - 1}{a + 2}$ 104. $\dfrac{\sqrt[3]{36}}{6}$

106. $-\frac{6}{5}(\sqrt{3} + \sqrt{8})$ 108. $\dfrac{3(\sqrt{x} - \sqrt{h})}{x - h}$

110. $\dfrac{2x + 5\sqrt{3x} + 9}{x - 3}$ 112. $\dfrac{4x - 9y}{x(2\sqrt{x} - 3\sqrt{y})}$

114. $\dfrac{4x - 27}{2x - 5\sqrt{3x} + 9}$ 116. 6.67×10^{-11}

118. 5.878×10^{12} 120. 1.89×10^{-13}

122. \$5,800,000,000

124. 0.000001 is already in decimal notation

126. 70,200,000,000,000,000

128. 1.5821708×10^{-5} yr   130. 2.19×10^9

132. $\approx 72 \text{ m}^2$ 134. $\approx 11{,}174.29$ m/sec

136. $\approx 3.8605 \text{ ft}^2$ 138. ≈ 346.34613 m/sec

Section 1.3, page 31

2. 0

4. No real solution

6. No real solution

8. $-5 \pm 2\sqrt{5}$ 10. $\dfrac{3 \pm \sqrt{5}}{2}$

12. $-1, -7$ 14. $2, -\frac{1}{2}$

16. No real solution

18. $\dfrac{5 \pm \sqrt{137}}{4}$ 20. $0, -\frac{6}{13}$ 22. $-\frac{1}{5}, \frac{1}{2}$

24. No real solution

26. $\dfrac{-35 \pm \sqrt{2105}}{20}$

28. $\dfrac{-2\sqrt{5} \pm \sqrt{20 + 5\sqrt{10}}}{5}$

30. $\dfrac{-13 \pm \sqrt{221}}{2}$ 32. $\frac{3}{4}$

34. $\dfrac{23 \pm \sqrt{521}}{2}$ 36. $\dfrac{30 \pm 15\sqrt{2}}{2}$ 38. $0.5, -0.1$

40. No real solution

42. 144, 1

44. No real solution

46. 4 ($y = -1$ rejected)

48. 117   50. $-\frac{7}{5}, 4, \pm\sqrt{5}$ 52. $\sqrt[3]{\frac{1}{4}}$

54. $\pm 2, \frac{4}{3}$ 56. 1   58. 6   60. $\frac{1}{2}$

62. $h \approx 7.4$m, $r \approx 3.7$m   64. 8 and 15 ft

66. a. 13 sec   68. 130
    b. $\frac{1}{2}, \frac{11}{2}$ sec

70. a. 808.8 m   72. 5%   74. 8%
    b. 100°C
    c. 98.9867°C
    d. 97.976°C

76. a. $h = \dfrac{\sqrt{3}a}{2}$ 78. a. 20
                                            b. 12
    b. $A = \dfrac{ha}{2}$ c. No
    c. $P = 3a$
    d. 30 cm

Section 1.4, page 34

2. (5,7]

4. $[-5, -3]$

6. $[5, 7)$

8. $(-\infty, 3)$

10. $\{x : -3 < x < -1\}$ 12. $\{x : x \le 5\}$
14. $\{x : x \ge 5\}$ 16. $t < 5; (-\infty, 5)$
18. $t > 3; (3, \infty)$ 20. $x < 2; (-\infty, 2)$
22. $x \le \frac{5}{3}; (-\infty, \frac{5}{3}]$ 24. $x \le \frac{3}{2}; (-\infty, \frac{3}{2}]$
26. $x > -\frac{5}{11}; (-\frac{5}{11}, \infty)$ 28. $3 > t \ge 2; [2, 3)$
30. $-2 \le x \le 5; [-2, 5]$ 32. $\frac{3}{2} \le y < \frac{11}{2}; [\frac{3}{2}, \frac{11}{2})$
34. $x \ge -\frac{3}{8}; [-\frac{3}{8}, \infty)$ 36. $0 \le t < \frac{32}{3}; [0, \frac{32}{3})$
38. $x \ge \frac{15}{4}; [\frac{15}{4}, \infty)$ 40. $x > \frac{4}{5}; (\frac{4}{5}, \infty)$
42. $(-4, 1)$ 44. $[-4, 5]$ 46. $\left(-\infty, -\frac{1}{\sqrt{3}}\right) \cup \left(\frac{1}{\sqrt{3}}, \infty\right)$
48. $(-\infty, -1) \cup (-1, \infty)$ 50. $(-\infty, -2] \cup [4, 5]$

52. $(-\infty, 0) \cup (0, 3) \cup (4, \infty)$ 54. $[-2, \frac{1}{2}] \cup [2, \infty)$
56. $(-5, 2)$ 58. $(-\frac{5}{3}, \frac{3}{2}]$ 60. $(-7, 2) \cup (3, \infty)$
62. $\varnothing$
64. $x > 23$ or $x < -23; (-\infty, -23) \cup (23, \infty)$
66. $-14 < x < 14; (-14, 14)$
68. $x \le -16$ or $x \ge -2; (-\infty, -16] \cup [-2, \infty)$
70. $12 \ge x \ge -9; [-9, 12]$

72. $-\dfrac{110}{7} < x < 14; \left(-\dfrac{110}{7}, 14\right)$

74. $[-3, -2] \cup [6, 7]$ 76. $\varnothing$
78. $[-\frac{1}{2}, \frac{7}{2}]$ 80. $[-4, 4]$ 82. $(-\infty, -\frac{1}{2})$
84. $(-\infty, -1) \cup (1, 4) \cup (6, \infty)$ 86. $[-2, 2]$
88. $p \le -12$ or $p \ge 12$ 90. $(-1, 1)$ 92. $(1, \infty)$
94. Not true; try $a = -3, b = -2$

96. $t \ge \frac{140}{13}$ hr

98. $t \ge 150$ innings

100.
    a. $\left(2 - \dfrac{\sqrt{3}}{2}, 2 + \dfrac{\sqrt{3}}{2}\right)$

    b. $\left[0, 2 - \dfrac{\sqrt{11}}{2}\right) \cup \left(2 + \dfrac{\sqrt{11}}{2}, \infty\right)$

102. $n \ge 20$ 104. $96.6 \le T \le 100.6$

Section 1.5, page 49

2.

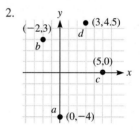

4. a. y-axis; b. quadrant II; c. x-axis; d. quadrant I
6. $\sqrt{74}$ 8. $6\sqrt{2}$ 10. 4
12. $\sqrt{x^2 + 4x + 13}$ 14. $2\sqrt{3 + \sqrt{5}}$
16. $2|b|$ 18. 3   20. $|b|$ 22. $(\frac{5}{2}, \frac{15}{2})$
24. $(0, 0)$ 26. $(-3, 6)$ 28. $\left(\dfrac{x - 2}{2}, -\dfrac{3}{2}\right)$
30. $\left(\dfrac{\sqrt{3}}{2}, \dfrac{\sqrt{5} - 2}{2}\right)$ 32. $(a, a - b)$
34. No such point exists.   36. $(0, \frac{29}{5})$
38. $(6, 12)$ 40. 36 units^2 42. $(-\frac{61}{6}, 0)$
44. $d(A, B) = \sqrt{106}$
    $d(A, C) = \sqrt{269}$ No, not collinear
    $d(B, C) = \sqrt{41}$

46. $d(A, B) = \sqrt{29}$

   $d(A, C) = \sqrt{145}$ $(\sqrt{29})^2 + (\sqrt{116})^2 = (\sqrt{145})^2$

   $d(B, C) = \sqrt{116}$ Yes, right triangle
48. $(x - h)^2 + (y - k)^2 = 25$

Section 1.6, page 59

2.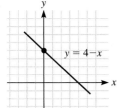

4.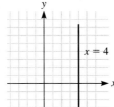

6.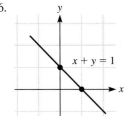

8.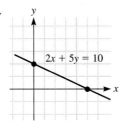

10.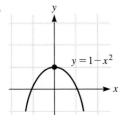

12.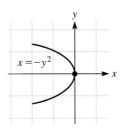

14.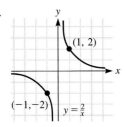

16.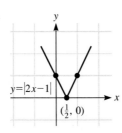

18.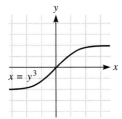

20.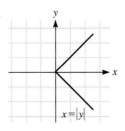

22.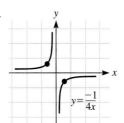

24.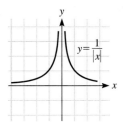

26.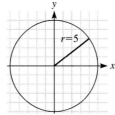

28.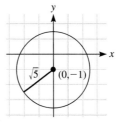

30.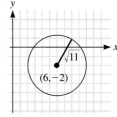

32.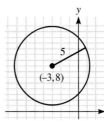

34.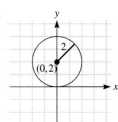

36. $(x - 3)^2 + (y + 4)^2 = 3$ 38. $x^2 + (y - a)^2 = 0.01$
40. $(x - 9)^2 + (y - 10)^2 = 100$
42. $\left(x + \frac{2}{3}\right)^2 + \left(y + \frac{2}{3}\right)^2 = \frac{4}{9}$
44. $\left(x + \frac{1}{2}\right)^2 + (y + 6)^2 = \frac{29}{4}$
46. $(x + 2)^2 + (y + 3)^2 = 121$
48. Center $= (-1.75, 2.1)$; radius $= \sqrt{17.4725} \approx 4.18$
50. Center $= (-\frac{2}{3}, \frac{3}{4})$; radius $= \dfrac{\sqrt{109}}{12}$
52. Yes   54. No   56. No   58. No
60. Equation of a line: $-20x + 2y - 31 = 0$

62. *a.* 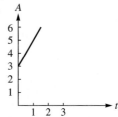   *b.* The area covered at time 0

Section 1.7, page 69

2. $\frac{2}{3}$ 4. $\frac{1}{4}$ 6. $m = -2$ 8. 7

10. *a.* Increases by 680
    b. Decreases by 132.6

12. $-\frac{1}{2}$

14. -3

16. Slope undefined

18. Slope undefined

20. $y = -\frac{1}{2}x$ 22. $y = -3$

24. $y = -4x$

26. $y + 6 = -\frac{2}{3}(x + 2)$ 28. $y = -6$

30. Slope $= -\frac{3}{2}$; y-intercept: $(0, 3)$

32. Slope $= \frac{10}{3}$; y-intercept: $\left(0, -\frac{1}{2}\right)$

34. Slope undefined. No y-intercept

36. Neither   38. Neither

40. Perpendicular   42. Parallel

44. $y = \frac{1}{3}x + 6$ 46. $x = 5$ 48. $y = 5$

50. $y = 1 - 2x$ 52. $y = 2$

54. $y = -\frac{9}{2}x - \frac{17}{2}$

56. Parallel: $y = 9x - 56$. Perpendicular: $y = -\frac{1}{9}x - \frac{4}{3}$

58. *a.* x-intercept: 4. y-intercept: -2
    b. x-intercept: 3. y-intercept: -4

60. $y + 2 = -\frac{2}{3}(x + 2)$

62. $\frac{1}{2}$

66. *a.* $N = .9Y + 2.2(1983 \leftrightarrow Y = 0)$
    b. 13.9 million

68. *a.* $P - 1.29 = -0.004(N - 60)$
    b. \$1.13. The predicted price is \$0.06 less than the listed price, \$1.19.

70. The slope represents the value the machine loses per year.

72. $S = 24,000 + 0.38x$, $S = $ salary, $x = $ sales

74. *a.* Rate of growth of cost per mile of driving a car with respect to time
    b. \$0.56 (increase)

76. *a.* Rate of change of Fahrenheit temperature with respect to Celsius temperature
    b. Increase by $\frac{9}{5}°$

Section 1.8, Cumulative Review Exercises, page 74

2. $-\dfrac{32b^7}{a^2}$ 4. $\frac{2}{5}xy$ 6. $\dfrac{9x^{18}z^{22}}{16y^8}$

8. $z^{4/3}$ 10. $(a^2 + b^2)^{3/2}$ 12. $\dfrac{b^3}{a^2}$

14. $\dfrac{1}{ab(a + b)}$ 16. $\dfrac{1}{a^2b^2 + ab + 1}$

18. $\dfrac{3a}{4}$ 20. $(x + y)^6 \sqrt{x + y}$

22. $16x^4y^6 \sqrt{\dfrac{\sqrt[5]{5^4 x^4 y^3}}{y}}$ 24. $\dfrac{1}{(x - 1)^{3n/2}}$

26. Cannot factor

28. $(3a + 8)(2a + 1)$

30. $(6x^a + 1)(x^a - 1)$

32. $7y(x - 2)(x + 2)$

34. $4(a^2 + 3b)(a^4 - 3a^2b + 9b^2)$

36. $(a - \frac{1}{2})(a^2 + \frac{1}{2}a + \frac{1}{4})$

38. $(x - 5y)^2$

40. $(a^{x+2} - b^{3x+5})(a^{x+2} + b^{3x+5})$

42. $\dfrac{1 - 2\sqrt{a} + a}{1 - a}$ 44. $\dfrac{\sqrt[3]{450}}{5}$

46. $\dfrac{1}{\sqrt{3 + h} + \sqrt{3}}$

48. Undefined (0 in denominator)

50. $\sqrt{1 - a^2} \big/ [1 + -a^2]$

52. $-\dfrac{-13\sqrt{7} - 8\sqrt{42} - 6\sqrt{3} + 22\sqrt{2}}{215}$

54. 8   56. $\frac{85}{7}$ 58. $-\frac{33}{7}$ 60. $-\frac{3}{4}, 1$

62. No real solutions

64. $-\frac{5}{2} \pm \frac{\sqrt{53}}{2}$ 66. $\dfrac{-\sqrt{3} \pm \sqrt{11}}{2}$

68. 62

70. No real solutions

72. 10   74. -5 76. $\frac{1}{3}, -\frac{1}{2}$ 78. 64

80. 64, 4   82. $x > 18; (18, \infty)$

84. 6, -1 86. $(\frac{1}{2}, -\frac{11}{2})$

88. $y = \frac{3}{5}x - \frac{29}{5}$

90. $y = \frac{1}{2}x + 4$; slope $= \frac{1}{2}$; $b = 4$

92. Neither   94. Neither

96. $\sqrt{47}$ 98.

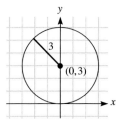

100. $(x - 2)^2 + (y + 1)^2 = 12$
     Center: $(2, -1)$;
     $r = \sqrt{12} = 2\sqrt{3}$

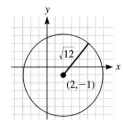

102. $x^2 + y^2 = 3$
Center: $(0, 0)$;
$r = \sqrt{3}$

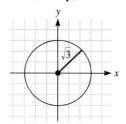

104. $3x + 4y = -23$ 106. $y - 3 = 3(x + 2)$
108. 3×10^{-5} 110. $24{,}000{,}000{,}000{,}000$
112. 1.118 cm   114. $s \approx 16$ cm

116. diameter $= \sqrt{\dfrac{150}{\pi}}$ inches, height $= 2\sqrt{\dfrac{150}{\pi}}$ inches

118. 200 km/hr   120. $\frac{5000}{17}$ ft^3 122. 9%
124. *a.* 5.5125 m
     b. 0.2485 m

c. $\dfrac{T^2}{-4\pi^2}(9.8) = L$

126. 2.374% 128. $-38.2° < F < 674.6°$
130. \$65,000 at 6%
     \$55,000 at 9%
132. *a.* 10 sec
     b. The ball will never reach a height of 1216 ft.

CHAPTER 2
Section 2.1, page 88
2. -13 4. $4(a + h) - 5$ 6. 1
8. $48t^2 - 4t - 1$ 10. 11 12. $|5 - (a + h)^2|$
14. $-\dfrac{5}{4}$ 16. $a^2 + a - 1$ 18. -28
20. $2a^3 + 6a^2h + 6ah^2 + 2h^3 - 3a^2 - 6ah - 3h^2$
22. 0 24. $1 + \dfrac{3}{t} + \dfrac{3}{t^2} + \dfrac{1}{t^3}$ 26. 0
28. $4t^2 + 8th + 4h^2 + 4t + 4h + 1$ 30. 12
32. 12 34. D: all real numbers; R: all real numbers
36. D: all real numbers; R: $\{-4\}$
38. D: all real numbers, $x \neq -\frac{7}{3}$;
    R: all real numbers, $y \neq 0$
40. D: $\left[-\frac{1}{4}, \infty\right)$; R: $[0, \infty)$
42. D: all real numbers; R: all real numbers
44. D: all real numbers; R: $[0, \infty)$
46. D: $[-\sqrt{3}, \sqrt{3}]$; R: $[0, \sqrt{3}]$
48. D: $\left[-\frac{3}{2}, \infty\right)$; R: $[0, \infty)$
50. $5, 10$ 52. $3a^2 + 3ah + h^2, 6a^2 + 2h^2$
54. $\dfrac{-2a - h}{a^2(a + h)^2}, -\dfrac{4a}{(a + h)^2(a - h)^2}$ 56. $-1, -2$
58. $5a^4 + 10a^3h + 10a^2h^2 + 5ah^3 + h^4$,
    $10a^4 + 20a^2h^2 + 2h^4$

60. D: $[0, \infty)$ 62. D: $[0, 4) \cup (4, \infty)$
64. D: all real numbers
66. D: all real numbers, $x \neq \frac{1}{2}, 1$
68. D: $[0, \infty)$ 70. D: all real numbers, $x \neq -\frac{9}{5}$
72. D: all real numbers, $x \neq 1$
74. $V(x) = x(20 - 2x)^2$
76. $R(x) = x^2 - 10x + 1200$ 78. $A(d) = \dfrac{\pi d^2}{4}$
80. $A(x) = \dfrac{x^2}{4\pi} + \left(7 - \dfrac{x}{4}\right)^2$
82. $t(w) = \dfrac{2080}{500 + w} + \dfrac{1440}{500 - w}$

Section 2.2, page 102
2. No   4. No
6. $f(-2) = 2$,
   $f(0) = 4$,
   $f(7) = 0$,
   $f(5) = -2$
8. $f(0) = -1$ $f(3) = 2$ $f(5) = 2$
10. D: $[-2, \infty]$ 12. R: $[-3, 2]$, D: $[-2, 5]$
14. D: $[-3, 3]$; R: $[-11, 1]$ 16. D: $[-2, 2]$; R: $[-3, 1]$

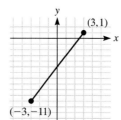
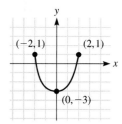

18. D: all real numbers;   20. D: $[-3, 3]$; R: $[-4, 5]$
    R: all real numbers

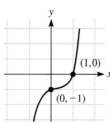
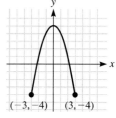

22. D: $(-\infty, -3] \cup [3, \infty)$;   24. D: $[-1, 3]$; R: $[0, 4]$
    R: $[0, \infty)$

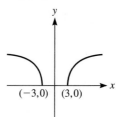
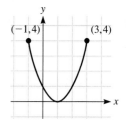

26.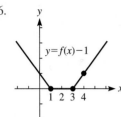

28.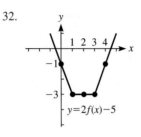

30.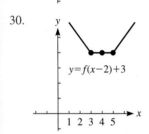

32.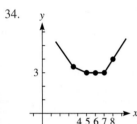

34.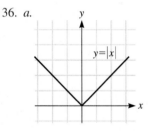

36. *a.* 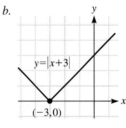 *b.*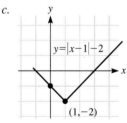

c. 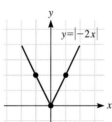 *d.*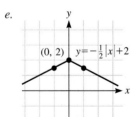

e. 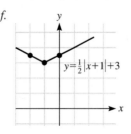 *f.*

38. Neither
40. Symmetrical with respect to y-axis
42. Symmetrical with respect to y-axis

44. Symmetrical with respect to y-axis
46. No   48. No   50. Yes   52. No
54. Odd   56. Even   58. Odd   60. Neither
62. Even   64. Even

66. $f(x) = \begin{cases} -1 & x < 0 \\ 0 & x = 0 \\ 1 & x > 0 \end{cases}$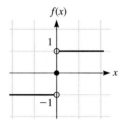

68. $f(x) = \begin{cases} |x| & x < 2 \\ -x^2 & 2 \le x < 3 \\ 2x - 1 & x \ge 3 \end{cases}$

70. $f(x) = \begin{cases} \dfrac{x^2 - 9}{x + 3} & x \ne -3 \\ 6 & x = -3 \end{cases}$

72. $f(x) = \begin{cases} -3 & x \notin I \\ 3 & x \in I \end{cases}$ 74.

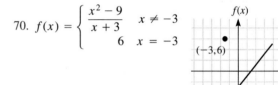

76. 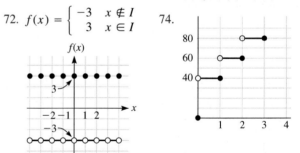   78.

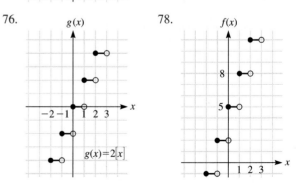

80. $-ax - by = c$ if symmetrical with respect to the origin:
$ax + by = c \Rightarrow c = 0$

Section 2.4, page 120

2. $f(x) = x^2 - 4$ D_f = all real numbers
     $g(x) = 2x + 3$ D_g = all real numbers
     $(f + g)(x) = x^2 + 2x - 1$ D_{f+g} = all real numbers
     $(f - g)(x) = x^2 - 2x - 7$ D_{f-g} = all real numbers
     $(fg)(x) = (x^2 - 4)(2x + 3)$ D_{fg} = all real numbers
     $(ff)(x) = (x^2 - 4)^2$ D_{ff} = all real numbers
     $\left(\dfrac{f}{g}\right)(x) = \dfrac{x^2 - 4}{2x + 3}$ $D_{f/g}$ = all real numbers,
                                    $x \neq -\frac{3}{2}$
     $\left(\dfrac{g}{f}\right)(x) = \dfrac{2x + 3}{x^2 - 4}$ $D_{g/f}$ = all real numbers,
                                    $x \neq \pm 2$
     $(f \circ g)(x) = (2x + 3)^2 - 4$
              $= 4x^2 + 12x + 5$ $D_{f \circ g}$ = all real numbers
     $(g \circ f)(x) = 2(x^2 - 4) + 3$
              $= 2x^2 - 5$ $D_{g \circ f}$ = all real numbers

4. $f(x) = \sqrt{x}$ $D_f = [0, \infty)$
     $g(x) = x^2$ D_g = all real numbers
     $(f + g)(x) = \sqrt{x} + x^2$ $D_{f+g} = [0, \infty)$
     $(f - g)(x) = \sqrt{x} - x^2$ $D_{f-g} = [0, \infty)$
     $(fg)(x) = x^{5/2}$ $D_{fg} = [0, \infty)$
     $(ff)(x) = \sqrt{x} \cdot \sqrt{x} = x$ $D_{ff} = [0, \infty)$
     $\left(\dfrac{f}{g}\right)(x) = \dfrac{\sqrt{x}}{x^2} = \dfrac{1}{x^{3/2}}$ $D_{f/g} = (0, \infty)$
     $\left(\dfrac{g}{f}\right)(x) = \dfrac{x^2}{\sqrt{x}} = x^{3/2}$ $D_{g/f} = (0, \infty)$
     $(f \circ g)(x) = \sqrt{x^2} = |x|$ $D_{f \circ g}$ = all real numbers
     $(g \circ f)(x) = (\sqrt{x})^2 = x$ $D_{g \circ f} = [0, \infty)$

6. 1   8. $x^2 + 2x + 2$ 10. 0

12. Not defined   14. $\dfrac{x^2 - 1}{2x + 3}$ 16. $x^4 - 2x^2$

18. $x^6 - 3x^4 + 2x^2$ 20. $2x^2 + 1$
22. $(f \circ g)(x) = x, (g \circ f)(x) = x$
24. $(f \circ g)(x) = x, (g \circ f)(x) = x$
26. $(f \circ g)(x) = x, (g \circ f)(x) = x$
28. $(f \circ g)(x) = |x|, (g \circ f)(x) = x$
30. $(f \circ g)(x) = \dfrac{13x^2 + 4x + 41}{-5x^2 - 44x + 9}$
     $(g \circ f)(x) = \dfrac{-3x^2 + 7}{7x^2 - 1}$
32. $(f \circ g)(x) = -4, (g \circ f)(x) = -11$
34. $f(x) = \sqrt[3]{x};\ \ g(x) = x^2 + 1$
36. $f(x) = \dfrac{1}{\sqrt{x}};\ \ g(x) = 7x + 2$
38. $f(x) = |x|;\ \ g(x) = 4x^2 - 3$
40. $f(x) = x^4;\ \ g(x) = \sqrt{x} + 3$
42. $f(x) = \sqrt{2 + x};\ \ g(x) = \sqrt{2 + x}$
44. $f(x) = x^{2/3};\ \ g(x) = x - 7$

46.
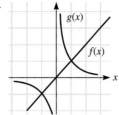
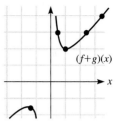

48.
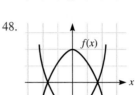
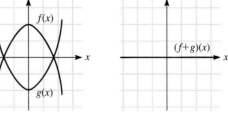

50.
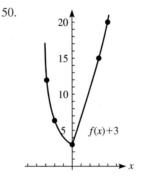

52.
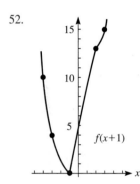

54.
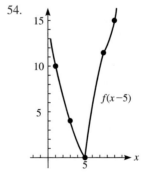

56.
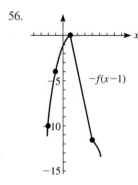

58.
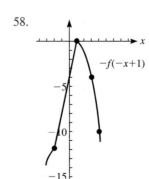

60.
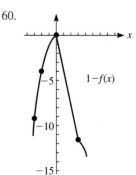

Section 2.5, page 131

2. Yes

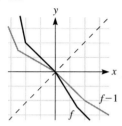

4. Yes

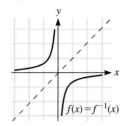

6. No

8. No

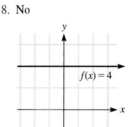

10. Yes; $f^{-1}(x) = 4x + 2$ 12. Yes; $f^{-1}(x) = x^2$, $x \geq 0$

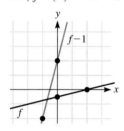

14. Yes;

$$f^{-1}(x) = \frac{1}{x}, \ x \neq 0$$

16. Yes;

$$f^{-1}(x) = -\sqrt{4 - x^2},$$
$$0 \leq x \leq 2$$

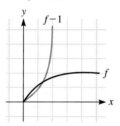

18. Yes   20. Yes   22. Yes   24. No

26. No   28. No   30. $\dfrac{x}{7} - \dfrac{2}{7}$

32. $\frac{1}{1.6}x + 4 = \frac{5}{8}x + 4$ 34. $\dfrac{-\sqrt{x}}{2}$

36. $\dfrac{x^2}{7} + \dfrac{2}{7}$, $x \geq 0$ 38. $\sqrt[3]{x + 2}$

40. $-\frac{7}{4}x$ 42. $\dfrac{2 + 3x}{8x - 7}$

44. $\sqrt{49 - x^2}$, $(0 \leq x \leq 7)$ 46. $2 - \dfrac{x^5}{3}$

48. $(f \circ f^{-1})(958.34) = 958.34$; $(f^{-1} \circ f)\left(\frac{765}{577}\right) = \frac{765}{577}$

50. $A \circ r(x) = r \circ A(x) = x$

52. n must be an odd integer.   54. $f(x) = \dfrac{1}{x}$, $g(x) = \dfrac{1}{x}$

56. Prove that if f is increasing over an interval, it has an inverse. $f(a) = f(b) \Leftrightarrow a = b$ for f, a strictly increasing function. Thus f is $1 - 1$ and has an inverse.

62. Yes; $f^{-1}(2) = \sqrt[3]{\dfrac{7.6}{3.9}} \approx 1.249$ 64. No

66. Yes

Section 2.6, page 137

2. *a.* $k = \frac{7}{4}$, $y = \frac{7}{4}x$; *b.* $y = \frac{28}{25}$

4. *a.* $k = 3.12$, $y = \dfrac{3.12}{x}$; *b.* $y = \dfrac{39}{8}$

6. *a.* $k = \frac{17}{100}$, $y = \frac{17}{100}xz$; *b.* $y = 721.14$

8. *a.* $k = 2$, $y = 2xz^2$; *b.* $y = 3.723408$

10. *a.* $k = 17.907433$, $y = \dfrac{17.907433xz}{wp}$;

b. $y = 43.418023$

12. *a.* $k = 80,000$, $y = \dfrac{80,000\sqrt{x}}{z^3w^2}$; *b.* $y = 15.625$

14. $A = \frac{1}{2}hb$; $\frac{1}{2}$ is variational constant; A varies jointly as h and b.

16. $I = Prt$; 1 is variational constant; I varies jointly as P, r and t.

18. 1.4×10^{-7} erg 20. 6.3430001 s   22. $4\sqrt{2}$ ft

24. Original force will be divided by 16

26. 681.3921 days

28. *a.* $P = 3.24$ watts; *b.* $P = .405$ watts;
c. fraction is $\frac{1}{8}$; *d.* $V = 20$ mph

30. $w = 1.14875 \times 10^{-38}$

32. $d = \dfrac{ab^3}{kc}$, d varies jointly as a and b^3 and inversely as c.

34. *a.* $x = \dfrac{yw}{kz^2}$; *b.* $w = \dfrac{kxz^2}{y}$; *c.* Variational constant for x is $1/k$. Variational constant for w is k. w varies jointly as x and z^2 and inversely as y.

Section 2.7, Cumulative Review Exercises, page 141

2. Yes   4. No   6. 36

8. $a^2 + 2ah + h^2 + 3a + 3h - 4$ 10. -1.1875

12. $\dfrac{3}{\sqrt{a + h} + \sqrt{a}}$ 14. $\dfrac{1}{a(a + h)}$

16. D: all real numbers; R: all real numbers

18. D: all real numbers, $x \neq -1$; R: all real numbers, $y \neq -2$

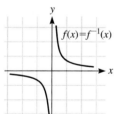

20. D: all real numbers, $x \neq 5, -6$; R: $(-\infty, 0) \cup \left[\frac{12}{121}, \infty\right)$
22. $(x + 8)(x - 3) \leq 0$; D: $[-8, 3]$
24. D: all real numbers
26. D: all real numbers, $x \neq 3$
28. D: $(1, \infty)$

30.

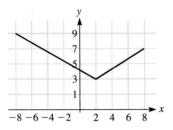

32.

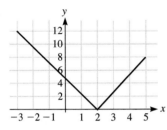

34. Symmetric with respect to origin
36. No symmetry
38. Symmetric with respect to origin
40. Neither
42. Even   44. Neither
46. Odd   48. All real numbers

50. a. all real numbers, $x \neq -\frac{3}{2}$; b. $\left(\frac{f}{g}\right)(x) = \frac{x^2 - 1}{2x + 3}$

52. a. all real numbers; b. $(f + g)(x) = x^2 + 2x + 2$
54. a. all real numbers; b. $(g \circ f)(x) = 2x^2 + 1$
56. $(f \circ g)(x) = x$; $(g \circ f)(x) = x$
58. $(f \circ g)(x) = x$; $(g \circ f)(x) = x$
60. $f(x) = x^{11}$; $g(x) = 3x^2 - 5x + 1$
62. $f(x) = x^5 + x^3 - 2x^2 - 7$; $g(x) = x^3 + 2$
64. $y = \frac{4}{3}x - \frac{11}{3}$ 66. $y = -\frac{1}{3}x + \frac{7}{3}$

68. $k = 10$; a. $y = \frac{10xw}{z^3}$; b. $y = 0.03$

70. $D_{f \circ g}$: $\varnothing$; $D_{g \circ f}$: all real numbers; $x \neq 4$
72. Yes   74. No   76. $f^{-1}(x) = \frac{1}{9}(x + 14)$
78. $f^{-1}(x) = \sqrt{16 - x^2}$, $0 \leq x \leq 4$
80. $(f \circ f^{-1})(78.8999) = 78.8999$;
    $(f^{-1} \circ f)(-2, 344, 789) = -2, 344, 789$
82. a. $S = kwh^2$ b. Choose figure with $h = 18$ cm.
84. a. For each increase in Fahrenheit temperature, Celsius temperature increases 5/9 of a degree;
    b. Celsius temperature drops 5/9 degree.

86. a. $V = \frac{k}{P}$; b. $k = 800$; $V = \frac{800}{15} = \frac{160}{3}$ ft^3

88. a. K changes by $1°$; b. $C = -273.15$; $F = -459.67$

90. $\frac{50}{b - 2} + \frac{50}{b + 2}$

92.
$$C(x) = \begin{cases} 12x & 0 \leq x \leq 50 \\ 10x & 50 < x \leq 200 \\ 9x & 200 < x \end{cases}$$

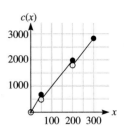

CHAPTER 3

Section 3.1, page 154

2. (f) 4. (a) 6. (e)

8. $f(x) = \frac{8}{3}x^2$ 10. $f(x) = \frac{5}{4}(x - 3)^2 + 2$

12. $f(2) = 3$; $f(5) = 7$, $f(x) = \frac{4}{9}(x - 2) + 3$

14. a. $(0, 0)$; b. Upward;   16. a. $(-1, 0)$;
    c. b. Downward;
                                      c.

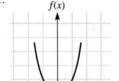

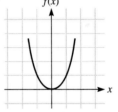   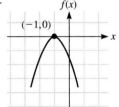

18. a. $(-2, 0)$;   20. a. $(4, -3)$; b. Upward;
    b. Downward; c.
    c.

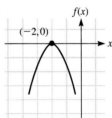

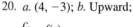

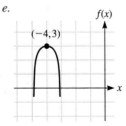

22. a. $f(x) = -(x + 4)^2 + 3$;   24. a. $f(x) = 2\left(x + \frac{1}{4}\right)^2 - \frac{9}{8}$;
    b. $(-4, 3)$; b. $\left(-\frac{1}{4}, -\frac{9}{8}\right)$;
    c. Downward; d. max: 3; c. Upward; d. min: $-\frac{9}{8}$;
    e. e.

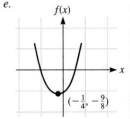

26. min: 0   28. min: 2   30. min: 0   32. max: 1
34. min: $-\frac{27}{32}$ 36. min: $4,173,333.33

38. Numbers are: $\frac{13}{2}, -\frac{13}{2}$.   40. h, $b = 72.5$ yds

42. *a.* $-3x^2 + 380x$; *b.* $\frac{190}{3}$; *c.* $\frac{36,100}{3}$ yd^2

44. Dimensions are $\frac{49'}{2} \times \frac{49'}{2}$; area is $\frac{2401}{4}$ ft^2

46. *a.* $P(x) = -x^2 + 280x - 5000$; *b.* 140 stereos;
    c. $P(140) = \$14,600$

48. *a.* $A(x) = -\frac{1}{2}\pi x^2 - 2x^2 + 28x$;
    b. $\frac{28}{\pi + 4} \approx 3.92$ ft; *c.* $\frac{392}{\pi + 4} \approx 54.89$ ft.^{2}

50. *a.* $S(t) = -16t^2 + 15.5t + 1800$; *b.* $\frac{31}{64}$
    c. $\frac{461,761}{256}$ ft ≈ 1804 ft

52. $P\left(\frac{200}{11}\right) = 1818.\overline{18}$

54. No real solution

56. Critical value: $\frac{3c}{2(1-c)}$; $f\left(\frac{3c}{2(1-c)}\right) = \frac{-9c^2}{4(1-c)} + 12$

58. $f(x) = 5(x+2)^2 + 3$

60.
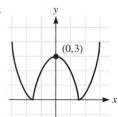

62.
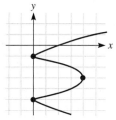

Section 3.2, page 163

2.
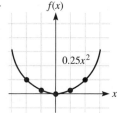

4.
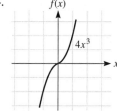

6.
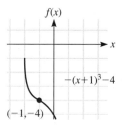

8.
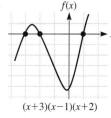

10.
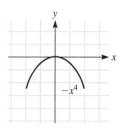

12.
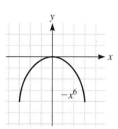

14.
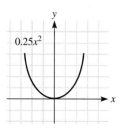

16.
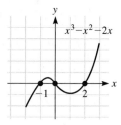

18.
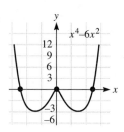

20.
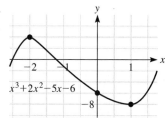

22.
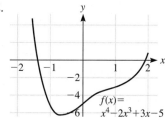

24.
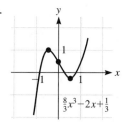

26.
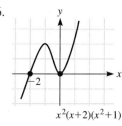

28. *a.*
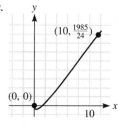

b. min $-\frac{1}{48}$ at $x = \frac{1}{2}$

30.
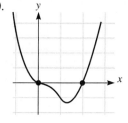

32.

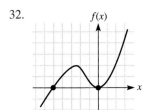

34. *a.*

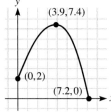

b. Maximum ≈ 7.4 at
$t \approx 3.9$
c. Estimate: $0 \le t \le 6.77$

36. (*d*)   38. (*a*)   40. (*c*)

Section 3.3, page 172

2. $2i$ 4. $3\sqrt{3}i$ 6. $-3\sqrt{2}i$ 8. $-5\sqrt{5}i$
10. $6 - 2\sqrt{21}i$ 12. $-2\sqrt{19}i + 5\sqrt{5}i$
14. $-i$ 16. $-i$ 18. 1 20. i
22. $-18 - i$ 24. $38i$ 26. i 28. 0
30. i 32. $(2\sqrt{3} + 36)i$ 34. $3 + i$
36. $\frac{1}{2}$ 38. $5 + i$ 40. 0 42. $4 - 6i$
44. $22 - 5i$ 46. $9 + 32i$ 48. 74 50. $-40i - 30$
52. $8 + 6i$ 54. $\frac{3}{5} + \frac{4}{5}i$ 56. $\frac{1}{3} + \frac{2}{3}\sqrt{2}i$
58. $-\sqrt{2}i - 1$ 60. $\frac{12}{109} - \frac{40}{109}i$ 62. $\frac{55}{338} - \frac{37}{338}i$
64. $\frac{1}{2} - \frac{1}{2}i$ 66. $\frac{-237}{425} - \frac{106}{425}i$ 68. $-\frac{164}{221} - \frac{129}{221}i$

70. $-\dfrac{1}{49} + \dfrac{4\sqrt{3}}{49}i$ 72. $\pm 3i$ 74. $\pm\sqrt{17}i$

76. $x^2 + 25$ 78. $x^2 - \dfrac{\sqrt{3}}{2}x + \dfrac{7}{16}$

80. $(x - 5i)(x + 5i)$ 82. $(x - \sqrt{5}i)(x + \sqrt{5}i)$
84. $(x + 4iy)(x - 4iy)$ 86. $x = -\frac{28}{5}, y = -\frac{11}{5}$
88. $x = 8, y = -2$ 98. $\overline{\alpha}^3 + \overline{\alpha}^2 + \overline{\alpha} + 1$
100. $5\overline{\alpha}^4 - 6\overline{\alpha}^3 + 2\overline{\alpha}^2 + 3\overline{\alpha} - 23$

Section 3.4, page 181

2. $q(x) = 2x^4 - 2x^3 - x^2 + 3x - 4; r(x) = 7$
4. $q(x) = 3x^3 - x^2 + 4x + 3; r(x) = -7$
6. $q(x) = \frac{1}{2}x^4 - \frac{5}{8}x^3 + \frac{1}{32}x^2 + \frac{59}{128}x - \frac{423}{512}; r(x) = \frac{3651}{512}$
8. $q(x) = 2x^3 - 5x + 2; r(x) = 4x + 1$
10. $q(x) = x^2 + x + 1; r(x) = 0$
12. $q(x) = 2x^2 - 3; r(x) = -x + 6$
14. $q(x) = x; r(x) = -x$
16. $q(x) = x^4 - bx^3 + b^2x^2 - b^3x + b^4; r(x) = 0$
18. $f(x) = -\frac{7}{12}(x - 5)(x + 2)$
20. $f(x) = -\frac{7}{2}[x - (2 - i)][x - (2 + i)]$ or $f(x)$
$= \frac{7}{2}(x^2 - 4x + 5)$
22. $f(x) = 12x(x + 1)(x - 2)$
24. No such polynomial
26. $f(x) = 10\left[x - \left(\dfrac{2 + \sqrt{3}}{5}\right)\right]\left[x - \left(\dfrac{2 - \sqrt{3}}{5}\right)\right]$
$\times \left[x - \left(\frac{1}{2} - \frac{3}{2}i\right)\right]\left[x - \left(\frac{1}{2} + \frac{3}{2}i\right)\right]$

28. $[x - (1 + \sqrt{5})]$ and $[x - (1 - \sqrt{5})]$ are factors.
30. *a.* $f(2) = 347$; *b.* 347; *c.* $f(-1) = 26$; *d.* 26
32. *a.* $f(-i) = -5$; *b.* -5
34. $q(x) = 2x^4 - 2x^3 - x^2 + 3x - 4; r(x) = 7$
36. $q(x) = 6x^3 + x^2 + 10x + 12; r(x) = 2$
38. $q(x) = x^2 + x + 1; r(x) = 0$
40. $q(x) = 4x^4 - 3x^3 + \frac{17}{4}x^2 - \frac{51}{16}x + \frac{89}{64}; r(x) = \frac{1013}{256}$
42. $q(x) = x^4 - bx^3 + b^2x^2 - b^3x + b^4; r(x) = 0$
44. $f(-2) = 28; f(0) = -2; f(2) = 0$
46. $f(-2) = -41; f(1) = -5; f(4) = 49$
48. $f(-3) = -470; f(20) = 2720775; f\left(-\frac{1}{2}\right) = -\frac{135}{32}$
50. -1: No; 2: Yes   52. i: No; 0: Yes
54. 3: No; $-\frac{1}{2}$: Yes   56. $\{-6, -2, 3\}$
58. $\{-1, 2, 3\}$ 60. $(-1, 2) \cup (3, \infty)$ 62. $k = 2$
64. $q(x) = x^2 - 4x + 11, r(x) = 4$
66. $q(x) = (2x^2 - 5x - 3)^2(x - 3), r(x) = 0$
68. Prove that $(x - a)$ is a factor of $x^n - a^n$ for all natural numbers n. Let $f(x) = x^n - a^n$. Then $f(a) = a^n - a^n = 0$; so, by the factor theorem, $(x - a)$ is a factor.
70. Prove that $(x + a)$ is a factor of $x^n - a^n$ for all even natural numbers n. Let $f(x) = x^n - a^n$. Then $f(-a) = (-a)^n - a^n = a^n - a^n = 0$; so, by the factor theorem, $(x + a)$ is a factor.
72. $q(x) \ne x^{n+1} + x^{n-2}a + x^{n-3}a^2 - \cdots - a^{n-1}$

Section 3.5, page 191

2. $3x^4(x + 1 - \sqrt{2})(x + 1 + \sqrt{2})$
4. $(x - 2i)(x + 2i)(x - 5)(x + 1)$
6. $(x + 2)(x - 1 - \sqrt{3}i)(x - 1 + \sqrt{3}i)$
8. $(x - 5)(x - \sqrt{7}i)(x + \sqrt{7}i)$

10. $\left(x - \dfrac{1 + \sqrt{21}}{2}\right)^2 \left(x - \dfrac{1 - \sqrt{21}}{2}\right)^2$

12. $\frac{3}{2}$, mult 1; 4, mult 2; -1, mult 2

14. $-\frac{1}{2}$, mult 3; 3, mult 3

16. 3, mult 2; -2, mult 2; -3, mult 2; 1, mult 2

18. 0, mult 2; -5, mult 1;
$\dfrac{-3 + \sqrt{11}i}{2}$, mult 2; $\dfrac{-3 - \sqrt{11}i}{2}$, mult 2

20. 0, mult 5; $\sqrt{5}i$, mult 3; $-\sqrt{5}i$, mult 3; 1, mult 1
22. 4   24. 7

26. $(t - 1)\left(t + \dfrac{1 + \sqrt{3}i}{2}\right)\left(t + \dfrac{1 - \sqrt{3}i}{2}\right)(t + 1)$
$\times \left(t - \dfrac{1 + \sqrt{3}i}{2}\right)\left(t - \dfrac{1 - \sqrt{3}i}{2}\right);$
zeros are 1, $\dfrac{-1 \pm \sqrt{3}i}{2}$, -1, $\dfrac{1 \pm \sqrt{3}i}{2}$

28. $(x + 1)\left(x - \dfrac{1 + \sqrt{3}i}{2}\right)\left(x - \dfrac{1 - \sqrt{3}i}{2}\right)$;

    zeros are $-1, \dfrac{1 \pm \sqrt{3}i}{2}$

30. $(x - 2)(x + 1 - \sqrt{3}i)(x + 1 + \sqrt{3}i)$;

    zeros are $2, -1 \pm \sqrt{3}i$

32. $(y - \sqrt{3})(y + \sqrt{3})(y - \sqrt{3}i)(y + \sqrt{3}i)$;

    zeros are $\pm\sqrt{3}, \pm\sqrt{3}i$

34. $p^2(p - i)^2(p + i)^2$; zeros are $0, \pm i$

36. $(x - 3i)(x + 3i)(x - 1 + \sqrt{3})(x - 1 - \sqrt{3})$;

    zeros are $\pm 3i, 1 \pm \sqrt{3}$

38. $(x - \sqrt{7})(x + \sqrt{7})(x - \sqrt{5}i)(x + \sqrt{5}i)$;

    zeros are $\pm\sqrt{7}, \pm\sqrt{5}i$

40. $(t - 4)(t + 2\sqrt{3})(t - 2\sqrt{3})$; zeros are $4, \pm 2\sqrt{3}$

42. Positive real zeros: at most 4; negative real zeros: none; nonreal: at least 2

44. Positive real zeros: at most 3; negative real zeros: 1; nonreal: at least 4

46. Positive real zeros: at most 4; negative real zeros: none; nonreal: at least 2

48. Positive real zeros: at most 2; negative real zeros: at most 2; nonreal: can't say

50. Positive real zeros: at most 3; negative real zeros: 1; nonreal: at least 4

52. Positive real zeros: at most 3; negative real zeros: at most 2; nonreal: can't say; 0 is a zero of mult 3

54. Remaining zeros are $-i$ and $-\frac{2}{3}$.

56. Remaining zeros are $1 - 2i, 1 \pm \sqrt{2}$.

58. Remaining zero is -1.

60. Remaining zeros are $\sqrt{2}i, \frac{2}{3}, \frac{1}{2}$.

Section 3.6, page 200

2. No rational zeros   4. 4

6. No rational zeros   8. ± 1

10. 0 is an upper bound; -3 is a lower bound.

12. 2 is an upper bound; -2 is a lower bound.

14. 2 is an upper bound; -2 is a lower bound.

16. 1 is an upper bound; -2 is a lower bound.

18. No rational zeros   20. No rational zeros

22. $\pm\frac{2}{3}$ 24. ± 1 are rational zeros

26. No rational zeros   28. $f(-2) < 0; f(-1) > 0$

30. $f(1) < 0; f(2) > 0$ 32. $f(3) < 0; f(4) > 0$

34. $f(-1) > 0; f(0) < 0; f(1) < 0; f(2) > 0$

36. -1.3247 38. 1.2134

40. $1.5174; -1.5174$ 42. $-0.7405; 3.2724$

44. $0.2626; 1.2014; 3.4050$ 46. $x \approx 1.07$ or $x = 3$

48. $x \approx 3.9236$ or $x \approx 15.8886$

50. $t \approx 0.6460; t \approx 3.9083$

52. Zeros are $-3, \dfrac{1 \pm \sqrt{11}i}{2}$ 54. Zeros are $-2, -1 \pm \sqrt{2}i$

56. Zeros are $\dfrac{-1 \pm \sqrt{7}i}{2}, \pm 1$

58. Zeros are $\pm\frac{1}{2}, \dfrac{1 \pm \sqrt{5}}{2}$ 60. Zeros are $-1, 2, -3$

62. Zeros are $\pm\sqrt{5 \pm \sqrt{2}}$

64. Solutions are $-\frac{1}{2}, \frac{3}{4}, -\frac{2}{3}$.

66. Solutions are $-\frac{3}{2}, -3 \pm 2\sqrt{2}$.

68. Solutions are $-2, 3$.

70. Consider $f(x) = x^2 - 5$. By Theorem 1, rational zeros are of the form c/d where c is a factor of 5 and d is a factor of 1. Thus possible rational zeros are $\pm 1, \pm 5$. But $\sqrt{5}$ is a zero of $f(x)$ and hence is irrational.

74. 3.1998

Section 3.7, page 212

2. $x = 0, y = 0$ 4. $x = 0, y = 0$

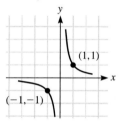

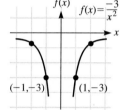

6. $x = 1, y = 0$ 8. $x = 3, y = 0$

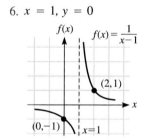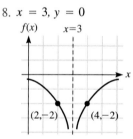

10. $x = 2, y = -1$ 12. $x = \frac{16}{5}, y = \frac{-2}{5}$

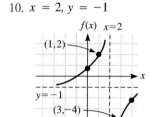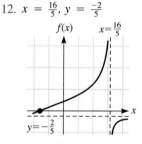

14. $y = 0$ 16. $x = -2, x = 2, y = 0$

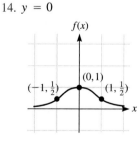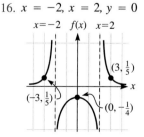

18. $x = -2$, $x = 3$, $y = 0$

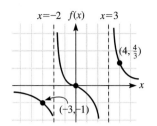

20. $x = -2$, $y = x - 2$

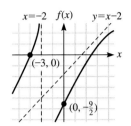

22. $x = 2$, $x = -3$, $y = 1$

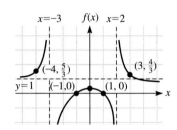

24. $x = -2$, $y = x - 3$

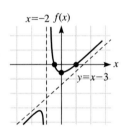

26. $x = 0$, $y = x$

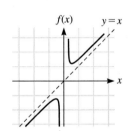

28. $x = -\frac{4}{3}$, $x = 2$, $y = \frac{2}{3}$

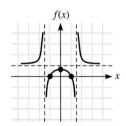

30. $x = 0$ and $y = x^2$

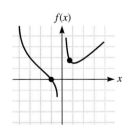

32. No asymptotes

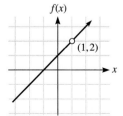

34. No asymptotes

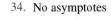
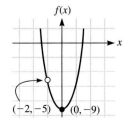

36. No asymptotes

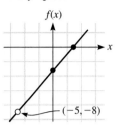

38. $x = 0$, $y = 1$

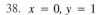
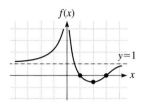

40. $x = \pm 3$, $y = x - 2$

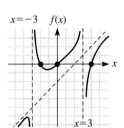

42. $x = 1$, $x = -2$, $y = x - 5$

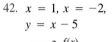
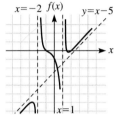

44. a. $t = \dfrac{617}{r}$

b.

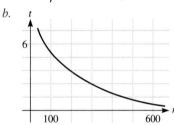

46. Asymptotes: $y = 5x + 22500$, $x = 0$

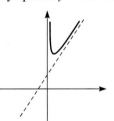

48. a. $v = -6.818$ mm; b. $v = \dfrac{uF}{(u - F)}$

50. Horizontal asymptote:
$y = 0$
No vertical asymptote

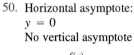

52.

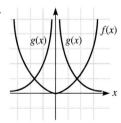

54.

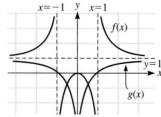

56.

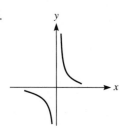

62. (*b*)   64. (*a*)   66. (*c*)

70.

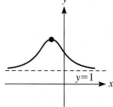

72.

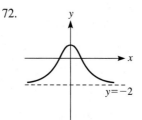

Section 3.8, Cumulative Review Exercises, page 216

2. *a.* $f(x) = 2(x - 1)^2 + 4$;
   b. Vertex: $(1, 4)$;
   c. Upward;
   d. Minimum: 4
   e.

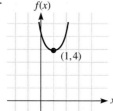

4. *a.* $f(x) = \frac{1}{2}(x + 2)^2 + 3$;
   b. Vertex: $(-2, 3)$;
   c. Upward;
   d. Minimum: 3
   e.

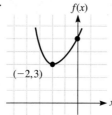

6. Minimum: $-\frac{569}{8}$ 8. Minimum: -13.814594

10. Numbers are 15.5 and 15.5.

12. *a.* $-x^2 + 80x - 700$; *b.* 40; *c.* 900

14.

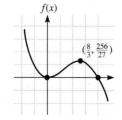

16.

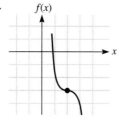

18.

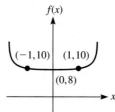

20.

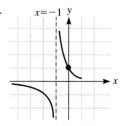

22.

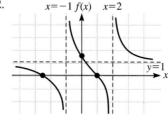

24.

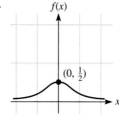

26.

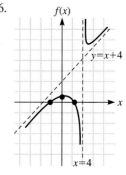

28.

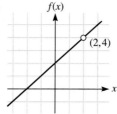

30.

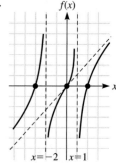

32. $q(x) = x^2 + 4x + 4$; $r(x) = -2x^2 - 10x - 19$

34. $f(x) = -\frac{5}{2}(x + 2 - \sqrt{3})(x + 2 + \sqrt{3})$

36. $f(x) = (x - 1 - \sqrt{2})(x - 1 + \sqrt{2})(x - 3)(x + 2)$

38. $q(x) = 4x^4 - 8x^3 + 16x^2 - 32x + 70$; $r(x) = -139$

40. $q(x) = \frac{1}{6}x - \frac{1}{3}$; $r(x) = 0$

42. $f(0) = -1$; $f(-1) = 4$; $f(i) = 0$

44. $f(2) = 3$; $f(3) = 0$; $f(-2) = -105$; 2: No; 3: Yes;
    -2: No

46. $x = \sqrt[3]{100}$ or $x = -\dfrac{100^{1/3} \pm \sqrt{3}i(100)^{1/3}}{2}$

48. 0; mult. 3; 4: mult. 2; $-\frac{3}{2}$: mult. 1

50. i: mult. 2; $-i$: mult. 2; 1: mult. 3; -1: mult. 3

52. At most 6

54. Pos. real: at most 3; neg. real: at most 3.

56. $k = -19$

58. Other zeros are $-1 \pm \sqrt{2}$.

60. Other zeros are $2 - 4i, \pm \sqrt{2}$.

62. Upper bound: 6; lower bound; -1.

64. Upper bound: 1; lower bound: -2.

66. No rational zeros

68. No rational zeros

70. 2, 3, 5

72. $1, \frac{1}{2}, \frac{1}{3}$

74. Solutions are: $\pm\sqrt{\dfrac{-1 \pm \sqrt{3}i}{2}}$

76. $f(2) < 0; f(3) > 0$

78. 2.0945

80. -1.1673

82. *a.* $R_1^3 + 2R_1^2 - 4R_1 - 6 = 0$;
    b. $R_1 = 1.8661, R_2 = 3.8661, R_3 = 4.8661$

84. *a.* 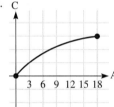 *b.*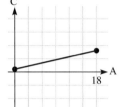

c. $A \approx 9.8$ and 1.2

86. No *x*-intercepts

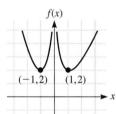

88. *x*-intercept: (0, 0)

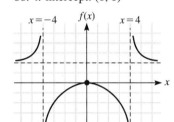

90. $a = -\dfrac{1}{702}$

92. Prove $\sqrt{3}$ is irrational.
Consider $f(x) = x^2 - 3$. Rational zeros are of the form
c/d where c is a factor of 3 and d is a factor of 1.
Thus potential rational zeros are $\pm 3, \pm 1$. Now $\sqrt{3}$
clearly is a zero and is thus irrational.

CHAPTER 4

Section 4.1, page 228

2. *a.* 5; *b.* 15.425847; *c.* 16.188929; *d.* 16.241123; *e.*
   16.24243; *f.* 16.242451; *g.* 16.24245083; *h.* ≈ 16.24245

4. *a.* $f(1) = \frac{1}{3}$; $f(2) = \frac{1}{9}$; $f(3) = \frac{1}{27}$; $f(0) = 1$; $f(-1) = 3$;
   $f(-2) = 9$; $f(-3) = 27$; $f\left(\dfrac{1}{2}\right) = \dfrac{1}{\sqrt{3}}$

b.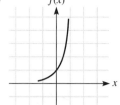

6.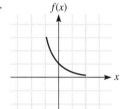

8.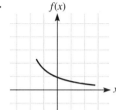

10.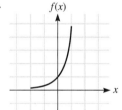

12.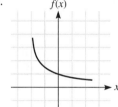

14.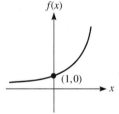

16.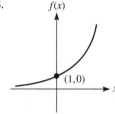

18.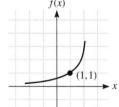

20.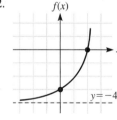

22.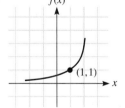

24.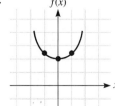

26.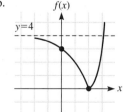

28.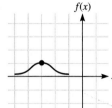

30.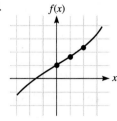

28. a.

t	$1 - e^{-0.18t}$
1	0.165
2	0.302
3	0.417
5	0.593
13	0.904

b.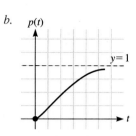

32. a. $Q(t) = [A_0(0.88741)^t]$

   b. $Q(\frac{1}{365}) = 3.999$ g, $Q(1) = 3.51$ g, $Q(2) = 3.079$ g,

     $Q(10.6) = 1$ g, $Q(20) = 0.292$ g

34. a. $N(t) = [5(1.0162503)^t]$

   b. $N(43) = 10$ billion, $N(86) = 20$ billion,

     $N(10) = 5.87$ billion, and $N(13) = 6.17$ billion

36. a. $N(t) = 100,000(1.0717735)^t$

   b. $N(5) = \$141,421.36$; $N(10) = \$200,000$;

     $N(20) = \$400,000$; $N(32) = \$918,958.68$;

   c. $t = 33.22$ yr

38. *(d)*   40. *(h)*

42. *(g)*   44. *(j)*   46. *(b)*

30. a.

t	$\dfrac{4000}{1 + 200e^{-2t}}$
5	3964
10	3999
20	3999
30	3999

b.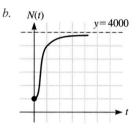

32. a.

t	$\dfrac{800}{1 + 132e^{-0.4t}}$
1	9
2	13
6	62
12	383
30	799
60	799

b.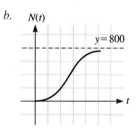

Section 4.2, page 237

2. $e^2(= 7.3890561)$ 4. $e^{-2}(= 0.1353353)$

6. $\dfrac{1}{e} + 4e^2(= 29.924104)$

8. $4 + 4e + e^2(= 22.262183)$

10. $16e^{-4x}$ 12. $1 + 4e^x + 4e^{2x}$

14. $e^{3x} + 3e^x + 3e^{-x} + e^{-3x}$

16. $e^{2x} - e^{-2x}$ 18. $e^{3x} - 1$ 20. e^{2x}

22. -301.85096

34. $\dfrac{2 + 2e^{-x}}{(1 + e^x)^2}$ 36. $\dfrac{e^x(e^h - 1)}{h}$

38. $\dfrac{(e^x - e^{-x}e^{-h})(e^h - 1)}{2h}$

Section 4.3, page 247

2. 3125 4. 729 6. $\frac{1}{25}$ 8. $4^t = Q$

10. $\log_{20} 1 = 0$ 12. $\log_{27} 9 = \frac{2}{3}$ 14. $\log_{27} 3 = \frac{1}{3}$

16. $\log_a c = b$ 18. 1 20. 1

22. 0 24. 0 26. $3x$ 28. k

30. 25 32. 4 34. $3x + 5$ 36. $x^2 + 3$

24. a.

t	$0.75(1 - e^{-0.05t})$
1	0.037
2	0.071
10	0.295
20	0.474
50	0.688

b.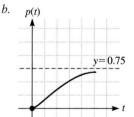

26. a.

t	$(1 - e^{-0.06t})$
1	0.058
2	0.113
10	0.451
20	0.699
50	0.95

b.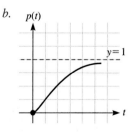

38. a.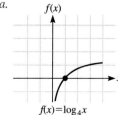

$f(x) = \log_4 x$

b.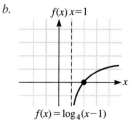

$f(x) = \log_4(x - 1)$

c.

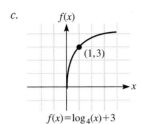

$f(x)=\log_4(x)+3$

40.

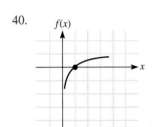

42. $x=-3$

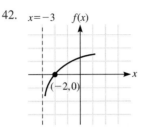

44.

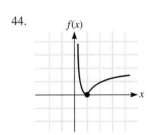

46.

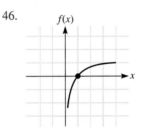

48.

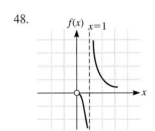

50. $1 + \log_{10} 5$

52. $\log_a 5x$ 54. $\log_a 2\sqrt{x}$ 56. 0

58. $\log_a \dfrac{1}{x+2}$ 60. $\log_a(m^3 + n^3)$

62. $3\log_c x + \frac{1}{4}\log_c y - \frac{1}{4}\log_c z$

64. $3\log_5 x + 4\log_5 y + 4\log_5 z$

66. $-2\log_2 x - \log_2 y - 3\log_2 z$

68. $\frac{1}{2}\log_a x - \log_a y - \frac{3}{2}\log_a z$

70. $2\log_6 x + \log_6 z - 2\log_6 y$

72. $\log_t(P - 2Q) + \log_t(P^2 + 2PQ + 4Q^2)$

74. $\frac{1}{2}\log_a(x - y)$

76. $\log_b(2x + y) + \log_b(3x + 2y)$

78.

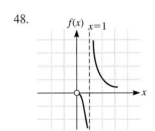

80. 0.9208 82. -0.2625 84. -0.4354 86. -2

88. 2.3125 90. 0.118733 92. -0.13125

94. -0.13938 96. 17 98. 19

100. $H = \dfrac{-10}{\log_2(0.2466)} (\approx 4.95109 \text{ days})$

102. $D = \dfrac{1}{\log_2\left(\frac{10,352,650}{10,000,000}\right)} (\approx 20 \text{ min})$

104. $\sqrt{3}$ 106. No solution

108. $\frac{1}{49}$ 110. 4 112. -2 114. 0

116. $x = -\frac{3}{5}$ 118. $x = \frac{1}{2}, -2$ 120. $x = \frac{3}{2}$

122. $x = \dfrac{\log_4(13) - 5}{3}$ 124. $x = \frac{18}{5}$

126. $x = -\frac{2}{3}$ 128. $x = 3$ 130. $\frac{1}{2}$

132. True   134. False   136. True   138. False

140. $0 < x < 1$ 142. $x \geq 5$ 144. $x \geq 0$

146. $x \geq -2$ 148. $4\log_b(a - b)$

150. Prove $\log_a\left(\dfrac{x + \sqrt{x^2 - 1}}{x - \sqrt{x^2 - 1}}\right) = 2\log_a(x + \sqrt{x^2 - 1})$

Note: $\dfrac{x + \sqrt{x^2 - 1}}{x - \sqrt{x^2 - 1}} \cdot \dfrac{x + \sqrt{x^2 - 1}}{x + \sqrt{x^2 - 1}}$

$$= \dfrac{(x + \sqrt{x^2 - 1})^2}{x^2 - (x^2 - 1)}$$

$$= (x + \sqrt{x^2 - 1})^2$$

Thus, $\log_a\left(\dfrac{x + \sqrt{x^2 - 1}}{x - \sqrt{x^2 - 1}}\right) = \log_a\left((x + \sqrt{x^2 - 1})^2\right)$

$$= 2\log_a\left(x + \sqrt{x^2 - 1}\right)$$

152. $f^{-1}(x) = 2^x$ 154. $f^{-1}(x) = \sqrt{\log_2 x}$

Section 4.4, page 255

2. 2 4. 4 6. -2 8. -4

10. 2.0958665 12. 0.9943172 14. -2.1090204

16. 8.6033609 18. 28.114611 20. -6.2553603

22. -18.004277 24. 135.1 decibels

26. $I = 10^{-4}$ watts/cm^2

28. $10 \log \left(\dfrac{I_1 I_2}{I_0^2} \right)$

30. $R(I_0) = 0$

32. One higher on the Richter scale

34. 8.9

36. $\log \left(\dfrac{I_1 \cdot I_2}{I_0^2} \right)$

38. *a.* 7.4685211; *b.* 10.79588; *c.* 3.39794

40. $10^{-pH} = H^+$ 42. $H^+ > 10^{-7}$ 44. 2

46. 1   48. t 50. $64e^3$

52. 4   54. 66   56. $\dfrac{1}{\ln 100} \cdot (10)^{2x}$

58. $\frac{1}{2}$ 60. x^2 62. $\ln 2$

64. $\frac{1}{2} \cdot 36 \cdot e - \frac{1}{2} \cdot 4 \cdot e = 16e$ 66. $\ln 19$

68. $\dfrac{-4^{1/x}(x + \ln 4)}{x^3}$ 70. $t \approx 4256.88$ yr

72. *a.* $k = \dfrac{\ln 2}{T}$; *b.* 77 yr; *c.* $k \approx 3.0136\%$

74. *a.* $P(1) = \$10,832.87; P(2) = \$11,735.11;$
   $P(5) = \$14,918.25; P(10) = \$22,255.41;$
   b. $P(1) = \$10,800; P(2) = \$11,664;$
   $P(5) = \$14,693.28;$
   $P(10) = \$21,589.25$
   c. $T = 8.664$ yr; *d.* $k = 7\%$

76. $A_0 = \$4740.76$

78. *a.* $P(1000) = 974.74072$ millibars;
   $P(2000) = 937.92642$ millibars;
   $P(10,000) = 689.29649$ millibars;
   $P(30,000) = 319.15328$ millibars;
   b. $a = 1250$ ft

80. *a.* $I(1) = 24.66\% I_0; I(2) = 6.08\% I_0; I(3) = 1.5\% I_0$
   b. $I(10) = 8.315 \times 10^{-5} \% I_0$

82. $D = 1082$ ft; $N = \frac{1}{2}; L \approx 1309$ ft

84. *a.* $k \approx 2.9168826$, $T(5) = 26°C$;
   b. $k = 0.9710156$, $T(10) = 25.998°C$

86. 2.2454485   88. 0.4102726   90. $\dfrac{\ln 2.347}{\ln 10}$

92. $\dfrac{\ln \sqrt{x}}{\ln 10}$ 94. $\dfrac{\log 78.56}{\log e}$ 96. $\dfrac{\log t^4}{\log e}$

98. $\dfrac{1}{2}(x^2 + 1)\ln(x^2 + 1) + \dfrac{x^2}{4}(2 \ln x - 3)$

100. 1.990208×10^{10} yd^2; 551.9 yr   102. -0.89698

104. 0.92797   106. *(b)*   108. *(d)*

Section 4.5, page 262

2. $x = \pm \sqrt{\dfrac{\ln 5}{\ln 8}} \approx \pm 0.879759$

4. $x = \dfrac{-4}{7 - 2\ln 5} \approx -1.0578864$

6. $x = \dfrac{3 + 2e^7}{2 - 5e^7} \approx -0.40069$

8. $t = \ln 60$ 10. $t = -50 \ln 0.06$

12. $k = \frac{1}{9} \ln \left(\frac{52}{25} \right)$ 14. $t = 20 \ln(5)$

16. $x = \dfrac{5 \pm \sqrt{29}}{2}$ 18. $x = -\frac{56}{3}$

20. $x = 10, (x \neq -1)$ 22. $x = 5, (x \neq -2)$

24. $x = \dfrac{3 + \sqrt{149}}{2} \left(x \neq \dfrac{3 - \sqrt{149}}{2} \right)$

26. No solution   28. $x = 1000, \frac{1}{100}$

30. $x = 2, (x \neq -1)$ 32. $x = 3, 4$

34. $x = 5$ 36. $x = 6$ 38. $x = \ln 2$

40. $x = \ln \left(\dfrac{4 + \sqrt{31}}{5} \right)$ 42. $x = \ln(3 \pm 2\sqrt{2})$

44. $x = \frac{1}{2} \ln \frac{3}{5}$ 46. $x = 32$ or -32

48. $x = \frac{4445}{1111}$ 50. $x = 10^{(1 \pm \sqrt{13})/2}$

52. $x = \frac{10}{13}$ 54. $t = \dfrac{1}{k} \ln \left(\dfrac{N_0}{N(t)} \right)$

56. $t = -\dfrac{1}{k} \ln \left[\dfrac{T - T_0}{|T_1 - T_0|} \right]$ 58. $t = \ln(x + \sqrt{x^2 + 1})$

60. $t = -\dfrac{1}{k} \ln \left[1 - \dfrac{v}{v_T} \right]$ 62. $t = \log_B \left(\dfrac{N}{G} \right)$

68. $f(x) = 5e^{x(3 - \ln 5)/2}$ 70. $f(x) = \frac{1}{5}e^{x \ln 5}$

72. $f(x) = 10e^{x \ln 100/10}$ 74. 113.219669

76. No solution   78. 0.21740968

Section 4.6, Cumulative Review Exercises, page 264

2. $10^{0.3010} = 2$ 4. $(0.25)^{1/2} = 0.5$ 6. $3^{3/2} = \sqrt{27}$

8. $\log_2 \frac{1}{2} = -1$ 10. $\log_{16} 4.015 = x$

12. *a.* 50 watts;
   b. $W(30) = 44.346$ watts; $W(365) = 11.612$ watts;
   c. $t \approx 173.3$ days;   *d.* 402.4 days

14.

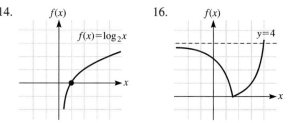

$f(x) = \log_2 x$

16.

$y = 4$

18.

20.

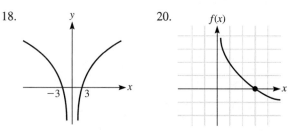

22. $x = \frac{3}{4}$ ⬜ 24. $x = e^4$ ⬜ 26. $x = \frac{1}{1000}$

28. True for all $x > 0$ ⬜ 30. $x = \frac{1}{9}$

32. $x = 10, (x \neq 0)$ ⬜ 34. $x = 100$

36. $x = \ln\left[\frac{-1 + \sqrt{22}}{3}\right], \left(x \neq \frac{-1 - \sqrt{22}}{3}\right)$

38. No solution ⬜ 40. $k = -\frac{1}{10}\ln\frac{13}{20}$

42. $x = 0.56$ ⬜ 44. 224.87508 ⬜ 46. $e^{3x} + 8$

48. a. $N(0) = 711$ ⬜ b.

t	$\dfrac{6400}{1 + 8e^{-0.23t}}$
0	711
1	870
2	1058
6	2124
12	4249
30	6349
60	6400

c.

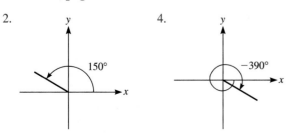

d. $t \approx 9.04$ min

50. $2(3x - 1)$ ⬜ 52. 0 ⬜ 54. $1 + \log_4 x$

56. $3x^2$ ⬜ 58. $\frac{5}{2}\log_a x$ ⬜ 60. $\log(1 - t)$

62. $\frac{3}{2}\log_a x + \log_a y - \frac{1}{2}\log_a z$ ⬜ 64. 2.099

66. $1 - \log_b(b + 1)$ ⬜ 68. -1.609 ⬜ 70. $4.7\sqrt{b}$

72. 19 ⬜ 74. $D \approx 109.7$ yr ⬜ 76. 110 decibels

78. a. $k \approx 0.5108256$; b. $A(t) = 5e^{-0.5108256t}$;
c. $t \approx 1.79$ hr

80. $D \approx 5.78$ days ⬜ 82. $t \approx 43.5$ yr; near the end of 1991

84. $t = \ln(\ln N)$ ⬜ 86. $t = \log_n\left(\frac{y}{a}\right)$

88. $t = \dfrac{-0.07}{\ln(0.02)}(\approx 0.0178936)$ ⬜ 90. $\dfrac{e^{-x}(e^{-h} - 1)}{h}$

92. Prove $(\log_x a)(\log_a b) = \log_x b$.

$\log_a b = \dfrac{\log_x b}{\log_x a}$ by change of base formula.

Thus $(\log_x a)(\log_a b) = \log_x b$.

94. $\log_x b = \dfrac{\log_b b}{\log_b x} = \dfrac{1}{\log_b x}$

CHAPTER 5

Section 5.1, page 276

2.

4.

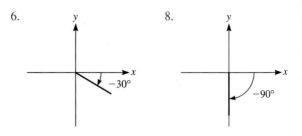

6.

8.

10. a. $11°49'$; b. $101°49'$ ⬜ 12. a. $17'37''$; b. $90°17'37''$

14. $78.1833°$ ⬜ 16. $89.7064°$ ⬜ 18. $-78.2167°$

20. $244.7694°$ ⬜ 22. $78°48'25''$ ⬜ 24. $-189°37'12''$

26. $780°15''$ ⬜ 28. $-222°13'20''$ ⬜ 30. $\dfrac{3\pi}{5}$ radians

32. $-\frac{125}{144}\pi$ radians ⬜ 34. $\dfrac{65\pi}{36}$ radians ⬜ 36. $\dfrac{-\pi}{2}$ radians

38. $612°$ ⬜ 40. $-270°$ ⬜ 42. $-107.315°$

44. $840°$ ⬜ 46. Angles measure $\dfrac{2\pi}{9}, \dfrac{2\pi}{9}$, and $\dfrac{5\pi}{9}$ radians

48. $\dfrac{2\pi}{5}$ radians; complement measures $18° = \dfrac{\pi}{10}$ radians

50. 0.2 m ⬜ 52. $L = 7.94$ cm ⬜ 54. $\frac{3}{5}$ radians

56. $\dfrac{200\pi}{25,753}$ ft ⬜ 58. 0.418879 in/sec

60. $\dfrac{225\pi}{143}$ mph ⬜ 62. 73107.692 radians/hr

66. 2.426056 radians ⬜ 68. 13.2500258 radians

70. $118°46'52''$ ⬜ 72. $3°17'3''$

Section 5.2, page 286

	t	$\sin t$	$\cos t$	$\tan t$	$\cot t$	$\sec t$	$\csc t$
2.	$\frac{3\pi}{4}$	$\frac{\sqrt{2}}{2}$	$-\frac{\sqrt{2}}{2}$	-1	-1	$\sqrt{2}$	$\sqrt{2}$
4.	$\frac{11\pi}{6}$	$-\frac{1}{2}$	$\frac{\sqrt{3}}{2}$	$-\frac{1}{\sqrt{3}}$	$-\sqrt{3}$	$\frac{2}{\sqrt{3}}$	-2
6.	$\frac{8\pi}{3}$	$\frac{\sqrt{3}}{2}$	$-\frac{1}{2}$	$-\sqrt{3}$	$-\frac{1}{\sqrt{3}}$	-2	$\frac{2}{\sqrt{3}}$
8.	$-\frac{\pi}{3}$	$-\frac{\sqrt{3}}{2}$	$\frac{1}{2}$	$-\sqrt{3}$	$-\frac{1}{\sqrt{3}}$	2	$-\frac{2}{\sqrt{3}}$
10.	$-\frac{7\pi}{3}$	$-\frac{\sqrt{3}}{2}$	$\frac{1}{2}$	$-\sqrt{3}$	$-\frac{1}{\sqrt{3}}$	2	$-\frac{2}{\sqrt{3}}$
12.	-6π	0	1	0	Undef.	1	Undef.
14.	$180°$	0	-1	0	Undef.	-1	Undef.
16.	$300°$	$-\frac{\sqrt{3}}{2}$	$\frac{1}{2}$	$-\sqrt{3}$	$-\frac{1}{\sqrt{3}}$	2	$-\frac{2}{\sqrt{3}}$
18.	$-90°$	-1	0	Undef.	0	Undef.	-1
20.	$150°$	$\frac{1}{2}$	$-\frac{\sqrt{3}}{2}$	$-\frac{1}{\sqrt{3}}$	$-\sqrt{3}$	$-\frac{2}{\sqrt{3}}$	2
22.	$570°$	$-\frac{1}{2}$	$-\frac{\sqrt{3}}{2}$	$\frac{1}{\sqrt{3}}$	$\sqrt{3}$	$-\frac{2}{\sqrt{3}}$	-2
24.	$-480°$	$-\frac{\sqrt{3}}{2}$	$-\frac{1}{2}$	$\sqrt{3}$	$\frac{1}{\sqrt{3}}$	-2	$-\frac{2}{\sqrt{3}}$
26.	$-\frac{25\pi}{4}$	$-\frac{\sqrt{2}}{2}$	$\frac{\sqrt{2}}{2}$	-1	-1	$\sqrt{2}$	$-\sqrt{2}$
28.	32π	0	1	0	Undef.	1	Undef.
30.	50π	0	1	0	Undef.	1	Undef.
32.		$-\frac{2}{5}$	$-\frac{4}{5}$	$-\frac{3}{4}$	$-\frac{4}{3}$	$-\frac{5}{4}$	$\frac{5}{3}$
34.		$-\frac{5}{13}$	$-\frac{12}{13}$	$\frac{5}{12}$	$\frac{12}{5}$	$-\frac{13}{12}$	$-\frac{13}{5}$

36. 0.3256   38. 0.2047   40. 29.3711   42. 0.9166
44. 0.4002   46. 1.1967   48. 0.0757   50. 0.6911
52. 2.6608   54. Not defined   56. 0.8415
58. 1.0459 or 59.9284°
60. 1.3429 or 76.9411°
62. 1.4239 or 81.5810°
64. 1.4783 or 84.7007°
66. 1.3694 or 78.4630°
68. 0.0417 or 2.3880°

Section 5.3, page 291

	$\sin\theta$	$\cos\theta$	$\tan\theta$	$\cot\theta$	$\sec\theta$	$\csc\theta$
2.	$\frac{18}{30}=\frac{3}{5}$	$\frac{24}{30}=\frac{4}{5}$	$\frac{18}{24}=\frac{3}{4}$	$\frac{4}{3}$	$\frac{5}{4}$	$\frac{5}{3}$
4.	$\frac{24}{25}$	$\frac{7}{25}$	$\frac{24}{7}$	$\frac{7}{24}$	$\frac{25}{7}$	$\frac{25}{24}$
6.	$\frac{5}{\sqrt{34}}$	$\frac{3}{\sqrt{34}}$	$\frac{5}{3}$	$\frac{3}{5}$	$\frac{\sqrt{34}}{3}$	$\frac{\sqrt{34}}{5}$
8.	$\frac{\sqrt{4-a^2}}{2}$	$\frac{a}{2}$	$\frac{\sqrt{4-a^2}}{a}$	$\frac{a}{\sqrt{4-a^2}}$	$\frac{2}{a}$	$\frac{2}{\sqrt{4-a^2}}$
10.	$\frac{42,163}{75,098}$	$\frac{62,145}{75,098}$	$\frac{42,163}{62,145}$	$\frac{62,145}{42,163}$	$\frac{75,098}{62,145}$	$\frac{75,098}{42,163}$

12. $\sin\theta = \dfrac{\sqrt{4-c^2}}{c}$ $\cos\theta = \dfrac{2}{a}$ $\tan\theta = \dfrac{\sqrt{4-c^2}}{2}$
   $\cot\theta = \dfrac{2}{\sqrt{4-c^2}}$ $\sec\theta = \dfrac{c}{2}$ $\csc\theta = \dfrac{c}{\sqrt{4-c^2}}$

14. $\sin\theta = \dfrac{\sqrt{6}}{\sqrt{7}}$ $\cos\theta = \dfrac{1}{\sqrt{7}}$ $\tan\theta = \sqrt{6}$
   $\cot\theta = \dfrac{1}{\sqrt{6}}$ $\sec\theta = \sqrt{7}$ $\csc\theta = \dfrac{\sqrt{7}}{\sqrt{6}}$

16. $\sin\theta = \dfrac{1}{\sqrt{10}}$ $\cos\theta = \dfrac{3}{\sqrt{10}}$ $\tan\theta = \dfrac{1}{3}$
   $\cot\theta = 3$ $\sec\theta = \dfrac{\sqrt{10}}{3}$ $\csc\theta = \sqrt{10}$

18. $\sin\theta = \dfrac{\sqrt{5}}{4}$ $\cos\theta = \dfrac{4}{\sqrt{11}}$ $\tan\theta = \sqrt{\dfrac{5}{11}}$
   $\cot\theta = \sqrt{\dfrac{11}{5}}$ $\sec\theta = \dfrac{\sqrt{11}}{4}$ $\csc\theta = \dfrac{4}{\sqrt{5}}$

20. $\sin\theta = \dfrac{1}{\sqrt{1+v^2}}$ $\cos\theta = \dfrac{v}{\sqrt{1+v^2}}$ $\tan\theta = \dfrac{1}{v}$
   $\cot\theta = v$ $\sec\theta = \dfrac{\sqrt{1+v^2}}{v}$ $\csc\theta = \sqrt{1+v^2}$

22. $\sin\theta = 0.9974$ $\cos\theta = 0.0721$ $\tan\theta = 13.8404$
   $\cot\theta = 0.0723$ $\sec\theta = 13.8765$ $\csc\theta = 1.0026$

24. $\sin\theta = 0.9922$ $\cos\theta = 0.125$ $\tan\theta = 7.9373$
   $\cot\theta = 0.126$ $\sec\theta = 8$ $\csc\theta = 1.0079$

26. $\sin\theta = \dfrac{20}{25} = \dfrac{4}{5}$ $\cos\theta = \dfrac{15}{25} = \dfrac{3}{5}$ $\tan\theta = \dfrac{20}{15} = \dfrac{4}{3}$
   $\cot\theta = \dfrac{3}{4}$ $\sec\theta = \dfrac{5}{3}$ $\csc\theta = \dfrac{5}{4}$

28. 3.6006636; from calculator: 5.7978837
30. $x \geq 0.001$

Section 5.4, page 295

2. $a = 294.3916$; $c = 330.4034$; $\alpha = 63°$
4. $b = 29.459$; $\beta = 71°15'$; $c = 31.1101$

6. $B = 14.84°$; $A = 75.16°$; $c = 12.1037$
8. $B = 34.1233°$; $A = 55.876°$; $a = 3.3941$
10. $a = 6.0407$; $c = 22.14$; $A = 15°50'$
12. $b = 297.6480$; $c = 452.1797$; $B = 41°,\ 10'$
14. $B = 67.744°$; $A = 22.256°$; $b = 95.7894$
16. $b = 38960.637$; $c = 45759.493$; $A = 31°38'$
18. $A = 55.15°$; $B = 34.85°$; $a = 11.4891$
20. $32.7352°$; $57.2648°$
22. 38.079 m, 66.8014°, 23.1986°
24. 21.8014° (angle with bench); 68.1986° (angle with leg);
   80.777 cm (length of support)
26. 100.045 m, 1.718°   28. 1433.94 ft
30. total amount of wire needed = 100.92 ft
32. 485.57 ft   34. $h = 19.7$ 36. yes, 38.7°
38. $5.6 \times 6 = 33.6$ m   40. 10,404.7 ft
42. South: 100 miles; west: 173.2 miles
44. 2442.76 ft   46. 356.3 ft
48. Distance from destroyer to carrier: 7.14 miles
   Distance from carrier to lighthouse: 12.3 miles
50. $A = \frac{1}{2}bc\sin\alpha$

Section 5.5, page 309

2. a. 2π; b. 1
   c.

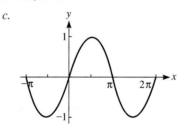

4. a. π; b. 1
   c.

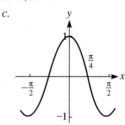

6. a. 2; b. 1
   c.

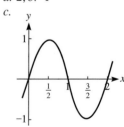

8. a. 8π; b. 1   c.

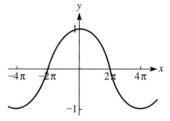

10. *a.* 2π; *b.* 3   *c.*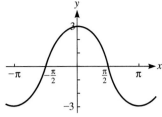

12. *a.* 2π; *b.* 2   *c.*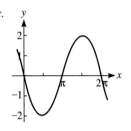

14. *a.* $\dfrac{2\pi}{3}$; *b.* 3   *c.*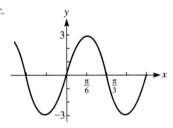

16. *a.* 2; *b.* 2   *c.*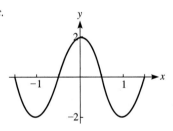

18. *a.* $\frac{1}{3}$; *b.* 0.25   *c.*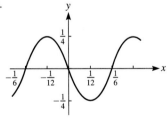

20. *a.* 8; *b.* 6   *c.*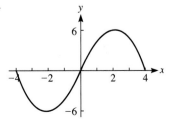

22. *a.* 2π; *b.* 1; *c.* $\dfrac{\pi}{4}$ left

d.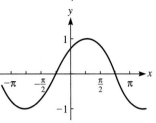

24. *a.* 2π; *b.* 2; *c.* π right

d.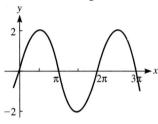

26. *a.* 2π; *b.* 2; *c.* $\dfrac{3\pi}{2}$ left   28. *a.* 2; *b.* 4; *c.* $\dfrac{2}{\pi}$ right

d. 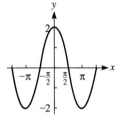   *d.*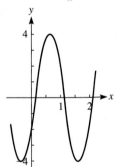

30. *a.* 1; *b.* $\frac{5}{2}$; *c.* $-\dfrac{1}{\pi}$

d.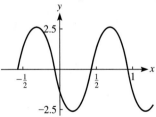

32. *a.* 2π; *b.* 2; *c.* None   34. *a.* π; *b.* $\dfrac{1}{2}$; *c.* $\dfrac{\pi}{2}$ right

d. 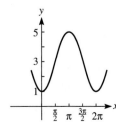   *d.*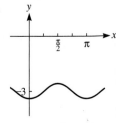

36. *a.* $\dfrac{2\pi}{3}$; *b.* $\frac{1}{3}$; *c.* $\dfrac{\pi}{4}$ left   38. *a.* π; *b.* $\frac{1}{6}$: *c.* $\dfrac{\pi}{3}$ left

40. *a.* $\frac{2}{3}$; *b.* 4; *c.* $\frac{5}{12}$ right

42. *a.*

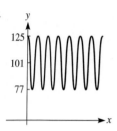

b. 80 beats/sec;
c. 77 min, 125 max

44. *a.* $\dfrac{2\pi}{\pi/45} = 90$; *b.* 5000 miles

c.

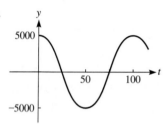

46. *a.*

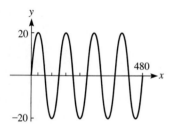

b. max = 20 m, min = −20 m; *c.* 120

48. *a.* $H(221) = 13.5$ hr; *b.* $H(335) = 9.8$ hr;

c.

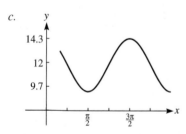

d. 365 days;
e. 171st day (longest), 354th day (shortest)

50. *b.* 110 rabbits/ mile²; 50 rabbits/ mile²
c. 13 lynx/ mile²; 9 lynx/ mile²

52. $p = \dfrac{2\pi}{\pi/4} = 8$, p.s.: 1 right

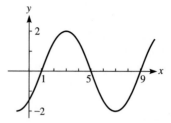

54. $3\sin\left(\dfrac{\pi}{6}x\right)$ 56. $2\cos\left(2x + \dfrac{\pi}{2}\right)$

58. $4.3\sin\left[\dfrac{2\pi}{7}x + \dfrac{20\pi}{7}\right]$

60. 0.9426   62. no zeros on $[-\pi, \pi]$
64. period = $\frac{8}{3}$, ampl. = 4, phase shift = 0
66. period = $\frac{2}{3}$, ampl. = 1.6, phase shift = 0
68. 0   70. (*b*)   72. (*c*)
74. (*h*)   76. (*d*)

Section 5.6, page 321

2.

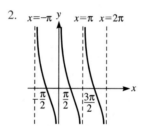

4.

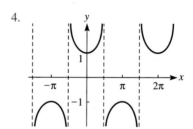

6.

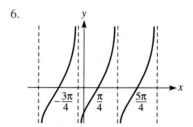

8.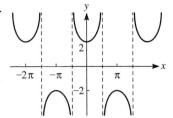

10.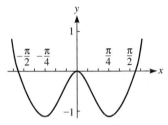

12.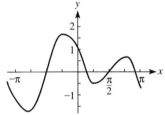

14.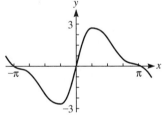

16.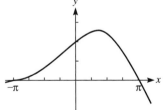

18.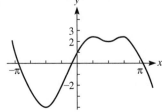

20.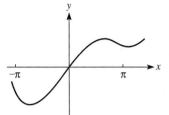

22.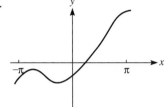

24.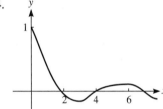

26.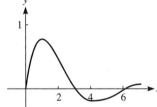

28.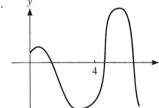

30.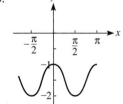

32.

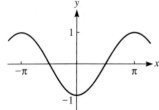

34.

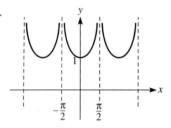

36.

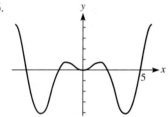

38.

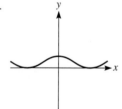

42. *a.*

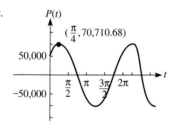

b. Revenue does not cover operating expenses.

48. (*e*)   **50.** (*a*)   **52.** (*f*)

Section 5.7, page 327

	$\sin\theta$	$\cos\theta$	$\tan\theta$	$\cot\theta$	$\sec\theta$	$\csc\theta$
2.	$\dfrac{1}{3}$	$-\dfrac{2\sqrt{2}}{3}$	$-\dfrac{1}{2\sqrt{2}}$	$-2\sqrt{2}$	$-\dfrac{3}{2\sqrt{2}}$	3
4.	$-\dfrac{\sqrt{55}}{8}$	$\dfrac{3}{8}$	$-\dfrac{\sqrt{55}}{3}$	$\dfrac{3}{\sqrt{55}}$	$\dfrac{8}{3}$	$-\dfrac{8}{\sqrt{55}}$

	$\sin\theta$	$\cos\theta$	$\tan\theta$	$\cot\theta$	$\sec\theta$	$\csc\theta$
6.	$-\dfrac{\sqrt{22}}{5}$	$-\dfrac{\sqrt{3}}{5}$	$\sqrt{\dfrac{22}{3}}$	$\sqrt{\dfrac{3}{22}}$	$-\dfrac{5}{\sqrt{3}}$	$-\dfrac{5}{\sqrt{22}}$
8.	$-\dfrac{2}{\sqrt{5}}$	$-\dfrac{1}{\sqrt{5}}$	2	$\dfrac{1}{2}$	$-\sqrt{5}$	$-\dfrac{\sqrt{5}}{2}$

10. $\sin t - 1$ **12.** $\sin t$ **14.** $\cos t + 1$

16. $1 - 3\cos^2 t + 3\cos^4 t - 2\cos^6 t$

18. $\csc^5 t - \csc^3 t$ **20.** $2\cos^2 t - 1\,(= 1 - 2\sin^2 t)$

22. $1 - 3\cos^2 t + 3\cos^4 t - 2\cos^6 t$

24. $\cos^3 t - \dfrac{1}{\cos^3 t}$ **26.** $2\sin^2 t - 1$

28. $\cos^2 t + 2 + \dfrac{1}{\cos^2 t}$ **30.** $\cot t$

32. $\dfrac{2(1 - \sin^2\theta)}{11\sin\theta}$ **34.** $\dfrac{1}{\cos y + 3}$ **36.** $\sin\beta - 6$

38. $\dfrac{2\sin^4 t - 2\sin^2 t + 1}{2\sin^2 t - 1}$ **40.** $3\cos p$

42. 1 **44.** $\dfrac{3(\sin x - 7)}{7}$ **46.** $\csc^4 t$

Section 5.8, Cumulative Review Exercises, page 328

2. *a.* $15°48'22''$; *b.* $105°48'22''$

4. $74.1939°$ **6.** $-22.3728°$ **8.** $-13°7''$

10. $56°13'48''$ **12.** $\dfrac{\pi}{180}$ radians

14. $\dfrac{11,723}{9000}\pi$, or 1.3026π radians   **16.** $660°$

18. $645.1505°$ **20.** $\dfrac{15}{\pi}$ cm   **22.** 0.785

24. angular speed is 248,914.29 radians/hr

	$\sin\theta$	$\cos\theta$	$\tan\theta$	$\cot\theta$	$\sec\theta$	$\csc\theta$
26.	$\dfrac{6}{2\sqrt{34}} = \dfrac{3}{\sqrt{34}}$	$-\dfrac{5}{\sqrt{34}}$	$-\dfrac{3}{5}$	$-\dfrac{5}{3}$	$-\dfrac{\sqrt{34}}{5}$	$\dfrac{\sqrt{34}}{3}$

28. -0.0872 **30.** -0.9615 **32.** 1.7661

34. Undefined

	$\sin\theta$	$\cos\theta$	$\tan\theta$	$\cot\theta$	$\sec\theta$	$\csc\theta$
36.	$\dfrac{2}{7}$	$\dfrac{3\sqrt{5}}{7}$	$\dfrac{2}{3\sqrt{5}}$	$\dfrac{3\sqrt{5}}{2}$	$\dfrac{7}{3\sqrt{5}}$	$\dfrac{7}{2}$
38.	$\dfrac{5}{\sqrt{26}}$	$\dfrac{1}{\sqrt{26}}$	5	$\dfrac{1}{5}$	$\sqrt{26}$	$\dfrac{\sqrt{26}}{5}$

40. 0.5973 radians, or $34.2219°$

42. 1.4706 radians, or $84.2608°$

	t	$\sin\theta$	$\cos\theta$	$\tan\theta$	$\cot\theta$	$\sec\theta$	$\csc\theta$
44.	0	0	1	0	undef.	1	undef.

t	$\sin\theta$	$\cos\theta$	$\tan\theta$	$\cot\theta$	$\sec\theta$	$\csc\theta$
46. $-\dfrac{5\pi}{6}$	$-\dfrac{1}{2}$	$-\dfrac{\sqrt{3}}{2}$	$\dfrac{1}{\sqrt{3}}$	$\sqrt{3}$	$-\dfrac{2}{\sqrt{3}}$	-2
48. $-300°$	$\dfrac{\sqrt{3}}{2}$	$\dfrac{1}{2}$	$\sqrt{3}$	$\dfrac{1}{\sqrt{3}}$	2	$\dfrac{2}{\sqrt{3}}$
50. $-\dfrac{5\pi}{4}$	$\dfrac{\sqrt{2}}{2}$	$-\dfrac{\sqrt{2}}{2}$	-1	-1	$-\dfrac{2}{\sqrt{2}}$	$\dfrac{2}{\sqrt{2}}$

52. $(2n-1)\pi$, n an integer

54. $\dfrac{\pi}{6} + 2k\pi$; $\dfrac{5\pi}{6} + 2k\pi$, k an integer

56. $20 + 9\cos t - 20\cos^2 t$

58. $\dfrac{\sin^4 t}{1 - \sin^2 t}$ 60. $\dfrac{-1}{\sec t(\sec t + 1)}$

62. $\cos^2 x$ 64. $\cos^2\theta$ 66. 0 68. $\cot^2 x$
70. $0, 1$ 72. ± 1

74. 1 76. $\dfrac{\sqrt{25 - \cos^2\theta}}{5 + \cos\theta}$

78. Domain of $\sin\theta$, $\cos\theta$: $\mathcal{R}$
   Domain of $\tan\theta$, $\sec\theta$: $\mathcal{R}$, $x \ne \dfrac{2k+1}{2}\pi$,
   $k = 0, \pm 1, \pm 2, \ldots$
   Domain of $\csc\theta$, $\cot\theta$: $\mathcal{R}$, $x \ne k\pi$,
   $k = 0, \pm 1, \pm 2, \ldots$

80. 1

82. a. 2π; b. $\dfrac{\pi}{2}$ left; c. 3 84. a. π; b. $\dfrac{\pi}{4}$ right; c. 3

d.
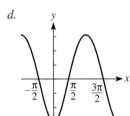

d.
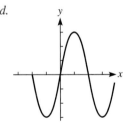

86.
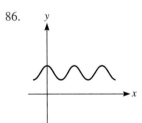

88.
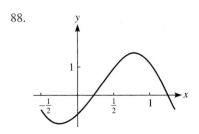

90.
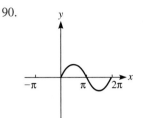

92. $-4\cos(\pi x)$ 94. $y = 0.3\sin\left(\dfrac{\pi}{6}x + \dfrac{2\pi}{3}\right)$

96. $\dfrac{\pi}{5}\left(2 + \sin^2\dfrac{7\pi}{12}\right)$ 98. 28 ft   100. 70.5°

102. Distance west: 75 miles; Distance north: 129.9 miles
104. 1.11 miles

106. a.

t	$14 - 8\cos(39\pi t)$	t	$14 - 8\cos(39\pi t)$
0	6 in.	1	22 in.
0.1	6.392 in.	2	6 in.
0.2	7.528 in.	3	22 in.
0.4	11.528 in.	4	6 in.
0.5	14 in.	10	6 in.

b.
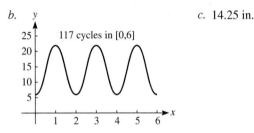

c. 14.25 in.

CHAPTER 6
Section 6.1, page 339

2. $\dfrac{1 - \sin^2 x \pm \sqrt{1 - \sin^2 x}}{\sin x}$ where $\cos x = \pm\sqrt{1 - \sin^2 x}$

4. $\dfrac{1}{1 - \sin^2 x}$ 6. $\dfrac{\sin x}{1 - \sin^2 x}$ 8. $\dfrac{1}{\sin x}$

10. $2\sin x$ 12. $\dfrac{1}{\cos x}$ 14. $\dfrac{\cos^2 x}{1 - \cos^2 x}$

16. $\dfrac{\cos^2 x}{1 - \cos^4 x}$ 18. $\dfrac{1}{1 - \cos^2 x}$ 20. $\tan x$

22. $\tan^2 x$ 24. $\dfrac{1}{\cos x}$, or $\sec x$

26. $\dfrac{1}{a\sec x}$ 28. $5\cos x$

30. $a^2\sin x$ 32. $64\tan^2 x \sec x$

34. $\cos^2\beta + 1 = \cos^2\beta + \cos^2\beta + \sin^2\beta = 2\cos^2\beta + \sin^2\beta$

36. $\tan x \sin x + \cos x = \dfrac{\sin x}{\cos x}\cdot\sin x + \dfrac{\cos^2 x}{\cos x} = \dfrac{1}{\cos x}$
$= \sec x$

38. $\dfrac{1+\sin\beta}{\cos\beta}\cdot\dfrac{1-\sin\beta}{1-\sin\beta}=\dfrac{1-\sin^2\beta}{\cos\beta(1-\sin\beta)}=\dfrac{\cos\beta}{1-\sin\beta}$

40. $\dfrac{1-\tan^2 a}{1-\sec^2 a}=\dfrac{1-(\sin^2 a/\cos^2 a)}{1-(1/\cos^2 a)}=\dfrac{\cos^2 a-\sin^2 a}{\cos^2 a-1}$

$\qquad=\dfrac{\cos^2 a-\sin^2 a}{-\sin^2 a}=1-\cot^2 a$

42. $\dfrac{\cos a}{\sec a-\tan a}=\dfrac{\cos^2 a}{1-\sin a}\cdot\dfrac{1+\sin a}{1+\sin a}=1+\sin a$

44. $1+\sec^2 a\tan^2 a=\sec^2 a-\tan^2 a+\sec^2 a\tan^2 a$

$\qquad=\sec^2 a+\tan^2 a(\sec^2 a-1)=\sec^2 a+\tan^4 a$

46. $\dfrac{\tan^2 a-1}{\tan a-\cot a}=\dfrac{(\sin^2 a-\cos^2 a)/\cos^2 a}{(\sin a/\cos a)-(\cos a/\sin a)}$

$\qquad=\dfrac{(\sin^2 a-\cos^2 a)/\cos^2 a}{(\sin^2 a-\cos^2 a)/(\cos a\sin a)}=\dfrac{\cos a\sin a}{\cos^2 a}=\tan a$

48. $\sin^3\beta\cos\beta-\sin^5\beta\cos\beta$

$\qquad=\cos\beta\sin^3\beta(1-\sin^2\beta)=\cos^3\beta\sin^3\beta$

50. $1-\dfrac{\cos x}{\sec x}=1-\cos^2 x=\sin^2 x=\dfrac{\sin x}{\csc x}$

52. $\dfrac{2\sin^3\alpha}{1-\cos\alpha}=\dfrac{2\sin\alpha(1-\cos\alpha)(1+\cos\alpha)}{1-\cos\alpha}$

$\qquad=2\sin\alpha(1+\cos\alpha)$

54. $\cos(-a)\csc(-a)=\dfrac{\cos a}{-\sin a}=-\cot a$

56. $\dfrac{\sin^3\theta+\cos^3\theta}{\sin\theta+\cos\theta}=\sin^2\theta-\sin\theta\cos\theta+\cos^2\theta$

$\qquad=1-\sin\theta\cos\theta$

58. $\dfrac{\sec x}{\sec x-1}\cdot\dfrac{\sec x+1}{\sec x+1}-\dfrac{\sec x+1}{\tan^2 x}$

$\qquad=\dfrac{\sec^2 x+\sec x-\sec x-1}{\tan^2 x}=\dfrac{\tan^2 x}{\tan^2 x}=1$

60. $\dfrac{\tan x+\sin x}{\tan x-\sin x}=\dfrac{1+\cos x}{1-\cos x}\cdot\dfrac{1+\cos x}{1+\cos x}=\dfrac{(1+\cos x)^2}{\sin^2 x}$

62. $\dfrac{1+\cot x}{1-\cot x}+\dfrac{\cot x+1}{\cot x-1}=0$

Section 6.2, page 350

2. $\dfrac{\sqrt{2}+\sqrt{6}}{4}$ **4.** $\dfrac{-\sqrt{2}-\sqrt{6}}{4}$ **6.** $\dfrac{\sqrt{3}+1}{1-\sqrt{3}}$

8. $\dfrac{\sqrt{2}-\sqrt{6}}{4}$ **10.** $\dfrac{\sqrt{2}-\sqrt{6}}{4}$ **12.** $\dfrac{1+\sqrt{3}}{1-\sqrt{3}}$

14. $\sin\theta$ **16.** $\dfrac{1}{2}\sin\theta-\dfrac{\sqrt{3}}{2}\cos\theta$

18. $\dfrac{\tan\theta+1}{1-\tan\theta}$ **20.** $\dfrac{\sqrt{3}}{2}\cos\theta-\dfrac{1}{2}\sin\theta$

22. $\dfrac{\sqrt{2}}{2}(\cos\theta+\sin\theta)$ **24.** $\dfrac{\tan\theta-1}{1+\tan\theta}$

26. $\dfrac{\sqrt{3}}{2}$ **28.** -1 **30.** 0 **32.** $\sqrt{3}$

34. $\dfrac{\sqrt{6}-\sqrt{2}}{4}$ **36.** $-\dfrac{1}{2}$ **38.** $\dfrac{120}{169}$

40. Undefined.   **42.** $-\dfrac{24}{25}$ **44.** 0

46. $\dfrac{24}{7}$ **48.** $\tan\theta=-\dfrac{7}{4}$

50. Perpendicular lines. Therefore the angle is $\dfrac{\pi}{2}$. Tangent undefined.

52. $\tan\theta=\dfrac{3}{11}$ **54.** 1.1284221 radians

60. $\sqrt{2}\sin\left(x+\dfrac{7\pi}{4}\right)$ **62.** $5\sin\left[x+\arcsin\left(-\dfrac{3}{5}\right)\right]$

64. $5\sin\left[x+\arcsin\left(\dfrac{4}{5}\right)\right]$ **66.** $5\sin\left(2x+\arcsin\dfrac{3}{5}\right)$

68. $\dfrac{\sqrt{2}}{2}$ **70.** -1

76. $\sin 2x=\sin(x+x)=\sin x\cos x+\sin x\cos x$

$\qquad=2\sin x\cos x$

78. $\sin(x+y)\sec(x+y)=\dfrac{\sin x\cos y+\cos x\sin y}{\cos x\cos y-\sin x\sin y}$

$\qquad=\dfrac{\tan x+\tan y}{1-\tan x\tan y}$

80. $\sin(x-y)\sin(x+y)$

$\qquad=[\sin x\cos y-\sin y\cos x][\sin x\cos y+\sin y\cos x]$

$\qquad=\sin^2 x\cos^2 y+\sin x\sin y\cos y\cos x$

$\qquad\quad-\sin y\cos x\sin x\cos y-\sin^2 y\cos^2 x$

$\qquad=\sin^2 x\cos^2 y-\sin^2 y\cos^2 x$

$\qquad\quad+\sin^2 x\sin^2 y-\sin^2 x\sin^2 y$

$\qquad=\sin^2 x(\cos^2 y+\sin^2 y)-\sin^2 y[\cos^2 x+\sin^2 x]$

$\qquad=\sin^2 x-\sin^2 y$

82. $\cos\left(\dfrac{\pi}{6}+x\right)\cos\left(\dfrac{\pi}{6}-x\right)-\sin\left(\dfrac{\pi}{6}+x\right)\sin\left(\dfrac{\pi}{6}-x\right)$

$\qquad=\left[\cos\dfrac{\pi}{6}\cos x-\sin\dfrac{\pi}{6}\sin x\right]$

$\qquad\quad\times\left[\cos\dfrac{\pi}{6}\cos x+\sin\dfrac{\pi}{6}\sin x\right]$

$\qquad\quad-\left[\sin\dfrac{\pi}{6}\cos x+\cos\dfrac{\pi}{6}\sin x\right]$

$\qquad\quad\times\left[\sin\dfrac{\pi}{6}\cos x-\cos\dfrac{\pi}{6}\sin x\right]$

$\qquad=\left(\dfrac{\sqrt{3}}{2}\right)^2\cos^2 x-\left(\dfrac{1}{2}\right)^2\sin^2 x-\left(\dfrac{1}{2}\right)^2\cos^2 x$

$\qquad\quad+\left(\dfrac{\sqrt{3}}{2}\right)^2\sin^2 x=\dfrac{3}{4}-\dfrac{1}{4}=\dfrac{1}{2}$

Section 6.3, page 362

2. *a.* 1; *b.* 0; *c.* Undefined.

4. *a.* -0.96; *b.* 0.28; *c.* -3.4285714

6. *a.* $\dfrac{\sqrt{3}}{2}$; *b.* $-\dfrac{1}{2}$; *c.* $-\sqrt{3}$ **8.** *a.* $-\dfrac{\sqrt{3}}{2}$; *b.* $-\dfrac{1}{2}$; *c.* $\sqrt{3}$

10. $\dfrac{3\tan x-\tan^3 x}{1-3\tan^2 x}$ **12.** $8\cos^4 x-8\cos^2 x+1$

14. *a.* $\dfrac{1}{\sqrt{10}}$; *b.* $-\dfrac{3}{\sqrt{10}}$; *c.* $-\dfrac{1}{3}$

16. a. $\sqrt{\dfrac{2+\sqrt{2}}{4}}$; b. $\sqrt{\dfrac{2-\sqrt{2}}{4}}$; c. $\dfrac{2+\sqrt{2}}{\sqrt{2}}$

18. a. $\sqrt{\dfrac{5-\sqrt{21}}{10}}$; b. $\sqrt{\dfrac{5+\sqrt{21}}{10}}$; c. $\sqrt{\dfrac{5-\sqrt{21}}{5+\sqrt{21}}}$

20. a. $\dfrac{5}{\sqrt{26}}$; b. $-\dfrac{1}{\sqrt{26}}$; c. -5

22. $\dfrac{\sqrt{11}}{3\sqrt{2}}$ 24. $\dfrac{1}{5}\sqrt{\dfrac{21}{2}}$ 26. $\dfrac{\sqrt{2-\sqrt{2}}}{2}$

28. $\dfrac{\sqrt{2}}{2+\sqrt{2}}$ 30. $\dfrac{2}{\sqrt{2-\sqrt{2}}}$ 32. $-\dfrac{\sqrt{2-\sqrt{2}}}{2}$

34. $\sin 100°$ 36. $\cos\dfrac{\pi}{4}$ 38. $\tan\dfrac{\pi}{10}$ 40. $\cos 65°$

42. $\tan 20°$ 44. $\sin(-30°)$ 46. -1 48. $\sin^3 x$

50. $\csc x$

52. $\cos^2\dfrac{x}{2} + 2\cos\dfrac{x}{2}\sin\dfrac{x}{2} + \sin^2\dfrac{x}{2} = 1 + \sin x$

54. $\dfrac{2\cos x}{\csc x - 2\sin x} = \dfrac{2\cos x\sin x}{1 - 2\sin^2 x} = \dfrac{\sin 2x}{\cos 2x} = \tan 2x$

56. $\left(\dfrac{\sin x}{1+\cos x}\right)^2 - 2\left(\dfrac{1}{1+\cos x}\right) + 1$

$= \dfrac{\sin^2 x - 2(1+\cos x) + (1+\cos x)^2}{(1+\cos x)^2}$

$= \dfrac{\sin^2 x - 2 - 2\cos x + 1 + 2\cos x + \cos^2 x}{(1+\cos x)^2} = 0$

58. $\dfrac{\sin 3x}{\sin x} = \dfrac{-4\sin^3 x + 3\sin x}{\sin x} = -4\sin^2 x + 3$

60. $\dfrac{2 + \sin 2x}{2} = 1 + \sin x\cos x$

$= \sin^2 x + \sin x\cos x + \cos^2 x$

$= \dfrac{\sin^3 x - \cos^3 x}{\sin x - \cos x}$

62. $\dfrac{1 + \tan^2 x}{2\tan x} + \dfrac{1}{2}[\cot x + \tan x] = \dfrac{1}{2}\left[\dfrac{\sin^2 x + \cos^2 x}{\cos x\sin x}\right]$

$= \csc 2x$

64. $\dfrac{\csc x - 2\sin x}{2\cos x} = \dfrac{1 - 2\sin^2 x}{2\sin x\cos x} = \dfrac{\cos 2x}{\sin 2x} = \cot 2x$

66. $\cot x\sin 2x = \dfrac{\cos x}{\sin x}\cdot 2\sin x\cos x = 2\cos^2 x$

$= \cos 2x + 1$

68. $\dfrac{2}{1+\cos x} - \dfrac{1+\cos x}{1+\cos x} = \dfrac{1-\cos x}{1+\cos x} = \tan^2\dfrac{x}{2}$

70. $\sin x\sec\dfrac{x}{2} = 2\sin\dfrac{x}{2}\cos\dfrac{x}{2}\sec\dfrac{x}{2} = 2\sin\dfrac{x}{2}$

72. $\dfrac{\sin x\cos x}{1 - 2\sin^2 x} = \dfrac{(1/2)\sin 2x}{\cos 2x} = \dfrac{1}{2}\tan 2x$

$= \dfrac{1}{\cot x - \tan x}$ (using exercise 63)

74. $\dfrac{\sin x}{\csc 3x} - \dfrac{\cos 3x}{\sec x} = (\sin x)(\sin 3x) - (\cos 3x)(\cos x)$

$= -\cos(3x + x) = -\cos 4x$

$= -[\cos^2 2x - \sin^2 2x]$

Section 6.4, page 366

2. $\dfrac{1}{2}\sin 3a + \dfrac{1}{2}\sin a$ 4. $\dfrac{1}{2}\cos(3x) - \dfrac{1}{2}\cos(7x)$

6. $\dfrac{1}{2}\cos(9y) + \dfrac{1}{2}\cos(y)$ 8. $\dfrac{1}{2}\cos(5y) - \dfrac{1}{2}\cos(y)$

10. $\dfrac{1}{2}\cos(5x) + \dfrac{1}{2}\cos(x)$ 12. $2\sin\dfrac{7\pi}{6} - 2\sin\dfrac{\pi}{2}$

14. $\dfrac{1}{2}\cos(2x) + \dfrac{1}{2}\cos(2y)$ 16. $2\sin 60°\sin 20°$

18. $2\sin 3x\cos 2x$ 20. $2\sin\dfrac{\pi}{6}\cos\dfrac{\pi}{3}$

22. $2\cos x\cos y$ 24. $2\sin(5x)\cos(3x)$

26. $2\sin 50°\cos 30°$

28. $\dfrac{1}{2}[\cos(x+y)+\cos(x-y)] = \dfrac{1}{2}[\cos x\cos y - \sin x\sin y$

$+ \cos x\cos y + \sin x\sin y] = \cos x\cos y$

30. Consider $\cos w\cos z = \dfrac{1}{2}[\cos(w+z)+\cos(z-w)]$. Let

$\left.\begin{array}{l} w + z = x \\ w - z = y \end{array}\right\} \Rightarrow w = \dfrac{x+y}{2}, \quad z = \dfrac{x-y}{2}$

Then $2\cos\left(\dfrac{x+y}{2}\right)\cos\left(\dfrac{x-y}{2}\right) = \cos x + \cos y$.

32. Consider $\sin w\cos z = \dfrac{1}{2}[\sin(w+z)+\sin(w-z)]$. Let

$\left.\begin{array}{l} w + z = 2x \\ w - z = 2y \end{array}\right\} \Rightarrow w = x + y, \quad z = x - y$

Then

$\sin(x+y)\cos(x-y) = \dfrac{1}{2}[\sin(2x) + \sin(2y)]$

34. $\dfrac{\cos x + \cos 9x}{\sin x + \sin 9x} = \dfrac{2\cos 5x\cos 4x}{2\sin 5x\cos 4x} = \cot 5x$

36. $\dfrac{\cos 5x + \cos 2x}{\sin 5x + \sin 2x} = \dfrac{2\cos(7x/2)\cos(3x/2)}{2\sin(7x/2)\cos(3x/2)} = \cot\left(\dfrac{7x}{2}\right)$

38. $\dfrac{\sin 5x + \sin 3x}{\cos 5x - \cos 3x} = \dfrac{2\sin 4x\cos x}{-2\sin 4x\sin x} = -\cot x$

40. $\cos(x+y)\cos(x-y) = \dfrac{1}{2}[\cos(2x) + \cos(2y)]$

$= \dfrac{1}{2}[1 - 2\sin^2 x + 2\cos^2 y - 1] = \cos^2 y - \sin^2 x$

42. $\dfrac{\cos 2x}{\sec x} - \dfrac{\sin x}{\csc 2x} = \cos x\cos 2x - \sin x\sin 2x = \cos 3x$

44. $\dfrac{\sin x + \sin y}{\cos x + \cos y} = \dfrac{2\sin[(x+y)/2]\cos(x-y)/2}{2\cos[(x+y)/2]\cos(x-y)/2}$

$= \tan\left(\dfrac{x+y}{2}\right)$

46. $\dfrac{\sin 4x + \sin 6x}{\cos 4x - \cos 6x} = \dfrac{2\sin 5x\cos(-x)}{-2\sin 5x\sin(-x)} = -\cot(-x)$

$= \cot x$

48. -1 50. -1 52. $\cot\dfrac{29\pi}{144}$

54. $-\cot\left(\dfrac{7\pi}{36}\right)$ 56. $\dfrac{\sqrt{3}+1}{4}$

Section 6.5, page 371

2. $-\dfrac{\pi}{3}$ 4. $\dfrac{\pi}{3}$ 6. 0 8. $\dfrac{5\pi}{4}$

10. $\dfrac{\pi}{3}$ 12. $\dfrac{\pi}{6}$ 14. $-\dfrac{\pi}{4}$ 16. $\dfrac{\pi}{4}$

18. $-\dfrac{\sqrt{3}}{2}$ 20. -1 22. $\dfrac{\pi}{2}$ 24. 1

26. $\dfrac{5}{13}$ 28. $\dfrac{5\pi}{12}$ 30. 1

40. $\cos^{-1}\left[\sin\left(x + \dfrac{\pi}{2}\right)\right] = x$ for $0 \le x \le \pi$

$\sin\left(x + \dfrac{\pi}{2}\right) = \sin x \cos \dfrac{\pi}{2} + \sin \dfrac{\pi}{2} \cos x = \cos x$

$\cos^{-1}(\cos x) = x$ for $0 \le x \le \pi$

42. $\sin(\tan^{-1} x) = \dfrac{2x}{1 + x^2}$

$\sin 2\theta = 2 \sin \theta \cos \theta$

$= 2 \cdot \dfrac{x}{\sqrt{1 + x^2}} \cdot \dfrac{1}{\sqrt{1 + x^2}} = \dfrac{2x}{1 + x^2}$

46. $\dfrac{\cos(y) + 1}{2}$ 48. $\cot\left(\dfrac{y}{3}\right) - 3$ 50. $2 - \sec 3y$

52. $\dfrac{3 - \sin\left(\dfrac{y}{4}\right)}{2}$ 54. $\dfrac{\tan(5y) - 1}{4}$ 56. $\sqrt{1 - x^2}$

58. $\dfrac{\sqrt{x^2 - 4}}{x}$ 60. $\dfrac{1 - x^2}{1 + x^2}$ 62. $\dfrac{3}{\sqrt{x^2 - 9}}$

64. $\sqrt{1 + x^2}$ 66. $1 - 2x^2$ 68. $2x\sqrt{1 - x^2}$

70. $\sin y = 2x$ $\quad$ $\csc y = \dfrac{1}{2x}$

$\cos y = \sqrt{1 - 4x^2}$ $\quad$ $\sec y = \dfrac{1}{\sqrt{1 - 4x^2}}$

$\tan y = \dfrac{2x}{\sqrt{1 - 4x^2}}$ $\quad$ $\cot y = \dfrac{\sqrt{1 - 4x^2}}{2x}$

72. $\sin y = \dfrac{\sqrt{4 - x^2}}{2}$ $\quad$ $\csc y = \dfrac{2}{\sqrt{4 - x^2}}$

$\cos y = \dfrac{x}{2}$ $\quad$ $\sec y = \dfrac{2}{x}$

$\tan y = \dfrac{\sqrt{4 - x^2}}{x}$ $\quad$ $\cot y = \dfrac{x}{\sqrt{4 - x^2}}$

74. $\sin y = \dfrac{\sqrt{x^2 - 1}}{x}$ $\quad$ $\csc y = \dfrac{x}{\sqrt{x^2 - 1}}$

$\cos y = \dfrac{1}{x}$ $\quad$ $\sec y = x$

$\tan y = \sqrt{x^2 - 1}$ $\quad$ $\cot y = \dfrac{1}{\sqrt{x^2 - 1}}$

76. $\sin y = \dfrac{1}{\sqrt{x^2 + 1}}$ $\quad$ $\csc y = \sqrt{x^2 + 1}$

$\cos y = \dfrac{x}{\sqrt{x^2 + 1}}$ $\quad$ $\sec y = \dfrac{\sqrt{x^2 + 1}}{x}$

$\tan y = \dfrac{1}{x}$ $\quad$ $\cot y = x$

78.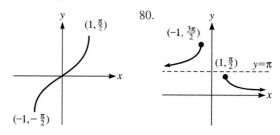

80.

82. $\sin(x + y) = -\dfrac{\sqrt{3}}{2}$; $\cos(x + y) = -\dfrac{1}{2}$;

$\sin(2x) = 0$; $\cos(2x) = -1$

84. $\sin(x + y) = \dfrac{3}{\sqrt{10}}$; $\cos(x + y) = \dfrac{-1}{\sqrt{10}}$;

$\sin(2x) = \frac{4}{5}$; $\cos(2x) = -\frac{3}{5}$

86. $\sin(x + y) = 1$; $\cos(x + y) = 0$; $\sin(2x) = \dfrac{\sqrt{3}}{2}$;

$\cos(2x) = -\frac{1}{2}$

88. Choose $x = 0$: $\cos^{-1} 0 = \dfrac{\pi}{2} \ne 2 \cos^{-1} 0 = \pi$

90. Take $x = 1$. $\qquad$ 92. Take $x = 1$.

Section 6.6, page 376

2. $x = \dfrac{(2n + 1)\pi}{2}$, $\quad n = 0, \pm 1, \pm 2, \ldots$

4. $x = \dfrac{(2n + 1)\pi}{2}$, $\quad n = 0, \pm 1, \pm 2, \ldots$

6. No solution

8. $x = \pi + 2\pi n$, $\quad n = 0, \pm 1, \pm 2, \ldots$

10. $x = \dfrac{3\pi}{4} + \pi n$, $\quad n = 0, \pm 1, \pm 2, \ldots$

12. $x = \dfrac{4\pi}{3} + 2\pi n$ or $x = \dfrac{5\pi}{3} + 2\pi n$; $\quad n = 0, \pm 1, \pm 2, \ldots$

14. $x = \dfrac{\pi}{3} + 2\pi n$ or $x = \dfrac{2\pi}{3} + 2\pi n$, $\quad n = 0, \pm 1, \pm 2, \ldots$

16. $x = \dfrac{2\pi}{3} + 2\pi n$ or $x = \dfrac{4\pi}{3} + 2\pi n$; $\quad n = 0, \pm 1, \pm 2, \ldots$

18. $x = 0.3965474 + 2\pi n$ or $x = 5.8866379 + 2\pi n$; $n = 0, \pm 1, \pm 2, \ldots$

20. $x = 0.4464875 + 2k\pi$ or $x = 2.6951052 + 2k\pi$; $k = 0, \pm 1, \pm 2, \ldots$

22. $x = 1.3497012 + k\pi$; $\quad k = 0, \pm 1, \pm 2, \ldots$

24. $-\dfrac{\pi}{3}, 0, \dfrac{\pi}{3}$ 26. $-\dfrac{\pi}{6}, \dfrac{\pi}{6}$ 28. $\dfrac{7\pi}{24}, -\dfrac{5\pi}{24}$

30. $-\dfrac{\pi}{6}$ 32. $\dfrac{11\pi}{12}, \dfrac{19\pi}{12}$ 34. $\dfrac{7\pi}{6}, \dfrac{3\pi}{2}, \dfrac{11\pi}{6}$

36. $0, \pi, \dfrac{\pi}{3}, \dfrac{4\pi}{3}$ 38. $\dfrac{\pi}{2}, \dfrac{3\pi}{2}$

In the box at top right:

$\cos y = \dfrac{x}{\sqrt{x^2 + 1}}$ $\qquad$ $\sec y = \dfrac{\sqrt{x^2 + 1}}{x}$

$\tan y = \dfrac{1}{x}$ $\qquad$ $\cot y = x$

40. $\dfrac{\pi}{6}, \dfrac{5\pi}{6}, \dfrac{7\pi}{6}, \dfrac{11\pi}{6}$

42. $0, \dfrac{\pi}{3}, \dfrac{5\pi}{3}$

44. $\dfrac{\pi}{6}, \dfrac{\pi}{2}, \dfrac{7\pi}{6}, \dfrac{3\pi}{2}$

46. $\dfrac{2\pi}{3}, \dfrac{4\pi}{3}$

48. No solution

50. $\dfrac{\pi}{4}, \dfrac{\pi}{2}, \dfrac{3\pi}{4}, \dfrac{5\pi}{4}, \dfrac{3\pi}{2}, \dfrac{7\pi}{4}$

52. 1.570796

54. 2.03444, 0.588, 5.176, 3.7296

56. No solution

58. $0, \dfrac{\pi}{4}, \dfrac{3\pi}{4}, \dfrac{5\pi}{4}, \dfrac{7\pi}{4}, \dfrac{\pi}{2}, \dfrac{3\pi}{2}, \pi$

60. $\dfrac{\pi}{4}, \dfrac{3\pi}{4}, \dfrac{5\pi}{4}, \dfrac{7\pi}{4}$

62. $\dfrac{\pi}{2}, \dfrac{3\pi}{2}, \dfrac{\pi}{6}, \dfrac{5\pi}{6}$

64. $\dfrac{\pi}{3}, \dfrac{2\pi}{3}, \dfrac{4\pi}{3}, \dfrac{5\pi}{3}$

66. $\dfrac{\pi}{12}, \dfrac{5\pi}{12}, \dfrac{7\pi}{12}, \dfrac{11\pi}{12}, \dfrac{13\pi}{12}, \dfrac{17\pi}{12}, \dfrac{19\pi}{12}, \dfrac{23\pi}{12}$

68. $\dfrac{\pi}{3}, \dfrac{5\pi}{3}$

70. 0

72. 1.9635, 5.10509, 0.3927, 3.53429

74. $\dfrac{3\pi}{4}, \dfrac{7\pi}{4}$

76. 1.24905, 4.39064

78. $(1 - 2\sin^2 x)\sec 2x = \dfrac{1 - 2\sin^2 x}{\cos 2x} = 1$

80. $\dfrac{\sin x}{\sin x \cot x + \cos x} = \dfrac{\sin x}{2\cos x} = \dfrac{\tan x}{2}$

82. $\sin^2 x = 1 - \cos^2 x$

84. $\dfrac{\tan x - \tan y - \tan z - \tan x \tan y \tan z}{1 + \tan x \tan y + \tan x \tan z - \tan y \tan z}$

86. $\sin 3x = -4\sin^3 x + 3\sin x$

Section 6.7, Cumulative Review Exercises, page 379

2. $\cot x$

4. $\tan x$

6. $\cos x$

	$\sin x$	$\cos x$	$\tan x$	$\cot x$	$\sec x$	$\csc x$
8.	$\dfrac{\sqrt{3}}{2}$	$-\dfrac{1}{2}$	$-\sqrt{3}$	$-\dfrac{1}{\sqrt{3}}$	-2	$\dfrac{2}{\sqrt{3}}$
10.	$-\dfrac{5}{13}$	$-\dfrac{12}{13}$	$\dfrac{5}{12}$	$\dfrac{12}{5}$	$-\dfrac{13}{12}$	$-\dfrac{13}{5}$

12. $\tan x$

18. $-\dfrac{(\sqrt{6} + \sqrt{2})}{4}$

20. $\dfrac{1 + \sqrt{3}}{\sqrt{3} - 1}$

22. $\dfrac{1 + \tan x}{1 - \tan x}$

24. $\dfrac{\sqrt{2}}{2}\cos x + \dfrac{\sqrt{2}}{2}\sin x$

26. $\dfrac{4\sqrt{3} + 3}{10}$

28. $-\dfrac{1}{2}$

30. $-\dfrac{120}{169}$

32. $\dfrac{24}{7}$

34. $\dfrac{7}{17}; \theta \approx 0.3906$

36. 1

38. 1

40. LHS: $\cos[(\pi + x) - (\pi - x)] = \cos 2x$

42. $\cot(x + y) = \dfrac{1 - \tan x \tan y}{\tan x + \tan y} = \dfrac{\cot x \cot y - 1}{\cot y + \cot x}$

44. *a.* $\sin 2x = -\dfrac{\sqrt{3}}{2}; b.\ \cos 2x = -\dfrac{1}{2};$

c. $\tan 2x = \sqrt{3}; d.\ \sin\dfrac{x}{2} = \dfrac{\sqrt{3}}{2};$

e. $\cos\dfrac{x}{2} = \dfrac{1}{2}; f.\ \tan\dfrac{x}{2} = \sqrt{3}$

46. *a.* $\sin 2x = -\dfrac{24}{25}; b.\ \cos 2x = -\dfrac{7}{25};$

c. $\tan 2x = \dfrac{24}{7}; d.\ \sin\dfrac{x}{2} = \dfrac{1}{\sqrt{5}};$

e. $\cos\dfrac{x}{2} = -\dfrac{2}{\sqrt{5}}; f.\ \tan\dfrac{x}{2} = -\dfrac{1}{2}$

48. $\dfrac{2\tan 7x}{1 - \tan^2 7x}$

50. $4\cos^3 x - 3\cos x$

52. $\sec 2x = \dfrac{1}{\cos 2x} = \dfrac{1}{1 - 2\sin^2 x}$

54. $2\cos^2\dfrac{x}{2} = 2\left(\sqrt{\dfrac{1 + \cos x}{2}}\right)^2 = 1 + \cos x$

56. $\sec\dfrac{x}{2} = \pm\sqrt{\dfrac{2}{1 + \cos x}} = \pm\dfrac{\sqrt{2(1 + \cos x)}}{1 + \cos x}$

58. $\tan x = -1.25356\left(\text{exact}: -\sqrt{\dfrac{11}{7}}\right)$

60. $\sin x = \dfrac{\sqrt{29}}{5\sqrt{2}}$

62. $\dfrac{2 - \sqrt{2}}{\sqrt{2}}$

64. $\dfrac{2}{\sqrt{\sqrt{2} + 2}}$

66. $\dfrac{\sqrt{2}}{2}$

68. $\cos\dfrac{\pi}{8}$

70. $\cos x$

72. $\sin 3x$

74. $\cos 8x - \cos 10x$

76. $\sin 8x - \sin 2x$

78. $2\sin 20° \cos 40°$

80. $-2\sin 2x \sin x$

82. $\dfrac{\cos 4x + \cos 2x}{\sin 4x - \sin 2x} = \dfrac{2\cos 3x \cos x}{2\sin x \cos 3x} = \cot x$

84. $\dfrac{\cos 2x - \cos 4x}{2\sin 3x} = \dfrac{-2\sin 3x \sin(-x)}{2\sin 3x} = \sin x$

86. $\dfrac{\pi}{6}$

88. $\dfrac{1}{2}$

90. 0

94. $\sin\left(\dfrac{y}{3}\right) - 1$

96. $1 - \cos\left(\dfrac{2y}{3}\right)$

98. $\dfrac{x}{\sqrt{x^2 + 9}}$

100. $\dfrac{\sqrt{25 - x^2}}{5}$

102. $\sin y = x \qquad \csc y = \dfrac{1}{x}$

$\cos y = \sqrt{1 - x^2} \qquad \sec y = \dfrac{1}{\sqrt{1 - x^2}}$

$\tan y = \dfrac{x}{\sqrt{1 - x^2}} \qquad \cot y = \dfrac{\sqrt{1 - x^2}}{x}$

104. $\sin y = -\dfrac{1}{x} \qquad \csc y = -x$

$\cos y = -\dfrac{\sqrt{x^2 - 1}}{x} \qquad \sec y = -\dfrac{x}{\sqrt{x^2 - 1}}$

$\tan y = \dfrac{1}{\sqrt{x^2 - 1}} \qquad \cot y = \sqrt{x^2 - 1}$

106. $\sin(x+y) = \dfrac{\sqrt{3}}{2}; \cos(x+y) = -\dfrac{1}{2}; \sin 2x = \dfrac{\sqrt{3}}{2};$

$\cos 2x = -\dfrac{1}{2}; \tan 2x = -\sqrt{3}$

108. $x = \dfrac{\pi}{6} + 2\pi k$ or $x = \dfrac{5\pi}{6} + 2\pi k;$ $\quad k = 0, \pm 1, \pm 2, \ldots$

110. $x = -1.0180407 + \pi k;$ $\quad k = 0, \pm 1, \pm 2, \ldots$

112. $\dfrac{\pi}{6} + \dfrac{\pi}{3}k; k = 0, \pm 1, \pm 2, \ldots$

114. $x = \dfrac{\pi}{3}, \dfrac{2\pi}{3}, \dfrac{4\pi}{3}, \dfrac{5\pi}{3}$

116. No solution   118. No solution

120. 1.77215, 4.51103, 1.318116, 4.96507

CHAPTER 7
Section 7.1, page 388

2. $21°; 138°; 112.03$

4. $22.7°; 123.8°; 752.8$ 6. $110°; 354.7; 687.5$

8. $\beta = 98°; a = 10.1; c = 15.9$

10. $\alpha = 113°; b = 3.61; a = 5.52$

12. $\gamma = 71°20'; b = 77.74; c = 77.81$

14. $\gamma = 2.57°; \alpha = 174.68°; a = 11.98$

16. $\beta = 9.43°; \gamma = 25.57°; c = 10.53$

18. $\gamma = 102.85°; b = 3.07; c = 13.33$

20. $\alpha = 30°; \beta = 92°; b = 99.93$

22. $\gamma = 24.3°$ or $155.7°$

If $\gamma = 24.3°$, $\alpha = 142.7°$ and $a = 700.41$

If $\gamma = 155.7°$, $\alpha = 11.3°$ and $a = 226.48$

24. $21.55°; \gamma = 122.45°; c = 34.46$

26. $a = 60.44$ km (closer tower)

28. The shortest distance is 2.56 km.

30. 17.93 m   32. 1714 m

34. Ground speed is 361.6 ft/second $= 246.5$ mph

36. 79.28 miles   38. 163 m

40. $\overline{CA} = 14.5$ km   42. 107.7

44. No solution   46. 44

Section 7.2, page 395

2. $c = 3.2; \alpha = 9.1°; \beta = 140.9°$

4. $a = 2.6; \beta = 19.1°; \gamma = 130.9°$

6. $a = 10.9; \beta = 28.2°; \gamma = 6.8°$

8. $b = 28.2; \alpha = 7.9°; \gamma = 112.1°$

10. $y = 39.9; \angle X = 38.8°; \angle Z = 29.2°$

12. No such triangle

14. $a = 15.8; \beta = 38°; \gamma = 69.2°$

16. $\alpha = \gamma = 42.7°; \beta = 94.7°$

18. $\alpha = 16.3°; \beta = 73.7°; \gamma = 90°$

20. $\alpha = 28.8°; \beta = 108.1°; \gamma = 43.2°$

22. $\alpha = 22.6°; \beta = 67.4°; \gamma = 90°$

24. $b = 0.9; \alpha = 86.2°; \gamma = 21.8°$

26. $\overline{BC} = 45.9$ m

28. Angles are $138.9°$ and $41.4°$.

30. 20.3 ft^3 32. 17.3 miles; speed is 132 mph

34. 1152.5 miles   36. 76.8 cm

38. Heading: $270° + 24.5° = 294.5°$ from north

40. Area $= 21$ 42. Area $= 20.1$ 44. Area $= 81.4$

46. Area $= \frac{1}{2}ab\sin C$

Consider the law of cosines: $c^2 = a^2 + b^2 - 2ab\cos C$. That is,

$$\cos C = \frac{a^2 + b^2 - c^2}{2ab}$$

Thus

$$\sin C = \sqrt{1 - \left(\frac{a^2 + b^2 - c^2}{2ab}\right)^2} \quad (\sin C \geq 0)$$

and

$$\text{area} = \frac{1}{2}ab\sqrt{1 - \left(\frac{a^2 + b^2 - c^2}{2ab}\right)^2}$$

48. Consider the law of cosines: $a^2 = b^2 + c^2 - 2bc\cos A$. That is,

$$\frac{a^2 - b^2 - c^2}{-2bc} = \cos A$$

Thus,

$$1 - \cos A = 1 + \frac{a^2 - b^2 - c^2}{2bc} = \frac{a^2 - (b^2 - 2bc + c^2)}{2bc}$$

$$= \frac{a^2 - (b - c)^2}{2bc} = \frac{(a - b + c)(a + b - c)}{2bc}$$

Section 7.3, page 403

2. 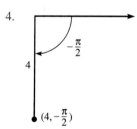 $(6, 45°)$

4.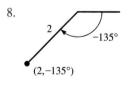

6. 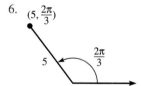 $\left(5, \frac{2\pi}{3}\right)$

8.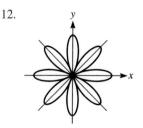

10.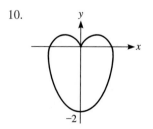

12.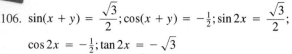

14.

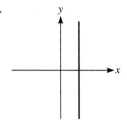

16.

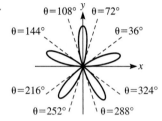

18.

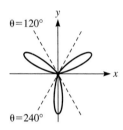

20.

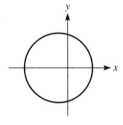

22.

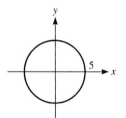

24.

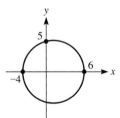

26.

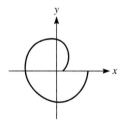

28.

30. $(9\sqrt{2}, 135°)$
34. $(4, 330°)$
38. $(2, 210°)$
42. $r = 2\csc\theta$
46. $r = \dfrac{4}{\cos\theta + \sin\theta}$
50. $r = \dfrac{5}{3\cos\theta + 4\sin\theta}$
54. $x = 2$

32. $(1, 315°)$
36. $(5, 143.1°)$
40. $(\sqrt{13}, 123.7°)$
44. $r = 25\cot\theta\csc\theta$
48. $r = 5\sec\theta$
52. $x^2 + y^2 - 2x - 3y = 0$
56. $x^2 - 4y - 4 = 0$

58. $x + y = 2$

60. $x^2 + y^2 + 2y + 2x = 0$

62.

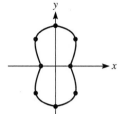

64.

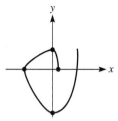

66. Points: $(2, k\pi)$, $k = 0, \pm1, \pm2, \dots$

Section 7.4, page 409

2. $\sqrt{29}$ 4. 5 6. $\sqrt{5}$ 8. $\sqrt{34}$

10. Choose $z = 1 + 3i$, $w = 1 - 3i$.
Then $|z| - |w| = 0$, $|z + w| = 2$. Therefore $|z| - |w| \neq |z + w|$

12. Let $z = a + bi$; then $\bar{z} = a - bi$.
Now $|z| = \sqrt{a^2 + b^2} = |\bar{z}|$. So $|z\bar{z}| = |(a + bi)(a - bi)|$
$= |a^2 + b^2| = a^2 + b^2 = |z| \cdot |\bar{z}|$

14. $3\sqrt{2}\left(\cos\dfrac{\pi}{4} + i\sin\dfrac{\pi}{4}\right)$ 16. $5\left(\cos\dfrac{\pi}{2} + i\sin\dfrac{\pi}{2}\right)$

18. $10(\cos\pi + i\sin\pi)$

20. $2\sqrt{2}\left(\cos\dfrac{7\pi}{4} + i\sin\dfrac{7\pi}{4}\right)$

22. $6\sqrt{2}\left(\cos\dfrac{3\pi}{4} + i\sin\dfrac{3\pi}{4}\right)$

24. $2\left(\cos\dfrac{4\pi}{3} + i\sin\dfrac{4\pi}{3}\right)$ 26. $2\left(\cos\dfrac{5\pi}{6} + i\sin\dfrac{5\pi}{6}\right)$

28. $8\left(\cos\dfrac{\pi}{3} + i\sin\dfrac{\pi}{3}\right)$ 30. $4(\cos 0 + i\sin 0)$

32. $2\left(\cos\dfrac{3\pi}{2} + i\sin\dfrac{3\pi}{2}\right)$

34.

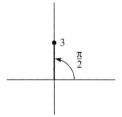

36.

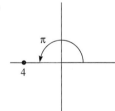

38.

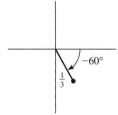

40.

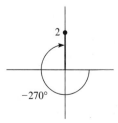

42.

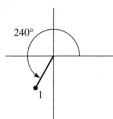

44.

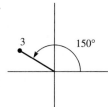

46.

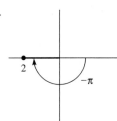

48.

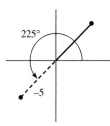

10. $\dfrac{\sqrt{2}}{2} + \dfrac{\sqrt{2}}{2}i$ 12. $2^{10}\left(-\frac{1}{2} + \frac{\sqrt{3}}{2}i\right)$

14. -2^{18} 16. $3^6\left(\frac{1}{2} - \frac{\sqrt{3}}{2}i\right)$

18. $16\left(\frac{1}{2} - \frac{\sqrt{3}}{2}i\right) = 8 - 8\sqrt{3}i$

20. 3^{20} 22. $2; -1 + \sqrt{3}i; -1 - \sqrt{3}i$

24. $t_j = 2^{1/6}\left(\cos\left[\dfrac{(5\pi/6) + 2\pi j}{6}\right] + i\sin\left[\dfrac{(5\pi/6) + 2\pi j}{6}\right]\right)$,
$j = 0, 1, 2, \ldots, 5$

26. $\dfrac{\sqrt{2}}{2} + \dfrac{\sqrt{2}}{2}i; -\dfrac{\sqrt{2}}{2} + \dfrac{\sqrt{2}}{2}i; -\dfrac{\sqrt{2}}{2} - \dfrac{\sqrt{2}}{2}i; \dfrac{\sqrt{2}}{2} - \dfrac{\sqrt{2}}{2}i$

28. $t_j = (\sqrt{2})^{1/5}\left(\cos\left[\dfrac{(7\pi/4) + 2\pi j}{5}\right]\right.$
$\left. + i\sin\left(\left[\dfrac{(7\pi/4) + 2\pi j}{5}\right]\right)\right)$, $j = 0, 1, 2, 3, 4$

30. $-2\sqrt{2} + 2\sqrt{2}i; 2\sqrt{2} - 2\sqrt{2}i$

32. $1 + \sqrt{3}i; -1 - \sqrt{3}i$

34. $t_j = 2\left[\cos\left(\dfrac{150° + 360°j}{5}\right) + i\sin\left(\dfrac{150° + 360°j}{5}\right)\right]$,
$j = 0, 1, 2, 3, 4$

36. $\dfrac{5\sqrt{2}}{2} + \dfrac{5\sqrt{2}}{2}i; 5[\cos(165°) + i\sin(165°)]$;
$[\cos(285°) + i\sin(285°)]$

38. $5\sqrt{3} - 5i; -5\sqrt{3} + 5i$

40. $2; 1 + \sqrt{3}i; -1 + \sqrt{3}i; -1 - \sqrt{3}i; 1 - \sqrt{3}i, -2$

42. $t_j = 3\left[\cos\left(\dfrac{\pi + 2\pi j}{5}\right) + i\sin\left(\dfrac{\pi + 2\pi j}{5}\right)\right]$,
$j = 0, 1, 2, 3, 4$

44. $t_j = 2\left(\cos\left[\dfrac{(5\pi/6) + 2\pi j}{3}\right] + i\sin\left[\dfrac{(5\pi/6) + 2\pi j}{3}\right]\right)$,
$j = 0, 1, 2$

46. $t_j \cos\left[\dfrac{(\pi/2) + 2\pi j}{5}\right] + i\sin\left[\dfrac{(\pi/2) + 2\pi j}{5}\right]$,
$j = 0, 1, 2, 3, 4$

48. $\sqrt{2} + \sqrt{2}i; -\sqrt{2} + \sqrt{2}i; -\sqrt{2} - \sqrt{2}i; \sqrt{2} - \sqrt{2}i$

50. $1; \frac{1}{2} + \frac{\sqrt{3}}{2}i; -\frac{1}{2} + \frac{\sqrt{3}}{2}i; -1; -\frac{1}{2} - \frac{\sqrt{3}}{2}i; \frac{1}{2} - \frac{\sqrt{3}}{2}i$

52. $t_j = 2^{1/5}\left(\cos\left[\dfrac{(7\pi/4) + 2\pi j}{5}\right] + i\sin\left[\dfrac{(7\pi/4) + 2\pi j}{5}\right]\right)$,
$j = 0, 1, 2, 3, 4$

54. $t_j = 2^{1/3}\left(\cos\left[\dfrac{\pi/3 + 2\pi j}{3}\right] + i\sin\left[\dfrac{\pi/3 + 2\pi j}{3}\right]\right)$,
$j = 0, 1, 2$

56. $-\dfrac{1}{512} + \dfrac{\sqrt{3}}{512}i$

50. $zw = 6(\cos 135° + i\sin 135°)$;
$\dfrac{z}{w} = \frac{2}{3}(\cos 225° + i\sin 225°)$

52. $zw = 12\left(\cos\frac{\pi}{4} + i\sin\frac{\pi}{4}\right)$; $\dfrac{z}{w} = \frac{4}{3}\left(\cos\frac{5\pi}{4} + i\sin\frac{5\pi}{4}\right)$

54. $zw = 6(\cos\pi + i\sin\pi)$; $\dfrac{z}{w} = \frac{3}{2}\left(\cos\frac{5\pi}{3} + i\sin\frac{5\pi}{3}\right)$

56. $zw = 32\left(\cos\frac{7\pi}{4} + i\sin\frac{7\pi}{4}\right)$; $\dfrac{z}{w} = 2\left(\cos\frac{\pi}{4} + i\sin\frac{\pi}{4}\right)$

58. $zw = 12(\cos 195° + i\sin 195°)$;
$\dfrac{z}{w} = \frac{1}{3}(\cos 285° + i\sin 285°)$

60. $zw = 20(\cos 150° + i\sin 150°)$; $\dfrac{z}{w} = 5(\cos 30° + i\sin 30°)$

62. $zw = 6\left(\cos\frac{13\pi}{12} + i\sin\frac{13\pi}{12}\right)$;
$\dfrac{z}{w} = \frac{1}{6}\left(\cos\frac{5\pi}{12} + i\sin\frac{5\pi}{12}\right)$

64. -5 66. $-\frac{3}{2} + \frac{3\sqrt{3}}{2}i$ 68. $1 - \sqrt{3}i$

70. $-1 + i$ 72. $-3\sqrt{3} + 3i$ 74. $-2i$

76. $\left(\frac{3}{2} + 3\sqrt{2}\right) + \left(\frac{3\sqrt{3}}{2} - 3\sqrt{2}\right)i$

78. $(-6\sqrt{2} - 3) - 6\sqrt{2}i$

80. $zw = 2.071 - 7.727i; \dfrac{z}{w} = 0.41 + 0.287i$

82. $zw = -14.863 - 13.383i; \dfrac{z}{w} = -4.973 + 0.523i$

Section 7.5, page 415

2. 1

4. $-2^{30}i$

6. $-648 + 648\sqrt{3}i$

8. $-\frac{1}{2} + \frac{\sqrt{3}}{2}i$

58. $16 + 16i$

60. $64 + 64i$

62. $-54 + 54\sqrt{3}i$

64. $-\dfrac{1}{2} - \dfrac{\sqrt{3}}{2}i$

Section 7.6, page 422

2.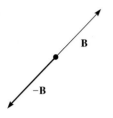

4.

6.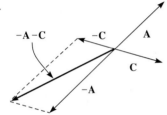

8.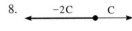

10.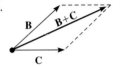

12. *a.* $2\sqrt{2}$; *b.* $-2i - 2j$ 14. *a.* $5\sqrt{2}$; *b.* $-5i + 5j$

16. *a.* 8; *b.* $4\sqrt{3}i + 4j$ 18. *a.* 4; *b.* $2i - 2\sqrt{3}j$

20. *a.* 13; *b.* $-5i + 12j$ 22. *a.* 10; *b.* $-6i - 8j$

24. *a.* $\sqrt{34}$; *b.* $5i + 3j$ 26. 100.37 lbs; 111.47 lbs

28. 17.36 lbs   30. 50 lbs

32. 6815.66 lbs   34. 44°

36. Plane is blown 9.1° west; speed = 253.2 mph

38. Course of ship is 9.5° south of east; 18.2 mph

40. 172.8°   42. Bearing is 230°, speed 305 knots

44. Speed = 40.22 km/hr

46. Bearing is 85.07°, speed = 174.6 mph

48. $\langle 6, -12 \rangle$ 50. $\langle -10, 10 \rangle$ 52. $\langle 19, -14 \rangle$

54. $\langle -8, 0 \rangle$ 56. $\langle 5, 14 \rangle$ 58. -2026

64. yes   66. no   68. $\cos^{-1} 0 = \dfrac{\pi}{2}$

70. $\cos^{-1}\left(-\dfrac{2}{\sqrt{5}}\right)$

Section 7.7, Cumulative Review Exercises, page 424

2. $b = 7.1; \alpha = 105°; a = 13.7$

4. $c = 24.79; \beta = 23.79°; \alpha = 126.21°$

6. 9.29 ft   8. 17.1 m

10. $d_2 = \sqrt{B^2 + 10^2 - 20B\cos 35°}$,

$\quad d_1 = \sqrt{B^2 + 10^2 - 20B\cos 145°}$

12. 744.6 miles

14. If $\alpha = 48.59°$, $\beta = 101.42°$ then $b = 39.21$.
If $\alpha = 131.41°$, $\beta = 18.59°$ then $b = 12.75$.

16. 96   18. 64.95

20. $3\sqrt{2}$ 22. 2

24. *a.* $3\sqrt{2}\left(\cos\dfrac{7\pi}{4} + i\sin\dfrac{7\pi}{4}\right)$; *b.* $3\sqrt{2}$; *c.* $\dfrac{7\pi}{4}$

26. *a.* $5\left(\cos\dfrac{\pi}{2} + i\sin\dfrac{\pi}{2}\right)$; *b.* 5; *c.* $\dfrac{\pi}{2}$

28.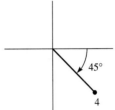

30.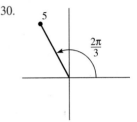

32. $zw = 6(\cos 30° + i\sin 30°)$,

$\quad \dfrac{z}{w} = \dfrac{2}{3}(\cos 270° + i\sin 270°)$

34. $zw = 8\left[\cos\left(\dfrac{5\pi}{6}\right) + i\sin\left(\dfrac{5\pi}{6}\right)\right]$,

$\quad \dfrac{z}{w} = 2\left[\cos\dfrac{\pi}{2} + i\sin\dfrac{\pi}{2}\right]$

36. $\dfrac{\sqrt{3}}{2} - \dfrac{3}{2}i$ 38. $\sqrt{3} + i$ 40. 1

42. $-27i$ 44. $-648 - 648\sqrt{3}i$

46. $-2\sqrt{2} - 2\sqrt{2}i; 2\sqrt{2} + 2\sqrt{2}i$

48. $t_j = 2\left(\cos\left[\dfrac{(\pi/3) + 2\pi j}{4}\right] + i\sin\left[\dfrac{(\pi/3) + 2\pi j}{4}\right]\right)$,
$\quad j = 0, 1, 2, 3$

50. $t_j = 5\left(\cos\left[\dfrac{(3\pi/4) + 2\pi j}{3}\right]\right.$
$\quad\quad\quad \left. + i\sin\left[\dfrac{(3\pi/4) + 2\pi j}{3}\right]\right), \quad j = 0, 1, 2$

52. $2; -1 + \sqrt{3}i; -1 - \sqrt{3}i$

54. $t_j = 2\left(\cos\left[\dfrac{(5\pi/6) + 2\pi j}{4}\right] + i\sin\left[\dfrac{(5\pi/6) + 2\pi j}{4}\right]\right)$,
$\quad j = 0, 1, 2, 3$

56.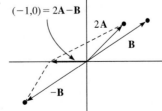

58.

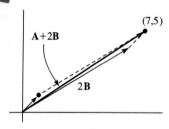

$A+2B$ (7,5) 2B

60. *a.* 2; *b.* $-i - \sqrt{3}j$

62. *a.* $3\sqrt{2}$;
    b. $3i - 3j$

64. 203.96 mph;
    Therefore, bearing $-78.7°$.

66. 48.7 lbs   68. $\langle 1, 0 \rangle$ 70. $\langle 8, 5 \rangle$

72.

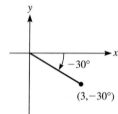

$-30°$ (3,−30°)

74.

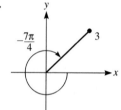

$-\frac{7\pi}{4}$ 3

76.

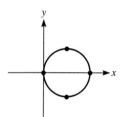

78.

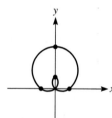

80. $\left(2, \dfrac{5\pi}{3}\right)$ 82. $\left(4\sqrt{2}, \dfrac{5\pi}{4}\right)$ 84. $r = 6\csc\theta$

86. $r^2 + 2r\sin\theta = 0$ 88. $x = 2$

90. $x\sqrt{x^2 + y^2} = y$ 92. $4y^2 - 9x^2 = 36$

CHAPTER 8

Section 8.1, page 436

2. $x = -\frac{14}{3}; y = \frac{9}{4}$ 4. $x = \frac{26}{11}; y = -\frac{28}{11}$

6. $x = \frac{5}{3}; y = \frac{2}{5}$ 8. $x = \frac{5}{2}; y = -\frac{3}{2}$

10. $(3, 0); \left(-\frac{9}{5}, \frac{12}{5}\right)$ 12. $x = 2; y = 2$ 14. $y = 1; x = 1$

16. $(\sqrt{3}, 1); (\sqrt{3}, -1); (-\sqrt{3}, 1); (-\sqrt{3}, -1)$

18. No real solution   20. 200 @ \$5; 75 @ \$2.50

22. $y = \frac{140}{29}x - \frac{530}{29}$

24. \$8,000 @ 8.5%, \$9,000 @ 9.5%

26. No solution   28. $\pm 5\sqrt{5}$ 30. 125

32. 12 paintings   34. 45 mph

36. *a.* 2.85 hr
    b. 6.9 mph (avg. speed of the jet stream);
         511.2 mph (avg. speed of plane)

Section 8.2, page 440

2. $x = 2, y = 1$ 4. $x = 0, y = -3$

6. $x = -\frac{13}{15}, y = \frac{2}{5}$ 8. $x = 5, y = -4$

10. $x = -8, y = 14$ 12. $x = 2, y = 3$

14. Width: 15 m; length: 20 m

16. 100 @\$4.50, 350 @\$6   18. 4 hrs, 10 hrs

20. 40 mph   22. $x = \dfrac{a}{b}, y = -\dfrac{a}{b}$

Section 8.3, page 450

2. $\left(-\frac{67}{4}, \frac{31}{2}, 15\right)$ 4. $(-8, 7, 2)$ 6. $(1, 0, -4)$

8. $\begin{bmatrix} 0 & -1 & 5 \\ -1 & 2 & 3 \\ 4 & 1 & 0 \end{bmatrix}$ 10. $\begin{bmatrix} -8 & -2 & 0 \\ -1 & 2 & 3 \\ 0 & -1 & 5 \end{bmatrix}$

12. $\begin{bmatrix} 4 & 1 & 0 \\ -2 & 0 & 26 \\ 0 & -1 & 5 \end{bmatrix}$ 14. $\begin{bmatrix} 7 & 5 & -2 \\ -1 & 2 & 3 \\ 0 & -1 & 5 \end{bmatrix}$

16. $\begin{bmatrix} 1 & -1 & 5 & -6 \\ 0 & 1 & -(\frac{8}{3}) & (\frac{14}{3}) \\ 0 & 0 & 1 & -1 \end{bmatrix}$

18. $\begin{vmatrix} 1 & 3 & 9 \\ 0 & 1 & \frac{11}{3} \end{vmatrix}$ 20. $\begin{bmatrix} 1 & 1 & 1 & 2 \\ 0 & 1 & 1 & 1 \\ 0 & 0 & 1 & -1 \end{bmatrix}$

22. $\begin{bmatrix} 1 & \frac{7}{2} & \frac{1}{2} \\ 0 & 1 & 1 \end{bmatrix}$ 24. $\begin{bmatrix} 1 & (\frac{1}{2}) & (\frac{3}{2}) & 6 \\ 0 & 1 & (\frac{6}{5}) & (\frac{14}{5}) \\ 0 & 0 & 1 & 4 \end{bmatrix}$

26. $\begin{bmatrix} 1 & 2 & -3 & 4 \\ 0 & 1 & -(\frac{8}{3}) & (\frac{7}{3}) \\ 0 & 0 & 1 & -(\frac{1}{2}) \end{bmatrix}$

28. $x = \frac{1}{18}, y = -\frac{3}{2}, z = \frac{11}{18}$

30. $x = 0, y = 1, z = 2$

32. $x = \frac{2}{11}z + \frac{12}{11}, y = \frac{7}{11}z + \frac{20}{11}, z = $ any number

34. $x = -2, y = 1, z = 3$

36. $x = 0, y = 2, z = -2, w = 1$ 38. No solution

40. $x = -\frac{2}{3}z + \frac{1}{6}, y = \frac{1}{3}z + \frac{1}{6}, z = $ any number

42. $(13, 8, 2, -5)$

44. $x = -\frac{2}{11}w + \frac{81}{11}, y = \frac{1}{22}w + \frac{10}{11}, z = \frac{4}{11}w - \frac{8}{11},$
    $w = $ any number

46. 150 mph, 30 mph   48. fifteen $33\frac{1}{3}$s, thirty 45s

Section 8.4, page 458

2. $y < 3x - 10$ 4. $\dfrac{x^2}{36} + \dfrac{y^2}{4} \le 1$

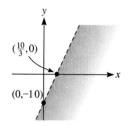

$\left(\frac{10}{3}, 0\right)$ (0,−10)

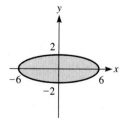

6. $x \geq -y^2 - 4$

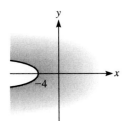

8. $y \leq \dfrac{16}{x}$

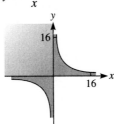

26. $\begin{cases} y \geq 2x - 6 \\ y > 6 - x \\ x \geq -1 \end{cases}$

28. $\begin{cases} y \leq 2x + 1 \\ y \geq -2x + 1 \\ x \leq 2 \end{cases}$

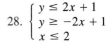
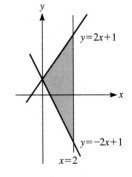

10. $(x + 1)^2 \geq (y + 1)^2$

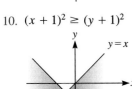

12. $x < y$

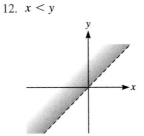

14. $y \geq -1 - x^2$

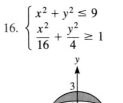

16. $\begin{cases} x^2 + y^2 \leq 9 \\ \dfrac{x^2}{16} + \dfrac{y^2}{4} \geq 1 \end{cases}$

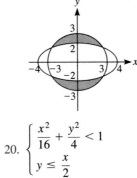

30. $\begin{cases} x \geq 0 \\ y \geq 0 \\ y \leq -2x + 4 \\ y \geq \frac{2}{3}x - 2 \end{cases}$

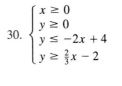

18. $\begin{cases} x^2 + y^2 \leq 36 \\ -4 \leq x \leq 4 \end{cases}$

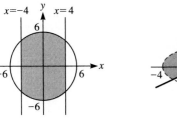

20. $\begin{cases} \dfrac{x^2}{16} + \dfrac{y^2}{4} < 1 \\ y \leq \dfrac{x}{2} \end{cases}$

32. $\begin{cases} x \geq 10 \\ y \geq 20 \\ y \leq -\frac{2}{3}x + \frac{100}{3} \\ y \leq -\frac{5}{4}x + 50 \end{cases}$

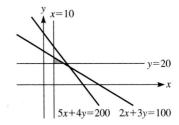

22. $\begin{cases} y \leq -(x - 3)^2 + 2 \\ y \geq -\frac{3}{5}x - 4 \end{cases}$

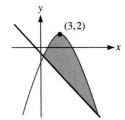

24. $\begin{cases} y > |x + 2| \\ |x| \leq 3 \end{cases}$

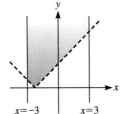

34. $\begin{cases} x \geq 0 \\ y \geq 0 \\ y \geq -\frac{4}{5}x + \frac{22}{25} \\ y \geq -\frac{4}{3}x + \frac{6}{5} \end{cases}$

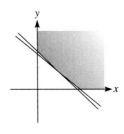

36. $\begin{cases} y \geq -2x + 4 \\ y \geq -\frac{3}{4}x + \frac{11}{4} \\ y \geq -\frac{1}{2}x + 2 \\ x, y \geq 0 \end{cases}$

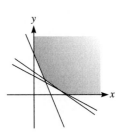

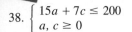

Answers to Even-Numbered Exercises

38. $\begin{cases} 15a + 7c \le 200 \\ a, c \ge 0 \end{cases}$

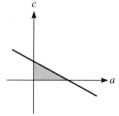

Section 8.5, page 461

2.

	$f(x, y) = x + 4y$
(0, 7)	28 ← max
(0, 0)	0 ← min
(6, 2)	14
(5, 4)	21

4.

	$F(x, y) = 2x - 3y$
(0, 5)	-15 ← min
(5, 0)	5 ← max
$\left(\frac{15}{8}, \frac{15}{8}\right)$	$-\frac{15}{8}$

6.

	$f(x, y) = -3x + y$
(6, 0)	-18
(8, 0)	-24 ← min
(6, 2)	-16
$\left(\frac{14}{3}, \frac{8}{3}\right)$	$-\frac{34}{3}$ ← max

8.

	$f(x, y) = x + 5y$
(1, 8)	41 ← max
(6, 7)	41 ← max
(4, 5)	29
(2, 3)	17 ← min

10.

	$f(x, y) = 3x + 8y$
(0, 9)	72 ← max
(4, 4)	44
(3, 0)	9
(0, 0)	0 ← min

12.

	$f(x, y) = 5x + 3y$
(3, 6)	33
(6, 5)	45 ← max
(8, 1)	43
(4, 1)	23
(1, 4)	17 ← min

14. Southeast Flint → Brighton   15 cars
Southeast Flint → Farmington   20 cars
Southwest Flint → Brighton   15 cars
Southwest Flint → Farmington   0 cars
Minimum cost = $695

16. 800 30-watt receivers, 300 50-watt receivers, max profit = $125,000

	$P(x, y) = 100x + 150y$
(0, 0)	0
(0, 700)	105,000
(300, 600)	120,000
(800, 300)	125,000 ← max
(1100, 0)	110,000

18.

	$P(x, y) = .08x + .075y$
(6000, 0)	$ 480
(6000, 30,000)	$2730
(10,000, 30,000)	$3050
(22,000, 18,000)	$3110
(22,000, 0)	$1760

Max: Invest $22,000 in stocks, $18,000 in bonds for an annual income of $3110.

20.

	$P(x, y) = 15x + 24y$
(5, 4)	171
$\left(\frac{5}{2}, 4\right)$	133.5
$\left(7, \frac{2}{5}\right)$	114.6 ← min
(7, 4)	171

22.

	$C(x, y) = 350x + 200y$
(0, 40)	8000
(4, 24)	6200 ← min
(10, 15)	6500
(40, 0)	14,000

Min: Use 4 carloads of A, 24 of B at a minimum cost of $6200.

24.

	$P(x, y) = 40x + 30y$
(0, 0)	0
(0, 240)	$7200
(40, 160)	$8000 ← max
(160, 0)	$6400

Max: 80 acres of corn, 160 acres of oats.

Section 8.6, Cumulative Review Exercises, page 464

2. $y = 5, x = -2$ 4. $(3, 0), \left(-\frac{9}{5}, \frac{12}{5}\right)$
6. $(-1, 2, -3)$
8. 835 @$20, 585 @$16
10. The number is 425.   12. $x = 0, y = 2, z = 3$
14. $(-1, 2)$ 16. $x = -1, y = -\frac{1}{2}$
18. $y = 20, x = 14$; dimensions: 14 m by 20 m
20.   22.

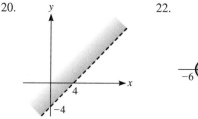

24.

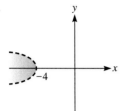

26.

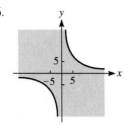

28.

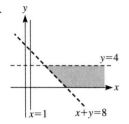

30.

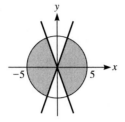

32.

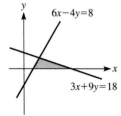

34.

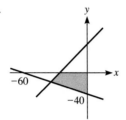

36.

	$f(x, y) = 4x + y$
$(2, 1)$	$9 \leftarrow$ min
$(1, 9)$	13
$(3, 5)$	$17 \leftarrow$ max
$(2, 7)$	15

38.

	$F(x, y) = \frac{3}{2}x + y$
$\left(3, \frac{1}{2}\right)$	$5 \leftarrow$ min
$(6, 2)$	$11 \leftarrow$ max
$(4, 0)$	6
$(6, 0)$	9

40.

	$f(x, y) = 3x + y$
$(1, 8)$	11
$(6, 9)$	27
$(9, 4)$	$31 \leftarrow$ max
$(5, 1)$	16
$(2, 2)$	$8 \leftarrow$ min

42. $y = 5, \quad x = 180$

	$P(x, y) = 10x + 25y$
$(0, 0)$	0
$(0, 50)$	1250
$(180, 5)$	$\$1925 \leftarrow$ max
$(187.5, 0)$	$\$1875$

Max: 180 car tires, 5 tractor tires

CHAPTER 9

Section 9.1, page 471

2. $\dfrac{3}{x - 2} + \dfrac{1}{x + 1}$

4. $\dfrac{1}{x} + \dfrac{3}{x + 2} - \dfrac{2}{x - 1}$

6. $\dfrac{2}{x - 3} - \dfrac{4}{3x - 2}$

8. $\dfrac{5}{2x - 1} + \dfrac{1}{x - 2} - \dfrac{2}{(x - 2)^2}$

10. $\dfrac{-2}{2x - 3} + \dfrac{3}{x + 2}$

12. $2x + 1 + \dfrac{\frac{12}{5}}{x - 3} - \dfrac{\frac{12}{5}}{x + 2}$

14. $2x + 1 + \dfrac{4}{2x - 1} + \dfrac{3}{x - 4}$

16. $\dfrac{5}{2x - 1} + \dfrac{3}{x + 1}$

18. $-\dfrac{4}{x} + \dfrac{1}{x + 1} - \dfrac{3}{(x + 1)^2}$

20. $\dfrac{6}{x - 4} + \dfrac{3}{x - 2} - \dfrac{4}{x + 1}$

22. $\dfrac{1}{x} - \dfrac{6}{x - 4} - \dfrac{4}{x + 1}$

24. $\dfrac{-1}{x + 1} + \dfrac{2}{(x + 1)^2} + \dfrac{3x + 1}{x^2 + 3}$

26. $\dfrac{-3}{x - 2} + \dfrac{2}{x + 2} + \dfrac{4}{x + 1}$

28. $\dfrac{1}{x - 2} + \dfrac{3}{(x - 2)^2} - \dfrac{2}{(x - 2)^3}$

30. $\dfrac{2}{2x - 3} - \dfrac{5}{x + 4}$

32. $\dfrac{2}{x - a} + \dfrac{3}{x^2 + ax + a^2}$

34. $y = \dfrac{3}{x - 2} + \dfrac{2}{x + 1}$

36. $y = \dfrac{-3}{x + 2} + \dfrac{3}{x - 1}$

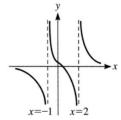

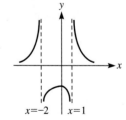

38. $x = 2, y = -2, z = -3$

Section 9.2, page 478

2. $\begin{bmatrix} 4 & -1 \\ 1 & 3 \end{bmatrix}$

4. $\begin{bmatrix} 12 & 6 \\ -3 & 0 \end{bmatrix}$

6. $\begin{bmatrix} 3 & -1 \\ -3 & -6 \end{bmatrix}$

8. $\begin{bmatrix} 6 & 6 \\ 8 & 28 \end{bmatrix}$

10. $\begin{bmatrix} 3 & -1 \\ 0 & 1 \end{bmatrix}$

12. $\begin{bmatrix} 0 & 5 \\ 0 & -4 \end{bmatrix}$

14. $\begin{bmatrix} -6 & 3 \\ 0 & 0 \end{bmatrix}$

16. $\begin{bmatrix} 0 & 0 \\ 0 & 0 \end{bmatrix}$

18. $\begin{bmatrix} -7 & -2 \\ 2 & 0 \end{bmatrix}$

20. $\begin{bmatrix} 10 & 30 \\ -7 & -21 \end{bmatrix}$

22. $\begin{bmatrix} -15 & -16 & 3 \\ -1 & 0 & 9 \\ 7 & 6 & 12 \end{bmatrix}$

24. $\begin{bmatrix} -9 \\ 19 \end{bmatrix}$

26. $\begin{bmatrix} 11 & 9 & -12 \\ 1 & 2 & -1 \end{bmatrix}$

28. $\begin{bmatrix} -4 & 0 & 1 \\ 28 & 0 & -7 \end{bmatrix}$

30. $\begin{bmatrix} 5 & 6 \\ -3 & -2 \end{bmatrix}$

32. *IF* is undefined.

34. $\begin{bmatrix} 1 \\ 7 \end{bmatrix}$

36. $A \times B$ is 4×2.

38. AB is undefined.

40. $A \times B$ is 5×3.

42. $x = 2, y = 7, z = 3, w = 3$

44. $x = 2, y = -3, w = 8, z = \frac{5}{3}$

46. Prove $A + B = B + A$.

$A + B = \begin{bmatrix} a_{11} & a_{12} \\ a_{21} & a_{22} \end{bmatrix} + \begin{bmatrix} b_{11} & b_{12} \\ b_{21} & b_{22} \end{bmatrix}$

$\quad = \begin{bmatrix} a_{11} + b_{11} & a_{12} + b_{12} \\ a_{21} + b_{21} & a_{22} + b_{22} \end{bmatrix}$

$\quad = \begin{bmatrix} b_{11} & b_{12} \\ b_{21} & b_{22} \end{bmatrix} + \begin{bmatrix} a_{11} & a_{12} \\ a_{21} & a_{22} \end{bmatrix} = B + A$

48. Prove $(A + B) + C = A + (B + C)$.

$(A + B) + C = \left(\begin{bmatrix} a_{11} & a_{12} \\ a_{21} & a_{22} \end{bmatrix} + \begin{bmatrix} b_{11} & b_{12} \\ b_{21} & b_{22} \end{bmatrix} \right) + \begin{bmatrix} c_{11} & c_{12} \\ c_{21} & c_{22} \end{bmatrix}$

$\quad = \begin{bmatrix} a_{11} + b_{11} + c_{11} & a_{12} + b_{12} + c_{12} \\ a_{21} + b_{21} + c_{21} & a_{22} + b_{22} + c_{22} \end{bmatrix}$

$A + (B + C) = \begin{bmatrix} a_{11} & a_{12} \\ a_{21} & a_{22} \end{bmatrix} + \left(\begin{bmatrix} b_{11} & b_{12} \\ b_{21} & b_{22} \end{bmatrix} + \begin{bmatrix} c_{11} & c_{12} \\ c_{21} & c_{22} \end{bmatrix} \right)$

$\quad = \begin{bmatrix} a_{11} + b_{11} + c_{11} & a_{12} + b_{12} + c_{12} \\ a_{21} + b_{21} + c_{21} & a_{22} + b_{22} + c_{22} \end{bmatrix}$

Section 9.3, page 488

2. 14   4. 2   6. -1 8. 10

10. -40 12. -9 14. -2 16. -3

18. 12   20. 7840   22. 20   24. -4

26. -88 28. $\left(-\frac{37}{7}, \frac{40}{7} \right)$ 30. $(1, 1)$

32. Cramer's rule not applicable.

34. $D = \begin{vmatrix} 3 & -6 \\ -2 & 4 \end{vmatrix} = 0$

    Cramer's rule not applicable.

36. $\left(\frac{1}{18}, -\frac{3}{2}, \frac{11}{18} \right)$ 38. $(0, 1, 2)$

40. $(1, -2, 3, -4)$ 42. $(13, 8, 2, -5)$

44. $\begin{vmatrix} x & y & 1 \\ 1 & 1 & 1 \\ 0 & -1 & 1 \end{vmatrix} = 0$ 46. $\begin{vmatrix} x & y & 1 \\ 10 & 0 & 1 \\ 0 & 8 & 1 \end{vmatrix} = 0$

48. Collinear   50. Not collinear   52. Area $= 33$

56. $x = 30$ 58. $m = 1, -5$ 60. 0

Section 9.4, page 495

2. $A^{-1} = \begin{bmatrix} -5 & 8 \\ 2 & -3 \end{bmatrix}$ 4. $A^{-1} = \begin{bmatrix} \frac{1}{6} & \frac{2}{3} \\ \frac{1}{6} & -\frac{1}{3} \end{bmatrix}$

6. $A^{-1} = \begin{bmatrix} 5 & 3 \\ 3 & 2 \end{bmatrix}$

8. $\begin{vmatrix} 1 & -4 & 8 \\ 1 & -3 & 2 \\ 2 & -7 & 10 \end{vmatrix} = 0.$ Inverse does not exist.

10. $A^{-1} = \begin{bmatrix} \frac{7}{4} & \frac{5}{2} & 3 \\ -\frac{1}{4} & -\frac{1}{2} & 0 \\ -\frac{1}{4} & -\frac{1}{2} & -1 \end{bmatrix}$

12. $A^{-1} = \begin{bmatrix} \frac{2}{3} & 0 & \frac{1}{6} \\ \frac{1}{3} & 0 & \frac{1}{3} \\ -2 & 1 & -\frac{1}{2} \end{bmatrix}$

14. $A^{-1} = \begin{bmatrix} 1 & 2 & 3 & 4 \\ 0 & 1 & 3 & -5 \\ 0 & 0 & 1 & -2 \\ 0 & 0 & 0 & 1 \end{bmatrix}$

16. Yes   18. No   20. Yes

22. $x = -1, y = 3$ 24. $x = 3, y = 0$

26. $x = -1, y = -\frac{2}{3}$ 28. $x = 2, y = 3$

30. $x = 124, y = 14, z = -51$

32. $x = -\dfrac{13}{4(k - 1)}, \quad y = -\dfrac{13}{4(k - 1)} - \dfrac{5}{2}$

    $z = \dfrac{17k - 4}{8(k - 1)} \quad (k \neq 1)$

34. $A = \begin{bmatrix} x & 0 & 0 \\ 0 & y & 0 \\ 0 & 0 & z \end{bmatrix} \quad (xyz \neq 0)$

    $A^{-1} = \begin{bmatrix} \frac{1}{x} & 0 & 0 \\ 0 & \frac{1}{y} & 0 \\ 0 & 0 & \frac{1}{z} \end{bmatrix}$

36. Let $A = \begin{bmatrix} a & b \\ c & d \end{bmatrix}$. If $\Delta = |A| = 0$, then A^{-1} does not

    exist because $A^{-1} = \begin{bmatrix} \frac{d}{\Delta} & -\frac{b}{\Delta} \\ -\frac{c}{\Delta} & \frac{a}{\Delta} \end{bmatrix}$

Section 9.5, Cumulative Review Exercises, page 498

2. $\dfrac{31}{y - 2} - \dfrac{49}{2y - 3}$ 4. $\dfrac{6}{x - 4} + \dfrac{3}{x - 2} - \dfrac{4}{x + 1}$

6. $EF = \begin{bmatrix} 7 \\ -15 \end{bmatrix}$ 8. $2E - 3A = \begin{bmatrix} -6 & 2 \\ -5 & 19 \end{bmatrix}$

10. $CH = \begin{bmatrix} 10 & 29 & 1 \\ 15 & 37 & 4 \end{bmatrix}$ 12. $CD = \begin{bmatrix} 7 & 17 & -4 \\ 9 & 22 & 0 \end{bmatrix}$

14. FG undefined.   16. AB is 3×3.

18. AB is undefined.   20. $x = 4, y = 2, w = 1$

22. -5 24. -6 26. 11   28. 152

30. -7 32. -210 34. $x = 23$

36. $A^{-1} = \begin{bmatrix} \frac{2}{11} & \frac{3}{11} \\ -\frac{1}{11} & \frac{4}{11} \end{bmatrix}$

38. $A^{-1} = \begin{bmatrix} \frac{1}{3} & 0 & \frac{1}{3} \\ -\frac{2}{5} & \frac{2}{5} & \frac{1}{5} \\ \frac{2}{15} & \frac{1}{5} & -\frac{1}{15} \end{bmatrix}$

40. A^{-1} doesn't exist because $|A| = 0$.

42. Not inverses **44.** $x = -\frac{1}{3}, y = 2$

46. $x = 3, y = -3, z = -2$

CHAPTER 10

Section 10.1, page 509

2. 1 **4.** 120 **6.** 56 **8.** 210

10. 1 **12.** 220 **14.** $56a^3b^5$

16. $-4^9 \cdot 715y^4$ **18.** $\frac{42}{3125} \cdot \frac{1}{a^2}$ **20.** $70 \cdot 5^4 7^4 a^8 b^{12}$

22. $a^4 + 4a^3b + 6a^2b^2 + 4ab^3 + b^4$

24. $4096a^6 - 6144a^5b + 3840a^4b^2 - 1280a^3b^3 + 240a^2b^4 - 24ab^5 + b^6$

26. $292 + 112\sqrt{6}$

28. $\frac{a^{12}}{b^6} - \frac{6a^{10}}{b^2} + 15a^8b^2 - 20a^6b^6 + 15a^4b^{10} - 6a^2b^{14} + b^{18}$

30. $243x^{10} + 810x^9y^3 + 1080x^8y^6 + 720x^7y^9 + 240x^6y^{12} + 32x^5y^{15}$

32. $\frac{1}{x^3} + \frac{6}{x^2} + \frac{15}{x} + 20 + 15x + 6x^2 + x^3$

34. 0 **36.** 0

38. $a^6 + 6\sqrt{5}a^5 + 75a^4 + 100\sqrt{5}a^3 + 375a^2 + 150\sqrt{5}a + 125$

40. $a^2 + 3a^{4/3}b^{1/3} + 3a^{2/3}b^{2/3} + b$

42. $512p^9 + 6912p^8q + 41,472p^7q^2 + 145,152p^6q^3$

44. $a^5 + 10a^4\sqrt{ab} + 45a^4b + 120a^3b\sqrt{ab} + 210a^3b^2$

46. $\frac{5}{729}a^6b^4 + \frac{10a^8b^{9/2}}{19683} + \frac{a^{10}b^5}{59049}$

48. There is no such term.

50. $\frac{100 \cdot 99 \cdot 98 \cdot 97}{4!}(0.91)^{96}(0.09)^4 \approx 0.03$

52. 0.99664 **54.** $\frac{3y}{x^{5/3}}$

56. $nx^{n-1} + \binom{n}{2}x^{n-2}h + \cdots + nxh^{n-2} + h^{n-1}$

58. $a^8 + 4a^6b^2 + 4a^6c^2 + 6a^4(b^4 + 2b^2c^2 + c^4) + 4a^2(b^6 + 3b^4c^2 + 3b^2c^4 + c^6) + b^8 + 4b^6c^2 + 6b^4c^4 + 4b^2c^6 + c^8$

60. $1 + 2x + 3x^2 + 4x^3 + \cdots + 8x^7$

62. $1 + \frac{1}{3}x - \frac{1}{9}x^2 + \frac{5}{81}x^3 - \frac{10}{243}x^4 + \frac{22}{729}x^5 - \frac{154}{6561}x^6 + \frac{374}{19,683}x^7$

64. $\frac{(3x + 6)}{n + 1}$

Section 10.2, page 515

2. Prove: $1 + 4 + 7 + \cdots + (3n - 2) = \frac{n(3n - 1)}{2}$.

For $n = 1$: $3 - 2 = \frac{1(2)}{2} = 1$. Assume true for n.

Consider: $1 + 4 + 7 + \cdots + (3n - 2) + 3(n + 1) - 2$

$= \frac{n(3n - 1)}{2} + 3n + 1$

$= \frac{3no^2 - n + 6n + 2}{2} = \frac{(3n + 2)(n + 1)}{2}$

$= \frac{[3(n + 1) - 1](n + 1)}{2}$

4. Prove: $1^2 + 3^2 + 5^2 + \cdots + (2n - 1)^2 = \frac{n(4n^2 - 1)}{3}$.

For $n = 1$: $(2 - 1)^2 = 1 = \frac{4 - 1}{3}$. Assume true for n.

Consider: $1^2 + 3^2 + \cdots + (2n - 1)^2 + [2(n + 1) - 1]^2$

$= \frac{n(4n^2 - 1)}{3} + (2n + 1)^2$

$= \frac{(2n + 1)(2n^2 - n + 6n + 3)}{3}$

$= \frac{(2n + 1)(2n^2 + 5n + 3)}{3}$

$= \frac{(2n + 1)(2n + 3)(n + 1)}{3}$

$= \frac{(n + 1)[2(n + 1) - 1][2(n + 1) + 1]}{3}$

6. Prove: $3 + 3^2 + 3^3 + \cdots + 3^n = \frac{3^{n+1} - 3}{2}$.

For $n = 1$: $3 = \frac{3^2 - 3}{2} = 3$. Assume true for n.

Consider: $3 + 3^2 + \cdots + 3^n + 3^{n+1}$

$= \frac{3^{n+1} - 3}{2} + 3^{n+1}$

$= \frac{3^{n+1} - 3 + 2 \cdot 3^{n+1}}{2}$

$= \frac{3 \cdot 3^{n+1} - 3}{2} = \frac{3^{n+2} - 3}{2}$

8. Prove: $\frac{1}{1 \cdot 2} + \frac{1}{2 \cdot 3} + \frac{1}{3 \cdot 4} + \cdots + \frac{1}{n(n + 1)} = \frac{n}{n + 1}$.

For $n = 1$: $\frac{1}{1(2)} = \frac{1}{1 + 1}$. Assume true for n.

Consider: $\frac{1}{1 \cdot 2} + \frac{1}{2 \cdot 3} + \cdots + \frac{1}{n(n + 1)} + \frac{1}{(n + 1)(n + 2)}$

$= \frac{n}{n + 1} + \frac{1}{(n + 1)(n + 2)}$

$= \frac{n^2 + 2n + 1}{(n + 1)(n + 2)} = \frac{n + 1}{n + 2}$

10. Prove: $2 + 2^2 + 2^3 + \cdots + 2^n = 2(2^n - 1)$.

For $n = 1$: $2 = 2(2^1 - 1)$. Assume true for n.

Consider: $2 + 2^2 + \cdots + 2^n + 2^{n+1}$
$$= 2(2^n - 1) + 2^{n+1}$$
$$= 2(2^n - 1 + 2^n)$$
$$= 2(2^{n+1} - 1).$$

12. Prove: $1 + 3 + 5 + \cdots + (2n - 1) = n^2$.
For $n = 1$: $2(1) - 1 = 1$. Assume true for n.
Consider: $1 + 3 + 5 + \cdots + 2n - 1 + 2(n + 1) - 1 = n^2 + 2n + 2 - 1 = (n + 1)^2$.

14. Prove: $1 + 2 + 3 + \cdots n < \dfrac{(2n + 1)^2}{8}$.

For $n = 1$: $1 < \dfrac{(2 + 1)^2}{8} = \dfrac{9}{8}$. Assume true for n.

Consider: $1 + 2 + \cdots + n + n + 1 < \dfrac{(2n + 1)^2}{8} + (n + 1)$

RHS: $\dfrac{(2n + 1)^2 + 8n + 8}{8} = \dfrac{4n^2 + 4n + 1 + 8n + 8}{8}$

$= \dfrac{4n^2 + 12n + 9}{8} = \dfrac{(2n + 3)^2}{8} = \dfrac{[2(n + 1) + 1]^2}{8}$.

16. Prove: $3^{2n} - 1$ is a multiple of 8.
For $n = 1$: $3^2 - 1 = 8$. Assume true for n: $3^{2n} - 1 = 8k$ where k is an integer.
Then $9(3^{2n} - 1) = 9 \cdot 8k$.
Further $9(3^{2n} - 1) + 8 = 9 \cdot 8k + 8$
$3^{2n+2} - 1 = 8(9k + 1)$, that is, $3^{2(n+1)} - 1$ is a multiple of 8.

18. Prove 4 is a factor of $5^n - 1$.
True for $n = 1$ because $5 - 1 = 4$. Assume true for n; that is, $5^n - 1 = 4k$ where k is an integer.
Then $5(5^n - 1) = 5 \cdot 4k$
Further $5(5^n - 1) + 4 = 5 \cdot 4k + 4$
So $5^{n+1} - 1 = 4(5k + 1)$,
That is, 4 is a factor of $5^{n+1} - 1$.

20. Prove $n^3 - n$ is divisible by 6.
True for $n = 1$ because $1 - 1 = 0$ is divisible by 6. Assume true for n; that is, $n^3 - n = 6k$ where k is an integer.
Consider: $(n + 1)^3 - (n + 1)$
$$= n^3 + 3n^2 + 3n + 1 - n - 1$$
$$= (n^3 - n) + (3n^2 + 3n)$$
$$= 6k + 3(n^2 + n)$$
$$= 6k + 3(2j)$$
$$= 6(k + j) \Leftarrow \text{ divisible by 6}$$

22. Prove 3 is a factor of $n^3 + 2n$.
True for $n = 1$ because $1 + 2 = 3$. Assume true for n; that is, $n^3 + 2n = 3k$ where k is an integer.
Consider: $(n + 1)^3 + 2(n + 1)$
$$= n^3 + 3n^2 + 3n + 1 + 2n + 2$$
$$= (n^3 + 2n) + 3(n^2 + n + 1)$$
$$= 3k + 3j \Leftarrow 3 \text{ is a factor of } (n + 1)\text{st term}$$

24. Prove $3^n < 3^{n+1}$.
Clear for $n = 1$ because $3 < 3^2 = 9$. Assume true for n.
Then $3 \cdot 3^n < 3 \cdot 3^{n+1} \Rightarrow 3^{n+1} < 3^{n+2}$.

26. Prove $2n < 2^n$ for $n \geq 3$.
True for $n = 3$ because $2 \cdot 3 = 6 < 2^3 = 8$. Assume true for n; that is, $2n < 2^n$.
Now $2n + 2 < 2^n + 2$, or $2(n + 1) < 2^n + 2^n = 2 \cdot 2^n$ (because $2 < 2^n$ for $n \geq 3$).
Therefore $2(n + 1) < 2^{n+1}$.

28. $f(n) = 3 + 7 + 11 + \cdots + (4n - 1)$

a.

n	1	2	3	4	5	6
$f(n)$	3	10	21	36	55	78

b. $f(7) = 27 + 78 = 105$
c. $f(n) = 3n + 2n(n - 1)$

30. $f(n) = 1 + 3 + 5 + \cdots + (2n - 1)$

a.

n	1	2	3	4	5	6
$f(n)$	1	4	9	16	25	36

b. $f(7) = 49$; c. $f(n) = n^2$

32. Prove that it is possible to pay any debt of \$8, \$9, \$10, $\ldots$, \$n and so on by using only \$3 and \$5 bills.
True for \$8 $= \$3 + \$5$. Assume true for n; that is, $n = 3j + 5k$ where j and k are nonnegative integers.
Consider $n + 1 = 3j + 5k + 1$. If $k \geq 1$, then:
$$n + 1 = 3j + 5k + 6 - 5$$
$$= 3(j + 2) + 5(k - 1) \Leftarrow \text{ true for } k \geq 1$$

If $k = 0$, then $n = 3j$, $j \geq 3$. So
$$n + 1 = 3j + 1 = 3(j - 3) + 9 + 1$$
$$= 3(j - 3) + 2 \cdot 5 \Leftarrow \text{ true here as well}$$

34. Prove that the measure of the angles in an n-sided convex polygon is $180°(n - 2)$ for $n \geq 3$. Clear for a triangle ($n = 3$) that has degree measure $180°$. Assume true for n. Show true for $n + 1$. Now, the addition of an edge to an n-sided convex polygon also contributes one vertex. Essentially we are going to "glue" a triangle onto the previous figure. Thus the degree measure of an $(n + 1)$–sided convex polygon is $180°(n - 2) + 180° = 180°(n - 1)$.

36. Prove DeMoivre's theorem: $[r(\cos\theta + i\sin\theta)]^n = r^n[\cos n\theta + i\sin n\theta]$.
Clear for $n = 1$. Assume true for n; show true for $n + 1$:
$[r(\cos\theta + i\sin\theta)]^n r(\cos\theta + i\sin\theta)$
$= r^n(\cos n\theta + i\sin n\theta) \cdot r(\cos\theta + i\sin\theta)$
$\Rightarrow [r(\cos\theta + i\sin\theta)]^{n+1}$
$= r^{n+1}[\cos n\theta \cos\theta - \sin n\theta \sin\theta$
$\quad + i\sin n\theta \cos\theta + i\sin\theta \cos n\theta]$
$= r^{n+1}[\cos(n\theta + \theta) + i\sin(n\theta + \theta)]$
$= r^{n+1}[\cos(\theta[n + 1]) + i\sin(\theta[n + 1])]$

38. Prove $\cos(\theta + n\pi) = (-1)^n \cos\theta$ for positive integers n.
True for $n = 1$ because $\cos(\theta + \pi) = \cos\theta \cos\pi - \sin\theta \sin\pi = -\cos\theta$. Assume true for n. Now
$\cos(\theta + (n + 1)\pi) = \cos[(\theta + n\pi) + \pi]$
$= \cos(\theta + n\pi)\cos\pi - \sin(\theta + n\pi)\sin\pi$
$= -\cos(\theta + n\pi)$
$= -[(-1)^n \cos\theta]$
$= (-1)^{n+1} \cos\theta$

40. Prove $\log n < n$ for $n \geq 1$.

True for $n = 1$ because $\log 1 = 0 < 1$. Assume true for n; that is, $\log n < n \Rightarrow n < 10^n$. Now

$$n + 1 < 10^n + 1 < 10^n + 9 \cdot 10^n \quad (\text{for } n \geq 1)$$

So $n + 1 < 10(10^n) = 10^{n+1} \Rightarrow \log(n+1) < \log 10^{n+1}$, as was to be shown.

42. Prove $(1 + x)^n \geq 1 + nx$ if $x \geq -1$ for $n \geq 1$.

True for $n = 1$ because $1 + x = 1 + x$. Assume true for n. Then $(1 + x)^n(1 + x) \geq (1 + nx)(1 + x)$ (note $x \geq -1$.) So

$$(1 + x)^{n+1} \geq 1 + nx + x + nx^2$$

$$(1 + x)^{n+1} \geq 1 + (n + 1)x + nx^2 \geq 1 + (n + 1)x.$$

44. Nothing

46. P_{2n-1} is true for every integer $n \geq 0$.

48. P_{4^n} is true for $n \geq 1$.

Section 10.3, page 520

2. $a_1 = -3; \quad a_2 = -2; \quad a_3 = -1; \quad a_8 = 4$

4. $a_1 = 0; \quad a_2 = \frac{1}{3}; \quad a_3 = \frac{2}{4} = \frac{1}{2}; \quad a_8 = \frac{7}{9}$

6. $a_1 = 3^{-3} = \frac{1}{27}; \quad a_2 = 3^{-2} = \frac{1}{9}; \quad a_3 = \frac{1}{3};$
$a_8 = 3^4 = 81$

8. $a_1 = 1; \quad a_2 = 8; \quad a_3 = 27; \quad a_8 = 512$

10. $a_1 = \sqrt{2}; \quad a_2 = -2; \quad a_3 = \sqrt{6}; \quad a_8 = -4$

12. $a_1 = 2^2 - 1 = 3; \quad a_2 = 2^3 - 2^1 = 6;$
$a_3 = 2^4 - 2^2 = 12; \quad a_8 = 2^9 - 2^7 = 384$

14. $a_1 = -2; \quad a_2 = \frac{8+1}{8-2} = \frac{9}{6} = \frac{3}{2};$
$a_3 = \frac{27+1}{27-2} = \frac{28}{25}; \quad a_8 = \frac{8^3+1}{8^3-2} = \frac{171}{170}$

16. $a_1 = 0; \quad a_2 = 0; \quad a_3 = 0; \quad a_8 = 0$

18. $a_1 = \cos\left(\frac{5\pi}{2}\right) = 0; \quad a_2 = \cos(5\pi) = -1;$

$a_3 = \cos\left(\frac{15\pi}{2}\right) = 0; \quad a_8 = \cos(20\pi) = 1$

20. $a_1 = -3; \quad a_2 = -6; \quad a_3 = -12; \quad a_4 = -24;$
$a_5 = -48$

22. $a_1 = 1; a_2 = \frac{1}{9}; a_3 = -\frac{1}{243}; a_4 = -\frac{1}{3^9}; a_5 = \frac{1}{3^{14}}$

24. $a_1 = 2; \quad a_2 = 9; \quad a_3 = 28; \quad a_4 = 65; \quad a_5 = 126$

26. $a_1 = \frac{1}{2}; \quad a_2 = \frac{1}{3!} = \frac{1}{6}; \quad a_3 = \frac{1}{4!} = \frac{1}{24};$

$a_4 = \frac{1}{5!} = \frac{1}{120}; \quad a_5 = \frac{1}{6!} = \frac{1}{750}$

28. $a_1 = 0; \quad a_2 = 2; \quad a_3 = 6; \quad a_4 = 12; \quad a_5 = 20$

30. $a_1 = -2; a_2 = \frac{1}{2}; a_3 = -2; a_4 = \frac{1}{2}; a_5 = -2$

32. $a_1 = a_2 = a_3 = a_4 = a_5 = 0$

34. 15   36. -22 38. 20   40. 45

42. 1   44. -199 46. -420 48. 3^n

50. $\frac{1}{n}(-1)^{n+1}$ 52. $13 + 4n$

54. $\sum_{n=1}^{6} 3 \cdot 2^{n-1}$ 56. $\sum_{n=1}^{8} (-2)^n$ 58. $\sum_{n=1}^{5} (4n - 3)$

60. $\frac{62}{39}$ 62. $x = 2$ 64. $x = 10$

Section 10.4, page 525

2. $d = 1$ 4. $d = 2$ 6. $d = 5$ 8. $d = -4$

10. 72   12. -191 14. -83 16. 17

18. 13.5   20. $d = -\frac{21}{2}, a_1 = \frac{455}{2}$

22. $d = 3, a_1 = -24$ 24. $d = \frac{1}{40}, a_1 = \frac{4}{10} = \frac{2}{5}$

26. $d = -2, a_1 = 7$ 28. $d = 4, a_1 = \frac{1}{2}$

30. $d = \frac{18}{10} = \frac{9}{5}$ 32. -61

34. $n = 30$ 36. $a_1 = 11$ 38. $n = 41$ 40. 1365

42. 1435   44. -1150 46. 675   48. 3822

50. -8544 52. 70 wpm   54. 51   56. $d = 2.5$

58. $d = 4$ 60. $\sum_{n=0}^{4} \left(3 - \frac{1}{2}n\right)$ 62. $\sum_{n=0}^{5} (-2 - 6n)$

64. $a_n = n^2, \quad b_n = a_n - a_{n-1}$
$b_{n+1} = a_{n+1} - a_n = (n + 1)^2 - n^2$
$b_{n+1} - b_n = (n + 1)^2 - n^2 - [n^2 - (n - 1)^2]$
$\phantom{b_{n+1} - b_n} = n^2 + 2n + 1 - n^2 - [n^2 - n^2 + 2n - 1]$
$\phantom{b_{n+1} - b_n} = 2$

Therefore, $b_1, b_2, b_3, \ldots$ is an arithmetic sequence.

66. $a_{n-1} = a_1 + (n - 2)d$
$a_n = a_1 + (n - 1)d$
$a_{n+1} = a_1 + nd$

$$\frac{a_{n-1} + a_{n+1}}{2} = \frac{2a_1 + (n - 2 + n)d}{2}$$
$$= a_1 + (n - 1)d = a_n$$

Section 10.5, page 532

2. $r = \frac{1}{2}$ 4. $r = 2$ 6. $r = -\frac{1}{2}$ 8. $\frac{512}{19,683}$

10. $\frac{-1}{2187}$ 12. $\frac{1}{512}$ 14. 262,144   16. 729

18. 9,765,625   20. $\frac{2}{27}$

22. The total earnings after winning 18 games is \$2,621,430.

24. \$295.24   26. \$19,293.71

28. 4 billion   30. 64   32. $n = 8$

34. $a_1 = \frac{1}{81}; a_2 = \frac{1}{9}; a_4 = 9; a_5 = 81$

36. $n = 7; S_n = 635$ 38. $a_1 = 2; a_9 = 512$

40. $a_5 = 810; S_5 = 1210$ 42. $a_1 = 64; S_n = 2059$

44. $a_6 = -\frac{243}{8}; S_6 = -\frac{133}{8}$

46. $a_1 = -\sqrt{5}, n = 9$ 48. $a_1 = 2; a_n = 2(\sqrt{3})^{n-1}$

50. $a_n = \frac{(\sqrt[3]{18})^n}{\sqrt[3]{12}}$ 52. \$1,489.35   54. \$27,874.86

56. $\frac{5}{6}$ 58. 3   60. $\frac{3}{2}$

62. Sum not defined.   64. $a_1 = 6$ 66. $r = \frac{4}{5}$

68. $r = \frac{1}{2}$ 70. 24 m   72. \$543,179.08

74. $n \approx 13.1$ (a little over 13 years)

76. Yes, if alternating or all terms are negative.

Section 10.6, page 540

2. 6,760,000   4. 24   6. 60

8. 362,880   10. 19,962,600   12. 840

14. $\dfrac{9!}{(9-P)!}$ **16.** 60,480 **18.** $n(n-1)(n-2)(n-3)$

20. 6 **22.** 120 **24.** 210 **26.** 362,880

28. 80,640 **30.** 36 **32.** 6 **34.** 495

36. $\dfrac{n(n-1)(n-2)(n-3)}{4!}$ **38.** 1

40. 635,013,559,600

42. 511 **44.** 210 **46.** 45 **48.** 83,160

52. *a.* 56; *b.* 252 **54.** Diagonals = 170; 10.

56. Prove $_{n+1}C_r = {_n}C_{r-1} + {_n}C_r$

$$_nC_{r-1} + {_n}C_r = \frac{n!}{(r-1)!(n-r+1)!} + \frac{n!}{r!(n-r)!}$$

$$= \frac{rn! + (n-r+1)n!}{r!(n-r+1)!} = \frac{(n+1)n!}{r!(n-r+1)!}$$

$$= \frac{(n+1)!}{r!(n-r+1)!} = {_{n+1}}C_r$$

58. $n = 100$ **60.** $n = 7$ **62.** $n = 8$

64. Prove $_5P_r = 5(_4P_{r-1})$

$$5(_4P_{r-1}) = 5 \cdot \frac{4!}{[4-(r-1)]!} = \frac{5!}{(5-r)!} = {_5}P_r$$

Section 10.7, page 545

2. $S = \{TA, TC, TK, AT, AC, AK, CT, CA, CK, KT,$
$KA, KC\}$

4. $S = \{GGGG, BGGG, GBGG, GGBG, GGGB,$
$BBGG, BGBG, BGGB, GBBG, GGBB,$
$GBGB, BBBG, BBGB, BGBB,$
$GBBB, BBBB\}$
4 points correspond to families with 3 boys and 1 girl.

6. $S = \{ABC, ABD, ABE, ACD, ACE, ADE, BCD,$
$BCE, BDE, CDE\}$

8. 6 points **10.** $_{48}C_4 = 194,580$ entries in sample space.

12. $\{HHHH, HHHT, HHTH, HTHH, THHH, HHTT,$
$HTTH, HTHT, TTHH, THTH, THHT, TTTH,$
$TTHT, THTT, HTTT, TTTT\}$

14. $\{GGG, GGR, GRG, RGG, GRR, RGR, RRG, RRR\}$

16. $\{RR, RF, FR, FF\}$ **18.** $\{MM, MW, WM, WW\}$

20. $_7P_3 = 210$ points in sample space consisting of sequences of three letters from word.

22. $P(\text{exactly one king}) = \dfrac{32}{221}$

24. $\dfrac{8}{27}$ **26.** $\dfrac{5}{8}$ **28.** $\dfrac{1}{2}$ **30.** $\dfrac{1}{595}$

32. $\dfrac{5}{33}$ **34.** $\dfrac{1}{66}$ **36.** $\dfrac{1}{21}$ **38.** $\dfrac{143}{380}$

40. $\dfrac{19}{35}$ **42.** $\dfrac{1}{945}$

Section 10.8, Cumulative Review Exercises, page 548

2. $27a^3b^6 - 108a^4b^5 + 144a^5b^4 - 64a^6b^3$

4. $a^3 - 6a^2 + 15a - 20 + \dfrac{15}{a} - \dfrac{6}{a^2} + \dfrac{1}{a^3}$

6. $-\dfrac{49152a^4}{125}$

8. Prove $1 + 2 + 3 + \cdots + n = \dfrac{n^2+n}{2}$.

True for $n = 1$ because $1 = \dfrac{1^2+1}{2}$. Assume true for n.

Thus $1 + 2 + 3 + \cdots + n + n + 1$

$$= \frac{n^2+n}{2} + n + 1$$

$$= \frac{n^2+n+2n+2}{2} = \frac{(n+1)^2+(n+1)}{2}$$

10. Prove $\dfrac{1}{2} + \dfrac{1}{2^2} + \dfrac{1}{2^3} + \cdots + \dfrac{1}{2^n} = 1 - \dfrac{1}{2^n}$.

True for $n = 1$ because $\frac{1}{2} = 1 - \frac{1}{2}$. Assume true for n.

Then $\dfrac{1}{2} + \dfrac{1}{2^2} + \dfrac{1}{2^3} + \cdots + \dfrac{1}{2^n} + \dfrac{1}{2^{n+1}}$

$$= 1 - \frac{1}{2^n} + \frac{1}{2^{n+1}}$$

$$= 1 + \frac{1}{2^n}\left[\frac{1}{2} - 1\right]$$

$$= 1 + \frac{1}{2^n}\left[-\frac{1}{2}\right] = 1 - \frac{1}{2^{n+1}}$$

14. $f(n) = 1 + 3 + 9 + \cdots + 3^{n-1}$

a.

n	1	2	3	4	5	6
$f(n)$	1	4	13	40	121	364

b. 9841; *c.* 9841; *d.* $f(n) = \dfrac{3^n-1}{2}$

18. $a_1 = 5$; $a_2 = 7$; $a_3 = 9$; $a_{10} = 23$

20. $a_1 = 1$; $a_2 = 4$; $a_3 = 9$; $a_{10} = 100$

22. $a_1 = 0$; $a_2 = 2$; $a_3 = 4$; $a_4 = 6$; $a_5 = 8$

24. $a_1 = -2$; $a_2 = -1$; $a_3 = -\frac{1}{2}$; $a_4 = -\frac{1}{4}$;
$a_5 = -\frac{1}{8}$

26. 36 **28.** 0 **30.** $a_{n+1} - a_n = -4$

32. $a_{n+1} - a_n = \frac{1}{2}$ **34.** 39 **36.** -7

38. $d = -3, a_1 = 7$ **40.** $d = \frac{45}{11}$, $a_1 = \frac{336}{11}$

42. $n = 21$ **44.** -56 **46.** 750 **48.** 725

50. 90 m **52.** 713 cm **54.** $r = \frac{1}{2}$ **56.** $r = -4$

58. $\dfrac{-4}{390,625}$ **60.** $\dfrac{32}{9}$ **62.** $a_n = (-3)\left(-\frac{2}{3}\right)^{n-1}$

64. $n = 8$ **66.** \$5392.73

68. $r = 2^{6/5}, S_n = \dfrac{(256)2^{1/5} - 2}{1 - 2^{6/5}}$

70. $a_n = 4(3)^{n-1}$ **72.** Sum not defined

74. $\dfrac{64}{7}$ **76.** $\dfrac{1200}{7}$ **78.** 80 **80.** 24

82. 720 **84.** 6 **86.** 3.9542426×10^{21}

88. 24 **90.** 7 **92.** 45

94. 2352 committees **96.** 4800

98. 20,358,520 points in sample space.

100. $\{(1, 1), (1, 2), (1, 3), (1, 4), (1, 5), (1, 6), (2, 1), \ldots,$
$(6, 1), (6, 2), (6, 3), (6, 4), (6, 5), (6, 6)\}$

102. $\dfrac{1}{5}$ **104.** $\dfrac{4}{663}$

CHAPTER 11
Section 11.1, page 556

2. vertex: $(0, 0)$; focus: $\left(\frac{1}{16}, 0\right)$; directrix: $x = -\frac{1}{16}$

4. vertex: $(0, 0)$; focus: $(0, -3)$; directrix: $y = 3$

6. vertex: $\left(-2, \frac{4}{3}\right)$; focus: $\left(-2, \frac{35}{24}\right)$; directrix: $y = \frac{29}{24}$

8. vertex: $\left(\frac{1}{2}, 0\right)$; focus: $\left(\frac{1}{2}, \frac{1}{16}\right)$; directrix: $y = -\frac{1}{16}$

10. vertex: $\left(-\frac{14}{3}, \frac{1}{2}\right)$; focus: $\left(-\frac{215}{48}, \frac{1}{2}\right)$;

    directrix: $x = -\frac{233}{48}$

12. vertex: $\left(-\frac{3}{8}, -\frac{3}{2}\right)$; focus: $\left(\frac{9}{8}, -\frac{3}{2}\right)$;

    directrix: $x = -\frac{15}{8}$

14. vertex: $\left(-\frac{3}{2}, \frac{49}{10}\right)$; focus: $\left(-\frac{3}{2}, \frac{171}{40}\right)$;

    directrix: $y = \frac{221}{40}$

16. $y^2 = 12x$

18. $(x - 1)^2 = -4(y - 1)$

20. $y^2 = -8x$

22. $(x - 4)^2 = -8(y - 2)$

24. $(x - 1)^2 = -6\left(y - \frac{7}{2}\right)$

26. $y^2 = 2\left(x - \frac{1}{2}\right)$

28. $x^2 = \frac{9}{10}y$

30. $(y - 1)^2 = -\frac{1}{2}(x - 2)$

32. $(x - 2600)^2 = \frac{2600^2}{300}y$

34. $x^2 = -240y$

36. $y - 4 = 4(x - 2) \Rightarrow y = 4x - 4$

38. $y = \frac{1}{8}x + \frac{45}{4}$

42. (f) 44. (a) 46. (e)

Section 11.2, page 564

2. $\dfrac{x^2}{25} + \dfrac{(y - 1)^2}{16} = 1$

4. $\dfrac{(x - 1)^2}{16} + \dfrac{(y - 2)^2}{12} = 1$

6. $\dfrac{x^2}{25} + \dfrac{y^2}{4} = 1$ 8. $\dfrac{(x - 5)^2}{9} + \dfrac{(y - 2)^2}{25} = 1$

10. $\dfrac{x^2}{9} + \dfrac{16y^2}{63} = 1$ 12. $\dfrac{x^2}{25} + \dfrac{y^2}{24} = 1$

14. $\dfrac{(x - 3)^2}{20} + \dfrac{(y - 1)^2}{36} = 1$

16. $\dfrac{x^2}{9} + \dfrac{y^2}{4} = 1$ 18. $\dfrac{4}{25}x^2 + \dfrac{y^2}{25} = 1$

20. $\dfrac{x^2}{4} + \dfrac{y^2}{9} = 1$

    $a = 3, b = 2, c = \sqrt{5}, e = \frac{\sqrt{5}}{3}$; foci: $(0, \pm\sqrt{5})$

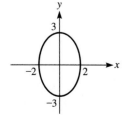

22. $\dfrac{x^2}{4} + \dfrac{y^2}{8} = 1$

    $a = 2\sqrt{2}, b = 2, c = 2, e = \frac{1}{\sqrt{2}}$; foci: $(0, \pm 2)$

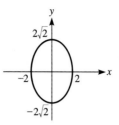

24. $\dfrac{(x + 2)^2}{16} + \dfrac{(y - 2)^2}{4} = 1$; $a = 4, b = 2, c = 2\sqrt{3}$;

    foci: $(-2 \pm 2\sqrt{3}, 2)$; $e = \frac{\sqrt{3}}{2}$

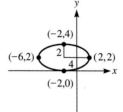

26. $\dfrac{(x - 1)^2}{18} + \dfrac{(y - 2)^2}{36} = 1$

    $a = 6, b = 3\sqrt{2}, c = 3\sqrt{2}, e = \frac{\sqrt{2}}{2}$

    center: $(1, 2)$; foci: $(1, 2 \pm 3\sqrt{2})$

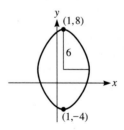

28. $\dfrac{(x + 1)^2}{6} + \dfrac{(y + 2)^2}{4} = 1$

    $a = \sqrt{6}, b = 2, c = \sqrt{2}, e = \frac{\sqrt{2}}{\sqrt{6}} = \frac{1}{\sqrt{3}}$

    center: $(-1, -2)$; foci: $(-1 \pm \sqrt{2}, -2)$

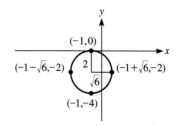

30. $\dfrac{x^2}{9} + y^2 = 1$

$a = 3, b = 1, c = 2\sqrt{2}, e = \dfrac{2\sqrt{2}}{3}$

center: $(0, 0)$;   foci: $(\pm 2\sqrt{2}, 0)$

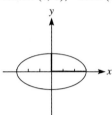

32. $\dfrac{(x + 1)^2}{25} + \dfrac{(y + 3)^2}{4} = 1$

$a = 5, b = 2, c = \sqrt{21}, e = \dfrac{\sqrt{21}}{5}$

center: $(-1, -3)$;   foci: $(-1 \pm \sqrt{21}, -3)$

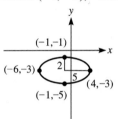

34. $\dfrac{(x + 1)^2}{25} + \dfrac{(y + 1)^2}{16} = 1$

$a = 5, b = 4, c = 3, e = \dfrac{3}{5}$

center: $(-1, -1)$;   foci: $(-4, -1), (2, -1)$

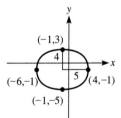

36. $\dfrac{(x + 1)^2}{\frac{4}{3}} + \dfrac{y^2}{4/3} = 1; a = b = \dfrac{2}{\sqrt{3}}, c = 0, e = 0;$

center: $(-1, 0)$;   foci: $= (-1, 0)$

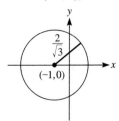

38. Show that the length of the latus recta are $2b^2/a$. Without loss of generality, assume the major axis lies along

the x-axis and the center of the ellipse is the origin. Now one endpoint of the latus recta is (c, y). Solve for y:

$$\dfrac{c^2}{a^2} + \dfrac{y^2}{b^2} = 1 \Rightarrow \dfrac{y^2}{b^2} = 1 - \dfrac{c^2}{a^2}$$

$$\Rightarrow y^2 = b^2 \left(\dfrac{a^2 - c^2}{a^2} \right)$$

$$= \dfrac{b^4}{a^2} \Rightarrow y = \pm \dfrac{b^2}{a} \qquad \text{(same for } x = -c)$$

Thus the length of the latus recta are: $\dfrac{2b^2}{a}$

40. $\dfrac{x^2}{(24,000)^2} + \dfrac{y^2}{(22,000)^2} = 1; e \approx 0.39965$

42. (b) 44. (c) 46. (e)

Section 11.3, page 571

2. $y^2 - x^2 = 1,$ $c = 0,$ $e = \sqrt{2}$

center: $(0, 0)$;   vertices: $(0, \pm 1)$;

foci: $(0, \pm \sqrt{2})$;   asymptotes: $y = \pm x$

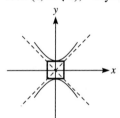

4. $\dfrac{y^2}{9} - \dfrac{x^2}{4} = 1,$ $c = \sqrt{13},$ $e = \dfrac{\sqrt{13}}{3}$

center: $(0, 0)$;   vertices: $(0, \pm 3)$;

foci: $(0, \pm \sqrt{13})$;   asymptotes: $y = \pm \dfrac{3}{2}x$

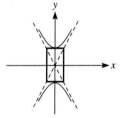

6. $y^2 - \dfrac{x^2}{3} = 1,$ $c = 2,$ $e = 2$

center: $(0, 0)$;   vertices: $(0, \pm 1)$;

foci: $(0, \pm 2)$;   asymptotes: $y = \pm \dfrac{1}{\sqrt{3}}x$

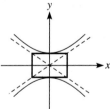

8. $\dfrac{(x + 3)^2}{9} - \dfrac{y^2}{4} = 1,$ $4 = c^2 - 9;$ center: $(-3, 0)$;

vertices: $(0, 0)$, $(-6, 0)$; foci: $(-3 \pm \sqrt{13}, 0)$;

asymptotes: $y = \frac{2}{3}x + 2$, $y = -\frac{2}{3}x - 2$, $e = \frac{\sqrt{13}}{3}$

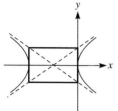

10. $\dfrac{x^2}{1/3} - \dfrac{y^2}{1/2} = 1$, $c^2 = \frac{5}{6}$, $e = \sqrt{\frac{5}{2}}$

center: $(0, 0)$;   vertices: $\left(\pm\frac{1}{\sqrt{3}}, 0\right)$;

foci: $\left(\pm\sqrt{\frac{5}{6}}, 0\right)$;   asymptotes: $y = \pm\sqrt{\frac{3}{2}}\,x$

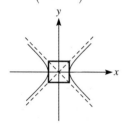

12. $\dfrac{x^2}{1/2} - \dfrac{y^2}{1/3} = 1$, $c^2 = \frac{5}{6}$, $e = \sqrt{\frac{5}{3}}$

center: $(0, 0)$;   vertices: $\left(\pm\frac{1}{\sqrt{2}}, 0\right)$;

foci: $\left(\pm\sqrt{\frac{5}{6}}, 0\right)$;   asymptotes: $y = \pm\sqrt{\frac{2}{3}}\,x$

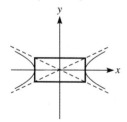

14. $\dfrac{y^2}{3} - \dfrac{(x + 2)^2}{10} = 1$, $c = \sqrt{13}$, $e = \sqrt{\frac{13}{3}}$

center: $(-2, 0)$;   vertices: $(-2 \pm \sqrt{3}, 0)$;

foci: $(-2 \pm \sqrt{13}, 0)$;   asymptotes: $y = \pm\sqrt{\frac{3}{10}}(x + 2)$

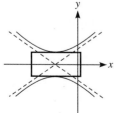

16. $\dfrac{(x - 2)^2}{95/24} - \dfrac{(y - 1/4)^2}{95/16} = 1$, $c^2 = \frac{475}{48}$

center: $\left(2, \frac{1}{4}\right)$;   vertices: $\left(2 \pm \sqrt{\frac{95}{24}}, \frac{1}{4}\right)$;

foci: $\left(2 \pm \sqrt{\frac{475}{48}}, \frac{1}{4}\right)$, $e = \sqrt{\frac{5}{2}}$;

asymptotes: $y = \frac{1}{4} \pm \sqrt{\frac{3}{2}}(x - 2)$

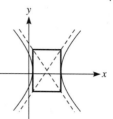

18. $\dfrac{(x - 3/2)^2}{23/8} - \dfrac{y^2}{23} = 1$, $c = \sqrt{\frac{207}{8}}$

center: $\left(\frac{3}{2}, 0\right)$;   vertices: $\left(\frac{3}{2} \pm \sqrt{\frac{23}{8}}, 0\right)$;

foci: $\left(\frac{3}{2} \pm \sqrt{\frac{207}{8}}, 0\right)$;   asymptotes: $y = \pm 2\sqrt{2}\left(x - \frac{3}{2}\right)$

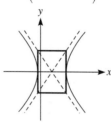

20. $\dfrac{(x + 1)^2}{3} - \dfrac{y^2}{300} = 1$, $c = \sqrt{303}$, $e = \sqrt{101}$

center: $(-1, 0)$;   vertices: $(-1 \pm \sqrt{3}, 0)$;

foci: $(-1 \pm \sqrt{303}, 0)$;   asymptotes: $y = \pm 10(x + 1)$

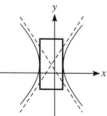

22. $\dfrac{x^2}{25} - \dfrac{y^2}{100} = 1$

24. $\dfrac{(y + 2)^2}{9/4} - \dfrac{(x + 1)^2}{1/4} = 1$

26. $\dfrac{x^2}{64} - \dfrac{y^2}{80} = 1$ 28. $\dfrac{(y - 1)^2}{9} - \dfrac{(x + 2)^2}{27} = 1$

30. $y^2 - 4x^2 = 1$ 32. $\dfrac{(y - 3/2)^2}{49/4} = \dfrac{(x - 1)^2}{49/30} = 1$

34. Parabola   36. Circle   38. Ellipse

40. $\frac{5}{48}$ miles away from the rescue team

42. Let $a_1 = at$, $b_1 = bt$. Assume horizontal transverse axis. (The proof is similar for vertical transverse axis.) Equation of hyperbola is

$$\frac{(x - h)^2}{a_1^2} - \frac{(y - k)^2}{b_1^2} = 1 = \frac{(x - h)^2}{(at)^2} - \frac{(y - k)^2}{(bt)^2}.$$

Asymptotes are: $y = k \pm \dfrac{b_1}{a_1}(x - h) = k \pm \dfrac{b}{a}(x - h)$.

(Same equations as original.)

44. An ellipse is more elongated.

46. (a) 48. (c) 50. (b)

Section 11.4, page 577

2. $(-4, 4)$ 4. $(6, 6)$ 6. $(0, -1)$

8. $(-2, -3)$ 10. $(-8, \frac{5}{2})$

12. $(X - 1)^2 - 3(X - 1) + 2(Y + 2) = 1$

14. $2(X - 5)^2 + 3(X - 5)(Y - 1) - 2(Y - 1) = 4$

16. $(X + 2)^2 - 3(X + 2)(Y - 3) + (Y - 3)^2 = 0$

18. $(X + 3)^2 + 3(X + 3)(Y + 2) - (Y + 2)^2 = 4$

20. $(X - 7) - 3(3Y + 2) = 6$

22. $\left(\dfrac{X}{2} - \dfrac{\sqrt{3}Y}{2}\right)^2 + \left(\dfrac{\sqrt{3}X}{2} + \dfrac{Y}{2}\right)^2 = 16$

24. $4\left(\dfrac{\sqrt{2}X}{2} - \dfrac{\sqrt{2}Y}{2}\right)^2 + 9\left(\dfrac{\sqrt{2}X}{2} + \dfrac{\sqrt{2}Y}{2}\right)^2 = 36$

26. $\left(-\dfrac{X}{2} + \dfrac{\sqrt{3}Y}{2}\right) - 1 = \dfrac{1}{4}\left(\dfrac{\sqrt{3}X}{2} + \dfrac{Y}{2} + 2\right)^2$

28. $4\left(-\dfrac{\sqrt{3}X}{2} + \dfrac{Y}{2}\right)^2 + 4\left(-\dfrac{\sqrt{3}X}{2} + \dfrac{Y}{2}\right)\left(-\dfrac{X}{2} - \dfrac{\sqrt{3}Y}{2}\right)$

$+ \left(-\dfrac{X}{2} - \dfrac{\sqrt{3}Y}{2}\right)^2 = 4$

30. $\left(-\dfrac{\sqrt{2}X}{2} + \dfrac{\sqrt{2}Y}{2}\right) = 3\left(\dfrac{\sqrt{2}X}{2} + \dfrac{\sqrt{2}Y}{2}\right)$

32. $\theta = 45°$ 34. $\theta = \frac{1}{2}\tan^{-1}(\frac{4}{5})$

36. $\theta = \frac{1}{2}\tan^{-1}(-\frac{5}{2})$ 38. $\theta = \frac{1}{2}\tan^{-1}(-\frac{6}{13})$

40. $\theta = 45°$

Section 11.5, page 582

2. $r = \dfrac{2}{1 + \sin\theta}$ 4. $r = \dfrac{1}{1 + \sin\theta}$

6. $r = \dfrac{6}{1 - \cos\theta}$ 8. $r = \dfrac{6}{1 - \sin\theta}$

10. $r = \dfrac{6}{3 + 2\sin\theta}$ 12. $r = \dfrac{2}{5 + \sin\theta}$

14. $r = \dfrac{2}{1 - 2\sin\theta}$ 16. $r = \dfrac{8}{3 - 5\sin\theta}$

18. parabola, focus: (0, 0) directrix $x = -3$

20. hyperbola, focus: (0, 0) directrix $x = -\frac{5}{3}$

22. ellipse, focus: (0, 0) directrix $x = 3$

24. hyperbola, $e = \frac{3}{2}$, focus: (0, 0) directrix $y = -2$

26. hyperbola, $e = \frac{5}{2}$, focus: (0, 0) directrix $y = \frac{1}{5}$

28. parabola, focus: (0, 0) directrix $y = -\frac{1}{3}$

30. $r = \dfrac{\sqrt{5}}{8 + 4\sqrt{3}\sin\theta}$

32. $r = \dfrac{25}{4 + \sqrt{41}\sin\theta}$

34. 13, 263.2 miles

Section 11.6, page 585

2. $x = 2t - 1$, $y = t + 1$,   4. $x = 2t$, $y = \ln t$,
     $-1 \le t \le 1$ $0 < t < 5$

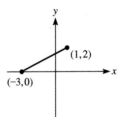   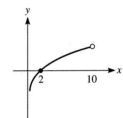

6. $x = 3 + 2\sin\theta$, $y = 1 + 2\cos\theta$, $0 \le \theta \le 2\pi$

θ	$x = 3 + 2\sin\theta$	$y = 1 + 2\cos\theta$
0	3	3
$\pi/4$	$3 + \sqrt{2}$	$1 + \sqrt{2}$
$\pi/2$	5	1
$3\pi/4$	$3 + \sqrt{2}$	$1 - \sqrt{2}$
π	3	-1
$5\pi/4$	$3 - \sqrt{2}$	$1 - \sqrt{2}$
$3\pi/2$	1	1
$7\pi/4$	$3 - \sqrt{2}$	$1 + \sqrt{2}$

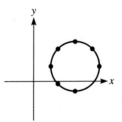

8.

10. $x = 3\sin\theta$,
$y = -3\cos\theta$,
$0 \le \theta \le 6\pi$

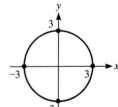

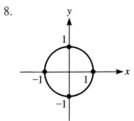

12. $x = 2 + \cos t$, $y = -1 + 3\sin t$, $0 \le t \le 2\pi$

t	$2 + \cos t$	$-1 + 3\sin t$
0	3	-1
$\pi/2$	2	2
π	1	-1
$3\pi/2$	2	-4

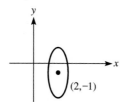

14. $x = \dfrac{1}{t^2 + 1}$,

$y = \dfrac{t}{t^2 + 1}$, for all real t

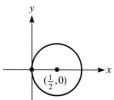

16. $x = \cos t$, $y = \sin t$,

$0 \le t \le \dfrac{\pi}{2}$

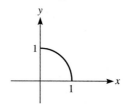

18. $x - 2y + 3 = 0$, $-3 \le x \le 1$

20. $y = \ln \frac{x}{2}$, $0 < x < 10$

22. $x^2 + y^2 - 6x - 2y + 6 = 0$, $1 \le x \le 5$

24. $x^2 + y^2 = 1$, $-1 \le x \le 1$

26. $x^2 + y^2 = 9$, $-3 \le x \le 3$

28. $9x^2 + y^2 - 36x - 2y + 28 = 0$, $1 \le x \le 3$

30. $y = x\left(\pm\sqrt{\dfrac{1}{x} - 1}\right) = \pm\sqrt{x - x^2}$, $0 < x \le 1$

32. $x^2 + y^2 = 1$, $0 \le x \le 1$

34. $\dfrac{(x - h)^2}{a^2} + \dfrac{(y - k)^2}{b^2} = \dfrac{a^2 \cos^2 \theta}{a^2} + \dfrac{b^2 \sin^2 \theta}{b^2}$

$= \cos^2 \theta + \sin^2 \theta = 1$

36. $x = 3 + 2\cos\theta$, $y = -1 + 2\sin\theta$, $0 \le \theta \le 2\pi$

38. $x = 2\cos\theta$, $y = 5\sin\theta$, $\theta = \dfrac{\pi}{2}$ to $\theta = \dfrac{-7\pi}{2}$

40. $x = 5 + 2\cos\theta$, $y = -2 + 3\sin\theta$ $(0 \le \theta \le 2\pi)$

42. $x^2 - y^2 = 1$, $x = \sec\theta$, $y = \tan\theta$, $\begin{cases} -\dfrac{\pi}{2} < \theta < \dfrac{\pi}{2} \\[2mm] \dfrac{\pi}{2} < \theta < \dfrac{3\pi}{2} \end{cases}$

46. Line segment from $(3, -1)$ to $(7, 1)$

48. Same graph   50. Reflection across line $y = x$

Section 11.7, Cumulative Review Exercises, page 588

2. vertex: $\left(0, \frac{3}{2}\right)$; focus: $(0, 2)$; directrix: $y = 1$

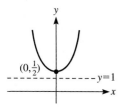

4. $\dfrac{y^2}{10} - \dfrac{x^2}{10/3} = 1$, $c = \dfrac{2\sqrt{30}}{3}$, $e = \dfrac{2\sqrt{3}}{3}$

center: $(0, 0)$; foci: $\left(0, \pm\frac{2\sqrt{30}}{3}\right)$;

transverse axis $= 2\sqrt{10}$, conjugate axis $= \dfrac{2\sqrt{10}}{3}$

asymptotes: $y = \pm\sqrt{3}x$; vertices: $(0, \pm\sqrt{10})$

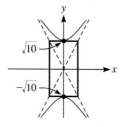

6. $\dfrac{(x - 1)^2}{\frac{25}{2}} + \dfrac{(y - 3)^2}{5} = 1$, $c^2 = \dfrac{15}{2}$

semimajor: $\frac{5}{\sqrt{2}}$; semiminor: $\sqrt{5}$;

foci: $\left(1 \pm \sqrt{\frac{15}{2}}, 3\right)$; $e = \dfrac{\sqrt{15/2}}{\sqrt{25/2}} = \sqrt{\frac{3}{5}}$

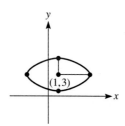

8. $\dfrac{(x - 1)^2}{3^3} + \dfrac{(y + 2)}{4^3} = 1$, $p = -\frac{27}{256}$

vertices: $(1, 62)$; foci: $\left(1, 62 - \frac{27}{256}\right)$;

directrix: $y = 62 + \frac{27}{256}$, $e = 1$

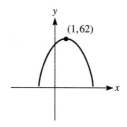

10. $\dfrac{x^2}{3} - \dfrac{y^2}{3} = 1$

center: $(0, 0)$; vertices: $(\pm\sqrt{3}, 0)$; foci: $(\pm\sqrt{6}, 0)$;

asymptotes: $y = \pm x$, $e = \sqrt{2}$

transverse axis = conjugate axis = $2\sqrt{3}$

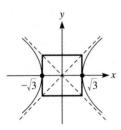

12. $\dfrac{(x - 3/10)^2}{39/100} + \dfrac{(y + 3/2)^2}{39/40} = 1$, $c^2 = \dfrac{234}{400} = \dfrac{117}{200}$

Therefore: center: $\left(\dfrac{3}{10}, -\dfrac{3}{2}\right)$; foci: $\left(\dfrac{3}{10}, -\dfrac{3}{2} \pm \sqrt{\dfrac{117}{200}}\right)$;

semiminor: $\dfrac{\sqrt{39}}{10}$; semimajor: $\sqrt{\dfrac{39}{40}}$;

$e = \dfrac{\sqrt{117/200}}{\sqrt{39/40}} = \sqrt{\dfrac{117}{39} \cdot \dfrac{1}{5}} = \sqrt{\dfrac{3}{5}}$

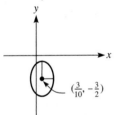

14. $\dfrac{(x + 5/2)^2}{21/4} + \dfrac{(y - 3)^2}{21/8} = 1$, $c = \sqrt{\dfrac{21}{8}}$

center: $\left(-\dfrac{5}{2}, 3\right)$; semimajor: $\dfrac{\sqrt{21}}{2}$; semiminor: $\sqrt{\dfrac{21}{8}}$;

$e = \dfrac{\sqrt{21/8}}{\sqrt{21/4}} = \dfrac{1}{\sqrt{2}}$, foci $= \left(-\dfrac{5}{2} \pm \sqrt{\dfrac{21}{8}}, 3\right)$

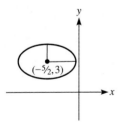

16. $(1, -1)$

18. $(-6, -1)$

20. $\left(\dfrac{1}{2}, 1\right)$

22. $(X + 2)^2 + 2(X + 2)(Y - 3) - 2(Y - 3) = 3$

24. $(X - 3)^2 - 3(X - 3)(Y - 2) - (Y - 2)^2 = 4$

26. $(X + 1)^2 + 3(X + 1)(Y - 3) = 6$

28. $-2XY = 25$

30. $4\left(\dfrac{\sqrt{3}X}{2} + \dfrac{Y}{2}\right)^2 - 9\left(-\dfrac{X}{2} + \dfrac{\sqrt{3}Y}{2}\right)^2 = 36$

32. $\left(-\dfrac{\sqrt{2}X}{2} - \dfrac{\sqrt{2}Y}{2}\right)^2 - 4\left(-\dfrac{\sqrt{2}X}{2} - \dfrac{\sqrt{2}Y}{2}\right)\left(\dfrac{\sqrt{2}X}{2} - \dfrac{\sqrt{2}Y}{2}\right) +$

$\left(\dfrac{\sqrt{2}X}{2} - \dfrac{\sqrt{2}Y}{2}\right)^2 = 4$

34. $\theta = 45°$

36. $\theta = 0°$

38. $r = \dfrac{3}{3 + \cos\theta}$

40. $r = \dfrac{6}{1 - 2\sin\theta}$

42. ellipse, $e = \dfrac{1}{2}$, focus: $(0, 0)$ directrix $x = -4$

44. hyperbola, $e = 3$ focus: $(0, 0)$ directrix $x = 2$

46. $x = t^2$, $y = -t$, $0 \le t \le 1$

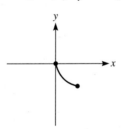

48. $x = \sin t$, $y = t$, $0 \le t \le \pi$

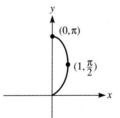

50. $y = \dfrac{1}{\sqrt{x - 1}}$, $x > 1$

52. $y = x + 6$, $-2 \le x \le 0$

54. $y = 3e^{x/5} + 4$, $x \le 5 \ln 4$